CW00739524

ISBN 978-0-265-61513-3
PIBN 10057546

A SYSTEM OF INSTRUCTION IN QUANTITATIVE CHEMICAL ANALYSIS.

BY

DR. C. REMIGIUS FRESENIUS,

PROFESSOR OF CHEMISTRY AND NATURAL PHILOSOPHY, WIESBADEN.

From the last English and German Editions.

EDITED BY

O. D. ALLEN, PH.D.,

PROFESSOR OF ANALYTICAL CHEMISTRY AND METALLURGY IN THE SHEFFIELD SCIENTIFIC SCHOOL YALE COLLEGE.

WITH THE CÖOPERATION OF

SAMUEL W. JOHNSON, M.A.,

PROFESSOR OF THEORETICAL AND AGRICULTURAL CHEMISTRY IN THE SHEFFIELD SCIENTIFIC SCHOOL.

NEW YORK:

JOHN WILEY & SONS,

15 ASTOR PLACE.

1881.

S. W. GREEN'S SON,
Printer, Electrotyper and Binder,
74 Beekman Street, New York.

EDITOR'S PREFACE

TO THE SECOND AMERICAN EDITION.

In the preparation of this edition of Fresenius' Quantitative Analysis, the general plan announced by the editor of the first American edition in the preceding preface has been followed. Although the original work, as it appears in the last foreign editions, has been somewhat abridged, it is believed that little which is useful to the student has been omitted from the present work. All processes which are described are given with the full details, and, with few exceptions, as far as practicable in the language used by the author.

The desired reduction of the bulk of the original treatise has been effected by the omission of one or more processes when several are given for the same purpose, or more rarely by the entire omission of a whole subject.

The subjects omitted, in addition to those mentioned in the preceding preface, are: "The Determination of the Equivalent of Organic Compounds," "The Assay of Silver Ores," and "The Assay of Gold Ores." On the other hand, many new processes and modifications of old processes appearing in the recently published first volume of the sixth German edition are included, and may be regarded as valuable additions to the General Part.

Additions made by the editors are usually distinguished by enclosure in brackets [ ].

The more important additions of this kind are in those chapters (in the Special Part) which treat of the analysis of products pertaining to the Metallurgy of Iron and to Commercial Fertilizers.

The entire chapter on the latter subject has been prepared by Professor S. W. Johnson and Dr. E. H. Jenkins, Chemist of the Connecticut Agricultural Experiment Station. It describes the methods and plans of analysis adopted in that institution after much experience and research.

The new system of chemical notation and nomenclature is employed throughout the book, although the old system is still retained even in the last foreign editions. It is confidently believed that this change, so long deferred for reasons perhaps sufficiently valid, can at the present time be made with advantage to the student and instructor.

The editor is under obligation to Messrs. W. J. Comstock and A. B. Howe, Ph.D., instructors in the Sheffield Laboratory, and to Professor W. G. Mixter, of the Sheffield Scientific School, for information and advice which their experience has enabled them to give regarding many processes, and for valuable assistance in various other ways.

The task of preparing this edition was undertaken and carried out with the generous co-operation of Professor S. W. Johnson. To him, therefore, most especially are due thanks from the editor and from those who may believe that they find any advantage in the possession of the book in its present form.

O. D. ALLEN.

SHEFFIELD LABORATORY OF YALE COLLEGE, Feb., 1881.

EDITOR'S PREFACE

TO THE FIRST AMERICAN EDITION.

IN preparing this edition of Fresenius' Quantitative Chemical Analysis, the editor has sought by various changes to adapt it to the wants of the American student.

The foreign editions have attained such encyclopedic dimensions as to occasion the beginner no little confusion and embarrassment. For this reason the bulk of the work has been considerably reduced. A few processes which the editor's experience has convinced him are untrustworthy, and many more that can well be spared because they are tedious or unnecessary, have been omitted. The entire chapter on Analysis of Mineral Waters, excellent as it is, has been suppressed on account of its length, and because the few who have occasion to make detailed investigations in that direction have access to the original sources of information.

The section on Organic Analysis has been reduced from sixty to thirty pages, mainly by the omission of processes which, from their antiquity or inferiority, are more curious than useful. The chapters on Acidimetry and Alkalimetry have been likewise greatly condensed, and all that especially relates to Soils and Ashes of Plants has been left out. The recent appearance of an excellent special treatise on "Agricultural Chemical Analysis" by Professor Caldwell, of Cornell University, justifies the last-mentioned omission.

On the other hand, some important matter has been added. Bunsen's invaluable new methods of treating precipitates are described in his own (translated) words. Various new methods of estimation and separation are incorporated in their proper places.

The editor thankfully acknowledges his indebtedness to several gentlemen for special contributions to this work, viz.: To Dr. J. Lawrence Smith, who has kindly furnished a manuscript account of his admirable method of fluxing silicates for the estimation of alkalies. To O. D. Allen, Esq., late chemist to the Freedom Iron Works, Lewistown, Pennsylvania, for copious notes of his extensive experience in the analyses of steel, iron, and iron ores, which have been freely employed in § 229. To Mr. William G. Mixter, chief assistant in the Sheffield Laboratory, for the account of the gold and silver assay. To Professor Brush, of Yale College, Professor Collier, of Vermont University, and B. S. Burton, Esq., of Philadelphia, for various important facts and suggestions. Just before going to press, Dr. Wolcott Gibbs has communicated an account of his new method of finding at once the total correction for temperature, pressure, and moisture in absolute determinations of nitrogen or other gases, which, from its simplicity, convenience, and accuracy, must prove of the highest service in chemistry. It will be found in the Appendix, p. 838.

The additions which have been made to the methods of examining ores, it is believed, adapt the work to meet all the ordinary requirements of the metallurgical and mining student.

The editor's additions are distinguished, in all important cases, by enclosure in brackets, [ ].

While fully recognizing the necessity of teaching the new notation and nomenclature of chemistry, the editor has in this book retained the old system, because it is identified with the chemical literature of the century, and cannot be speedily forgotten by practical men. At a time when the most elementary text-books are framed on the "modern" system, it is important to keep the student exercised in the language of the old masters of the science, which is still, and must for some time remain, a part of the vernacular of the physician, the apothecary, the metallurgist, and the manufacturer.

SAMUEL W. JOHNSON.

SHEFFIELD LABORATORY OF YALE COLLEGE, Dec., 1869.

CONTENTS.

SECTION II.

SECTION III.

SECTION VI.

PART II.

SPECIAL PART.

PART III.

APPENDIX.

INTRODUCTION.

As we have already seen in the "Manual of Qualitative Analysis,"—to which the present work may be regarded as the sequel,—Chemical Analysis comprises two branches, viz.: *qualitative analysis* and *quantitative analysis*, the object of the former being to ascertain the *nature*, that of the latter to determine the *amount*, of the several component parts of any compound.

By QUALITATIVE ANALYSIS we convert the *unknown* constituents of a body into certain *known* forms and combinations; and we are thus enabled to draw correct inferences respecting the nature of these unknown constituents. Quantitative analysis attains its object, according to circumstances, often by very different ways; the two methods most widely differing from each other, are *analysis by weight*, or *gravimetric analysis*, and *analysis by measure*, or *volumetric analysis*.

GRAVIMETRIC ANALYSIS has for its object to convert the *known* constituents of a substance into forms or combinations which will admit of the most exact determination of their weight, and of which, moreover, the composition is accurately known. These new forms or combinations may be either *educts* from the analyzed substance, or they may be *products*. In the former case the ascertained weight of the eliminated substance is the direct expression of the amount in which it existed in the compound under examination; whilst in the latter case, that is, when we have to deal with *products*, the quantity in which the eliminated constituent was originally present in the analyzed compound, has to be deduced by calculation from the quantity in which it exists in its new combination.

The following example will serve to illustrate these points:—Suppose we wish to determine the quantity of mercury contained in the chloride of that metal; now, we may do this, either by precipitating the metallic mercury from the solution of the chloride, say by means of stannous chloride; or we may attain our object by precipitating the solution by sulphuretted hydrogen, and weighing the precipitated mercuric sulphide. 100 parts of mercuric chloride consist of 73·82 of mercury and 26·18 of chlorine; consequently, if the process is conducted with absolute accuracy, the precipitation of mercury in 100 parts of mercuric chloride by stannous chloride will yield 73·82 parts of metallic mercury. With equally exact manipulation the other method yields 85·634 parts of mercuric sulphide.

Now, in the former case we find the number 73·82 directly; in the latter case we have to deduce it by calculation:—(100 parts of mercuric sulphide contain 86·207 parts of mercury; how much mercury do 85·634 parts contain?)

$$100 : 85{\cdot}634 :: 86{\cdot}207 : x - x = 73{\cdot}82.$$

As already hinted, it is absolutely indispensable that the forms into which bodies are converted for the purpose of estimation by weight should fulfil two conditions: first, they must be capable of being weighed exactly; secondly, they must be of known composition,—for it is quite obvious, on the one hand, that accurate quantitative analysis must be altogether impossible if the substance the quantity of which it is intended to ascertain, does not admit of correct weighing; and on the other hand, it is equally evident that if we do not know the exact composition of a new product, we lack the necessary basis of our calculation.

VOLUMETRIC ANALYSIS is based upon a very different principle from that of gravimetric analysis; viz., it effects the quantitative determination of a body, by converting it from a certain definite state to another equally definite state, by means of a fluid of accurately known power of action, and under circumstances which permit the analyst to mark with rigorous precision the exact point when the conversion is accomplished. The following example will serve to illustrate the principle of this method:—Potassium permanganate added to a solution of ferrous sulphate, acidified with sulphuric acid, immediately converts the ferrous sulphate into ferric sulphate; the permanganic acid, which is characterized by its intense color, yielding up oxygen and forming with the free sulphuric acid present colorless manganous sulphate. If, therefore, to an acidified fluid containing a ferrous salt we add, drop by drop, a solution of potassium permanganate, its red color continues for some time to disappear upon stirring; but at last a point is reached when the coloration imparted to the fluid by the last drop added remains: this point marks the termination of the conversion of the ferrous salt into a ferric salt.

If now we convert a known weight of iron into a ferrous sulphate by dissolving it in dilute sulphuric acid, and ascertain by suitable measuring apparatus the volume of a solution of potassium permanganate required to convert the ferrous sulphate to ferric sulphate, we can by means of this permanganate solution determine unknown quantities of ferrous iron in a solution. This is accomplished by adding the permanganate solution until the above described reaction is completed, and noting the volume used. The amount of iron present can now be calculated by comparing the volume used with that used when a known quantity of iron was present, as the weight of iron must in both cases be proportional to volume of permanganate used.

To this brief intimation of the general purport and object of quantitative analysis, and the general mode of proceeding in analytical researches, I have to add that certain qualifications are essential to those who would devote themselves successfully to the pursuit of this branch. These qualifications are, 1, theoretical knowledge; 2, skill in manipulation; and 3, strict conscientiousness.

The preliminary *knowledge* required consists in an acquaintance with qualitative analysis, the stoïchiometric laws, and simple arithmetic. Thus prepared, we shall understand the method by which bodies are separated and determined, and we shall be in a position to perform our calculations, by which, on the one hand, the formulæ of compounds are deduced from the analytical results, and, on the other hand, the correctness of the adopted methods is tested, and the results obtained are controlled. To this *knowledge* must be joined the *ability of performing the necessary practical operations.* This axiom generally holds good for all applied sciences, but if it is true of one more than another, quantitative analysis is that one. The most extensive and solid theoretical acquirements will not enable us, for instance, to determine the amount of common salt present in a solution, if we are without the requisite dexterity to transfer a fluid from one vessel to another without the smallest loss by spirting, running down the side, &c. The various operations of quantitative analysis demand great aptitude and manual skill, which can be acquired only by practice. But even the possession of the greatest practical skill in manipulation, joined to a thorough theoretical knowledge, will still prove insufficient to insure a successful pursuit of quantitative researches, unless also combined with *a sincere love of truth and a firm determination to accept none but thoroughly confirmed results.*

Every one who has been engaged in quantitative analysis knows that cases will sometimes occur, especially when commencing the study, in which doubts may be entertained as to whether the result will turn out correct, or in which even the operator is *positively convinced* that it *cannot* be *quite* correct. Thus, for instance, a small portion of the substance under investigation may be spilled; or some of it lost by decrepitation; or the analyst may have reason to doubt the accuracy of his weighing; or it may happen that two analyses of the same substance do not exactly agree. In all such cases it is indispensable that the operator should be conscientious enough to repeat the whole process over again. He who is not possessed of this self-command—who shirks trouble where truth is at stake—who would be satisfied with mere assumptions and guess-work, where the attainment of positive certainty is the object, must be pronounced just as deficient in the necessary qualifications for quantitative analytical researches as he who is wanting in knowledge or skill. He, therefore, who cannot fully trust his work—who cannot swear to the correctness of his results, may indeed occupy himself with quantitative analysis by way of practice, but he ought on no account to publish or use his results as if they were positive, since such proceeding could not conduce to his own advantage, and would certainly be mischievous as regards the science.

The domain of quantitative analysis may be said to extend over all matter—that is, in other words, anything corporeal may become the object of quantitative investigation. The present work, however, is intended to embrace only the substances used in pharmacy, arts, trades, and agriculture.

Quantitative analysis may be subdivided into two branches, viz., analysis of *mixtures*, and analysis of *chemical compounds*. This division may appear at first sight of very small moment, yet it is necessary that we should establish and maintain it, if we would form a clear conception of the value and utility of quantitative research. The quantitative analysis of mixtures, too, has not the same aim as that of chemical compounds; and the method applied to secure the correctness of the results in the former case is different from that adopted in the latter. The quantitative analysis of chemical compounds also rather subserves the purposes of the science, whilst that of mixtures belongs to the practical purposes of life. If, for instance, I analyze the salt of an acid, the result of the analysis will give me the constitution of that acid, its combining proportion, saturating capacity, &c.; or, in other words, the results obtained will enable me to answer a series of questions of which the solution is important for the theory of chemical science: but if, on the other hand, I analyze gunpowder, alloys, medicinal mixtures, ashes of plants, &c., &c., I have a very different object in view; I do not want in such cases to apply the results which I may obtain to the solution of any theoretical question in chemistry, but I want to render a practical service either to the arts and industries, or to some other science. If in the analysis of a chemical compound I wish to control the results obtained, I may do this in most cases by means of calculations based on stoïchiometric data, but in the case of a mixture a second analysis is necessary to confirm the correctness of the results afforded by the first.

The preceding remarks clearly show the immense importance of quantitative analysis. It may, indeed, be averred that chemistry owes to this branch its elevation to the rank of a science, since quantitative researches have led us to discover and determine the laws which govern the combinations and transpositions of the elements. Stoïchiometry is entirely based upon the results of quantitative investigations; all rational views respecting the constitution of compounds rest upon them as the only safe and solid basis.

Quantitative analysis, therefore, forms the strongest and most powerful lever for chemistry as a science, and not less so for chemistry in its applications to the practical purposes of life, to trades, arts, manufactures, and likewise in its application to other sciences. It teaches the mineralogist the true nature of minerals, and suggests to him principles and rules for their recognition and classification. It is an indispensable auxiliary to the physiologist; and agriculture has already derived much benefit from it; but far greater benefits may be predicted. We need not expatiate here upon the advantages which medicine, pharmacy, and every branch of industry derive, either directly or indirectly, from the practical application of its results. On the other hand, the benefit thus bestowed by quantitative analysis upon the various sciences, arts, etc., has been in a measure reciprocated by some of them. Thus whilst stoïchiometry owes its establishment to quantitative analysis, the stoïchiometric laws afford us the means of controlling the results of our analyses so accurately as to justify the reliance which we now generally place on them. Again, whilst quantitative analysis has advanced the progress of arts and industry, our manufacturers in return supply us with the most perfect platinum, glass, and porcelain vessels, and with articles of india-rubber, without which it would be next to impossible to conduct our analytical operations with the minuteness and accuracy which we have now attained.

Although the aid which quantitative analysis thus derives from stoïchiometry, and the arts and manufactures, greatly facilitates its practice, and although many determinations are considerably abbreviated by volumetric analysis, it must be admitted, notwithstanding, that the pursuit of this branch of chemistry requires considerable expenditure of time. This remark applies especially to those who are commencing the study, for they must not allow their attention to be divided upon many things at one time, otherwise the accuracy of their results will be more or less injured. I would therefore advise every one desirous of becoming an analytical chemist to arm himself with a considerable share of patience, reminding him that it is not at one bound, but gradually, and step by step, that the student may hope to attain the necessary certainty in his work, the indispensable self-reliance which can alone be founded on one's own results. However mechanical, protracted, and tedious the operations of quantitative analysis may appear to be, the attainment of accuracy will amply compensate for the time and labor bestowed upon them; whilst, on the other hand, nothing can be more disagreeable than to find, after a long and laborious process, that our results are incorrect or uncertain. Let him, therefore, who would render the study of quantitative analysis agreeable to himself, from the very outset endeavor, by strict, nay, scrupulous adherence to the conditions laid down, to attain correct results, at any sacrifice of time. I scarcely know a better and more immediate reward of labor than that which springs from the attainment of accurate results and perfectly corresponding analyses. The satisfaction enjoyed at the success of our efforts is surely in itself a sufficient motive for the necessary expenditure of time and labor, even without looking to the practical benefits which we may derive from our operations.

The following are the substances treated of in this work:—

I. METALLOIDS, OR NON-METALLIC ELEMENTS.

Oxygen, Hydrogen, Sulphur, [*Selenium,*] *Phosphorus, Chlorine, Iodine, Bromine, Fluorine, Nitrogen, Boron, Silicon, Carbon.*

II. METALS.

Potassium, Sodium, [*Lithium,*] *Barium, Strontium, Calcium, Magnesium, Aluminium, Chromium,* [*Titanium,*] *Zinc, Manganese, Nickel, Cobalt, Iron,* [*Uranium,*] *Silver, Mercury, Lead, Copper, Bismuth, Cadmium,* [*Palladium,*] *Gold, Platinum, Tin, Antimony, Arsenic,* [*Molybdenum*].

(The elements enclosed within brackets are considered in supplementary paragraphs, and more briefly than the rest.)

I have divided my subject into three parts. In the first, I treat of quantitative analysis generally; describing the execution of analysis. In the second, I give a detailed description of several special analytical processes. And in the third, a number of carefully selected examples, which may serve as exercises for the groundwork of the study of quantitative analysis.

The following table will afford the reader a clear and definite notion of the contents of the whole work:—

I. GENERAL PART.

1. Operations.
2. Reagents.
3. Forms and combinations in which bodies are separated from others, or in which their weight is determined.
4. Determination of bodies in simple compounds.
5. Separation of bodies.
6. Organic elementary analysis.

II. SPECIAL PART.

1. Analysis of waters.
2. Analysis of such minerals and technical products as are most frequently brought under the notice of the chemist; including methods for ascertaining their commercial value.
3. Analysis of atmospheric air.

III. EXERCISES FOR PRACTICE.

APPENDIX.

1. Analytical experiments.
2. Calculation of analyses.
3. Tables for calculation.

PART I.

GENERAL PART.

THE EXECUTION OF ANALYSIS.

SECTION I.

OPERATIONS.

§ 1.

Most of the operations performed in quantitative research are the same as in qualitative analysis, and have been accordingly described in my work on that branch of analytical science. With respect to such operations I shall, therefore, confine myself here to pointing out any modifications they may require to adapt them for application in the quantitative branch; but I shall, of course, give a full description of such as are resorted to exclusively in quantitative investigations. Operations forming merely part of certain specific processes will be found described in the proper place, under the head of such processes.

I. Determination of Quantity.

§ 2.

The quantity of solids is usually determined by *weight;* the quantity of gases and fluids, in many cases by *measure;* upon the care and accuracy with which these operations are performed, depends the value of all our results; I shall therefore dwell minutely upon them.

§ 3.

1. Weighing.

To enable us to determine with precision the correct weight of a substance, it is indispensable that we should possess, 1st, a good balance, and 2d, accurate weights.

a. THE BALANCE.

Fig. 1 represents a form of balance well adapted for analytical purposes. There are several points respecting the construction and properties of a good balance, which it is absolutely necessary for every chemist to understand. The usefulness of this instrument depends upon two points: 1st, its *accuracy*, and 2d, its *sensibility* or *delicacy*.

§ 4.

The ACCURACY of a balance depends upon the following conditions :—

α. The fulcrum or the point on which the beam rests must lie above the centre of gravity of the balance.

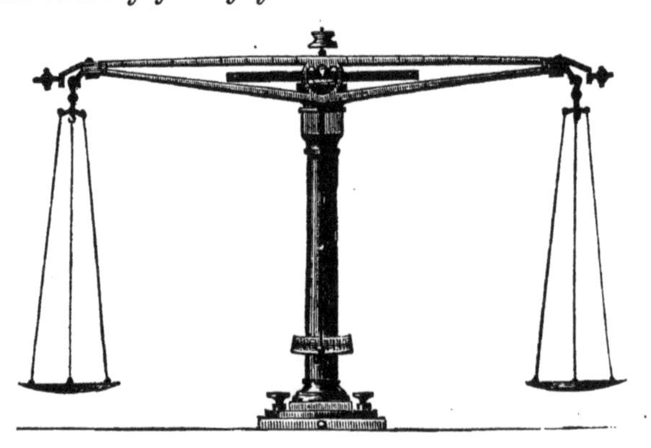

Fig. 1.

This is in fact a condition essential to every balance. If the fulcrum were placed *in* the centre of gravity of the balance, the beam would not oscillate, but remain in any position in which it is placed, assuming the scales to be equally loaded. If the fulcrum be placed *below* the centre of gravity, the balance will be overset by the slightest impulse.

When the fulcrum is above the centre of gravity the balance represents a pendulum, the length of which is equal to that of the line uniting the fulcrum with the centre of gravity, and this line forms right angles with the beam, in whatever position the latter may be placed. Now if we impart an impetus to a ball suspended by a thread, the ball, after having terminated its vibrations, will invariably rest in its original perpendicular position under the point of suspension. It is the same with a properly adjusted balance—impart an impetus to it, and it will oscillate for some time, but it will invariably return to its original position; in other words, its centre of gravity will finally fall back into its perpendicular position under the fulcrum, and the beam must consequently reassume the horizontal position.

But to judge correctly of the force with which this is accomplished, it must be borne in mind that a balance is not a simple pendulum, but a compound one, *i. e.*, a pendulum in which not one, but many material points move round the turning point. The inert mass to be moved is accordingly equal to the sum of these points, and the moving force is equal to the excess of the material points below, over those above the fulcrum.

β. *The points of suspension of the scales must be on an exact level with the fulcrum.* If the fulcrum be placed below the line joining the points of suspension, increased loading of the scales will continually tend to raise the centre of gravity of the whole system, so as to bring it nearer and nearer the fulcrum; the weight which presses upon the scales combining in the relatively high-placed points of suspension; at last, when the scales have been loaded to a certain degree, the centre of gravity will shift altogether to the fulcrum, and the balance will consequently cease to vibrate—any further addition of weight will finally overset the beam by placing the centre of gravity above the fulcrum. If, on the other hand, the fulcrum be placed above the line joining the points of suspension, the centre of gravity will become more and more depressed in proportion as the loading of the scales is increased; the line of the pendulum will consequently be lengthened, and a greater force will be required to produce an equal turn; in other words, the balance will grow less sensitive the greater the load. But when the three edges are in one plane, increased loading of the scales will, indeed, continually tend to raise the centre of gravity towards the fulcrum, but the former can in this case never *entirely* reach the latter, and consequently the balance will never altogether cease to vibrate upon the further addition of weight, nor will its sensibility be lessened; on the contrary—speaking theoretically—a greater degree of sensibility is imparted to it. This increase of sensibility is, however, compensated for by other circumstancas. (*See* § 5.)

γ. The beam must be sufficiently rigid to bear without bending the greatest weight that the construction of the balance admits of; since the bending of the beam would of course depress the points of suspension so as to place them below the fulcrum, and this would, as we have just seen, tend to diminish the sensibility of the balance in proportion to the increase of the load. It is, therefore, necessary to avoid this fault by a proper construction of the beam. The form best adapted for beams is that of an isosceles obtuse-angled triangle, or of a rhombus.

δ. The arms of the balance must be of equal length, i. e., *the points of suspension must be equidistant from the fulcrum*, for if the arms are of unequal length the balance will not be in equilibrium, supposing the scales to be loaded with equal weights, but there will be preponderance on the side of the longer arm.

§ 5.

The SENSIBILITY of a balance depends principally upon the three following conditions:—

α. The friction of the edges upon their supports must be as slight as possible. The greater or less friction of the edges upon their supports depends upon both the form and material of those parts of the balance. The edges *must* be made of good steel, the supports *may* be made of the same material; it is better, however, that the centre edge at least should rest upon an agate plane. To form a clear conception of how necessary it is that even the end edges should have as little friction as possible, we need simply reflect upon what would happen were we to fix the scales immovably to the beam by means of rigid rods. Such a contrivance would at once altogether annihilate the sensibility of a balance, for if a weight were placed upon one scale, this certainly would have a tendency to sink; but at the same time the connecting rods being compelled to form constantly a right angle with the beam, the weighted scale would incline inwards, whilst the other scale would turn outwards, and thus the arms would become unequal, the shorter arm being on the side of the weighted scale, whereby the tendency of the latter to sink would be immediately compensated for. The more considerable the friction becomes at the end edges of a balance, the more the latter approaches the state just now described, and consequently the more is its sensibility impaired.

β. The centre of gravity must be as near as possible to the ful-

crum. The nearer the centre of gravity approaches the fulcrum, the shorter becomes the pendulum. If we take two balls, the one suspended by a short and the other by a long thread, and impart the same impetus to both, the former will naturally swing at a far greater angle from its perpendicular position than the latter. The same must of course happen with a balance; the same weight will cause the scale upon which it is placed to turn the more rapidly and completely, the shorter the distance between the centre of gravity and the fulcrum. We have seen above, that in a balance where the three edges are on a level with each other, increased loading of the scales will continually tend to raise the centre of gravity towards the fulcrum. A good balance will therefore become more delicate in proportion to the increase of weights placed upon its scales; but, on the other hand, its sensibility will be diminished in about the same proportion by the increment of the mass to be moved, and by the increased friction attendant upon the increase of load; in other words, the delicacy of a good balance will remain the same, whatever may be the load placed upon it. The nearer the centre of gravity lies to the fulcrum, the slower are the oscillations of the balance. Hence in regulating the position of the centre of gravity we must not go too far, for if it approaches the fulcrum too nearly, the operation of weighing will take too much time.

γ. The beam must be as light as possible. The remarks which we have just now made will likewise show how far the weight of the beam may influence the sensibility of a balance. We have seen that if a balance is not actually to become less delicate on increased loading, it must on the one hand have a tendency to become more delicate by the continual approach of the centre of gravity to the fulcrum. Now it is evident, that the more considerable the weight of the beam is, the less will an equal load placed upon both scales alter the centre of gravity of the whole system, the more slowly will the centre of gravity approach the fulcrum, the less will the increased friction be neutralized, and consequently the less sensibility will the balance possess. Another point to be taken into account here is, that the moving forces being equal, a lesser mass or weight is more readily moved than a greater. (§ 4 *α.*)

§ 6.

We will now proceed, first, to give the student a few general rules to guide him in the purchase of a balance intended for the purposes of quantitative analysis; and, secondly, to point out the best method of testing the accuracy and sensibility of a balance.

1. A balance able to bear 70 or 80 grammes in each scale, suffices for most purposes.

2. The balance must be enclosed in a glass case to protect it from dust. This case ought to be sufficiently large, and, more especially, its sides should not approach too near the scales. It must be constructed in a manner to admit of its being opened and closed with facility, and thus to allow the operation of weighing to be effected without any disturbing influence from currents of air. Therefore, either the front part of the case should consist of three parts, viz., a fixed centre part and two lateral parts, opening like doors; or, if the front part happens to be made of one piece, and arranged as a sliding-door, the two sides of the case must be provided each with a door.

3. The balance must be provided with a proper contrivance to render it immovable whilst the weights are being placed upon the scale. This is most commonly effected by an arrangement which enables the operator to lift up the beam and thus to remove the middle edge from its support, whilst the scales remain suspended.

It is highly advisable to have the case of the balance so arranged that the contrivances for lifting the beam and fixing the scales can be worked while the case remains closed, and consequently from without.

4. It is necessary that the balance should be provided with an index to mark its oscillations; this index is appropriately placed at the bottom of the balance.

5. The balance must be provided with a spirit level, to enable the operator to place the three edges on an exactly horizontal level; it is best also for this purpose that the case should rest upon three screws.

6. It is very desirable that the beam should be graduated into tenths, so as to enable the operator to weigh the milligramme and its fractions with a centigramme "rider."*

7. The balance must be provided with a screw to regulate the centre of gravity, and likewise with two screws to regulate the

* [Becker's later balances have beams graduated to twelfths, and a rider weighing 12 mgrs. This enables the operator to use nearly the whole of the graduation.]

equality of the arms, and finally with screws to restore the equilibrium of the scales, should this have been disturbed.

§ 7.

The following experiments serve to test the accuracy and sensibility of a balance.

1. The balance is, in the first place, accurately adjusted, if necessary, either by the regulating screws, or by means of tinfoil, and a milligramme weight is then placed in one of the scales. A good and practically useful balance must turn very distinctly with this weight; a delicate chemical balance should indicate the $\frac{1}{10}$ of a milligramme with perfect distinctness.

2. Both scales are loaded with the maximum weight the construction of the balance will admit of—the balance is then *accurately* adjusted, and a milligramme added to the weight in the one scale. This ought to cause the balance to turn to the same extent as in 1. In most balances, however, it shows somewhat less on the index. It follows from § 5 β that the balance will oscillate more slowly in this than in the first experiment.

3. The balance is accurately adjusted (should it be necessary to establish a perfect equilibrium between the scales by loading the one with a minute portion of tinfoil, this tinfoil must be left remaining upon the scale during the experiment); both scales are then equally loaded, say, with fifty grammes each, and, if necessary, the balance is again adjusted (by the addition of small weights). The load of the two scales is then interchanged, so as to transfer that of the right scale to the left, and *vice versâ.* A balance with perfectly equal arms must maintain its absolute equilibrium upon this interchange of the weights of the two scales.

4. The balance is accurately adjusted; it is then arrested and again set in motion; the same process should be repeated several times. A good balance must invariably reassume its original equilibrium. A balance the end edges of which afford too much play to the hook resting upon them, so as to allow the latter slightly to alter its position, will show perceptible differences in different trials. This fault, however, is possible only with balances of defective construction.

A balance to be practically useful for the purposes of quantitative analysis *must* stand the first, second, and last of these tests. A slight inequality of the arms is of no great consequence, as the error that it would occasion may be completely prevented by the manner of weighing.

As the sensibility of a balance will speedily decrease if the steel edges are allowed to get rusty, delicate balances should never be kept in the laboratory, but always in a separate room. It is also advisable to place within the case of the balance a vessel half filled with calcined carbonate of potassa, to keep the air dry. I need hardly add that this salt must be re-calcined as soon as it gets moist.

§ 8.

b. The Weights.

1. The French gramme is the best standard for calculation. A set of weights ranging from fifty grammes to one milligramme may be considered sufficient for all practical purposes. With regard to the set of weights, it is generally a matter of indifference for scientific purposes whether the gramme, its multiples and fractions, are really and perfectly equal to the accurately adjusted *normal* weights of the corresponding denominations;* but it is *absolutely necessary* that they should agree *perfectly* with each other, *i. e.*, the centigramme weight must be exactly the one hundredth part of the gramme weight of the set, etc. etc.

2. The whole of the set of weights should be kept in a suitable, well-closing box; and it is desirable likewise that a distinct compartment be appropriated to every one even of the smaller weights.

3. As to the shape best adapted for weights, I think that of short frusta of cones inverted, with a handle at the top, the most convenient and practical form for the large weights; square pieces of foil, turned up at one corner, are best adapted for the small weights. The foil used for this purpose should not be too thin, and the compartments adapted for the reception of the several smaller weights in the box, should be large enough to admit of their contents being taken out of them with facility, or else the smaller weights will soon get cracked, bruised, and indistinct. Every one

---

* Still it would be desirable that mechanicians who make gramme-weights intended for the use of the chemist, should endeavor to procure *normal* weights. It is very inconvenient, in many cases, to find notable differences between weights of the same denomination, but coming from different makers; as I myself have often had occasion to discover.

of the weights (with the exception of the milligramme) should be distinctly marked.

4. With respect to the material most suitable for the manufacture of weights, we commonly rest satisfied with having the smaller weights only, from 1 or 0·5 gramme downwards, made of platinum or aluminium foil, using brass weights for all the higher denominations. Brass weights must be carefully shielded from the contact of acid or other vapors, or their correctness will be impaired; nor should they ever be touched with the fingers, but always with small pincers. But it is an erroneous notion to suppose that weights slightly tarnished are unfit for use. It is, indeed, hardly possible to prevent weights for any very great length of time from getting slightly tarnished. I have carefully examined many weights of this description, and have found them as exactly corresponding with one another in their relative proportions as they were when first used. The tarnishing coat, or incrustation, is so extremely thin, that even a very delicate balance will generally fail to point out any perceptible difference in the weight.

The following is the proper way of *testing the weights:*

One scale of a delicate balance is loaded with a one-gramme weight, and the balance is then completely equipoised by taring with small pieces of brass, and finally tinfoil (not paper, since this absorbs moisture). The weight is then removed, and replaced successively by the other gramme weights, and afterwards by the same amount of weight in pieces of lower denominations.

The balance is carefully scrutinized each time, and any deviation from the exact equilibrium marked. In the same way it is seen whether the two-gramme piece weighs the same as two single grammes, the five-gramme piece the same as three single grammes and the two-gramme piece, &c. In the comparison of the smaller weights thus among themselves, they must not show the least difference on a balance turning with $\frac{1}{10}$ of a milligramme. In comparing the larger weights with all the small ones, differences of $\frac{1}{10}$ to $\frac{2}{10}$ of a milligramme may be passed over. If you wish them to be more accurate, you must adjust them yourself. In the purchase of weights chemists ought always to bear in mind that an accurate weight is truly valuable, whilst an inaccurate one is absolutely worthless. It is the safest way for the chemist to test every weight he purchases, no matter how high the reputation of the maker.

§ 9.

c. The Process of Weighing.

We have two different methods of determining the weight of substances; the one might be termed *direct weighing*, the other is called *weighing by substitution*.

In *direct weighing*, the substance is placed upon one scale, and the weight upon the other. If we possess a balance, the arms of which are of equal length, and the scales in a perfect state of equilibrium, it is indifferent upon which scale the substance is placed in the several weighings required during an analytical process; *i.e.*, we may weigh upon the right or upon the left side, and change sides at pleasure, without endangering the accuracy of our results. But if, on the contrary, the arms of our balance are not perfectly equal, or if the scales are not in a state of perfect equilibrium, we are compelled to weigh invariably upon the same scale, otherwise the correctness of our results will be more or less materially impaired.

Suppose we want to weigh one gramme of a substance, and to divide this amount subsequently into two equal parts. Let us assume our balance to be in a state of perfect equilibrium, but with unequal arms, the left being 99 millimetres, the right 100 millimetres long; we place a gramme weight upon the left scale, and against this, on the right scale, as much of the substance to be weighed as will restore the equilibrium of the balance.

According to the axiom, "masses are in equilibrium upon a lever, if the products of their weights into their distances from the fulcrum are equal," we have consequently upon the right scale 0·99 grm. of substance, since $99 \times 1·00 = 100 \times 0·99$. If we now, for the purpose of weighing one half the quantity, remove the whole weight from the left scale, substituting a 0·5 grm. weight for it, and then take off part of the substance from the right scale, until the balance recovers its equilibrium, there will remain 0·495 grm.; and this is exactly the amount we have removed from the scale: we have consequently accomplished our object with respect to the relative weight; and as we have already remarked, the absolute weight is not generally of so much importance in scientific work. But if we attempted to halve the substance which we have on the right scale, by first removing both the weight and the substance from the scales, and placing subsequently a 0·5 grm. weight upon the *right* scale, and part of the substance upon the *left*, until the balance recovers its equilibrium, we should have 0·505 of substance upon the left scale, since $100 \times 0·500 = 99 \times 0.505$; and cousequently, instead of exact halves, we should have one part of the substance amounting to 0·505, the other only to 0·485.

If the scales of our balance are not in a state of absolute equilibrium, we are obliged to weigh our substances in vessels to insure accurate results (although the arms of the balance be perfectly equal). It is self-evident that the weights in this case must likewise be invariably placed upon one and the same scale, and that the difference between the two scales must not undergo the slighest variation during the whole course of a series of experiments.

From these remarks result the two following rules:—

1. It is, under all circumstances, advisable to place the substance invariably upon one and the same scale—most conveniently upon the left.

2. If the operator happens to possess a balance for his own private and *exclusive* use, there is no need that he should adjust it at the commencement of every analysis; but if the balance be used in common by several persons, it is absolutely necessary to ascertain, before every operation, whether the state of absolute equilibrium may not have been disturbed.

Weighing by substitution yields not only *relatively*, but also *absolutely* accurate results; no matter whether the arms of the balance be of exactly equal lengths or not, or whether the scales be in perfect equipoise or not.

The process is conducted as follows: the material to be weighed—say a platinum crucible—is placed upon one scale, and the other scale is accurately counterpoised against it. The platinum crucible is then removed, and the equilibrium of the balance restored by substituting weights for the removed crucible. It is perfectly obvious that the substituted weights will invariably express the real weight of the crucible with absolute accuracy. We weigh by substitution whenever we require the greatest possible accuracy; as, for instance, in the determination of atomic weights. The process may be materially shortened by first placing a tare (which must of course be heavier than the substance to be weighed) upon one scale, say the left, and loading the other scale with weights until equilibrium is produced. This tare is always retained on the left scale. The weights after being noted are removed. The substance is placed on the right scale, together with the smaller weights requisite to restore the equilibrium of the balance. The sum of the weights added is then subtracted from the noted weight of the counterpoise: the remainder will at once indicate the absolute weight of the substance. Let us suppose, for instance, we have on the left scale a tare requiring a weight of fifty grammes to counterpoise it. We place a platinum crucible on the right scale, and find that it requires an addition of weight to the extent of 10 grammes to counterpoise the tare on the left. Accordingly, the crucible weighs 50 *minus* 10=40 grammes.

§ 10.

The following *rules* will be found useful in performing the process of weighing:—

1. The safest and most expeditious way of ascertaining the exact weight of a substance, is to avoid trying weights at random; instead of this, a strictly systematic course ought to be pursued in counterpoising substances on the balance. Suppose, for instance, we want to weigh a crucible, the weight of which subsequently turns out to be 6.627 grammes; well, we place 10 grammes on the other scale against it, and we find this is too much; we place the weight next in succession, *i.e.*, 5 grammes, and find this too little; next 7, too much; 6, too little; 6·5, too little; 6·7, too much; 6·6, too little; 6·65, too much; 6·62, too little; 6·63, too much; 6·625, too little; 6·627, right.

I have selected here, for the sake of illustration, a most complicated case; but this systematic way of laying on the weights will in most instances lead to the desired end, in half the time required when weights are tried at random. After a little practice a few minutes will suffice to ascertain the weight of a substance to within the $\frac{1}{10}$ of a milligramme, provided the balance does not oscillate too slowly.

2. The milligrammes and fractions of milligrammes are determined by a centigramme rider (to be placed on or between the divisions on the beam) far more expeditiously and conveniently than by the use of the weights themselves, and at the same time with equal accuracy.

3. Particular care and attention should be bestowed on entering the weights in the book. The best way is to write down the weights first by inference from the blanks, or gaps in the weight box, and to control the entry subsequently by removing the weights from the scale, and replacing them in their respective compartments in the box. The student should from the commencement make it a rule to enter the number to be deducted in the *lower line;* thus, in the upper line, the weight of the crucible + the substance; in the lower line, the weight of the empty crucible.

4. The balance ought to be arrested every time any change is contemplated, such as removing weights, substituting one weight for another, &c. &c., or it will soon get spoiled.

5. Substances (except, perhaps, pieces of metal, or some other bodies of the kind) must never be placed *directly* upon the scales, but ought to be weighed in appropriate vessels of platinum, silver, glass, porcelain, &c., never on paper or card, since these, being liable to attract moisture, are apt to alter in weight. The most common method is to weigh in the first instance the vessel by itself, and to introduce subsequently the substance into it; to weigh again, and subtract the former weight from the latter. In many instances, and more especially where several portions of the same substance are to be weighed, the united weight of the vessel and of its contents is first ascertained; a portion of the contents is then shaken out, and the vessel weighed again; the loss of weight expresses the amount of the portion taken out of the vessel.

6. Substances liable to attract moisture from the air, must be weighed invariably in closed vessels (in covered crucibles, for instance, or between two watch-glasses, or in a closed glass tube); fluids are to be weighed in small bottles closed with glass stoppers.

7. A vessel ought never to be weighed whilst warm, since it will in that case invariably weigh lighter than it really is. This is owing to two circumstances. In the first place, every body condenses upon its surface a certain amount of air and moisture, the quantity of which depends upon the temperature and hygroscopic state of the air, and likewise on its own temperature. Now suppose a crucible has been weighed cold at the commencement of the operation, and is subsequently weighed again whilst hot, together with the substance it contains, and the weight of which we wish to determine. If we subtract for this purpose the

weight of the cold crucible, ascertained in the former instance, from the weight found in the latter, we shall subtract too much, and consequently we shall set down less than the real weight for the substance. In the second place, bodies at a high temperature are constantly communicating heat to the air immediately around them; the heated air expands and ascends, and the denser and colder air, flowing towards the space which the former leaves, produces a current which tends to raise the scale, making it thus appear lighter than it really is.

8. If we suspend from the end edges of a correct balance respectively 10 grammes of platinum and 10 grammes of glass, by wires of equal weight, the balance will assume a state of equilibrium; but if we subsequently immerse the platinum and glass completely in water, this equilibrium will at once cease, owing to the different specific gravity of the two substances; since, as is well known, substances immersed in water lose of their weight a quantity equal to the weight of their own bulk of water. If this be borne in mind, it must be obvious to every one that weighing in the air is likewise defective, inasmuch as the bulk of the substance weighed is not the same with that of the weight. This defect, however, is so very insignificant, owing to the trifling specific gravity of the air in proportion to that of solid substances, that we may generally disregard it altogether in analytical experiments. In cases, however, where *absolutely* accurate results are required, the bulk both of the substance examined, and of the weight, must be taken into account, and the weight of the corresponding volume of air added respectively to that of the substance and of the weight, making thus the process equivalent to weighing *in vacuo*.

§ 11.

2. Measuring.

The process of measuring is confined in analytical researches mostly to gases and liquids. The method of measuring gases has been brought to such perfection that it may be said to equal in accuracy the method of weighing. However, such accurate measurements demand an expenditure of time and care, which can be bestowed only on the nicest and most delicate scientific investigations.*

The measuring of liquids in analytical investigations was resorted to first by DESCROIZILLES ("Alkalimeter," 1806). GAY-LUSSAC materially improved the process, and indeed brought it to the highest degree of perfection (measuring of the solution of chloride of sodium in the assay of silver in the wet way). More recently F. MOHR† has bestowed much care and ingenuity upon the production of appropriate and convenient measuring apparatus, and has added to our store the eminently practical *compression stop-cock burette.* The process is now resorted to even in most accurate scientific investigations, since it requires much less time than the process of weighing.

The accuracy of all measurings depends upon the proper construction of the measuring vessels, and also upon the manner in which the process is conducted.

§ 12.

a. THE MEASURING OF GASES.

We use for the measuring of gases graduated tubes of greater or less capacity, made of strong glass, and closed by fusion at one end, which should be rounded. The following tubes will be found sufficient for all the processes of gas measuring required in organic elementary analyses.

1. A bell-glass capable of holding from 150 to 250 c. c., and about 4 centimetres in diameter; divided into cubic centimetres.

2. Five or six glass tubes, about 12 to 15 millimetres in diameter in the clear, and capable of holding from 30 to 40 c. c. each, divided into $\frac{1}{5}$ c. c.

The sides of these tubes should be pretty thick, otherwise they will be liable to break, especially when used to measure over mercury. The sides of the bell-glass should be about 3, of the tubes about 2 millimetres thick.

The most important point, however, in connection with meas-

* [The student who will practise the accurate measurement of gases in any but the simplest cases, must refer for all details to Bunsen's "Gasometry" (translated by Roscoe), and Russell, Jour. Chem. Soc., 1868, p. 128, as the subject is too extensive for the limits of this volume.]

† "Lehrbuch der Titrirmethode," by Dr. Fr. Mohr. Brunswick, 1855.

uring instruments is that they be correctly graduated, since upon this of course depends the accuracy of the results. For the method of graduating I refer to GREVILLE WILLIAMS' "Chemical Manipulation."*

In testing the measuring tubes we have to consider three things.

1. Do the divisions of a tube correspond with each other?

2. Do the divisions of each tube correspond with those of the other tubes?

3. Do the volumes expressed by the graduation lines correspond with the weights used by the analyst?

These three questions are answered by the following experiments:

a. The tube which it is intended to examine is placed in a perpendicular position, and filled gradually with accurately measured small quantities of mercury, care being taken to ascertain with the utmost precision whether the graduation of the tube is proportionate to the equal volumes of mercury poured in. The measuring-off of the mercury is effected by means of a small glass tube, sealed at one end, and ground perfectly even and smooth at the other. This tube is filled to overflowing by immersion under mercury, care being taken to allow no air bubbles to remain in it; the excess of mercury is then removed by pressing a small glass plate down on the smooth edge of the tube.†

b. Different quantities of mercury are successively measured off in one of the smaller tubes, and then transferred into the other tubes. The tubes may be considered in perfect accordance with each other, if the mercury reaches invariably the same divisional point in every one of them.

Such tubes as are intended simply to determine the relative volume of different gases, need only pass these two experiments; but in cases where we want to calculate the *weight* of a gas from its *volume*, it is necessary also to obtain an answer to the third question. For this purpose—

c. One of the tubes is accurately weighed and then filled with

* [See also Cary Lea, Am. Jour. Sci. and Arts, 2d ser., vol. 42, p. 375.]

† As warming the metal is to be carefully avoided in this process, it is advisable not to hold the tube with the hand in immersing it in the mercury, but to fasten it in a small wooden holder.

distilled water of a temperature of 16° to the last mark of the graduated scale; the weight of the water is then accurately determined. If the tube agrees with the weights, every 100 c. c. of water of 16° must weigh 99·9 grm. But should it not agree, no matter whether the error lie in the graduation of the tube or in the adjustment of the weights, we must apply a correction to the volume observed before calculating the weight of a gas therefrom. Let us suppose, for instance, that we find 100 c. c. to weigh only 99·6 grm.: assuming our weights to be correct, the c. c. of our scale are accordingly too small; and to convert 100 of these c. c. into normal c. c. we say:—

$$99{\cdot}9 : 99{\cdot}6 :: 100 : x.$$

In the measuring of gases we must have regard to the following points:—

1. Correct reading-off. 2. The temperature of the gas. 3. The degree of pressure operating upon it. And 4. The circumstance whether it is dry or moist. The three latter points will be readily understood, if it be borne in mind that any alteration in the temperature of a gas, or in the pressure acting upon it, or in the tension of the admixed aqueous vapor, involves likewise a considerable alteration in its volume.

§ 13.

1. Correct Reading-off.

This is rather difficult, since mercury in a cylinder has a convex surface (especially observable with a narrow tube), owing to its own cohesion; whilst water, on the other hand, under the same circumstances has a concave surface, owing to the attraction which the walls of the tube exercise upon it. The cylinder should invariably be placed in a perfectly perpendicular position, and the eye of the operator brought to a level with the surface of the fluid.

In reading-off over water, the middle of the dark zone formed by that portion of the liquid that is drawn up around the inner walls of the tube, is assumed to be the real surface; whilst when operating with mercury, we have to place the real surface in a plane exactly in the middle between the highest point of the surface of the mercury, and the points at which the latter is in actual contact with the walls of the tube. However, the results obtained in this way are only approximate.

Absolutely accurate results cannot be arrived at, in measuring over water or any other fluid that adheres to glass. But over mercury they may be arrived at if the error of the meniscus be determined and the mercury be read off at the highest point. The determination of the error of the meniscus is performed for each tube, once for all, in the following manner: some mercury is poured into the tube, and its height read-off right on a level with the top of the convex surface exhibited by it; a few drops of solution of chloride of mercury are then poured on the top of the metal; this causes the convexity to disappear; the height of the mercury in the tube is now read-off again and the difference noted. In the process of graduation, the tube stands upright, in that of measuring gases, it is placed upside down; the difference observed must accordingly be doubled, and the sum added to each volume of gas read off.

§ 14.

2. Influence of Temperature.

The temperature of gases to be measured is determined either by making it correspond with that of the confining fluid, and ascertaining the latter, or by suspending a delicate thermometer by the side of the gas to be measured, and noting the degree which it indicates.

If the construction of the pneumatic apparatus permits the total immersion of the cylinder in the confining fluid, uniformity of temperature between the latter and the gas which it is intended to measure, is most readily and speedily obtained; but in the reverse case, the operator must always, after every manipulation, allow half an hour or, in operations combined with much heating, even an entire hour to elapse, before proceeding to observe the state of the mercury in the cylinder, and in the thermometer.

Proper care must also be taken, after the temperature of the gas has been duly adjusted, to prevent re-expansion during the reading-off; all injurious influences in this respect must accordingly be carefully guarded against, and the operator should, more especially, avoid laying hold of the tube with his hand (in pressing it down, for instance, into the confining fluid); making use, instead, of a wooden holder.

§ 15.

3. Influence of Pressure.

With regard to the third point, the gas is under the actual pressure of the atmosphere if the confining fluid stands on an exact level both in and outside the cylinder; the degree of pressure exerted upon it may therefore at once be ascertained by consulting the barometer. But if the confining fluid stands *higher* in the cylinder than outside, the gas is under *less* pressure,—if *lower*, it is under *greater* pressure than that of the atmosphere; in the latter case, the perfect level of the fluid inside and outside the cylinder may readily be restored by raising the tube; if the fluid stands higher in the cylinder than outside, the level may be restored by depressing the tube; this however can only be done in cases where we have a trough of sufficient depth. When operating over water, the level may in most cases be readily adjusted; when operating over mercury, it is, more especially with wide tubes, often impossible to bring the fluid to a perfect level inside and outside the cylinder.

§ 16.

4. Influence of Moisture.

In measuring gases saturated with aqueous vapor, it must be taken into account that the vapor, by virtue of its tension, exerts a pressure upon the confining fluid. The necessary correction is simple, since we know the respective tension of aqueous vapor for the various degrees of temperature. But before this correction can be applied, it is, of course, necessary that the gas should be actually saturated with the vapor. It is, therefore, indispensable in measuring gases to take care to have the gas thoroughly saturated with aqueous vapor, or else absolutely dry.

It is quite obvious from the preceding remarks, that volumes of gases can be compared only if measured at the same temperature, under the same pressure, and in the same hygroscopic state. They are generally reduced to 0°, 0·76 met. barometer, and absolute dryness. How this is effected, as well as the manner in which we deduce the weight of gases from their volume, will be found in the chapter on the calculation of analyses.

§ 17.

b. The Measuring of Fluids.

In consequence of the vast development which volumetric analysis has of late acquired, the measuring of fluids has become an operation of very frequent occurrence. According to the different objects in view, various kinds of measuring vessels are employed. The operator must, in the case of every measuring vessel, carefully distinguish whether it is graduated for *holding* or for *delivering* the exact number of c. c. marked on it. If you have made use of a vessel of the former description in measuring off 100 c. c. of a fluid, and wish to transfer the latter completely to another vessel, you must, after emptying your measuring vessel, rinse it, and add the rinsings to the fluid transferred; whereas, if you have made use of a measuring vessel of the latter description, there must be no rinsing.

α. Measuring vessels graduated for holding the exact measure of fluid marked on them.

aa. Measuring vessels which serve to measure out one definite quantity of fluid.

We use for this purpose—

§ 18.

1. *Measuring Flasks.*

Fig. 2 represents a measuring flask of the most practical and convenient form.

Measuring flasks of various sizes are sold in the shops, holding respectively 200, 250, 500, 1000, 2000, &c., c. c. As a general rule, they have no ground-glass stoppers; it is, however, very desirable, in certain cases, to have measuring flasks with ground stoppers. The flasks must be made of well-annealed glass of uniform thickness, so that fluids may be heated in them. The line-mark should be placed within the lower third, or at least within the lower half, of the neck.

Measuring flasks, before they can properly be employed in analytical operations, must first be carefully tested. The best and simplest way of effecting this is to proceed thus:—Put the flask, perfectly dry inside and outside, on the one scale of a sufficiently delicate balance, together with a weight of 1000 grm. in the case of a litre flask, 500 grm. in the case of a half-litre flask, &c., restore the equilibrium by placing the requisite quantity of shot and tinfoil on the other scale, then remove the flask and the weight from the balance, put the flask on a perfectly level surface, and pour in distilled water of 16°,* until the lower border of the dark zone formed by the top of the water around the inner walls corresponds with the line-mark. After having thoroughly dried the neck of the flask above the mark, replace it upon the scale: if this restores the perfect equilibrium of the balance, the water in the flask weighs, in the case of a litre measure, exactly 1000 grm. If the scale bearing the flask sinks, the water in it weighs as much above 1000 grm. as the additional weights amount to which you have to put in the other scale to restore the equilibrium; if it rises, on the other hand, the water weighs as much less as the weights amount to which you have to put in the scale with the flask to effect the same end.

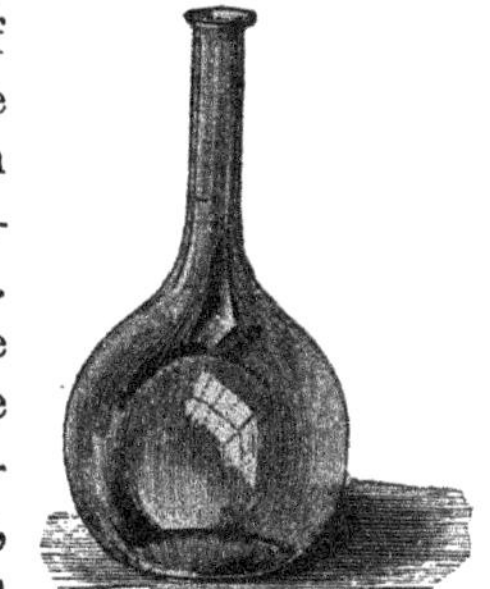

Fig. 2.

* To use water in the state of its highest density, viz., of 4°, 1 c. c. of which weighs exactly 1 grm., and, accordingly, 1 litre, exactly 1000 grms., is less practical, as the operations must in that case be conducted in a room as cold; since, in a warmer room, the outside of the flask would immediately become covered with moisture, in consequence of the air cooling below dew-point. Nor can I recommend F Mohr's suggestion to make litre-flasks, and measuring vessels in general, upon a plan to make the litre-flask, for instance, hold, not 1000 grm. water at 4°, but 1000 grm. at 16°, since in an arrangement of the kind proper regard is not paid to the actual meaning of the term "litre" in the scientific world; and measuring vessels of the same nominal capacity, made by different instrument-makers, are thus liable to differ to a greater or less extent. One litre-flask, according to Mohr, holds 1001·2 standard c. c. I consider it impractical to give to the c. c. another signification in vessels intended for measuring fluids than in vessels used for the measuring of gases, which latter demand strict adhesion to the standard c. c., as it is often required to deduce the weight of a gas by calculating from the volume.

If the water in the litre measure weighs 999 grm.,* in the half-litre measure, 499·5 grm., &c., the measuring flasks are correct. Differences up to 0·100 grm., in the litre measure, up to 0.070 grm. in the half-litre measure, and up to 0·050 grm. in the quarter-litre measure, are not taken into account, as one and the same measuring flask will be found to offer variation to the extent indicated, in repeated consecutive weighings, though filled each time exactly up to the mark with water of the same temperature.

Though a flask should, upon examination, turn out not to hold the exact quantity of water which it is stated to contain, it may yet possibly agree with the other measuring vessels, and may accordingly still be perfectly fit for use for most purposes. Two measuring vessels agree among themselves if the marked Nos. of c. c. bear the same proportion to each other as the weights found; thus, for instance, supposing your litre-measure to hold 998 grm. water of 16°, and your 50 c. c. pipette to deliver 49·9 grm. water of the same temperature, the two measures agree, since

$$1000 : 50 = 998 : 49 \cdot 9.$$

To prepare or correct a measuring flask, tare the dry litre, half-litre, or quarter-litre flask, and then weigh into it, by substitution, (§ 9) 999 grm., or, as the case may be, the half or quarter of that quantity of distilled water of 16°. Put the flask on a perfectly horizontal support, place your eye on an exact level with the surface of the water, and mark the lower border of the dark zone by two little dots made on the glass with a point dipped into thick asphaltum varnish, or some other substance of the kind. Now pour out the water, place the flask in a convenient position, and cut with a diamond a fine distinct line into the glass from one dot to the other.

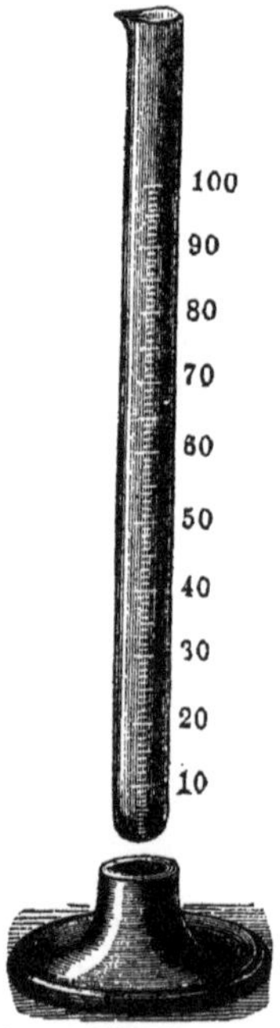

100
90
80
70
60
50
40
30
20
10

Fig. 3.

bb. Measuring vessels which serve to measure out any quantities of fluid at will.

---

* With absolute accuracy, 998·981 grm.

§ 19.

2. *The Graduated Cylinder.*

This instrument, represented in fig. 3, should be from 2 to 3 cm. wide, of a capacity of 100—300 c. c., and divided into single c. c. It must be ground at the top, that it may be covered quite close with a ground-glass plate. The measuring with such cylinders is not quite so accurate as with measuring flasks, as in the latter the volume is read off in a narrower part. The accuracy of measuring cylinders may be tested in the same way as in the case of measuring flasks, viz., by weighing into them water of 16° ; or, also, very well, by letting definite quantities of fluid flow into the cylinder from a correct pipette, or burette graduated for delivering, and observing whether or not they are correctly indicated by the scale of the cylinder.

β. MEASURING VESSELS GRADUATED FOR DELIVERING THE EXACT MEASURE OF FLUID MARKED ON THEM (graduated *à l'écoulement*).

aa. Measuring vessels which serve to measure out one definite quantity of fluid.

§ 20.

3. *The Graduated Pipette.*

This instrument serves to take out a definite volume of a fluid from one vessel, and to transfer it to another; it must accordingly be of a suitable shape to admit of its being freely inserted into flasks and bottles.

We use pipettes of 1, 5, 10, 20, 50, 100, 150, and 200 c. c. capacity. The proper shape for pipettes up to 20 c. c. capacity is represented in fig. 4; fig. 5 shows the most practical form for larger ones. To fill a pipette suction is applied to the upper aperture, either directly with the lips or through a caoutchouc tube, until the fluid stands above the mark; the upper orifice (which is somewhat narrowed and ground) is then closed with the first finger of the right hand (the point of which should be a little moist); the outside is then wiped dry, if required, and, the pipette being held in a perfectly vertical direction, the fluid is made to drop out, by lifting the finger a little, till it has fallen to the required level; the loose drop is carefully wiped off, and the contents of the tube are then finally transferred to the other vessel. In this process it is found that the fluid does not run out completely, but that a small portion of it remains adhering to the glass in the point of the pipette; after a time, as this becomes increased by other minute particles of fluid trickling down from the upper part of the tube, a drop gathers at the lower orifice, which may be allowed to fall off from its own weight, or may be made to drop off by a slight shake. If, after this, the point of the pipette be laid against a moist portion of the inner side of the vessel, another minute portion of fluid will trickle out, and, lastly, another trifling droplet or so may be got out by blowing into the pipette. Now, supposing the operator follows no fixed rule in this respect, letting the fluid, for instance, in one operation simply run out, whilst in another operation he lets it drain afterwards, and in a third blows out the last particles of it from the pipette, it is evident that the respective quantities of fluid delivered in the several operations cannot be quite equal. I prefer in all cases the second method, viz., to lay the point of the pipette, whilst draining, finally against a moist portion of the side of the vessel, which I have always found to give the most accurately corresponding measurements.

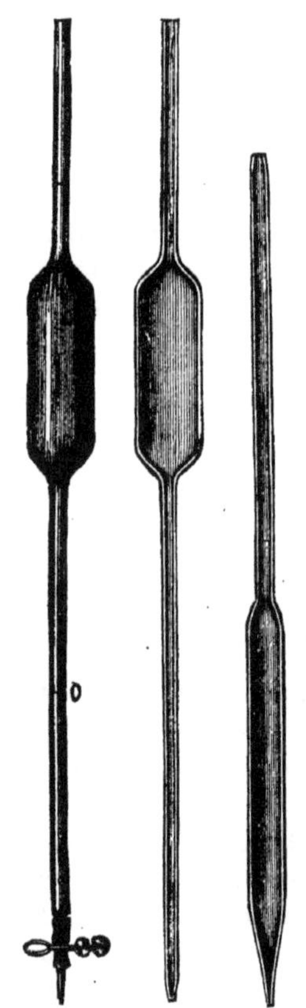

Fig. 4. Fig. 5. Fig. 6.

The correctness of a pipette is tested by filling it up to the mark with distilled water of 16°, letting the water run out, in the manner just stated, into a tared vessel, and weighing; the pipette may be pronounced correct if 100 c. c. of water of 16° weigh 99·9 grm.

Testing in like manner the accuracy of the measurements made with a simple hand pipette, we find that one and the same pipette will in repeated consecutive weighings of the contents, though filled and emptied each time with the minutest care, show differences up to 0·010 grm. for 10 c. c. capacity, up to 0·040 grm. for 50 c. c. capacity.

The accuracy of the measurements made with a pipette may be heightened by giving the instrument the form and construction shown in fig. 6, and fixing it to a holder.

It will be seen from the drawing that these pipettes are emptied only to a certain mark in the lower tube, and that they are provided with a *compression stop-cock*, a contrivance which we shall have occasion to describe in detail when on the subject of burettes. This contrivance reduces the differences of measurements with one and the same 50 c. c. pipette to 0·005 grm.

Pipettes are used more especially in cases where it is intended to estimate different constituents of a substance in separate portions of the same: for instance, 10 grm. of the substance under examination are dissolved in a 250 c. c. flask, the solution is diluted up to the mark, shaken, and 2, 3, or 4 several portions are then taken out with a 50 c. c. pipette. Each portion consists of $\frac{1}{5}$ part of the whole, and accordingly contains 2 grm. of the substance. Of course the pipette and the flask must be in perfect harmony. Whether they are may be ascertained by, for instance, emptying the 50 c. c. pipette 5 times into the 250 c. c. flask, and observing if the lower edge of the dark zone of fluid coincides with the mark. If it does not, you may make a fresh mark, which, no matter whether it is really correct or not, will bring the two instruments in question into conformity with each other.

Cylindrical pipettes, graduated throughout their entire length, may be used also to measure out any given quantities of liquid; however, these instruments can properly be employed only in processes where minute accuracy is not indispensable, as the limits of error in reading off the divisions in the wider part of the tube are not inconsiderable. For smaller quantities of liquid this inaccuracy may be avoided by making the pipettes of tubes of uniform width, having a small diameter only, and narrowed at both ends. (Fr. Mohr's measuring pipettes.)

When a fluid runs out of a pipette, drops sometimes remain here and there adhering to the tube; this arises from a film of fat on the inside; it may be removed by keeping the instrument some time filled with a solution of bichromate of potassa mixed with sulphuric acid.

bb. Measuring vessels which serve to measure out quantities of fluid at will.

4. *The Burette.*

Of the various forms and dispositions of this instrument, the following appear to me the most convenient:—

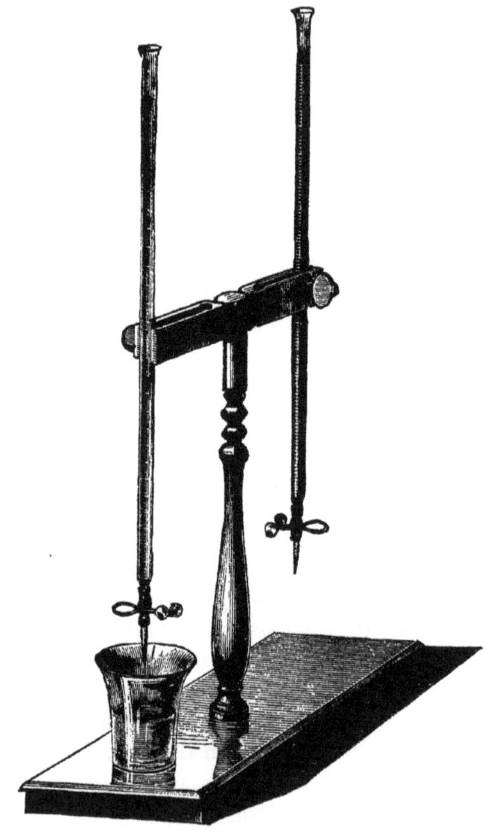

Fig. 7.

§ 21.

I. *Mohr's Burette*, (Compression cock burette).

For this excellent measuring apparatus, which is represented in fig. 7, we are indebted to FR. MOHR. It consists of a cylindrical tube, narrower towards the lower end for about an inch, with a slight widening, however, at the extreme point, in order that the caoutchouc connector may take a firm hold. I only use burettes of two sizes, viz., of 30 c. c., divided into $\frac{1}{10}$ c. c.; and of 50 c. c., divided into $\frac{1}{2}$ c. c. The former I employ principally in scientific, the latter chiefly in technical investigations. The usual length of my 30 c. c. burette is about 50 cm.; the graduated portion occupies about 49 cm. The diameter of the tube is accordingly about 10 mm. in the clear; the upper orifice is, for the convenience of filling, widened in form of a funnel, measuring 20 mm. in diameter; the width of the lower orifice is 5 mm. For very delicate processes, the length of the graduated portion may be extended to 50 or 52 cm., leaving thus intervals of nearly 2 mm. between the small divisional lines. In my 50 c. c. burettes the graduated portion of the tube is generally 40 cm. long.

To make the instrument ready for use, the narrowed lower end of the tube is warmed a little, and greased with tallow; a caoutchouc tube, about 30 mm. long, and having a diameter of 3 mm. in the clear, is then drawn over it; into the other end of this is inserted a tube of pretty thick glass, about 40 mm. long, and drawn out to a tolerably fine point; it is advisable to slightly widen the upper end of this tube also, and to cover it with a thin coat of tallow; and also to tie linen-thread, or twine, round both ends of the connector, to insure perfect tightness.

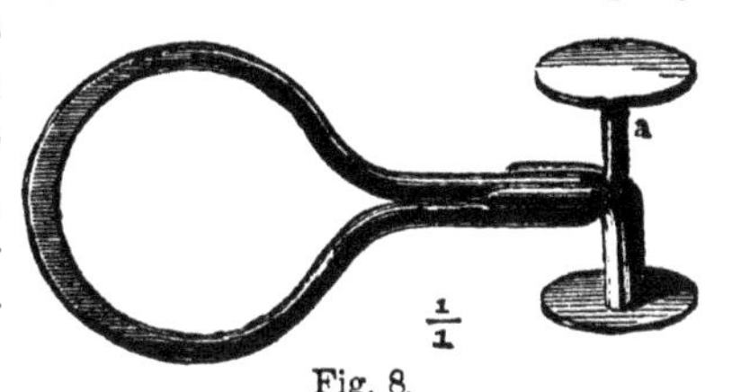

Fig. 8.

The space between the lower orifice of the burette and the upper orifice of the small delivery tube should be about 15 mm. The India rubber tube is now pressed together between the ends of the tubes by the compression-cock (or clip). This latter instrument is usually made outof brass wire; the form represented in fig. 8 was given by MOHR.

A good clip must pinch so tight that not a particle of fluid can make its way through the connector when compressed by it; it must be so constructed that the analyst may work it with perfect facility and exactness, so as to regulate the outflow of the liquid with the most rigorous accuracy, by bringing a higher or less degree of pressure to bear upon it.

For supporting MOHR's burettes, I use the holder represented in fig. 7; this instrument, whilst securely confining the tube, permits its being moved up and down with perfect freedom, and also its being taken out, without interfering with the compression-cock. The position of the burette must be strictly perpendicular, to insure which, care must be taken to have the grooves of the cork lining, which are intended to receive the tube, perfectly vertical, with the lower board of the stand in a horizontal position.

To charge the burette for a volumetrical operation, the point of the instrument is immersed in the liquid, the compression-cock opened, and a little liquid, sufficient at least to reach into the burette tube, sucked up by applying the mouth to the upper end; the cock is then closed, and the liquid poured into the burette until it reaches up to a little above the top mark. The burette having, if required, been duly adjusted in the proper vertical position, the liquid is allowed to drop out to the exact level of the top mark. The instrument is now ready for use. When as much liquid has flowed out as is required to attain the desired object, the analyst, before proceeding to read off the volume used, has to wait a few minutes, to give the particles of fluid adhering to the sides of the emptied portion of the tube proper time to run down. This is an indispensable part of the operation in accurate measurements, since, if neglected, an experiment in which the standard liquid in the burette is added slowly to the fluid under examination (in which, accordingly, the minute particles of fluid adhering to the glass have proper time afforded them during the operation itself to run down), will, of course, give slightly different results from those arrived at in another experiment, where the larger portion of the standand fluid is applied rapidly, and the last few drops alone are added slowly.

The *way* in which the *reading-off* is effected, is a matter of great importance in volumetric analysis; the first requisite is to bring the eye to a level with the top of the fluid. We must consequently settle the question—What is to be considered the top?

If you hold a burette, partly filled with water, between the eye and a strongly illumined wall, the surface of the fluid presents the appearance shown in fig. 10; if you hold close behind the tube a sheet of white paper, with a strong light falling on it, the surface of the fluid presents the appearance shown in fig. 9.

In the one as well as in the other case, you have to read off at the lower border of the dark zone, this being the most distinctly marked line. Fr. Mohr recommends the following device for reading-off:—Paste on a sheet of very white paper a broad strip of black paper, and, when reading-off, hold this close behind the burette, in a position to place the border line between white and black from 2 to 3 mm. below the lower border of the dark zone, as shown in fig. 11; read-off at the lower border of the dark zone.

Great care must be taken to hold the paper invariably in the same position, since, if it be held lower down, the lower border of the black zone will move higher up.

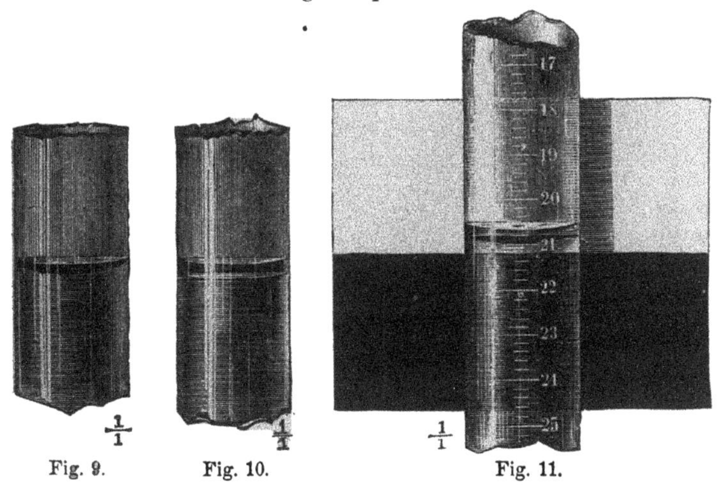

Fig. 9. Fig. 10. Fig. 11.

I prefer to read-off in a light which causes the appearance represented in fig. 9.

By the use of Erdmann's float * all uncertainties in reading-off may be avoided. Fig. 12 represents a burette thus provided. In this case we always read off the degree of the burette which coincideswith the circle in the middle of the float. The float must beso fitted to the width of the burette that when placed in the filled burette, it will, on allowing the fluid to run out gradually, sink down with the same without wavering, and when it has been pressed down into the fluid of the closed burette, it will slowly rise again. The weight of the float must, if necessary, be so regu-

* Journ. f. prakt. Chem. 71, 194.

lated by mercury that when placed in the filled tube it may cut the fluid with its top uniformly all round. A further important condition of the float is that its axis should coincide as nearly as possible with that of the burette tube, so that the division-mark on the burette may be always parallel with the circular line on the float.

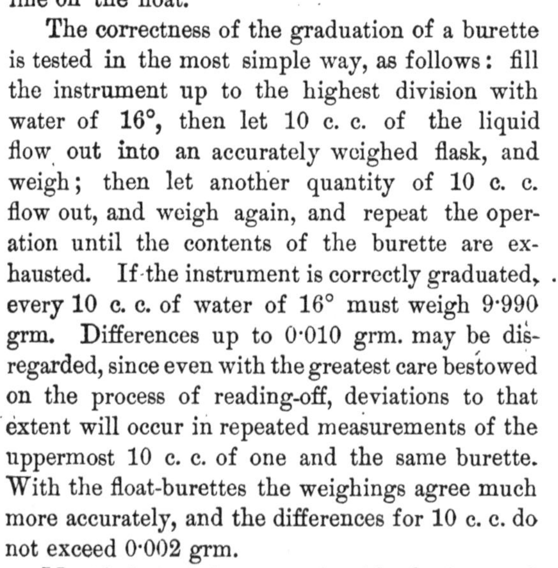
Fig. 12.

The correctness of the graduation of a burette is tested in the most simple way, as follows: fill the instrument up to the highest division with water of 16°, then let 10 c. c. of the liquid flow out into an accurately weighed flask, and weigh; then let another quantity of 10 c. c. flow out, and weigh again, and repeat the operation until the contents of the burette are exhausted. If the instrument is correctly graduated, every 10 c. c. of water of 16° must weigh 9·990 grm. Differences up to 0·010 grm. may be disregarded, since even with the greatest care bestowed on the process of reading-off, deviations to that extent will occur in repeated measurements of the uppermost 10 c. c. of one and the same burette. With the float-burettes the weighings agree much more accurately, and the differences for 10 c. c. do not exceed 0·002 grm.

Mohr's burette is unquestionably the best and most convenient instrument of the kind, and ought to be employed in the measurement of all liquids which are not injuriously affected by contact with caoutchouc. Of the standard solutions used at present in volumetric analysis, that of permanganate of potassa alone cannot bear contact with caoutchouc.

§ 22.

II. *Gay-Lussac's Burette.*

Fig. 13 represents this instrument in, as I believe, its most practical form.

I make use of two sizes, one of 50 c. c. divided into ½ c. c., the other of 30 c. c. divided into $\frac{1}{10}$ c. c. The former is about 33 cm. long; the graduated portion occupies about 25 cm.; the internal diameter of the wide tube measures 15 mm.; that of the narrow tube 4 mm., which in the upper bent end gradually decreases to 2 mm. The graduated portion of the smaller burette is about 28 cm. long, and has accordingly an internal diameter of about 11 mm.

The stand which I make use of to rest my burettes in, consists of a disk of solid wood, from 5 to 6 cm. high, and from 10 to 12 cm. in diameter, with holes made with the auger and chisel, of proper size to receive the bottom part of the burettes.

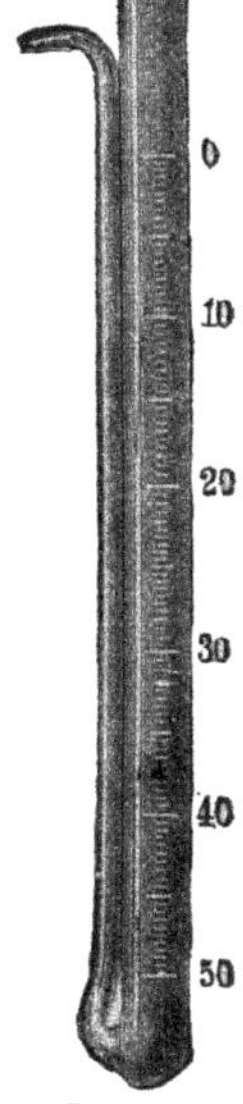

Fig. 13.

To complete the instrument, MOHR suggests the use of a perforated cork, bearing a short glass tube bent at aright angle. The cork being inserted into the mouth of the wide tube, a piece of caoutchouc is drawn over the short glass tube; by blowing into this with greater or less force, the outflow of the liquid from the spout of the slightly slanting burette may be regulated at pleasure.

The reading-off of the height of the liquid is effected in the same way as explained in § 21. I prefer, however, placing the burette firmly against a perpendicular partition, either a strongly illumined door, or the pane of a window, to insure the vertical position of the instrument. It is only when operating with more highly concentrated, and accordingly opaque solutions of permanganate of potassa, that the method of reading-off requires modification; in that case, the upper border of the liquid is noted; and the best way is to place the burette against a white background, and read off by reflected light.

§ 23.

III. *Geissler's Burette.*

In this instrument, which is represented in fig. 14, the narrow tube is placed inside the wide tube instead of outside, as in GAY-LUSSAC's burette. The part of the inner tube projecting beyond the wide tube is thick in the glass; whilst the part inside, which is of the same inside width, is made of very thin glass.

This is a very convenient instrument, and less liable to fracture than GAY-LUSSAC's burette.

II. Preliminary Operations.—Preparation of Substances for the Processes of Quantitative Analysis.

§ 24.

1. The Selection of the Sample.

Before the analyst proceeds to make the quantitative analysis of a body, he cannot too carefully consider whether the desired result is fully attained if he simply knows the respective quantity of every individual constituent of that body. This primary point is but too frequently disregarded, and thus false impressions are made, even by the most careful analysis. This remark applies both to scientific and to technical investigations.

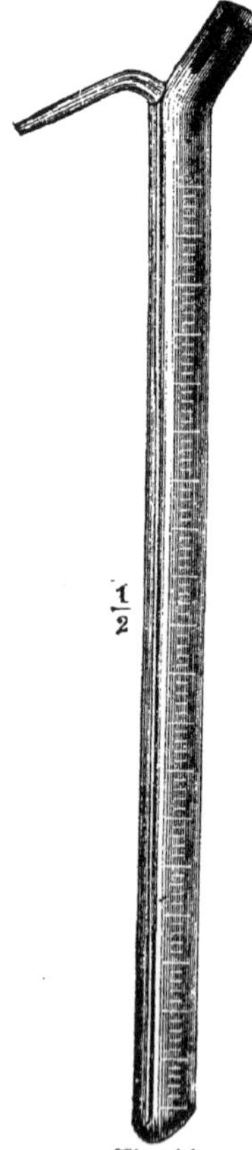

Fig. 14.

Therefore, if you have to determine the constitution of a mineral, take the greatest possible care to remove in the first place every particle of gangue, and disseminated impurities; remove any adherent matter by wiping or washing, then wrap the substance up in a sheet of thick paper, and crush it to pieces on a steel anvil; and pick out with a pair of small pincers the cleanest pieces. Crystalline substances, prepared artificially, ought to be purified by recrystallization; precipitates by thorough washing, &c., &c.

In technical investigations,—when called upon, for instance, to determine the amount of peroxide presentin a manganese ore, or the amount of iron present in an iron ore,—the first point for consideration ought to be whether the samples selected correspond as much as possible to the average quality of the ore. What would it serve, indeed, to the purchaser of a manganese mine to know the amount of peroxide present in a select, possibly particularly rich, sample?

These few observations will suffice to show that no universally applicable and valid rules to guide the analyst in the selection of the sample can be laid down; he must in every individual case, on the one hand, examine the substance carefully, and more particularly also under the microscope, or through a lens; and, on the other hand, keep clearly in view the *object* of the investigation, and then take his measures accordingly.

§ 25.

2. MECHANICAL DIVISION.

In order to prepare a substance for analysis, *i.e.*, to render it accessible to the action of solvents or fluxes, it is generally indispensable, in the first place, to divide it into minute parts, since this will create abundant points of contact for the solvent, and will counteract, and, as far as practicable, remove the adverse influences of the power of cohesion, thus fulfilling all the conditions necessary to effect a complete and speedy solution.

The means employed to attain this object vary according to the nature of the different bodies we have to operate upon. In many cases, simple crushing or pounding is sufficient; in other cases it is necessary to reduce the powder to the very highest degree of fineness, by sifting or by elutriation.

The operation of powdering is conducted in mortars; the first and most indispensable condition is, that the material of the mortar be considerably harder than the substance to be pulverized, so as to prevent, as far as practicable, the latter from being contaminated with any particles of the former. Thus, for pounding salts and other substances possessing no very considerable degree of hardness, porcelain mortars may be used, whilst the pounding of harder substances (of most minerals, for instance,) requires vessels of agate, chalcedony, or flint. In such cases, the larger pieces are first reduced to a coarse powder; this is best effected by wrapping them up in several sheets of writing-paper, and striking them with a hammer upon a steel or iron plate; the coarse powder thus obtained is then pulverized, in small portions at a time, in an agate mortar, until it is reduced to the state of an impalpable powder. If we have but a small portion of a mineral to operate upon, and indeed in all cases where we are desirous of avoiding loss, it is advisable to use a steel mortar (fig. 15) for the preparatory reduction of the mineral to coarse powder.

ab and *cd* represent the two parts of the mortar; these may be readily taken asunder. The substance to be crushed (having, if practicable, first been broken into small pieces), is placed in the cylindrical chamber *ef*; the steel cylinder, which fits somewhat loosely into the chamber, serves as pestle. The mortar is placed upon a solid support, and perpendicular blows are repeatedly struck upon the pestle with a hammer until the object in view is attained.

Fig. 15.

Minerals which are very difficult to pulverize should be strongly ignited, and then suddenly plunged into cold water, and subsequently again ignited. This process is of course applicable only to minerals which lose no essential constituent on ignition, and are perfectly insoluble in water.

In the purchase of agate mortars, especial care ought to be taken that they have no palpable cracks or indentations; very slight cracks, however, that cannot be felt, do not render the mortar useless, although they impair its durability.

Minerals insoluble in acids, and which consequently require fusing, must especially be finely divided, otherwise we cannot calculate upon complete decomposition. This object may be obtained either by triturating the pounded mineral with water, or by elutriation, or by sifting; the two former processes, however, can be resorted to only in the case of substances which are not attacked by water. It is quite clear that analysts must in future be much more cautious in this point than has hitherto been the case, since we know now that many substances which are usually held to be insoluble in water are, when in a state of minute division, strongly affected by that solvent; thus, for instance, water, acting upon some sorts of finely pulverized glass, is found to rapidly dissolve from 2 to 3 per cent. of powder even in the cold. (PELOUZE.*) Thus, again, finely divided feldspar, granite, trachyte and porphyry give up to water both alkali and silica. (H. LUDWIG.†)

Trituration with water (*levigation*). Add a little water to the pounded mineral in the mortar, and triturate the paste until all crepitation ceases, or, which is a more expeditious process, transfer

* Compt. Rend., t. xliii. pp. 117–123. † Archiv der Pharm. 91, 147.

the mineral paste from the mortar to an agate or flint slab, and triturate it thereon with a muller. Rinse the paste off, with the washing bottle, into a smooth porcelain basin of hemispheric form, evaporate the water on the water-bath, and mix the residue most carefully with the pestle. (The paste may be dried also in the agate mortar, but at a very gentle heat, since otherwise the mortar might crack.)

To perform the process of *elutriation*, the pasty mass, having first been very finely triturated with water, is washed off into a beaker, and stirred with distilled water; the mixture is then allowed to stand a minute or so, after which the supernatant turbid fluid is poured off into another beaker. The sediment, which contains the coarser parts, is then again subjected to the process of trituration, etc., and the same operation repeated until the whole quantity is elutriated. The turbid fluid is allowed to stand at rest until the minute particles of the substance held in suspension have subsided, which generally takes many hours. The water is then finally decanted, and the powder dried in the beaker.

The process of *sifting* is conducted as follows: a piece of fine, well-washed, and thoroughly dry linen is placed over the mouth of a bottle about 10 cm. high, and pressed down a little into the mouth, so as to form a kind of bag; a portion of the finely triturated substance is put into the bag, and a piece of soft leather stretched tightly over the top by way of cover. By drumming with the finger on the leather cover, a shaking motion is imparted to the bag, which makes the finer particles of the powder gradually pass through the linen. The portion remaining in the bag is subjected again to trituration in an agate mortar, and, together with a fresh portion of the powder, sifted again; and the same process is continued until the entire mass has pass through the bag into the glass.

When operating on substances consisting of different compounds it would be a grave error indeed to use for analysis the powder resulting from the first process of elutriation or sifting, since this will contain the more readily pulverizable constituents in a greater proportion to the more resisting ones than is the case with the original substance.

Great care must, therefore, also be taken to avoid a loss of substance in the process of elutriation or sifting, as this loss is likely to be distributed unequally among the several component parts.

In cases where it is intended to ascertain the average composition of a heterogeneous substance, of an iron ore for instance, a large average sample is selected, and reduced to a coarse powder; the latter is thoroughly intermixed, a portion of it powdered more finely, and mixed uniformly, and finally the quantity required for analysis is reduced to the finest powder. The most convenient instrument for the crushing and coarse pounding of large samples of ore, &c., is a steel anvil and hammer. The anvil in my own laboratory consists of a wood pillar, 85 cm. high and 26 cm. in diameter, into which a steel plate, 3 cm. thick and 20 cm. in diameter, is let to the depth of one-half of its thickness. A brass ring, 5 cm. high, fits round the upper projecting part of the steel plate. The hammer, which is well steeled, has a striking surface of 5 cm. diameter. An anvil and hammer of this kind afford, among others, this advantage, that their steel surfaces admit most readily of cleaning. To convert the coarse powder into a finer, a smooth-turned steel mortar of about 130 mm. upper diameter and 74 mm. deep is used—the final trituration is conducted in an agate mortar.

§ 26.

3. Drying.

Bodies which it is intended to analyze quantitatively must be, when weighed, in a definite state, in a condition in which they can be always obtained again.

Now, the essential constituents of a substance are usually accompanied by an unessential one, viz., a greater or less amount of water, enclosed either within its lamellæ, or adhering to it from the mode of its preparation, or absorbed by it from the atmosphere. It is perfectly obvious that to estimate correctly the quantity of a substance, we must, in the first place, remove this variable amount of water. *Most solid bodies*, therefore, *require to be dried before they can be quantitatively analyzed.*

The operation of drying is of the very highest importance for the correctness of the results; indeed it may safely be averred that many of the differences observed in analytical researches proceed entirely from the fact that substances are analyzed in different states of moisture.

Many bodies contain, as is well known, water which is proper to them either as inherent in their constitution or as so-called water of crystallization. In contradistinction to this, we will employ the term *moisture* to designate that variable adherent or mechanically enclosed water, with the removal of which the operation of drying in the sense here in view is alone concerned.

In the drying of substances for quantitative analysis, our object is to remove all moisture, without interfering in the slightest degree with combined water or any other constituent of the body. To accomplish this object, it is absolutely requisite that we should know the properties which the substance under examination manifests in the dry state, and whether it loses water or other constituents at a red heat, or at 100°, or in dried air, or even simply in contact with the atmosphere. These data will serve to guide us in the selection of the process of desiccation best suited to each substance.*

The following classification may accordingly be adopted:—

a. Substances which lose water even in simple contact with the atmosphere; such as sodium sulphate, crystallized sodium carbonate, etc. Substances of this kind turn dull and opaque when exposed to the air, and finally crumble wholly or partially to a white powder. They are more difficult to dry than many other bodies. The process best adapted for the purpose, is to press the pulverized salts with some degree of force between thick layers of fine white blotting-paper, repeating the operation with fresh paper until the last sheets remain *absolutely dry.*

It is generally advisable in the course of this operation to repowder the salt.

b. Substances which do not yield water to the atmosphere (*unless it is perfectly dry*), *but effloresce in artificially dried air;* such as magnesium sulphate, sodium potassium tartrate (Rochelle salt), &c. Salts of this kind are reduced to powder, which, if it be very moist, is pressed between sheets of blotting-paper, as in *a;* after this operation, it must be allowed to remain for some time spread in a thin layer upon a sheet of blotting-paper, effectually protected against dust, and shielded from the direct rays of the sun.

* The dried substance should always at once be transferred to a well-closed vessel; glass tubes, sealed at one end, and of sufficiently thick glass to bear the firm insertion of tight-fitting smooth corks—weighing-tubes—are usually employed for this purpose.

§ 27.

c. Substances which undergo no alteration in dried air, but lose water at 100°; calcium tartrate, for instance. These are finely pulverized; the powder is put in a thin layer into a watch-glass or shallow dish, and the latter placed inside a chamber in which the air is kept dry by means of sulphuric acid. This process is usually conducted in one of the following apparatuses, which are termed *desiccators*, and subserve still another purpose besides that of drying, viz., that of allowing hot crucibles, dishes, etc., to cool in dry air.

In fig. 16, *a* represents a glass plate (ground-glass plates answer the purpose best), *b*, a bell jar with ground rim, which is greased with tallow; *c* is a glass basin with sulphuric acid; *d*, a round iron

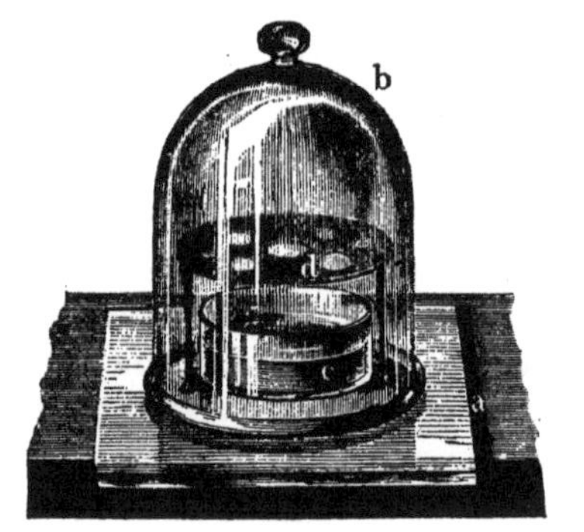

Fig. 16.

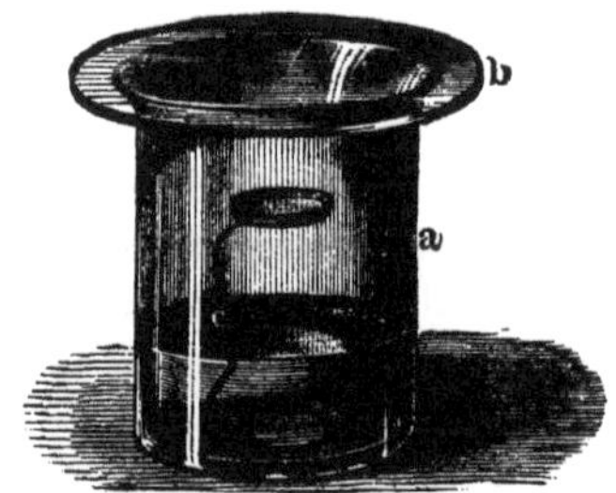

Fig. 17.

plate, supported on three feet, with circular holes of various sizes, for the reception of the watch-glasses, crucibles, etc., containing the substance.

In fig. 17, *a* represents a beaker with ground and greased rim, and filled to one-fourth or one-third with concentrated sulphuric acid; *b* is a ground-glass plate; *c* is a bent wire of lead, which serves to support the watch-glass containing the substance.

Fig. 18 represents a readily portable desiccator, used more particularly to receive crucibles in course of cooling, and carry them to the balance. The instrument consists of a box made of strong glass; the lid must be ground to shut air-tight; the place on which it joins is greased with tallow. The outer diameter of my boxes is 105 mm.; the sides are 6 mm. thick. The aperture has a diameter of 80 mm.; the box up to the small part is 65 mm. high; the lid has the same height; the small part itself is 15 mm. high, and ground to a slightly conical shape. A brass ring, with rim, fits exactly into the aperture; the rim must not project beyond the glass. The ring bears a triangle of iron, or, better, platinum wire, intended for the reception of crucibles, &c.

Fig. 18.

The body which it is intended to dry is kept exposed to the action of the dry air in the glass, until it shows no further diminution of weight. Substances upon which the oxygen of the air exercises a modifying influence are dried in a similar manner, under the exhausted receiver of an air-pump. Substances which, though losing no water in dry air, yet give off ammonia, are dried over quicklime, mixed with some chloride of ammonium in powder, and consequently in an anhydrous ammoniacal atmosphere.

§ 28.

d. Substances which at 100° *completely lose their moisture, without suffering any other alteration*, such as hydrogen potassium tartrate, sugar, etc. These are dried in the water-bath; in the case of slow-drying substances, or where it is wished to expedite the operation, with the aid of a current of dry air.

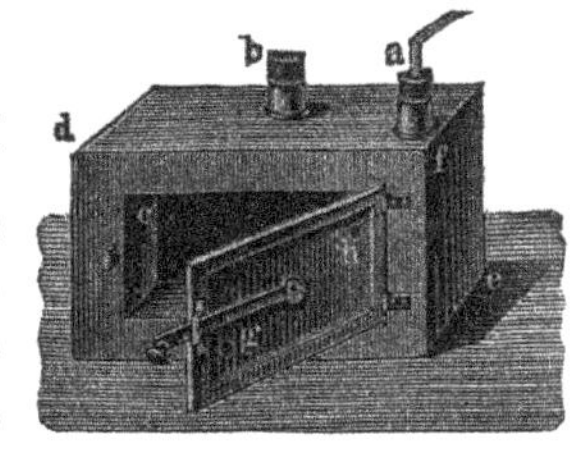

Fig. 19.

Fig. 19 represents the water-bath most commonly used. It is made of sheet copper. The engraving renders a detailed description unnecessary. The inner chamber, *c*, is surrounded on five sides by the outer case or jacket, *d e*, without communicating with it. The object of the apertures *g* and *h* is to effect change of air, which purpose they answer sufficiently well. When it is intended to use the apparatus, the outer case is filled to about one-half with rain-water, and the aperture *a* is closed with a perforated cork, into which a glass tube is fitted; the aperture *b* is entirely closed. If the apparatus is intended to be heated over charcoal, it should have a length of about 20 cm. from *d* to *f*; but if over a gas-, spirit-, or oil-lamp, it should be only about 13 cm. long. In the former case, the inner chamber is 17 cm. deep, 14 cm. broad, and 10 cm. high; in the latter case, it is 10 cm. deep, 9 cm. broad, and 6 cm. high. The temperature in the inner chamber never quite reaches 100°; to bring it up to 100°, F. ROCHLEDER has suggested to close *b* with a double-limbed tube, the outer longer limb of which dips into a cylinder filled with water; *a* is in that case closed with a perforated cork bearing a sufficiently tall funnel tube, which fits air-tight in the cork. The lower end of this tube reaches down to one inch from the bottom.

In large analytical laboratories water is usually kept boiling all day long, for the production of distilled water. The boilers used in my own laboratory have the shape of somewhat oblong square boxes, about 120 cm. long, 60 cm. broad, and 24 cm. high; the front of the boiler has soldered into it, one above the other, two rows of drying chambers, of the kind shown in fig. 19. This gives so many ovens that almost every student may have one for his special use. Most of these ovens are from 11 to 12 cm. deep and broad, and 8 cm. high; some of them, however, are 16 cm. deep and broad, to enable them to receive large-sized dishes. The substances to be dried are usually put on double watch-glasses, laid one within the other, which are placed in the oven, and the door is then closed. In the subsequent process of weighing, the upper glass, which contains the substance, is covered with the lower one. The glasses must be quite cold before they are placed on the scale. In cases where we have to deal with hygroscopic substances, the reabsorption of water upon cooling is prevented by the selection of close-fitting glasses, which are held tight together by a clasp (fig. 20), and allowed to cool with their contents under a bell-glass over sulphuric acid (see fig. 16). These latter instructions

Fig. 20.

apply equally to the process of drying conducted in other apparatus.

The clasp used for keeping the watch-glasses pressed together —and which in all cases where it is intended to ascertain the loss of weight which a substance suffers on desiccation, is to be looked upon as belonging to the glasses, and must accordingly be weighed with them—is constructed of two strips of thin brass plate, about 10 cm. long, and 1 cm. wide, which are laid the one over the other, and soldered together at the ends, to the extent of 5 to 6 mm.

The following apparatus (fig. 21) serves for drying substances in a current of air :—

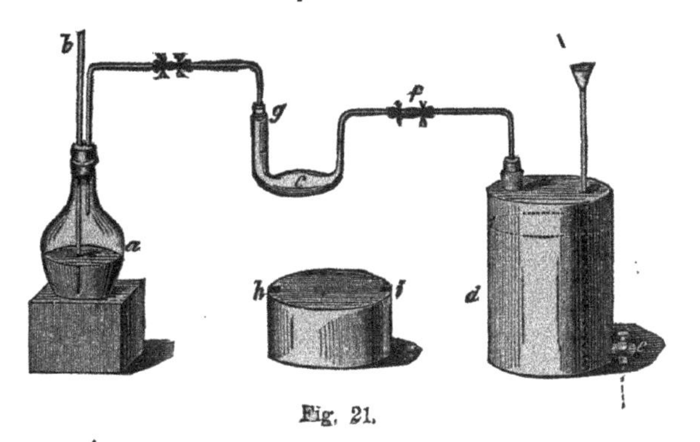

Fig. 21.

a represents a flask filled to one-third with concentrated sulphuric acid; *c* a glass vessel (commonly called a Liebig's drying-tube), and *d* a tin vessel provided with a stop-cock at *e*, and arranged in other respects as the cut shows.

h, *i*, represents a small tin vessel, containing water and covered with a lid; two apertures are cut into the border of the latter, to receive the ascending limbs of *c*.

The tube *c* is first weighed with the substance, then placed in the water-bath, *h*, *i*, which is placed over a spirit- or gas-lamp; the aspirator *d* is then filled with water, and *c* connected with the flask *a* by the perforated cork *g*, and with *d* by means of a caoutchouc tube *f*. If the stop-cock *e* be now opened so as to cause the water to drop from *d*, the air will pass through the tube *b*, and after being dehydrated by the sulphuric acid, will pass over the heated substance in *c*. After the operation has been continued for some time, it is interrupted for the purpose of weighing the tube c and its contents, and then resumed again, and continued until the weight of c (and its contents) remains stationary. The current of cold air exercising its constant cooling action upon the substance, the latter never really reaches 100°. It is, therefore, sometimes advisable to substitute for the water in the bath a saturated solution of common salt.

With this substitution, the apparatus represented in fig. 21 will be found to effect its purpose the most expeditiously. It is not adapted, however, for drying such substances as have a tendency to fuse or agglutinate at 100°.

§ 29.

e. Substances which persistently retain moisture at 100°, *or become completely dry only after a very long time; but which are decomposed by a red heat.*

The desiccation of such substances is effected in the air-bath or oil-bath, the temperature being raised to 110–120°, and still higher, and, according to circumstances, with or without application of a current of air, carbon dioxide, or hydrogen.

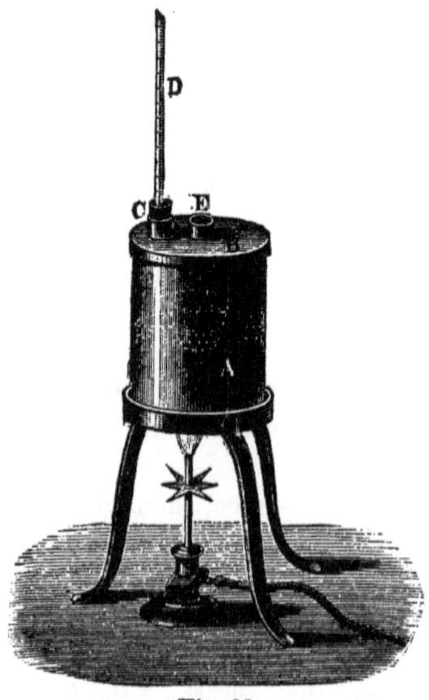

Fig. 22.

Figs. 22 and 23 represent two air-baths of simple construction; the former (fig. 22) adapted for the desiccation of a single substance, the latter suited for the simultaneous drying of several substances.

In fig. 22, A is a box of strong sheet copper, about 11 cm. high, and 9 cm. in diameter. The box is closed with the loose-fitting cover B, which is provided with a narrow rim, and has two apertures, C and E; C is intended to receive the thermometer D, which is fitted into it by a perforated cork, E affords an exit to the aqueous vapors, and is, according to circumstances, either left open, or loosely closed. In the interior of the box, about half-way up, are fixed three pins, supporting a triangle of moderately stout wire, upon which the crucible with the substance is placed uncovered. The bulb of the thermometer approaches the crucible as closely as possible, but without touching the triangle. The heating is effected by means of a gas- or spirit-lamp. When the apparatus has cooled sufficiently to allow its being laid hold of without inconvenience, the lid is removed, the crucible, which is still warm, taken out, covered, and allowed to cool in a desiccator; and weighed when cold.

In fig. 23' *a b* is a case of strong sheet copper, with riveted or locked joints, of a width and depth of 15 to 20 cm., and corresponding height. The aperture *c* is intended to receive a perforated cork, into which is fixed a thermometer, *d*, which reaches into the interior of the case; within is a shelf, on which are placed the watch-glasses with the substances to be dried. The case is heated by means of a gas-, spirit-, or oil-lamp. When the temperature has once reached the intended point, it is easy to maintain it pretty constant, by regulating the flame.* In order to limit as much as possible the cooling from without, it is advisable to put over the whole apparatus a pasteboard hood with a movable front.

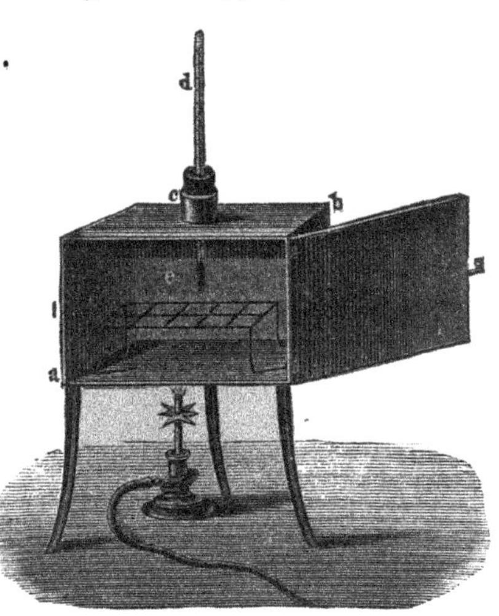

Fig. 23.

[The air-bath, fig. 23, by a slight alteration, may serve for desiccating in a stream of dry air. For this purpose, cut a circular orifice, 35 mm. wide, in each end of the copper chamber, and rivet over each orifice a copper tube or ring of corresponding diameter, and 25 mm. long. Fit a glass tube of 20 mm. diameter, by means of perforated corks, into these openings, so that it shall traverse the chamber and project 40–50 mm. beyond the corks at each end.

* With a gas-lamp, Kemp's regulator improved by Bunsen, may advantageously be used to obtain constant temperatures.

The copper tubes should be so adjusted that the glass tube shall stand horizontally in the chamber, at the same height as the thermometer bulb and just behind it. To produce the current of dry air one of the projecting ends of the wide tube is connected by a narrow glass tube and perforated cork, with an aspirator as in fig. 21, the other with a large calcium chloride tube; the water of the aspirator is allowed to run off somewhat rapidly at first, more slowly afterwards. The end of the tube that delivers the air into the wide tube is recurved, so that the substance within shall not be carried away in the current.

The substance to be dried is weighed out in a tray of platinum or porcelain, fig. 24, which is pushed within the wide glass tube by help of a wire. When the substance is hygroscopic, the tray is placed horizontally within a test-tube, which is corked while the weight is being ascertained. The substance and tray, after drying, may be cooled in the same test-tube; in that case just before putting on the balance, the cork should be removed momentarily to allow the tube to fill with air.]

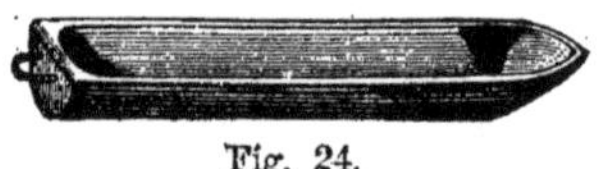

Fig. 24.

§ 30.

The copper apparatus represented in fig. 19, when made with brazed joints, can be employed also as a paraffine-bath; when used for that purpose, the outer case is filled to two-thirds with paraffine. To note the temperature, a thermometer is inserted, by means of a perforated cork, in the aperture *a*; with the bulb reaching nearly to the bottom, or, at all events, entirely immersed in the paraffine.

Many organic substances, when dried at a somewhat high temperature, suffer alteration by the action of the atmospheric oxygen. In the desiccation of such substances, oxygen must accordingly be excluded.

[The drying of such bodies is conducted as just described in the modified air-bath, but in a stream of dried and purified hydrogen or carbonic acid (see § 29). The gas is evolved from a self-regulating generator (see fig. 50), § 108.

§ 31.

f. Substances which suffer no alteration at a red heat, such as barium sulphate, pearlash, etc., are very readily freed from moisture. They need simply be heated in a platinum or porcelain crucible over a gas or spirit-lamp until the desired end is attained. The crucible, having first been allowed to cool a little, is put, still hot, under a desiccator, and finally weighed when cold.

III. General Procedure in Quantitative Analyses.

§ 32.

It is important, in the first place, to observe that we embrace in the following general analytical method only the separation and determination of the metals and their combinations with the metalloids, and of the inorganic acids and salts. With respect to the quantitative analysis of other compounds, it is not easy to lay down a universally applicable method, except that their constituents usually require to be converted first into acids or bases, before their separation and estimation can be attempted; this is the case, for instance, with phosphorus sulphide, sulphur chloride, iodine chloride, nitrogen sulphide, &c.

The quantitative analysis of a substance presupposes an accurate knowledge of the properties of the same, and of the nature of its several constituents. These data will enable the operator at once to decide whether the direct estimation of each individual constituent is necessary; whether he need operate only on one portion of the substance, or whether it would be advantageous to determine each constituent in different portions. Let us suppose, for instance, we have a mixture of sodium chloride and anhydrous sodium sulphate, and wish to ascertain the proportion in which these two substances are mixed. Here it would be superfluous to determine each constituent directly, since the determination either of the quantity of the chlorine, or of the sulphuric acid, is quite sufficient to answer the purpose; still the estimation of both the chlorine and the sulphur trioxide will afford us an infallible control for the correctness of our analysis; since the united weights of these two substances, added to the sodium and soda respectively equivalent to them, must be equal to the weight of the substance taken.

These estimations may be made, either in one and the same portion of the mixture, by first precipitating the sulphuric acid with barium nitrate, and subsequently the hydrochloric acid from the filtrate with solution of silver nitrate; or a separate portion of the mixture may be appropriated to each of these two operations. Unless there is some objection to its use (*e.g.*, deficiency or heterogeneousness of substance), the latter method is more convenient and generally yields more accurate results; since, in the former method, the unavoidable washing of the first precipitate swells the amount of liquid so considerably that the analysis is thereby delayed, and, moreover, loss of substance less easily guarded against.

Before beginning all analyses, at least those of a more complex nature, the student should write out an exact plan, and accurately note on paper, during the entire process, everything that he does. It is in the highest degree unwise to rely on the memory in a complicated analysis. When students, who imagine they can do so, come, a week or a fortnight after they have begun their analysis, to work out the results, they find generally too late that they have forgotten much, which now appears to them of importance to know. The intelligent pursuit of chemical analysis consists in the projecting and accurate testing of the plan; acuteness and the power of passing in review all the influencing chemical relations must here support each other. He who works without a thoroughly thought-out plan, has no right to say he is practising chemistry; for a mere unthinking stringing together of a series of filtrations, evaporations, ignitions, and weighings, howsoever well these several operations may be performed, is not chemistry.

We will now proceed to describe the various operations constituting the process of quantitative analysis.

§ 33.

1. Weighing the Substance.

The amount of matter required for the quantitative analysis of a substance depends upon the nature of its constituents; it is, therefore, impossible to lay down rules for guidance on this point. Half a gramme of sodium chloride, and even less, is sufficient to effect the estimation of the chlorine. For the quantitative analysis of a mixture of common salt and anhydrous sodium sulphate, 1 gramme will suffice; whereas, in the case of ashes of plants, complex minerals, &c., 3 or 4 grammes, and even more, are required. 1 to 3 grm. can therefore be indicated as the average quantity suitable in most cases. For the estimation of constituents present in very minute proportions only, as, for instance, sodium and potassium in limestones, phosphorus or sulphur in cast-iron, &c., much greater quantities are often required—10, 20, or 50 grammes.

The greater the amount of substance taken the more accurate will be the analysis; the smaller the quantity, the sooner, as a rule, will the analysis be finished. We would advise the student to endeavor to combine accuracy with economy of time. The less substance he takes to operate upon, the more carefully he ought to weigh; the larger the amount of substance, the less harm can result from slight inaccuracies in weighing. Somewhat large quantities of substance are generally weighed to 1 milligramme; minute quantities, to $\frac{1}{10}$ of a milligramme.

If *one* portion of a substance is to be weighed off, we first weigh two watch-glasses which fit on each other, or else an empty platinum crucible with lid, then we put some substance in, and weigh again; the difference between the two weighings gives the weight of the substance taken.

If *several* quantities of a substance are to be operated upon, the best way is to weigh off the several portions successively; which may be accomplished most readily by weighing in a glass tube, or other appropriate vessel, the whole amount of substance, and then shaking out of the tube the quantities required one after another into appropriate vessels, weighing the tube after each time.

The work may often also be materially lightened, by weighing off a larger portion of the substance, dissolving this to $\frac{1}{4}$, $\frac{1}{2}$ or 1 litre, and taking out for the several estimations aliquot parts, with the 50 or 100 c.c. pipette. The first and most essential condition of this proceeding, of course, is that the pipettes must accurately correspond with the measuring flasks (§§ 18 and 20).

§ 34.

2. Estimation of the Water.

If the substance to be examined—after having been freed from moisture by a suitable drying process (§§ 26–32)—contains water, it is usual to begin by determining the amount of this water. This operation is generally simple; in some instances, however, it has its difficulties. This depends upon various circumstances, viz., whether the compounds intended for analysis yield their water readily or not; whether they can bear a red heat without suffering decomposition; or whether, on the contrary, they give off other volatile substances, besides water, even at a lower temperature.

The correct knowledge of the constitution of a compound depends frequently upon the accurate estimation of the water contained in it; in many cases—for instance, in the analysis of the salts of known acids—the estimation of the water contained in the analyzed compound suffices to enable us to deduce the formula. The estimation of the water contained in a substance is, therefore, one of the most important, as well as most frequently occurring operations of quantitative analysis. The proportion of water contained in a substance may be determined in two ways, viz., *a*, from the diminution of weight consequent upon the expulsion of the water; *b*, by weighing the amount of water expelled.

§ 35.

a. Estimation of the Water from the Loss of Weight.

This method, on account of its simplicity, is most frequently employed. The *modus operandi* depends upon the nature of the substance under examination.

α. The substance bears ignition without losing other Constituents besides Water, and without absorbing Oxygen.

The substance is weighed in a platinum or porcelain crucible, and placed over the gas- or spirit-lamp; the heat should be very gentle at first, and gradually increased. When the crucible has been maintained some time at a red heat, it is allowed to cool a little, put still warm under the desiccator, and finally weighed when cold. The ignition is then repeated, and the weight again ascertained. If no further diminution of weight has taken place, the process is at at end, the desired object being fully attained. But if the weight is less than after the first heating, the operation must be repeated until the weight remains constant.

In the case of silicates, the heat must be raised to a very high degree, since many of them (*e.g.* talc, steatite, nephrite) only begin at a red heat to give off water, and require a yellow heat for the complete expulsion of that constituent. (TH. SCHEERER.*) Such bodies are therefore ignited over a blast-lamp.

In the case of substances that have a tendency to puff off, or to spirt, a small flask or retort may sometimes be advantageously substituted for the crucible. Care must be taken to remove the last traces of aqueous vapor from the vessel, by suction through a glass tube.

Decrepitating salts (sodium chloride, for instance) are put—finely pulverized, if possible—in a small covered platinum crucible, which is then placed in a large one, also covered; the whole is weighed, then heated, gently at first for some time, then more strongly; finally, after cooling, weighed agaiu.

β. The substance loses on ignition other Constituents besides Water (Boracic Acid, Sulphuric Acid, Silicon Fluoride, &c.).

Here the analyst has to consider, in the first place, whether the water may not be expelled at a lower degree of heat, which does not involve the loss of other constituents. If this may be done, the substance is heated either in the water-bath, or where a higher temperature is required, in the air-bath or oil-bath, the temperature being regulated by the thermometer. The expulsion of the water may be promoted by the co-operation of a current of air (compare §§ 29 and 30); or by the addition of pure dry sand to the substance, to keep it porous.† The process must be continued under these circumstances also, until the weight remains constant.

In cases where, for some reason or other, such gentle heating is insufficient, the analyst has to consider whether the desired end may not be attained at a red heat, by adding some substance that will retain the volatile constituent whose loss is apprehended. Thus, for instance, the crystallized sulphate of alumina loses at a red heat, besides water, also sulphuric acid; now, the loss of the latter constituent may be guarded against by adding to the sulphate an excess (about six times the quantity) of finely pulverized, recently ignited, pure lead oxide. But the addition of this substance will not prevent the escape of silicon fluoride from silicates when exposed to a red heat (LIST‡).

* Jahresber. von Liebig u. Kopp, 1851, 610.
† Ann. d. Chem. u. Pharm. 53, 233. ‡ Ibid. 81, 189.

Thus again, the amount of water in commercial iodine may be determined by triturating the iodine together with eight times the quantity of mercury, and drying the mixture at 100° (BOLLEY*).

γ. *The substance contains several differently combined quantities of Water which require different Degrees of Temperature for Expulsion.*

Substances of this nature are heated first in the water-bath, until their weight remains constant; they are then exposed in the oil- or air-bath to 150°, 200°, or 250°, &c., and finally, when practicable, ignited over a gas- or spirit-lamp. [In such experiments, it is best to proceed as described, § 29, p. 53, viz., to heat in a current of dried air, hydrogen, or carbon dioxide.]

In this manner differently combined quantities of water may be distinguished, and their respective amounts correctly estimated. Thus, for instance, crystallized sulphate of copper contains 28·87 per cent. of water, which escapes at a temperature below 140°, and 7·22 per cent., which escapes only at a temperature between 220° and 260°.

δ. *When the substance has a tendency to absorb oxygen* (from the presence of ferrous compounds, for instance) the water is better determined in the direct way, than by the loss. (§ 36.)

§ 36.

b. ESTIMATION OF WATER BY DIRECT WEIGHING.

This method is resorted to by way of control, or in the case of substances which, upon ignition, lose, besides water, other constituents, which cannot be retained even by the addition of some other substance (*e.g.*, carbon dioxide, oxygen), or in the case of substances containing bodies inclined to oxidation (*e.g.*, ferrous compounds). The principle of the method is to expel the water by the application of a red heat, so as to admit of the condensation of the aqueous vapor, and the collection of the condensed water in an appropriate apparatus, partly physically, partly by the agency of some hygroscopic substance. The increase in the weight of this apparatus represents the quantity of the water expelled.

The operation may be conducted in various ways; the following is one of the most appropriate:—

* Dingler's Polyt. Journ., 126, 39.

B, fig. 25 represents a gasometer filled with air; *b* a flask half-filled with concentrated sulphuric acid; *c* and *a o* are calcium chloride tubes; *d* is a bulb-tube.

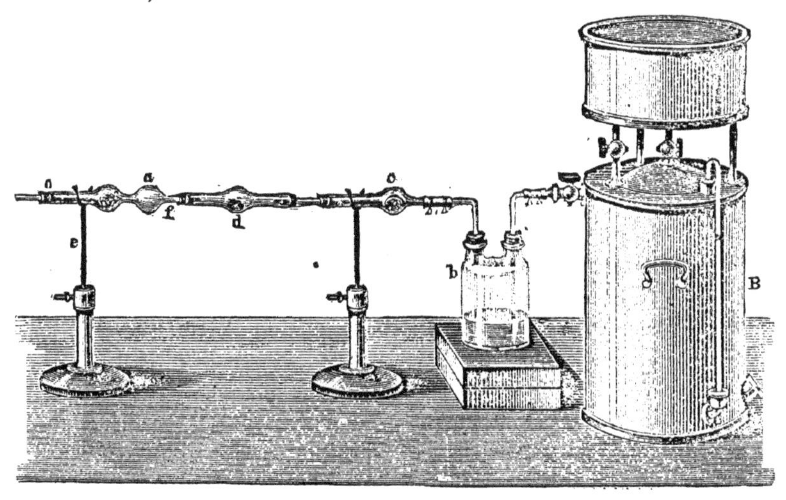

Fig. 25.

The substance intended for examination is weighed in the perfectly dry tube *d*,* which is then connected with *c* and the weighed calcium chloride tube *a o*, by means of sound and well-dried perforated corks.

The operation is commenced by opening the stop-cock of the gasometer a little, to allow the air, which loses all its moisture in *b* and *c*, to pass slowly through *d*; the tube *d* is then heated to beyond the boiling-point of water, by holding a lamp towards *f*, taking care not to burn the cork; and finally, the bulb which contains the substance is exposed to a low red heat, the temperature at *f* being maintained all the while at the point indicated. When the expulsion of the water has been accomplished, a slow current of air is still kept up till the bulb-tube is cold; the apparatus is then disconnected, and the calcium chloride tube *a o*, weighed. The increase in the weight of this tube represents the quantity of water originally present in the substance examined.

* [It is usually better to weigh off the substance into a tray or boat of porcelain or platinum, and place this within a straight tube of hard glass and ignite by means of a tube furnace.]

The empty bulb *a*, in which the greater portion of the water collects, has not only for its object to prevent the liquefaction of the calcium chloride, but enables the analyst also to test the condensed water as to its reaction and purity.

The apparatus may, of course, be modified in various ways; thus, the chloride of calcium tubes may be U-shaped; a U-tube, filled with pieces of pumice-stone saturated with sulphuric acid, may be substituted for the flask with sulphuric acid; and the gasometer may be replaced by an aspirator (fig. 21) joined to *o*.

The expulsion of the aqueous vapor from the tube containing the substance under examination, into the calcium chloride tube, may be effected also by other means than a current of air supplied by a gasometer or aspirator; viz., the substance under examination may be heated to redness in a perfectly dry tube, together with lead carbonate, since the carbon dioxide escaping from the latter at a red heat, serves here the same purpose as a stream of air. This method is principally applied in cases where it is desirable to retain an acid which otherwise would volatilize together with the water; thus, it is applied, for instance, for the direct estimation of the water contained in acid potassium sulphate.

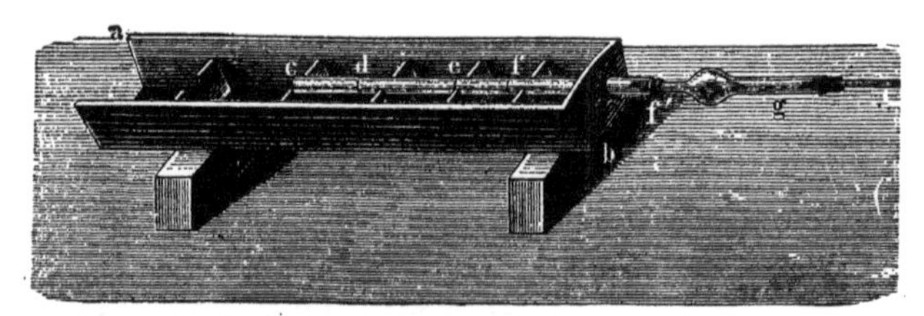

Fig. 26.

Fig. 26. represents the disposition of the apparatus.

a b is a common combustion furnace; *c f'* a tube filled as follows:—from *c* to *d* with lead carbonate,* from *d* to *e* the substance intimately mixed with lead carbonate, and from *e* to *f* pure lead carbonate. The calcium chloride tube *g*, being accurately weighed, is connected with the tube *c f'*, by means of a well-dried perforated cork, *f'*.

The operation is commenced by surrounding the tube with red-

* The lead carbonate must have been previously ignited to incipient decomposition, and cooled in a closed tube.

hot charcoal, advancing from f' toward c; the fore part of the tube which protrudes from the furnace should be maintained at a degree of heat which barely permits the operator to lay hold of it with his fingers. All further particulars of this operation will be found in the chapter on organic elementary analysis. The mixing is performed best in the tube with a wire. The tube $c f'$ may be short and moderately narrow.

The volatilization of an acid cannot in all cases be prevented by lead oxide; thus, for instance, we could not determine the water in crystallized boracic acid by the above process. This could readily be done, however, by igniting the acid mixed with excess of dry sodium carbonate in a glass tube drawn out behind in the form of a beak, receiving the water in a calcium chloride tube, and transferring the final residue of aqueous vapor into the Ca Cl, tube by suction, after the point of the beak has been broken off. (See Organic Analysis.)

The foregoing methods for the direct estimation of water do not, however, yet embrace all cases in which those described in § 35 are inapplicable; since they can be employed only if the substances escaping along with the water are such as will not wholly or partly condense in the calcium chloride tube (or in a tube containing fused potassa, or one filled with pumice-stone saturated with sulphuric acid, which might be used instead). Thus they are perfectly well adapted for determining the water in the basic zinc carbonate, but they cannot be applied to determine the water in sodium ammonium sulphate. With substances like the latter, we must either have recourse to the processes of organic elementary analysis, or we must rest satisfied with the indirect estimation of the water.

§ 37.

3. Solution of Substances.

Before pursuing the analytical process further, it is in most cases necessary to obtain a solution of the substance. This operation is simple where the body may be dissolved by direct treatment with water, or acids, or alkalies, &c.; but it is more complicated in cases where the body requires fluxing as an indispensable preliminary to solution.

When we have mixed substances to operate upon, the component parts of which behave differently with solvents, it is not by any means necessary to dissolve the whole substance at first; on the contrary, the separation may, in such cases, be often effected, in the most simple and expeditious manner, by the solvents themselves. Thus, for instance, a mixture of potassium nitrate, calcium carbonate, and barium sulphate may be readily and accurately analyzed by dissolving out, in the first place, the potassium nitrate with water, and subsequently the calcium carbonate by hydrochloric acid, leaving the insoluble barium sulphate.

§ 38.

a. Direct Solution.

The direct solution of substances is effected, according to circumstances, in beakers, flasks, or dishes, and may, if necessary, be promoted by the application of heat; for which purpose the water-bath will be found most convenient. In cases where an open fire, or the sand-bath, or an iron-plate is resorted to, the analyst must take care to guard against actual ebullition of the fluid, since this would render a loss of substance from spirting almost unavoidable, especially in cases where the process is conducted in a dish. Fluids containing a sediment, either insoluble, or, at least, not yet dissolved, will, when heated over the lamp, often bump and spirt even at temperatures far short of the boiling-point.

In cases where the solution of a substance is attended with evolution of gas, the process is conducted in a flask, placed in a sloping position, so that the spirting drops may be thrown against the walls of the vessel, and thus secured from being carried off with the stream of the evolved gas; or it may be conducted in a beaker, covered with a large-sized watch-glass, which, after the solution is effected, and the gas expelled by heating on the water-bath, must be thoroughly rinsed with the washing-bottle.

In cases where the solution has to be effected by means of concentrated volatile acids (hydrochloric acid, nitric acid, aqua regia), the operation should never be conducted in a dish, but always in a flask covered with a watch-glass, or placed in a slanting position, and the application of too high a temperature must be avoided. The operation should always be conducted also under a hood, with proper draught, to carry off the escaping acid vapors. In my own laboratory, I use for the latter purpose the following simple contrivance: a leaden pipe, permanently fixed in a convenient position, leads from the working table through the wall or the window-frame into the open air. The end in the laboratory is connected with one of the mouths of a two-necked bottle which contains a little water. The other mouth of the bottle is closed with a perforated cork, bearing a firmly-fixed glass tube bent at a right angle; the portion of the tube which enters the bottle must not dip into the water. The solution-flask being now closed with a perforated cork, or an india-rubber cap, bearing a glass tube, connected by means of india-rubber with the bent tube in the double-necked bottle, the vapors evolved are carried out of the laboratory without the least inconvenience to the operator; moreover, no receding of fluid upon cooling need be apprehended. Instead of conveying the vapors away through a tube leading into the open air, a conical glass-tube filled with pieces of broken glass, moistened with water or solution of sodium carbonate, may be fixed on the second mouth of the double-necked bottle. I, however, prefer the other method. In some cases, it is advisable also to conduct the escaping vapors into a little water, and, when solution has been effected, make the water recede by withdrawing the lamp, since this will, at the same time, serve to dilute the solution; care must be taken, however, to guard against a premature receding of the water in consequence of an accidental cooling of the solution flask.

It is often necessary, in conducting a process of solution, to guard against the action of the atmospheric oxygen; in such cases, a slow stream of carbon dioxide is transmitted through the solution-flask; in some cases it is sufficient to expel the air, by simply first putting a little hydrogen sodium carbonate into the flask, containing an excess of acid, before introducing the substance.

§ 39.

b. Solution, preceded by Fluxing.

Substances insoluble in water, acids, or aqueous alkalies, usually require decomposition by fluxing, to prepare them for analysis. Substances of this kind are often met with in the mineral kingdom; most silicates, the sulphates of the alkali-earth metals, chrome ironstone, &c., belong to this class.

The object and general features of the process of fluxing have already been treated of in the qualitative part of the present work.

The special methods of conducting this important operation will be described hereafter under "The analysis of silicates," and in the proper places; as a satisfactory description of the process, with its various modifications, cannot well be given without entering more minutely into the particular circumstances of the several special cases.

Decomposition by fluxing often requires a higher temperature than is attainable with a spirit-lamp with double draught, or with a common gas-lamp. In such cases, the glass-blower's lamp, fed with gas, is used with advantage.

§ 40.

4. Conversion of the dissolved Substance into a weighable Form.

The conversion of a substance in a state of solution into a form adapted for weighing may be effected either by *evaporation* or by *precipitation.* The former of these operations is applicable only in cases where the substance, the weight of which we are desirous to ascertain, either exists already in the solution in the form suitable for the determination of its weight, or may be converted into such form by evaporation in conjunction with some reagent. The solution must, moreover, contain the substance unmixed, or, at least, mixed only with such bodies as are expelled by evaporation or at a red-heat. Thus, for instance, the amount of sodium sulphate present in an aqueous solution of that substance may be ascertained by simple evaporation; whilst the potassium carbonate contained in a solution would better be converted into potassium chloride, by evaporating with solution of ammonium chloride.

Precipitation may always be resorted to, whenever the substance in solution admits of being converted into a combination which is insoluble in the menstruum present, provided that the precipitate is fit for determination, which can never be the case unless it can be washed and is of constant composition.

§ 41.

a. Evaporation.

In processes of evaporation for pharmaceutical or technico-chemical purposes the principal object to be considered is saving of time and fuel; but in evaporating processes in quantitative analytical researches this is merely a subordinate point, and the analyst has to direct his principal care and attention to the means of guarding against loss or contamination of the substance operated upon.

The simplest case of evaporation is when we have to *concentrate a clear fluid, without carrying the process to dryness.* To effect this object, the fluid is poured into a basin, which should not be filled to more than two-thirds. Heat is then applied by placing the basin either on a water-bath, sand-bath, common stove, or heated iron plate, or over the flame of a gas- or spirit-lamp, care being taken always to guard against actual ebullition, as this invariably and unavoidably leads to loss from small drops of fluid spirting out. Evaporation over a gas- or spirit-lamp, when conducted with proper care, is an expeditious and cleanly process. BUNSEN's gas-lamp may be used most advantageously in operations of this kind; a little wire-gauze cap, loosely fitted upon the tube of the lamp, is a material improvement. By means of this simple arrangement it is easy to produce even the smallest flame, without the least apprehension of ignition of the gas within the tube.

If the evaporation is to be effected on the water-bath, and the operator happens to possess a BEINDORF, or other similarly-constructed steam apparatus, the evaporating-dish may be placed simply into an opening corresponding in size. Otherwise recourse must be had to the water-bath, illustrated by fig. 27.

Fig. 27.

It is made of strong sheet copper, and when used is half filled with water, which is kept boiling over a gas-, spirit-, or oil-lamp. The breadth from *a* to *b* should be from 12 to 18 cm. Various flat rings of the same outside diameter as the top of the bath, and adapted to receive dishes and crucibles of different sizes, are essential adjuncts to the bath. These rings when required are simply laid on the bath.

It will occasionally happen that the water in the bath completely evaporates; in such cases, residues are heated to a higher degree than is desirable, concentrated solutions spirt, &c. To avoid these inconveniences, water-baths have been devised with an arrangement for maintaining a constant level of water.

If the operator can conduct his processes of evaporation in a room set apart for the purpose, where he may easily guard against any occurrence tending to suspend dust in the air, he will find it no very difficult task to keep the evaporating fluid clean; in this case it is best to leave the dishes uncovered. But in a large laboratory, frequented by many people, or in a room exposed to draughts of air, or in which coal fires are burning, the greatest caution is required to protect the evaporating fluid from contamination by dust or ashes.

For this purpose the evaporating dish is either covered with a sheet of filtering-paper turned down over the edges, or a glass rod twisted into a triangular shape (fig. 28) is laid upon it, and a sheet of filtering-paper spread over it, which is kept in position by a glass rod laid across, the latter again being kept from rolling down by the slightly turned up ends, *a* and *b*, of the triangle.

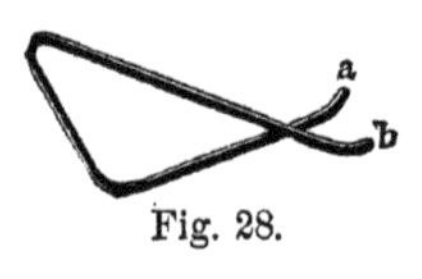

Fig. 28.

The best way, however, is the following:—Take two small thin wooden hoops (fig. 29), one of which fits loosely in the other; spread a sheet of blotting-paper over the smaller one, and push the other over it. This forms a cover admirably adapted to the purpose; and whilst in no way interfering with the operation, it completely protects the evaporating fluid from dust, and may be readily taken off; the paper cannot dip into the fluid; the cover lasts a long time, and may, moreover, at any time be easily renewed.

Fig. 29.

It must be borne in mind, however, that the common filtering-paper contains always certain substances soluble in acids, such as lime, ferric oxide, &c., which, were covers of the kind just described used over evaporating dishes containing a fluid evolving acid vapors, would infallibly dissolve in these vapors, and the solution dripping down into the evaporating fluid, would speedily contaminate it. Care must be taken, therefore, in such cases, to use only such filtering-paper as has been freed by washing from substances soluble in acids.

Evaporation for the purpose of concentration may be effected also in flasks; these are only half filled, and placed in a slanting position. The process may be conducted on the sand-bath, or over a gas- or spirit-lamp, or even, and with equal propriety, over a charcoal fire. In cases where the operation is conducted over a lamp or a charcoal fire, it is the safest way to place the flasks on wire gauze. Gentle ebullition of the fluid can do no harm here, since the slanting position of the flask guards effectively against risk of loss from the spirting of the liquid. Still better than in flasks, the object may be attained by evaporating in tubulated retorts with open tubulure and neck directed obliquely upwards. The latter acts as a chimney, and the constant change of air thus effected is extremely favorable to evaporation.

The evaporation of *fluids containing a precipitate* is best conducted on the water-bath; since on the sand-bath, or over the lamp, it is next to impossible to guard against loss from bumping. This

Fig. 30.

bumping is occasioned by slight explosions of steam, arising from the sediment impeding the uniform diffusion of the heat. Still there remains another, though less safe way, viz., to conduct the evaporation in a crucible placed in a slanting position, as illustrated in fig. 30. In this process, the flame is made to play upon the crucible above the level of the fluid.

Where a fluid has to be evaporated to dryness, as is so often the case, the operation should always, if possible, be terminated on the water-bath. In cases where the nature of the dissolved substance precludes the application of the water-bath, the object in view may often be most readily attained by heating the contents of the dish from the top, which is effected by placing the dish in a proper position in a drying closet, whose upper plate is heated by a flame (that of the water- or sand-bath) passing over it. If the substance is in a covered platinum dish or crucible, place the gas-lamp in such a position that the flame may act on the cover from above.

In cases where the heat has to be applied from the bottom, a method must be chosen which admits of an equal diffusion and ready regulation of the heat.

An air-bath is well adapted for this purpose, *i.e.*, a dish of iron plate, in which the porcelain or platinum dish is to be placed on a wire triangle, so that the two vessels may be at all points ¼ to ½ inch distant from each other. The copper apparatus, fig. 27, may also serve as an air-bath, although I must not omit to mention that this mode of application will in the end seriously injure it. If the operation has to be conducted over a lamp, the dish should be placed high above the flame; best on wire gauze, since this will greatly contribute to an equal diffusion of the heat. The use of the sand-bath is objectionable here, because with that apparatus we cannot reduce the heat so speedily as may be desirable. An iron plate heated by gas may perhaps be used with advantage. But no matter which method be employed, *this* rule applies equally to all of them; that the operator must watch the process, from the moment that the residue begins to thicken, in order to prevent spirting, by reducing the heat, and breaking the pellicles which form on the surface, with a glass rod, or a platinum wire or spatula.

Saline solutions that have a tendency, upon their evaporation, to creep up the sides of the vessel, and may thus finally pass over the brim of the latter, thereby involving the risk of a loss of substance, should be heated from the top, in the way just indicated; since by that means the sides of the vessel will get heated sufficiently to cause the instantaneous evaporation of the ascending liquid, preventing thus its overflowing the brim. The inconvenience just alluded to may, however, be obviated also, in most cases, by covering the brim, and the uppermost part of the inner side of the vessel, with a very thin coat of tallow, thus diminishing the adhesion between the fluid and the vessel.

In the case of liquids evolving gas-bubbles upon evaporating, particular caution is required to guard against loss from spirting. The safest way is to heat such liquids in an obliquely-placed flask, or in a beaker covered with a large watch-glass; the latter is removed as soon as the evolution of gas-bubbles has ceased, and the fluid that may have spirted up against it is carefully rinsed into the glass, by means of a washing-bottle. If the evaporation has to be conducted in a dish, a rather capacious one should be selected, and a very moderate degree of heat applied at first, and until the evolution of gas has nearly ceased.

If a fluid has to be evaporated *with exclusion of air*, the best way is to place the dish under the bell of an air-pump, over a vessel with sulphuric acid, and to exhaust; or a tubulated retort may be used through whose tubulure hydrogen or carbon dioxide is passed by the acid of a tube not quite reaching to the surface of the fluid.

The material of the evaporating vessels may exercise a much greater influence on the results of an analysis than is generally believed. Many rather startling phenomena that are observed in analytical processes may arise simply from a contamination of the evaporated liquid by the material of the vessel; great errors may also spring from the same source.

The importance of this point has induced me to subject it to a searching investigation (see Appendix, Analytical Experiments, 1—4), of which I will here briefly intimate the results.

Distilled water kept boiling for some length of time in glass (flasks of Bohemian glass) dissolves very appreciable traces of that material. This is owing to the formation of soluble silicates; the particles dissolved consist chiefly of potassa, or soda and lime, in combination with silicic acid. A much larger proportion of the glass is dissolved by water containing caustic or carbonated alkali; boiling solution of ammonium chloride also strongly attacks glass vessels. Boiling dilute acids, with the exception, of course, of hydrofluoric and hydrofluosilicilic acids, exercise a less powerful solvent action on glass than pure water. Porcelain (Berlin dishes) is much less affected by water than glass; alkaline liquids also exercise a less powerful solvent action on porcelain than on glass; the quantity dissolved is, however, still notable. Solution of ammonium chloride acts on porcelain as strongly as on glass; dilute acids, though exercising no very powerful solvent action on porcelain, yet attack that material more strongly than glass. It results from these data, that in analyses pretending to a high degree of accuracy, platinum or platinum-iridium or silver dishes should always be preferred. The former may be used in all cases where no free chlorine, bromine, or iodine is present in the fluid, or can be formed during evaporation. Fluids containing caustic alkalies may safely be evaporated in platinum, but not to the point of fusion of the residue. Silver vessels should never be used to evaporate acid fluids nor liquids containing alkaline sulphides; but they are admirably suited for solutions of alkali hydroxides and carbonates, as well as of most normal salts.

§ 42.

We come now to *weighing the residues remaining upon the evaporation of fluids.* We allude here simply to such as are soluble in water; those which are separated by filtration will be treated of afterwards. Residues are generally weighed in the same vessel in which the evaporation has been completed, for which purpose platinum dishes, from 4 to 8 cm. in diameter, provided with light covers, or large platinum crucibles, are best adapted, since they are lighter than porcelain vessels of the same capacity.

Fig. 31.

However, in most cases, the amount of liquid to be evaporated is too large for so small a vessel, and its evaporation in portions would occupy too much time. The best way, in cases of this kind, is to concentrate the liquid first in a larger vessel, and to terminate the operation afterwards in the smaller weighing vessel. In transferring the fluid from the larger to the smaller vessel, the lip of the former is slightly greased, and the liquid made to run down a glass rod. (See fig. 31.)

Finally the large vessel is carefully rinsed with a washing-bottle, until a drop of the last rinsing leaves no longer a residue upon evaporation on a platinum knife. When the fluid has thus been transferred to the weighing-vessel, the evaporation is completed on the water-bath and the residuary substance finally ignited, provided, of course, it will admit of this process. For this purpose the dish is covered with a lid of thin platinum (or a thin glass plate), and then placed high over the flame of a lamp, and heated very gently until all the water which may still adhere to the substance is expelled; the dish is now exposed to a stronger, and finally to a red heat. (Where a glass plate is used, this must, of course, be removed before igniting.) In this case it is also well to make the flame play obliquely on the cover from above, so as to run as little risk as possible of loss by spirting. After cooling in a desiccator, the covered dish is weighed with its contents. When operating upon substances which decrepitate, such as sodium chloride, for instance, it is advisable to expose them—after their removal from the water-bath, and previously to the application of a naked flame—to a temperature somewhat above 100°, either in the air-bath, or on a sand-bath, or on a common stove.

If the residue does not admit of ignition, as is the case, for instance, with organic substances, ammonium salts, &c., it is dried at a temperature suited to its nature. In many cases, the temperature of the water-bath is sufficiently high for this purpose, for the drying of ammonium chloride, for instance; in others, the air or oil-bath must be resorted to. (See §§ 29 and 30.) Under any circumstances, the desiccation must be continued until the substance ceases to suffer the slightest diminution in weight, after renewed exposure to heat for half an hour. The dish should invariably be covered during the process of weighing.

If, as will frequently happen, we have to deal with a fluid containing a small quantity of a potassium or sodium salt, the weight of which we want to ascertain, in presence of a comparatively large amount of an ammonium salt, which has been mixed with it in the course of the analytical process, I prefer the following method: The saline mass is thoroughly dried, in a large dish, on the water-bath, or, towards the end of the process, at a temperature somewhat exceeding 100°. The dry mass is then, with the aid of a platinum spatula, transferred to a small glass dish, which is put aside for a time in a desiccator. The last traces of the salt left adhering to the sides and bottom of the large dish are rinsed off with a little water into the small dish, or the large crucible, in which it is intended to weigh the salt; the water is then evaporated, and the dry contents of the glass dish are added to the residue: the ammonium salts are now expelled by ignition, and the residuary fixed salts finally weighed. Should some traces of the saline mass adhere to the smaller glass dish, they ought to be removed and transferred to the weighing vessel, with the aid of a little pounded ammonium chloride, or some other ammonium salt, as the moistening again with water would involve an almost certain loss of substance.

§ 43

b. PRECIPITATION.

Precipitation is resorted to in quantitative analysis far more frequently than evaporation, since it serves not merely to convert substances into forms adapted for weighing, but also, and more especially, to separate them from one another. The principal intention in precipitation, for the purpose of quantitative estimations, is to convert the substance in solution into a form in which it is insoluble in the menstruum present. The result will, therefore, *cæteris paribus*, be the more accurate, the more the precipitated body deserves the epithet insoluble, and in cases where precipitates are of the same degree of solubility, that one will suffer the least loss which comes in contact with the smallest amount of solvent.

Hence it follows, first, that in all cases where other circumstances do not interfere, it is preferable to precipitate substances in their most insoluble form; thus, for instance, barium had better be precipitated as sulphate than as carbonate; secondly, that when we have to deal with precipitates that are not quite insoluble in the menstruum present, we must endeavor to remove that menstruum, as far as practicable, by evaporation; thus a dilute solution of strontium should be concentrated, before proceeding to precipitate the strontium with sulphuric acid; and, thirdly, that when we have to deal with precipitates slightly soluble in the liquid present, but insoluble in another menstruum, into which the former may be converted by the addition of some substance or other, we ought to endeavor to bring about this modification of the menstruum. Thus, for instance, alcohol may be added to water, to induce complete precipitation of ammonium platinic chloride, lead chloride, calcium sulphate, &c.; thus again, ammonium magnesium phosphate may be rendered insoluble in an aqueous menstruum by adding ammonia to the latter, &c.

Precipitation is generally effected in beakers. In cases, however, where we have to precipitate from fluids in a state of ebullition, or where the precipitate requires to be kept boiling for some time with the fluid, flasks or dishes are substituted for beakers, with due regard always to the material of which they are made (see Evaporation, § 41, at the end).

The separation of precipitates from the fluid in which they are suspended, is effected either by *decantation* or *filtration*, or by both these processes jointly. But, before proceeding to the separation of the precipitate by any of these methods, the operator must know whether the precipitant has been added in sufficient quantity, and whether the precipitate is completely formed. To determine the latter point, an accurate knowledge of the properties of the various precipitates must be attained, which we shall endeavor to supply in the third section. To decide the former question, it is usually sufficient to add to the fluid (after the precipitate has settled) cautiously a fresh portion of the precipitant, and to note if a further turbidity ensues. This test, however, is not infallible, when the precipitate has not the property of forming immediately; as, for instance, is the case with ammonium phospho-molybdate. When this is apprehended, pour out (or transfer with a pipette) a small quantity of the clear supernatant fluid into another vessel, add some of the precipitant, warm if necessary; and after some time look and see whether a fresh precipitate has formed. As a general rule, the precipitated liquid should be allowed to stand at rest for several hours, before proceeding to the separation of the precipitate. This rule applies more particularly to crystalline, pulverulent, and gelatinous precipitates, whilst curdy and flocculent precipitates, more particularly when the precipitation was effected at a boiling temperature, may often be filtered off immediately. However, we must observe here, that all general rules, in this respect, are of limited application.

§ 44.

α. Separation of Precipitates by Decantation.

When a precipitate subsides so completely and speedily in a fluid that the latter may be decanted off perfectly clear, or drawn off with a syphon, or removed by means of a pipette, and that the washing of the precipitate does not require a very long time, decantation is often resorted to for its separation and washing; this is the case, for instance, with chloride of silver, metallic mercury, &c.

Decantation will always be found a very expeditious and accurate method of separation, if the process be conducted with due care; it is necessary, however, in most cases, to promote the speedy and complete subsidence of the precipitate; and it may be laid down as a general rule, that heating the precipitate with the fluid will produce the desired effect. Nevertheless, there are instances in which the simple application of heat will not suffice; in some cases, as with silver chloride, for instance, agitation must be resorted to; in other cases, some reagent or other is to be added—hydrochloric acid, for instance, in the precipitation of mercury, &c. We shall have occasion, subsequently, in the fourth section, to discuss this point more fully, when we shall also mention the vessels best adapted for the application of this process to the various precipitates.

After having washed the precipitate repeatedly with fresh quantities of the proper fluid, until there is no trace of a dissolved substance to be detected in the last rinsings, it is placed in a crucible or dish, if not already in a vessel of that description; the fluid still adhering to it is poured off as far as practicable, and the precipitate is then, according to its nature, either simply dried, or heated to redness.

A far larger amount of water being required for washing precipitates by decantation than on filters, the former process can be expected to yield *accurate* results only where the precipitates are *absolutely insoluble.* For the same reason, decantation is not ordinarily resorted to in cases where we have to determine other constituents in the decanted fluid.

The decanted fluid must be allowed to stand at rest from twelve to twenty-four hours, to make quite sure that it contains no particles of the precipitate; if, after the lapse of this time, no precipitate is visible, the fluid may be thrown away; but if a precipitate has subsided, this had better be estimated by itself, and the weight added to the main amount; the precipitate may, in such cases, be separated from the supernatant fluid by decantation, or by filtration.

§ 45.

β. Separation of Precipitates by Filtration.

This operation is resorted to whenever decantation is impracticable, and, consequently, in the great majority of cases; provided always the precipitate is of a nature to admit of its being completely freed, by mere washing on the filter, from all foreign substances. Where this is not the case, more particularly, therefore, with gelatinous precipitates, aluminium hydroxide for instance, a combination of decantation and filtration is resorted to (§ 48).

aa. Filtering Apparatus.

Filtration, as a process of quantitative analysis, is almost exclusively effected by means of paper.

Plain circular filters are most generally employed; plaited filters are only occasionally used. Much depends upon the quality of the paper. Good filtering paper must possess the three following properties:—1. It must completely retain the finest precipitates; 2. It must filter rapidly; and 3. It must be as free as possible from any admixture of inorganic bodies, but more especially from such as are soluble in acid or alkaline fluids.

It is a matter of some difficulty, however, to procure paper fully answering these conditions. The *Swedish filtering paper*, with the water-mark J. H. Munktell, is considered the best, and, consequently, fetches the highest price; but even this answers only the first two conditions, being by no means sufficiently pure for very accurate analyses, since it leaves upon incineration about 0·3 per cent. of ash,* and yields to acids perceptible traces of lime, magnesia, and ferric oxide. For exact experiments it is, consequently, necessary first to extract the paper with dilute hydrochloric acid, then to wash the acid completely out with water, and finally to dry the paper. In the case of very fine filtering paper, the best way to perform this operation is to place the ready-cut filters, several together, in a funnel, exactly the same way as if intended for immediate filtration; they are then moistened with a mixture of one part of ordinary pure hydrochloric acid with two parts of water, which is allowed to act on them for about ten minutes; after this all traces of the acid are carefully removed by washing the filters in the funnel repeatedly with warm water. The funnel being then covered with a piece of paper, turned over the edges, is put in a warm place until the filters are dry. Compare the instruction given in the "Qual. Anal.," Am. Ed., p. 8, on the preparation of washed filters. Filter paper containing lead, and which is consequently blackened by sulphuretted hydrogen, should be rejected.

---

* Plantamour found the ash of Swedish filtering paper to consist of 63·23 silicic acid, 12·83 lime, 6·21 magnesia, 2·94 alumina, and 13·92 ferric oxide, in 100 parts.

Ready-cut filters of various sizes should always be kept on hand. Filters are either cut by circular patterns of pasteboard or tin, or, still better, by Mohr's filter-patterns, fig. 32. This little apparatus is made of tin-plate, and consists of two parts. *B* is a quadrant fitting in *A*, whose straight edges are turned up, and which is slightly smaller than *B*. The sheets of filter-paper are first cut up into squares, which are folded in quarters, and placed in *A*, then *B* is placed on the top, and the free edge of the paper is cut off with scissors. Filters cut in this way are perfectly circular, and of equal size.

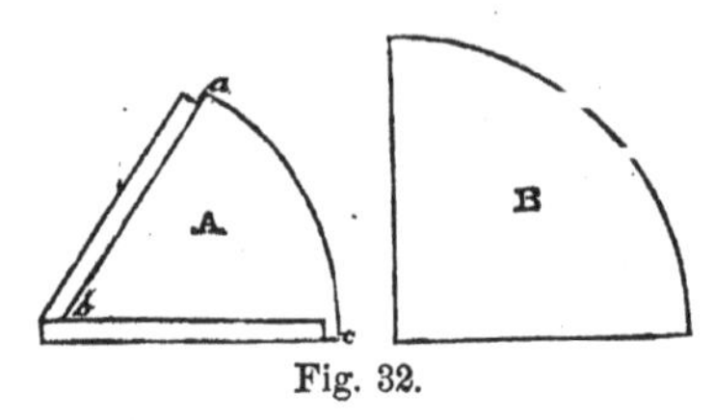

Fig. 32.

Several pairs of these patterns of various sizes (3, 4, 5, 6, 6·5, and 8 cm. radius) should be procured. In taking a filter for a given operation, you should always choose one which, after the fluid has run through, will not be more than half filled with the precipitate.

As to the funnels, they should be inclined at the angle of 60°, and not bulge at the sides. Glass is the most suitable material for them.

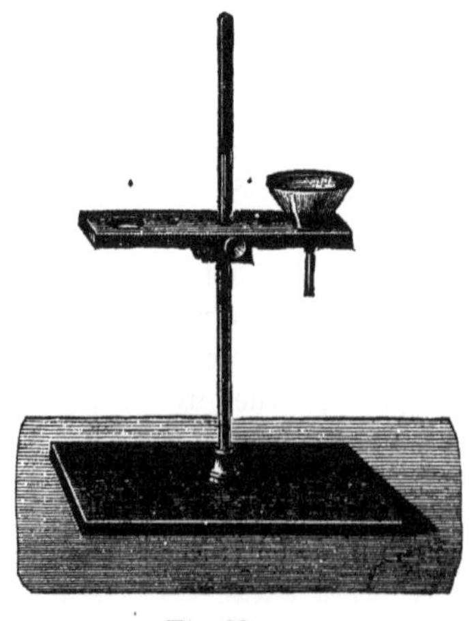

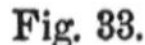

Fig. 33.

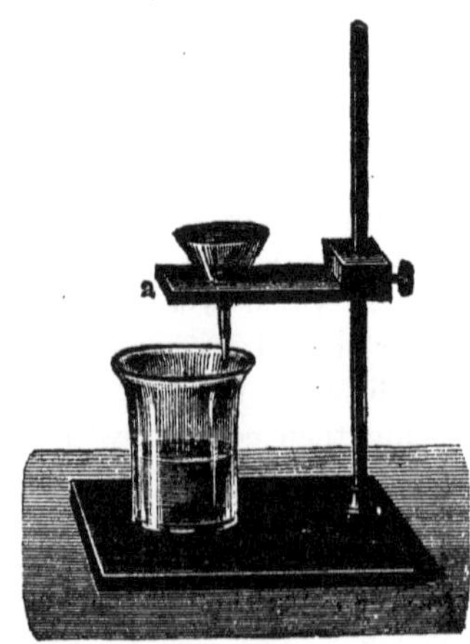

Fig. 34.

The filter should never protrude beyond the funnel. It should come up to one or two lines from the edge of the latter.

The filter is firmly pressed into the funnel, to make the paper fit closely to the side of the latter; it is then moistened with water; any extra water is not poured out, but allowed to drop through.

The stands shown in figs. 33 and 34 complete the apparatus for filtering.

The stands are made of hard wood. The arm holding the funnel or funnels must slide easily up and down, and be fixable by the screw. The holes for the funnels must be cut conically, to keep the funnels steadily in their place.

These stands are very convenient, and may be readily moved about without interfering with the operation.

§ 46.

bb. Rules to be observed in the Process of Filtration.

In the case of curdy, flocculent, gelatinous, or crystalline precipitates there is no danger of the fluid passing turbid through the filter. But with fine pulverulent precipitates it is generally *necessary*, and always *advisable*, to let the precipitate subside, and then filter the supernatant liquid, before proceeding to place the precipitate upon the filter. We generally proceed in this way also with other kinds of precipitates, especially with those that require to stand long before they completely separate. Precipitates which have been thrown down hot, are most properly filtered off before cooling (provided always there be no objections to this course), since hot fluids run through the filter more speedily than cold ones. Some precipitates have a tendency to be carried through the filter along with the fluid; this may be prevented in some instances by modifying the latter. Thus barium sulphate, when filtered from an aqueous solution, passes rather easily through the filter—the addition of hydrochloric acid or ammonium chloride prevents this in a great measure.

If the operator finds, during a filtration, that the filter would be much more than half filled by the precipitate, he would better use an additional filter, and thus distribute the precipitate over the two; for, if the first were too full, the precipitate could not be properly washed.

The fluid ought never to be poured directly upon the filter, but always down a glass rod, and the lip or rim of the vessel from which the fluid is poured should always be slightly greased with tallow.* The stream ought invariably to be directed towards the sides of the filter, never to the centre, since this might occasion loss by splashing. In cases where the fluid has to be filtered off, with the least possible disturbance of the precipitate, the glass rod must not be placed, during the intervals, in the vessel containing the precipitate; but it may conveniently be put into a clean glass, which is finally rinsed with the wash-water.

The filtrate is received either in flasks, beakers, or dishes, according to the various purposes for which it may be intended. Strict care should be taken that the drops of fluid filtering through glide down the side of the receiving vessel; they should never be allowed to fall into the centre of the filtrate, since this again might occasion loss by splashing. The best method is that shown in fig. 34, viz., to rest the point of the funnel against the upper part of the inside of the receiving vessel.

If the process of filtration is conducted in a place perfectly free from dust, there is no necessity to cover the funnel, nor the vessel receiving the filtrate; however, as this is but rarely the case, it is generally indispensable to cover both. This is best effected with round plates of sheet-glass. The plate used for covering the receiving vessel should have a small U-shaped piece cut out of its edge, large enough for the funnel-tube to go through. The effect desired may be produced by cautiously chipping out the glass bit by bit with the aid of a key. Plates perforated in the centre are worthless as regards the object in view.

After the fluid and precipitate have been transferred to the filter, and the vessel which originally contained them has been rinsed repeatedly with water, it happens generally that small particles of the precipitate remain adhering to the vessel, which cannot be removed with the glass rod. From beakers or dishes these particles may be readily removed by means of a feather prepared for the purpose by tearing off nearly the whole of the plumules, leaving only a small piece at the end which should be cut perfectly straight. From flasks, minute portions of heavy precipitates which are not adherent, are readily removed by blowing a jet of water into the flask, held inverted over the funnel; this is effected

* The tallow may be kept under the edge of the work-table at a convenient point, where it will adhere by a little pressure. The best way of applying the tallow to the lip of a vessel is with the greased finger.

by means of the washing-bottle shown in fig. 36. If the minute adhering particles of a precipitate cannot be removed by mechanical means, solution in an appropriate menstruum must be resorted to, followed by re-precipitation. Bodies for which we possess no solvent, such as barium sulphate, for instance, must not be precipitated in flasks.

§ 47.

cc. Washing of Precipitates.

After having transferred the precipitate completely to the filter, we have next to perform the operation of washing; this is effected by means of one of the well-known washing-bottles, of which I prefer the one represented in fig. 35 in every respect. The doubly perforated stoppers are of vulcanized rubber.

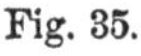

Fig. 35.

Fig. 36.

Fig. 37.

Care must always be taken to properly regulate the jet, as too impetuous a stream of water might occasion loss of substance.

In cases where a precipitate has to be washed with great caution, the apparatus illustrated in fig. 37 will be found to answer very well.

The construction of this apparatus requires no explanation. When the flask is inverted, it supplies a fine continuous jet of water.

Precipitates requiring washing with water, are washed most expeditiously with hot water, provided always there be no special reason against its use. The washing-bottle shown in fig. 35 is particularly well adapted for this purpose. The cork which is fastened to the neck of the flask with wire serves to facilitate holding it.

It is a rule in washing precipitates not to add fresh wash-water to the filter till the old has quite run through. In applying the jet of water you have to take care on the one hand that the upper edge of the filter is properly washed, and on the other hand that no canals are formed in the precipitate, through which the fluid runs off, without coming in contact with the whole of the precipitate. If such canals have formed and cannot be broken up by the jet, the precipitate must be stirred cautiously with a small platinum knife or glass rod.

The washing may be considered completed when all soluble matter that is to be removed has been got rid of. The beginner who devotes proper attention to the completion of this operation shuns one of the rocks which he is most likely to encounter. Whether the precipitate has been completely washed may generally be ascertained by slowly evaporating a drop of the last washings upon a platinum knife, and observing if a residue is left. But in cases where the precipitate is not altogether insoluble in water (strontium sulphate, for instance), recourse must be had to more special tests, which we shall have occasion to point out in the course of the work. The student should never discontinue the washing of a precipitate because he simply *imagines* it is finished —he must be certain.

§ 48.

γ. Separation of Precipitates by Decantation and Filtration combined.

In the case of precipitates which, from their gelatinous nature, or from the firm adhesion of certain coprecipitated salts, oppose insuperable, or, at all events, considerable obstacles to perfect washing on the filter, the following method is resorted to: Let the precipitate subside as far as practicable, pour the nearly clear supernatant liquid on the filter, stir the precipitate up with the washing fluid (in certain cases, where such a course is indicated, heat to boiling), let it subside again, and repeat this operation until the precipitate is almost thoroughly washed. Transfer it now to the filter, and complete the operation with the washing-bottle (see § 47). This method is highly to be recommended; there are many precipitates that can be thoroughly washed only by its application.

In cases where it is not intended to weigh the precipitate washed by decantation, but to dissolve it again, the operation of washing is entirely completed by decantation, and the precipitate not even transferred to the filter. The re-solution of the bulk of the precipitate being effected in the vessel containing it, the filter is placed over the latter, and the solvent passed through it. Although the termination of the operation of washing may be usually ascertained by testing a sample of the washings for one of the substances originally present in the solution which has to be removed (for hydrochloric acid, for instance, with nitrate of silver), still there are cases in which this mode of proceeding is inapplicable. In such cases, and indeed in processes of washing by decantation generally, BUNSEN's method will be found convenient —viz., to continue the process of washing until the fluid which had remained in the beaker, after the first decantation, has undergone a ten thousand-fold dilution. To effect this, measure with a slip of paper the height from the bottom of this beaker to the surface of the fluid remaining in it, together with the precipitate, after the first decantation; then fill the beaker with water, if possible, boiling, and measure the entire height of the fluid; divide the length of the second column by that of the first. Go through the same process each time you add fresh water, and always multiply the quotient found with the number obtained in the preceding calculation, until you reach 10000.

§ 49.

FURTHER TREATMENT OF PRECIPITATES.

Before proceeding to weigh a precipitate, it still remains to convert it into a form of accurately known composition. This is done either by igniting or by drying. The latter proceeding is more protracted and tedious than the former, and is, moreover, apt to give less accurate results. The process of drying is, therefore, as a general rule, applied only to precipitates which cannot bear exposure to a red heat without undergoing total or partial volatilization; or whose residues left upon ignition have no constant composition; thus, for instance, drying is resorted to in the case of mercuric sulphide, arsenious sulphide, and other metallic sulphides; and also in the case of silver cyanide, potassium platinic chloride, etc.

But whenever the nature of the precipitate (*e.g.*, barium sulphate, lead sulphate, and many other compounds) leaves the operator at liberty to choose between drying and heating to redness, the process is almost invariably preferred.

§ 50.

aa. Drying of Precipitates.

When a precipitate has been collected, washed, and dried on a filter, minute particles of it adhere so firmly to the paper that it is found impossible to remove them. The weighing of dried precipitates involves, therefore, in all accurate analyses, the drying and weighing of the filter also. To obtain accurate results, it is necessary to dry and weigh the filter before using it; the temperature at which the filter is dried must be the same as that to which it is intended subsequently to expose the precipitate. Another condition is that the filtering-paper must not contain any substance liable to be dissolved by the fluid passing through it.

The drying is conducted either in the water-, air-, or oil-bath, according to the degree of heat required. The weighing is performed in a closed vessel, mostly between two clasped watch-glasses (fig. 20), or in a platinum crucible. When the filter appears dry, it is placed between the warm watch-glasses, or in a warm crucible, allowed to cool under a bell-glass, over sulphuric acid, and weighed. The reopened crucible or watch-glasses, together with the filter, are then again exposed for some time to the required degree of heat, and, after cooling, weighed once more. If the weight does not differ from that found at first, the filter may be considered dry, and we have simply to note the collective weight of the watch-glasses, clasp, and filter, or of the crucible and filter.

After the washing of the precipitate has been concluded and the water allowed to run off as far as possible, the filter with the precipitate is taken off the funnel, folded up, and placed upon blotting-paper, which is then kept for some time in a moderately warm place protected from dust; it is finally put into one of the watch-glasses, or into the uncovered platinum crucible, with which it was first weighed, and exposed to the appropriate degree of heat, either in the water-, air-, or oil-bath. When it is judged that the precipitate is dry, the second watch-glass, or the lid of the crucible is put on (with the clasp pushed over the two in the former case), and the whole, after cooling in the desiccator, is weighed. The filter and the precipitate are then again exposed, in the same way, to the proper drying temperature, allowed to cool, and weighed again, the same process being repeated until the weight remains constant or varies only to the extent of a few deci-milli-grammes. By subtracting from the weight found the tare of the crucible or watch-glasses and filter, we obtain the weight of the dry precipitate. [The filter must not be dried too long, as it slowly loses weight, and even becomes brown from decomposition when heated to 100° for days together.]

Fig. 38.

It happens sometimes that the precipitate nearly fills the filter, or retains a considerable amount of water; or sometimes the paper is so thin that its removal from the funnel cannot well be effected without tearing. In all such cases, the best way is to let the filter and precipitate get nearly dry in the funnel, which may be effected readily by covering the latter with a piece of blotting paper * to keep out the dust, and placing it, supported on a broken beaker (fig. 38), or some other vessel of the kind, on the steam-apparatus or sand-bath, or stove, or on a heated iron plate. For support to a funnel while drying a hollow frustum of a cone open both ends, made of sheet zinc (fig. 39), is convenient. Two sizes may be used, 10 cm. and 12 cm. high respectively. The lower diameter should be from 7 to 8, the upper from 4 to 6 cm.

Fig. 39.

§ 51.

bb. Ignition of Precipitates.

In this process it is necessary to burn the filter and substract the weight of the filter ash from the total weight found.

If care be taken to make the filters always of the same paper,

* Turned down over the rim. Or more neatly as follows:—Wet a common cut filter, stretch it over the ground top of the funnel, and then gently tear off the superfluous paper. The cover thus formed continues to adhere after drying with some force.

and to cut every size by a pattern, the quantity of ash which each size yields upon incineration may be readily determined. It is necessary, however, to determine separately the quantity of ash left by ordinary filters, and that left by filters which have been washed with hydrochloric acid and water; on an average the latter leave about half as much ash as the former. To determine the filter ash take ten filters (or an equal weight of cuttings from the same paper), burn them in an obliquely-placed platinum crucible, and ignite until every trace of carbon is consumed; then weigh the ash, and divide the amount found by ten; the quotient expresses, with sufficient precision, the average quantity of ash which every individual filter leaves upon incineration.

In the ignition of precipitates, the following four points have to be more particularly regarded:

1. No loss of substance must be incurred;
2. The ignited precipitates must really be the bodies they are represented to be in the calculation of the results;
3. The incineration of the filters must be complete;
4. The crucibles must not be attacked.

The following two methods seem to me the simplest and most appropriate of all that have as yet been proposed. The selection of either depends upon certain circumstances, which I shall immediately have occasion to point out. But no matter which method is resorted to, the precipitate must always be thoroughly dried, before it can properly be exposed to a red heat. The application of a red heat to moist precipitates, more particularly to such as are very light and loose in the dry state (silicic acid, for instance), involves always a risk of loss from the impetuously escaping aqueous vapors carrying away with them minute particles of the substance. Some other substances, as aluminium hydroxide or ferric hydroxide, for instance, form small hard lumps; if such lumps are ignited while still moist within they are liable to fly about with great violence. The best method of drying precipitates as a preliminary to ignition is as described in § 50, the last paragraph.

Respecting the ignition, the *degree* of heat to be applied and the *duration* of the process must, of course, depend upon the nature of the precipitate and upon its deportment at a red heat. As a general rule, a moderate red heat, applied for about five minutes, is found sufficient to effect the purpose; there are, however, many exceptions to this rule which will be indicated where-ever they occur.

Whenever the choice is permitted between porcelain and platinum crucibles, the latter are always preferred, on account of their comparative lightness and infrangibility, and because they are more readily heated to redness. The crucible selected should always be of sufficient capacity, as the use of crucibles deficient in size involves the risk of loss of substance. The proper size, in most cases, is 4 cm. in height, and 3·5 cm. in diameter. That the crucible must be perfectly clean, both inside and outside, need hardly be mentioned. The analyst should acquire the habit of cleaning and polishing the platinum crucible always after using it. This should be done by friction with moist sea-sand whose grains are all round and do not scratch. The sand is rubbed on with the finger, and the desired effect is produced in a few minutes. The adoption of this habit is attended with the pleasure of always working with a bright crucible and the profit of prolonging its existence. This mode of cleaning is all the more necessary, when one ignites over gas-lamps, since at this high temperature crucibles soon acquire a gray coating, which arises from a superficial loosening of the platinum. A little burnishing with sea-sand readily removes the appearance in question, without causing any notable diminution of the weight of the crucible. The foregoing remarks on platinum crucibles refer equally to those of iridium-platinum—which, by the by, are now much used, and very highly to be recommended—only the restoration of the polish is somewhat more difficult with the latter, on account of the greater hardness of the alloy. If there are spots on the platinum or iridium-platinum crucibles, which cannot be removed by the sand without wearing away too much of the metal, a little potassium disulphate is fused in the crucible, the fluid mass shaken about inside, allowed to cool, and the crucible finally boiled with water. There are two ways of cleaning crucibles soiled outside; either the crucible is placed in a larger one, and the interspace filled with potassium disulphate, which is then heated to fusion; or the crucible is placed on a platinum-wire triangle, heated to redness, and then sprinkled over with powdered potassium disulphate. Instead of the sulphate you may use borax. Never forget at last to polish the crucible with sea-sand again.

When the crucible is clean, it is placed upon a clean platinum-wire triangle (fig. 40), ignited, allowed to cool in the desiccator, and weighed. This operation, though not indispensable, is still always advisable, that the weighing of the empty and filled crucible may be performed under as nearly as possible the same circumstances. The empty crucible may of course be weighed *after* the ignition of the precipitate; however, it is preferable in most cases to weigh it *before*. The ignition is effected with a BERZELIUS spirit-lamp or a gas-lamp, or else in a muffle. In igniting reducible substances over lamps, the analyst must always be on his guard against the contact of unconsumed hydrocarbons even in covered crucibles. When gas-lamps are used there is especial need of caution in this respect. Reduction will be avoided if the flame is made no larger than necessary, if the crucible is supported in the upper part of the flame, and if, when the crucible is in a slanting position, it is heated from behind.

Fig. 40.

We pass on now to the description of the special methods.

§ 52.

FIRST METHOD. (*Ignition of the Precipitate with the Filter.*)

This method is resorted to in cases where there is no danger of a reduction of the precipitate by the action of the carbon of the filter. The mode of proceeding is as follows:—

The perfectly dry filter, with the precipitate, is removed from the funnel, and its sides are gathered together at the top, so that the precipitate lies enclosed as in a small bag. The filter is now put into the crucible, which is then covered and heated over a spirit-lamp with double draught, or over gas very gently, to effect the slow charring of the filter; the cover is now removed, the crucible placed obliquely, and a stronger degree of heat applied, until complete incineration of the filter is effected; the lid, which had in the meantime best be kept on a porcelain plate, or in a porcelain crucible, is put on again, and a red heat applied for some time longer, if needed; the crucible is now allowed to cool a little, and is then, while still hot, though no longer red hot,* taken off

* Taking hold of a *red hot* crucible with brass tongs might cause the formation of black rings round it.

with a pair of tongs of brass or polished iron (fig. 41), and put in the desiccator, where it is left to cool; it is finally weighed.

The combustion of the carbon of the filter may be promoted, in cases where it proceeds too slowly, by pushing the non-consumed particles, with a smooth and rather stout platinum wire, within the focus of the strongest action of the heat and air. And the operator may also increase the draught of air by leaning the lid of the crucible against the latter in the manner illustrated in fig. 42.

It will occasionally happen that particles of the carbon of the filter obstinately resist incineration. In such cases the operation may be promoted by putting a small lump of fused, dry ammonium

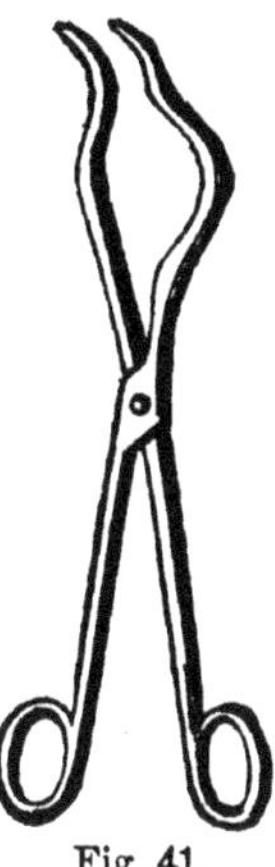

Fig. 41.

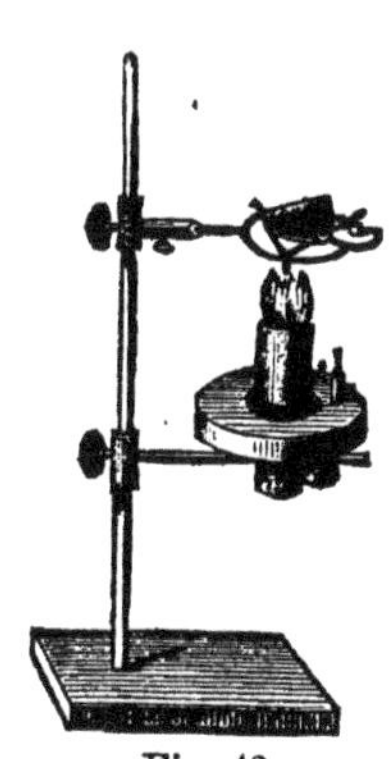

Fig. 42.

nitrate into the crucible, placing on the lid and applying a gentle heat at first, which is gradually increased. However, as this way of proceeding is apt to involve some loss of substance, its application should not be made a general rule.

In cases where the bulk of the precipitate is easily detached from the filter, the preceding method is occasionally modified in this, that the precipitate is put into the crucible, and the filter, with the still adhering particles, folded loosely together, and laid over the precipitate. In other respects, the operation is conducted in the manner above described.

§ 53.

SECOND METHOD. (*Ignition of the Precipitate apart from the Filter.*)

This method is resorted to in cases where a reduction of the precipitate from the action of the carbon of the filter is apprehended; and also where the ignited precipitate is required for further examination, which the presence of the filter ash might embarrass. It may be employed also, instead of the first method, in all cases where the precipitate is easily detached from the filter. The mode of proceeding is as follows:—

The crucible intended to receive the precipitate is placed upon a sheet of glazed paper; the perfectly dry filter with the precipitate is taken out of the funnel, and gently pressed together over the paper, to detach the precipitate from the filter; the precipitate is now shaken into the crucible, and the particles still adhering to the filter are removed from it, as far as practicable, by further pressing or gentle rubbing together of the folded filter, and are then also transferred to the crucible. The filter is now spread open upon the sheet of glazed paper, and then folded in form of a little square box, enclosed on all sides by the parts turned up; any minute particles of the precipitate that may have dropped on the glazed paper are brushed into this little box, with the aid of a small feather; the box is closed again, rolled up, and one end of a long platinum wire spirally wound round it. The crucible being placed on or above a porcelain plate, the little roll is lighted, and, during its combustion, held over the crucible, so that the falling particles of the precipitate or filter ash may drop into it, or, at least, into the porcelain plate. In this way, and by occasionally holding the little roll again in or against the flame, the incineration of the filter is readily and safely effected. When the operation is terminated, a slight tap will suffice to drop the ash and the remaining particles of the precipitate into the crucible, which is then covered, and the ignition completed as in § 52. Where it is intended to keep the ash separate from the precipitate, it is made to drop into the lid of the crucible, in which case it is better to ignite the crucible with the principal portion of the precipitate first. This method of incinerating the filter, devised by BUNSEN, is preferable to the method formerly in use, in which the filter, freed, as far as practicable, from the precipitate, was burnt either whole or cut up into little bits on the lid of the crucible, the operation being promoted when necessary by gently pressing the still unconsumed particles with a platinum wire, or platinum spatula, against the red-hot lid. No matter which method of incineration is resorted to, the operation must always be conducted in a spot entirely protected from draughts.

Certain precipitates suffer some essential modification in their properties, in their solubility, for instance, from ignition. In cases where a portion of a substance of the kind is required, after the weighing, for some other purpose with which the effects of a red heat would interfere, the two operations of drying and igniting may be combined in the following way:—The precipitate is collected on a filter dried at 100°; it is then also dried, at 100°, and weighed (§ 50). A portion of the dry precipitate is put into a tared crucible, and its exact weight ascertained; it is then exposed to a red heat, allowed to cool in the usual way, and weighed again; the diminution of weight which it has undergone is calculated on the whole amount of the precipitate.

§ 53, a.

BUNSEN'S METHOD OF RAPID FILTRATION.*

A precipitate is washed either by filtration or by decantation: in the former case the portion of liquid not mechanically retained is allowed to drain from the precipitate; in the latter it is separated by simply pouring it away, the foreign substances contained in the precipitate being then removed by the repeated addition of some washing-fluid, in each successive portion of which the precipitate is, as far as possible, uniformly suspended, this process being continued until the amount of impurity becomes so minute that its presence may be entirely disregarded.

In the process of filtration as hitherto conducted, the time required is so long and the quantity of wash-water needed so great that some simplification of this continually recurring operation is in the highest degree desirable. The following method, which depends not upon the removal of the impurity by simple attenuation, but upon its displacement by forcing the wash-water through the

* Ann. der Chem. und Pharm., vol. cxlviii. p. 269; Am. Jour. Sci., xlvii. p. 321.

precipitate, appears to me to combine all the requisite conditions and therefore to satisfy the need.

The rapidity with which a liquid filters depends, *cæteris paribus*, upon the difference which exists between the pressure upon its upper and lower surfaces. Supposing the filter to consist of a solid substance, the pores of which suffer no alteration by pressure, or by any other influence, then the volume of liquid filtered in the unit of time is nearly proportional to the difference in pressure: this is clearly shown by the following experiments, made with pure water and a filter consisting of a thin plate of artificial pumice-stone. The thin plate of pumice was hermetically fastened into a funnel consisting of a graduated cylindrical glass vessel, the lower end of which was connected with a large thick flask by means of a tightly fitting caoutchouc cork. The pressure in the flask was then reduced by rarefying the air by means of a method to be described upon another occasion; and for each difference of pressure p, measured by a mercury column, the number of seconds t was observed which a given quantity of water occupied in passing through the filter. The following are the results:—

I.

p. metre.	t. "	pt.
0·179	91·7	16·4
0·190	81·0	15·4
0·282	52·9	14·9
0·472	33·0	15·6

In the ordinary process of filtration, p on the average amounts to no more than 0·004 to 0·008 metre. The advantage gained, therefore, is easily perceived when we can succeed by some simple, practicable, and easily attainable method in multiplying this difference in pressure one or two hundred times, or, say, to an entire atmosphere, without running any risk of breaking the filter. The solution of this problem is very easy: an ordinary glass funnel has only to be so arranged that the filter can be completely adjusted to its side even to the very apex of the cone. For this purpose a glass funnel is chosen possessing an angle of 60°, or as nearly 60° as possible, the walls of which must be completely free from inequalities of every description; and into it is placed a second funnel made of exceedingly thin platinum-foil, and the sides of which possess exactly the same inclination as those of the glass funnel. An ordinary paper filter is then introduced into this compound funnel in the usual manner; when carefully moistened and so adjusted that no air-bubbles are visible between it and the glass, this filter, when filled with a liquid, will support the pressure of an extra atmosphere without ever breaking.

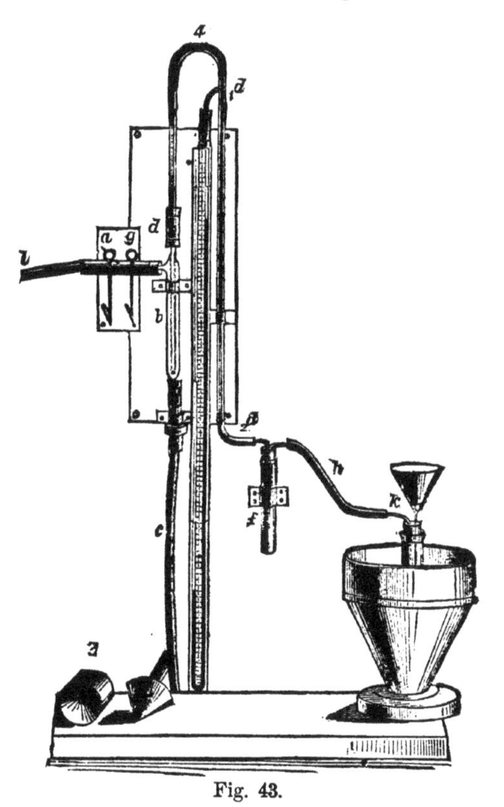

Fig. 43.

The platinum funnel is easily made from thin platinum-foil in the following manner:—In the carefully chosen glass funnel is placed a *perfectly accurately fitting* filter made of writing-paper; this is kept in position by dropping a little melted sealing-wax between its upper edge and the glass; the paper is next saturated with oil and filled with liquid plaster of Paris, and before the mixture solidifies a small wooden handle is placed in the centre.

After an hour or so the plaster cone with the adhering paper filter can be withdrawn by means of the handle from the funnel, to which it accurately corresponds. The paper on the outside of the cone is again covered with oil, and the whole carefully inserted into liquid plaster of Paris contained in a small crucible 4 or 5 centims. in height. After the mixture has solidified, the cone may be easily withdrawn; the adhering paper filter is then detached, and any small pieces of paper still remaining removed by gentle rubbing with the finger. In this manner a solid cone is obtained accurately fitting into a hollow cone, and of which the angle of inclination perfectly corresponds with that of the glass funnel.

Fig. 43, 1, represents the cones. By their help the small platinum funnel is made. A piece of platinum (shown three-fourths of the natural size in fig. 44)* is cut from foil of such a thickness that one square centimetre weighs about 0·154 grm., and from the centre *a* a vertical incision is made by the scissors to the edge *c b d*. The small piece of foil is next rendered pliable by being heated to redness, and is placed upon the solid cone in such a manner that its centre *a* touches the apex of the latter; the side *a b d* is then closely pressed upon the plaster, and the remaining portion of the platinum wrapped as equally and as closely as possible around the cone. On again heating the foil to redness, pressing it once more upon the cone, and inserting the whole into the hollow cone, and turning it round once or twice under a gentle pressure, the proper shape is completed. The platinum funnel, which should not allow of the transmission of light through its extreme point, even now possesses such stability that it may be immediately employed for any purpose. If desired, it may be made still stronger by soldering down the overlapping portion in one spot only to the upper edge of the foil by means of a grain or two of gold and borax; in general, however, this precaution is unnecessary. If the shape has in any degree altered during this latter process, it is simply necessary to drop the platinum funnel into the hollow cone and then to insert the solid cone, when by one or two turns of the latter the proper form

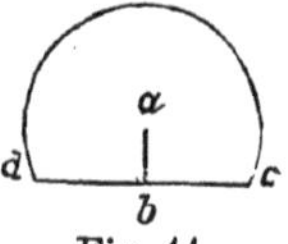

Fig. 44.

* The diameter of *a* in the original drawing is 2·5 centimetres. Perforated platinum cones admirably adapted for use with the BUNSEN filtering apparatus can now be purchased of dealers in chemical apparatus, or of the manufacturer, Mr. J. Bishop, Sugartown, Chester Co., Pa.

may be immediately restored. The platinum funnel is placed in the bottom of the glass funnel, the dry paper filter then introduced in the ordinary manner, moistened, and freed from all adhering air-bubbles by pressure with the finger. A filter so arranged and in perfect contact with the glass, when filled with a liquid will support the pressure of an entire atmosphere without the least danger of breaking; and the interspace between the folds of the platinum foil is perfectly sufficient to allow of the passage of a continuous stream of water.

In order to be able to produce the additional pressure of an atmosphere, the filtered liquid is received in a strong glass flask instead of in beakers.* This flask is closed by means of a doubly perforated caoutchouc cork, through one of the holes of which the neck of the glass funnel is passed to a depth of *from* 5 *to* 8 *centimetres* (fig. 43, *k*); through the other is fitted a narrow tube open at both ends, the lower end of which is brought *exactly to the level of the lower surface of the cork*, to the other is adapted the caoutchouc tube connected with the apparatus destined to produce the requisite difference in pressure: this apparatus will be described immediately. The flasks are placed in a metallic or porcelain vessel, in the conical contraction of which several strips of cloth are fastened. This method of supporting the flask has the advantage that, in one and the same vessel, flasks varying in size from 0·5 to 2·5 litres stand equally well and that by simply laying a cloth over the mouth of the vessel, the consequences of an explosion (which through inexperience or carelessness is possible) are rendered harmless.

It is impossible to employ any of the air-pumps at present in use to create the difference in pressure, since the filtrate not unfrequently contains chlorine, sulphurous acid, hydric sulphide, and other substances which would act injuriously upon the metallic portions of these instruments. I therefore employ a *water* air-pump constructed on the principle of SPRENGEL's mercury-pump, and which appears to me preferable to all other forms of air-pump for chemical purposes, since it effects a rarefaction to within 6 or 12 millimetres pressure of mercury.

Fig. 43 shows the arrangement of this pump. On opening the pinch-cock *a*, water flows from the tube *l* into the enlarged glass

* These flasks must be somewhat thicker than those ordinarily used, in order to prevent the possibility of their giving way under the atmospheric pressure.

vessel *b*, and thence down the leaden pipe *c*. This pipe has a diameter of about 8 millims., and extends downward to a depth of 30 or 40 feet, and ends in a sewer or other arrangement serving to convey the water away. The lower end of the tube *d* possesses a narrow opening; it is hermetically sealed into the wider tube *b*, and reaches nearly to the bottom of the latter. A manometer is attached to the upper continuation of this tube *d* by means of a side tube at d^1; at d^2 is attached a strong thick caoutchouc tube possessing an internal diameter of 5 millims. and an external diameter of 12 millims.; this leads to the flask which is to be rendered vacuous, and is connected with it by means of the short narrowed tube *k*. Between the air-pump and the flask is placed the small thick glass vessel *f*, in which, when one washes with hot water, the steam which may be carried over is condensed. All the caoutchouc joinings are made with very thick tubing, the internal diameter of which amounts to about 5 millims., the external diameter to about 17 millims. The entire arrangement is screwed down upon a board fastened to the wall, in such a manner that each separate piece of the apparatus is held by a single fastening only, in order to prevent the tubes being strained and broken by the possible warping of the board. On releasing the pinchcock *a*, water flows from the conduit *l* down the tube *c* to a depth of more than 30 feet, carrying with it the air which it sucks through the small opening of the tube *d* in the form of a continuous stream of bubbles. No advantage is gained by increasing the rapidity of the flow, since the friction exerted by the water upon the sides of the leaden pipe acts directly as a counter-pressure, and a comparatively small increase in the rapidity of the flow is accompanied by a great increase in the amount of this friction. Accordingly at *g* is a second pinchcock, by which the stream can be once for all so regulated that, on completely opening the cock *a*, the friction, on account of the diminished rate of flow, is rendered sufficiently small to allow of the maximum degree of rarefaction. Such an apparatus, when properly regulated once for all by means of the cock *g*, exhausts in a comparatively short time the largest vessels to within a pressure of mercury equal to the tension of aqueous vapor at the temperature possessed by the stream.* The tension

* The time required to obtain the above degree of exhaustion in a flask of from 1 to 3 litres capacity ranges from six to ten minutes; the quantity of water necessary amounts to about 40 or 50 litres.

exerted by the water-stream in my laboratory, in which six of these pumps are used, amounts to about 7 millims. in winter and 10 millims. in summer. The filtration is made in the following manner: The flask standing in the metallic or porcelain vessel is connected by means of the slightly drawn-out tube *k* with the caoutchouc tube *h* attached to the pump, the cock *a* having been previously opened and the properly fitted moistened filter filled with the liquid to be filtered. As usual, the clear supernatant fluid is first poured upon the filter; in a moment or two the filtrate runs through in a continuous stream, often so rapidly that one must hasten to keep up the supply of liquid, since it is advisable to maintain the filter as full as possible. After the precipitate has been entirely transferred, the filtrate passes through drop by drop, and the manometer not unfrequently now shows a pressure of an extra atmosphere. The filter may be filled (in fact this is to be recommended) with the precipitate to within a millimetre of its edge, since the precipitate, in consequence of the high pressure to which it is subjected, becomes squeezed into a thin layer broken up by innumerable fissures. As soon as the liquid has passed through and the first traces of this breaking up become evident, the precipitate will be found to have been so firmly pressed upon the paper, that on cautiously pouring water over it it remains completely undisturbed. The washing is effected by carefully pouring water down the side of the funnel to within a centimetre *above* the rim of the filter: the washing flask for this purpose is not applicable; the water must be poured from an open vessel. After the filter has in this manner been replenished four times with water and allowed to drain for a few minutes, it will be found to be already so far dried, in consequence of the high pressure to which it has been subjected, that without any further desiccation it may be withdrawn, together with the precipitate, from the funnel, and immediately ignited, with the precautions to be presently given, in the crucible.

§ 53, b.

BUNSEN'S SIMPLIFIED EXHAUSTING APPARATUS.

It is not necessary to use a pump as powerful as that described, since a fall of 10 or 15 feet is sufficient to filter a precipitate according to the above described method, and so far to dry it that it can be immediately ignited in the crucible. The simple arrangement represented in fig. 45 answers this purpose. It consists of two equal-sized bottles, *a* and *a'*, of from 2 to 4 litres capacity, each of which is provided near the bottom with a small stopcock designed to regulate the flow of water. Suppose *a* filled with water and placed upon a shelf as high above the ground as possible, and *a'* placed empty on the floor, and the two stopcocks connected by means of caoutchouc tubing *c*, then on allowing water to flow down the tube the air in the upper bottle becomes somewhat rarefied; and in order to employ the consequent difference in pressure (amounting to a column of mercury about 0·2 metre in height) for the purpose of filtration, it is only necessary to connect the mouth of the upper bottle with the tube of the filter-flask. When the water has ceased to flow, the position of the bottle is reversed, when the operation recommences. So small a pressure as 0·2 metre suffices to render the filter and its contents so far dry that they may be immediately withdrawn from the funnel and ignited without any other preliminary desiccation.

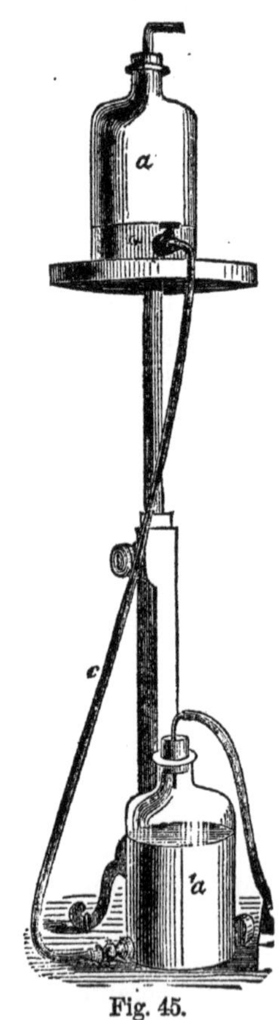

Fig. 45.

§ 53, c.

Bunsen's Method of Drying and Igniting Precipitates.

If a precipitate be heated in a platinum crucible immediately after filtration by the older process, a portion will inevitably be projected out of the crucible. Hitherto, therefore, it has been necessary to dry the filter and precipitate before ignition. Now to dry a quantity of hydrated chromium sesquioxide containing 0·2436 grm. Cr^2O^3 in a water-bath at 100° C. requires at least five hours; and, moreover, bringing the dried precipitate into the crucible, burning the filter, and gradually igniting the mass is in the highest degree tedious and troublesome. All this expenditure of time and labor may be saved by employing the new method. By its means a precipitate is as completely dried upon the filter in from 1 to 5 minutes as if it had been exposed from 5 to 8 hours in a drying chamber; and it can immediately, filter and all, be thrown into a platinum or porcelain crucible and ignited without the slightest fear of its spurting. By operating in the following manner the filter burns quietly without flame or smoke; this phenomenon, although remarkable, easily admits of an explanation. The portion of filter-paper free from precipitate is tightly wrapped round the remainder of the filter in such a manner that the precipitate is enveloped in from four to six folds of clean paper. The whole is then dropped into the platinum or porcelain crucible lying obliquely upon a triangle over the lamp, and pushed down against its sides with the finger. The cover is then supported against the mouth of the crucible in the ordinary way, and the ignition commenced by heating the portion of the crucible in contact with the cover. When the flame has the proper size and position, the filter carbonizes quietly without any appearance of flame or considerable amount of smoke. When the carbonization proceeds too slowly, the flame is moved a little toward the bottom of the crucible. After some time the precipitate appears to be surrounded only by an extremely thin envelope of carbon, possessing exactly the form (of course diminished in size) of the original filter; the flame is then increased, and the crucible maintained at a bright-red heat until the carbon contained in this envelope is consumed. The combustion proceeds so quietly that the resulting ash surrounding the precipitate possesses, even to the smallest fold, the exact form of the original filter. If the ash shows here and there a dark color, it is simply necessary to heat the crucible over a blast-lamp for a few minutes to effect the complete removal of the trace of carbon. This method of burning a filter is extremely convenient and accurate; it is only necessary to give a little attention at first to the slow carbonization of the paper, after which the further progress of the operation may be left to itself.

Gelatinous, finely divided, granular, and crystalline precipitates, such as alumina, calcium oxalate, barium sulphate, silica, &c., may with equal facility be treated in this manner; so that even in this particular the work, in comparison with the method generally adopted, is considerably shortened and simplified. [This method should not, of course, be used when the substance to be ignited is such as to be injuriously affected by the reducing action of filter-paper.]

§ 53, d.

USE OF ASBESTOS FILTERS WITH BUNSEN'S FILTERING APPARATUS.

A method of filtering, washing, and igniting precipitates without the use of paper filters, which in many cases possesses great advantages, has been devised by F. A. GOOCH, and is described as follows.* *First.* White, silky, anhydrous asbestos is scraped to a fine short down with an ordinary knife-blade, boiled with hydrochloric acid to remove traces of iron or other soluble matter, washed by decantation, and set aside for use.

Fig. 46.

Secondly. A platinum crucible of ordinary size, preferably of the broad low pattern (fig. 46), is chosen, and the bottom (fig. 47) perforated with fine holes (the more numerous and the finer the better) by means of a steel point; or better still, the bottom may be made of fine platinum gauze. Next, a Bunsen funnel of the proper size is selected, and over the top a short piece of rubber tubing † is stretched and drawn down until the portion above the funnel arranges itself at right angles to the stem. Within the opening in the rubber, the perforated crucible is fitted as shown in fig. 48, and the funnel is connected with the receiver of a Bunsen pump or other exhausting apparatus in the ordinary way.

Fig. 47.

Fig. 48.

To make the asbestos filter, the pressure of the pump is applied, and a little of the asbestos prepared as described, and suspended in water, is poured into the crucible. The rubber and the crucible

* Proceedings of Am. Acad. Arts and Sciences, 1878, p. 342.

† If suitable rubber tubing is not at hand for fitting the crucible to the funnel, a piece of strong glass tube, preferably tapering slightly, may be used in place of a funnel. The diameter of the tube should exceed that of the crucible. One end is drawn down to size of a common funnel stem; the crucible is then fitted to the large end by means of a short section of large rubber tubing, or a bored rubber stopper.

are held together by the exhaustion of the vacuum pump with sufficient force to make an air-tight joint; the water is drawn through and the asbestos is deposited almost instantly in a close compact layer on the perforated bottom; more asbestos (if necessary) in suspension as before being poured upon the first until the layer becomes sufficiently thick for the purpose for which it is intended. Finally a little distilled water is drawn through the apparatus to wash away any filaments which might cling to the under side, and the filter is ready for use; the whole process occupying less time than is required to fold and fit an ordinary paper filter to a funnel.

To prepare the filter for the weighing of a precipitate, the crucible with the felt of asbestos undisturbed is removed from the funnel and ignited. In case the precipitate to be subsequently collected must be heated to high temperature for a long time, it is better to enclose the perforated crucible with its felt within another crucible; because in such cases asbestos felt is apt to curl at the edges, and without such precaution some of the precipitate might drop through the perforations and be lost. For drying at low temperatures, however, and even for ordinary ignition, a second crucible is unnecessary; but, during the ignition of an easily reducible substance care must be taken to avoid contact of unburnt gas with the perforated bottom.

To perform the filtration, the crucible is replaced in the funnel, the pressure applied, and the process conducted precisely as in ordinary filtration by the Bunsen pump. It is necessary to observe that the vacuum pump be started before pouring the liquid upon the filter. The final drying or ignition, as the case may be, of the precipitate and filter is made without difficulty, or need of extra precaution.

For turbid liquids, or gelatinous precipitates, instead of the perforated crucible a platinum cone may be used, the upper part being made of foil, the lower part of gauze. This process is recommended not only for such precipitates as have heretofore usually been collected upon weighed paper filters, but also for many other precipitates which are usually ignited, but whose proper ignition is more or less interfered with by the presence of carbon.

§ 54.

5. Analysis by Measure (Volumetric Analysis).

The principle of volumetric analysis has been explained already in the "Introduction," where we have seen how the quantity of iron present in a fluid as a ferrous salt may be determined by means of a solution of potassium permanganate, the value of which has been previously ascertained by observing the quantity required to convert a known amount of iron from a ferrous to a ferric salt.

Solutions of accurately known composition or strength, used for the purposes of volumetric analysis, are called *standard solutions.* They may be prepared in two ways, viz., (*a*) by dissolving a weighed quantity of a substance in a definite volume of fluid; or (*b*), by first preparing a suitably concentrated solution of the reagent required, and then determining its exact strength by a series of experiments made with it upon weighed quantities of the body for the determination of which it is intended to be used.

In the preparation of standard solutions by method *a*, the weight of the reagent taken for 1000 c.c. may, if desired, be a weight exactly *equivalent* to 1 gramme of hydrogen (see § 192, *c*, *δ*). In the case of standard solutions prepared by method *b*, this may also be easily done, by diluting to the required degree the still somewhat too concentrated solution, after having accurately determined its strength; however, as a rule, this latter process is only resorted to in technical analyses, where it is desirable to avoid all calculation. Fluids which contain the eq. number of grammes of a substance in one litre, are called *normal solutions;* those which contain $\frac{1}{10}$ of this quantity, *decinormal solutions.*

The determination of a standard solution intended to be used for volumetric analysis is obviously a most important operation; since any error in this will, of course, necessarily falsify every analysis made with it. In scientific and accurate researches it is, therefore, always advisable, whenever practicable, to examine the standard solution—no matter whether prepared by method *a*, or by method *b*, with subsequent dilution to the required degree—by experimenting with it upon accurately weighed quantities of the body for the determination of which it is to be used.

In the previous remarks I have made no difference between fluids of known composition and those of known power; and this has hitherto been usual. But by accepting the two expressions as synonymous, we take for granted that a fluid exercises a chemical action exactly corresponding to the amount of dissolved substance it contains—that, for instance, a solution of sodium chloride containing 1 mol. Na Cl will precipitate exactly 1 at. silver. This presumption, however, is very often not absolutely correct, as will be shown with reference to this very example, § 115, *b*, 5. In such cases, of course, it is not merely advisable, but even absolutely necessary, to determine the strength of the fluid by experiment, although the amount of the reagent it contains may be exactly known, for the power of the fluid can be inferred from its composition only approximately and not with perfect exactness. If a standard solution keeps unaltered, this is a great advantage, as it dispenses with the necessity of determining its strength before every fresh analysis.

That particular change in the fluid operated upon by means of a standard solution which marks the completion of the intended decomposition, is termed the FINAL REACTION. This consists either in a *change of color*, as is the case when a solution of potassium permanganate acts upon an acidified solution of ferrous salt, or a solution of iodine upon a solution of sulphuretted hydrogen mixed with starch paste; or in the *cessation of the formation of a precipitate* upon further addition of the standard solution, as is the case when a standard solution of sodium chloride is used to precipitate silver from its solution in nitric acid; or in *incipient precipitation*, as is the case when a standard solution of silver is added to a solution of hydrocyanic acid mixed with an alkali; or in *a change in the action of the examined fluid upon a particular reagent*, as is the case when a solution of sodium arsenite is added, drop by drop, to a solution of chloride of lime, until the mixture no longer imparts a blue tint to paper moistened with potassium iodide and starch-paste, &c.

The more sensitive a final reaction is, and the more readily, positively, and rapidly it manifests itself, the better is it calculated to serve as the basis of a volumetric method. In cases where it is an object of great importance to ascertain with the greatest practicable precision the exact moment when the reaction is completed, the analyst may sometimes prepare, besides the actual standard solution, another, ten times more dilute, and use the latter to finish the process, carried nearly to completion with the former.

But a good final reaction is not of itself sufficient to afford a safe basis for a good volumetric method; this requires, as the first and most indispensable condition, that the particular decomposition which constitutes the leading point of the analytical process should —at least under certain known circumstances—remain unalterably the same. Wherever this is not the case—where the action varies with the greater or less degree of concentration of the fluid, or according as there may be a little more or less free acid present; or according to the greater or less rapidity of action of the standard solution; or where a precipitate formed in the course of the process has not the same composition throughout the operation—the basis of the volumetric method is fallacious, and the method itself, therefore, of no value.

SECTION II.

REAGENTS.

§ 55.

For general information respecting reagents, I refer the student to my volume on "Qualitative Analysis."

The instructions given here will be confined to the preparation, testing, and most important uses of those chemical substances which subserve principally and more exclusively the purposes of quantitative analysis. Those reagents which are employed in qualitative investigations, having been treated of already in the volume on the qualitative branch of the analytical science, will therefore be simply mentioned here by name.

The reagents used in quantitative analysis are properly arranged under the following heads:—

A. Reagents for gravimetric analysis in the wet way.
B. Reagents for gravimetric analysis in the dry way.
C. Reagents for volumetric analysis.
D. Reagents used in organic analysis.

The mode of preparing the fluids used in volumetric analysis, will be found where we shall have occasion to speak of their application.

A. REAGENTS FOR GRAVIMETRIC ANALYSIS IN THE WET WAY.

I. SIMPLE SOLVENTS.

§ 56.

1. Distilled Water (see "Qual. Anal.").

Water intended for quantitative investigations must be perfectly pure. Water distilled from glass vessels leaves a residue upon evaporation in a platinum vessel (see experiment No. 5), and is therefore inapplicable for many purposes; as, for instance, for the determination of the exact degree of solubility of sparingly soluble substances. For certain uses it is necessary to free the water by ebullition from atmospheric air and carbonic acid.

2. Alcohol (see "Qual. Anal.").

a. Absolute alcohol. *b.* Common alcohol of various degrees of strength.

3. Ether.

The application of ether as a solvent is very limited. It is more frequently used mixed with alcohol, in order to diminish the solvent power of the latter for certain substances, *e.g.*, ammonium platinic chloride. The ordinary ether of the shops will answer the purpose.

4. Carbon Disulphide (see "Qual. Anal.").

II. ACIDS AND HALOGENS.

a. Oxygen Acids.

§ 57.

1. Sulphuric Acid.

a. Concentrated sulphuric acid of the shops.
b. Concentrated pure sulphuric acid.
c. Dilute sulphuric acid.
See "Qual. Anal."

2. Nitric Acid.

a. Pure nitric acid of 1·2 sp. gr. (see "Qual. Anal.").
b. Red fuming nitric acid (concentrated nitric acid containing some hyponitric acid).

Preparation.—Two parts of pure, dry potassium nitrate are introduced into a capacious retort, and one part of concentrated sulphuric acid is added either through the tubulure of the retort, or if a common non-tubulated one is used, through the neck by means of a long funnel-tube bent at the lower end, carefully avoiding soiling the neck of the retort. The latter being put into a vessel filled with sand, or, better still, with iron turnings, is then connected with a receiver, but not quite air-tight. The distillation is conducted at a gradually increased heat, and carried to dryness. The cooling of the receiver must be properly attended to during the distillation. In the preparation of small quantities, the retort is placed on a piece of wire-gauze, and heated with charcoal; in this process it is always advisable to coat the retort by repeated application of a thin paste made of clay and water; a little borax or sodium carbonate should be added to the water used for making the paste.

Tests.—Red fuming nitric acid must be in a state of the greatest possible concentration, and perfectly free from sulphuric acid. In order to detect minute traces of the latter, evaporate a few c. c. of the specimen in a porcelain dish nearly to dryness, dilute the residue with water, add some barium chloride, and observe whether a precipitate forms on standing.

Uses.—A powerful oxidizing agent and solvent; it serves more especially to convert sulphur and metallic sulphides into sulphuric acid and sulphates respectively.

3. Acetic Acid (see "Qual. Anal.").

4. Tartaric Acid (see "Qual. Anal.").

b. Hydrogen Acids and Hologens.

§ 58.

1. Hydrochloric Acid.

a. Pure hydrochloric acid of 1·12 sp. gr. (see "Qual. Anal.").

b. Pure fuming hydrochloric acid of about 1·18 sp. gr.

Preparation.—As in "Qual. Anal." § 26, with this modification, however, that only 3 or 4 parts of water, instead of 6, are put into the receiver, to 4 parts of sodium chloride in the retort. The greatest care must be taken to keep the receiver cool, and to change it as soon as the tube through which the gas is conducted into it begins to get hot, since it is now no longer hydrochloric acid gas which passes over, but an aqueous solution of the gas, in form of vapor, which would simply weaken the fuming acid, if it were allowed to mix with it.

Tests.—The fuming acid must, for many purposes, be perfectly free from chlorine and sulphurous acid. For the mode of testing for these impurities, see "Qual Anal." loc. cit. Test for sulphuric acid as under Nitric Acid, above.

Uses.—Fuming hydrochloric acid has a much more energetic action than the dilute acid; it is, therefore, used instead of the latter in cases where a more rapid and energetic action is desirable.

2. HYDROFLUORIC ACID.

This is employed for the decomposition of silicates and borates, sometimes in the gaseous form, sometimes in the condition of aqueous solution. In the first case, the substance to be decomposed is introduced into the leaden box, in which the hydrofluoric gas is being generated; in the latter case, we must first prepare the aqueous acid. The raw material employed is fluor spar or kryolite (LUBOLDT*). Both are first finely powdered, and then treated with concentrated sulphuric acid. To 1 part kryolite, 2½ parts sulphuric acid are used; to 1 part fluor spar, 2 parts sulphuric acid are used. If the latter is employed, allow the mixture to stand in a dry place for several days, stirring every now and then, so that the silicic acid (which is generally contained in fluor spar) may first escape in the form of fluosilicic gas. Convenient distillatory apparatus have been described by LUBOLDT (loc. cit.) and by H. BRIEGLEB.† The latter commends itself especially on account of its relatively small cost. It consists of a leaden retort, with a movable leaden top, which can be luted on. The receiver belonging to it is a box of lead, with a tubulure at the side, into which the neck of the retort just enters. The cover of the receiver is raised conical, and is provided at the top with an exit tube of lead. In the receiver a platinum dish containing water is placed, all joints are luted, and the retort is carefully heated in a sand-bath. The aqueous hydrofluoric acid found at the end of the operation in the platinum dish is perfectly pure. The small quantity of impure hydrofluoric acid which collects on the bottom of the receiver is thrown away. The hydrofluoric acid must entirely volatilize when heated in a platinum dish on a water-bath. The pure acid gives no precipitate when neutralized with potash, while potassium silicofluoride separates if the acid contains hydrofluosilicic acid. The acid is best preserved in gutta-percha bottles, as recommended by STÄDELER. The greatest caution must be observed in preparing this acid, since, whether in the fluid or gaseous condition, it is one of the most injurious substances.

3. CHLORINE AND CHLORINE-WATER (see "Qual. Anal.").

4. NITRO-HYDROCHLORIC ACID (see "Qual. Anal.").

5. HYDROFLUOSILICIC ACID (see "Qual. Anal.").

---

* Journ. für prakt. Chem., 76, 330.

† Annal. d. Chem. u. Pharm., 111, 380.

c. Sulphur Acids.

1. HYDROSULPHURIC ACID (see "Qual. Anal.").

III. BASES AND METALS.

a. Oxygen Bases and Metals.

§ 59.

α. Alkali Bases.

1. POTASSIUM HYDROXIDE OR POTASSA AND SODIUM HYDROXIDE OR SODA (see "Qual. Anal.").

All the four sorts of the caustic alkalies mentioned in the qualitative part are required in quantitative analysis, viz., common solution of soda, potassa purified with alcohol, solution of potassa prepared with baryta, and absolutely pure soda. Pure solution of potassa may be obtained also by heating to redness for half an hour in a copper crucible, a mixture of 1 part of potassium nitrate, and 2 or 3 parts of thin sheet copper cut into small pieces, treating the mass with water, allowing the oxide of copper to subside in a tall vessel, and removing the supernatant clear fluid by means of a syphon (WÖHLER).*

2. AMMONIA (see "Qual. Anal.").

β. Alkali-earth Bases.

1. BARIUM HYDROXIDE OR BARYTA (see "Qual. Anal.").

2. CALCIUM HYDROXIDE OR LIME.

Finely divided calcium hydroxide mixed with water (milk of lime), is used more particularly to effect the separation of magnesium, &c., from the alkali metals. Milk of lime intended to be used for that purpose must, of course, be perfectly free from alkalies. To insure this the slaked lime should be thoroughly washed, by repeated boiling with fresh quantities of distilled water. This operation is conducted best in a silver dish. When cold, the milk of lime so prepared is kept in a well-stoppered bottle.

* Sodium hydroxide, made by acting on pure water by pure sodium and fusing in silver vessels, is to be had cheaply of the Magnesium Metal Company, Salford, Manchester, England.

γ. *Heavy Metals, and their Oxides.*

§ 60.

1. Zinc.

Zinc has of late been much used as a reagent in quantitative analysis. It serves more especially to effect the reduction of ferric to ferrous salts, and also the precipitation of copper from solutions of its salts. Zinc intended to be used for the former purpose must be free from iron, for the latter free from lead, copper, and other metals which remain undissolved upon treating the zinc with dilute acids.

To procure zinc which leaves no residue upon solution in dilute sulphuric acid, there is commonly no other resource but to re-distil the commercial article.

This is effected in a retort made of the material of Hessian or black-lead crucibles. The operation is conducted in a wind-furnace with good draught. The neck of the retort must hang down as perpendicular as possible. Under the neck is placed a basin or small tub, filled with water. The distillation begins as soon as the retort is at a bright red heat. As the neck of the retort is very liable to become choked up with zinc, or oxide of zinc, it is necessary to keep it constantly free by means of a pipe-stem. The zinc obtained by this re-distillation is nearly or quite free from lead.

Tests.—The following is the simplest way of testing the purity of zinc: dissolve the metal in dilute sulphuric acid in a small flask provided with a gas-evolution tube, place the outer limb of the tube under water, and when the solution is completed, let the water entirely or partly recede into the flask; after cooling, add to the fluid, drop by drop, a sufficiently dilute solution of potassium permanganate. If a drop of that solution imparts the same red tint to the zinc solution as to an equal volume of water, the zinc may be considered free from iron. I prefer this way of testing the purity of zinc to other methods, as it affords, at the same time, an approximate, or, if the zinc has been weighed, and the permanganate solution (which, in that case, must be considerably diluted) measured, an accurate and precise knowledge of the quantity of iron present. If lead or copper are present, these metals remain undissolved upon solution of the zinc.

2. Lead Oxide.

Precipitate pure lead nitrate or acetate with ammonium carbonate, wash the precipitate, dry, and ignite gently to complete decomposition.

Lead oxide is often used to fix an acid, so that it is not expelled even by a red heat.

b. Sulphur Bases.

1. Ammonium Sulphide (see "Qual. Anal.").

We require both the colorless monosulphide, and the yellow polysulphide.

2. Sodium Sulphide (see "Qual. Anal.").

IV. SALTS.

a. Salts of the Alkalies.

§ 61.

1. Potassium Sulphate (see "Qual. Anal.").
2. Ammonium Oxalate (see "Qual. Anal.").
3. Sodium Acetate (see "Qual. Anal.").
4. Ammonium Succinate.

Preparation.—Saturate succinic acid, which has been purified by dissolving in nitric acid and recrystallizing, with dilute ammonia. The reaction of the new compound should be rather slightly alkaline than acid.

Uses.—This reagent serves occasionally to separate ferric iron from other metals.

5. Sodium Carbonate (see "Qual. Anal.").

This reagent is required both in solution and in pure crystals; in the latter form to neutralize an excess of acid in a fluid which it is desirable not to dilute too much.

6. Ammonium Carbonate (see "Qual. Anal.").
7. Sodium Hydrogen Sulphite (see "Qual. Anal.").
8. Sodium Thiosulphate (Hyposulphite), $N_2S_2O_3$.

This salt occurs in commerce. It should be dry, clear, well crystallized, completely and with ease soluble in water. The solution must give with silver nitrate at first a white precipitate, must not effervesce with acetic acid, and when acidified must give no precipitate with barium chloride, or at most, only a slight turbidity. The acidified solution must, after a short time, become milky from separation of sulphur.

Uses.—Sodium thiosulphate is used for the precipitation of several metals, as sulphides, particularly in separations, for instance, of copper from zinc; it also serves as solvent for several salts (silver chloride, calcium sulphate, &c.); lastly, it is employed in volumetric analysis, its use here depending on the reaction $2(Na_2S_2O_3) + 2I = 2NaI + Na_2S_4O_6$.

9. POTASSIUM NITRITE (see "Qual. Anal.").
10. POTASSIUM DICHROMATE (see "Qual. Anal.").
11. AMMONIUM MOLYBDATE (see "Qual. Anal.").
12. AMMONIUM CHLORIDE (see "Qual. Anal.").
13. POTASSIUM CYANIDE (see "Qual. Anal.").

b. Salts of the Alkali-earth Metals.

§ 62.

1. BARIUM CHLORIDE (see "Qual. Anal.").

The following process gives a very pure barium chloride, free from calcium and strontium:—Transmit through a concentrated solution of impure barium chloride hydrochloric gas, as long as a precipitate continues to form. Nearly the whole of the barium chloride present is by this means separated from the solution, in form of a crystalline powder. Collect this on a filter, let the adhering liquid drain off, wash the powder repeatedly with small quantities of pure hydrochloric acid, until a sample of the washings, diluted with water, and precipitated with sulphuric acid, gives a filtrate which, upon evaporation in a platinum dish, leaves no residue. The hydrochloric mother-liquor serves to dissolve fresh portions of witherite. I make use of the barium chloride so obtained, principally for the preparation of perfectly pure barium carbonate, which is often required in quantitative analyses.

2. BARIUM ACETATE.

Preparation.—Dissolve pure barium carbonate in moderately dilute acetic acid, filter, and evaporate to crystallization.

Tests.—Dilute solution of barium acetate must not be rendered turbid by solution of silver nitrate. See also "Qual. Anal.," *Barium* chloride, the same tests being also used to ascertain the purity of the acetate.

Uses.—Barium acetate is used instead of barium chloride, to effect the precipitation of sulphuric acid, in cases where it is desirable to avoid the introduction of a chloride into the solution, or to convert the base into an acetate. As the reagent is seldom required, it is best kept in crystals.

3. BARIUM CARBONATE (see "Qual. Anal.").

4. STRONTIUM CHLORIDE.

Preparation.—Strontium chloride is prepared from strontianite or celestine, by the same processes as barium chloride. The pure crystals obtained are dissolved in alcohol of 96 per cent., the solution is filtered, and kept for use.

Uses.—The alcoholic solution of strontium chloride is used to effect the conversion of alkali sulphates into chlorides, in cases where it is desirable to avoid the introduction into the fluid of a salt insoluble in alcohol.

5. CALCIUM CHLORIDE (see "Qual. Anal.").

6. MAGNESIUM CHLORIDE OR MAGNESIUM MIXTURE.

Dissolve 11 parts crystallized magnesium chloride ($MgCl_2 + 6 H_2O$) and 28 parts ammonium chloride in 130 parts water, add 70 parts dilute ammonia solution (sp. gr. 0·96). Allow the mixture to stand one or two days and filter. This solution, commonly called "magnesia mixture," is used to precipitate phosphoric acid. An excess is required to effect complete precipitation. Prepared as here described, about 10 c. c. should be used in ordinary cases for every 0·1 gramme P_2O_5.

A solution containing the same per cent. (approximately) of magnesium chloride and other constituents may also be prepared from common calcined magnesia (MgO), provided it is free from the other alkali-earth metals, as follows:—Add to 11 parts magnesia sufficient hydrochloric acid to effect solution, next add a slight excess of magnesia and boil to separate traces of iron; filter, and add 140 parts ammonium chloride and 350 parts dilute ammonia. Dilute with water until volume equals 1000 c. c. for every 11 grammes of MgO used. Allow the mixture to stand two or three days, and filter if necessary.

c. Salts of the Heavy Metals.

§ 63.

1. FERROUS SULPHATE (see "Qual. Anal.").
2. FERRIC CHLORIDE (see "Qual. Anal.").
3. URANIC ACETATE.

Heat finely powdered pitchblende with dilute nitric acid, filter the fluid from the undissolved portion, and treat the filtrate with hydrosulphuric acid to remove the lead, copper, and arsenic; filter again, evaporate to dryness, extract the residue with water, and filter the solution from the oxides of iron, cobalt, and manganese. Uranic nitrate crystallizes from the filtrate; purify this by recrystallization, and then heat the crystals until a small portion of uranic oxide is reduced. Warm the yellowish-red mass thus obtained with acetic acid, filter and let the filtrate crystallize. The crystals are uranic acetate, and the mother-liquor contains the undecomposed nitrate (WERTHEIM).

Tests.—Solution of uranic acetate after acidification with hydrochloric acid must not be altered by hydrosulphuric acid; ammonium carbonate must produce in it a precipitate, soluble in an excess of the precipitant.

Uses.—Uranic acetate may serve, in many cases, to effect the separation and determination of phosphoric acid.

4. SILVER NITRATE (see "Qual. Anal.").
5. LEAD ACETATE (see "Qual. Anal.").
6. MERCURIC CHLORIDE (see "Qual. Anal.").
7. STANNOUS CHLORIDE (see "Qual. Anal.").
8. PLATINIC CHLORIDE (see "Qual. Anal.").
9. SODIUM PALLADIO-CHLORIDE (see "Qual. Anal.").

B. REAGENTS FOR GRAVIMETRIC ANALYSIS IN THE DRY WAY.

§ 64.

1. SODIUM CARBONATE, pure anhydrous (see "Qual. Anal.").
2. MIXED SODIUM AND POTASSIUM CARBONATES (see "Qual. Anal.").
3. BARIUM HYDROXIDE OR BARYTA (see "Qual. Anal." and § 59).
4. POTASSIUM NITRATE (see "Qual. Anal.").
5. SODIUM NITRATE (see "Qual. Anal.").
6. BORAX (fused).

Preparation.—Heat crystallized borax (see "Qual. Anal.) in a platinum or porcelain dish, until there is no further intumescence; reduce the porous mass to powder, and heat this in a platinum crucible until it is fused to a transparent mass. Pour the semi-fluid, viscid mass upon a fragment of porcelain. A better way is to fuse the borax in a net of platinum gauze, by making the gas blowpipe-flame act upon it. The drops are collected in a platinum dish. The vitrified borax obtained is kept in a well-stoppered bottle. But as it is always necessary to heat the vitrified borax previous to use, to make quite sure that it is perfectly anhydrous, the best way is to prepare it only when required.

Uses.—Vitrified borax is used to effect the expulsion of carbonic acid and other volatile acids, at a red heat.

7. Potassium Disulphate.

Preparation.—Mix 87 parts of normal potassium sulphate (see "Qual. Anal."), in a platinum crucible, with 49 parts of concentrated pure sulphuric acid, and heat to gentle redness until the mass is in a state of uniform and limpid fusion. Pour the fused salt on a fragment of porcelain, or into a platinum dish standing in cold water. After cooling, break the mass into pieces, and keep for use.*

Uses.—This reagent serves as a flux for certain native compounds of alumina and chromic oxide. Potassium disulphate is used also, as we have already had occasion to state, for the cleansing of platinum crucibles; for this latter purpose, however, the salt which is obtained in the preparation of nitric acid will be found sufficiently pure.

8. Ammonium Carbonate (solid).

Preparation.—See "Qual. Anal."—This reagent serves to convert the acid alkali sulphates into normal salts. It must completely volatilize when heated in a platinum dish.

9. Ammonium Nitrate.

Preparation.—Neutralize pure ammonium carbonate with pure nitric acid, warm, and add ammonia to slightly alkaline reaction; filter, if necessary, and let the filtrate crystallize. Fuse the crystals in a platinum dish, and pour the fused mass upon a piece of porcelain; break into pieces whilst still warm, and keep in a well-stoppered bottle.

Tests.—Ammonium nitrate must leave no residue when heated in a platinum dish.

* [J. Lawrence Smith advises the use of sodium disulphate for fluxing aluminous compounds, as the fused mass is much more readily soluble in water.]

Uses.—Ammonium nitrate serves as an oxidizing agent; for instance, to convert lead into lead oxide, or to effect the combustion of carbon, in cases where it is desired to avoid the use of fixed salts.

10. Ammonium Chloride.

Preparation and Tests.—See "Qual. Anal."

Uses.—Ammonium chloride is often used to convert metallic oxides and acids, *e.g.*, lead oxide, zinc oxide, stannic oxide, arsenic acid, antimonic acid, &c., into chlorides (ammonia and water escape in the process). Many metallic chlorides being volatile, and others volatilizing in presence of ammonium chloride fumes, they may be completely removed by igniting them with ammonium chloride in excess, and thus many compounds, *e.g.*, alkali antimonates, may be easily and expeditiously analyzed. Ammonium chloride is also used to convert various salts of other acids into chlorides, *e.g.*, small quantities of alkali sulphates.

11. Hydrogen Gas.

Preparation.—Hydrogen gas is evolved when dilute sulphuric acid is added to granulated zinc. It may be purified from traces of foreign gases either by passing first through mercuric chloride solution, then through potash solution, or as recommended by Stenhouse, by passing through a tube filled with pieces of charcoal. If the gas is desired dry, pass through sulphuric acid or a calcium chloride tube.

Tests.—Pure hydrogen gas is inodorous. It ought to burn with a colorless flame, which, when cooled by depressing a porcelain dish upon it, must deposit nothing on the surface of the dish except pure water (free from acid reaction).

Uses.—Hydrogen gas is frequently used, in quantitative analysis, to reduce oxides, chlorides, sulphides, &c., to the metallic state.

12. Chlorine.

Preparation.—See "Qual. Anal."—Chlorine gas is purified and dried by transmitting it through concentrated sulphuric acid, or a calcium chloride tube.

Uses.—Chlorine gas serves principally to produce chlorides, and to separate the volatile from the non-volatile chlorides; it is also used to displace and indirectly determine bromine and iodine.

C. REAGENTS USED IN VOLUMETRIC ANALYSIS.

§ 65.

Under this head are arranged the most important of those substances, which serve for the preparation and testing of the fluids required in volumetric analysis, and have not been given *sub A* and *B*.

1. Pure Crystallized Oxalic Acid, $H_2C_2O_4 + 2H_2O$.

The introduction of crystallized oxalic acid as a basis for alkalimetry and acidimetry is due to Fr. Mohr. It is also employed to determine the strength of, or to *standardize*, a solution of potassium permanganate, 1 molecule of potassium permanganate being required, in the presence of free sulphuric acid, to convert 5 molecules of oxalic acid into carbon dioxide and water ($K_2Mn_2O_8 + 5H_2C_2O_4 + 3H_2SO_4 = K_2SO_4 + 2MnSO_4 + 8H_2O + 10CO_2$). We use in most cases the pure crystallized acid which has the formula $H_2C_2O_4 + 2H_2O$, and of which the molecular weight is accordingly 126.

Preparation.—See "Qual. Anal.," under Ammonium Oxalate.

Tests.—The crystals of oxalic acid must not show the least sign of efflorescence (to which they are liable even at 20° in a dry atmosphere); they must dissolve in water to a perfectly clear fluid; when heated in a platinum dish, they must leave no fixed and incombustible residue (calcium carbonate, potassium carbonate, &c.). If the acid obtained by a first crystallization fails to satisfy these requirements, it must be recrystallized.

2. Tincture of Litmus.

Preparation.—Digest 1 part of litmus of commerce with 6 parts of water, on the water-bath, for some time, filter, divide the blue fluid into 2 portions, and saturate in one half the free alkali, by stirring repeatedly with a glass rod dipped in very dilute nitric acid, until the color just appears red; add the remaining blue half, together with 1 part of strong spirit of wine, and keep the tincture which is now ready for use, in a small open bottle, not quite full, in a place protected from dust. In a stoppered bottle the tincture would speedily lose color.

Tests.—Litmus tincture is tested by coloring with about 100 cubic centimetres of water distinctly blue, dividing the fluid into two portions, and adding to the one the least quantity of a dilute acid, to the other a trace of solution of soda. If the one portion acquires a distinct red, the other a distinct blue tint, the litmus tincture is fit for use, as neither acid nor alkali predominates.

3. POTASSIUM PERMANGANATE.

Preparation.—Mix 8 parts of very finely powdered pure pyrolusite, or manganese binoxide, with 7 parts of potassium chlorate, put the mixture into a shallow cast-iron pot, and add 37 parts of a solution of potassa of 1·27 specific gravity (the same solution as is used in organic analysis *); evaporate to dryness, stirring the mixture during the operation; put the residue before it has absorbed moisture, into an iron or Hessian crucible, and expose to a dull-red heat, with frequent stirring with an iron rod or iron spatula, until no more aqueous vapors escape and the mass is in a faint glow. Remove the crucible now from the fire, and transfer the friable mass to an iron pot. Reduce to coarse powder, and transfer this, in small portions at a time, to an iron vessel containing 100 parts of boiling water; keep boiling, replacing the evaporating water, and passing a stream of carbon dioxide through the fluid (MULDER †). The originally dark-green solution of potassium manganate soon changes, with separation of hydrated manganese binoxide, to the deep violet-red of the permanganate. When it is considered that the conversion is complete, allow to settle, take out a small quantity of the clear liquid, boil and pass carbon dioxide through it. If a precipitate forms, the conversion is not yet complete.

The solution may be filtered through gun-cotton. Evaporate, crystallize, and dry the crystals on a porous tile.

The pure salt is now to be obtained in commerce.

4. AMMONIUM FERROUS SULPHATE.

$$FeSO_4.(NH_4)_2SO_4 + 6H_2O.$$

FR. MOHR has proposed to employ this double salt, which is not liable to efflorescence and oxidation, as an agent to determine the strength of the permanganate solution.

Preparation.—Take two equal portions of dilute sulphuric

---

* Or instead of the solution, use 10 parts of the hydroxide KOH. In this case fuse the potash and the chlorate together first; and then project the manganese into the crucible.

† Jahresbericht von Kopp und Will, 1858, 581.

acid, and warm the one with a moderate excess of small iron nails free from rust, until the evolution of hydrogen gas has altogether or very nearly ceased; neutralize the other portion exactly with ammonium carbonate, and then add to it a few drops of dilute sulphuric acid. Filter the solution of the ferrous sulphate into that of the ammonium sulphate, evaporate the mixture a little, if necessary, and then allow the salt to crystallize. Let the crystals, which are hard and of a pale-green color, drain in a funnel, then wash them in a little water, dry thoroughly on blotting-paper in the air, and keep for use.

The molecular weight of the salt (392) is exactly 7 times the atomic weight of iron (56). The solution of the salt in water which has been just acidified with sulphuric acid must not become red on the addition of potassium sulphocyanate.

5. AMMONIA-IRON-ALUM.

$$(NH_4)_2SO_4 \cdot Fe_2(SO_4)_3 + 24H_2O.$$

Preparation.—Bring into a large porcelain dish 58 grms. of pure crystallized ferrous sulphate (see Fresenius' "Qual. Anal." Am. ed., p. 73), together with a quantity of oil of vitriol equivalent to 8·3 grms. of sulphuric anhydride (SO_3), (see Table, § 191). Heat upon a sand-bath, adding nitric acid from time to time, in small portions, until the iron has all passed into ferric sulphate, or until a drop of the solution gives no blue coloration with potassium ferricyanide. Heat further, and evaporate until the excess of nitric acid is expelled, then add 14 grms. of ammonium sulphate,* and, if need be, hot water sufficient to bring the salt into solution; filter into a porcelain capsule and set aside, under cover, to crystallize.

The iron-alum separates in cubo-octahedrons, which may be yellowish, lilac, or colorless. If dark in color, dissolve in warm water, add a few drops of oil of vitriol, and crystallize again. Rinse the pale or colorless crystals, after separation from the mother-liquor, with cold water, wrap up closely in filter paper, and allow them to dry at the ordinary temperature.†

* If not on hand, this salt may be prepared by saturating oil of vitriol with ammonium carbonate and evaporating to dryness. 30 grammes of oil of vitriol give somewhat more than is required above.

† Examinations of iron-alum thus prepared show that the variations in the

The yield should be about 80 grms. The dry salt should be pulverized, pressed between folds of paper until freed from mechanically adhering water, and preserved in a well-stoppered bottle.

Uses.—Ammonia-iron-alum furnishes the best means of obtaining a definite quantity of iron in a ferric salt for making standard solutions, being easily obtained pure and inalterable if kept away from acid vapors. Its purity may be readily controlled by ascertaining the loss on careful ignition, which should leave a residue of 16·6 per cent. of ferric oxide of iron, corresponding to 11·62 per cent. of metallic iron.

6. Pure Iodine.

Preparation.—Triturate iodine of commerce with $\frac{1}{8}$ part of its weight of potassium iodide, dry the mass in a large watch-glass with ground rim, warm this gently on a sand-bath, or on an iron plate, and as soon as violet fumes begin to escape, cover it with another watch-glass of the same size. Continue the application of heat until all the iodine is sublimed, and keep in a well-closed glass bottle. The chlorine or bromine, which is often found in iodine of commerce, combines, in this process, with the potassium, and remains in the lower watch-glass, together with the excess of potassium iodide.

Tests.—Iodine purified by the process just now described, must leave no fixed residue when heated on a watch-glass. But, even supposing it should leave a trace on the glass, it would be of no great consequence, as the small portion intended for use has to be resublimed immediately before weighing.

---

color of the salt, from colorless to rose, are not connected with appreciable differences of composition.

J. H. Grove, of the Sheffield Laboratory, obtained the following results in the examination of ammonia-iron-alum crystals, the ferric oxide being estimated by ignition :—

	Fe_2O_3
1st	16·59
	16·55
	16·59
2d	16·53
3d	16·57
4th	16·57
5th	16·58
6th	16·50
	16·56
7th	16·55
Calculated	16·60

Uses.—Pure iodine is used to determine the amount of iodine contained in the solution of iodine in potassium iodide, employed in many volumetric processes.

7. Potassium Iodide.

Small quantities of this article may be procured cheaper in commerce than prepared in the laboratory. For the preparation of potassium iodide intended for analytical purposes I recommend Baup's method, improved by Frederking, because the product obtained by this process is free from iodic acid.

Tests.—Put a sample of the salt in dilute sulphuric acid. If the iodide is pure, it will dissolve without coloring the fluid; but if it contain potassium iodate, the fluid will acquire a brown tint, from the presence of free iodine, the sulphuric acid setting free iodic and hydriodic acids which react on each other ($HIO_3 + (HI)_5 = (H_2O)_3 + I_6$) with liberation of iodine which remains in solution. Mix the solution of another sample with silver nitrate, as long as a precipitate continues to form; add solution of ammonia in excess, shake the mixture, filter, and supersaturate the filtrate with nitric acid. The formation of a white, curdy precipitate indicates the presence of chloride in the potassium iodide. Presence of potassium sulphate is detected by means of solution of barium chloride, with addition of some hydrochloric acid.

Uses.—Potassium iodide is used as a solvent for iodine in the preparation of standard solutions of iodine; it is employed also to absorb free chlorine. In the latter case every atom of chlorine liberates an atom of iodine, which is retained in solution by the agency of the excess of potassium iodide. The potassium iodide intended for these uses must be free from potassium iodate and carbonate; the presence of trifling traces of potassium chloride or potassium sulphate is of no consequence.

8. Arsenious Oxide (As_2O_3).

The arsenious oxide sold in the shops in large pieces, externally opaque, but often still vitreous within, is generally quite pure. The purity of the article is tested by moderately heating it in a glass tube, open at both ends, through which a feeble current of air is transmitted. Pure arsenious oxide must completely volatilize in this process; no residue must be left in the tube upon the expulsion of the sublimate from it. If a non-volatile residue is left which, when heated in a current of hydrogen gas, turns black, the arsenious oxide contains antimony teroxide, and is unfit for use in analytical processes. Dissolve about 10 grms. of the arsenious oxide to be tested in soda, and add 1—2 drops lead acetate. If a brownish color is produced, the arsenious oxide contains arsenious sulphide and cannot be used. Arsenious oxide dissolves in a solution of sodium carbonate forming sodium arsenite which is used to determine hypochlorous acid, free chlorine, iodine, &c.

9. SODIUM CHLORIDE.

Perfectly pure rock-salt is best suited for analytical purposes. It must dissolve in water to a clear fluid; ammonium oxalate, sodium phosphate, and barium chloride must not trouble the solution. Pure sodium chloride may be prepared also by MARGUERITTE'S process, viz., conduct into a concentrated solution of common salt hydrochloric gas to saturation, collect the small crystals of sodium chloride which separate on a funnel, let them thoroughly drain, wash with hydrochloric acid, and dry the sodium chloride finally in a porcelain dish, until the hydrochloric acid adhering to it has completely evaporated. The mother-liquor contains the small quantities of calcium sulphate, magnesium chloride, &c., originally present in the salt.

Uses.—Sodium chloride serves as a volumetric precipitating agent in the determination of silver, and also to determine the strength of solutions of silver intended for the estimation of chlorine. We usually fuse it before weighing. The operation must be conducted with caution, and must not be continued longer than necessary; for if the gas-flame acts on the salt, hydrochloric acid escapes, while sodium carbonate is formed.

10. METALLIC SILVER.

The silver obtained by the proper reduction of the pure chloride of the metal alone can be called chemically pure. The silver precipitated by copper is never absolutely pure, but contains generally about $\frac{1}{1000}$ of copper.

Chemically pure silver is only used in small quantity to prepare the dilute solution employed for the determination of silver. The solution of silver required for the estimation of chlorine need not be made with absolutely pure silver, as the strength of this solution had always best be determined *after* the preparation, by means of pure sodium chloride.

D. REAGENTS USED IN ORGANIC ANALYSIS.

§ 66.

1. Cupric Oxide.

Preparation.—Stir pure* copper scales (which should first be ignited in a muffle) with pure nitric acid in a porcelain dish to a thick paste; after the effervescence has ceased, heat gently on the sand-bath until the mass is perfectly dry. Transfer the green basic salt produced to a Hessian crucible, and heat to a moderate redness, until no more fumes of hyponitric acid escape; this may be known by the smell, or by introducing a small portion of the mass into a test tube, closing the latter with the finger, heating to redness, and then looking through the tube lengthways. The uniform decomposition of the salt in the crucible may be promoted by stirring the mass from time to time with a hot glass rod. When the crucible has cooled a little, reduce the mass, which now consists of pure cupric oxide, to a tolerably fine powder, by triturating it in a brass or porcelain mortar; pass through a metal sieve, and keep in a well-stoppered bottle for use. It is always advisable to leave a small portion of the oxide in the crucible, and to expose this again to an intense red heat. This agglutinated portion is not pounded, but simply broken into small fragments.

Another method is to dissolve pure copper in pure nitric acid, evaporate to dryness in a porcelain dish, ignite the copper nitrate thus obtained in a Hessian crucible until no fumes arise on stirring the top of the mass with a rod. A portion in the bottom of the crucible will be sintered if a proper heat has been applied, while the upper part will be pulverulent. Treat the sintered portion as above, and reserve each separately. This method gives a reliable product.

Tests.—Pure cupric oxide is a compact, heavy, deep-black powder, gritty to the touch; upon exposure to a red heat it must evolve no hyponitric acid fumes, nor carbon dioxide; the latter would indicate presence of fragments of charcoal, or particles of dust. It must contain nothing soluble in water. That portion of the oxide which has been exposed to an intense red heat should be hard, and of a grayish-black color.

* If the scales contain lime, digest them with water, containing a little nitric acid, for a long time; wash, and then proceed as above.

Uses.—Cupric oxide serves to oxidize the carbon and hydrogen of organic substances, yielding up its oxygen wholly or in part, according to circumstances. That portion of the oxide which has been heated to the most intense redness is particularly useful in the analysis of volatile fluids.

N.B. The cupric oxide, after use, may be regenerated by oxidation with nitric acid, and subsequent ignition. Should it have become mixed with alkali salts in the course of the analytical process, it is first digested with very dilute cold nitric acid, and washed afterwards with water. To purify cupric oxide containing chloride, E. ERLENMEYER recommends to ignite it in a tube, first in a stream of moist air, and finally, when the escaping gas ceases to redden litmus paper, in dry air. By these operations any oxides of nitrogen that may have remained are also removed.

2. LEAD CHROMATE.

Preparation.—Precipitate a clear filtered solution of lead acetate, slightly acidulated with acetic acid, with a small excess of potassium dichromate; wash the precipitate by decantation, and at last on a linen strainer; dry, put in a Hessian crucible, and heat to bright redness until the mass is fairly in fusion. Pour out upon a stone slab or iron plate, break, pulverize, pass through a fine metallic sieve, and keep the tolerably fine powder for use.

Tests.—Lead chromate is a heavy powder, of a dirty yellowish-brown color. It must evolve no carbon dioxide upon the application of a red heat; the evolution of carbon dioxide would indicate contamination with organic matter, dust, &c. It must contain nothing soluble in water.

Uses.—Lead chromate serves, the same as cupric oxide, for the combustion of organic substances. It is converted, in the process of combustion, into chromic oxide and basic lead chromate. It suffers the same decomposition, with evolution of oxygen, when heated by itself above its point of fusion. The property of lead chromate to fuse at a red heat renders it preferable to cupric oxide as an oxidizing agent, in cases where we have to act upon difficultly combustible substances.

N.B. Lead chromate may be used a second time. For this purpose it is fused again (being first roasted, if necessary), and then powdered. After having been twice used it is powdered, moistened with nitric acid, dried, and fused. In this way the lead chromate may be used over and over again indefinitely (VOHL*).

3. OXYGEN GAS.

Preparation.—Triturate 100 grammes of potassium chlorate with 5 grammes of finely pulverized manganese binoxide, and introduce the mixture into a plain retort, which must not be more than half full; expose the retort over a charcoal fire or a gas-lamp, at first to a gentle, and then to a gradually increased heat. As soon as the salt begins to fuse, shake the retort a little, that the contents may be uniformly heated. The evolution of oxygen speedily commences, and proceeds rapidly at a relatively low temperature, provided the above proportions be adhered to. As soon as the air is expelled from the retort, connect the glass tube fixed in the neck of the retort by means of a tight-fitting cork, with an india-rubber tube inserted in the lower orifice of the gasometer; the glass tube must be sufficiently wide, and there must be sufficient space left around the india-rubber to permit the free efflux of displaced water. Continue the application of heat to the retort till the evolution of gas has ceased. 100 grammes of potassium chlorate give about 27 litres of oxygen.

The oxygen produced by this process is moist, and may contain traces of carbon dioxide, and also of chlorine. These impurities must be removed and the oxygen thoroughly dried, before it can be used in organic analysis. The gas is therefore passed from the gasometer first through a solution of potassa of 1·27 sp. gr., then through U tubes containing granulated soda lime, and finally, according to circumstances, through U tubes containing calcium chloride or pumice-stone moistened with sulphuric acid.

Tests.—A chip of wood which has been kindled and blown out so as to leave a spark at the extremity must immediately burst into flame in oxygen gas. The gas must not render lime-water or a solution of silver nitrate turbid when transmitted through these fluids.

4. SODA-LIME.

Preparation.—Take solution of soda (NaOH), ascertain its specific gravity, weigh out a certain quantity, calculate the weight of sodium hydroxide present, add twice this latter weight of the best quick-lime, allow the lime to slake, and then evaporate to dryness

* Annalen d. Chem. u. Pharm., 106, 127.

in an iron vessel. Heat the residue in an iron or Hessian crucible; keep for some time at a low red heat. Break up while still warm in an iron mortar, and pass the whole through a sieve with meshes about 3 mm. wide. Reject the finest portion (removing it with a fine sieve) and keep the granulated product in a well-closed bottle.

Use.—Granulated soda-lime prepared as above described forms an excellent absorbent for carbon dioxide. It was formerly also used for nitrogen determination instead of the following:

5. Soda-lime for Nitrogen Determinations.*

Preparation.—Equal weights of sal-soda in clean (washed) large crystals and of good white promptly slaking quick-lime are separately so far pulverized as to pass through holes of $\frac{1}{16}$ inch, then well mixed together, placed in an iron pot which should not be more than half filled, and gently heated, at first without stirring. The lime soon begins to combine with the crystal water of the sodium carbonate, the whole mass heats strongly, swells up, and in a short time yields a fine powder, which may then be stirred to effect intimate mixture and to drive off the excess of water so that the mass is not perceptibly moist and yet short of the point at which it rises in dust on handling. When cold it is secured in well-closed bottles or fruit jars, and is ready for use.

6. Metallic Copper.

Metallic copper serves, in the analysis of nitrogenous substances, to effect the reduction of the nitric oxide gas that may form in the course of the analytical process.

It is used either in the form of turnings, or copper scales reduced by hydrogen; or of small rolls made of fine copper wire gauze. A length of from 7 to 10 centimetres is given to the spirals or rolls, and just sufficient thickness to admit of their being inserted into the combustion tube. To have it perfectly free from dust, oxide, &c., it is first heated to redness in the open air, in a crucible, until the surface is oxidized; it is then put into a glass or porcelain tube, through which an uninterrupted current of dry hydrogen gas is transmitted; and when all atmospheric air has been expelled from the evolution apparatus and the tube, the latter is in its whole length heated to redness. The operator should

* S. W. Johnson. Report of the Conn. Agr. Expr. Station, 1878, p. 111.

make sure that the atmospheric air has been thoroughly expelled, before he proceeds to apply heat to the tube; neglect of this precaution may lead to an explosion.

7. Potassium Hydroxide or Potassa.

a. Solution of Potassa.

Solution of potassa is prepared from the carbonate, with the aid of milk of lime, in the way described in the "Qualitative Analysis," for the preparation of solution of soda. The proportions are—1 part of potassium carbonate to 12 parts of water, and $\frac{2}{3}$ part of lime, slaked to paste with three times the quantity of warm water.

The decanted clear solution is evaporated, in an iron vessel, over a strong fire, until it has a specific gravity of 1·27; it is then, whilst still warm, poured into a bottle, which is well closed, and allowed to stand at rest until all solid particles have subsided. The clear solution is finally drawn off from the deposit, and kept for use.

b. Fused Potassa (common).

The commercial potassa in sticks (impure KOH usually combined with more or less H_2O) will answer the purpose. If you wish to prepare it, evaporate solution of potassa (*a*) in a silver vessel, over a strong fire, until the residuary hydroxide flows like oil, and white fumes begin to rise from the surface. Pour the fused mass out on a clean iron plate, and break it up into small pieces. Keep in a well-stoppered bottle for use.

c. Potassa (purified with alcohol), see "Qual. Anal.," p. 43.

Uses.—Solution of potassa serves for the absorption, and at the same time for the estimation of carbon dioxide. In many cases, a tube filled with fragments of fused potassa is used, in addition to the apparatus filled with solution of potassa. Potassa purified with alcohol, which is perfectly free from potassium sulphate, is employed for the determination of sulphur in organic substances.

8. Calcium Chloride.

a. Pure Calcium Chloride.

Preparation.—Dissolve marble in commercial hydrochloric acid diluted with four or five times its volume of water. (The waste solution resulting from the preparation of carbon dioxide may be used.) Add to this solution with stirring lime, slaked with sufficient water to give it the consistency of thin paste until it gives an alkaline reaction and a pellicle of calcium carbonate forms on the surface on standing exposed to the air. Iron, manganese, and especially magnesium are usually present in such a solution, and are precipitated by the calcium hydroxide—the iron, however, not completely. After a few hours, filter and pass hydrogen sulphide through the alkaline solution until a filtered portion is no longer blackened by this reagent. Let the solution stand for twelve hours, then filter from the iron sulphide. Add next hydrochloric acid to strongly acid reaction to convert the calcium sulphide and calcium oxychloride which may be present into chloride. Concentrate in a porcelain dish. If sulphur separates, after a short time filter again, and continue the evaporation to dryness with addition of a little more hydrochloric acid toward the end of the process. Finally expose the residue to a tolerably strong heat about (200°) on the sand-bath, until it is changed throughout to a white porous perfectly opaque mass, which point can be ascertained by breaking up a piece detached from the top. The product is $CaCl_2 + (H_2O)_2$. Reduce while still hot to granules of the proper size ($\frac{1}{8}$ to $\frac{1}{16}$ of an inch) by means of suitable sieves and a mortar previously warmed, and keep in well-closed bottles.

b. Crude fused Calcium Chloride.

Preparation.—Neutralize the alkaline solution obtained in *a* (without separating the little iron present with H_2S) exactly with hydrochloric acid, and evaporate to dryness in an iron pan; fuse the residue in an iron or Hessian crucible, pour out the fused mass, and break into pieces. Preserve it in well-stoppered bottles.

Uses.—The crude fused calcium chloride serves to dry moist gases; the pure chloride is used in elementary organic analysis for the absorption and estimation of water formed by the hydrogen contained in the analyzed substance. A solution of the pure calcium chloride should not show an alkaline reaction. A calcium chloride tube filled with it should not gain weight when a very slow current of perfectly dry carbon dioxide is passed through it an hour.

9. Potassium Dichromate.

Bichromate of potassa of commerce is purified by repeated recrystallization, until barium chloride produces, in the solution of a sample of it in water, a precipitate which completely dissolves in hydrochloric acid. Potassium dichromate thus perfectly free from sulphuric acid is required more particularly for the oxidation of organic substances with a view to the estimation of the sulphur contained in them. Where the salt is intended for other purposes, *e.g.*, to determine the carbon of organic bodies, by heating them with potassium dichromate and sulphuric acid, one recrystallization is sufficient.

SECTION III.

FORMS AND COMBINATIONS IN WHICH SUBSTANCES ARE SEPARATED FROM EACH OTHER, OR IN WHICH THEIR WEIGHT IS DETERMINED.

§ 67.

THE quantitative analysis of a compound substance requires, as the first and most indispensable condition, a correct and accurate knowledge of the composition and properties of the new combinations into which it is intended to convert its several individual constituents, for the purpose of separating them from one another, and determining their several weights. Regarding the properties of the new compounds, we have to inquire more particularly, in the first place, how they behave with solvents; secondly, what is their deportment in the air; and, thirdly, what is their behavior on ignition? It may be laid down as a general rule, that compounds are the better adapted for quantitative determination the more insoluble they are, and the less alteration they undergo upon exposure to air or to a high temperature.

With respect to the composition of a compound, it is better adapted to the quantitative determination of a body the less it contains relatively of that body; since any error in weighing or loss of the compound to be weighed will have the less influence on the accuracy of the results the less the percentage it contains of the substance to be determined.

In this section those combinations of the several bodies which are best adapted for their quantitative determination are enumerated and described. The description given of the external form and appearance of the new compounds relates more particularly to the state in which they are obtained in our analyses. With regard to the properties of the new compounds, we shall confine ourselves to the enumeration of those which bear upon the special objects we have more immediately in view.

[The percentage compositions of these compounds are stated in connection with their description. For this purpose the symbols of the constituent elements of the compounds in many cases (viz.: when they are oxygen salts) are grouped in a manner different from that used to express their chemical constitution. This grouping constitutes a kind of formulæ differing from either the empirical or rational in ordinary use in modern text-books of chemistry, but identical with that formerly in general use (the old system). These formulæ are based upon the fact that in all oxygen salts, whether normal, acid, basic, ortho-, meta-, or pyro-salts, there is just enough oxygen to form with the radicals present, both basic and acid, their corresponding oxides or anhydrides, and with hydrogen, if present, water. They represent oxides (and water) jointly equivalent in weight to the radicals, hydrogen, and remaining oxygen, which rational formulæ represent as existing in oxygen salts.

EXAMPLES.

Potassium sulphate, $SO_2 < \begin{matrix} OK \\ OK \end{matrix} = K_2O,SO_3.$

Hydrogen potassium sulphate,

$$2\left(SO_2 < \begin{matrix} OH \\ OK \end{matrix}\right) = K_2O,H_2O,2SO_3.$$

Potassium disulphate,

$$O < \begin{matrix} SO_2 - OK \\ SO_2 - OK \end{matrix} = K_2O,2SO_3.$$

Ammonium magnesium phosphate,

$$2\left(PO \begin{matrix} \diagup O\,NH_4 \\ - O \\ \diagdown O \end{matrix} \begin{matrix} \\ > Mg \\ \end{matrix}\right) = 2MgO,(NH_4)_2O,P_2O_5.$$

Magnesium pyrophosphate,

$$O \begin{matrix} \diagup PO < \begin{matrix} O \\ O \end{matrix} > Mg \\ \diagdown PO < \begin{matrix} O \\ O \end{matrix} > Mg \end{matrix} = 2MgO,P_2O_5.$$

Most analytical chemists prefer to present the results of analyses of oxygen salts in percentages of oxides (or anhydrides) and water on account of the simplicity of computations required. Accordingly, in the following section, the percentage composition of oxygen salts is given in this manner, accompanied by corresponding formulæ and molecular weights. These formulæ are in every case preceded by rational formulæ constructed in accordance with the theory of the constitution of oxygen salts which is now generally accepted.]

A. FORMS IN WHICH THE BASIC RADICALS ARE WEIGHED OR PRECIPITATED.

BASIC RADICALS OF THE FIRST GROUP.

§ 68.

1. POTASSIUM.

The combinations best suited for the weighing of potassium are POTASSIUM SULPHATE, POTASSIUM CHLORIDE, and POTASSIUM PLATINIC CHLORIDE.

a. Potassium sulphate crystallizes usually in small, hard, straight, four-sided prisms, or in double six-sided pyramids; in the analytical process it is obtained as a white crystalline mass. It dissolves with some difficulty in water (1 part requiring 10 parts of water of 12°), it is almost absolutely insoluble in pure alcohol, but slightly more soluble in alcohol containing sulphuric acid (Expt. No. 6). It does not affect vegetable colors; it is unalterable in the air. The crystals decrepitate strongly when heated, yielding at the same time a little water, which they hold mechanically confined. The decrepitation of crystals that have been kept long drying is less marked. At a good red heat the salt fuses without volatilizing or decomposing. At a white heat a little of the salt volatilizes and also some sulphuric acid, so that the residue possesses an alkaline reaction (AL. MITSCHERLICH,* BOUSSINGAULT†). When exposed to a red heat, in conjunction with ammonium chloride, potassium sulphate is partly, and, upon repeated application of the process, wholly converted, with effervescence, into potassium chloride (H. ROSE).

---

* Journ. f. prakt. Chem. 83, 486. † Zeitschr. f. anal. Chem. 7, 244.

COMPOSITION.

$SO_2 < {OK \atop OK} =$	K_2O . . .	94·26	54·09
	SO_3 . . .	80·00	45·91
		174·26	100·00

The acid potassium sulphate ($KHSO_4$), which is produced when the normal salt is evaporated to dryness with free sulphuric acid, is readily soluble in water, and fusible even at a moderate heat. At a red heat it loses sulphuric acid, and is converted into normal potassium sulphate, but not readily—the complete conversion of the acid into the normal salt requiring the long-continued application of an intense red heat. However, when heated in an atmosphere of ammonium carbonate—which may be readily procured by repeatedly throwing into the faint red-hot crucible containing the acid sulphate small lumps of pure ammonium carbonate, and putting on the lid—the acid salt changes readily and quickly to the normal sulphate. The transformation may be considered complete as soon as the salt, which was so readily fusible before, is perfectly solid at a faint red heat.

b. Potassium chloride crystallizes usually in cubes, often lengthened to columns; rarely in octahedra. In analysis we obtain it either in the former shape, or as a crystalline mass. It is readily soluble in water, but much less so in dilute hydrochloric acid; in absolute alcohol it is nearly insoluble, and but slightly soluble in common alcohol. It does not affect vegetable colors, and is unalterable in the air. When heated, it decrepitates, unless it has been kept long drying, with expulsion of a little water mechanically confined in it. At a moderate red heat, it fuses unaltered and without diminution of weight; when exposed to a higher temperature, it volatilizes in white fumes; this volatilization proceeds the more slowly the more effectually the access of air is prevented (Expt. No. 7). When repeatedly evaporated with solution of oxalic acid in excess, it is converted into potassium oxalate. When evaporated with excess of nitric acid, it is converted readily and completely into nitrate. On ignition with ammonium oxalate, potassium carbonate and potassium cyanide are formed in noticeable quantities.

COMPOSITION.

K	39·13	52·46
Cl	35·46	47·54
	74·59	100·00

c. Potassium platinic chloride presents either small reddish-yellow octahedra, or a lemon-colored powder. It is difficultly soluble in cold, more readily in hot water; nearly insoluble in absolute alcohol, and but sparingly soluble in common alcohol—one part requiring for its solution, respectively, 12083 parts of absolute alcohol, 3775 parts of alcohol of 76 per cent. and 1053 parts of alcohol of 55 per cent. (Expt. No. 8, *a.*) Presence of free hydrochloric acid sensibly increases the solubility (Expt. No. 8, *b*). In caustic potassa it dissolves completely to a yellow fluid. It is unalterable in the air, and at 100°. On exposure to an intense red heat, four atoms of chlorine escape, metallic platinum and potassium chloride being left; but even after long-continued fusion, there remains always a little potassium platinic chloride which resists decomposition. Complete decomposition is easily effected, by igniting the double salt in a current of hydrogen gas, or with some oxalic acid.

According to ANDREWS, potassium platinic chloride, even though dried at a temperature considerably exceeding 100°, retains still ·0055 of its weight of water.

COMPOSITION.

$(KCl)_2$ . . .	149·18	30·56	K_2 . . .	78·26	16·03
$PtCl_4$ . . .	339·02	69·44	Pt . . .	197·18	40·39
			Cl_6 . . .	212·76	43·58
	488·20	100·00		488·20	100·00

d. Potassium silicofluoride is obtained on mixing a solution of a potassium salt with hydrofluosilicic acid in the form of a translucent iridescent precipitate, which increases and completely separates, when an equal volume of strong alcohol is added to the fluid. After being filtered off, washed with weak alcohol and dried, it is a soft white powder. It is difficultly soluble in cold water, far more readily in boiling water, not at all or in merest traces soluble in a mixture of water and strong alcohol in equal parts, but it is decidedly more soluble in the presence of any considerable quantity of free acid, especially hydrochloric or sulphuric acid. When potassa is added to the boiling aqueous solution of the salt the following change takes place: $(KF)_2SiF_4 + 4KOH = 6KF + Si(OH)_4$, the solution turning from acid to neutral (principle of STOLBA's volumetric method of estimating potassium). As soon as it is ignited the salt fuses, gives off silicon fluoride and leaves potassium fluoride.

§ 69.

2. SODIUM.

Sodium is usually weighed as SODIUM SULPHATE, SODIUM CHLORIDE, or SODIUM CARBONATE. It is separated from potassium in the form of SODIUM PLATINIC CHLORIDE, from other bodies occasionally in the form of sodium silicofluoride.

a. Anhydrous normal *sodium sulphate* is a white powder or a white very friable mass. It dissolves readily in water; but is sparingly soluble in absolute alcohol; presence of free sulphuric acid slightly increases its solubility in that menstrum; it is somewhat more readily soluble in common alcohol (Expt. No. 9). It does not affect vegetable colors; upon exposure to moist air, it slowly absorbs water (Expt. No. 10). At a gentle heat it is unaltered, at a strong red heat it fuses without decomposition or loss of weight. At a white heat it loses weight by volatilization of sodium sulphate and also of sulphuric acid (AL. MITSCHERLICH, BOUSSINGAULT). When ignited with ammonium chloride, it behaves like potassium sulphate.

COMPOSITION.

$SO_2 < \begin{matrix} ONa \\ ONa \end{matrix} =$			
	Na_2O	62·08	43·69
	SO_3	80·00	56·31
		142·08	100·00

The acid sodium sulphate (sodium hydrogen sulphate, $NaHSO_4$) which is always produced upon the evaporation of a solution of the normal salt with sulphuric acid in excess, fuses even at a gentle heat; it may be readily converted into the normal salt in the same manner as the acid potassium sulphate (see § 68, *a*).

b. Sodium chloride crystallizes in cubes, octahedra, and hollow four-sided pyramids. In analysis it is frequently obtained as an amorphous mass. It dissolves readily in water, but is much less soluble in hydrochloric acid; it is nearly insoluble in absolute alcohol, and but sparingly soluble in common alcohol; 100 parts of alcohol of 75 per cent. dissolve, at a temperature of 15°, ·7 part (WAGNER). It is neutral to vegetable colors. Exposed to a somewhat moist atmosphere, it slowly absorbs water (Expt. No. 12). Crystals of this salt that have not been kept drying a considerable time decrepitate when heated, yielding a little water, which they hold mechanically confined. The salt fuses at a red heat without decomposition; at a white heat, and in open vessels even at a bright red heat, it volatilizes in white fumes (Expt. No. 13). If a carburetted hydrogen flame acts on fusing sodium chloride, hydrochloric acid escapes, and some sodium carbonate is formed. On evaporation with oxalic or nitric acid as well as by ignition with ammonium oxalate, it behaves like the corresponding potassium salt.

COMPOSITION.

Na	23·04	39·38
Cl	35·46	60·62
	58·50	100·00

c. Anhydrous *sodium carbonate* is a white powder or a white very friable mass. It dissolves readily in water, but much less so in solution of ammonia (MARGUERITTE); it is insoluble in alcohol. Its reaction is strongly alkaline. Exposed to the air, it absorbs water slowly. On moderate ignition to incipient fusion it scarcely loses weight; on long fusion, however, it volatilizes to a considerable extent (Comp. Expt. 14).

COMPOSITION.

$CO < \begin{matrix} ONa \\ ONa \end{matrix} = \begin{matrix} Na_2O \\ CO_2 \end{matrix}$ . . .	62·08 44·00	58·52 41·48
	106·08	100·00

d. Sodium platinic chloride crystallizes with 6 mol. water $(NaCl)_2.\ PtCl_4 + 6\,H_2O$, in light yellow, transparent, prismatic crystals which dissolve readily both in water and in common alcohol.

e. Sodium silicofluoride is similar in properties to the corresponding potassium salt. It has an analogous composition, and is decomposed in the same way by alkalies. It is, however, considerably more soluble in water and in diluted alcohol.

§ 70.

3. AMMONIUM.

Ammonium is most appropriately weighed as AMMONIUM CHLORIDE, or as AMMONIUM PLATINIC CHLORIDE, or it may be estimated from the weight of the PLATINUM in the latter compound.

Under certain circumstances ammonium may also be estimated from the volume of the NITROGEN GAS eliminated from it; and it is frequently estimated by alkalimetry.

a. Ammonium chloride crystallizes in cubes and octahedra, but more frequently in feathery crystals. In analysis we obtain it uniformly as a white mass. It dissolves readily in water, but is difficultly soluble in common alcohol. It does not alter vegetable colors, and remains unaltered in the air. Solution of ammonium chloride, when evaporated on the water-bath, loses a small quantity of ammonia, and becomes slightly acid. The diminution of weight occasioned by this loss of ammonia is very trifling (Expt. No. 15). At 100° ammonium chloride loses nothing, or very little of its weight (comp. same Expt.). At a higher temperature it volatilizes readily, and without undergoing decomposition.

COMPOSITION.

NH_4	18·04	33·72	NH_3	17·04	31·85
Cl	35·46	66·28	HCl	36·46	68·15
	53·50	100·00		53·50	100·00

100 parts of ammonium chloride correspond to 48·67 parts of ammonium oxide.

b. Ammonium platinic chloride occurs either as a heavy, lemon-colored powder, or in small, hard octahedral crystals of a bright yellow color. It is difficultly soluble in cold, but more readily in hot water. It is very sparingly soluble in absolute alcohol, but more readily in common alcohol—1 part requiring of absolute alcohol, 26535 parts; of alcohol of 76 per cent., 1406 parts; of alcohol of 55 per cent., 665 parts. The presence of free acid sensibly increases its solubility (Expt. No. 16). It remains unaltered in the air, and at 100°. It loses a little water between 100° and 125°. Upon ignition chlorine and ammonium chloride escape, leaving the metallic platinum as a porous mass (spongy platinum). However, if due care be not taken, in this process, to apply the heat gradually, the escaping fumes will carry off particles of platinum, which will coat the lid of the crucible. For properties of metallic platinum, see § 89, *a*.

COMPOSITION.

$(NH_4Cl)_2$ . . .	107·00	23·99	$(NH_4)_2$ . . .	36·08	8·09
$PtCl_4$ . . .	339·02	76·01	Pt	197·18	44·21
			Cl_6 . . .	212·76	47·70
	446·02	100·00		446·02	100·00

N_2 . . .	28·08	6·295	$(NH_3)_2$ . .	34·08	7·64
H_8 . . .	8·00	1·794			
Pt . . .	197·18	44·209	$(HCl)_2$ . .	72·92	16·35
Cl_6 . . .	212·76	47·702	$PtCl_4$. . .	339·02	76·01
	446·02	100·000		446·02	100·00

100 parts of ammonium platinic chloride correspond to 11·677 parts of ammonium oxide.

c. Nitrogen gas is colorless, tasteless, and inodorous; it mixes with air, without producing the slightest coloration; it does not affect vegetable colors. Its specific gravity is ·97137 (REGNAULT). One litre weighs at 0°, and ·76 metre bar., 1·25617 grm. It is difficultly soluble in water, 1 volume of water absorbing, at 0°, and ·76 pressure, ·02035 vol.; at 10°, ·01607 vol.; at 15°, ·01478 vol. of nitrogen gas (BUNSEN).

BASIC RADICALS OF THE SECOND GROUP.

§ 71.

1. BARIUM.

Barium is weighed as BARIUM SULPHATE, BARIUM CARBONATE, and BARIUM SILICOFLUORIDE.

a. Artificially prepared *barium sulphate* presents the appearance is of a fine white powder. When recently precipitated, it difficult to obtain a clear filtrate, especially if the precipitation was effected in the cold, and the solution contains neither hydrochloric acid nor ammonium chloride. It is as good as insoluble in cold and in hot water. (1 part of the salt requires more than 400,000 parts of water for solution.) It has a great tendency, upon precipitation, to carry down with it other substances contained in the solution from which it separates, more particularly barium nitrate, nitrates and chlorates of the alkali metals, ferric oxide, &c. Several of the impurities, such, for instance, as potassium or sodium chlorates, may be removed by igniting the barium sulphate, moistening with hydrochloric acid, evaporating the latter off and exhausting the residue with water; other impurities again, such as potassium or sodium nitrates, cannot be removed by this treatment. Even the precipitate obtained from a solution of barium chloride by means of sulphuric acid in excess contains traces of barium chloride, which it is impossible to remove, even by washing with boiling water, but which are dissolved by nitric acid (SIEGLE). Cold dilute acids dissolve trifling, yet appreciable traces of barium sulphate; for instance, 1000 parts of nitric acid of 1·032 sp. gr. dissolve ·062 parts (CALVERT), 1000 parts of hydrochloric acid containing 3 per cent. dissolve ·06 parts.* Cold concentrated acids dissolve considerably more; thus, 1000 parts of nitric acid of 1·167 sp. gr. dissolve 2 parts (CALVERT). Boiling hydrochloric acid also dissolves appreciable traces; thus 230 c.c. hydrochloric acid of 1·02 sp. gr. were found, after a quarter of an hour's boiling with ·679 grm. barium sulphate, to have dissolved of it ·048 grm. Acetic acid dissolves less barium sulphate than the other acids; thus, 80 c.c. acetic acid of 1·02 sp. gr. were found, after a quarter of an hour's boiling with ·4 grm., to have dissolved only ·002 grm. (SIEGLE). Free chlorine considerably increases its solubility (O. L. ERDMANN). Several salts more particularly interfere with the precipitation of barium by sulphuric acid. I observed this some time ago with magnesium chloride, but ammonium nitrate (MITTENTZWEY), alkali nitrates generally,* and more particularly alkali citrates (SPILLER), possess this property in a high degree. In the last case the precipitate appears on the addition of hydrochloric acid. If a fluid contains metaphosphoric acid, barium cannot be completely precipitated out of it by means of sulphuric acid; the resulting precipitate too contains phosphoric acid (SCHEERER, RUBE). Barium

* Zeitschr. f. anal. Chem. 9, 62.

sulphate dissolves in tolerable quantity in concentrated sulphuric acid, but separates again on dilution. It is as good as insoluble in a boiling solution of ammonium sulphate (1 in 4). Barium sulphate remains quite unaltered in the air, at 100°, and even at a red heat. At a strong white heat it loses sulphuric acid (BOUSSINGAULT).* On ignition with charcoal, or under the influence of reducing gases, it is converted comparatively easily, but as a rule only partially, into barium sulphide. On ignition with ammonium chloride, barium sulphate undergoes partial decomposition. It is not affected, or affected but very slightly, by cold solutions of the hydrogen carbonates of the alkali metals or of ammonium carbonate; solutions of normal sodium and potassium carbonates when cold have only a slight decomposing action upon it; but when boiling, and upon repeated application, they effect at last the complete decomposition of the salt (H. ROSE). By fusion with sodium or potassium carbonate, barium sulphate is readily decomposed.

COMPOSITION.

$SO_2<{O \atop O}>Ba =$			
	BaO	153	65·67
	SO_3	80	34·33
		233	100·00

b. Artificially prepared *barium carbonate* presents the appearance of a white powder. It dissolves in 14137 parts of cold, and in 15421 parts of boiling water (Expt. No. 17). It dissolves far more readily in solutions of ammonium chloride or ammonium nitrate; from these solutions it is, however, precipitated again, though not completely, by caustic ammonia. In water containing free carbonic acid, barium carbonate dissolves to an acid carbonate. In water containing ammonia and ammonium carbonate, it is nearly insoluble, one part requiring about 141000 parts (Expt. No. 18). Its solution in water has a very faint alkaline reaction. Alkali citrates and metaphosphates impede the precipitation of barium by ammonium carbonate. It is unalterable in the air, and at a red heat. When exposed to the strongest heat of a blast-furnace, it slowly yields up the whole of its carbonic acid; this expulsion of the carbonic acid is promoted by the simultaneous action of aqueous vapor. Upon heating it to redness with charcoal, caustic baryta is formed, with evolution of carbon monoxide.

* Zeitschr. f. anal. Chem. 7, 244.

COMPOSITION.

$$CO<^{O}_{O}>Ba = \begin{array}{lrr} BaO \;\ldots\ldots & 153 & 77\cdot67 \\ CO_2 \;\ldots\ldots & 44 & 22\cdot33 \\ \hline & 197 & 100\cdot00 \end{array}$$

c. Barium silicofluoride forms small, hard, and colorless crystals, or (more generally) a crystalline powder. It dissolves in 3800 parts of cold water; in hot water it is more readily soluble (Expt. No. 19). The presence of free hydrochloric acid increases its solubility considerably (Expt. No. 20). Ammonium chloride acts also in the same way (1 part silicofluoride of barium dissolves in 428 parts of saturated, and 589 parts of dilute solution of ammonium chloride. J. W. MALLET). In common alcohol it is almost insoluble. It is unalterable in the air, and at 100°; when ignited, it is decomposed into silicon fluoride, which escapes, and barium fluoride, which remains.

COMPOSITION.

BaF_2 . . .	175	62·72	Ba . . .	137	49·10
SiF_4 . . .	104	37·28	Si . . .	28	10·04
	279	100·00	F_6 . . .	114	40·86
				279	100·00

§ 72.

2. STRONTIUM.

Strontium is weighed either as STRONTIUM SULPHATE, or as STRONTIUM CARBONATE.

a. Strontium sulphate, artificially prepared, is a white powder, sometimes dense and crystalline, sometimes loose and bulky. It dissolves in 6895 parts of cold, and 9638 parts of boiling water (Expt. No. 21). In water containing sulphuric acid, it is still more difficultly soluble, requiring from 11000 to 12000 parts (Expt. No. 22). Of cold hydrochloric acid of 8·5 per cent., it requires 474 parts; of cold nitric acid of 4·8 per cent., 432 parts; of cold acetic acid of 15·6 per cent. of $HC_2H_3O_2$, as much as 7843 parts (Expt. No. 23). It dissolves in solutions of potassium chloride and magnesium chloride, in quantity which increases with the concentration, also in solutions of sodium chloride and calcium chloride in greatest quantity when the solutions are of medium concentration (A. VIRCK*); it is precipitated from these solutions by sulphuric acid. Metaphosphoric acid (SCHEERER, RUBE), and also alkali citrates, but not free citric acid (SPILLER), impede the precipitation of strontium by sulphuric acid. It is as good as insoluble in absolute alcohol, in common alcohol, and in a boiling solution of ammonium sulphate (1 in 4). It does not alter vegetable colors; and remains unaltered in the air, and at a red heat. When exposed to a most intense red heat, it fuses with loss of a small quantity of sulphuric acid (M. DARMSTADT †); all the sulphuric acid will escape on very strong ignition continued for a length of time (BOUSSINGAULT ‡). When ignited with charcoal, or under the influence of reducing gases, it is converted into strontium sulphide. Solutions of acid and normal carbonates of potassium, sodium, and ammonium decompose strontium sulphate completely at the common temperature, even when considerable quantities of alkali sulphates are present (H. ROSE). Boiling promotes the decomposition.

COMPOSITION.

$SO_2<^{O}_{O}>Sr =$			
	SrO . . .	103·5	56·40
	SO_3 . . .	80·0	43·60
		183·5	100·00

b. Strontium carbonate, artificially prepared, is a white, soft, loose powder. It dissolves, at the common temperature, in 18045 parts of water (Expt. No. 24): presence of ammonia diminishes its solubility (Expt. No. 25). It dissolves pretty readily in solutions of ammonium chloride and ammonium nitrate, but is precipitated again from these solutions by ammonia and ammonium carbonate, and more completely than barium carbonate under similar circumstances. Water impregnated with carbonic acid dissolves it as an acid carbonate. Its reaction is very feebly alkaline. Alkali citrates and metaphosphates impede the precipitation of strontium by alkali carbonates. Ignited with access of air it is infusible, but when exposed to a most intense heat, it fuses and gradually loses its carbonic acid. On ignition with charcoal, strontium oxide is formed, with evolution of carbon monoxide gas.

* Zeitschr. f. anal. Chem. 1, 473. † *Ib.* 6, 376. ‡ *Ib.* 7, 244.

COMPOSITION.

$CO < {O \atop O} > Sr =$			
	SrO . . .	103·50	70·17
	CO_2 . . .	44·00	29·83
		147·50	100·00

§ 73.

3. CALCIUM.

Calcium is weighed either as CALCIUM SULPHATE, CALCIUM CARBONATE, or CALCIUM OXIDE; to convert it into the latter forms, it is first usually precipitated as calcium oxalate.

a. Artificially prepared anhydrous *calcium sulphate* is a loose, white powder. It dissolves, at the common temperature, in 430 parts, at 100°, in 460 parts of water (POGGIALE). Presence of hydrochloric acid, nitric acid, ammonium chloride, sodium sulphate, or sodium chloride, increases its solubility. It dissolves with comparative ease, especially on gently warming, in aqueous solution of sodium thiosulphate (DIEHL), and also in a boiling solution of ammonium sulphate (1 in 4). The aqueous solution of calcium sulphate does not alter vegetable colors. In alcohol of 90 per cent or stronger it is almost absolutely insoluble. Exposed to the air, it slowly absorbs water. It remains unaltered at a dull-red heat. Heated to intense bright redness, it fuses, losing weight considerably from loss of sulphuric acid (AL. MITSCHERLICH *). On long ignition at a white heat all the sulphuric acid escapes (BOUSSINGAULT†). On ignition with charcoal, or under the influence of reducing gases, it is converted into calcium sulphide. Solutions of normal and acid carbonates of the alkali metals decompose calcium sulphate more readily still than strontium sulphate.

COMPOSITION.

$SO_2 < {O \atop O} > Ca =$			
	CaO	56	41·18
	SO_3	80	58·82
		136	100·00

b. Calcium carbonate artificially produced by the precipitation of a calcium salt with ammonium carbonate is at first loose and

* Jour. f. prakt. Chem. 83, 485.

† Zeitschr. f. anal. Chem. 7, 244.

amorphous, but after some time becomes a white, fine, crystalline powder, which under the microscope has sometimes the form of calcite, sometimes that of aragonite. It is very slightly soluble in water. By protracted boiling 1 litre of water dissolves ·034 grm., according to A. W. HOFMANN, or ·036 grm. according to C. WELTZIEN; so one part requires 28500 parts of water for solution. The solution has a barely-perceptible alkaline reaction. In water containing ammonia and ammonium carbonate the crystallized salt dissolves much more sparingly (Expt. No. 27), one part requiring about 65000 parts; this solution is not precipitated by ammonium oxalate. Amorphous calcium carbonate is also much more insoluble in water containing ammonia than in pure water (DIVERS*). Presence of ammonium chloride and of ammonium nitrate increases the solubility of calcium carbonate; but the salt is precipitated again from these solutions by ammonia and ammonium carbonate, and more completely than barium carbonate under similar circumstances. Normal salts of potassium and sodium, and also normal calcium and magnesium salts (HUNT), likewise increase its solubility, the precipitation of calcium by the alkali carbonates is completely prevented or considerably interfered with by the presence of alkali citrates (SPILLER) or metaphosphates (RUBE). Water impregnated with carbonic acid dissolves calcium carbonate as acid carbonate. Calcium carbonate remains unaltered in the air at 100°, and even at a low red heat; but upon the application of a stronger heat, more particularly with free access of air, it gradually loses its carbonic acid. By means of a gas blowpipe-lamp, calcium carbonate (about ·5 grm.), in an open platinum crucible, is without difficulty reduced to calcium oxide; attempts to effect complete reduction over a spirit lamp with double draught have, however, failed (Expt. No. 28). It is decomposed far more readily when ignited with charcoal, giving off its carbonic acid in the form of carbon monoxide.

COMPOSITION.

$CO < {O \atop O} > Ca =$			
	CaO	56	56·00
	CO_2	44	44·00
		100	100·00

* Jour. Chem. Soc. 1870, 362.

c. Calcium oxalate, precipitated from hot or concentrated solutions, is a fine white powder consisting of infinitely minute indistinct crystals, and almost absolutely insoluble in water. The salt has the formula, $CaC_2O_4 + H_2O$. When precipitated from cold, extremely-dilute solutions, the salt presents a more distinctly crystalline appearance, and consists of a mixture of $CaC_2O_4 + H_2O$ and $CaC_2O_4 + 3H_2O$ (SOUCHAY and LENSSEN). Presence of free oxalic acid and acetic acid slightly increases the solubility of calcium oxalate. The stronger acids (hydrochloric acid, nitric acid) dissolve it readily; from these solutions it is precipitated again unaltered, by alkalies, and also (provided the excess of acid be not too great) by alkali oxalates or acetates added in excess. Calcium oxalate does not dissolve in solutions of potassium chloride, sodium chloride, ammonium chloride, barium chloride, calcium chloride, and strontium chloride, even though these solutions be hot and concentrated; but, on the other hand, it dissolves readily and in appreciable quantities, in hot solutions of the salts belonging to the magnesium group. From these solutions it is reprecipitated by an excess of alkali oxalate (SOUCHAY and LENSSEN). Alkali citrates (SPILLER) and metaphosphates (RUBE) impede the precipitation of lime by alkali oxalates. When treated with solutions of many of the heavy metals, *e.g.*, with solution of cupric chloride, silver nitrate, &c., calcium oxalate suffers decomposition, a soluble calcium salt being formed, and an oxalate of the heavy metal, which separates immediately, or after some time (REYNOSO). Calcium oxalate is unalterable in the air, and at 100°. Dried at the latter temperature, it has invariably the following composition (Expt. No. 28, also SOUCHAY and LENSSEN *).

$$\begin{matrix} CO-O \\ | \\ CO-O \end{matrix} \Big\rangle Ca + H_2O = \begin{matrix} CaO & \ldots & 56 & 38 \cdot 36 \\ C_2O_3 & \ldots & 72 & 49 \cdot 32 \\ H_2O & \ldots & 18 & 12 \cdot 32 \\ & & \overline{146} & \overline{100 \cdot 00} \end{matrix}$$

At 205° calcium oxalate loses its water, without undergoing decomposition; at a somewhat higher temperature, still scarcely reaching dull redness, the anhydrous salt is decomposed, without actual separation of carbon, into carbon monoxide and calcium carbonate. The powder, which was previously of snowy whiteness,

* Anal. d. Chem. und Pharm. 100, 322.

transiently assumes a gray color in the course of this process, even though the oxalate be perfectly pure. Upon continued application of heat this gray color disappears again. If the calcium oxalate is heated in small, coherent fragments, such as are obtained upon drying the precipitated salt on a filter, the commencement and progress of the decomposition can be readily traced by this transient appearance of gray. If the process of heating be conducted properly, the residue will not contain a trace of calcium oxide. Hydrated calcium oxalate exposed suddenly to a dull-red heat, is decomposed with considerable separation of carbon. By ignition over the gas blowpipe calcium oxalate is converted into calcium oxide.

d. Calcium oxide obtained by continued strong ignition of the oxalate or carbonate appears as a white, infusible powder, unalterable by ignition. By standing in the air it attracts water and carbonic acid, but not rapidly enough to interfere with accurate weighing. By treatment with a little water calcium hydroxide is formed with evolution of much heat; on igniting again the water of hydration is readily and completely removed. Pure calcium oxide dissolves in dilute hydrochloric acid with evolution of heat, but without effervescence.

§ 74.

4. Magnesium.

Magnesium is weighed as MAGNESIUM SULPHATE, MAGNESIUM PYROPHOSPHATE, or MAGNESIUM OXIDE. To convert it into the pyrophosphate, it is precipitated as NORMAL AMMONIUM MAGNESIUM PHOSPHATE.

a. Anhydrous *magnesium sulphate* presents the appearance of a white, opaque mass. It dissolves readily in water. It is nearly altogether insoluble in absolute alcohol, but it is somewhat soluble in common alcohol.

It does not alter vegetable colors. Exposed to the air it absorbs water rapidly. At a moderate red heat, it remains unaltered; but when heated to intense redness, it undergoes partial decomposition, losing part of its acid, after which it is no longer perfectly soluble in water. By means of a gas blowpipe it as tolerably easy to expel the whole of the sulphuric acid from small quantities of magnesium sulphate (Expt. No. 30). Ignited with ammonium chloride magnesium sulphate is not decomposed.

COMPOSITION.

$SO_2 <^{O}_{O}> Mg =$			
	MgO	40	33·33
	SO_3	80	66·67
		120	100·00

b. Ammonium magnesium phosphate is a white crystalline powder. It dissolves, at the common temperature, in 15293 parts of cold water (Expt. No. 31). In water containing ammonia, it is much more insoluble. 1000 grm. of a mixture of 3 parts water and 1 part ammonia solution, dissolved only a quantity corresponding to ·004 grm. pyrophosphate (KISSEL*); the salt was considerably more soluble when ammonium chloride was also present; thus, in one of KISSEL's experiments a quantity corresponding to ·011 grm. pyrophosphate was dissolved by 1000 grm. fluid containing 18 grm. ammonium chloride. Presence of excess of magnesium sulphate diminishes the solubility in dilute ammonia, even in the presence of ammonium chloride, to such an extent that the quantity dissolved by 1000 grm. fluid cannot be estimated (KISSEL); the precipitate, under these circumstances, is liable, especially in the absence of much ammonium chloride, and when a large excess of magnesium sulphate is present, to contain some magnesium hydroxide or basic magnesium sulphate (KUBEL,† KISSEL). Sodium phosphate also diminishes (to about the same extent as magnesium sulphate) the solubility of the salt in water containing ammonium chloride and ammonia (W. HEINTZ ‡). It dissolves readily in acids, even in acetic acid. Its composition is expressed by the formula $NH_4MgPO_4 + 6H_2O$. 5 mol. of water escape at 100°, the remaining water together with ammonia are expelled, at a red heat, leaving $Mg_2P_2O_7$. On the application of a stronger heat the mass passes through a state of incandescence, if the salt were pure; the weight of the residue is not affected. The incandescence may not take place at all in the presence of small quantities of calcium salts,

* Zeitschr. f. anal. Chem. 8, 173. † *Ib.* 8, 125. ‡ *Ib.* 9, 16.

of other magnesium salts, or of silicic acid. It is occasioned not by the passage of the orthophosphate into the pyrophosphate, but by the passage from the crystalline to the amorphous condition (O. Popp *). If ammonium magnesium phosphate is dissolved in dilute hydrochloric or nitric acid and ammonia be then added to the solution, the salt is reprecipitated completely, or more correctly, only so much remains in solution as corresponds to its ordinary solubility in water containing ammonia and ammonium salt.

c. Magnesium pyrophosphate presents the appearance of a white mass, often slightly inclining to gray. It is barely soluble in water, but readily so in hydrochloric acid, and in nitric acid. It remains unaltered in the air, and at a red heat; at a very intense heat it fuses unaltered. Exposed at a white heat to the action of hydrogen, $Mg_3(PO_4)_2$ is formed, while PH_3, P and P_2O_5 escape. $3(Mg_2P_2O_7) = 2(Mg_3(PO_4)_2 + P_2O_5$ (Struve†). It leaves the color of moist turmeric-, and of reddened litmus-paper unchanged. If we dissolve it in hydrochloric or nitric acid, add water to the solution, boil for some time, and then precipitate with ammonia in excess, we obtain a precipitate of ammonium magnesium phosphate which, after ignition, affords less $Mg_2P_2O_7$, than was originally employed. Weber‡ gives the loss as from 1·3 to 2·3 per cent. By long-continued fusion with mixed potassium and sodium carbonates, magnesium pyrophosphate is completely decomposed, the pyrophosphoric acid being re-converted into orthophosphoric. If, therefore, we treat the fused mass with hydrochloric acid, and then add water and ammonia, we re-obtain on igniting the precipitate the whole quantity of the salt used. If the solution of magnesium pyrophosphate in nitric acid is evaporated to dryness a white residue is left; if this is heated more strongly hyponitric acid is liberated, and the residue turns the color of cinnamon; on cooling it is yellowish-white. By heating still more strongly to incipient redness, rapid decomposition sets in, more hyponitric acid is evolved, and pure-white magnesium pyrophosphate is left. Unless the heat is applied with care the evolution of gas may be so rapid as to carry away particles of the substance (E. Luck).

---

* Zeitschr. f. anal. Chem. 13, 305. † Jour. f. prakt. Chem. 79, 349.

‡ Pogg. Ann. 73, 146.

COMPOSITION.

$$O \left\langle \begin{matrix} PO < {}^{O}_{O} > Mg \\ PO < {}^{O}_{O} > Mg \end{matrix} \right. = \begin{matrix} 2MgO \\ P_2O_5 \end{matrix}$$

$2MgO$ . . .	80	36·04
P_2O_5 . . .	142	63·96
	222	100·00

d. Magnesium oxide is a white, light, loose powder. It dissolves in 55,368 parts of cold, and in the same proportion of boiling water (Expt. No. 37). Its aqueous solution has a very slightly alkaline reaction. It dissolves in hydrochloric and in other acids, without evolution of gas. Magnesium oxide dissolves readily and in quantity, in solutions of normal ammonium salts, and also in solutions of potassium chloride and sodium chloride (Expt. No. 38) and potassium sulphate and sodium sulphate (R. WARINGTON, Jr.) it is more soluble than in water. Exposed to the air, it slowly absorbs carbonic acid and water. Magnesium oxide is highly infusible, remaining unaltered at a strong red heat, and fusing superficially only at the very highest temperature.

COMPOSITION.

Mg	24	60
O	16	40
	40	100

BASIC RADICALS OF THE THIRD GROUP.

§ 75.

1. ALUMINIUM.

Aluminium is usually precipitated as HYDROXIDE, occasionally as BASIC ACETATE or BASIC FORMATE, and always weighed as ALUMINIUM OXIDE.

a. Aluminium hydroxide, recently precipitated from a solution of an aluminium salt by an alkali is translucent, and when dried at 100° has the formula, $Al_2(OH)_6$. The precipitate invariably retains a minute proportion of the acid with which the aluminium was previously combined, as well as of the alkali which has served as the precipitant; it is freed with difficulty from these admixtures by repeated washing. It is insoluble in pure water;

but it readily dissolves in soda, potassa, and ethylamine (SONNENSCHEIN); it is sparingly soluble in ammonia, and insoluble in ammonium carbonate; presence of ammonium salts greatly diminishes its solubility in ammonia (Expt. No. 39). The correctness of this statement of mine in the first edition of the present work, has been amply confirmed since by MALAGUTI and DUROCHER;* and also by experiments made by my former assistant, Mr. J. FUCHS. The former chemists state also that, when a solution of aluminium is precipitated with ammonium sulphide, the fluid may be filtered off five minutes after, without a trace of aluminium in it. FUCHS did not find this to be the case (Expt. No. 40). Aluminium hydroxide, recently precipitated, dissolves readily in hydrochloric or nitric acid; but after filtration, or after having remained for some time in the fluid from which it has been precipitated, it does not dissolve in these acids without considerable difficulty, and long digestion. Aluminium hydroxide shrinks considerably on drying, and then presents the appearance of a hard, translucent, yellowish, or of a white, earthy mass. When ignited, it loses water, and this loss is frequently attended with slight decrepitation, and invariably with considerable diminution of bulk. Aluminium hydroxide precipitated from a solution of aluminium in potassa or soda by ammonium chloride is milk-white, denser, easier to wash, and much less soluble in ammonia than the variety above described. When dried at 100°, it has the formula $Al_2O_3 + (H_2O)_2$ (J. LÖWE†).

b. Aluminium oxide or alumina, prepared by heating the hydroxide to a moderate degree of redness, is a loose and soft mass; but upon the application of a very intense degree of heat, it concretes into small, hard lumps. At the most intense white heat, it fuses to a clear glass. Ignited alumina is dissolved by dilute acids with very great difficulty; in fuming hydrochloric acid, it dissolves upon long-continued digestion in a warm place, slowly, but completely. It dissolves tolerably easily and quickly by first heating with a mixture of 8 parts of concentrated sulphuric acid and 3 parts of water, and then adding water (A. MITSCHERLICH‡). Ignition in a current of hydrogen gas leaves it unaltered. By fusion with potassium disulphate, it is rendered soluble in water. Upon igniting alumina with ammonium chloride, aluminium

* Ann. de Chim. et de Phys. 3 Sér. 17, 421.

† Zeitschr. f. anal. Chem. 4, 350.

‡ Journ. f. prakt. Chem. 81, 110.

chloride escapes; but the process fails to effect complete volatilization of the alumina (H. ROSE). When alumina is fused at a very high temperature, with ten times its quantity of sodium carbonate, sodium aluminate is formed, which is soluble in water (R. RICHTER). Placed upon moist red litmus-paper, pure alumina does not change the color to blue.

COMPOSITION.

Al_2	55·00	53·40
O_3	48·00	46·60
	103·00	100·00

c. If to the solution of a salt of aluminium, sodium carbonate or ammonium carbonate be added, till the resulting precipitate only just redissolves on stirring, and then sodium acetate or ammonium acetate poured in in abundance and the mixture boiled some time, the aluminium is precipitated almost completely as basic acetate in the form of translucent flocks, so that if the filtrate be boiled with ammonium chloride and ammonia, only unweighable traces of aluminium hydroxide separate. If the quantity of sodium acetate employed be too small, the precipitate appears more granular, the filtrate would then contain a larger amount of aluminium. The precipitate cannot be very conveniently filtered and washed. In washing it is best to use boiling water, containing a little sodium acetate or ammonium acetate. The precipitate is readily soluble in hydrochloric acid.

d. If, instead of the acetates mentioned in *c*, the corresponding formates be used, a flocculent voluminous precipitate of basic aluminium formate is obtained, which may be very readily washed (FR. SCHULZE*).

§ 76.

2. CHROMIUM.

Chromium is usually precipitated as CHROMIC HYDROXIDE, and always weighed as chromic oxide.

a. Chromic hydroxide recently precipitated from a green solution, is greenish-gray, gelatinous, insoluble in water: it dissolves readily, in the cold, in solutions of potassa or soda, to a dark green fluid; it dissolves also in the cold, but rather sparingly, in solution

* Chem Centralbl. 1861, 3.

of ammonia, to a light violet red fluid. In acids it dissolves readily, with a dark green color. Presence of ammonium chloride exercises no influence upon the solubility of the hydroxide in ammonia. Boiling effects the complete separation of the hydroxide from its solutions in potassa, or ammonia (Expt. No. 41). The dried hydroxide is a greenish-blue powder; it is converted into oxide with loss of water at a gentle red heat.

b. CHROMIC OXIDE, produced by heating the hydroxide to dull redness, is a dark green powder; upon the application of a higher degree of heat, it assumes a lighter tint, but suffers no diminution of weight; the transition from the darker to the lighter tint is marked by a vivid incandescence of the powder. The feebly ignited oxide is difficultly soluble in hydrochloric acid, and the strongly ignited oxide is altogether insoluble in that acid. It remains unaltered when ignited with ammonium chloride, or in a current of hydrogen. By fusion with sodium carbonate and potassium nitrate, potassium chromate is formed.

COMPOSITION.

Cr_2	104·96	68·62
O_3	48·00	31·38
	152·96	100·00

BASIC RADICALS OF THE FOURTH GROUP.

§ 77.

1. ZINC.

Zinc is weighed in the form of OXIDE or SULPHIDE; it is precipitated as BASIC CARBONATE, or as SULPHIDE.

a. Basic zinc carbonate, recently precipitated, is white, flocculent, nearly insoluble in water—(one part requiring 44600 parts, Expt. No. 42)—but readily soluble in potassa, soda, ammonia, ammonium carbonate, and acids. The solutions in soda or potassa, if concentrated, are not altered by boiling; but if dilute, nearly all the zinc present is thrown down as a white precipitate. From the solutions in ammonia and ammonium carbonate, especially if they are dilute, zinc is likewise separated upon boiling. When a neutral solution of zinc is precipitated with sodium carbonate or potassium carbonate, carbonic acid is set free, since the precipitate formed is not

$ZnCO_3$, but consists of a compound of zinc hydroxide, with normal carbonate in proportions varying according to the concentration of the solution, and to the mode of precipitation. Owing to the presence and action of this carbonic acid, part of the zinc remains in solution; if filtered cold, therefore, the filtrate gives a precipitate with ammonium sulphide. But if the solution is precipitated boiling, and kept at that temperature for some time, the precipitation of the zinc is complete to the extent that the filtrate is not rendered turbid by ammonium sulphide; still, if the filtrate, mixed with ammonium sulphide, be allowed to stand at rest for many hours, minute and almost unweighable flakes of zinc sulphide will separate from the fluid. The precipitate of zinc carbonate, obtained in the manner just described, may be completely freed from all admixture of alkali by washing with hot water. If ammonium salts be present, the precipitation is not complete till every trace of ammonia is expelled. If the solution of a zinc salt is mixed with potassium or sodium carbonate in excess, the mixture evaporated to dryness, at a gentle heat, and the residue treated with cold water, a perceptible proportion of the zinc is obtained in solution as double carbonate of zinc and potassium or sodium; but if the mixture is evaporated to dryness, at a boiling heat, and the residue treated with hot water, the whole of the zinc, with the exception of an extremely minute proportion, as we have already had occasion to observe, is obtained as zinc carbonate. The dried basic zinc carbonate is a brilliant, white, loose powder; exposure to a red heat converts it into oxide.

b. Zinc oxide, produced from the carbonate by ignition, is a white light powder, with a slightly yellow tint. When heated, it acquires a yellow color, which disappears again on cooling. Upon ignition with charcoal, carbon monoxide and zinc fumes escape. By igniting in a rapid current of hydrogen, metallic zinc is produced; whilst by igniting in a feeble current of hydrogen, crystallized zinc oxide is obtained (St. Claire Deville). In the latter case, too, a portion of the metal is reduced and volatilized. Zinc oxide is insoluble in water. Placed on moist turmeric paper, it does not change the color to brown. In acids, zinc oxide dissolves readily and without evolution of gas. Ignited with ammonium chloride, fused zinc chloride is produced which volatilizes with very great difficulty if the air is excluded: but readily and completely, with free access of air, and with ammonium chloride fumes. Mixed with a sufficiency of powdered sulphur and ignited in a stream of hydrogen, the corresponding amount of sulphide is obtained (H. Rose).

COMPOSITION.

Zn	65·06	80·26
O	16·00	19·74
	81·06	100·00

c. Zinc sulphide, recently precipitated, is a white, loose hydrate. The following facts should here be mentioned with regard to its precipitation.* Colorless ammonium sulphide precipitates dilute solutions of zinc, but only slowly; yellow ammonium sulphide does not precipitate dilute solutions of zinc (1 : 5000) at all. Ammonium chloride favors the precipitation considerably. Free ammonia acts so as to keep the precipitate somewhat longer in suspension, otherwise it exerts no injurious influence. If the conditions which I shall lay down are strictly observed, zinc may be precipitated by ammonium sulphide from a solution containing only $\frac{1}{1000000}$. Hydrated zinc sulphide on account of its slimy nature easily stops up the pores of the filter, and cannot therefore be washed without difficulty on a filter. The washing is best performed by using water containing ammonium sulphide, and continually diminished quantities of ammonium chloride (at last none) (see Expt. No. 43). The hydrate is insoluble in water, in caustic alkalies, alkali carbonates, and the monosulphides of the alkali metals. It dissolves readily and completely in hydrochloric and in nitric, but only very sparingly in acetic acid. When dried, the precipitated zinc sulphide is a white powder; when air-dried its composition is $3ZnS + 2H_2O$; dried at 100°, $2ZnS + H_2O$; at 150°, $4ZnS + H_2O$ (A. Souchay†). On ignition it loses the whole of its water. During the latter process some hydrogen sulphide escapes, and the residue contains some oxide. By roasting in the air, and intense ignition, small quantities of zinc sulphide may be readily converted into the oxide. On igniting the dried zinc sulphide, mixed with powdered sulphur, in a stream of hydrogen, the pure anhydrous sulphide is obtained (H. Rose). The latter suffers no loss of weight worth mentioning by ignition for five minutes over the gas blowpipe; but if such ignition is

* Jour. f. prakt. Chem. 82, 263. † Zeitschr. f. anal. Chem. 7, 78.

very protracted the loss of weight becomes considerable (AL. CLASSEN*).

COMPOSITION.

Zn	65·06	67·03
S	32·00	32·97
	97·06	100·00

§ 78.

2. MANGANESE.

Manganese is weighed either as PROTOSESQUIOXIDE (MANGANOSO-MANGANIC OXIDE), as SULPHIDE, as MANGANOUS SULPHATE, or as PYROPHOSPHATE. With the view of converting it into these forms, it is precipitated as MANGANOUS CARBONATE, MANGANOUS HYDROXIDE, MANGANESE DIOXIDE, or AMMONIUM MANGANESE PHOSPHATE.

a. Manganese carbonate, recently precipitated, is white, flocculent, nearly insoluble in pure water, but somewhat more soluble in water impregnated with carbonic acid. Presence of sodium carbonate or potassium carbonate does not increase its solubility. Recently precipitated manganese carbonate dissolves pretty readily in ammonium chloride: it is owing to this property that a solution of manganese cannot be completely precipitated by potassium or sodium carbonate, in presence of ammonium chloride (or any other ammonium salt), until the latter is completely decomposed. If the precipitate, while still moist, is exposed to the air, or washed with water impregnated with air, especially if it is in contact with alkali carbonate, it slowly assumes a dirty brownish-white color, part of it becoming converted into hydrated protosesquioxide. Even long-continued washing will not remove the last traces of alkali salt from the precipitate. The wash-water often comes through turbid. If the filtrate and wash-water are evaporated to dryness and the residue is treated with boiling water, the small traces of manganous carbonate which were partly dissolved and partly suspended will remain behind in the form of hydrated protosesquioxide. Dried by pressure the precipitate is white, and consists of $MnCO_3 + H_2O$; dried in a vacuum it consists of $2(MnCO_3) + H_2O$ (E. PRIOR†); when dried with free access of air, the powder is of a dirty-white color. When strongly heated with access of air,

* Zeitschr. f. anal. Chem. 4, 421. † *Ib.* 8, 428.

this powder first turns black, and changes subsequently to brown protosesquioxide of manganese. However, this conversion takes some time, and must never be held to be completed until two weighings, between which the precipitate has been ignited again with free access of air, give perfectly corresponding results. On igniting the manganous carbonate, mixed with powdered sulphur, in a stream of hydrogen, manganese sulphide is obtained (H. ROSE).

b. Manganous hydroxide recently thrown down forms a white, flocculent precipitate, barely soluble in water and alkalies, but soluble in ammonium chloride; it immediately absorbs oxygen from the air, and turns brown, owing to the formation of hydrated protosesquioxide. On drying it in the air, a brown powder is obtained which, when heated to intense redness, with free access of air, is converted into protosesquioxide, and on ignition with sulphur, in a stream of hydrogen, is converted into sulphide.

c. Protosesquioxide of manganese, artificially produced, is a brown powder. All the oxides of manganese are finally converted into this by strong ignition in the air. Each time it is heated it assumes a darker color, but its weight remains unaltered. It is insoluble in water, and does not alter vegetable colors. If ignited with ammonium chloride, it is converted into the manganous chloride. When heated with concentrated hydrochloric acid, it dissolves to chloride with evolution of chlorine ($Mn_3O_4 + 8HCl = 3MnCl_2 + 2Cl + 4H_2O$). On ignition with sulphur in a stream of hydrogen it is converted into sulphide (H. ROSE). On ignition in oxygen it is converted into manganic oxide (SCHNEIDER). On ignition in hydrogen it is converted into manganous oxide.

COMPOSITION.

Mn_3	165·00	72·05
O_4	64·00	27·95
	229·00	100·00

d. Manganese dioxide is occasionally produced in analysis by exposing a concentrated solution of manganous nitrate to a gradually increased temperature. At 140° brown flakes separate, at 155° much nitrous acid is disengaged, and the whole of the manganese separates as anhydrous dioxide. It is brownish-black, and is deposited on the sides of the vessel, with metallic lustre. It is insoluble in weak nitric acid, but dissolves to a small amount in hot and concentrated nitric acid (DEVILLE). In hydrochloric acid it dissolves with evolution of chlorine, in concentrated sulphuric acid with liberation of oxygen. The dioxide is also sometimes obtained in the hydrated condition in analytical separations, thus when we precipitate a solution of a manganous salt with sodium hypochlorite, or, after addition of sodium acetate, with bromine or chlorine in the heat. The brownish-black flocculent precipitate thus obtained, contains alkali, from which it cannot be well freed by washing.

e. Manganese sulphide, prepared in the wet way, generally forms a flesh-colored precipitate. I must make a few remarks with reference to its precipitation.* This is effected but incompletely if we add to a pure manganous solution only ammonium sulphide, no matter whether it be colorless or yellow, while it is perfectly effected if ammonium chloride be used in addition. A large quantity even of ammonium chloride does not impede the precipitation. Ammonia in small quantity is not injurious, but in large quantity it interferes with complete precipitation, especially in the presence of ammonium polysulphide (A. CLASSEN†). In all cases we must allow to stand at least 24 hours, and with very dilute solutions 48 hours, before filtering. Colorless or slightly yellow ammonium sulphide is the most appropriate precipitant. In the presence of ammonium chloride even a large excess of ammonium sulphide is uninjurious. If the precipitation is conducted as directed, the manganese can be precipitated from solutions which contain an amount equivalent to only $\frac{1}{400000}$ of the manganous oxide. If the flesh-colored hydrated sulphide remains some time under the fluid, from which it was precipitated, it sometimes becomes converted into the green anhydrous sulphide.‡ This conversion is more likely to take place when a large excess of ammonium sulphide has been used; heating favors it, ammonium chloride hinders it. The conversion is occasionally rapid. The green sulphide thus obtained consists of eight-sided tables distinctly visible under the microscope (F. MUCK§). In acids (hydrochloric, sulphuric, acetic, &c.) the hydrated sulphide dissolves with evolution of hydrogen sulphide. If the precipitate, while still moist, is exposed to the air, or washed with water impregnated with air, it changes to brown, hydrated protosesquioxide of manganese

* Journ. f. prakt. Chem. 82, 265.
† Zeitschr. f. anal. Chem. 8, 370.
‡ Journ. f. prakt. Chem. 82, 268.
§ Zeitschr. f. Chem. N. F. 6, 6.

being formed, together with a small portion of manganous sulphate. Hence in washing the hydrate we always add some ammonium sulphide to the wash-water, and keep the filter as full as possible with the same. We guard against the filtrate running through turbid, by adding gradually decreasing quantities of ammonium chloride to the wash-water (at last none). (Expt. No. 44.) On igniting the precipitate mixed with sulphur in a stream of hydrogen the anhydrous sulphide remains. If we have gently ignited during this process, the product is light green; if we have strongly ignited, it is dark green to black. Neither the green nor the black sulphide attracts oxygen or water quickly from the air (H. ROSE). The anhydrous sulphide is also readily soluble in dilute acids.

COMPOSITION.

Mn	55·00	63·22
S	32·00	36·78
	87·00	100·00

f. Anhydrous manganous sulphate, produced by exposing the crystallized salt to the action of heat, is a white, friable mass, readily soluble in water. It resists a very faint red heat; but upon exposure to a more intense red heat, it suffers more or less complete decomposition—oxygen, sulphur dioxide, and sulphur trioxide being evolved, and protosesquioxide of manganese remaining behind. Ignited with sulphur in a stream of hydrogen it is transformed into sulphide (H. ROSE).

COMPOSITION.

$SO_2 < {O \atop O} > Mn =$

MnO . .	71·00	47·02
SO_3 . . .	80·00	52·98
	151·00	100·00

g. Ammonium manganese phosphate.—GIBBS* says that this precipitate is insoluble in boiling water, but I have not found this to be the case. My results are that 1 part dissolves in 32092 parts of cold water, in 20122 parts boiling water, and 17755 parts of water containing $\frac{1}{10}$ of ammonium chloride. It has the formula

* SILLIM. Amer. Journ. (ii.) 44, 216.

$NH_4MnPO_4 + H_2O$. It presents pale pink scales of pearly lustre, which sometimes turn reddish on the filter. On ignition it is converted into manganese pyrophosphate.

h. Manganese pyrophosphate is the white residue left on the ignition of the preceding.

COMPOSITION.

$$O \left\langle \begin{array}{l} PO < {O \atop O} > Mn \\ PO < {O \atop O} > Mn \end{array} \right. = \begin{array}{l} 2MnO \\ P_2O_5 \end{array}$$

2MnO	142	50
P_2O_5	142	50
	284	100

§ 79.

3. NICKEL.

Nickel is precipitated as HYDROXIDE, and as SULPHIDE. It is weighed in the form of NICKELOUS OXIDE, of METALLIC NICKEL, or of anhydrous NICKELOUS SULPHATE.

a. Nickelous hydroxide forms an apple-green precipitate, almost absolutely insoluble in water. When precipitated from a solution of the chloride or sulphate, it retains some of the acid even after long washing (TEICHMANN*). It is also very difficult to remove the last traces of alkali. It dissolves with some difficulty in ammonia and ammonium carbonate, far more readily in the presence of an ammonium salt. From these solutions it is completely precipitated by excess of potassa or soda; application of heat promotes the precipitation. It is unalterable in the air; on ignition, it passes into nickelous oxide.

b. Nickelous oxide is a dirty grayish-green powder. When obtained by heating the nitrate to redness, it always contains some nickelic oxide, and requires very strong and protracted ignition for conversion into the pure green nickelous oxide (W. J. RUSSELL). It is insoluble in water, but readily soluble in hydrochloric acid. It does not affect vegetable colors. It suffers no variation of weight upon ignition with free access of air. Mixed with ammonium chloride and ignited, it is reduced to metallic nickel (H. ROSE); it is also easily reduced by ignition in hydrogen or carbon monoxide.

* Annal. d. Chem. u. Pharm. 156, 17.

COMPOSITION.

Ni	59	78·67
O	16	21·33
	75	100·00

c. Metallic nickel obtained by the reduction of nickelous oxide with hydrogen has the form of a gray powder, or if the heat has been very strong, and it has melted, it is lustrous and white like silver. It is unaltered in weight by ignition in hydrogen, when ignited in the air it is superficially oxidized. It is attracted by the magnet. It is dissolved slowly by hydrochloric acid and dilute sulphuric acid, and readily by moderately strong nitric acid.

d. Anhydrous nickelous sulphate obtained by evaporating a solution of the chloride, nitrate, &c., with sulphuric acid is yellow, soluble in water to a green fluid. The hydrous salt may be rendered anhydrous without loss of acid by cautious heating in a platinum dish, but at low redness it begins to blacken at the edges and loses acid (F. GAUHE*).

e. Hydrated nickelous sulphide, prepared in the wet way, forms a black precipitate, insoluble in water. I must make some observations on its precipitation.† In order to precitate the nickel from a pure solution completely and with ease, ammonium chloride must be present; it is not enough to add ammonium sulphide alone. A large quantity even of ammonium chloride produces no injurious effect. In the presence of free ammonia, on the contrary, some nickel remains in solution. In this case, the supernatant fluid appears brown. As precipitant, colorless or light-yellow ammonium sulphide containing no free ammonia should be used, a large excess must be avoided. If the directions given are adhered to—allowing to stand 48 hours—the nickel may be precipitated by means of ammonium sulphide, from solutions containing only $\frac{1}{800000}$ of the oxide. As the precipitate is liable to take up oxygen from the air, being transformed into sulphate, a little ammonium sulphide is mixed with the wash-water, to which also it is advisable to add ammonium chloride (less and less—at last none); the filter should be kept full (Expt. No. 45). Brown filtrates, containing nickel sulphide in solution, may be freed from the latter by acidulation with acetic acid, and boiling some time.

* Zeitschr. f. anal. Chem. 4, 190. † Journ. f. prakt. Chem. 82, 257.

The sulphide falls down, and may now be filtered off. It is very sparingly soluble in concentrated acetic acid, somewhat more soluble in hydrochloric acid. It is more readily soluble still in nitric acid, but its best solvent is nitro-hydrochloric acid. It loses its water upon the application of a red heat; when ignited in the air, it is transformed into a basic compound of nickelous oxide with sulphuric acid. Mixed with sulphur and ignited in a stream of hydrogen, a fused mass remains, of pale yellow color and metallic lustre. This consists of Ni_2S, but its composition is not perfectly constant (F. GAUHE*). [Nickel may be precipitated as a sulphide, dense in form, easy to wash, and not readily oxidizing by contact with air, by proceeding as follows: To the solution, which should be concentrated and contain a liberal quantity of ammonium salts, add ammonia (if necessary) to alkaline reaction, then acetic acid to *slight* acid reaction, also ammonium or sodium acetate, and heat to boiling. Transmit H_2S gas through the boiling solution. Since *much* free acetic acid prevents complete precipitation, it is necessary sometimes when much nickel is present to partially neutralize once or twice the acid set free during the process.]

Nickel sulphide may be converted into nickelous sulphate by dissolving in nitric acid and evaporating with sulphuric acid.

§ 80.

4. COBALT.

Cobalt is weighed in the PURE METALLIC state, or as COBALTOUS SULPHATE. Besides the properties of these substances, we have to study also those of COBALTOUS HYDROXIDE, of the SULPHIDE, and of the TRIPOTASSIUM COBALTIC NITRITE.

a. Cobaltous hydroxide.—Upon precipitating a solution of a cobaltous salt with potassa, a blue precipitate (a basic salt) is formed at first, which, upon boiling with potassa in excess, excluded from contact of air, changes to light red cobaltous hydroxide; if, on the contrary, this process is conducted with free access of air, the precipitate becomes discolored, and finally black, part of the cobaltous hydroxide being converted into cobaltic hydroxide. But the hydroxide prepared in this way, retains always a certain quantity of the acid, and, even after the most thorough washing

* Zeitschr. f. anal. Chem. 4, 191.

with hot water, also a small amount of the alkaline precipitant. The latter, however, is not enough to spoil the accuracy of the results (H. ROSE, F. GAUHE*). Cobaltous hydroxide is insoluble in water, and also in dilute potassa; it is somewhat soluble in very concentrated potassa, and readily in ammonium salts. When dried in the air, it absorbs oxygen, and acquires a brownish color. By strong ignition it is converted into cobaltous oxide (even if some higher oxide had formed from boiling or drying in the air); if cooled with exclusion of air, as in a current of carbon dioxide, pure light-brown cobaltous oxide will be left; if cooled, on the contrary, with access of air, it is more or less changed to black protosesquioxide (cobaltoso-cobaltic oxide) (W. J. RUSSELL†). By ignition in a current of hydrogen, metallic cobalt is left, from which any traces of alkali may now be almost completely removed by boiling water.

b. The *metallic cobalt* obtained according to *a*, or by igniting the chloride or the protosesquioxide (produced by igniting the nitrate) in hydrogen is a grayish-black powder, which is attracted by the magnet, and is more difficultly fusible than gold. If the reduction has been effected at a faint heat, the finely divided metal burns in the air to protosesquioxide of cobalt, which is not the case if the reduction has been effected at an intense heat. Cobalt does not decompose water, either at the common temperature, or upon ebullition—except sulphuric acid be present, in which case decomposition will ensue. Heated with concentrated sulphuric acid, it forms cobaltous sulphate, with evolution of sulphur dioxide. In nitric acid it dissolves readily to cobaltous nitrate.

c. Cobalt sulphide, produced in the wet way, forms a black precipitate, insoluble in water, alkalies, and alkali sulphides. With regard to its precipitation,‡—this is effected but slowly and imperfectly by ammonium sulphide alone; in the presence of ammonium chloride however, it takes place quickly and completely. Free ammonia is not injurious; it is all one, whether colorless or yellow ammonium sulphide is employed. If the directions given are observed, cobalt may be precipitated from a solution containing no more than $\frac{1}{800000}$ of the protoxide. In the moist condition, exposed to the air, it oxidizes to sulphate. In washing it, therefore, water containing ammonium sulphide is employed, and

* Zeitschr. f. anal. Chem. 4, 54. † Journ. Chem. Soc. (2) 1, 51.

‡ Journ. f. prakt. Chem. 82, 262.

the filter is kept full. It is advisable also to mix a little ammonium chloride with the wash-water, but its quantity should be gradually decreased, and the last water used must contain none. It is but sparingly soluble in acetic acid and in dilute mineral acids, more readily in concentrated mineral acids, and most readily in warm nitro-hydrochloric acid. Mixed with sulphur and ignited in a stream of hydrogen, we obtain a product which varies in composition according to the temperature employed. The residue is therefore not suited for the determination of cobalt (H. ROSE). Cobalt can be precipitated as sulphide completely in the presence of a very small amount of free acetic acid by hydrogen sulphide in the same manner as nickel (see § 79, *e*). Cobalt sulphide may be converted into cobaltous sulphate by heating in the air, moistening with nitric acid, evaporating with sulphuric acid and igniting.

d. Cobaltous sulphate crystallizes, in combination with 7 aq., slowly in oblique rhombic prisms of a fine red color. The crystals yield the whole of the water, at a moderate heat, and are converted into a rose-colored anhydrous salt, which bears the application of a low red heat without losing acid. At a stronger heat the edges become black and some sulphuric acid escapes (F. GAUHE*). It dissolves rather difficultly in cold, but more readily in hot water.

COMPOSITION.

$SO_2 < {O \atop O} > Co =$			
	CoO	75	48·39
	SO_3	80	51·61
		155	100·00

e. Tripotassium cobaltic nitrite.—If a solution of a cobalt salt (not too dilute) is mixed with excess of potassa and then with acetic acid till the precipitate is redissolved, and a concentrated solution of potassium nitrite previously acidified with acetic acid is added, first a dirty, brownish precipitate forms which gradually turns yellow and crystalline, especially on the application of a gentle heat (N. W. FISCHER†). The composition of this precipitate corresponds to the formula $(KNO_2)_6Co_2(NO_2)_6$ + aq. x (SADTLER). Dried at 100° its composition is somewhat variable (STROMEYER, ERDMANN‡). It is decidedly soluble in water, less in potassium acetate whether neutral or acidified with acetic acid,

* Zeitschr. f. anal. Chem. 4, 55. † Pogg. Ann. 72, 477.
‡ Journ. f. prakt. Chem. 97, 385.

not in potassium acetate to which some potassium nitrite has been added, not in potassium nitrite, nor in alcohol of 80 per cent. On washing with water or solution of potassium acetate, unless potassium nitrite is added, nitric oxide is constantly evolved in small quantities. It is decomposed with separation of brown cobaltic hydroxide, with difficulty by solution of potassa, with ease by soda or baryta. On being moistened with sulphuric acid and ignited (finally with addition of ammonium carbonate) it leaves $2(CoSO_4) + 3(K_2SO_4)$, but there is a difficulty in driving off all the excess of acid without decomposing the cobaltous sulphate. The yellow salt is soluble in hydrochloric acid, potassa precipitates the whole of the cobalt from this solution as hydroxide.

§ 81.

5. Ferrous Iron; and 6. Ferric Iron.

Iron is usually weighed in the form of FERRIC OXIDE, occasionally as SULPHIDE. We have to study also the FERRIC HYDROXIDE, the FERRIC SUCCINATE, the FERRIC ACETATE, and the FERRIC FORMATE.

a. Ferric hydroxide, recently prepared, is a reddish-brown precipitate, insoluble in water, in dilute alkalies, and in ammonium salts, but readily soluble in acids; it shrinks very greatly on drying. When dry, it presents a brown, hard mass, with shining conchoidal fracture. If the precipitant alkali is not used in excess, the precipitate contains basic salt; on the other hand, if the alkali has been used in excess, a portion of it is invariably carried down in combination with the ferric hydroxide,—on which account ammonia alone can properly be used in analysis for this purpose. Under certain circumstances, for instance, by protracted heating of a solution of ferric acetate on the water-bath (which turns the solution from blood-red to brick-red, and makes it appear turbid by reflected light), and subsequent addition of some sulphuric acid or salt of an alkali, a reddish-brown hydrated ferric oxide is produced, which is insoluble in cold acids, even though concentrated, and is not attacked even by boiling nitric acid (L. PEAN DE ST. GILLES*).

Closely allied to ferric hydroxide are the highly basic salts obtained by mixing dilute cold solutions of ferric salts, best ferric chloride, with much ammonium chloride, cautiously adding am-

* Journ. f. prakt. Chem. 66, 137.

monium carbonate till the fluid on standing in the cold instead of becoming clear turns more turbid if anything, and then boiling. The precipitates, thus produced in the fluid which still retains its acid reaction, contain the whole of the iron present and play an important part in analytical separations. They should be washed with boiling water containing ammonium chloride, being soluble to a slight extent in pure water. They are not suitable for ignition, as ferric chloride might occasionally escape from them.

b. Ferric hydroxide is, upon ignition, converted into *ferric oxide*. If the hydroxide has been superficially dried only, the violent escape of steam from the lumps is likely to occasion loss; but if it has been dried as much as possible by suction and still remains moist, it may be ignited without fear of loss. Pure ferric oxide, when placed upon moist reddened litmus-paper, does not change the color to blue. It dissolves slowly in dilute, but more rapidly in concentrated hydrochloric acid; the application of a moderate degree of heat effects this solution more readily than boiling. With a mixture of 8 parts concentrated sulphuric acid and 3 parts water, it behaves in the same manner as alumina. The weight of ferric oxide does not vary upon ignition in the air; when ignited with ammonium chloride, ferric chloride escapes. Ignition with charcoal, in a closed vessel, reduces it more or less. Strongly ignited with sulphur in a stream of hydrogen, it is transformed into ferrous sulphide.

COMPOSITION.

Fe_2	112	70·00
O_3	48	30·00
	160	100·00

c. Ferrous sulphide, produced in the wet way, forms a black precipitate. The following facts are to be noticed with regard to its precipitation.† Ammonium sulphide used alone, whether colorless or yellow, precipitates pure neutral solutions of ferrous salts, but slowly and imperfectly. Ammonium chloride acts very favorably; a large excess even is not attended with inconvenience. Ammonia has no injurious action. It is all the same whether the ammonium sulphide be colorless or light yellow. If the direc-

† *Ib.*, 82, 268.

tions given are observed, iron may be precipitated by means of ammonium sulphide, from solutions containing only $\frac{1}{1000000}$ of ferrous oxide. In such a case, however, it is necessary to allow to stand forty-eight hours. Since the precipitate rapidly oxidizes in contact with air, ammonium sulphide is to be added to the wash-water, and the filter kept full. It is well also to mix a little ammonium chloride with the wash-water, but the quantity should be continually reduced, and the last water used should contain none. In mineral acids, even when very dilute, the hydrated sulphide dissolves readily. Mixed with sulphur, and strongly ignited in a stream of hydrogen, anhydrous ferrous sulphide remains (H. Rose).

COMPOSITION.

Fe	56	63·64
S	32	36·36
	88	100·00

d. When a neutral solution of a ferric salt is mixed with a neutral solution of an alkali succinate, a cinnamon-colored precipitate of a brighter or darker tint of a *basic ferric succinate* is formed, succinic acid being set free. The free succinic acid does not exercise any perceptible solvent action upon the precipitate in a cold and highly dilute solution, but it redissolves the precipitate a little more readily in a warm solution. The precipitate must therefore be filtered cold, if we want to guard against re-solution. Formerly the precipitate was erroneously supposed to consist of a normal salt, decomposable by hot water into an insoluble basic and a soluble acid compound. Basic ferric succinate is insoluble in cold, and but sparingly soluble in hot water. It dissolves readily in mineral acids. Ammonia, especially if warm, deprives it of the greater portion of its acid, leaving compounds which are highly basic ferric succinates (Döpping).

e. If to a solution of a ferric salt, sodium carbonate be added in the cold, till the fluid contains no more free acid, and in consequence of the formation of basic salt has become deep red, but remains still perfectly clear, and then sodium acetate be poured in and the mixture boiled, the whole of the iron will be precipitated as *basic ferric acetate.*

f. Instead of the sodium or ammonium acetate used in *e*, the corresponding formates may be used. The *basic ferric formate* here obtained is more easily washed than the basic acetate (F. SCHULZE*).

BASIC RADICALS OF THE FIFTH GROUP.

§ 82.

1. SILVER.

Silver may be weighed in the METALLIC state, as CHLORIDE, SULPHIDE, or CYANIDE.

a. Metallic silver, obtained by the ignition of salts of silver with organic acids, &c., is a loose, white, glittering mass of metallic lustre; but, when obtained by reducing silver chloride, &c., in the wet way, by zinc, it is a dull-gray powder. It fuses at about 1000°. Its weight is not altered by moderate ignition. It may, however, be distilled by the heat of the oxyhydrogen flame (CHRISTOMANOS†). It dissolves readily and completely in dilute nitric acid.

b. Silver chloride, recently precipitated, is white and curdy. On shaking, the large spongy flocks combine with the smaller particles, so that the fluid becomes perfectly clear. This result is, however, only satisfactorily effected when the flocks have been recently precipitated in presence of excess of silver solution (compare G. J. MULDER‡). Silver chloride is in a very high degree insoluble in water, and in dilute nitric acid; strong nitric acid, on the contrary, does dissolve a trace. Hydrochloric acid, especially if concentrated and boiling, dissolves it very perceptibly. According to PIERRE, 1 part of silver chloride requires for solution 200 parts of strong hydrochloric acid and 600 parts of a dilute acid, composed of 1 part strong acid and 2 parts water. On sufficiently diluting such a solution with cold water the silver chloride is precipitated so completely that the filtrate is not colored by hydrogen sulphide. Silver chloride is insoluble, or very nearly so, in concentrated sulphuric acid; in the dilute acid it is as insoluble as in water. In a solution of tartaric acid silver chloride dissolves perceptibly on warming; on cooling, however, the solution deposits the whole, or, at all events, the greater part of it. Aqueous solutions of chlorides (of sodium, potassium, ammonium, calcium, zinc,

* Chem. Centralblatt, 1861, 3. † Zeitschr. f. anal. Chem. 7, 299.

‡ Die Silberprobirmethode, translated into German by D. Chr. Grimm, pp. 19 and 311. Leipzig: J. J. Weber. 1859.

&c.) all dissolve appreciable quantities of silver chloride, especially if they are hot and concentrated. On sufficient dilution with cold water the dissolved portion separates so completely that the filtrate is not colored by hydrogen sulphide. The solutions of alkali and alkali-earth nitrates also dissolve a little silver chloride. The solubility in the cold is trifling; in the heat, on the contrary, it is very perceptible. A strong solution of silver nitrate dissolves it slightly, especially in the heat; but I have found it insoluble in a moderately dilute cold solution of lead nitrate. The action of mercuric salts upon it is remarkable. When well washed and treated with a very dilute solution of mercuric chloride it becomes white if previously a little blackened by light, is easily diffused in the fluid, and is but tardily deposited. This depends upon the mercuric salt being taken up; if the silver salt is washed the mercuric salt will be removed. Mercuric nitrate acts in a similar way, but a certain quantity of silver passes at the same time into solution. Silver chloride is much more difficultly dissolved by mercuric acetate than by mercuric nitrate; therefore, if you have a solution of mercuric nitrate containing silver chloride, if the mercuric salt is not present in enormous quantity, the silver may be almost absolutely thrown down by addition of an alkali acetate (H. DEBRAY*). Solutions of potash and soda decompose silver chloride, even at the ordinary temperature, more readily on boiling; silver oxide separates, and chloride of the alkali metal is formed. Solution of sodium or potassium carbonate decomposes silver chloride only very imperfectly even on boiling; after long boiling decided traces of chlorine are found in the filtrate. Silver chloride dissolves readily in aqueous ammonia, and also in the solution of potassium cyanide and that of sodium thiosulphate. According to WALLACE and LAMONT† 1 part of silver chloride dissolves in 12·88 parts of strong aqueous ammonia of ·89 sp. gr. Under the influence of light silver chloride soon changes to violet, finally black, losing chlorine, and passing partly into Ag_2Cl. The change is quite superficial, but the loss of weight resulting is very appreciable (MULDER, *op. cit.*, p. 21). If silver chloride that has become violet or black from the influence of light be treated with aqueous ammonia, it dissolves with separation of a very small quantity of metallic silver, Ag_2Cl gives AgCl and Ag (WITTSTEIN). On long contact (say for 24

* Zeitschr. f. Chem. 13, 348. † Chem. Gaz. 1859, 137.

hours) with water, especially of 75°, silver chloride, although removed from the influence of light, becomes gray, and, it appears, decomposed; the precipitate is found to contain silver oxide, and the water hydrochloric acid (MULDER). On digestion with excess of solution of potassium bromide or iodide, silver chloride is completely transformed into silver bromide or iodide, as the case may be (FIELD*). On drying, silver chloride becomes pulverulent; on heating it turns yellow; at 260° it fuses to a transparent yellow fluid; at a very high heat it volatilizes without decomposition. On cooling after fusion it presents a colorless or pale yellowish mass. Fused in chlorine gas, it absorbs some chlorine; on cooling, this escapes, but not completely. If it is to be completely expelled, and, in very delicate experiments this must be done, we pass carbon dioxide before allowing to cool (STAS †). Ignition with charcoal fails to effect its reduction to the metallic state; but it may be readily so reduced in a current of hydrogen, carburetted hydrogen, or carbon monoxide.

COMPOSITION.

Ag	107·93	75·27
Cl	35·46	24·73
	143·39	100·00

c. Silver sulphide, prepared in the wet way, is a black precipitate, insoluble in water, dilute acids, alkalies, and alkali sulphides. It is unalterable in the air; after being allowed to subside, it is filtered and washed with ease, and may be dried at 100° without decomposition. It dissolves in concentrated nitric acid, with separation of sulphur. Solution of potassium cyanide dissolves it with difficulty, if it has been precipitated from a very dilute solution with less difficulty; the quantity of potassium cyanide, too, has great influence on the effect. For instance, if silver cyanide, is dissolved in a bare sufficiency of potassium cyanide and hydrogen sulphide, or ammonium sulphide is added, silver sulphide is thrown down; if, on the other hand, a large excess of potassium

* Quart. Journ. Chem. Soc. 10, 234.

† Recherches sur less rapports réciproques des poids atomiques, p. 37. Bruxelles, 1860. The loss of weight which about 100 grm. chloride of silver suffered, by the expulsion of the absorbed chlorine, was from 7 to 13 mgrm.

cyanide is present, no precipitate will be produced. If silver sulphide is dissolved in a concentrated solution of potassium cyanide, it will generally separate at once on addition of much water (BECHAMP*). Ignited in a current of hydrogen, it passes readily and completely into the metallic state (H. ROSE).

COMPOSITION.

Ag_2	215·86	87·09
S	32·00	12·91
	247·86	100·00

d. Silver cyanide, recently thrown down, forms a white curdy precipitate insoluble in water and dilute nitric acid, soluble in potassium cyanide and also in ammonia; exposure to light fails to impart the slightest tinge of black to it; it may be dried at 100° without decomposition. Upon ignition, it is decomposed into cyanogen, which escapes, and metallic silver, which remains, mixed with a little paracyanide of silver. By boiling with a mixture of equal parts of sulphuric acid and water, it is, according to GLASSFORD and NAPIER, dissolved to silver sulphate, with liberation of hydrocyanic acid.

COMPOSITION.

Ag	107·93	80·56
CN	26·04	19·44
	133·97	100·00

§ 83.

2. LEAD.

Lead is weighed as OXIDE, SULPHATE, CHROMATE, CHLORIDE, and SULPHIDE. Besides these compounds, we have also to study the CARBONATE and the OXALATE.

a. Normal lead carbonate forms a heavy, white, pulverulent precipitate. It is but very slightly soluble in perfectly pure (boiled) water, one part requiring 50550 parts (see Expt. No. 47, *a*); but it dissolves somewhat more readily in water containing ammonia and ammonium salts (comp. Expt. No. 47, *b* and *c*), and also in

* Journ. f. prakt. Chem. 60, 64.

water impregnated with carbonic acid. It loses its carbonic acid when ignited.

b. *Lead oxalate* is a white powder, very sparingly soluble in water. The presence of ammonium salts slightly increases its solubility (Expt. No. 48). When heated in close vessels, it leaves lead suboxide; but when heated with access of air, the yellow oxide.

c. *Lead oxide*, produced by igniting the carbonate or oxalate, is a lemon-yellow powder, inclining sometimes to a reddish-yellow, or to a pale yellow. When this yellow lead oxide is heated, it assumes a brownish-red color, without the slightest variation of weight. It fuses at an intense red heat. Ignition with charcoal reduces it. When exposed to a white heat, it rises in vapor. Placed upon moist red litmus paper, it changes the color to blue. When exposed to the air, it slowly absorbs carbonic acid. Mixed with ammonium chloride and ignited, it is converted into lead chloride. Lead oxide in a state of fusion readily dissolves silicic acid and the earthy bases with which the latter may be combined.

COMPOSITION.

Pb	207	92·83
O	16	7·17
	223	100·00

d. *Lead sulphate* is a heavy white powder. It dissolves, at the common temperature, in 22800 parts of pure water (Expt. No. 49*); it is less soluble in water containing sulphuric acid (1 part requiring 36500 parts—Expt. No. 50); it is far more readily soluble in water containing ammonium salts; from this solution it may be precipitated again by adding sulphuric acid in excess (Expt. No. 51). It is almost entirely insoluble in common alcohol. Of the ammonium salts, the nitrate, acetate, and tartrate are more especially suited to serve as solvents for lead sulphate: the two latter salts are made strongly alkaline by addition of ammonia, previous to use (WACKENRODER). Lead sulphate dissolves in concentrated hydrochloric acid, upon heating. In nitric acid it dissolves the more readily, the more concentrated and hotter the acid; water fails to precipitate it from its solution in nitric acid; but the addition of a copious amount of dilute sulphuric acid causes its precipi-

* According to G. F. RODWELL 1 part dissolves in 31696 parts water at 15° (Chem. News, 1866, 50).

tation from this solution. The more nitric acid the solution contains, the more sulphuric acid is required. It dissolves sparingly in concentrated sulphuric acid, and the dissolved portion precipitates again upon diluting with water (more completely upon addition of alcohol). A moderately concentrated solution of sodium thiosulphate dissolves lead sulphate completely even if cold, more readily if warmed. On boiling, the solution becomes black, from separation of a small quantity of lead sulphide (J. Löwe*). The solutions of alkali carbonates and alkali hydrogen carbonates convert lead sulphate, even at the common temperature, completely into lead carbonate. The solutions of the normal alkali carbonates, but not those of the alkali hydrogen carbonates, dissolve some lead oxide in this process (H. Rose†). Lead sulphate dissolves readily in hot solutions of potassa or soda. It is unalterable in the air, and at a gentle red heat; when exposed to a full red heat, it fuses without decomposition (Expt. No. 52), provided always reducing gases be completely excluded—for, if this is not the case, the weight will continually diminish, owing to reduction to sulphide (Erdmann‡). At a white heat the whole of the sulphuric acid gradually escapes (Boussingault§). When it is ignited with charcoal, lead sulphide is formed at first; if the heat be raised, this sulphide reacts on undecomposed sulphate, metallic lead and sulphur dioxide being produced. Fusion with potassium cyanide reduces the whole of the lead to the metallic state. Lead sulphate mixed with sulphur and exposed to intense ignition in a current of hydrogen yields the sulphide, but loss can scarcely be avoided (compare *f*).

COMPOSITION.

$SO_2<{O \atop O}>Pb =$			
	PbO	223	73·60
	SO_3	80	26·40
		303	100·00

e. Lead chloride obtained by precipitation is a white crystalline powder. It separates in needles from a hot solution containing a certain quantity of hydrochloric acid; occasionally it presents wedge-shaped crystals, or when separated from a strong hydrochloric solution, hexagonal tables. At 17°·7 water dissolves ·946

* Journ. f. prakt. Chem, 74, 348.
† Pogg. Annal. 95, 426.
‡ Journ. f. prakt. Chem. 62, 381.
§ Zeitschr. f. anal. Chem. 7, 244.

per cent.; a fluid containing 15 per cent. of hydrochloric acid of 1·162 sp. gr. dissolves ·090; a fluid containing 20 per cent. acid dissolves ·111 per cent.; a fluid containing 80 per cent. acid dissolves 1·498 per cent. Pure hydrochloric acid of the above strength dissolves 2·900 per cent. (J. CARTER BELL*). Lead chloride is less soluble in water containing nitric acid than in water (1 part requires 1636 parts, BISCHOF). It is extremely sparingly soluble in alcohol of 70 to 80 per cent., and altogether insoluble in absolute alcohol. It is unalterable in the air. It fuses at a temperature below red heat, without loss of weight. When exposed to a higher temperature, with access of air, it volatilizes slowly, being partially decomposed: chlorine gas escapes, and a mixture of lead oxide and chloride remains.

COMPOSITION.

Pb	207·00	74·48
Cl_2	70·92	25·52
	277·92	100·00

f. Lead sulphide, prepared in the wet way, is a black precipitate, insoluble in water, dilute acids, alkalies, and alkali sulphides. In precipitating it from a solution containing free hydrochloric acid, it is necessary to dilute plentifully, otherwise the precipitation will be incomplete. Even if a fluid only contain 2·5 per cent. HCl, the whole of the lead will not be precipitated (M. MARTIN †). It is unalterable in the air; it cannot be dried at 100° without decomposition. According to H. ROSE it increases perceptibly in weight by oxidation; in the case of long-protracted drying even becoming a few per-cents heavier.‡ I have confirmed his statement (see Expt. No. 53). If lead sulphide mixed with sulphur is heated gently in a current of hydrogen, so that the lower quarter of the crucible is red hot, lead sulphide is left without loss of weight. By continuing a gentle heat the weight gradually diminishes; by strong ignition the loss is rapid. This loss is partly owing to volatilization of lead sulphide, but mainly to escape of sulphur in the form of hydrogen sulphide and formation of Pb_2S, or even of lead (A. SOUCHAY §). It dissolves in concentrated hot

* Jour. Chem. Soc. (2) 6, 355.

† Journ. f. prakt. Chem. 67, 374.

‡ Pogg. Annal. 91, 110; and 110, 134.

§ Zeitschr. f. anal. Chem. 4, 63.

hydrochloric acid, with evolution of hydrogen sulphide. In moderately strong nitric acid lead sulphide dissolves, upon the application of heat, with separation of sulphur;—if the acid is rather concentrated, a small portion of lead sulphate is also formed. Fuming nitric acid acts energetically upon lead sulphide, and converts it into sulphate without separation of sulphur.

COMPOSITION.

Pb	207	86·61
S	32	13·39
	239	100·00

g. For the composition and properties of *lead chromate*, see *Chromic acid*, § 93.

§ 84.

3. Mercury in Mercurous Compounds; and 4. Mercury in Mercuric Compounds.

Mercury is weighed either in the METALLIC STATE, as MERCUROUS CHLORIDE, or as SULPHIDE, or occasionally as MERCURIC OXIDE.

a. Metallic mercury is liquid at the common temperature; it has a tin-white color. When pure, it presents a perfectly bright surface. It is quite unalterable in the air at the common temperature. It boils at 360°. It evaporates, but very slowly, at the ordinary temperature of summer. Upon long-continued boiling with water, a small portion of mercury volatilizes, and traces escape along with the aqueous vapor, whilst a very minute proportion remains suspended (not dissolved) in the water (comp. Expt. No. 54). This suspended portion of mercury subsides completely after long standing. When mercury is precipitated from a fluid, in a very minutely divided state, the small globules will readily unite to a large one if the mercury be perfectly pure; but even the slightest trace of extraneous matter, such as fat, etc., adhering to the mercury will prevent the union of the globules. Mercury does not dissolve in hydrochloric acid, even in concentrated; it is barely soluble in dilute cold sulphuric acid, but dissolves readily in nitric acid.

b. Mercurous chloride, prepared in the wet way, is a heavy white powder. It is almost absolutely insoluble in cold water; in boiling water it is gradually decomposed, the water taking up chlorine and mercury; upon continued boiling, the residue acquires a gray color. Highly dilute hydrochloric acid fails to dissolve it at the common temperature, but dissolves it slowly at a higher temperature; upon ebullition, with access of air, the whole of the mercurous chloride is gradually dissolved; the solution contains mercuric chloride ($Hg_2Cl_2 + 2HCl + O = 2HgCl_2 + H_2O$). When acted upon by boiling concentrated hydrochloric acid, it is rather speedily decomposed into mercury, which remains undissolved, and mercuric chloride, which dissolves. Boiling nitric acid dissolves it to mercuric chloride and nitrate. Chlorine water and nitrohydrochloric acid dissolve it to mercuric chloride, even in the cold. Solutions of ammonium chloride, sodium chloride, and potassium chloride, decompose it into metallic mercury and mercuric chloride, which latter dissolves; in the cold, this decomposition is but slight; heat promotes the action. It is soluble in hot solution of mercurous nitrate, and still more in that of mercuric nitrate; on cooling it crystallizes out almost completely (DEBRAY*). It does not affect vegetable colors; it is unalterable in the air, and may be dried at 100°, without loss of weight; when exposed to a higher degree of heat, though still below redness, it volatilizes completely, without previous fusion.

COMPOSITION.

Hg_2	400·00	84·94
Cl_2	70·92	15·06
	470·92	100·00

c. Mercuric sulphide, prepared in the wet way, is a black powder, insoluble in water. Dilute hydrochloric acid and dilute nitric acid fail to dissolve it, hot concentrated nitric acid scarcely attacks it, boiling hydrochloric acid has no action on it. By prolonged heating with red fuming nitric acid it is finally converted into a white compound, $2HgS + Hg(NO_3)_2$, which is insoluble, or barely soluble, in nitric acid. It dissolves readily in nitrohydrochloric acid. From a solution of mercuric chloride containing much free hydrochloric acid, the whole of the metal cannot be precipitated as

* Compt. Rend. 70, 995.

sulphide by means of hydrogen sulphide, until the solution is properly diluted. Should such a solution be very concentrated, mercurous chloride and sulphur are precipitated (M. MARTIN*). Solution of potassa, even boiling, fails to dissolve it. It dissolves in potassium sulphide, but readily only in presence of free alkali. It is insoluble in potassium hydrosulphide and in the corresponding sodium compound, and is therefore precipitated from its solution in potassium or sodium sulphide by hydrogen sulphide or by ammonium hydrosulphide (C. BARFOED†). Small but distinctly perceptible traces dissolve on cold digestion with yellowish or yellow ammonium sulphide, but after hot digestion it is scarcely possible to detect any traces in solution.‡ Potassium cyanide and sodium sulphite do not dissolve it. On account of the solubility of mercuric sulphide in potassium sulphide, it is impossible to precipitate mercury by means of ammonium sulphide completely from solutions containing potassium or sodium hydroxides or carbonates. Such solutions may occur, for instance, when a solution of mercuric chloride contains much potassium chloride, or sodium chloride, for, in this case, no mercuric oxide would be precipitated on the addition of potassa or soda (H. ROSE§). In the air it is unalterable, even in the moist state, and at 100°. When exposed to a higher temperature, it sublimes completely and unaltered.

COMPOSITION.

Hg	200	86·21
S	32	13·79
	232	100·00

d. Mercuric oxide, prepared in the dry way, is a crystalline brick-colored powder, which, when exposed to the action of heat, changes to the color of cinnabar, and subsequently to a violet-black tint. It bears a tolerably strong heat without decomposition; but when heated to incipient redness, it is decomposed into mercury and oxygen; perfectly pure mercuric oxide leaves no residue upon ignition. Its escaping fumes also should not redden litmus-paper. Water takes up a trace of mercuric oxide, acquiring thereby a very weak alkaline reaction. Hydrochloric or nitric acid dissolves it readily.

* Journ. f. prakt. Chem. 67, 376.
† Zeitschr. f. anal. Chem. 4, 436.
‡ *Ib.* 3, 140.
§ Pogg. Annal. 110, 141.

COMPOSITION.

Hg	200	92·59
O	16	7·41
	216	100·00

§ 85.

5. COPPER.

Copper is usually weighed in the METALLIC STATE, or in the form of CUPRIC OXIDE, or of CUPROUS SULPHIDE. Besides these forms, we have to examine CUPRIC SULPHIDE, CUPROUS OXIDE, and CUPROUS SULPHOCYANATE.

a. Copper, in the pure state, is a metal of a peculiar well-known color. It fuses only at a white heat. Exposure to dry air, or to moist air, free from carbon dioxide, leaves the fused metal unaltered; but upon exposure to moist air impregnated with carbon dioxide, it becomes gradually tarnished and coated with a film, first of a blackish-gray, finally of a bluish-green color. Precipitated finely divided copper, in contact with water and air, oxidizes far more quickly, especially at an elevated temperature. On igniting copper in the air, it oxidizes superficially to a varying mixture of cuprous and cupric oxide. In hydrochloric acid, in the cold, it does not dissolve if air be excluded; in the heat it dissolves but slightly if the metal is in a compact state. Finely divided copper on the contrary dissolves slowly when heated with strong hydrochloric acid, hydrogen being evolved and cuprous chloride being formed (WELTZIEN*). Copper dissolves readily in nitric acid. In ammonia it dissolves slowly if free access is given to the air; but it remains insoluble if the air is excluded. Metallic copper brought into contact in a closed vessel with solution of cupric chloride in hydrochloric acid, reduces the cupric to cuprous chloride, an atom of metal being dissolved for every molecule of chloride.

b. Cupric oxide.—If a dilute, cold, aqueous solution of a cupric salt is mixed with solution of potassa or soda in excess, a light blue precipitate of cupric hydroxide, $Cu(OH)_2$, is formed, which it is found difficult to wash. If the precipitate be left in the fluid

* Ann. d. Chem. u. Pharm. 136, 109.

from which it has been precipitated, it will, even at a summer heat, gradually change to brownish-black, passing, with separation of water, into $6CuO + H_2O$ (SOUCHAY). This transformation is immediate upon heating the fluid nearly to boiling. The fluid filtered off from the black precipitate is free from copper. It follows from this that the black precipitate is insoluble in dilute potassa. Concentrated potassa or soda on the contrary dissolves the hydroxide, and on long warming even the black oxide (O. Löw*). The resulting blue solutions remain clear on boiling, even if mixed with some water; but if boiled after being much diluted the whole of the copper will separate as black oxide. If a solution of a cupric salt contains non-volatile organic substances, the addition of alkali in excess will, even upon boiling, fail to precipitate the whole of the copper. The hydrated cupric oxide, $6CuO + H_2O$, precipitated with potassa or soda from hot dilute solutions obstinately retains a portion of the precipitant; it may, however, be completely freed from this by washing with boiling water. The precipitated oxide after ignition, or the oxide prepared by decomposing cupric carbonate or nitrate by heat, is a brownish-black, or black powder, the weight of which remains unaltered even upon strong ignition over the gas- or spirit-lamp, provided all reducing gases be excluded (Expt. No. 56). If cupric oxide is exposed to a heat approaching the fusing point of metallic copper, it fuses, yields oxygen, and becomes Cu_6O_5 (FAVRE and MAUMENE). It is very readily reduced by ignition with charcoal, or under the influence of reducing gases; heated in the air for a long time, the reduced metallic copper re-oxidizes. Mixed with sulphur and ignited in a current of hydrogen, towards the end strongly, cupric oxide passes into cuprous sulphide (Cu_2S—H. ROSE). Cupric oxide, in contact with the atmosphere, absorbs water; less rapidly after being strongly ignited (Expt. No. 57). It is nearly insoluble in water; but it dissolves readily in hydrochloric acid, nitric acid, &c.; less readily in ammonia. It does not affect vegetable colors.

COMPOSITION.

Cu	63·40	79·85
O	16·00	20·15
	79·40	100·00

* Zeitschr. f. anal. Chem. 9, 463.

c. Cupric sulphide, prepared in the wet way, is a brownish-black, or black precipitate, almost absolutely insoluble in water.* When exposed to the air in a moist state, it acquires a greenish tint and the property of reddening litmus paper, cupric sulphate being formed. Hence the sulphide must be washed with water containing hydrogen sulphide. It dissolves readily in boiling nitric acid, with separation of sulphur. Hydrochloric acid dissolves it with difficulty. This is the reason why hydrogen sulphide precipitates copper entirely from solutions which contain even a very large amount of free hydrochloric acid (GRUNDMANN†). Only when we dissolve a copper salt directly in pure hydrochloric acid of 1·1 sp. gr. does any copper remain unprecipitated (M. MARTIN‡). It does not dissolve in solutions of potassa and of potassium sulphide, particularly if these solutions be boiling; it dissolves perceptibly in colorless, and much more readily in hot yellow ammonium sulphide. Potassium cyanide dissolves the freshly precipitated sulphide readily and completely. Upon intense ignition in a current of hydrogen it is converted into pure Cu_2S.

d. If the blue solution which is obtained upon adding to solution of copper tartaric acid and then soda in excess, is mixed with solution of grape sugar or sugar of milk, and heat applied, an orange-yellow precipitate of cuprous hydroxide is formed, which contains the whole of the copper originally present in the solution, and after a short time, more particularly upon the application of a stronger heat, turns red, owing to the conversion of the hydroxide into anhydrous cuprous oxide (Cu_2O). The precipitate, which is insoluble in water, retains a portion of alkali with considerable tenacity. When treated with dilute sulphuric acid, it gives cupric sulphate which dissolves, and metallic copper which separates.

e. Cuprous sulphocyanate, $Cu_2(CNS)_2$, which is always formed when potassium sulphocyanate is added to a solution of copper, mixed with sulphurous or hypophosphorous acid, is a white precipitate insoluble in water, as well as in dilute hydrochloric or sulphuric acid. Dried at 115°, the salt retains from 1 to 3 per cent. of water, which is driven off only by heating to incipient decomposition; it is, therefore, not well adapted for direct weighing. When

* In some experiments that I made when examining the Weilbach water, I found that about 950000 parts of water are required to dissolve 1 part of CuS.

† Journ. f. prakt. Chem. 73, 241.

‡ *Ib.* 67, 375.

ignited with sulphur, with exclusion of air, it changes to Cu_2S (RIVOT*). When heated with hydrochloric acid and potassium chlorate, or with sulphuric acid and nitric acid, it is dissolved and suffers decomposition. Solutions of potassa and soda separate hydrated cuprous oxide, with formation of sulphocyanate of the alkali metal.

f. Cuprous sulphide, produced by heating CuS in a current of hydrogen or $Cu_2(CNS)_2$ with sulphur, is a grayish-black crystalline mass, which may be ignited and fused without decomposition if the air is excluded.

COMPOSITION.

Cu_2	126·80	79·85
S	32·00	20·15
	158·80	100·00

§ 86.

6. BISMUTH.

Bismuth is weighed as OXIDE, as METAL, or as CHROMATE $(Bi_2O,2CrO_4)$. Besides these compounds, we have to study here the BASIC CARBONATE, the BASIC NITRATE, the BASIC CHLORIDE, and the SULPHIDE.

a. Bismuth trioxide, prepared by igniting the carbonate or nitrate, is a pale lemon-yellow powder which, under the influence of heat, assumes transiently a dark yellow or reddish-brown color. When heated to intense redness, it fuses, without alteration of weight. Ignition with charcoal, or in a current of carbon monoxide, reduces it to the metallic state. Fusion with potassium cyanide also effects its complete reduction (H. ROSE†). It is insoluble in water, and does not affect vegetable colors. It dissolves readily in those acids which form soluble salts with it. When ignited with ammonium chloride it gives metallic bismuth, the reduction being attended with deflagration.

COMPOSITION.

Bi_2	416	89·655
O_3	48	10·345
	464	100·000

* *Ib.* 62, 252. † Journ. f. prakt. Chem. 61, 188.

b. Metallic bismuth is white, with a reddish tinge, moderately hard, brittle, with a tendency to crystallize. It fuses at 264°, and at a low white heat volatilizes. It does not oxidize in the air at the ordinary temperature, but with the co-operation of water it oxidizes slowly, more speedily on fusion. It dissolves in dilute nitric acid.

c. Bismuth carbonate.—Upon adding ammonium carbonate in excess to a solution of bismuth, free from hydrochloric acid, a white precipitate of basic bismuth carbonate ($Bi_2O_2CO_3$) is immediately formed; part of this precipitate, however, redissolves in the excess of the precipitant. But if the fluid with the precipitate be heated before filtration, the filtrate will be free from bismuth. (Potassium carbonate likewise precipitates solutions of bismuth completely; but the precipitate in this case invariably contains traces of potassium, which it is very difficult to remove by washing. Sodium carbonate precipitates solutions of bismuth less completely.) The precipitate is easily washed; it is practically insoluble in water, but dissolves readily, with effervescence, in hydrochloric and nitric acids. Upon ignition it leaves the oxide.

d. The *basic bismuth nitrate*, which is obtained by mixing with water a solution of the nitrate containing little or no free acid, presents a white, crystalline powder. It cannot be washed with pure cold water without suffering a decided alteration. It becomes more basic, while the washings show an acid reaction, and contain bismuth. If the basic salt, however, be washed with cold water containing $\frac{1}{500}$ of ammonium nitrate, no bismuth passes through the filter. The solution of ammonium nitrate must not be warm. These remarks only apply in the absence of free nitric acid (J. LÖWE*). On ignition the basic nitrate passes into the oxide.

e. Basic bismuth chloride, formed by adding much water to solution of bismuth containing hydrochloric acid or sodium chloride, is a brilliant white powder (BiOCl after drying at 100°). It is insoluble in water, but dissolves in concentrated hydrochloric or nitric acid. Fused with potassium cyanide it gives metallic bismuth.

f. Bismuth chromate ($Bi_2O_3,2CrO_3$), which is produced by adding potassium dichromate, slightly in excess, to a solution of

---

* *Ib.* 74, 341.

bismuth nitrate as neutral as possible, is an orange-yellow, dense, readily-subsiding precipitate, insoluble in water, even in presence of some free chromic acid, but soluble in hydrochloric acid and nitric acid. It may be dried at 100°–112° without decomposition (Löwe*).

COMPOSITION.

$$O \left< \begin{matrix} Bi < \begin{matrix} O \\ O \end{matrix} > CrO_2 \\ Bi < \begin{matrix} O \\ O \end{matrix} > CrO_2 \end{matrix} \right. =$$

Bi_2O_3 . .	464·00	69·78
$2CrO_3$ . .	200·96	30·22
	664·96	100·00

g. Bismuth trisulphide, prepared in the wet way, is a brownish black, or black precipitate, insoluble in water, dilute acids, alkalies, alkali sulphides, sodium sulphite, and potassium cyanide. In moderately concentrated nitric acid it dissolves, especially on warming, to nitrate, with separation of sulphur. Hence in precipitating bismuth from a nitric acid solution, care should be taken to dilute sufficiently. Hydrochloric acid impedes the precipitation by hydrogen sulphide only when a very large excess is present, and the fluid is quite concentrated. The sulphide does not change in the air. Dried at 100°, it continually takes up oxygen and increases slightly in weight; if the drying is protracted this increase may be considerable (Expt. No. 58). Fused with potassium cyanide, it is completely reduced (H. Rose). Reduction takes place more slowly by ignition in a current of hydrogen.

COMPOSITION.

Bi_2	416	81·25
S_3	96	18·75
	512	100·00

§ 87.

7. Cadmium.

Cadmium is weighed either as oxide or as sulphide. Besides these substances, we have to examine cadmium carbonate.

a. Cadmium oxide, produced by igniting the carbonate or nitrate, is a yellowish-brown or reddish-brown powder. The appli-

* Journ. f. prakt. Chem. 67, 291.

cation of a white heat fails to fuse, volatilize, or decompose it; it is insoluble in water, but dissolves readily in acids; it does not alter vegetable colors. Ignition with charcoal, or in a current of hydrogen, carbon monoxide, or carburetted hydrogen, reduces it readily, the metallic cadmium escaping in the form of vapor.

COMPOSITION.

Cd	112	87·50
O	16	12·50
	128	100·00

b. Cadmium carbonate is a white precipitate, insoluble in water and the fixed alkali carbonates, and extremely sparingly soluble in ammonium carbonate. It loses its water completely upon drying. Ignition converts it into oxide.

c. Cadmium sulphide, produced in the wet way, is a lemon-yellow to orange-yellow precipitate, insoluble in water, dilute acids, alkalies, alkali sulphides, sodium sulphite, and potassium cyanide (Expt. No. 59). It dissolves readily in concentrated hydrochloric acid, with evolution of hydrogen sulphide. In precipitating, therefore, with hydrogen sulphide, a cadmium solution should not contain too much hydrochloric acid, and should be sufficiently diluted. The sulphide dissolves readily in dilute sulphuric acid on heating. It dissolves in moderately concentrated nitric acid, with separation of sulphur. It may be washed, and dried at 100° or 105°, without decomposition. Even on gentle ignition in a current of hydrogen, it volatilizes in appreciable amount (H. Rose*), partially unchanged, partially as metallic vapor.

COMPOSITION.

Cd	112	77·78
S	32	22·22
	144	100·00

* Pogg. Annal. 110, 134.

METALS OF THE SIXTH GROUP.

§ 88.

1. GOLD.

Gold is always weighed in the metallic state. Besides METALLIC GOLD, we have to consider the TRISULPHIDE OR AURIC SULPHIDE.

a. Metallic gold, obtained by precipitation, presents a blackish-brown powder, destitute of metallic lustre, which it assumes, however, upon pressure or friction; when coherent in a compact mass, it exhibits the well-known bright yellow color peculiar to it. It fuses only at a white heat, and resists, accordingly, all attempts at fusion over a spirit-lamp. It remains wholly unaltered in the air and at a red heat, and is not in the slightest degree affected by water, nor by any simple acid. Nitrohydrochloric acid dissolves it to trichloride. Hot concentrated sulphuric acid containing a little nitric acid dissolves gold, especially if in a finely divided condition, to a yellow fluid, from which it is thrown down again by water (J. SPILLER†).

b. Auric sulphide.—When hydrogen sulphide is transmitted through a cold dilute solution of auric chloride, the whole of the gold separates as auric sulphide, Au_2S_3, in form of a brownish-black precipitate. If this precipitate is left in the fluid, it is gradually transformed into metallic gold and free sulphuric acid. Upon transmitting hydrogen sulphide through a warm solution of auric chloride, aurous sulphide Au_2S precipitates, with formation of sulphuric and hydrochloric acids.

Auric sulphide is insoluble in water, hydrochloric acid, and nitric acid, but dissolves in nitrohydrochloric acid. Colorless ammonium sulphide fails to dissolve it; but it dissolves almost entirely in yellow ammonium sulphide, and completely upon addition of potassa. It dissolves in potassa, with separation of gold. Yellow potassium sulphide dissolves it completely. It dissolves in potassium cyanide. Exposure to a moderate heat reduces it to the metallic state.

§ 89.

2. PLATINUM.

Platinum is invariably weighed in the METALLIC STATE; it is

† Chem. News, 14, 256.

generally precipitated as AMMONIUM PLATINIC CHLORIDE, or as POTASSIUM PLATINIC CHLORIDE, rarely as PLATINIC SULPHIDE.

a. Metallic platinum, produced by igniting ammonium platinic chloride, or potassium platinic chloride, presents the appearance of a gray, lustreless, porous mass (spongy platinum). The fusion of platinum can be effected only at the very highest degrees of heat. It remains wholly unaltered in the air, and in the most powerful furnaces. It is not attacked by water, or simple acids, and scarcely by aqueous solutions of the alkalies. Nitrohydrochloric acid dissolves it to platinic chloride.

b. The properties of *potassium platinic chloride*, and those of *ammonium platinic chloride*, have been given already in §§ 68 and 70 respectively.

c. Platinic sulphide.—When a concentrated solution of platinic chloride is mixed with hydrogen sulphide water, or when hydrogen sulphide gas is transmitted through a rather dilute solution of the chloride, no precipitate forms at first; after standing some time, however, the solution turns brown, and finally a precipitate subsides. But if the mixture of solution of platinic chloride, with hydrogen sulphide in excess, is gradually heated (finally to ebullition), the whole of the platinum separates as platinic sulphide (free from any admixture of platinic chloride). Platinic sulphide is insoluble in water and in simple acids; but it dissolves in nitrohydrochloric acid. It dissolves partly in caustic alkalies, with separation of platinum, and completely in alkali sulphides, especially the polysulphides if used in sufficient excess. When hydrogen sulphide is transmitted through water holding minutely divided platinic sulphide in suspension, the sulphide, absorbing hydrogen sulphide, acquires a light grayish-brown color; the hydrogen sulphide thus absorbed, separates again upon exposure to the air. When moist platinic sulphide is exposed to the air, it is gradually decomposed, being converted into metallic platinum and sulphuric acid. Ignition in the air reduces platinic sulphide to metallic platinum.

§ 90.

3. Antimony.

Antimony is weighed as ANTIMONIOUS SULPHIDE, as ANTIMONY TETROXIDE (or ANTIMONIOUS ANTIMONATE), or more rarely in the METALLIC state.

a. Upon transmitting hydrogen sulphide through a solution of antimonious chloride mixed with tartaric acid, an orange precipitate of amorphous *antimonious sulphide* is obtained, mixed at first with a small portion of basic antimony chloride. However, if the fluid is thoroughly saturated with hydrogen sulphide, and a gentle heat applied, the chloride mixed with the precipitate is decomposed, and pure antimonious sulphide obtained. Antimonious sulphide is insoluble in water and dilute acids; it dissolves in concentrated hydrochloric acid, with evolution of hydrogen sulphide. In precipitating with hydrogen sulphide, therefore, antimony solutions should not contain too much free hydrochloric acid, and should be sufficiently diluted. The amorphous antimonious sulphide dissolves readily in dilute potassa, ammonium sulphide, and potassium sulphide, sparingly in ammonia, very slightly in ammonium carbonate, and not at all in hydrogen potassium sulphite. The amorphous sulphide, dried in the desiccator at the ordinary temperature, loses very little weight at 100°; if kept for some time at this latter temperature its weight remains constant. But it still retains a little water, which does not perfectly escape even at 190°, but at 200° the sulphide becomes anhydrous, turning black and crystalline (H. ROSE* and Expt. No. 60). Ignited gently in a stream of carbon dioxide, the weight of this anhydrous sulphide remains constant; at a stronger heat a small amount volatilizes. The amorphous sulphide, if long exposed to the action of air, in presence of water, slowly takes up oxygen, so that on treatment with tartaric acid it yields a filtrate containing antimony.

Antimonic sulphide is insoluble in water, also in water containing hydrogen sulphide. It dissolves completely in ammonia, especially on warming; traces only dissolve in ammonium carbonate. On heating dried antimonic sulphide in a current of carbon dioxide 2 atoms of sulphur escape, black crystalline antimonious sulphide remaining.

On treating antimonious or antimonic sulphide with fuming nitric acid violent oxidation sets in. We obtain first antimonic acid and pulverulent sulphur; on evaporating to dryness antimonic acid and sulphuric acid; and lastly on igniting antimony tetroxide. The same antimony tetroxide is obtained by igniting the sulphide

---

* Journ. f. prakt. Chem. 59, 331.

with 30 to 50 times its amount of mercuric oxide (BUNSEN*). [According to later investigations of BUNSEN,† the temperature necessary to reduce Sb_2O_5 to Sb_2O_4 lies so near that which reduces Sb_2O_4 to Sb_2O_3 that it is not easy to bring antimony into Sb_2O_4 for weighing. It is possible only by using a large covered platinum or rather large open porcelain crucible (by suitable choice of size of crucible and intensity of flame) and heating with a gas blast lamp so that the bottom only of the crucible reaches a strong red heat, to drive off exactly one atom of oxygen from Sb_2O_5.] Ignition in a current of hydrogen converts the sulphides of antimony into the metallic state.

COMPOSITION.

Sb_2	244·00	71·77
S_3	96·00	28·23
	340·00	100·00

b. Antimony tetroxide is a white powder, which, when heated, acquires transiently a yellow tint; it is infusible; it loses weight when ignited intensely in a small platinum crucible with a gas blast flame (BUNSEN†). It is almost insoluble in water, and dissolves in hydrochloric acid with very great difficulty. It undergoes no alteration on treatment with ammonium sulphide. It manifests an acid reaction when placed upon moist litmus-paper.

COMPOSITION.

Sb_2	244	79·22
O_4	64	20·78
	308	100·00

c. Metallic antimony, produced in the wet way, by precipitation, presents a lustreless black powder. It may be dried at 100° without alteration. It fuses at a moderate red heat. Upon ignition in a current of gas, *e.g.*, hydrogen, it volatilizes, without formation of antimonetted hydrogen. Hydrochloric acid has very little action on it, even when concentrated and boiling. Nitric acid converts it into antimonious oxide, mixed with more or less

* Annal. de Chem. u. Pharm. 106, 3. † Zeitschr. f. anal. Chem. 1879, 268.

antimony tetroxide, according to the concentration of the nitric acid.

§ 91.

4. Tin in Stannous Compounds; and 5. Tin in Stannic Compounds.

Tin is generally weighed in the form of STANNIC OXIDE; besides stannic oxide, we have to examine stannous sulphide and stannic sulphide.

a. Stannic oxide.—If a solution of an alkali, sodium sulphate or ammonium nitrate is added to a solution of stannic chloride, *stannic acid* (H_2SnO_3) is precipitated. This precipitate is soluble in excess of soda, and does not separate again even on the addition of a large quantity of soda (C. F. Barfoed*). It is also readily soluble in hydrochloric acid.

By the action of nitric acid on metallic tin, or by evaporating a solution of tin with an excess of nitric acid, a white residue is obtained which is *metastannic acid* ($Sn_5H_{10}O_{15}$?). This residue is insoluble in water, but very slightly soluble in nitric acid, or sulphuric acid. By heating with hydrochloric acid it does not dissolve, but is changed to metastannic chloride, which is soluble in water after removal of the excess of hydrochloric acid. Soda added to a solution of metastannic chloride precipitates sodium metastannate, which is insoluble in excess of soda and in weak alcohol. Upon intense ignition, both stannic and metastannic acids are converted into stannic oxide. Mere heating to redness is not sufficient to expel all the water (Dumas†).

Stannic oxide is a straw-colored powder, which under the influence of heat, transiently assumes a different tint, varying from bright yellow to brown. It is insoluble in water and acids, and does not alter the color of litmus-paper. Mixed with ammonium chloride in excess, and ignited, it volatilizes completely as stannic chloride. If stannic oxide is fused with potassium cyanide, all the tin is obtained in form of metallic globules, which may be completely, and without the least loss of metal, freed from the adhering slag, by extracting with dilute alcohol, and rapidly decanting the fluid from the tin globules (H. Rose‡).

---

* Zeitschr. f. anal. Chem. 7, 260. † Annal. d. Chem. u. Pharm. 105, 104.
‡ Journ. f. prakt. Chem. 61, 189.

COMPOSITION.

Sn	118	78·67
O_2	32	21·33
	150	100·00

b. Hydrated stannous sulphide forms a brown precipitate, insoluble in water, hydrogen sulphide water, and dilute acids. In precipitating tin from stannous solutions by means of hydrogen sulphide, free hydrochloric acid must not be present in too large amount, and the solution must be diluted sufficiently. Ammonia fails to dissolve it; but it dissolves pretty readily in yellow ammonium sulphide, and in yellow potassium sulphide; it dissolves readily in hot concentrated hydrochloric acid. Heated, with exclusion of air, it loses its water, and is rendered anhydrous; when exposed to the continued action of a gentle heat, with free access of air, it is converted into sulphur dioxide, which escapes, and stannic oxide, which remains.

c. Hydrated stannic sulphide, precipitated by acids from the solution of its alkali sulphur salts, is a light-yellow precipitate. In washing with pure water, it is inclined to yield a turbid filtrate and to stop up the pores of the filter; this annoyance is got over by washing with water containing sodium chloride, ammonium acetate, or the like (BUNSEN). On drying, the precipitate assumes a darker tint. It is insoluble in water; it dissolves with difficulty in ammonia, but readily in potassa, alkali sulphides, and hot concentrated hydrochloric acid. It is insoluble in hydrogen potassium sulphite. In precipitating tin from stannic solutions by hydrogen sulphide, the solution should not contain too much free hydrochloric acid, and should be sufficiently diluted. According to C. F. BARFOED* the precipitates thus produced are not pure hydrated stannic sulphide, but a mixture of this with stannic or metastannic acid, as the case may be. The precipitate thrown down from ordinary stannic chloride keeps its yellow color even after long standing in the fluid, and dissolves completely in excess of soda; that thrown down from the metastannic chloride is first white and becomes gradually yellow, it turns brown on standing in the fluid and dissolves in excess of soda, leaving, however, a considerable residue of sodium metastannate. When heated, with

* Zeitschr. f. anal. Chem. 7, 261.

exclusion of air, stannic sulphide loses its water of hydration, and, at the same time, according to the degree of heat, one-half or one-fourth of its sulphur, becoming converted either into stannous sulphide or the sesquisulphide of tin; when heated very slowly, with free access of air, it is converted into stannic oxide, with disengagement of sulphur dioxide.

§ 92.

6. Arsenic of Arsenious Compounds; and 7. Arsenic of Arsenic Compounds.

Arsenic is weighed either as LEAD ARSENATE, as ARSENIOUS SULPHIDE, as AMMONIUM MAGNESIUM ARSENATE, as MAGNESIUM PYROARSENATE, or as URANYL PYROARSENATE; besides these forms, we have here to examine also ARSENIO-MOLYBDATE OF AMMONIUM.

a. Lead arsenate, in the pure state, is a white powder, which agglutinates when exposed to a gentle red heat, at the same time transitorily acquiring a yellow tint; it fuses when exposed to a higher degree of heat. When strongly ignited, it suffers a slight diminution of weight, losing a small proportion of arsenic acid, which escapes as arsenious oxide and oxygen. In analysis we have never occasion to operate upon the pure lead arsenate, but upon a mixture of it with lead oxide.

b. Arsenious sulphide forms a precipitate of a rich yellow color; it is insoluble in water,* and also in hydrogen sulphide water. When boiled with water, or left for several days in contact with that fluid, it undergoes a very trifling decomposition: a trace of arsenious acid dissolves in the water, and a minute proportion of hydrogen sulphide is disengaged. This does not in the least interfere, however, with the washing of the precipitate. The precipitate may be dried at 100°, without decomposition; the whole of the water which it contains is expelled at that temperature. When exposed to a stronger heat, it transitorily assumes a brownish-red color, fuses, and finally rises in vapor, without decomposition. It dissolves readily in alkalies, alkali carbonates,

* In some experiments which I had occasion to make, in the course of an analysis of the springs of Weilbach (Chemische Untersuchung der wichtigsten Nassauischen Mineralwasser von Dr. Fresenius, V. Schwefelquelle zu Weilbach. Weisbaden, Kreidel und Niedner. 1856), I found that one part of As_2S_3 dissolves in about one million parts of water.

alkali sulphides, potassium hydrogen sulphite, and nitrohydrochloric acid; but it is scarcely soluble in boiling concentrated hydrochloric acid. Red fuming nitric acid converts it into arsenic acid and sulphuric acid. It is insoluble in carbon disulphide.

COMPOSITION.

As_2	150	60·98
S_3	96	39·02
	246	100·00

c. Ammonium magnesium arsenate forms a white, somewhat transparent, finely crystalline precipitate, which when dried in a desiccator has the formula $NH_4MgAsO_4 + 6H_2O$. After drying at 100°, its composition is $(NH_4MgAsO_4)_2 + H_2O$. At a higher temperature, say 105°—110°, more water escapes, and at 130° this loss is considerable (PULLER*). Upon ignition it loses water and ammonia, and changes to magnesium pyroarsenate, $Mg_2As_2O_7$. On rapid ignition the escaping ammonia has a reducing action on the arsenic acid, and a notable loss is occasioned (H. ROSE); by raising the heat very gradually reduction may be avoided (H. ROSE, WITTSTEIN,† PULLER), or by passing a current of dry oxygen during the ignition. Ammonium magnesium arsenate dissolves very sparingly in water, one part of the salt dried at 100°, requiring 2656, one part of the anhydrous salt, 2788 parts of water of 15°. It is far less soluble in ammoniated water, one part of the salt dried at 100° requiring 15038, one part of the anhydrous salt, 15786 parts of a mixture of one part of solution of ammonia (·96 sp. gr.), and 3 parts of water at 15°. In water containing ammonium chloride, it is much more readily soluble, one part of the anhydrous salt requiring 886 parts of a solution of one part of ammonium chloride in 7 parts of water. Presence of ammonia diminishes the solvent capacity of the ammonium chloride; one part of the anhydrous salt requires 3014 parts of a mixture of 60 parts of water, 10 of solution of ammonia (·96 sp. gr.) and one of ammonium chloride.‡ A solution of ammonium chloride, ammonia and magnesium sulphate dissolves much less of the salt than ammoniated water; thus, PULLER (loc. cit.) found that one

* Zeitschr. f. anal. Chem. 10, 62. † *Ib.* 2, 19.

‡ Zeitschr. f. anal. Chem. 3, 206. PULLER obtained almost the same numbers (*Ib.* 10, 53).

part of the anhydrous salt dissolved in 32827 parts of a fluid containing $\frac{1}{20}$ of magnesia mixture (p. 113). Excess of alkali arsenate still more diminishes the solubility of the salt in water containing ammonia and ammonium chloride (Puller).

COMPOSITION OF AMMONIUM MAGNESIUM ARSENATE DRIED AT 100°.

$$2\left(AsO \begin{matrix} \diagup ONH_4 \\ - O \\ \diagdown O \end{matrix} > Mg\right) + H_2O =$$

$2MgO$ . . .	80·00	21·05
$(NH_4)_2O$ . .	52·08	13·68
As_2O_5 . . .	230·00	60·53
H_2O . . .	18·00	4·74
	380·08	100·00

d. Magnesium pyroarsenate, obtained by careful ignition of the preceding salt, is white, infusible by ignition in a porcelain crucible even over the blowpipe, but agglutinating at a still higher temperature, and finally fusing. After ignition in a porcelain crucible it dissolves readily in hydrochloric acid: ammonia precipitates ammonium magnesium arsenate from the solution in a crystalline form.

COMPOSITION.

$$O\begin{matrix} \diagup AsO < \begin{matrix} O \\ O \end{matrix} > Mg \\ \diagdown AsO < \begin{matrix} O \\ O \end{matrix} > Mg \end{matrix} =$$

$2MgO$ . . .	80	25·81
As_2O_5 . . .	230	74·19
	310	100·00

e. Uranyl pyroarsenate.—If a solution of arsenic acid is mixed with potash in slight excess, then with acetic acid to strongly acid reaction, and finally with uranyl acetate, the whole of the arsenic is thrown down as $UO_2HAsO_4 + 4H_2O$. In the presence of salts of ammonia the precipitate also contains the whole of the arsenic, and consists of $UO_2NH_4AsO_4$ + water. Both precipitates are pale yellowish-green, slimy, insoluble in water, acetic acid and saline solutions, such as ammonium chloride, soluble in mineral acids. Boiling favors the separation of the precipitate, addition of a few drops of chloroform will help it to settle, the washing is to be effected by boiling up and decanting. Both precipitates give $(UO_2)_2As_2O_7$ on ignition. The latter is a light yellow residue; if it has turned greenish from the action of reducing gases, it may be restored to its proper color by moistening with nitric acid and re-igniting. On igniting the ammonium uranyl arsenate, the ammonia must first be expelled by cautious heating, or a current of oxygen must be passed during the ignition, otherwise the arsenic acid will be partially reduced, and arsenic will be lost (Puller*).

COMPOSITION.

$$O \left\langle \begin{array}{l} AsO < \begin{smallmatrix} O \\ O \end{smallmatrix} > UO_2 \\ AsO < \begin{smallmatrix} O \\ O \end{smallmatrix} > UO_2 \end{array} \right. =$$

$2UO_2O$ . .	571·2	71·29
As_2O_5 . .	230·0	28·71
	801·2	100·00

f. Arsenio-molybdate of ammonium.—If a fluid containing arsenic acid is mixed with excess of the nitric acid solution of ammonium molybdate, the fluid remains clear in the cold, but on heating a yellow precipitate of arsenio-molybdate of ammonium separates. This precipitate comports itself with solvents like the analogous compound of phosphoric acid; it is, like the latter, insoluble in water, nitric acid, dilute sulphuric acid and salts, provided an excess of solution of ammonium molybdate, mixed with acid in moderate excess, be present. Hydrochloric acid or metallic chlorides, when present in large quantity, interfere with the thoroughness of the precipitation. Seligsohn† found it to be composed of 87·666 per cent. of molybdic acid, 6·308 arsenic acid, 4·258 ammonia, and 1·768 water.

B. FORMS IN WHICH THE ACID RADICALS ARE WEIGHED OR PRECIPITATED.

ACIDS OF THE FIRST GROUP.

§ 93.

1. Arsenious Acid and Arsenic Acid.—See § 92.

2. Chromic Acid.

Chromic acid is weighed either as chromic oxide, or as lead chromate, or barium chromate. We have also to consider mercurous chromate.

a. Chromic oxide.—See § 76.

b. Lead chromate obtained by precipitation forms a bright-yel-

* Zeitschr. f. anal. Chem. 10, 72.   † Journ. f. prakt. Chem. 67, 481.

low precipitate, insoluble in water and acetic acid, barely soluble in dilute nitric acid, readily in solution of potassa. When lead chromate is boiled with concentrated hydrochloric acid, it is readily decomposed, lead chloride and chromic chloride being formed. Addition of alcohol tends to promote this decomposition. Lead chromate is unalterable in the air. It dries thoroughly at 100°. Under the influence of heat it transitorily acquires a reddish-brown tint; it fuses at a red heat; when heated beyond its point of fusion, it loses oxygen, and is transformed into a mixture of chromic oxide and basic lead chromate. Heated in contact with organic substances, it readily yields oxygen to the latter.

COMPOSITION.

$CrO_2<^{O}_{O}>Pb =$			
	PbO . . .	223·00	68·94
	CrO_3 . . .	100·48	31·06
		323·48	100·00

c. Barium chromate is obtained as a light-yellow precipitate on mixing a solution of an alkali chromate with barium chloride. It dissolves in hydrochloric and in nitric acid, but not in acetic acid. On washing with pure water, the latter begins to dissolve it slightly, as soon as all soluble salts are removed, to such an extent that the washings run off yellow. The precipitate is insoluble in saline solutions. Hence it is best to use a solution of ammonium acetate for washing (PEARSON and RICHARDS*). It is not decomposed by moderate ignition.

COMPOSITION.

$CrO_2<^{O}_{O}>Ba =$			
	BaO . . .	153·00	60·36
	CrO_3 . . .	100·48	39·64
		253·48	100·00

d. Mercurous chromate obtained by adding mercurous nitrate to an alkali chromate is a brilliant-red precipitate, which turns black by the action of light. It dissolves very slightly in cold water, more in boiling water, being partially converted into a mercuric salt; it dissolves slightly in dilute nitric acid. For washing,

* Zeitschr. f. anal. Chem. 9, 108.

it is best to use a dilute solution of mercurous nitrate containing but little free acid; in this solution it is insoluble (H. ROSE*).

3. SULPHURIC ACID.

Sulphuric acid is determined best in the form of BARIUM SULPHATE, for the properties of which see § 71.

4. PHOSPHORIC ACID.

The principal forms into which phosphoric acid is converted are as follows:—LEAD PHOSPHATE, MAGNESIUM PYROPHOSPHATE, MAGNESIUM PHOSPHATE $Mg_3(PO_4)_2$, FERRIC PHOSPHATE, URANYL PYROPHOSPHATE, STANNIC PHOSPHATE, and SILVER PHOSPHATE. Besides these compounds, we have to examine MERCUROUS PHOSPHATE and PHOSPHO-MOLYBDATE OF AMMONIUM.

a. The *lead phosphate* obtained in the course of analysis is rarely pure, but is generally mixed with free lead oxide. In this mixture we have accordingly the normal lead phosphate $Pb_3(PO_4)_2$; in the pure state, this presents the appearance of a white powder; it is insoluble in water, acetic acid, and ammonia. It dissolves readily in nitric acid. When heated it fuses without decomposition.

b. Magnesium pyrophosphate.—See § 74.

c. Magnesium phosphate ($Mg_3(PO_4)_2$).—A mixture of this compound with excess of magnesia is produced by mixing a solution of an alkali phosphate, containing ammonium chloride, with magnesia, evaporating, heating until the ammonium chloride is expelled, and finally treating with water. It is practically insoluble in water and in solutions of salts of the alkalies (FR. SCHULZE†).

d. Ferric phosphate.—If a solution of phosphoric acid or of calcium phosphate in acetic acid is carefully precipitated with a solution of ferric acetate, or with a mixture of iron-alum and sodium acetate, so that the iron salt may just predominate, the precipitate always contains 1 mol. P_2O_5 to 1 mol. Fe_2O_3 corresponding to the formula of normal ferric phosphate, $Fe_2(PO_4)_2$ (RÄWSKY, WITTSTEIN, E. DAVY‡); if, on the other hand, the ferric acetate is in larger excess, the precipitate is more basic. WITTSTEIN obtained, by using a considerable excess of ferric acetate, a precipitate containing $3P_2O_5$ to $4Fe_2O_3$. Precipitates obtained with a small excess of the precipitant possess a composition varying between the above-

* Pogg. Ann. 53, 124. † Journ. f. prakt. Chem. 63, 440.
‡ Phil. Mag. 19, 181.

mentioned limits. RAMMELSBERG obtained $Fe_2(PO_4)_2 + 4H_2O$, and WITTSTEIN subsequently the same compound (with $8H_2O$ instead of 4) upon mixing ferric sulphate with sodium phosphate in excess; with an insufficient quantity of sodium phosphate the latter chemist obtained a more yellowish precipitate which had a composition corresponding to the formula $3Fe_2(PO_4)_2 + Fe_2(OH)_6 + 8H_2O$. If an acid fluid containing a *considerable* excess of phosphoric acid is mixed with a small quantity of a ferric solution, and an alkali acetate is added, a precipitate of the formula, $Fe_2(PO_4)_2$ + water, is invariably obtained, which accordingly leaves upon ignition $Fe_2(PO_4)_2 = Fe_2O_3 + P_2O_5$ (WITTSTEIN). Fresh experiments which I have made upon this subject have convinced me of the perfect correctness of this statement. MOHR obtained the same results.* The precipitate is insoluble in a fluid containing salts, but when washing, as soon as the soluble salts are nearly removed, the precipitate begins to dissolve. The filtrate has an acid reaction, and contains iron and phosphoric acid. The precipitate, under these circumstances, alters in composition, and this explains why different results were obtained in the analysis of precipitates which had been washing for different lengths of time (FR. MOHR).

COMPOSITION.

$$\left.\begin{array}{l} PO\begin{cases} -O- \\ -O- \\ -O- \end{cases}Fe \\ PO\begin{cases} -O- \\ -O- \\ -O- \end{cases}Fe \end{array}\right\} =$$

P_2O_5 . . .	142	47·02
Fe_2O_3 . . .	160	52·98
	302	100·00

If we dissolve ferric phosphate in hydrochloric acid, supersaturate the solution with ammonia, and apply heat, we obtain more basic salts, viz., $3Fe_2O_3,2P_2O_5$ (RAMMELSBERG); $2Fe_2O_3,P_2O_5$ (WITTSTEIN—after long washing). In WITTSTEIN's experiment, the wash-water contained phosphoric acid. The white ferric phosphate does not dissolve in acetic acid, but it dissolves in a solution of ferric acetate. Upon boiling the latter solution (of ferric phosphate in ferric acetate), the whole of the phosphoric acid precipitates, with basic ferric acetate, as hyperbasic ferric phosphate. Similar extremely basic combinations are invariably obtained (often mixed with ferric hydroxide), upon precipitating with ammonia or barium

* Zeitschr. f. anal. Chem. 2, 250.

carbonate, a solution containing phosphoric acid and an excess of a ferric salt. The precipitate obtained by barium carbonate can be conveniently filtered off and washed, the filtrate is perfectly free from either iron or phosphoric acid; on the contrary, the precipitate obtained by ammonia, especially if the latter were much in excess, is slimy, and therefore difficult to wash, and the filtrate always contains small traces of both iron and phosphoric acid.

e. Uranyl pyrophosphate.—If the hot aqueous solution of a phosphate soluble in water or acetic acid is mixed, in presence of free acetic acid, with uranyl acetate, a precipitate of uranyl hydrogen phosphate is immediately formed. If the fluid contains much ammonium salt, the precipitate contains also uranyl ammonium phosphate. The same precipitate forms also if aluminium or ferric salts are present; but in that case it is always mixed with more or less aluminium or ferric phosphate. Presence of potassium or sodium salts, on the contrary, or of salts of the alkali-earth metals, has no influence on the composition of the precipitate. Ammonium uranyl phosphate ($UO_2NH_4PO_4 + xH_2O$) is a somewhat gelatinous, whitish-yellow precipitate, with a tinge of green. The best way of washing it, at least so far as the principal part of the operation is concerned, is by boiling with water and decanting. If, after having allowed the fluid in which the precipitate is suspended to cool a little, a few drops of chloroform are added, and the mixture is shaken or boiled up, the precipitate subsides much more readily than without this addition.

The precipitate is insoluble in water and in acetic acid; but it dissolves in mineral acids; ammonium acetate, added in sufficient excess, completely re-precipitates it from this solution, upon application of heat. Upon igniting the precipitate, no matter whether containing ammonium or not, uranyl pyrophosphate of the formula $(UO_2)_2P_2O_7$ is produced. This has the color of the yolk of an egg. If the precipitate is ignited in presence of charcoal or of some reducing gas, partial reduction to uranous phosphate ensues, owing to which the ignited mass acquires a greenish tint; however, upon warming the greenish residue with some nitric acid, the green uranous salt is readily reconverted into the yellow uranyl salt. Uranyl pyrophosphate is not hygroscopic, and may therefore be ignited and weighed in an open platinum dish (A. Arendt and W. Knop*).

---

* Chemisches Centralblatt, 1856, 769, 803; and 1857, 177.

$$O\begin{cases}PO<^{O}_{O}>UO_2\\PO<^{O}_{O}>UO_2\end{cases}=\begin{array}{llr}2\,UO_2O & \ldots & 571\cdot2 & 80\cdot09\\P_2O_5 & \ldots & 142\cdot0 & 19\cdot91\\ & & \overline{713\cdot2} & \overline{100\cdot00}\end{array}$$

The one-fifth part of the precipitate may accordingly be calculated as phosphoric anhydride in ordinary analyses.*

f. Stannic phosphate is never obtained in the pure state in the analytical process, but contains always an admixture of hydrated metastannic acid in excess, which, upon ignition, changes to metastannic acid. It has, generally speaking, the same properties as hydrated metastannic acid, and is more particularly, like the latter, insoluble in nitric acid. Upon heating with concentrated solution of potassa, potassium phosphate and metastannate are formed.

g. Normal silver phosphate is a yellow powder; it is insoluble in water, but readily soluble in nitric acid, and also in ammonia. In ammonium salts, it is difficultly soluble. It is unalterable in the air. Upon ignition, it acquires transiently a reddish-brown color; at an intense red heat, it fuses without decomposition.

$$2\left(PO\begin{cases}OAg\\OAg\\OAg\end{cases}\right)=\begin{array}{llr}3Ag_2O & \ldots & 695\cdot82 & 83\cdot05\\P_2O_5 & \ldots & 142\cdot00 & 16\cdot95\\ & & \overline{837\cdot82} & \overline{100\cdot00}\end{array}$$

h. Mercurous phosphate.—This compound is employed for the purpose of effecting the separation of phosphoric acid from many bases, after H. Rose's method.

Mercurous phosphate presents the appearance of a white crystalline mass, or of a white powder. It is insoluble in water, but dissolves in nitric acid. The action of a red heat converts it into fused mercuric phosphate, with evolution of vapor of mercury. Upon fusion with alkali carbonates, alkali phosphates are produced, and mercury, oxygen, and carbon dioxide escape.

i. Phospho-molybdate of ammonium.—This compound also

---

* The atomic weight of uranium is here taken as 237·6, according to Ebelmen. If we take it according to Péligot, as 240, the ignited phosphate would contain 80·22 UO_3, and 19·78 P_2O_5. W. Knop and Arendt found in four experiments 20·13, 20·06, 20·04, and 20·04 respectively (in another 20·77). It will be seen that these numbers agree better with the composition as reckoned from Ebelmen's than from Péligot's atomic weight.

serves to effect the separation of phosphoric acid from other bodies; it is of the utmost importance in this respect.

Phospho-molybdate of ammonium forms a bright yellow, readily subsiding precipitate. Dried at 100°, it has, according to SELIGSOHN, the following (average) composition:—

MoO_3	90·744
P_2O_5	3·142
$(NH_4)_2O$	3·570
H_2O	2·544
	100·000*

In the pure state, it dissolves but sparingly in cold water (1 in 10000—EGGERTZ); but it is soluble in hot water. It is readily soluble even in the cold, in caustic alkalies, alkali carbonates and phosphates, ammonium chloride, and ammonium oxalate. It dissolves sparingly in ammonium sulphate, potassium nitrate, and potassium chloride; and very sparingly in ammonium nitrate.

It is soluble in potassium sulphate and sodium sulphate, sodium chloride and magnesium chloride, and sulphuric, hydrochloric and nitric acids (concentrated and dilute). Water, containing 1 per cent. of common nitric acid, dissolves $\frac{1}{6600}$ (EGGERTZ). Application of heat does not check the solvent action of these substances. Presence of ammonium molybdate totally changes its deportment with acid fluids. Dilute nitric or sulphuric acid containing ammonium molybdate does not dissolve it; but much hydrochloric acid, even in the presence of ammonium molybdate, has a solvent action, and this acid consequently interferes with the complete precipitation of phosphoric acid by nitric acid solution of ammonium molybdate. The solution of the phospho-molybdate of ammonium in acids is probably attended, in all cases, with decomposition and separation of the molybdic acid, which cannot take place in the presence of ammonium molybdate (J. CRAW)†. Tartaric acid and similar organic substances entirely prevent the

* From the varying results of different analysts it is plain that the precipitate, prepared under apparently the same circumstances, has not always exactly the same composition. SONNENSCHEIN (Journ. f. prakt. Chem. 53, 342) found in the precipitate dried at 120°, 2·93—3·12 % P_2O_5; LIPOWITZ (Pogg. Annal. 109, 135), in the precipitate dried at from 20° to 30°, 3·607 % P_2O_5; EGGERTZ (Journ. f. prakt. Chem. 79, 496), 3·7 to 3·8 %.

† Chem. Gaz. 1852, 216.

precipitation of the phospho-molybdate of ammonium (EGGERTZ). In the presence of an iodide instead of a yellow precipitate, a green precipitate or a green fluid is formed, resulting from the reducing action of the hydriodic acid on the molybdic acid (J. W. BILL*). Other substances which reduce molybdic acid have of course a similar action.

5. BORACIC ACID.

POTASSIUM BOROFLUORIDE is the best form to convert boracic acid into for the purpose of the direct estimation of the acid. This compound is produced by mixing the solution of an alkali borate, in presence of a sufficient quantity of potassa, with hydrofluoric acid in excess, in a silver or platinum dish, and evaporating to dryness. The gelatinous precipitate which forms in the cold, dissolves upon application of heat, and separates from the solution subsequently, upon evaporation, in small, hard, transparent crystals. The compound has the formula KF,BF_3. It is soluble in water and also in dilute alcohol; but strong alcohol fails to dissolve it; it is insoluble also in concentrated solution of potassium acetate. It may be dried at 100°, without decomposition (AUG. STROMEYER†).

COMPOSITION.

K	39·13	31·02
B	11·00	8·72
F_4	76·00	60·26
	126·13	100·00

6. OXALIC ACID.

When oxalic acid is to be directly determined it is usually precipitated in the form of CALCIUM OXALATE; and its weight is inferred from the CALCIUM CARBONATE or CALCIUM OXIDE produced from the oxalate by ignition. For the properties of these bodies see § 73.

7. HYDROFLUORIC ACID.

The *direct* estimation of hydrofluoric acid is usually effected by weighing the acid in the form of CALCIUM FLUORIDE.

Calcium fluoride forms a gelatinous precipitate, which it is found difficult to wash. If digested with ammonia, previous to

* Sillim. Journ., July, 1858. † Annal. d. Chem. u. Pharm. 100, 82.

filtration, it is rendered denser and less gelatinous. It is not altogether insoluble in water; aqueous solutions of the alkalies fail to decompose it. It is very slightly soluble in dilute, but more readily in concentrated hydrochloric acid. When acted upon by sulphuric acid, it is decomposed, and calcium sulphate and hydrofluoric acid are formed. Calcium fluoride is unalterable in the air, and at a red heat. Exposed to a very intense heat, it fuses. Upon intense ignition in moist air, it is slowly and partially decomposed into calcium oxide and hydrofluoric acid. Mixed with ammonium chloride, and exposed to a red heat, calcium fluoride suffers a continual loss of weight; but the decomposition is incomplete.

COMPOSITION.

Ca	40	51·28
F_2	38	48.72
	78	100·00

We often determine fluorine, more particularly in presence of silicic acid, by converting it into silicon fluoride (SiF_4). This is a colorless gas, fuming in the air, with suffocating odor, of sp. gr. 3·574, which decomposes when mixed with water forming silica and hydrofluosilicic acid thus: $3SiF_4 + 2H_2O = 2H_2SiF_6 + SiO_2$.

8. Carbonic Acid.

The *direct* estimation of carbonic acid—which, however, is only rarely resorted to—is usually effected by weighing the acid in the form of calcium carbonate. For the properties of the latter substance, see § 73.

9. Silicic Acid.*

When silicic acid is separated by acids from aqueous solutions of alkali silicates, it is at first perfectly soluble in water. It becomes insoluble or rather difficultly soluble when it coagulates. Coagulation is a permanent change and is furthered by concentration and by elevation of temperature. Silicic acid solution containing 10 or 12 per cent. of SiO_2 coagulates at the ordinary temperature in a few hours, and immediately if heated. A solution of

* Free silicic acid in solution is assumed to have the composition expressed by the formula $Si(OH)_4$. Silicic anhydride (SiO_2) is usually called "silica." Compounds of SiO_2 with less water than corresponds to the formula $Si(OH)_4 = SiO_2(H_2O)_2$ are here called "hydrates of silica."

5 per cent. may be preserved without coagulating for five or six days, one of 2 per cent. for two or three months, and one of 1 per cent. for several years, and solutions containing $\frac{1}{10}$ per cent. or less are not appreciably altered by time. Solid matter in powder such as graphite, hastens coagulation, alkali salts induce it rapidly. Aqueous solutions of silicic acid may, on the contrary, be mixed with hydrochloric acid, nitric acid, acetic acid, tartaric acid and alcohol without coagulating. The gelatinous silicic acid produced by coagulation may contain more or less water, and it appears to be the more difficultly soluble in water, the less water it contains; thus a jelly of silicic acid containing 1 per cent. of silica (SiO_2) gives a solution with cold water containing 1 part of silica in about 5000 parts, a jelly of 5 per cent. gives a solution containing 1 part of silica in about 10000 parts of water. A jelly containing less water is still less soluble, and when the jelly is dried up to a gummy mass it is barely soluble at all; this is also the case with the pulverulent hydrate of silica obtained in the analysis of silicates by drying a jelly containing much salts at 100° (GRAHAM*). The hydrated silica dried at 100° dissolves but very slightly in acids (with the exception of hydrofluoric acid); it dissolves, however, in solutions of fixed alkalies and alkali carbonates, especially on heating. Aqueous ammonia dissolves the jelly in tolerable quantity and the dry hydrate in very notable quantity (PRIBRAM)†. Regarding the amount of water in the hydrate dried at given temperatures chemists do not agree.‡

On ignition all the hydrates pass into anhydrous silica. As the vapor escapes small particles of the extremely fine powder are liable to whirl up. This may be avoided by moistening the hydrate in the crucible with water, evaporating to dryness on a water bath, and then applying at first a slight and then a gradually increased heat.

The silica obtained by igniting the hydrate appears in the amorphous condition, with a sp. gr. of 2·2 to 2·3. It forms a

* Pogg. Annal. 123, 529. † Zeitschr. f. anal. Chem. 6, 119.

‡ DOVERI (Annal. de Chim. et de Phys. 21, 40; Annal. d. Chem. u. Pharm. 64, 256) found in the air-dried hydrate 16·9 to 17·8 % water; J. FUCHS (Annal. d. Chem. u. Pharm. 82, 119 to 123), 9·1 to 9·6; G. LIPPERT, 9·28 to 9·95. DOVERI found in the hydrate dried at 100°, 8·3 to 9·4; J. FUCHS, 6·63 to 6·96; G. LIPPERT 4·97 to 5·52. H. ROSE (Pogg. Annal. 108. 1; Journ. für prakt. Chem. 81, 227) found in the hydrate obtained by digesting stilbite with concentrated hydrochloric acid, and dried at 150°, 4·85 % water.

white powder insoluble in water, and acids (hydrofluoric excepted), soluble in solutions of the fixed alkalies and their carbonates, especially in the heat. Hydrofluoric acid readily dissolves amorphous silica; the solution leaves no residue on evaporation in platinum, if the silica was pure. The amorphous silica, when heated with ammonium fluoride in a platinum crucible, readily volatilizes. The ignited amorphous silica, exposed to the air, eagerly absorbs water, which it will not give up at from 100° to 150° (H. Rose). The lower the heat during ignition the more hygroscopic is the residue (Souchay*). Silica fuses at the strongest heat; the mass obtained being vitreous and amorphous. Amorphous silica ignited with ammonium chloride, at first loses weight, and then, when the ignition has rendered it denser, the weight remains constant.

The amorphous silica must be distinguished from the crystallized or crystalline variety, which occurs as rock crystal, quartz, sand, &c. This has a sp. gr. of 2·6 (Schaffgotsch), and is far more difficultly, and in far less amount, dissolved by potash solution or solution of fixed alkali carbonates; it is also more slowly attacked by hydrofluoric acid, or ammonium fluoride. Crystallized silica is not hygroscopic, whether strongly or gently ignited (Souchay). Vegetable colors are not changed either by silica or its hydrates.

COMPOSITION.

Si	28·00	46·67
O_2	32·00	53·33
	60·00	100·00

ACID RADICALS OF THE SECOND GROUP.

§ 94.

1. Hydrochloric Acid.

Hydrochloric acid is almost invariably weighed in the form of silver chloride—for the properties of which see § 82.

2. Hydrobromic Acid.

Hydrobromic acid is always weighed in the form of silver bromide.

* Zeitschr. f. anal. Chem. 8, 423.

Silver bromide, prepared in the wet way, forms a yellowish-white precipitate. It is wholly insoluble in water and in nitric acid, tolerably soluble in ammonia, readily soluble in sodium thiosulphate and potassium cyanide. Concentrated solutions of potassium, sodium, and ammonium chlorides and bromides dissolve it to a very perceptible amount, while in very dilute solutions of these salts it is entirely insoluble. Traces only dissolve in the alkali nitrates. It dissolves abundantly in a concentrated warm solution of mercuric nitrate. On digestion with excess of potassium iodide solution it is completely converted into silver iodide (FIELD). On ignition in a current of chlorine silver bromide is transformed into chloride; on ignition in a current of hydrogen it is converted into metallic silver. Exposed to the light it gradually turns gray, and finally black. Under the influence of heat, it fuses to a reddish liquid, which, upon cooling, solidifies to a yellow, horn-like mass. Brought into contact with zinc and water, it is decomposed: a spongy mass of metallic silver forms, and the solution contains zinc bromide.

COMPOSITION.

Ag	107·93	57·45
Br	79·95	42·55
	187·88	100·00

3. HYDRIODIC ACID.

Hydriodic acid is usually determined in the form of SILVER IODIDE, and occasionally also in that of PALLADIOUS IODIDE.

a. Silver iodide, produced in the wet way, forms a light-yellow precipitate, insoluble in water, and in dilute nitric acid, and very slightly soluble in ammonia. One part dissolves, according to WALLACE and LAMONT* in 2493 parts of aqueous ammonia sp. gr. ·89, according to MARTINI, in 2510 parts of ·96 sp. gr. It is copiously taken up by concentrated solution of potassium iodide, but it is insoluble in very dilute; it dissolves readily in sodium thiosulphate and in potassium cyanide; traces only are dissolved by alkali nitrates. In concentrated warm solution of mercuric nitrate it is copiously soluble. Hot concentrated nitric and sulphuric acids convert it, but with some difficulty, into silver nitrate and sulphate respectively, with expulsion of the iodine. Silver iodide acquires a

* Chem. Gaz. 1859, 137.

black color when exposed to the light. When heated, it fuses without decomposition to a reddish fluid, which, upon cooling, solidifies to a yellow mass, that may be cut with a knife. Under the influence of excess of chlorine in the heat it is completely converted into silver chloride; ignition in hydrogen reduces it but incompletely to the metallic state. When brought into contact with zinc and water, it is decomposed but incompletely: zinc iodide is formed, and metallic silver separates.

COMPOSITION.

Ag	107·93	45·97
I	126·85	54·03
	234·78	100·00

b. Palladious iodide, produced by mixing an alkali iodide with palladious chloride, is a deep brownish-black, flocculent precipitate, insoluble in water, and in dilute hydrochloric acid, but slightly soluble in saline solutions (sodium chloride, magnesium chloride, calcium chloride, &c.). It is unalterable in the air. Dried simply in the air it retains one molecule of water=5·05 per cent. Dried long *in vacuo*, or at a rather high temperature (70° to 80°), it yields the whole of this water, without the least loss of iodine. Dried at 100°, it loses a trace of iodine; at from 300° to 400°, the whole of the iodine is expelled. It may be washed with hot water, without loss of iodine.

COMPOSITION.

Pd	106·58	29·58
I_2	253·70	70·42
	360·28	100·00

4. Hydrocyanic Acid.

Hydrocyanic acid, if determined gravimetrically and directly, is always converted into SILVER CYANIDE—for the properties of which compound see § 82.

5. Hydrosulphuric Acid.

The forms into which the sulphur in hydrogen sulphide or metallic sulphides, is converted for the purpose of being weighed, are ARSENIOUS SULPHIDE, SILVER SULPHIDE, COPPER SULPHIDE, and BARIUM SULPHATE.

For the properties of the sulphides named, see §§ 82, 85, 92; for those of barium sulphate, see § 71.

ACID RADICALS OF THE THIRD GROUP.

§ 95.

1. Nitric Acid; and 2. Chloric Acid.

These two acids are never determined directly—that is to say, in compounds containing them, but always in an indirect way; generally volumetrically.

SECTION IV.

THE DETERMINATION (OR ESTIMATION) OF RADICALS.

§ 96.

In the preceding Section we have examined the composition and properties of the various forms and combinations in which radicals are separated from each other, or in which they are weighed. We have now to consider the special means and methods of converting them into such forms and combinations.

For the sake of greater clearness and simplicity, we shall, in the present Section, confine our attention to the various methods applied to effect the *determination of single radicals*, deferring to the next Section the consideration of the means adopted for separating them from each other.

As in the "Qualitative Analysis," the acids of arsenic will be treated of among the bases, on account of their behavior to hydrogen sulphide.

In the quantitative analysis of a compound we have to study first, the most appropriate method of dissolving it; and, secondly, the modes of determining the quantity of one or more of its constituents.

With regard to the latter point, we have to turn our attention, first, to the *performance;* and secondly, to the *accuracy* of the methods.

It happens very rarely in quantitative analyses that the amount of a substance, as determined by the analytical process, corresponds exactly with the amount theoretically calculated or actually present; and if it does happen, it is merely by chance.

It is of importance to inquire what is the reason of this fact, and what are the limits of inaccuracy in the several methods.

The cause of this almost invariably occurring discrepancy between the quantity present and that actually found, is to be ascribed either exclusively to *the execution*, or it lies partly in the *method itself*.

The *execution* of the analytical processes and operations can never be *absolutely* accurate, even though the greatest care and attention be bestowed on the most trifling minutiæ. To account for this, we need only bear in mind that our weights and measures are never *absolutely* correct, nor our balances *absolutely* accurate, nor our reagents *absolutely* pure; and, moreover, that we do not weigh *in vacuo;* and that, even if we deduce the weight *in vacuo* from the weight we *actually* obtain by weighing in the air, the very volumes on which the calculation is based are but approximately known;—that the hygroscopic state of the air is liable to vary between the weighing of the empty crucible and of the crucible + the substance;—that we know the weight of a filter ash only *approximately;*—that we can never succeed in *completely* keeping off dust, &c.

With regard to the *methods*, many of them are not entirely free from certain unavoidable *sources of error;*—precipitates are not *absolutely* insoluble; compounds which require ignition are not *absolutely* fixed; others, which require drying, have a slight tendency to volatilize; the final reaction in volumetric analyses is usually produced only by a small excess of the standard fluid, which is occasionally liable to vary with the degree of dilution, the temperature, &c.

Strictly speaking, no method can be pronounced quite free from defect; it should be borne in mind, for example, that even *barium sulphate* is not *absolutely* insoluble in water. Whenever we describe any method as free from sources of error, we mean, that no causes of considerable inaccuracy are inherent in it.

We have, therefore, in our analytical processes, invariably to contend against certain sources of inaccuracy which it is impossible to overcome entirely, even though our operations be conducted with the most scrupulous care and with the utmost attention to established rules. It will be readily understood that several defects and sources of error may, in some cases, *combine* to vitiate the results; whereas, in other cases, they may *compensate* one another, and thus enable us to attain a higher degree of accuracy. The *comparative* accuracy of the results attainable by an analytical method oscillates between two points—these points are called the limits of error. In the case of methods free from sources of error, these limits will closely approach each other; thus, for instance, in the determination of chlorine, with great care one will always be able to obtain between 99·9 and 100·1 for the 100 parts of chlorine actually present.

Less perfect methods will, of course, exhibit far greater discrepancies; thus, in the estimation of strontium by means of sulphuric acid, the most attentive and skilful operator may not be able to obtain more than 99·0 (and even less) for the 100 parts of strontium actually present. I may here incidentally state that the numbers occasionally given in this manner, in the course of the present work, to denote the degree of accuracy of certain methods, refer invariably to the substance estimated (chlorine, nitrogen, baryta, for instance), and not to the combination in which that substance may be weighed (silver chloride, ammonium platinic chloride, barium sulphate, for instance); otherwise the accuracy of various methods would not be comparable.

The occasional attainment of results *exactly* corresponding with the numbers calculated does not always justify the assumption, on the part of the student, that his operations, to have led to such a result, must have been conducted with the utmost precision and accuracy. It may sometimes happen, in the course of the analytical process, that one error serves to compensate another; thus, for instance, the analyst may, at the commencement of his operations, spill a minute portion of the substance to be analyzed; whilst, at a later stage of the process, he may recover the loss by an imperfect washing of the precipitate. As a general rule, results showing a trifling deficiency of substance may be looked upon as better proof of accurate performance of the analytical process than results exhibiting an excess of substance.

As not the least effective means of guarding against error and inaccuracies in *gravimetric analyses*, I would most strongly recommend the analyst, *after weighing a precipitate, &c., to compare its properties (color, solubility, reaction, &c.) with those which it should possess*, and which have been amply described in the preceding Section.

In my own laboratory, I insist upon all substances that are weighed in the course of an analysis being kept between watch-glasses, until the whole affair is concluded. This affords always a chance of testing them once more for some impurity, the presence of which may become suspected in the after-course of the process.

I. DETERMINATION OF BASIC RADICALS IN SIMPLE SALTS.

First Group.

POTASSIUM—SODIUM—AMMONIUM—(LITHIUM).

§ 97.

1. Potassium.

a. Solution.

Potassa and potassium salts of those inorganic acids which we have to consider here, are dissolved in water, in which menstruum they dissolve readily, or at all events, pretty readily.

Potassium salts of organic acids it is most convenient to convert into potassium sulphate. See p. 211.

b. Determination.

Potassium is weighed either as *potassium sulphate*, as *potassium chloride*, or as *potassium platinic chloride* (see § 68). It may also be determined volumetrically. For the alkalimetric estimation of potassa or potassium carbonate, see §§ 195 and 196.

We may convert into

1. Potassium Sulphate.

Potassium salts of strong volatile acids; *e.g.*, potassium chloride, potassium bromide, potassium nitrate, &c., and salts of organic acids.

2. Potassium Chloride.

In general, caustic potassa and potassium salts of weak volatile acids; also, and more particularly, potassium sulphate, chromate, chlorate, and silicate.

3. Potassium Platinic Chloride.

Potassium salts of non-volatile acids soluble in alcohol; *e.g.*, potassium phosphate, potassium borate.

The potassium in potassium borate may be determined also as sulphate (§ 136); and the potassium in the phosphate, as potassium chloride (§ 135).

The form of potassium platinic chloride may also be resorted to in general, for the estimation of potassium in all potassium salts of those acids which are soluble in alcohol. This form is, moreover, of especial importance, as that in which the separation of potassium from sodium, &c., is effected.

4. Potassium Silicofluoride.

Potassium salts of those acids which are soluble in weak alcohol, except borate.

1. *Determination as Potassium Sulphate.*

Evaporate the aqueous solution of the potassium sulphate to dryness, ignite the residue in a platinum crucible or dish, and weigh (§ 42). The residue must be thoroughly dried before you proceed to ignite it; the heat applied for the latter purpose must be moderate at first, and very gradually increased to the requisite degree; the crucible or dish must be kept well covered—neglect of these precautionary rules involves always a loss of substance from decrepitation. If free sulphuric acid is present, we obtain, upon evaporation, acid potassium sulphate; in such cases the acid salt is to be converted into the normal by igniting first alone (here it is best to place the lamp so that the flame may strike the dish-cover obliquely from above), then with ammonium carbonate. See § 68.

For properties of the residue, see § 68. Observe more particularly that the residue must dissolve to a clear fluid, and that the solution must be neutral. Should traces of platinum remain behind (the dish not having been previously weighed), these must be carefully determined, and their weight subtracted from that of the ignited residue.

With proper care and attention, this method gives accurate results.

To convert the above-mentioned salts (potassium chloride, &c.) into potassium sulphate, add to their aqueous solution a quantity of pure sulphuric acid more than sufficient to form normal sulphate with the whole of the potassium, evaporate the solution to dryness, ignite the residue, and convert the resulting acid potassium sulphate into the normal, by treating with ammonium carbonate (§ 68).

As the expulsion of a large quantity of sulphuric acid is a very disagreeable process, avoid adding too great an excess. Should too little of the acid have been used, which you may infer from the non-evolution of sulphuric acid fumes on ignition, moisten the residue with dilute sulphuric acid, evaporate, and again ignite. If you have to deal with a small quantity only of potassium chloride, &c., proceed at once to treat the dry salt, cautiously, with dilute sulphuric acid in the platinum crucible; provided the latter be capacious enough. In the case of potassium bromide and iodide, the use of platinum vessels must be avoided.

[Potassium salts of organic acids are directly converted into potassium sulphate by first carbonizing them at the lowest possible temperature, and after cooling adding some crystals of pure ammonium sulphate and a little water to the mass. The crucible being covered, the water is evaporated by heating the crucible cover, and the whole is afterwards heated to dull redness, until the excess of ammonium sulphate is destroyed. If the carbon is not fully consumed by this operation, add a little ammonium nitrate and repeat the ignition. Kämmerer.*] It is usually advisable to ignite finally in an atmosphere of ammonium carbonate.

2. *Determination as Potassium Chloride.*

General method the same as described in 1. The residue of potassium chloride must, previously to ignition, be treated in the same way as potassium sulphate, and for the same reason. The salt must be heated in a well-covered crucible or dish, and only to dull redness, as the application of a higher degree of heat is likely to cause some loss by volatilization. No particular regard need be had to the presence of free acid. For properties of the residue, see § 68. This method, if properly and carefully executed, gives very accurate results. The potassium chloride may, instead of being weighed, be determined volumetrically by § 141, *b*. This method, however, has no advantage in the case of single estimations, but saves time when a series of estimations has to be made.

In determining potassium in the carbonate it is sometimes desirable to avoid the effervescence occasioned by treatment with hydrochloric acid, as, for instance, in the case of the ignited residue of a potassium salt of an organic acid, which is contained in the crucible. This may be effected by treating the carbonate with solution of ammonium chloride in excess, evaporating and igniting, when ammonium carbonate and the excess of ammonium chloride will escape, leaving potassium chloride behind.

The methods of converting the potassium compounds specified above into potassium chloride, will be found in Part II. of this Section, under the respective heads of the acids which they contain.

[* Fres. Zeit. VII. 222.]

3. *Determination as Potassium Platinic Chloride.*

α. Potassium salts of volatile acids (nitric acid, acetic acid, &c.).

Mix the solution with hydrochloric acid, evaporate to dryness, dissolve the residue in a little water, add a concentrated solution of platinic chloride, as neutral as possible, in excess, and evaporate in a porcelain dish, on the water-bath, nearly to dryness, taking care not to heat the water-bath quite to boiling. Add alcohol of about 80 per cent. by volume to the residue and let it stand for some time, pour the alcoholic solution through a small filter, and treat the residue if necessary a few times with small quantities of alcohol of the same strength, until it appears to be pure potassium platinic chloride. Bring this upon the filter and wash completely by applying repeatedly small quantities of the same alcohol. Dry next the filter and its contents in the funnel, for it is necessary that the alcohol should be completely volatilized. Transfer the contents of the filter carefully to a watch-glass, and place the filter back into the funnel and dissolve and wash out the small quantity of adhering potassium platinic chloride with hot water. Evaporate the yellow solution thus obtained to dryness in a weighed platinum vessel. Then bring the chief quantity of the precipitate into the platinum dish and dry the whole to a constant weight at 130° C.

If the quantity of potassium platinic chloride obtained is very small, the whole may be dissolved from the filter, evaporated and dried in the same manner.*

β. Potassium salts of non-volatile acids (phosphoric acid, boracic acid, &c.).

Make a concentrated solution of the salt in water, add some hydrochloric acid, and platinic chloride in excess, mix with a tolerable quantity of the strongest alcohol, let the mixture stand 24 hours; after which filter, and proceed as directed in *α*.

Properties of the precipitate, § 68. This method, if properly executed, gives satisfactory results. Still there is generally a trifling loss of substance, potassium platinic chloride not being absolutely insoluble even in strong alcohol. In accurate analyses, therefore, the alcoholic washings must be evaporated, with addition of a little pure sodium chloride, at a temperature not exceeding 75°, nearly to dryness, and the residue treated once more with

* When many successive determinations are to be made, especially in technical analyses, much time can be saved by using GOOCH's apparatus (see p. 100) for washing and weighing the K_2PtCl_6.

alcohol of 80 per cent. A trifling additional amount of potassium platinic chloride is thus obtained, which is either added to the principal precipitate or collected on a separate small filter, and weighed by dissolving from the filter and evaporating to dryness as above described. The object of the addition of a little sodium chloride to the platinic chloride is to obviate the decomposition to which pure platinic chloride is more liable, upon evaporation in alcoholic solution alone, than it is when mixed with sodium platinic chloride. The atmosphere of a laboratory often contains ammonia, which might give rise to the formation of some ammonium platinic chloride, and to a consequent increase of weight in the potassium salt.

4. *Volumetric determination after conversion into Potassium Silicofluoride.*

To the moderately concentrated solution of the potassium salt in a beaker add a sufficiency of hydrofluosilicic acid,* and then an equal volume of pure strong alcohol. If the potassium salt was difficultly soluble (such as potassium platinic chloride), warm it with the hydrofluosilicic acid before adding the spirit. The potassium silicofluoride will separate as a translucent precipitate; when it has settled, filter, wash out the beaker with a mixture of equal parts strong alcohol and water, and wash the precipitate with the same mixture till the washings are no longer acid to litmus paper. Put the filter and precipitate into the beaker previously used, treat with water, add some tincture of litmus, heat to boiling, and add standard potash solution (§ 192) till the fluid is just blue, and remains so after continued boiling. The reaction is as follows: $(KF)_2SiF_4 + 4KOH = 6KF + Si(OH)_4$, consequently 2 atoms potassium in the standard solution correspond to 1 at. potassium originally present and precipitated as potassium silicofluoride (FR. STOLBA†).

If the solution of the potassium salt contains much free acid, particularly sulphuric acid, this is to be removed by heat before adding the hyrofluosilicic acid. Small quantities of ammonium salts are of no influence, large quantities should be removed. It need hardly be mentioned that other metals precipitable by

* W. KNOP and W. WOLF use hydrofluosilicate of aniline instead.—Zeitschr. f. anal. Chem. 1, 471.

† Zeitschr. f. anal. Chem. 3, 298.

hydrofluosilicic acid must be absent. The results are satisfactory. STOLBA obtained 99·2 to 100 per cent. Potassium platinic chloride may be easily converted into potassium silicofluoride; hence, in technical analyses, the potassium may be separated in the first form, and then titrated as the latter (STOLBA, *loc. cit.*).

§ 98.

2. SODIUM.

a. Solution.

See § 97, *a*—solution of potassa sodium—all the directions given in that place applying equally to the solution of soda and sodium salts.

b. Determination.

Sodium is determined either as *sodium sulphate*, as *sodium chloride*, or as *sodium carbonate* (§ 69). For the alkalimetric estimation of caustic soda and sodium carbonate, see §§ 195 and 196.

We may convert into

1. SODIUM SULPHATE; 2. SODIUM CHLORIDE.

In general the sodium salts corresponding to the potassium salts specified under the analogous potassium compounds, § 97.

3. SODIUM CARBONATE.

Caustic soda, sodium hydrogen carbonate, and sodium salts of organic acids, also sodium nitrate and sodium chloride.

In sodium borate the sodium is estimated best as sodium sulphate (§ 136); in the phosphate, as sodium chloride, or sodium carbonate (§ 135).

Sodium salts of organic acids are determined either, like the corresponding potassium compounds, as chloride, or—by preference —as carbonate. (This latter method is not so well adapted for potassium salts.) The analyst must here bear in mind, that when carbon acts on fusing sodium carbonate, carbon monoxide escapes, and caustic soda in not inconsiderable quantity is formed.

1. *Determination as Sodium Sulphate.*

If alone and in aqueous solution, evaporate to dryness, ignite and weigh the residue in a covered platinum crucible (§ 42). The process does not involve any risk of loss by decrepitation, as in the case of potassium sulphate. If free sulphuric acid happens to be present, this is removed in the same way as in the case of potassium sulphate.

With regard to the conversion of sodium chloride, &c., into sodium sulphate, see § 97, *b*, 1. For properties of the residue, see § 69. The method is easy, and gives accurate results.

2. Determination as Sodium Chloride.

Same method as described in 1. The rules given and the observations made in § 97, *b*, 2, apply equally here. For properties of the residue, see § 69.

The methods of converting sodium sulphate, chromate, chlorate, and silicate into sodium chloride, will be found in Part II. of this Section, under the respective heads of the acids which these salts contain.

3. Determination as Sodium Carbonate.

Evaporate the aqueous solution, ignite moderately, and weigh. The results are perfectly accurate. For properties of the residue, see § 69.

Caustic soda is converted into the carbonate by adding to its aqueous solution ammonium carbonate in excess, evaporating at a gentle heat, and igniting the residue.

Sodium hydrogen carbonate, if in the dry state, is converted into the normal carbonate by ignition. The heat must be very gradually increased, and the crucible kept well covered. If in aqueous solution, it is evaporated to dryness, in a capacious silver or platinum dish, and the residue ignited.

Sodium salts of organic acids are converted into the carbonate by ignition in a covered platinum crucible, from which the lid is removed after a time. The heat must be increased very gradually. When the mass has ceased to swell, the crucible is placed obliquely, with the lid leaning against it (see § 52, fig. 42), and a dull red heat applied until the carbon is consumed as far as practicable. The contents of the crucible are then warmed with water, and the fluid is filtered off from the residuary carbon, which is carefully washed. The filtrate and rinsings are evaporated to dryness with the addition of a little ammonium carbonate, and the residue is ignited and weighed. The ammonium carbonate is added, to convert any caustic soda that may have been formed into carbonate. The method, if carefully conducted, gives accurate results; however, a small loss of soda on carbonization is not to be avoided.

Sodium nitrate, or sodium chloride, may be converted into carbonate, by adding to their aqueous solution perfectly pure oxalic acid in moderate excess, and evaporating several times to dryness, with repeated renewal of the water. All the nitric acid of the sodium nitrate escapes in this process (partly decomposed, partly undecomposed); and equally so all the hydrochloric acid in the case of sodium chloride. If the residue is now ignited until the excess of oxalic acid is removed, sodium carbonate is left.

§ 99.

3. Ammonium.

a. Solution.

Ammonia is soluble in water, as are all ammonium salts of those acids which claim our attention here. It is not always necessary, however, to dissolve ammonium salts for the purpose of determining the amount of ammonium contained in them.

b. Determination.

Ammonium is weighed, as stated § 70, either in the form of *ammonium chloride*, or in that of *ammonium platinic chloride*. Into these forms it may be converted either *directly* or *indirectly* (*i.e.*, after expulsion as ammonia, and re-combination with an acid). Ammonium is also frequently determined by volumetric analysis, and its quantity is sometimes inferred, from the volume of nitrogen.

We convert directly into

1. Ammonium Chloride.

Ammonia gas and its aqueous solution, and also ammonium salts of weak volatile acids (ammonium carbonate, ammonium sulphide, &c.).

2. Ammonium Platinic Chloride.

Ammonium salts of acids soluble in alcohol, such as ammonium sulphate, ammonium phosphate, &c.

3. The methods based on the expulsion of ammonia from ammonium compounds, and also that of inferring the amount of ammonium from the volume of nitrogen eliminated in the dry way, are equally applicable to all ammonium salts.

The expulsion of ammonia in the dry way (by ignition with soda-lime), and its estimation from the volume of nitrogen eliminated in the dry way, being effected in the same manner as the estimation of the nitrogen in organic compounds, I refer the student to the Section on organic analysis. Here I shall only give the methods based upon the expulsion of ammonia and of nitrogen in the wet way. For the alkalimetric estimation of free ammonia, see §§ 195 and 196.

1. *Determination as Ammonium Chloride.*

Evaporate the aqueous solution of the ammonium chloride on the water-bath, and dry the residue at 100° until the weight remains constant (§ 42). The results are accurate. The volatilization of the chloride is very trifling. A direct experiment gave 99·94 instead of 100. (See Expt. 15.) The presence of free hydrochloric acid makes no difference; the conversion of caustic ammonia into ammonium chloride may accordingly be effected by supersaturating with hydrochloric acid. The same applies to the conversion of the carbonate, with this addition only, that the process of supersaturation must be conducted in an obliquely-placed flask, and the mixture heated in the same, till the carbonic acid is driven off. In the analysis of ammonium sulphide we proceed in the same way, taking care simply, after the expulsion of the hydrogen sulphide, and before proceeding to evaporate, to filter off the sulphur which may have separated. Instead of weighing the ammonium chloride, its quantity may be inferred by the determination of its chlorine according to § 141, *b*. (Comp. potassium chloride, § 97, *b*, 3.)

2. *Determination as Ammonium Platinic Chloride.*

α. Ammoniacal salts with volatile acids.

Same method as described in § 97, *b*, 3, *α* (potassium platinic chloride).

β. Ammonium salts of non-volatile acids.

Same method as described § 97, *b*, 3, *β* (potassium platinic chloride). The results obtained by these methods are accurate.

If you wish to control the results,* ignite the double chloride, wrapped up in the filter, in a covered crucible, and calculate the amount of ammonium from that of the residuary platinum. The results must agree. The heat must be increased very gradually.†

* If the ammonium platinic chloride is pure, which may be known by its color and general appearance, this control may be dispensed with.

† The best way is to continue the application of a moderate heat for a long time, then to remove the lid, place the crucible obliquely, with the lid leaning against it, and burn the charred filter at a gradually-increased heat (H. Rose).

Want of due caution in this respect is apt to lead to loss, from particles of the double salt being carried away with the ammonium chloride. Very small quantities of ammonium platinic chloride are collected on an unweighed filter, dried, and at once reduced to platinum by ignition.*

3. *Estimation by Expulsion of the Ammonia in the Wet Way.*

This method, which is applicable in all cases, may be effected in two different ways, viz.:

a. EXPULSION OF THE AMMONIA BY DISTILLATION WITH SOLUTION OF POTASSA, OR SODA, OR WITH MILK OF LIME.—Applicable in all cases where no nitrogenous organic matters from which ammonia might be evolved upon boiling with solution of potassa, &c., are present with the ammonium salts.

Weigh the substance under examination in a small glass tube, three centimetres long and one wide, and put the tube, with the substance in it, into a flask containing a suitable quantity of moderately concentrated solution of potassa or soda, or milk of lime, from which every trace of ammonia has been removed by protracted ebullition, but which has been allowed to get thoroughly cold again; place the flask in a slanting position on wire-gauze, and immediately connect it by means of a glass tube bent at an obtuse angle, with the glass tube of a small cooling apparatus. Connect the lower end of this tube, by means of a tight-fitting perforated cork, with a sufficiently large tubulated receiver which is in its turn connected with a U-tube by means of a bent tube passing through its tubulure.

If you wish to *determine volumetrically the quantity of ammonia expelled*, introduce the larger portion of a measured quantity of standard solution of acid (sulphuric, hydrochloric, or oxalic, § 192), into the receiver, the remainder into the U-tube; add to the portion of fluid in the latter a little water, and color the liquids in the receiver and U-tube red with 1 or 2 c. c. of tincture of litmus. The cooling-tube must not dip into the fluid in the receiver; the fluid in the U-tube must completely fill the lower part, but it must not rise high, as otherwise the passage of air bubbles might

* In a series of experiments to get the platinum from pure and perfectly anhydrous ammonium platinic chloride, by very cautious ignition, Mr. LUCIUS, one of my pupils, obtained from 44·1 to 44·3 per cent. of the metal, instead of 44·3.

easily occasion loss by spirting. The quantity of acid used must of course be more than sufficient to fix the whole of the ammonia expelled.

When the apparatus is fully arranged, and you have ascertained that all the joints are perfectly tight, heat the contents of the flask to gentle ebullition, and continue the application of the same degree of heat until the drops, as they fall into the receiver, have for some time altogether ceased to impart the least tint of blue to the portion of the fluid with which they first come in contact. Loosen the cork of the flask, allow to stand half an hour, pour the contents of the receiver and U-tube into a beaker, rinsing out with small quantities of water, determine finally with a standard solution of alkali the quantity of acid still free, which, by simple subtraction, will give the amount of acid which has combined with the ammonia; and from this you may now calculate the amount of the latter (§ 192). Results accurate.*

If you wish to *determine by the gravimetric method the quantity of ammonia expelled*, receive the ammonia evolved in a quantity of hydrochloric acid more than sufficient to fix the whole of it, and determine the ammonium chloride formed, either by simple evaporation, after the directions of 1, or as ammonium platinic chloride, after the directions of 2.

b. EXPULSION OF THE AMMONIA BY MILK OF LIME, WITHOUT APPLICATION OF HEAT.—This method, recommended by SCHLÖSING, is based upon the fact that an aqueous solution containing free ammonia gives off the latter completely, and in a comparatively short time, when exposed in a shallow vessel to the air, at the common temperature. It finds application in cases where the presence of organic nitrogenous substances, decomposable by boiling alkalies, forbids the use of the method described in 3, *a*; thus, for instance, in the estimation of the ammonia in urine, manures, &c.

The fluid containing the ammonia, the volume of which must not exceed 35 c. c., is introduced into a shallow flat-bottomed vessel from 10 to 12 centimetres in diameter; this vessel is put on a plate filled with mercury. A tripod, made of a massive glass rod, is placed in the vessel which contains the solution of the ammonium salt, and a saucer or shallow dish with 10 c. c. of the normal solution of oxalic or sulphuric acid (§ 192) put on it. A beaker is now

* [In thus estimating minute quantities of ammonia, the condensing tube must be of tin, since glass yields a sensible amount of alkali to hot-water vapor.]

inverted over the whole. The beaker is lifted up on one side as far as is required, and a sufficient quantity of milk of lime added by means of a pipette (which should not be drawn out at the lower end): The beaker is then rapidly pressed down, and weighted with a stone slab. After forty-eight hours the glass is lifted up, and a slip of moist reddened litmus paper placed in it; if no change of color is observable, this is a sign that the expulsion of the ammonia is complete; in the contrary case, the glass must be replaced. Instead of the beaker and plate with mercury, a bell-jar, with a ground and greased rim, placed air-tight on a level glass plate, may be used. A bell-jar, having at the top a tubular opening, furnished with a close-fitting glass stopper, answers the purpose best, as it permits the introduction of a slip of red litmus paper suspended from a thread; thus enabling the operator to see whether the combination of the ammonia with the acid is completed, without the necessity of removing the bell-jar. According to SCHLÖSING, forty-eight hours are always sufficient to expel 0·1 to 1 gramme of ammonia from 25 to 35 c. c. of solution. However, I can admit this statement only as regards quantities up to 0·3 grm.; quantities above this often require a longer time. I, therefore, always prefer operating with quantities of substance containing no more than 0·3 grm. ammonia at the most.

When all the ammonia has been expelled, and has entered into combination with the acid, the quantity of acid left free is determined by means of standard solution of alkali, and the amount of the ammonia calculated from the result (§ 192).

4. *Estimation by Expulsion of the Nitrogen in the Wet Way.*

A process for determining ammonium by means of the azotometer has been given by W. KNOP.* It depends on the separation of the nitrogen by a bromized and strongly alkaline solution of sodium hypochlorite.†

[The simplest azotometer is that described by RUMPF.‡ It

---

* Chem. Centralbl. 1860, 244.

† This is prepared as follows:—Dissolve 1 part of sodium carbonate in 15 parts of water, cool the fluid with ice, saturate perfectly with chlorine, keeping cold all the while, and add strong soda solution (of 25 per cent.) till the mixture on rubbing between the fingers makes the skin slippery. Before using, add to the quantity required for the series of experiments bromine in the proportion of 2–3 grm. to the litre, and shake.

‡ Fres. Zeit., VI. 398.

consists of a burette of 50 or 100 c. c. stationed in a glass cylinder nearly filled with mercury, and connected by a stout caoutchouc tube with a small bottle, *a*, fig. 49, to which is fitted a soft, thrice-perforated caoutchouc stopper. The stopper carries a thermometer and two short glass tubes, one of which joins it to the burette, and the other has attached a short bit of caoutchouc tubing and a pinch-cock, *e*. The weighed ammonium salt (not more than 0·4 grm.) is placed in the tube, *f*, with 10 c. c. of water, and 50 c. c. of the bromized hypochlorite solution are brought into the bottle, *a*. The cock, *e*, being open, the stopper is firmly fixed in its place, and the burette is depressed in the mercury until its uppermost degree exactly coincides with the surface of the metal. The cock is then closed, and the bottle is inclined to bring the two substances in contact. The ammonium salt is speedily decomposed. When no further evolution of gas takes place the burette is so adjusted that the level of the mercury without and within it shall nearly coincide, and the operator waits 10–20 minutes, or until the thermometer in *a* indicates the same temperature as the surrounding air. Then the adjustment of the burette to exact coincidence of the mercury level, within and without, is effected, and the volume of the gas is read off. The stand of the thermometer and barometer are also noted, and the recorded volume of nitrogen is corrected by use of the tables on pp. 223 and 224–225, by DIETRICH.

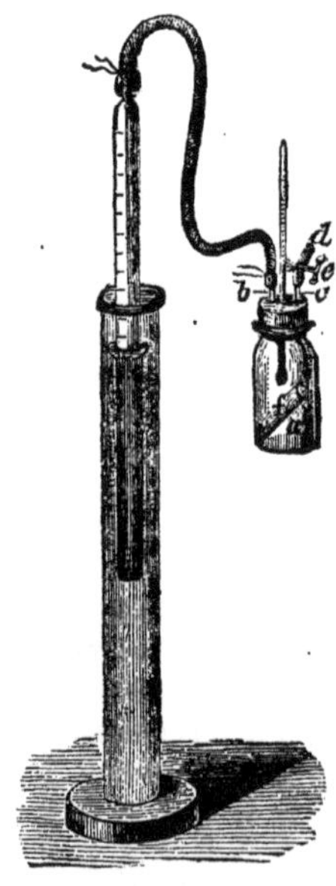

Fig. 49.

The first table gives a correction for the nitrogen which is absorbed by the 60 c. c. of liquid in the bottle *a*. The amount varies with the relative volumes of air and nitrogen, and is determined empirically by decomposing known quantities of ammonia and noting the difference between the obtained and the theoretical volume of nitrogen. The correction holds strictly, of course, only for a solution of such strength as that employed by DIETRICH and at the mean temperatures.

The second table serves to spare the labor of calculation. The weight of 1 c. c. of nitrogen, measured e. g. at 754 mm. of barome-

I. TABLE OF THE ABSORPTION OF NITROGEN GAS

in 60 *c. c. of liquid* (50 *c. c. of bromized hypochlorite and* 10 *c. c. of water*), *the hypochlorite having a sp. gr. of* 1.1, *and* 50 *c. c. evolving* 200 *mm. of nitrogen.*

Evolved........	1	2	3	4	5	6	7	8	9	10	11	12	13	14	15	16	17	18	19	20
Absorbed.......	0.06	0.08	0.11	0.13	0.16	0.18	0.21	0.23	0.26	0.28	0.31	0.33	0.36	0.38	0.41	0.43	0.46	0.48	0.51	0.53
Evolved........	21	22	23	24	25	26	27	28	29	30	31	32	33	34	35	36	37	38	39	40
Absorbed.......	0.56	0.58	0.61	0.63	0.66	0.68	0.71	0.73	0.76	0.78	0.81	0.83	0.86	0.88	0.91	0.93	0.96	0.98	1.01	1.03
Evolved........	41	42	43	44	45	46	47	48	49	50	51	52	53	54	55	56	57	58	59	60
Absorbed.......	1.06	1.08	1.11	1.13	1.16	1.18	1.21	1.23	1.26	1.28	1.31	1.33	1.36	1.38	1.41	1.43	1.46	1.48	1.51	1.53
Evolved........	61	62	63	64	65	66	67	68	69	70	71	72	73	74	75	76	77	78	79	80
Absorbed.......	1.56	1.58	1.61	1.63	1.66	1.68	1.71	1.73	1.76	1.78	1.81	1.83	1.86	1.88	1.91	1.93	1.96	1.98	2.01	2.03
Evolved........	81	82	83	84	85	86	87	88	89	90	91	92	93	94	95	96	97	98	99	100
Absorbed.......	2.06	2.08	2.11	2.13	2.16	2.18	2.21	2.23	2.26	2.28	2.31	2.33	2.36	2.38	2.41	2.43	2.46	2.48	2.51	2.53

II. TABLE OF THE WEIGHT OF

In Milligrammes for Pressures from 720 *to* 770 *m*

MILLIMETRES.

TEMPERATURE CELSIUS.	720	722	724	726	728	730	732	734	736	738	740		
10°	1.13380	1.13699	1.14018	1.14337	1.14656	1.14975	1.15294	1.15613	1.15932	1.16251	1.16570	1.16889	1.1720
11°	1.12881	1.13199	1.13517	1.13835	1.14153	1.14471	1.14789	1.15107	1.15424	1.15742	1.16060	1.16378	1.1669
12°	1.12376	1.12693	1.13010	1.13326	1.13643	1.13960	1.14277	1.14593	1.14910	1.15227	1.15543	1.15860	1.1617
13°	1.11875	1.12191	1.12506	1.12822	1.13138	1.13454	1.13769	1.14085	1.14401	1.14716	1.15032	1.15348	1.1566
14°	1.11369	1.11684	1.11999	1.12313	1.12628	1.12942	1.13257	1.13572	1.13886	1.14201	1.14515	1.14830	1.1514
15°	1.10859	1.11172	1.11486	1.11799	1.12113	1.12426	1.12739	1.13053	1.13366	1.13680	1.13993	1.14306	1.1462
16°	1.10346	1.10658	1.10971	1.11283	1.11596	1.11908	1.12220	1.12533	1.12845	1.13158	1.13470	1.13782	1.1409
17°	1.09828	1.10139	1.10450	1.10761	1.11073	1.11384	1.11695	1.12006	1.12317	1.12629	1.12940	1.13251	1.1356
18°	1.09304	1.09614	1.09924	1.10234	1.10544	1.10854	1.11165	1.11475	1.11785	1.12095	1.12405	1.12715	1.1302
19°	1.08744	1.09083	1.09392	1.09702	1.10011	1.10320	1.10629	1.10938	1.11248	1.11557	1.11866	1.12175	1.1248
20°	1.08246	1.08554	1.08862	1.09170	1.09478	1.09786	1.10094	1.10402	1.10710	1.11018	1.11327	1.11635	1.1194
21°	1.07708	1.08015	1.08322	1.08629	1.08936	1.09243	1.09550	1.09857	1.10165	1.10472	1.10779	1.11086	1.1139
22°	1.07166	1.07472	1.07778	1.08084	1.08390	1.08696	1.09002	1.09308	1.09614	1.09921	1.10227	1.10533	1.1083
23°	1.06616	1.06921	1.07226	1.07531	1.07836	1.08141	1.08446	1.08751	1.09056	1.09361	1.09666	1.09971	1.1027
24°	1.06061	1.06365	1.06669	1.06973	1.07277	1.07581	1.07885	1.08189	1.08493	1.08796	1.09100	1.09404	1.0970
25°	1.05499	1.05801	1.06104	1.06407	1.06710	1.07013	1.07316	1.07619	1.07922	1.08225	1.08528	1.08831	1.091
	720	722	724	726	728	730	732	734	736	738	740	74	

MILLIMETRES.

CUBIC CENTIMETRE OF NITROGEN.

of Mercury, and for Temperatures from 10° *to* 25° *C.*

MILLIMETRES.

746	748	750	752	754	756	758	760	762	764	766	768	770	TEMPERATURE CELSIUS.
1.17527	1.17846	1.18165	1.18484	1.18803	1.19122	1.19441	1.19760	1.20079	1.20398	1.20717	1.21036	1.21355	10°
1.17014	1.17332	1.17650	1.17168	1.18286	1.18603	1.18921	1.19239	1.19557	1.19875	1.20193	1.20511	1.20829	11°
1.16493	1.16810	1.17127	1.17444	1.17760	1.18077	1.18394	1.18710	1.19027	1.19344	1.19660	1.19977	1.20294	12°
1.15979	1.16295	1.16611	1.16926	1.17242	1.17558	1.17873	1.18189	1.18505	1.18820	1.19136	1.19452	1.19768	13°
1.15459	1.15774	1.16088	1.16403	1.16718	1.17032	1.17347	1.17661	1.17976	1.18291	1.18605	1.18920	1.19234	14°
1.14933	1.15247	1.15560	1.15873	1.16187	1.16500	1.16814	1.17127	1.17440	1.17754	1.18067	1.18381	1.18694	15°
1.14407	1.14720	1.15032	1.15344	1.15657	1.15969	1.16282	1.16594	1.16906	1.17219	1.17531	1.17844	1.18156	16°
1.13873	1.14185	1.14496	1.14807	1.15118	1.15429	1.15741	1.16052	1.16363	1.16674	1.16985	1.17297	1.17608	17°
1.13335	1.13645	1.13955	1.14266	1.14576	1.14886	1.15196	1.15506	1.15816	1.16126	1.16436	1.16746	1.17056	18°
1.12794	1.13103	1.13412	1.13721	1.14030	1.14340	1.14649	1.14958	1.15267	1.15576	1.15886	1.16195	1.16504	19°
1.12251	1.12559	1.12867	1.13175	1.13483	1.13791	1.14099	1.14408	1.14716	1.15024	1.15332	1.15640	1.15948	20°
1.11700	1.12007	1.12314	1.12621	1.12928	1.13236	1.13543	1.13850	1.14157	1.14464	1.14771	1.15078	1.15385	21°
1.11145	1.11451	1.11757	1.12063	1.12369	1.12675	1.12982	1.13288	1.13594	1.13900	1.14206	1.14512	1.14818	22°
1.10581	1.10886	1.11191	1.11496	1.11801	1.12106	1,12411	1.12716	1.13021	1.13326	1.13631	1.13936	1.14241	23°
1,10012	1.10316	1.10620	1.10924	1.11228	1.11532	1.11835	1.12139	1.12443	1.12747	1.13051	1.13355	1.13659	24°
1,09437	1.09740	1.10043	1.10346	1.10649	1.10952	1.11255	1.11558	1.11861	1.12164	1.12467	1.12770	1.13073	25°
746	748	750	752	754	756	758	760	762	764	766	768	770	

MILLIMETRES.

ter and 15° C., is found at the intersection of the vertical column 754 with the horizontal column 15°, is, viz., 1·16187.

To the observed volume of nitrogen add the amount absorbed as per Table I., and correct the total by Table II. It scarcely requires to be mentioned that good results can only be obtained in an apartment where the temperature is uniform, and when care is exercised to avoid warming the apparatus in handling. See DIETRICH's papers.*

§ 100.

Supplement to the First Group.

LITHIUM.

In the absence of other bases, lithium may, like potassium and sodium, be converted into anhydrous SULPHATE, and weighed in that form (Li_2SO_4). As lithium forms no acid sulphate, the excess of sulphuric acid may be readily removed by simple ignition. LITHIUM CARBONATE also, which is difficultly soluble in water, and fuses at a red heat without suffering decomposition, is well suited for weighing; whilst lithium chloride, which deliquesces in the air, and is by ignition in moist air converted into hydrochloric acid and lithium oxide, is unfit for the estimation of lithium.

In presence of other alkali metals, lithium is best converted into LITHIUM PHOSPHATE (Li_3PO_4), and weighed in that form. This is effected by the following process: add to the solution a sufficient quantity of sodium phosphate (which must be perfectly free from phosphates of the alkali-earth metals), and enough soda to keep the reaction alkaline, and evaporate the mixture to dryness; pour water over the residue, in sufficient quantity to dissolve the soluble salts with the aid of a gentle heat, add an equal volume of solution of ammonia, digest at a gentle heat, filter after twelve hours, and wash the precipitate with a mixture of equal volumes of water and solution of ammonia. Evaporate the filtrate and first washings to dryness, and treat the residue in the same way as before. If some more lithium phosphate is thereby obtained, add this to the principal quantity. The process gives, on an average, 99·61 for 100 parts of lithium oxide.

* Fres. Zeit. III. 162; IV. 141, and V. 36.

If the quantity of lithium present is relatively very small, the larger portion of the potassa or soda compounds should first be removed by addition of absolute alcohol to the most highly concentrated solution of the salts (chlorides, bromides, iodides, or nitrates, but not sulphates); since this, by lessening the amount of water required to effect the separation of the lithium phosphate from the soluble salts, will prevent loss of lithium (W. MAYER).*

The precipitated normal lithium phosphate has the formula $2Li_3PO_4 + H_2O$. It dissolves in 2539 parts of pure, and 3920 parts of ammoniated water; at 100°, it completely loses its water; if pure, it does not cake at a moderate red heat (MAYER).

The objections raised by RAMMELSBERG† to MAYER's method of estimating lithia I find to be ungrounded. According to my own experience, it appears that the filtrate and wash-water must be evaporated in a platinum dish not only once, but at least twice—in fact, till a residue is obtained which is completely soluble in dilute ammonia. Lithium phosphate may be dried at 100°, or ignited according to § 53, before being weighed. In the latter case, care must be taken to free the filter as much as possible from the precipitate before proceeding to incinerate it. I have thus obtained,‡ instead of 100 parts lithium carbonate, by drying at 100°, 99·84, 99·89, 100·41,—by igniting 99·66 and 100·05. The lithium phosphate obtained was free from sodium.

Second Group.

BARIUM—STRONTIUM—CALCIUM—MAGNESIUM.

§ 101.

1. BARIUM.

a. Solution.

Caustic baryta is soluble in water, as are many barium salts. Barium salts which are insoluble in water are, with almost the single exception of the sulphate, readily dissolved by dilute hydrochloric acid. The solution of the sulphate is effected by fusion with sodium carbonate, &c. (See § 132.)

---

* Annal. der Chem. u. Pharm. 98, 193, where Mayer has also demonstrated the non-existence of a sodium lithium phosphate of fixed composition (Berzelius), or of varying composition (Rammelsberg).

† Pogg. Annal. 102, 443.

‡ Zeitschr. f. Analyt. Chem. 1, 42.

b. Determination.

Barium is weighed either as *sulphate* or as *carbonate*, rarely (in the separation from strontia) as *barium silico-fluoride* (§ 71). Barium oxide or hydroxide, also barium carbonate, may also be determined by the volumetric (alkalimetric) method. Comp. § 198.

We may convert into

1. Barium Sulphate.

a. By Precipitation.

All barium compounds without exception.

b. By Evaporation.

All barium salts of volatile acids, if no other non-volatile body is present.

2. Barium Carbonate.

a. All barium salts soluble in water.

b. Barium salts of organic acids.

Barium is both precipitated and weighed, by far the most frequently as sulphate, the more so as this is the form in which it is most conveniently separated from other bases. The determination by means of evaporation (1, *b*) is, in cases where it can be applied, and where we are not obliged to evaporate large quantities of fluid, very exact and convenient. Barium is determined as carbonate in the wet way, when from any reason it is not possible or not desirable to precipitate it as sulphate. If a fluid or dry substance contains bodies which impede the precipitation of barium as sulphate or carbonate (alkali citrates, metaphosphoric acid, see § 71, *a* and *b*), such bodies must of course be got rid of, before proceeding to precipitation.

1. *Determination as Barium Sulphate.*

a. By Precipitation.

Heat the moderately dilute solution of barium, which must not contain too much free acid (and must, therefore, if necessary, first be freed therefrom by evaporation or addition of sodium carbonate), in a platinum or porcelain dish, or in a glass vessel, to incipient ebullition, add dilute sulphuric acid, as long as a precipitate forms, keep the mixture for some time at a temperature very near the boiling point, stirring if not on a water-bath, and allow the precipitate to subside; decant the almost clear supernatant fluid on a filter, boil the precipitate once with water and a little dilute sulphuric acid, then three or four times with water, then transfer it to the filter, and wash with boiling water, until the filtrate is no longer rendered turbid by barium chloride. Dry the precipitate, and treat it as directed in § 53. If the precipitate has been properly washed in the manner here directed, it is perfectly pure. In the presence of alkali salts, however, the precipitate will still contain small quantities of alkali sulphate.

b. By Evaporation.

Add to the solution, in a weighed platinum dish, pure sulphuric acid very slightly in excess, and evaporate on the water-bath; expel the excess of sulphuric acid by cautious application of heat, and ignite the residue.

For the properties of barium sulphate, see § 71.

Both methods, if properly and carefully executed, give almost absolutely accurate results.

2. *Determination as Barium Carbonate.*

a. In Solutions.

Mix the moderately dilute solution of the barium salt in a beaker with ammonia, add ammonium carbonate in slight excess, and let the mixture stand several hours in a warm place. Filter, wash the precipitate with water mixed with a little ammonia, dry, and ignite (§ 53).

For the properties of the precipitate, see § 71. This method involves a trifling loss of substance, as barium carbonate is not absolutely insoluble in water. The direct experiment, No. 62, gave 99·79 instead of 100.

If the solution contains a notable quantity of ammonium salts, the loss incurred is much more considerable, since the presence of such salts greatly increases the solubility of barium carbonate.

b. In Barium Salts of Organic Acids.

Heat the salt slowly in a covered platinum crucible, until no more fumes are evolved; place the crucible obliquely, with the lid leaning against it, and ignite, until the whole of the carbon is consumed, and the residue presents a perfectly white appearance: moisten the residue with a concentrated solution of ammonium carbonate, evaporate, ignite gently, and weigh. The results obtained by this method are quite satisfactory. A direct experiment, No. 63, gave 99·61 instead of 100. The loss of substance which almost invariably attends this method is owing to particles of the salt being carried away with the fumes evolved upon ignition, and is accordingly the less considerable, the more slowly and gradually the heat is increased. Omission of the moistening of the residue with ammonium carbonate would involve a further loss of substance, as the ignition of barium carbonate in contact with carbon is attended with formation of some caustic baryta, carbon monoxide gas being evolved.

§ 102.

2. Strontium.

a. Solution.

See the preceding paragraph (§ 101, *a.*—Solution of baryta and barium salts), the directions there given apply equally here.

b. Determination.

Strontium is weighed either as *strontium sulphate* or as *strontium carbonate* (§ 72). Strontium in the form of oxide, hydroxide, or carbonate, may be determined also by the volumetric (alkalimetric) method. Comp. § 198.

We may convert into

1. Strontium Sulphate.

a. By Precipitation.

All compounds of strontium without exception.

b. By Evaporation.

All strontium salts of volatile acids, if no other non-volatile body is present.

2. Strontium Carbonate.

α. All strontium compounds soluble in water.

β. Strontium salts of organic acids.

The method based on the precipitation of strontium with sulphuric acid yields accurate results only in cases where the fluid from which the strontium is to be precipitated may be mixed, without injury, with alcohol. Where this cannot be done, and where the method based on the evaporation of the solution of strontium with sulphuric acid is equally inapplicable, the conversion into the carbonate ought to be resorted to in preference, if admissible. As in the case of barium, so here, we have to be on our guard against the presence of substances which would impede precipitation.

1. *Determination as Strontium Sulphate.*

a. By Precipitation.

Mix the solution of the strontium salt (which must not be too dilute, nor contain much free hydrochloric or nitric acid) with dilute sulphuric acid in excess, in a beaker, and add at least an equal volume of alcohol; let the mixture stand twelve hours, and filter; wash the precipitate with dilute alcohol, dry and ignite (§ 53).

If the circumstances of the case prevent the use of alcohol, the fluid must be precipitated in a tolerably concentrated state, allowed to stand in the cold for at least twenty-four hours, filtered, and the precipitate washed with cold water, until the last rinsings manifest no longer an acid reaction, and leave no perceptible residue upon evaporation. If traces of free sulphuric acid remain adhering to the filter, the latter turns black on drying, and crumbles to pieces; too protracted washing of the precipitate, on the other hand, tends to increase the loss of substance.

Care must be taken that the precipitate be thoroughly dry, before proceeding to ignite it; otherwise it will be apt to throw off fine particles during the latter process. The filter, which is to be burnt apart from the precipitate, must be as clean as possible, or some loss of substance will be incurred; as may be clearly seen from the depth of the carmine tint of the flame with which the filter burns if the precipitate has not been properly removed.

For the properties of the precipitate, see § 72. When alcohol is used and the directions given are properly adhered to, the results are very accurate; when the sulphate of strontium is precipitated from an aqueous solution, on the contrary, a certain amount of loss is unavoidable, as strontium sulphate is not absolutely insoluble in water. The direct experiments, No. 64, gave only 98·12 and 98·02 instead of 100. However, the error may be rectified, by calculating the amount of strontium sulphate dissolved in the filtrate and the wash-water, basing the calculation upon the known degree of solubility of strontium sulphate in pure and acidified water. See Expt. No. 65, which, with this correction, gave 99·77 instead of 100. The necessity for making the correction may be obviated by washing with 1 part sulphuric acid mixed with 20 parts water till all substances precipitable by alcohol are removed, then with alcohol till all the sulphuric acid is removed. Strontium sulphate also carries down sulphates of other strong bases in small quantities.

b. By Evaporation.

The same method as described for barium, § 101, 1, *b*.

2. *Determination as Strontium Carbonate.*

a. In Solutions.

The same method as described § 101, 2, *a*. For the properties of the precipitate, see § 72. The method gives very accurate results, as strontium carbonate is nearly absolutely insoluble in water containing ammonia and ammonium carbonate. A direct experiment, No. 66, gave 99·82 instead of 100. Presence of ammonium salts exercises here a less adverse influence than the precipitation of barium carbonate.

b. In Salts of Organic Acids.

The same method as described § 101, 2, *b*. The remarks made there, respecting the accuracy of the results, apply equally here.

§ 103.

3. Calcium.

a. Solution.

See § 101, *a*.—Solution of barium. Calcium fluoride is, by means of sulphuric acid, converted into calcium sulphate, and the latter again, if necessary, decomposed by boiling or fusing with an alkali carbonate (§ 132). [Calcium sulphate dissolves readily in moderately dilute hydrochloric acid. It is much less soluble in strong hydrochloric acid.]

b. Determination.

Calcium is weighed either as *calcium sulphate*, as *calcium carbonate*, or *calcium oxide* (§ 73). Calcium in the form of oxide, hydroxide, or carbonate, may be determined also by the volumetric (alkalimetric) method. Comp. § 198.

We may convert into

1. Calcium Sulphate.

a. By Precipitation.

All calcium salts of acids soluble in alcohol, provided no other substance insoluble in alcohol be present.

b. By Evaporation.

All calcium salts of volatile acids, provided no non-volatile body be present.

2. Calcium Carbonate.

a. By Precipitation with Ammon Carbonate.

All calcium salts soluble in water.

b. By Precipitation with Ammonium Oxalate.

All calcium salts soluble in water or in hydrochloric acid without exception.

c. By Ignition.

Calcium salts of organic acids.

Of these several methods, 2, *b* (precipitation with ammonium oxalate) is the one most frequently resorted to. This, and the method 1, *b*, give the most accurate results. The method, 1, *a*, is usually resorted to only to effect the separation of calcium from other basic radicals; 2, *a*, generally only to effect the separation of calcium together with the other alkali-earth metals from the alkalies. As many bodies (alkali citrates, and metaphosphates) interfere with the precipitation of calcium by the precipitants given, these, if present, must be first removed.

1. Determination of Calcium Sulphate.

a. By Precipitation.

Mix the solution of calcium salt in a beaker, with dilute sulphuric acid in excess, and add twice the volume of alcohol; let the mixture stand twelve hours, filter, and *thoroughly* wash the precipitate with alcohol, dry, and ignite moderately (§ 53). For the properties of the precipitate, see § 73. The results are very accurate. A direct experiment, No. 67, gave 99·64 instead of 100.

b. By Evaporation.

The same method as described § 101, 1, *b*.

2. Determination as Calcium Carbonate or Calcium Oxide.

a. By Precipitation with Ammonium Carbonate.

The same method as described § 101, 2, *a*. The precipitate can be most conveniently weighed as calcium carbonate. It must be exposed only to a very gentle red heat, but this must be continued for some time. For the properties of the precipitate, see § 73.

This method gives very accurate results, the loss of substance incurred being hardly worth mentioning.

If the solution contains ammonium chloride or similar ammonium salts in considerable proportion, the loss of substance incurred is far greater. The same is the case if the precipitate is washed with pure instead of ammoniacal water. A direct experiment, No. 68, in which pure water was used, gave 99·17 instead of 100 parts of lime.

b. By Precipitation with Ammonium Oxalate.

α. The Calcium Salt is soluble in Water.

To the hot solution in a beaker, add ammonium oxalate in moderate excess, and then ammonia sufficient to impart an ammoniacal smell to the fluid; cover the glass, and let it stand in a warm place until the precipitate has completely subsided, which will require twelve hours, at least. Pour the clear fluid gently and cautiously, so as to leave the precipitate undisturbed, on a filter; wash the precipitate two or three times by decantation with hot water; lastly, transfer the precipitate also to the filter, by rinsing with hot water, taking care, before the addition of a fresh portion, to wait until the fluid has completely passed through the filter. Small particles of the precipitate, adhering firmly to the glass, are removed with a feather. If this fails to effect their complete removal, they should be dissolved in a few drops of highly dilute hydrochloric acid, ammonia added to the solution, and the oxalate obtained added to the first precipitate. Deviations from the rules laid down here will generally give rise to the passing of a turbid fluid through the filter. After having washed the precipitate, dry it on the filter in the funnel, and transfer the dry precipitate to a platinum crucible, taking care to remove it as completely as possible from the filter; burn the filter on a piece of platinum wire, letting the ash drop into the hollow of the lid; put the latter, now inverted, on the crucible, so that the filter ash may not mix with the precipitate; heat at first very gently, then more strongly, until the bottom of the crucible is heated to very faint redness. Keep it at that temperature from ten to fifteen minutes, removing the lid from time to time. I am accustomed during this operation to move the lamp backwards and forwards under the crucible with the hand, since, if you allow it to stand, the heat may very easily get too high. Finally allow to cool in the desiccator and weigh. After weighing, moisten the contents of the crucible, which must be perfectly white, or barely show the least tinge of gray, with a little water, and test this after a time with a minute slip of turmeric paper. Should the paper turn brown—a sign that the heat applied was too strong—rinse off the fluid adhering to the paper with a little water into the crucible, throw in a small lump of pure ammonium carbonate, evaporate to dryness (best in the water-bath), heat to very faint redness, and weigh the residue. If the weight has increased, repeat the same operation until the weight remains constant. This method gives nearly absolutely accurate results; and if the application of heat is properly managed, there is no need of the tedious evaporation with ammonium carbonate. A direct experiment, No. 69, gave 99·99 instead of 100.

If a gas blowpipe is at hand, or any other arrangement by means of which a platinum crucible may be raised to a white heat, the calcium oxalate may be converted into CAUSTIC LIME with results almost equally accurate; and I believe that this method, which requires less patience than the other, is more certain to yield good results in the hands of many persons. The calcium oxalate and the filter ash are transferred to a moderate-sized platinum crucible, which is ignited first over the BUNSEN, and then over the blowpipe. The crucible is then weighed, and ignited again over the blowpipe. The second ignition over the blowpipe should not reduce the weight. The duration of the ignition necessary varies from 5 to 15 or more minutes, according to intensity of heat and quantity of the precipitate. It is well to weigh the empty crucible again at the end of the operation, as platinum sometimes loses weight after violent and prolonged ignition. The results obtained by FRITZSCHE, COSSA,* and SOUCHAY scarcely differ from the calculated numbers. For properties of calcium oxide, see § 73.

The calcium oxalate may also be converted into SULPHATE. SCHRÖTTER ignites in a covered platinum crucible with pure ammonium sulphate. Or you may ignite in a covered platinum dish till the precipitate is for the most part converted into oxide, add a little water, then hydrochloric acid to effect solution, then pure sulphuric acid in excess, evaporate and ignite moderately. This process is also quite accurate.

Instead of converting calcium oxalate into carbonate or oxide for weighing, the quantity of calcium present in the salt may be determined also by two different volumetric methods.

* FRITZSCHE (Zeitschr. f. anal. Chem. 3, 179) and A. COSSA (*Ib.* 8, 141).

a. Ignite the oxalate, converting it thus into a mixture of calcium carbonate and oxide, and determine the quantity of the calcium by the alkalimetric method described in § 198; or,

b. Determine the oxalic acid in the well-washed but still moist calcium oxalate by means of potassium permanganate (§ 137).

With proper care, both these volumetric methods give as accurate results as those obtained by weighing. (Comp. Expt. No. 71.) They deserve to be recommended more particularly in cases where an entire series of quantitative estimations of calcium has to be made. Under certain circumstances it may also prove advantageous to precipitate calcium with a measured quantity of a standard solution of oxalic acid, filter, and determine the excess of oxalic acid in the filtrate, or an aliquot part of the same. (KRAUT.*)

β. The Salt is insoluble in Water.

Dissolve the salt in dilute hydrochloric acid. If the acid of the calcium salt is of a nature to escape in this operation (*e.g.*, carbonic acid), or to admit of its separation by evaporation (*e.g.*, silicic acid), proceed, after the removal of the acid, as directed in *α*. But if the acid cannot thus be readily got rid of (*e.g.*, phosphoric acid), proceed as follows: Add ammonia until a precipitate begins to form, re-dissolve this with a drop of hydrochloric acid, add ammonium oxalate in excess, and finally sodium acetate; allow the precipitate to subside, and proceed for the remainder of the operation as directed in *α*. In this process the free hydrochloric acid present reacts on the sodium acetate and ammonium oxalate, forming sodium and ammonium chlorides, with liberation of a corresponding amount of oxalic and acetic acids in which calcium oxalate is nearly insoluble. The method yields accurate results. A direct experiment, No. 72, gave 99·78 instead of 100.

c. By Ignition.

The same method as described § 101, 2, *b* (barium). The residue remaining upon evaporation with ammonium carbonate (which operation it is advisable to perform twice) must be ignited very gently. The remarks made in § 101, 2, *b*, in reference to the accuracy of the results, apply equally here. By way of control, the calcium carbonate may be converted into oxide or into calcium sulphate (see *b*, *α*), or it may be determined alkalimetrically (§ 198).

---

* Chem. Centralblatt, 1856, 316.

§ 104.

4. Magnesium.

a. Solution.

Many magnesium salts are soluble in water; those which are insoluble in that menstruum dissolve in hydrochloric acid, with the exception of some silicates and aluminates.

b. Determination.

Magnesium is weighed (§ 74) either as *sulphate* or as *pyrophosphate*, or as *magnesium oxide*. In the form of oxide or carbonate, it may be determined also by the alkalimetric method described in § 198.

We may convert into

1. Magnesium Sulphate.

a. Directly.

All magnesium salts of volatile acids, provided no other non-volatile substance be present.

b. Indirectly.

All magnesium salts soluble in water, and also those which, insoluble in that menstruum, dissolve in hydrochloric acid, with separation of their acid (provided no ammonium salts be present).

2. Magnesium Pyrophosphate.

All magnesium compounds without exception.

3. Magnesium Oxide.

a. Magnesium salts of organic acids, or of readily volatile inorganic oxygen acids.

b. Magnesium chloride, and magnesium compounds convertible into that salt.

The direct determination as magnesium sulphate is highly recommended in all cases where it is applicable. The indirect conversion into the sulphate serves only in the case of certain separations, and is hardly ever had recourse to where it can possibly be avoided. The determination as pyrophosphate is most generally resorted to; especially also in the separation of magnesium from other bases. The method based on the conversion of magnesium chloride into oxide is usually resorted to only to effect the separation of magnesium from the alkali metals. Magnesium phosphates are analyzed as § 135 directs.

1. *Determination as Magnesium Sulphate.*

Add to the solution excess of pure dilute sulphuric acid, evaporate to dryness, in a weighed platinum dish, on the water-bath; then heat at first cautiously, afterwards, with the cover on more strongly—here it is advisable to place the lamp so that the flame may play obliquely on the cover from above—until the excess of sulphuric acid is completely expelled; lastly, ignite gently over the lamp for some time; allow to cool, and weigh. Should no fumes of hydrated sulphuric acid escape upon the application of a strongish heat, this may be looked upon as a sure sign that the sulphuric acid has not been added in sufficient quantity, in which case, after allowing to cool, a fresh portion of sulphuric acid is added. The method yields very accurate results. Care must be taken not to use a very large excess of sulphuric acid. The residue must be exposed to a moderate red heat only, and weighed rapidly. For the properties of the residue, see § 74.

2. *Determination as Magnesium Pyrophosphate.*

The solution of the magnesium salt is mixed, in a beaker, with ammonium chloride, and ammonia added in slight excess. Should a precipitate form upon the addition of ammonia, this may be considered a sign that a sufficient amount of ammonium chloride has not been used; a fresh amount of that salt must consequently be added, sufficient to effect the re-solution of the precipitate formed. The clear fluid is then mixed with a solution of sodium phosphate or sodium ammonium phosphate* in excess, and the mixture stirred, taking care to avoid touching the sides of the beaker with the stirring-rod; otherwise particles of the precipitate are apt to adhere so firmly to the rubbed parts of the beaker, that it will be found difficult to remove them; the beaker is then covered, and allowed to stand at rest for twelve hours, without warming; after that time the fluid is filtered, and the precipitate collected on the filter, the last particles of it being rinsed out of the glass with a portion of the filtrate, with the aid of a feather; when the fluid has completely passed through, the precipitate is washed with a mixture of 3 parts of water, and 1 part of solution of ammonia of 0·96 sp. gr., the

* According to MOHR $(NaNH_4H)PO_4$ is preferable to $(Na_2H)PO_4$ as a precipitant. (See Zeitschr. f. Anal. Chem. 12, 36.)

operation being continued until a few drops of the fluid passing through the filter mixed with nitric acid and a drop of silver nitrate show only a very slight opalescence.

The precipitate is now thoroughly dried, and then transferred to a platinum crucible (§ 53); the latter, with the lid on, is exposed for some time to a very gentle heat, which is finally increased to intense redness. The filter, as clean as practicable, is incinerated in a spiral of platinum wire, and the ash transferred to the crucible, which is then once more exposed to a red heat, allowed to cool, and weighed. If the magnesium pyrophosphate is dark colored, moisten with a few drops of nitric acid, warm till dry, and ignite again.

For the properties of the precipitate and residue, see § 74.

This method, if properly executed, yields most accurate results. The precipitate must be washed completely, but not over-washed, and the washing water must always contain the requisite quantity of ammonia.

Direct experiments, No. 73, *a* and *b*, gave respectively 100·43 and 100·30 instead of 100.

3. *Determination as Magnesium Oxide.*

a. In Magnesium Salts of Organic or Volatile Inorganic Acids.

The magnesium salt is gently heated in a covered platinum crucible, increasing the temperature gradually, until no more fumes escape; the lid is then removed, and the crucible placed in an oblique position, with the lid leaning against it. A red heat is now applied, until the residue is perfectly white. For the properties of the residue, see § 74. The method gives the more accurate results the more slowly the salt is heated from the beginning. Some loss of substance is usually sustained, owing to traces of the salt being carried off with the empyreumatic products. Magnesium salts of readily volatile oxygen acids (carbonic acid, nitric acid), may be transformed into magnesium oxide in a similar way, by simple ignition. Even magnesium sulphate loses the whole of its sulphuric acid when exposed, in a platinum crucible, to the heat of the gas blowpipe-flame (SONNENSCHEIN). As regards small quantities of magnesium sulphate, I can fully confirm this statement.

b. Conversion of Magnesium Chloride into Magnesium Oxide. See § 153, 4, γ.

THIRD GROUP OF BASIC RADICALS.

ALUMINIUM—CHROMIUM—(TITANIUM).

§ 105.

1. ALUMINIUM.

a. Solution.

Aluminium compounds which are insoluble in water, dissolve, for the most part, in hydrochloric acid. Native crystallized aluminium oxide (sapphire, ruby, corundum, &c.), and many native aluminium compounds, and also artificially produced aluminium oxide after intense ignition, require fusing with sodium carbonate, caustic potassa, or barium hydroxide, as a preliminary step to their solution in hydrochloric acid. Many aluminium compounds which resist the action of concentrated hydrochloric acid, may be decomposed by protracted heating with moderately concentrated sulphuric acid, or by fusion with potassium disulphate; *e.g.*, common clay.

b. Determination.

Aluminium is invariably weighed as *aluminium oxide* (§ 75). The several aluminium salts are converted into aluminium oxide, either by precipitation as aluminium hydroxide, and subsequent ignition, or by simple ignition. Precipitation as basic acetate or basic formate is resorted to only in cases of separation.

We may convert into

ALUMINIUM OXIDE.

a. By Precipitation.	*b. By Heating or Ignition.*
All aluminium salts soluble in water, and those which, insoluble in that menstruum, dissolve in hydrochloric acid, with separation of their acid.	*α.* All aluminium salts of readily volatile oxygen acids (*e.g.*, aluminium nitrate). *β.* All aluminium salts of organic acids.

With regard to the method *a*, it must be remembered that the solution must contain no organic substances, which would interfere with the precipitation—*e.g.*, tartaric acid, sugar, &c. Should such be present, the solution must be mixed with sodium carbonate and potassium nitrate, evaporated to dryness in a platinum dish, the residue fused, then softened with water, transferred to a beaker, digested with hydrochloric acid, and the solution filtered, and then, but not before, precipitated.

The methods *b*, *α* and *β*, are applicable only in cases where no other fixed substances are present. The methods of determining aluminium in its combinations with phosphoric, boracic, silicic, and chromic acids, will be found in Part II. of this Section, under the heads of these several acids.

Determination as Aluminium Oxide.

a. By Precipitation.

Mix the moderately dilute hot solution of the aluminium salt, in a beaker or dish, with a tolerable quantity of ammonium chloride, if that salt is not already present; add ammonia *slightly* in excess, boil gently till the fluid gives a neutral or barely alkaline reaction (the fluid adhering to the test paper must be washed back). The fluid must not be heated too long, or it may become acid through decomposition of ammonium chloride, and some of the precipitate may redissolve; allow to settle; then decant the clear supernatant fluid on to a filter, taking care not to disturb the precipitate; pour boiling water on the latter in the beaker, stir, let the precipitate subside, decant again, and repeat this operation of washing by decantation a second and a third time; transfer the precipitate now to the filter, finish the washing with boiling water, dry thoroughly, ignite (§ 52), and weigh. The heat applied should be very gentle at first, and the crucible kept well covered, to guard against the risk of loss of substance from spirting, which is always to be apprehended if the precipitate is not *thoroughly* dry; towards the end of the process the heat should be raised to intense redness. In the case of aluminium *sulphate* the foregoing process is apt to leave some sulphuric acid in the precipitate, which, of course, vitiates the result. To insure the removal of this sulphuric acid, the precipitate should be exposed for 5–10 min. to the heat of the gas blowpipe flame. If there are difficulties in the way, preventing this proceeding, the precipitate, either simply washed or moderately ignited, must be re-dissolved in hydrochloric acid (which requires protracted warming with strong acid), and then precipitated again with ammonia; or the sulphate must first be converted into nitrate by decomposing it with lead nitrate, added in very slight excess, the excess of lead removed by means of hydrosulphuric acid, and the further process conducted according to the directions of *a* or *b*. For the properties of aluminium hydroxide and ignited aluminium oxide, see § 75. The method, if properly executed, gives very accurate results. But if a considerable excess of ammonia is used, more particularly in the absence of ammonium salts, and the liquid is filtered without boiling or long standing in a warm place to remove the ammonia, no trifling loss may be incurred. This loss is the greater, the more dilute the solution, and the larger the excess of ammonia. The precipitate cannot well be sufficiently washed on the filter on account of its gelatinous nature; on the other hand, if it be entirely washed by decantation, a very large quantity of wash-water must be used, hence it is advisable to combine the two methods, as directed.* In case the BUNSEN filtering apparatus is used for washing aluminium hydroxide, for which operation it is particularly desirable, the precipitate may be brought into the filter without washing by decantation, and may be ignited without previous drying. See § 53, *b*.

b. By Ignition.

α. Aluminium Salts of Volatile Oxygen Acids.

Ignite the salt (or the residue of the evaporated solution) in a platinum crucible, gently at first, then gradually to the very highest degree of intensity, until the weight remains constant. For the properties of the residue, see § 75. Its purity must be carefully tested. There are no sources of error.

β. Aluminium Salts of Organic Acids.

The same method as described § 104, 3, *a* (Magnesium).

§ 106.

2. CHROMIUM.

a. Solution.

Many chromic salts are soluble in water. Chromic hydroxide, and most of the salts insoluble in water, dissolve in hydrochloric acid. Ignition renders chromic oxide and many chromium salts insoluble in acids; this insoluble modification must be prepared for

* [When a solution of aluminium hydroxide in potassium or sodium hydroxide is boiled with excess of ammonium chloride, the aluminium separates completely as a hydrated oxide with two mol. of water, which may be washed with comparative ease. In certain cases, as where aluminium is separated from ferric iron by boiling their hydroxides with soda, this fact may be taken advantage of. LÖWE, Fres. Zeitschrift, IV. 355.]

solution in hydrochloric acid, by fusing with 3 or 4 parts of potassa. In the process of fusing a small quantity of potassium chromate is formed by the action of air; this, however, can be decomposed by heating with hydrochloric acid with formation of chromic chloride. Addition of alcohol greatly promotes the reduction to chromic chloride. Instead of this fusing with potassa, we frequently prefer to adopt a treatment, whereby the chromium is at once oxidized and converted into an alkali chromate (see 2). For the solution of chromic iron, see § 160.

b. Determination.

Chromium is always, when directly determined, weighed *as chromic oxide.* It is brought into this form either by precipitation as hydroxide and ignition, or by simple ignition. It may, however, also be estimated, by conversion into chromic acid, and determination as such.

We may convert into

1. Chromic Oxide.

a. By Precipitation.

All chromic salts soluble in water, and also those which, insoluble in that menstruum, dissolve in hydrochloric acid, with separation of their acid. Provided always that no organic substances (such as tartaric acid, oxalic acid, &c.) which interfere with the precipitation be present.

b. By Ignition.

α. All chromic salts of volatile oxygen acids, provided no non-volatile substances be present.

β. Chromic salts of organic acids.

2. Chromic acid, or, more correctly speaking, alkali chromate.
Chromic oxide and all chromic salts.

The methods of analyzing chromic phosphates, borates, silicates, and chromic chromate, will be found in Part II. of this Section, under the heads of the several acids of these compounds.

1. *Determination as Chromic Oxide.*

a. By Precipitation.

The solution, which must not be too highly concentrated, is heated to 100° in a platinum or porcelain dish. If the precipitation is effected in a glass vessel, considerable error is caused by contamination of the precipitate with silica. If porcelain is used, this error is slight. Ammonia is then added slightly in excess, and the mixture exposed to a temperature approaching boiling, until the fluid over the precipitate is perfectly colorless, presenting no longer the least shade of red; let the solid particles subside, wash three times by decantation, and lastly on the filter, with hot water, dry thoroughly, and ignite (§ 52). The heat in the latter process must be increased gradually, and the crucible kept covered, otherwise some loss of substance is likely to arise from spirting upon the incandescence of the chromic oxide which marks the passing of the soluble into the insoluble modification. For the properties of the precipitate and residue, see § 76. This method, if properly executed, gives accurate results.

b. By Ignition.

a. Chromic salts of Volatile Oxygen Acids.

The same method as described, § 105, *b*, *α* (Aluminium).

b. Chromic salts of Organic Acids.

The same method as described, § 104, 3, *a* (Magnesium).

2. CONVERSION OF CHROMIUM IN CHROMIC COMPOUNDS INTO ALKALI CHROMATE.

(For the estimation of chromic acid, see § 130.)

The following methods have been proposed with this view:—

a. The solution of the chromic salt is mixed with solution of potassa or soda in excess, until the chromic hydroxide, which forms at first, is redissolved. Chlorine gas is then conducted into the cold fluid until it acquires a yellowish-red tint; it is then mixed with potassa or soda in excess, and the mixture evaporated to dryness; the residue is ignited in a platinum crucible. The whole of the potassium (or sodium) chlorate formed is decomposed by this process, and the residue consists, therefore, now of an alkali chromate and potassium (or sodium) chloride.—(VOHL.)

b. Potassium hydroxide is heated in a silver crucible to calm fusion; the heat is then somewhat moderated, and the perfectly dry chromic compound projected into the crucible. When the substance is thoroughly moistened with the potassa, small lumps of fused potassium chlorate are added. A lively effervescence ensues, from the escape of oxygen; at the same time the mass acquires a more and more yellow color, and finally becomes clear and transparent. Loss of substance must be carefully guarded against (H. SCHWARZ).

c. Dissolve chromic hydroxide in solution of potassa or soda, add lead dioxide in sufficient excess, and warm. The yellow fluid produced contains all the chromium as lead chromate in alkaline solution. Filter from the excess of lead dioxide, add to the filtrate acetic acid to acid reaction, and determine the weight of the precipitated lead chromate (G. CHANCEL*).

[*d.* Render the solution of chromic salt nearly neutral by a solution of sodium carbonate, add sodium acetate in excess, heat and add chlorine water, or pass in chlorine gas, keeping the solution nearly neutral by occasional addition of sodium carbonate. The oxidation proceeds readily. Boil off excess of chlorine, when the chromic acid may be precipitated as lead chromate or barium chromate (W. GIBBS†).]

§ 107.

Supplement to the Third Group.

TITANIUM.

Titanium is always weighed as titanic oxide (TiO_2), *i.e.*, the oxide or anhydride corresponding to titanic acid ($Ti(OH)_4$). Titanic acid is precipitated with an alkali or by boiling its dilute acid solution. In *precipitating* acid solutions of titanic acid ammonia is employed; take care to add the precipitating agent only in slight excess, let the precipitate formed, which resembles aluminium hydroxide, deposit, wash, first by decantation, then completely on the filter, dry, and ignite (§ 52). If the solution contained sulphuric acid, put some ammonium carbonate into the crucible, after the first ignition, to secure the removal of every remaining trace of that acid. Lose no time in weighing the ignited titanic oxide, as it is slightly hygroscopic. Occasionally it is more convenient to precipitate titanic acid from its acid solutions by nearly neutralizing with ammonia, adding sodium acetate and boiling. The precipitate thus obtained is easily filtered and washed. If we have titanic acid dissolved in sulphuric acid, as for instance occurs when we fuse it with potassium disulphate and treat the mass with cold water, we may, by largely diluting, and long boiling, with renewal of the evaporating water, fully precipitate the titanic acid. If much free acid is present it must be nearly neu-

* Comp. rend. 43, 927. † [Am. Journ. Sci. 2 Ser. 39, 58.]

tralized with ammonia before boiling. In the process of igniting the dried precipitate, some ammonium carbonate is added. From dilute hydrochloric acid solutions of titanic acid, the latter separates completely only upon evaporating the fluid to dryness; and if the precipitate in that case were washed with pure water, the filtrate would be milky; acid must, therefore, be added to the water.

Titanic acid precipitated in the cold, washed with cold water, and dried without elevation of temperature, is completely soluble in hydrochloric acid; otherwise it dissolves only incompletely in that acid. The *metatitanic* acid thrown down from dilute acid solutions by boiling, is not soluble in dilute acids. Titanic oxide resulting from ignition of titanic or metatitanic acid does not dissolve even in concentrated hydrochloric acid, but it does dissolve by long heating with tolerably concentrated sulphuric acid. The easiest way of effecting its solution is to fuse it for some time with potassium disulphate, and treat the fused mass with a large quantity of cold water. Upon fusing with sodium carbonate, sodium titanate is formed, which, when treated with water, leaves acid sodium titanate, which is soluble in hydrochloric acid. Titanic oxide (TiO_2) consists of 60·98 per cent. of titanium, and 39·02 per cent. of oxygen. By fusing titanic oxide with three times it quantity of potassium hydrogen fluoride, potassium titanium fluoride is formed, which readily dissolves in very dilute hydrochloric acid (of sp. gr. 1·015) in the heat. On fusing a very low heat must be applied at first, till the excess of hydrofluoric acid has escaped, then the heat is quickly raised till the mass melts and the titanic oxide is just dissolved (MARIGNAC*). On heating with hydrofluoric and sulphuric acids practically no titanium fluoride escapes, but by heating with hydrofluoric acid some loss does occur (RILEY†).

---

* Zeitschr. f. anal. Chem. 7, 112. † *Ib.* 2, 71.

FOURTH GROUP OF BASIC RADICALS.

ZINC—MANGANESE—NICKEL—COBALT—FERROUS IRON—FERRIC IRON —(URANIUM AND URANYL).

§ 108.

1. Zinc.

a. Solution.

Many of the zinc salts are soluble in water. Metallic zinc, zinc oxide, and the salts, which are insoluble in water, dissolve in hydrochloric acid.

b. Determination.

Zinc is weighed either as *oxide* or as *sulphide* (§ 77). The conversion of zinc salts into the oxide is effected either by precipitation as basic zinc carbonate or sulphide, or by direct ignition. Besides these gravimetric methods, several volumetric methods are in use.

We may convert into

1. Zinc Oxide.

a. By Precipitation as Zinc Carbonate.

All zinc salts which are soluble in water, and all zinc salts of organic volatile acids; also those salts of zinc which, insoluble in water, dissolve in hydrochloric acid, with separation of their acid.

b. By Precipitation as Zinc Sulphide.

All compounds of zinc without exception.

c. By direct Ignition.

Zinc salts of volatile inorganic oxygen acids.

2. Zinc Sulphide.

All compounds of zinc without exception.

The method 1, *c*, is to be recommended only, as regards the more frequently occurring compounds of zinc, for the carbonate and the nitrate. The methods 1, *b*, or 2, are usually only resorted to in cases where 1, *a*, is inadmissible. They serve more especially to separate zinc from other basic radicals. Zinc salts of organic acids cannot be converted into the oxide by ignition, since this process would cause the reduction and volatilization of a small portion of the metal. If the acids are volatile, the zinc may be determined at once, according to method 1, *a*: if, on the contrary, the acids are non-volatile, the zinc is best precipitated as sulphide. For the analysis of zinc chromate, phosphate, borate, and silicate, look to the several acids. The volumetric methods are chiefly employed for technical purposes; see Special Part.

1. *Determination as Zinc Oxide.*

a. By Precipitation as Zinc Carbonate.

Heat the moderately dilute solution nearly to boiling in a *capacious* vessel,—a glass vessel is poorly adapted for this purpose, porcelain is better, and platinum best;—add, drop by drop, sodium carbonate till the fluid shows a strong alkaline reaction; boil a few minutes; allow to subside, decant through a filter, and boil the precipitate three times with water, decanting each time; then transfer the precipitate to the filter, wash completely with hot water, dry, and ignite as directed § 53, taking care to have the filter as clean as practicable, before proceeding to incinerate it. Should the solution contain ammonium salts, the ebullition must be continued until, upon a fresh addition of sodium carbonate, the escaping vapor no longer imparts a brown tint to turmeric paper. If the quantity of ammonium salts present is considerable, the fluid must be evaporated *boiling* to dryness. It is, therefore, in such cases more convenient to precipitate the zinc as sulphide (see *b*).

The presence of a great excess of acid in the solution of zinc must be as much as possible guarded against, that the effervescence from the escaping carbonic acid gas may not be too impetuous. The filtrate must always be tested with ammonium sulphide (with addition of ammonium chloride) to ascertain whether the whole of the zinc has been precipitated; a *slight* precipitate will indeed *invariably* form upon the application of this test; but, if the process has been properly conducted, this is so insignificant that it may be altogether disregarded, being limited to some exceedingly slight and imponderable flakes, which moreover make their appearance only after many hours' standing. If the precipitate is more considerable, however, it must be treated as directed in *b*, and the weight of the zinc oxide obtained added to that resulting from the first process. For the properties of the precipitate and residue, see § 77. This method yields pretty accurate results, though they are in most cases a little too low, as the precipitation is never *absolutely* complete, and as particles of the precipitate will always and unavoidably adhere to the filter, which exposes them to the chance of reduction and volatilization during the process of ignition. On the other hand, the results are sometimes too high; this is owing to defective washing, as may be seen from the alkaline reaction which the residue manifests in such cases. It is advisable also to ascertain whether the residue will dissolve in hydrochloric acid without leaving silica; this latter precaution is indispensable in cases where the precipitation has been effected in a glass vessel.

[It is often better, especially in presence of ammonium salts, to heat the dry zinc salt with excess of sodium carbonate in a platinum dish cautiously to near redness, then treat with hot water and wash as directed.]

b. By Precipitation as Zinc Sulphide.

Mix the solution, contained in a not too large flask and sufficiently diluted, with ammonium chloride, then add ammonia, till the reaction is just alkaline, and then colorless or slightly yellow ammonium sulphide in moderate excess. If the flask is not now quite full up to the neck, make it so with water, cork, allow to stand 12 to 24 hours in a warm place, wash the precipitate, if considerable, first by decantation, then on the filter with water containing ammonium sulphide and also less and less ammonium chloride (finally none). In decanting do not pour the fluid through the filter, but at once into a flask. After thrice decanting, filter the fluid that was poured off, and then transfer the precipitate to the filter, finishing the washing as directed. The funnel is kept covered with a glass plate. If the zinc is not to be determined according to 2, then put the moist filter with the precipitate in a beaker, and pour over it moderately dilute hydrochloric acid slightly in excess. Put the glass now in a warm place, until the solution smells no longer of hydrogen sulphide; dilute the fluid with a little water, filter, wash the original filter with hot water, and proceed with the solution of zinc chloride obtained as directed in *a*.

The following method also effects a practically complete precipitation of zinc from acid solution. Add sodium carbonate, at last drop by drop till a lasting precipitate forms, dissolve the latter by a drop of hydrochloric acid, pass hydrogen sulphide till the precipitate ceases to increase perceptibly, add sodium acetate, and again pass the gas. After washing with water containing hydrogen sulphide (which when the zinc sulphide had been thrown down by hydrogen sulphide from acetic acid solution, is easily done), treat as above directed.

From a solution of zinc acetate the metal may be precipitated completely, or nearly so, with hydrogen sulphide gas, even in presence of an excess of acetic acid, provided always no other free acid be present (Expt. No. 74). The precipitated zinc sulphide is washed with water impregnated with hydrogen sulphide, and, for the rest, treated exactly like the zinc sulphide obtained by precipitation with ammonium sulphide.

Small quantities of zinc sulphide may also be converted directly into the oxide, by heating in an open platinum crucible, to gentle redness at first, then, after some time, to most intense redness.

c. By direct Ignition.

The salt is exposed, in a covered platinum crucible, first to a gentle heat, finally to a most intense heat, until the weight of the residue remains constant. The action of reducing gases is to be avoided.

2. *Determination as Zinc Sulphide.*

The precipitated zinc sulphide, obtained as in 1, *b*, may be ignited in hydrogen and weighed. H. Rose,* who has lately recommended the process, employs the apparatus represented by fig. 50.

a contains concentrated sulphuric acid, *b*, calcium chloride. The porcelain crucible has a perforated porcelain or platinum cover, into the opening of which fits the porcelain or platinum tube, *d*. The latter is provided with an annular projection which rests on the cover, the tube itself extends some distance into the crucible. When the zinc sulphide has dried in the filter, it is transferred to the weighed porcelain crucible, the filter ashes added, powdered sulphur is sprinkled over the contents of the crucible, the cover is placed on, and hydrogen is passed in a moderate stream, a gentle heat is applied at first, which is afterwards raised for five minutes to intense redness; finally the crucible is allowed to cool with continued transmission of the gas, and the zinc sulphide is weighed.

* Pogg. Anal. 110, 128.

[Instead of the porcelain tube and perforated cover, a common tobacco-pipe may be employed, the bowl of the latter being inverted over or within a porcelain crucible. Hydrogen sulphide may be advantageously substituted for hydrogen.]

OESTEN'S experiments, which were adduced by ROSE in support of the accuracy of this method, were highly satisfactory.

Zinc sulphate, carbonate, and oxide may be converted into sulphide in the manner just described. They must, however, be mixed with an excess of powdered sulphur, otherwise you will lose some zinc from the reducing action of the hydrogen (H. ROSE.)

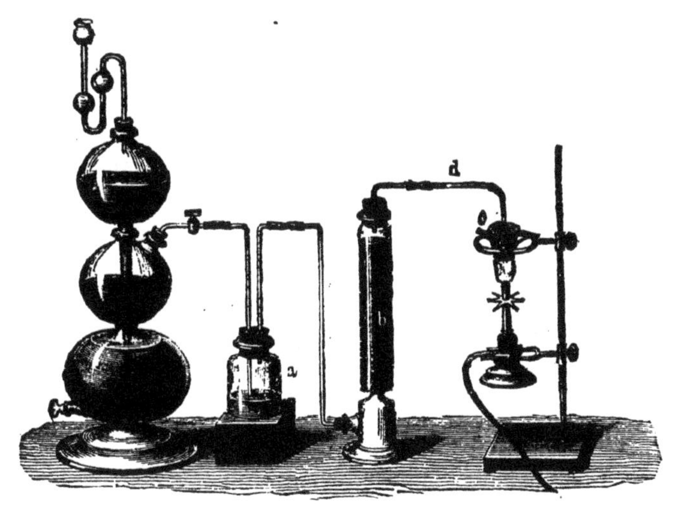

Fig. 50.

The properties of the hydrated and anhydrous zinc sulphide are given § 77; the results are accurate. Loss occurs only when the ignition is performed over the gas blowpipe (which is quite unnecessary), and continued longer than five minutes. Compare § 77, *c*.

§ 109.

2. MANGANESE.

a. Solution.

Many manganous salts are soluble in water. The manganous salts which are insoluble in that menstruum, dissolve in hydrochloric acid, which dissolves also all oxides of manganese. The solution of the higher oxides is attended with evolution of chlorine—equivalent to the amount of oxygen which the oxide under examination contains, more than manganous oxide (MnO)—and the fluid, after application of heat, is found to contain manganous chloride.

b. Determination.

Manganese is weighed either as *protosesquioxide*, as *sulphide*, or as *pyrophosphate* (§ 78). Into the form of protosesquioxide it is converted either by precipitation as manganous carbonate, or as manganous hydroxide, sometimes preceded by precipitation as manganous sulphide, or as manganese dioxide; or, finally, by direct ignition. [When determined as pyrophosphate it is precipitated as ammonium manganous phosphate.]

Manganese may be determined volumetrically in two different ways, one being applicable to any manganous solution, provided it be free from any other substance which exerts a reducing action on alkaline solution of potassium ferricyanide, the other being only admissible, when we have manganese in the condition of a perfectly definite higher oxide, and free from other bodies, which evolve chlorine on boiling with hydrochloric acid.

We may convert into

1. Manganese Protosesquioxide.

a. By Precipitation as Manganous Carbonate.

All soluble manganous salts of inorganic acids, and all manganous salts of volatile organic acids; also those manganous salts which, insoluble in water, dissolve in hydrochloric acid with separation of their acid.

b. By Precipitation as Manganese Hydroxide.

All the compounds of manganese, with the exception of its salts of non-volatile organic acids.

c. By Precipitation as Manganese Sulphide.

All compounds of manganese without exception.

d. By Separation as Manganese Dioxide.

All compounds of manganese in a slightly acid solution, especially manganous acetate and nitrate.

e. By direct Ignition.

All manganese oxides; manganous salts of readily volatile acids, and organic acids.

2. Manganese Sulphide.

All compounds of manganese without exception.

3. Manganese Pyrophosphate.

All the oxides of manganese and many manganous salts.

The method 1, *e*, is simple and accurate, but seldom admissible. The method 1, *a*, is the most usually employed; if one's choice is free, it is to be preferred to 1, *b*. The methods 1, *c*, and 2, are generally used, when the methods 1, *a*, or *b*, cannot be adopted—say on account of the presence of a non-volatile organic substance, and also when we have to separate manganese from other metals. The latter object may be attained also by the method 1, *d*. The process 3, is very convenient and accurate in absence of alkali-earth metals, and heavy metals. Manganous phosphate and borate are treated, either according to the method 1, *b*, as the salts precipitated from acid solution by potassa are completely decomposed upon boiling with excess of potassa, or according to the method 2. In silicates the manganese is determined after the separation of the silicic acid (§ 140), according to 1, *a*, or 3; for the analysis of manganous chromate, see § 130 (chromic acid). The volumetric method by reduction of potassium ferricyanide is especially suited for technical work, in which the highest degree of accuracy is not required. The estimation of manganese from the quantity of chlorine disengaged upon boiling the oxides with hydrochloric acid, is resorted to, more particularly, to determine the degrees of oxidation of manganese, and permits also the estimation of manganese in presence of other metals (see Section V.).

1. *Determination as Protosesquioxide of Manganese.*

a. By Precipitation as Manganous Carbonate.

The precipitation and washing are effected in exactly the same way as directed § 108, 1, *a* (determination of zinc as oxide, by precipitation as carbonate). If the filtrate is not absolutely clear, stand it in a warm place for twelve to twenty-four hours. A slight precipitate will then separate, which is collected on another small filter. The precipitate is dried, and then ignited as directed § 53. The lid is removed from the crucible, and a strong heat maintained until the weight of the residue remains constant. Care must be taken to prevent reducing gases finding their way into the crucible. For the properties of the precipitate and residue, see § 78. This method, if properly executed, gives accurate results.

The principal point is to continue the application of a sufficiently intense heat long enough to effect the object in view. It is necessary also to ascertain whether the residue has not an alkaline reaction, and having removed it from the platinum crucible, whether it dissolves in hydrochloric acid without leaving silica.

b. By Precipitation as Manganous Hydroxide.

The solution should not be too concentrated, and it is best to have it in a platinum dish. Precipitate with solution of pure soda or potassa, and proceed in all other respects as in *a*.

If phosphoric acid is present, or boracic acid, the fluid must be kept boiling for some time with an excess of alkali. For the properties of the precipitate, see § 78.

c. By Precipitation as Manganese Sulphide.

The solution contained in a comparatively small flask and not too dilute is first mixed with ammonium chloride (if an ammonium salt is not already present in sufficient quantity), then—if the fluid is acid—with ammonia, till it reacts neutral or very slightly alkaline; now add yellow ammonium sulphide, in moderate excess, if the flask is not already quite full up to the neck, add water till it is, cork, stand it in a warm place for at least twenty-four hours, wash the precipitate if at all considerable, first by decantation, then on the filter, using water containing ammonium sulphide, and also gradually-diminished quantities of ammonium chloride (finally none). In decanting, pour the fluid in a flask, not on the filter. After decanting three times, filter the fluids that have been poured off, transfer the precipitate to the filter, and finish the washing as above directed, without interruption. Keep the funnel covered with a glass plate. If you do not prefer to determine according to 2, proceed as follows:—Put the moist filter with the precipitate into a beaker, add hydrochloric acid, and warm until the mixture smells no longer of hydrogen sulphide; filter, wash the residuary paper carefully, and precipitate the filtrate as directed in *a*. The results are satisfactory, compare § 78, *e*.

Tartaric acid retards the precipitation, but does not render it less complete; citric acid prevents precipitation, or at least makes it incomplete.

d. By Separation as Manganese Dioxide.

Heat the solution of manganous acetate or some other manganous salt containing but little free acid, after addition of a sufficient quantity of sodium acetate, to from 50° to 60°, and transmit chlorine gas through the fluid, or add bromine (KÄMMERER,* WAAGE†). The whole of the manganese present falls down as dioxide (SCHIEL, RIVOT, BEUDANT, and DAGUIN). Wash, first by decantation, then upon the filter; dry, transfer the precipitate to a flask, add the filter ash, heat with hydrochloric acid, filter, and precipitate as directed in *a*. If the sodium acetate is deficient, and especially if hydrochloric acid is present, it may happen that the precipitation of the manganese by chlorine or bromine is not quite complete; it is therefore well, after filtering off the dioxide, to treat the filtrate with more sodium acetate, and again pass chlorine or add bromine. The separation of manganese as dioxide, by evaporating its solution in nitric acid to dryness, and heating the residue, finally to 155°, is given in Section V.

e. By direct Ignition.

The manganese compound under examination is introduced into a platinum crucible, which is kept closely covered at first, and exposed to a gentle heat; after a time the lid is taken off, and replaced loosely on the crucible, and the heat is increased to the highest degree of intensity, with careful exclusion of reducing gases; the process is continued until the weight of the residue remains constant. The conversion of the higher oxides of manganese into protosesquioxide of manganese requires more protracted and intense heating than the conversion of manganous oxide. In fact, it can hardly be effected without the use of a gas blowpipe. In the case of manganous salts of organic acids, care must always be taken to ascertain whether the whole of the carbon has been consumed; and should the contrary turn out to be the case the residue must either be dissolved in hydrochloric acid, and the solution precipitated as directed in *a*, or 3, or it must be repeatedly evaporated with nitric acid, until the whole of the carbon is oxidized. The method, if properly executed, gives accurate results. On the other hand, if the directions are not carefully attended to, one must not be surprised at considerable differences. In the ignition of manganous salts of organic acids, minute particles of the salt are generally carried away with the empyreumatic products evolved in the process, which, of course, tends to reduce the weight a little.

* Ber. der deutsch. Chem. Gesellsch. 4. 218.
† Zeitschr. f. anal. Chem 10. 206

2. *Determination as Manganous Sulphide.*

The sulphide precipitated as in 1, *c*, may be determined in this form, as follows: Dry, transfer the precipitate to a crucible, burn the filter, add the ashes, strew some sulphur on the top, ignite strongly in hydrogen (till it becomes black) and weigh as anhydrous manganous sulphide (H. ROSE*), compare the analogous process for zinc, § 108, 2.

The results obtained by OESTEN, and cited by ROSE, are perfectly satisfactory.

This method is shorter and more convenient than dissolving the moist sulphide in hydrochloric acid, and precipitating with sodium carbonate.

Manganous sulphate and all the oxides of manganese may be subjected to this process with the same result.

[3. *Determination as Manganous Pyrophosphate.*

To the solution of the manganous salt, which may contain ammonium or alkali salts, sodium phosphate is added in large excess above what is needful to convert the manganese into phosphate. The white precipitate which is formed unless considerable free acid is already present is then redissolved in sulphuric or chlorhydric acid, the liquid is heated to boiling, best in a platinum dish, and ammonia added in excess. The boiling is continued 10—15 minutes, whereby the white, semi-gelatinous precipitate first formed is converted into rose-colored, pearly scales. If one is obliged to precipitate in a glass beaker the precipitate may be converted into the crystalline form more safely by heating on the water-bath 1 or 2 hours, as it is likely to be thrown out of the beaker by boiling. The whole is kept hot for an hour longer, then filtered and washed with water containing a little ammonia. The precipitate of ammonium manganous phosphate is dried, separated from the filter, and converted by ignition into pyrophosphate. See § 78 (GIBBS,† HENRY‡).]

It is advantageous to use the Bunsen filtering apparatus for washing the precipitate on account of its slight solubility in water. (See § 78, *g*.) For the same reason when great accuracy is required it is recommended to evaporate the filtrate to dryness, redissolve with water and hydrochloric acid, make alkaline with ammonia,

* Pogg. Anal. 110, 122. † Am. Jour. Sci. 2d Ser. 44, p. 216.
‡ *Ib.* 47, p. 130.

and boil to precipitate and recover the small amount of manganese which may have passed into the filtrate.

4. *Volumetric determination by the Reduction of Ferricyanide of Potassium* (E. LENSSEN*).

The method is grounded on the fact that if a solution of a manganous salt is acted on by excess of alkaline solution of potassium ferricyanide at a boiling temperature in the presence of a sufficient amount of a ferric salt, all the manganese is precipitated as dioxide, while a corresponding quantity of potassium ferrocyanide is formed. By determining the latter, the amount of manganese present is obtained.

$$K_6Fe_2Cy_{12}+2K_2O+MnSO_4=2K_4FeCy_6+K_2SO_4+MnO_2.$$

Accordingly 1 at. manganese gives rise to 2 mol. potassium ferrocyanide. Of course all other reducing substances must be absent, and the manganese must be present entirely in the form of a manganous salt. If the solution contains no ferric salt, the precipitate is a combination of much dioxide, with little manganous oxide, not always in the same proportions. In performing the process, mix first with the acid solution of the manganous salt so much ferric chloride that you may be sure of having at least 1 mol. Fe_2Cl_6 to 1 atom Mn, and add the mixture gradually to a boiling solution of potassium ferricyanide, previously rendered strongly alkaline with potassa or soda. After boiling together a short time the brownish-black precipitate becomes granular and less bulky. Allow to cool *completely*, filter off and wash the precipitate, acidify the filtrate with hydrochloric acid, and estimate the potassium ferrocyanide with permanganate, according to § 147, II., *g. α.* If the liquid is filtered hot, the results are too high, as the filter in this case has a reducing action. The method may be shortened, as follows: After boiling, transfer the solution, together with the precipitate, to a measuring flask, allow to cool, fill up to the mark with water, shake, and allow to settle. Filter through a dry filter, take out a certain quantity with a pipette, and determine the ferrocyanide in this. A slight source of error is here introduced by disregarding the volume of the precipitate. The results adduced by LENSSEN are very satisfactory. I have myself repeatedly tested this method, and I have to remark as follows:—

* Journ. f. prakt. Chem. 80, 408.

a. If potassium ferricyanide is long boiled with pure potassa, a small quantity of ferrocyanide is invariably produced.

b. The potassa must be quite free from organic substances, and should therefore, if there is any doubt on this point, be fused in a silver dish before use, otherwise the error alluded to in *a* may be considerably increased.

c. The complete washing of the voluminous precipitate is attended with so much difficulty and loss of time as to render the method more troublesome than a gravimetric analysis.

d. The abridged method, on the other hand, may be of great service in certain cases, especially when a series of manganese determinations have to be made, the manganese not being in too minute quantities, and the highest degree of accuracy not being required. In my laboratory, by employing a slight excess of ferric salt, 97·9—100·12—98·21—98·99, and 100·4 were obtained, instead of 100. The inaccuracy increases on using a large excess of the iron.*

5. *Volumetric determination by boiling the higher oxides with hydrochloric acid, and estimating the chlorine evolved.*

The methods here employed will be found all together in the Special Part under "Valuation of Manganese Ores."

§ 110.

3. NICKEL.

a. Solution.

Many nickelous salts are soluble in water. Those which are insoluble, as also nickelous oxide, in its common modification, dissolve, without exception, in hydrochloric acid. The peculiar modification of nickelous oxide, discovered by GENTH, which crystallizes in octahedra, does not dissolve in acids, but is rendered soluble by fusion with potassium disulphate. Metallic nickel dissolves slowly, with evolution of hydrogen gas, when warmed with dilute hydrochloric or sulphuric acid; in nitric acid, it dissolves with great readiness. Nickel sulphide is but sparingly soluble in hydrochloric acid, but it dissolves readily in nitrohydrochloric acid. Nickelic oxide (Ni_2O_3) dissolves in hydrochloric acid, upon the application of heat, to nickelous chloride, with evolution of chlorine.

* Zeitschr. f. Anal. Chem. 3, 209.

b. Determination.

Nickel is best weighed as metal; it may be weighed also as nickelous oxide, or sulphate. The compounds of nickel are converted into nickelous oxide, usually by precipitation as nickelous hydroxide, preceded, in some instances, by precipitation as nickel sulphide, or by ignition.

We may convert into

1. Nickelous Oxide.

a. By Precipitation as Nickelous Hydroxide.

All nickel salts of inorganic acids which are soluble in water, and all its salts of volatile organic acids; likewise all salts of nickel which, insoluble in water, dissolve in the stronger acids, with separation of their acid.

b. By Precipitation as Nickel Sulphide.

All compounds of nickel without exception.

c. By Ignition.

Nickel salts of readily volatile oxygen acids, or of such oxygen acids as are decomposed at a high temperature (carbonic acid, nitric acid).

2. Metallic nickel: Nickelous oxide, also nickel chloride, bromide, and iodide.

3. Nickel sulphate: Nickel salts, whose acids are entirely expelled by heating and evaporating with sulphuric acid.

The method 1, *c*, is very good, but seldom admissible. The method 1, *a*, is most frequently employed. In the presence of sugar, or other non-volatile organic substance, it cannot be used. In this case we must either ignite and thereby destroy the organic matter before precipitating, or we must resort to the method *b*, which otherwise is hardly used except in separations. By whatever method nickelous oxide is obtained, it is best to convert it into metallic nickel (by method 2) before weighing. The conversion into nickel sulphate (method 3) is quickly executed, but it requires the greatest care to obtain trustworthy results. Nickel salts of chromic, phosphoric, boracic, and silicic acids are analyzed according to the methods given under the several acids.

1. *Determination as Nickelous Oxide.*

a. By Precipitation as Nickelous Hydroxide.

Mix the solution with pure solution of potassa or soda in excess, heat for some time nearly to ebullition, decant 3 or 4 times, boiling up each time, filter, wash the precipitate *thoroughly* with hot water, dry and ignite strongly (avoiding reducing gases if the oxide is to be weighed) (RUSSELL*) (§ 53). The precipitation is best effected in a platinum dish; in presence of nitrohydrochloric acid, or, if the operator does not possess a sufficiently capacious dish of the metal, in a porcelain dish; glass vessels do not answer the purpose so well. Presence of ammoniacal salts, or of free ammonia, does not interfere with the precipitation. For the properties of the precipitate and residue, see § 79. Instead of weighing the oxide it is better to reduce it to metal according to § 110, 2. The thorough washing of the precipitate is a most essential point. It is necessary also to ascertain whether the weighed metal (or oxide) has not an alkaline reaction, and whether it dissolves completely in nitric acid (or hydrochloric in case oxide is weighed).

b. By Precipitation as Sulphide of Nickel.

[*α*. Add to the solution, which should be concentrated, a large quantity of ammonium chloride. The precipitation is effected more readily if enough ammonium chloride is present to make the solution nearly saturated when cold. Make the solution neutral or better slightly acid by addition of ammonia or hydrochloric acid as the case demands. Heat to boiling in a flask and add drop by drop ammonium sulphide (which should be more or less yellow and contain no free ammonia), not fast enough to check the boiling. Use the least possible excess of ammonium sulphide. Ascertain when enough has been added by stopping the boiling long enough for the nickel sulphide to settle, and adding a drop to the clear surface of the solution. If more is required, raise the heat to boiling before adding it. When further addition of ammonium sulphide produces no more precipitate boil a few minutes longer, and add enough acetic acid to give a decided acid reaction. Add next a little hydrogen sulphide solution and filter, washing with a dilute solution of hydrogen sulphide. Test the filtrate by neutralizing with ammonia and adding one or two drops of ammonium

* Journ. Chem. Soc. 16, 58.

sulphide. If this causes a blackening of the fluid, boil with addition a slight excess of acetic to separate the nickel.

β. Prepare the solution as above described (1, *b*, *α*). Add sodium or ammonium acetate, if acetates are not already present; heat to boiling; transmit H_2S gas through the boiling solution about ten minutes. The precipitated nickel sulphide settles readily. Ascertain whether nickel has been completely separated by adding a drop of ammonium sulphide to the clear surface of the liquid. If no blackening, or only a white cloud of sulphur appears, add a little cold strong solution of hydrogen sulphide in water, filter, wash the precipitate and test further the filtrate for nickel as above described. If, on the contrary, the drop of ammonium sulphide causes a black coloration (nickel sulphide) an incomplete separation of nickel is indicated, which may be due to the presence of too much free acetic acid. Add, therefore, a few drops of ammonia, leaving the solution however still slightly acid, heat again to boiling and pass hydrogen sulphide, and so proceed till complete precipitation is effected. It should be borne in mind that a large amount of free acetic prevents precipitation of nickel as sulphide, while a small amount does not. The slight quantity of acetic present throughout the operation prevents the formation of ammonium sulphide which is a solvent for NiS, and also prevents the precipitation of the alkali-earth metals if they are present.]

Dry the washed nickel sulphide in the funnel, and transfer from the filter to a beaker; the filter is incinerated in a porcelain crucible and added to the dry precipitate. The precipitate is now treated with concentrated nitrohydrochloric acid, and the mixture digested at a gentle heat, until the whole of the nickel sulphide is dissolved, and the undissolved sulphur appears of a pure yellow; the fluid is then diluted, filtered, and the filtrate precipitated as directed in 1, *a*, and the nickel oxide thus obtained is reduced to metal according to directions in 2.

c. By direct Ignition.

The same method as described § 109, 1, *e*. (Manganese.)

2. *Determination as metallic Nickel.*

Ignite the oxide or chloride to be reduced in a porcelain crucible in a slow stream of hydrogen, at first gently, then more strongly till the weight is constant. For properties of the residue, see § 79, *c*. If on dissolving the metal in nitric acid any silica remains, this must be weighed and deducted.

3. *Determination as Nickel Sulphate.*

The nickel solution should be free from other non-volatile salts. Evaporate with a slight excess of pure sulphuric acid in a platinum dish to dryness and heat for 15 or 20 minutes moderately, so as just to drive off the excess of sulphuric acid without blackening the yellow sulphate at the edges. It is difficult to be sure of hitting the exact point, hence we cannot place dependence on this method nor on that of GIBBS, which consists in dissolving the sulphide in nitric acid and evaporating the solution with sulphuric acid. For the properties of the residue, see § 79, *d*.

§ 111.

4. COBALT.

a. Solution.

Cobalt and its compounds behave with solvents like the corresponding compounds of nickel. The protosesquioxide of cobalt obtained by SCHWARZENBERG in microscopic octahedra does not dissolve in boiling hydrochloric acid, or nitric acid, or nitrohydrochloric acid; but it dissolves in concentrated sulphuric acid, and in fusing potassium disulphate.

b. Determination.

Cobalt is determined in the metallic state or as sulphate, being usually first precipitated as cobaltous hydroxide, sulphide or tripotassium cobaltic nitrate.

We may convert into

1. METALLIC COBALT:

a. By direct reduction. All salts of cobalt, which can be immediately reduced by hydrogen (chloride, nitrate, carbonate, &c.).

b. By precipitation as cobaltous hydroxide. All salts soluble in water of inorganic acids, and insoluble salts of such acids as may be removed by solution. All salts of volatile organic acids.

c. By precipitation as sulphide. All compounds of cobalt without exception.

d. By precipitation as tripotassium cobaltic nitrite. All compounds of cobalt soluble in water or dilute acetic acid.

2. Cobalt sulphate:

a. By simple evaporation and ignition.—The oxygen compounds of cobalt and all cobaltous salts of acids which may be completely expelled by evaporation and ignition with sulphuric acid.

b. By precipitation as sulphide.—All compounds of cobalt without exception.

The method 1, *a*, is preferable to all others when it can be applied; it is quick and gives exact results. The method 1, *b*, gives better results than it used to be credited with. The direct conversion of suitable cobalt compounds into sulphate is also quite satisfactory. The precipitations as sulphide and as tripotassium cobaltic nitrate are rarely used except in separations.

1. *Determination as metallic Cobalt.*

a. By direct reduction.

Evaporate the solution of cobaltous chloride, or nitrate (which must be free from sulphuric acid and alkali), in a weighed crucible, to dryness, cover the crucible with a lid having a small aperture in the middle, conduct through this a moderate current of pure dry hydrogen, and then apply a gentle heat, which is to be increased gradually to intense redness. When the reduction is considered complete, allow to cool in the current of hydrogen, and weigh; ignite again in the same way and repeat the process until the weight remains constant. The results are accurate. For the properties of cobalt, see § 80.

As regards the apparatus to be employed, see § 108, 2.

b. By precipitation as cobaltous hydroxide.

The best material for the precipitating vessel is platinum, porcelain may also be used, but not glass. First remove any large excess of acid which may be present by evaporation. Heat nearly to boiling, add pure potash in slight excess, and continue heating till the precipitate is brownish-black. Pour the supernatant fluid through a filter, wash the precipitate by decantation with boiling water repeatedly, transfer it to the filter, and continue the washing with boiling water till the washings are free from any trace of dissolved substance. Dry, ignite in a porcelain crucible (§ 52) till the filter is thoroughly burnt, reduce in a current of hydrogen, wash the metal several times with boiling water, dry, ignite again in hydrogen and weigh. Test the weighed cobalt by dissolving in nitric acid. If any silica remains, this must be weighed and deducted. Mix the solution with ammonium chloride and ammonium carbonate, if a small precipitate (aluminium or ferric hydroxide) forms, ignite and weigh this too and deduct it. The results are excellent; the amount of alkali which remains with the metal when the work is done properly being exceedingly minute. Compare § 80, *a*.

c. By precipitation as sulphide.

Put the solution in a flask, add ammonium chloride, then ammonia just in excess, then ammonium sulphide as long as a precipitate is produced, fill up to the neck with water, cork and allow to stand 12 or 24 hours in a warm place. Decant, filter, and wash as directed § 109, 2. Finally, dry and proceed as directed § 110, *b*, *β*, to redissolve the cobalt sulphide. Determine the cobalt according to *b*. There are no sources of error in the precipitation with ammonium sulphide. For the properties of cobalt sulphide, see § 80. It cannot be brought into a weighable form by ignition in hydrogen, as the residue is a variable mixture of different sulphides (H. Rose). Cobalt may also be thrown down as sulphide by the other methods given under *Nickel*. The thorough precipitation of cobalt is much easier than that of nickel.

d. By precipitation as tripotassium cobaltic nitrate.

To the moderately concentrated solution of the cobalt salt add potash in excess, then acetic acid till the precipitate is just redissolved, then a concentrated solution of potassium nitrite previously just acidified with acetic acid, and allow to stand 24 hours at a gentle heat. Filter, wash with solution of potassium acetate (1 in 10) containing some potassium nitrite, till all foreign substances are removed, dry, dissolve with the filter ash in hydrochloric acid, filter and determine the cobalt according to 1, *b*. This method was introduced by A. Stromeyer;* the present modification, first suggested by H. Rose, and improved by Fr. Gauhe, is the surest to yield good results (Gauhe†). For the properties of the precipitate, see § 80, *e*.

2. *Determination as sulphate.*

a. By direct conversion.

Add to the solution a little more sulphuric than will suffice to form cobaltous sulphate with all the cobalt present. Evaporate,

* Annal. d. Chem. u. Pharm. 96, 218. † Zeitschr. f. anal. Chem. 4, 60.

using a platinum dish or platinum crucible, at all events, to finish the operation. Heat the residue cautiously over the lamp, gradually increasing the temperature to dull redness, and maintain at this point for 15 minutes. Should the edges blacken, moisten with dilute sulphuric acid, dry, and ignite again with greater caution. Properties of the precipitate, § 80. Results quite satisfactory.*

b. With previous precipitation as sulphide.

Precipitate the cobalt as sulphide according to 1, *c*, dissolve it as directed, evaporate with excess of sulphuric acid in a porcelain dish to dryness, take up the residue with water, transfer the solution to a weighed platinum dish and proceed according to 2, *a*.

§ 112.

5. Ferrous Iron.

a. Solution.

Many ferrous compounds are soluble in water. Those which are insoluble in water dissolve almost without exception in hydrochloric acid; the solutions, if not prepared with perfect exclusion of air, and with solvents absolutely free from air, contain invariably more or less ferric chloride. In cases where it is wished to avoid the chance of oxidation, the solution of the ferrous compound is effected in a small flask, through which a slow current of carbonic acid gas is passed, the transmission of the gas being continued until the solution is cold. Many native ferrous compounds cannot be thus dissolved. They are, indeed, rendered soluble by fusing with sodium carbonate, but in this process ferric oxide is formed. It is therefore advisable to heat such substances (in the finest powder) with a mixture of 3 parts concentrated sulphuric acid and 1 part water in a strong sealed tube of Bohemian glass for 2 hours at about 210°, or—in the case of silicates—to warm them with a mixture of 2 parts hydrochloric acid and 1 part strong hydrofluoric acid in a covered platinum dish (A. Mitscherlich†. See also Cooke's method of solution, § 160, 84). Metallic iron dissolves in hydrochloric acid, and in dilute sulphuric acid, with evolution of hydrogen, as ferrous chloride or sulphate respectively; in warm nitric acid it dissolves as ferric nitrate, and in nitrohydrochloric acid as ferric chloride.

---

* Compare Gauhe, Zeitschr. f. anal. Chem. 4, 55.

† Journ. f. prakt. Chem. 81, 116.

b. Determination.

Ferrous iron may be estimated 1, by dissolving, converting into ferric iron, and determining the latter gravimetrically or volumetrically; 2, by precipitating as sulphide, and weighing it as such, or determining it after conversion into a ferric salt; 3, by a direct volumetric method.

The methods 1 and 2 are, of course, only applicable when no ferric compound is present; the method 2 is scarcely ever used except for separations. The methods included under 3 are adapted to most cases, and, in absence of other reducing substances, are especially worthy of recommendation.

As the determination of iron as ferric oxide belongs to § 113, and as the process for precipitating ferrous iron as sulphide is the same as that for precipitating ferric iron in this form, nothing remains for us here but to describe the methods of converting ferrous into ferric salts and the processes included under 3.

1. *Methods of converting Ferrous into Ferric Iron.*

a. Methods, applicable in all cases.

Heat the solution of the ferrous salt with hydrochloric acid and add small portions of potassium chlorate, till the fluid, even after warming for some time, still smells strongly of chlorine. Our object may be also attained by passing chlorine gas or—in the case of small quantities—by addition of chlorine water, or very conveniently by adding solution of bromine in hydrochloric acid. If the solution is required to be free from excess of chlorine or bromine, it is finally heated, till all odor of chlorine or bromine has disappeared.

b. Methods which are only suitable when the iron is to be subsequently precipitated by ammonia, as ferric hydroxide.

Mix the solution of the ferrous salt in a flask with a little hydrochloric acid, if it does not already contain any; add some nitric acid, and heat the mixture for some time to incipient ebullition. The color of the fluid will show whether the nitric acid has been added in sufficient quantity. Though an excess of nitric acid does no harm, still it is better to avoid adding too much on account of the subsequent precipitation. In concentrated solutions, the addition of nitric acid produces a dark-brown color, which disappears upon heating. This color is owing to the nitrogen dioxide

(N_2O_3) formed dissolving in the portion of the solution which still contains ferrous salt.

c. Methods which can be employed only when the ferric iron is to be determined volumetrically.

Add to the hydrochloric solution small quantities of artificially prepared iron-free manganese dioxide, till the solution is of a dark olive-green color from the formation of manganic chloride; boil till this coloration and the odor of chlorine have disappeared (Fr. Mohr); or you may add pure potassium permanganate (in crystals or concentrated solution) till the fluid is just red and then boil, till the red color and chlorine-odor have vanished. These methods present the advantage of permitting complete conversion of ferrous into ferric salts without the use of any considerable excess of the oxidizing agent.

2. *Volumetric Determination.*

a. Marguerite's *Method.*

If we add to a solution of ferrous salt, containing an excess of sulphuric acid, potassium permanganate, the former is converted into a ferric salt by the oxidizing action of the latter ($10FeSO_4 + 8H_2SO_4 + K_2Mn_2O_8 = 5Fe_2(SO_4)_3 + K_2SO_4 + 2MnSO_4 + 8H_2O$). Now if we possess a solution of potassium permanganate, and know how much iron 100 c.c. of it can convert from the ferrous to the ferric condition, we can, with this, readily determine an unknown quantity of iron; we have simply, for this purpose, to dissolve the iron in acid, in the form of a ferrous salt, to oxidize the solution accurately, and note how many c.c. of the solution of potassium permanganate have been used to accomplish that object.

In the presence of hydrochloric acid (see γ), the change is not exactly in accordance with the above equation (Löwenthal and Lenssen*).

α. Titration of the Solution of Potassium Permanganate.

Dissolve 5 grm. (roughly weighed) of pure crystallized potassium permanganate in distilled water by the aid of heat, dilute to 1 litre, and preserve in a stoppered bottle. Action of direct sunlight on the solution should be avoided. The solution, if carefully kept, does not alter, but still it is well to titrate it afresh occasionally.

* Zeitschr. f. anal. Chem. 1, 329.

aa. Titration by Metallic Iron.

Weigh off accurately about 1 grm. thin soft iron wire, previously cleaned with emery paper, transfer to a $\frac{1}{4}$ litre measuring flask, containing 100 c.c. dilute sulphuric acid (1 to 5), add about 1 grm. sodium bicarbonate, to produce carbonic acid and expel the air, and then close the flask with an india-rubber stopper, provided with an evolution tube, as shown in fig. 51 ; *c* contains 20 or 30 c.c. water. Heat the flask at first gently, finally to gentle boiling till the iron is dissolved. The clip *b* is open, and the hydrogen escapes through the water in *c*. Meanwhile boil about 300 c.c. distilled water, to drive out all the air it contains, and then allow it to cool. As soon as the iron is entirely dissolved, remove the lamp and close the evolution tube with the clip. When the iron solution has cooled a little loose the clip, and allow the water in *c* to recede, pour the boiled water into *c*, and allow this also to recede till the solution nearly reaches the mark. Take out the evolution tube and close the flask with an unperforated stopper, allow to cool to the temperature of the room, fill with water to the mark, shake and allow to stand, so that the particles of carbon usually present may deposit. Now take out with a pipette 50 c.c. of the clear and nearly colorless fluid (containing $\frac{1}{5}$ of the iron weighed off), transfer to a 400 c.c. beaker, and dilute till the beaker is half full. Place the beaker on a sheet of white paper, or better, on a sheet of glass, with white paper underneath.

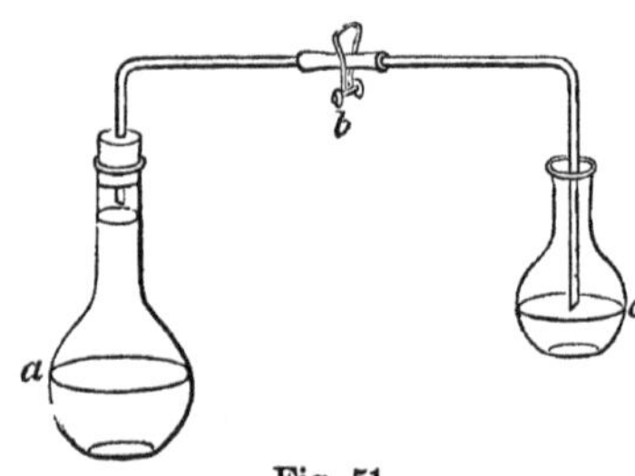

Fig. 51.

Fill a GAY-LUSSAC's or GEISSLER's burette of 30 c.c. capacity, divided into $\frac{1}{10}$ c.c. (see p. 41, figs. 13 and 14), up to zero, with solution of potassium permanganate, of which take care to have ready a sufficient quantity, perfectly clear and uniformly mixed.

Now add the permanganate to the ferrous solution, stirring the latter all the while with a glass rod. At first the red drops disappear very rapidly, then more slowly. The fluid, which at first was nearly colorless, gradually acquires a yellowish tint. From the instant the red drops begin to disappear more slowly, add the permanganate with more caution and in single drops, until the last drop imparts to the fluid a faint, but unmistakable reddish color, which remains on stirring. A little practice will enable you readily to hit the right point. As soon as the fluid in the burette has sufficiently collected again read off, and mark the number of c.c. used. The reading off must be performed with the greatest exactness (see § 22); the whole error should not amount to $\frac{1}{10}$ c.c.

The amount of permanganate solution used should be about 20 c.c. Repeat the experiment with another 50 c.c. of the iron solution. The difference between the permanganate used in the two cases should not be more than ·1 c.c.; if it is, make one more experiment and when the results are sufficiently near take the mean. Now calculate what quantity of iron is represented by 100 c.c. of the permanganate. To this end first divide the iron weighed off by 5, and then multiply by ·996, since soft iron wire contains on the average ·4 per cent. carbon, &c.; this gives the quantity of pure iron contained in 50 c.c. of the solution. Suppose we took 1·050 grm. iron wire, and used a mean of 21·3 c.c. permanganate, $\frac{1 \cdot 050}{5} = \cdot 210$, $\cdot 210 \times \cdot 996 = \cdot 20916$. And then by rule of three:—

$$21 \cdot 3 : \cdot 20916 :: 100 : x \text{——} x = \cdot 98197;$$

therefore, 100 c.c. permanganate = ·98197 pure iron.

If there is a deficiency of free acid in the solution of iron, the fluid acquires a brown color, turns turbid, and deposits a brown precipitate (manganese dioxide and ferric hydroxide). The same may happen also if the solution of potassium permanganate is added too quickly, or if the proper stirring of the iron solution is omitted or interrupted. Experiments attended with abnormal manifestations of the kind had always better be rejected. That the fluid reddened by the last drop of solution of potassium permanganate added, loses its color again after a time, need create no surprise or uneasiness; this decolorization is, in fact, quite inevitable, as a dilute solution of free permanganic acid cannot keep long undecomposed.

bb. Titration by Ammonium Ferrous Sulphate.

Weigh off, with the greatest accuracy, about 1·4 grm. of the pure salt prepared according to the directions given in § 65, 4, dissolve in about 200 c.c. distilled water, previously mixed with about 20 c.c. dilute sulphuric acid, and proceed as in *aa*.

By dividing the amount of salt weighed off by 7·0014 (or where great accuracy is not required by 7) we obtain the quantity of iron corresponding.

If the salt is not pure, if, for instance, it contains basic radicals isomorphous with ferrous iron (manganese, magnesium, &c.); or if it contains ferric iron, or is moist, the result will of course be too high.

cc. Titration by Oxalic Acid.

If solution of potassium permanganate is added to a warm solution of oxalic acid, mixed with sulphuric acid, the liberated permanganic acid oxidizes the oxalic acid to carbon dioxide and water [$5H_2C_2O_4 + K_2Mn_2O_8 + 3H_2SO_4 = K_2SO_4 + 2MnSO_4 + 10CO_2 + 8H_2O$]. For the oxidation of 1 mol. oxalic acid ($H_2C_2O_4$) and 2 at. iron (in the ferrous state) equal quantities of permanganic acid are accordingly required; therefore, 126 parts (1 mol.) of crystallized oxalic acid correspond, in reference to the oxidizing action of permanganic acid, to 112 parts (2 at.) of iron.

A solution of oxalic acid is altered by the action of light; it is, therefore, well only to dissolve as much as will be required for immediate use. Dissolve 1 to 1·2 grm. pure acid prepared by § 65, 1, to 250 c.c.; 50 c.c. of this solution are introduced into a beaker, diluted with about 100 c.c. water, from 6 to 8 c.c. conc. sulphuric acid added, and the fluid heated to about 60°. The beaker is then placed on a sheet of white paper, and permanganate added from the burette, with stirring. The red drops do not disappear at first very rapidly, but when once the reaction has fairly set in, they continue for some time to vanish instantaneously. As soon as the red drops begin to disappear more slowly, the solution of potassium permanganate must be added with great caution; if proper care is taken in this respect, it is easy to complete the reaction with a single drop of permanganate; this completion of the reaction is indicated with beautiful distinctness in the colorless fluid. To find the iron corresponding to the permanganate used, multiply the amount of crystallized oxalic acid in the 50 c.c. by 8 and divide by 9.

If the oxalic acid was not perfectly dry, or not quite pure, the result of the experiment will, of course, lead to fixing the strength of the solution of potassium permanganate too high. Instead of pure oxalic acid, SAINT-GILLES has proposed to use crystallized oxalate of ammonium $(NH_4)_2C_2O_4 + H_2O$). This can easily be prepared in the pure state,[1] keeps well, and can be weighed with accuracy. 142·08 parts of the crystallized salt correspond to 112 parts iron.

Of the foregoing three methods of standardizing solution of potassium permanganate, the first is the one originally proposed by MARGUERITE. Ammonium ferrous sulphate was first proposed by FR. MOHR, and oxalic acid by HEMPEL, as agents suitable for the purpose. With absolutely pure and thoroughly dry reagents, and proper attention, all three methods give correct results.

For myself, I prefer the first method, as the most direct and positive, the only doubtful point about it being the question whether the assumption that the iron wire contains 99·6 per cent. of chemically pure iron is quite correct; this, however, is of very trifling importance, as the error could not exceed $\frac{1}{10}$ or $\frac{2}{10}$ per cent.* The other two methods are, as may readily be seen, somewhat more convenient, but they are not so trustworthy unless you can insure the purity and dryness of the preparations.

For the analysis of very dilute solutions of iron, *e.g.*, chalybeate water, in which the amount of iron may be very approximately determined with great expedition, by direct oxidization with permanganate, a very dilute standard solution must be prepared. Such a solution may be made by diluting the previous solution with 9 parts of water or by dissolving ·5 grm crystals of potassium permanganate in 1 litre of water. It is to be directly standardized with correspondingly small quantities of iron, ferrous salt, or oxalic acid.

In experiments of this kind, the fact that a certain quantity of permanganate is required to impart a distinct color to pure acidified water (which is of no consequence in operations where the concentrated solution is used) must be taken into consideration; for where the solution used is so highly dilute, it takes indeed a measurable quantity of it to impart the desired reddish tint to the amount of water employed. In such cases, the volume of the solution of iron used for standardizing the permanganate and the volume of the weak ferruginous solution subjected to analysis should be the same, and either the two solutions should contain about the same quantity of iron, or by means of a special experiment, it is ascertained how many $\frac{1}{10}$ c.c. of the permanganate are required to

---

* If you are often making iron determinations, you may of course procure a quantity of wire and determine the amount of the foreign matter in it.

impart the desired pale red color to the same volume of acidified water. In the latter case, these $\frac{1}{10}$ c.c. will be deducted from the amount of permanganate used in the regular experiments.

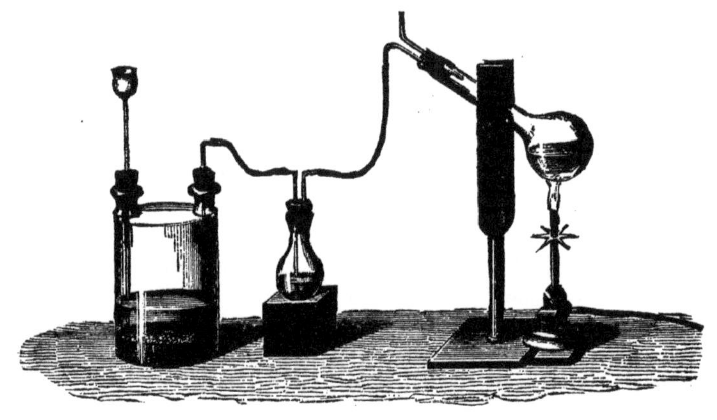

Fig. 52.

β. Performance of the Analytical Process.

This has been fully indicated in *α*. The compound to be examined is dissolved, preferably with application of a current of carbon dioxide* (see fig. 52). in dilute sulphuric acid, allowed to cool in the current of carbon dioxide, and suitably diluted (if practicable, the solution of a substance containing about ·2 grm. iron should be diluted to about 200 c.c.); if free acid is not present in sufficient quantity, dilute sulphuric acid is added till about 20 c. c. are present altogether, and then standard permanganate from the burette, to incipient reddening of the fluid. The volume of standard solution used is then read off. The strength of the solution of permanganate being known, the quantity of iron present in the examined fluid is found by a very simple calculation. Suppose 100 c. c. of solution of potassium permanganate to correspond to ·98 grm. iron, and 25 c.c. of the solution to have been used to effect the oxidation of the ferrous compound examined, then

$$100 : 25 :: \cdot 98 : x; \quad x = \cdot 245.$$

* If commercial hydrochloric acid is used for the preparation of CO_2 by action on marble, it must be free from *sulphurous acid*—an impurity which it often contains.

The quantity of ferrous iron originally present amounted accordingly to ·245 grm.

For the method of determining the total amount of iron present in a solution containing both ferrous and ferric salts, I refer to § 113; for that of determining the amount in each condition separately, to Section V.

γ. Process to be used with hydrochloric solutions of Iron.

In titrating hydrochloric acid solutions of iron with permanganate, it is essential that the standardizing of the reagent and the actual analysis be performed under the same circumstances as regards dilution, amount of acid, and temperature. Besides the proper reaction $10FeCl_2 + K_2Mn_2O_8 + 16HCl = 5Fe_2Cl_6 + 2KCl + 2MnCl_2 + 8H_2O$, the collateral reaction $K_2Mn_2O_8 + 16HCl = 2KCl + 2MnCl_2 + 8H_2O + 10Cl$ also takes place, in consequence of which a little chlorine is liberated. This chlorine does not combine with the ferrous chloride to form ferric chloride in the case of considerable dilution, but there occurs a condition of equilibrium in the fluid containing ferrous chloride, chlorine, and hydrochloric acid, which is destroyed by addition of a further quantity of either body (Löwenthal and Lenssen*). But since it is difficult to observe the above conditions of obtaining correct results, the determination in presence of hydrochloric acid is always less trustworthy than it is in sulphuric acid solutions.

The following method I have, however, found † to give the best results:—

Standardize the permanganate by means of iron dissolved in dilute sulphuric acid, make the iron solution to be tested up to ¼ litre, add 50 c.c. to a large quantity of water acidified with sulphuric acid (about 1 litre), titrate with permanganate, then again add 50 c.c. of the iron solution, and titrate again, &c. &c. The numbers obtained at the third and fourth time are taken. These are constant, while the number obtained the first time, and sometimes also the second time, differs. The result multiplied by 5 gives exactly the quantity of permanganate proportional to the amount of ferrous iron present.

b. Penny's *Method* (recommended subsequently by Schabus).

If potassium dichromate is added to a solution of a ferrous salt in presence of a strong free acid, the ferrous salt is converted into

* Zeitschr. f. anal. Chem. 1. 329. † *Ib.* 1, 361.

ferric salt, whilst a potassium, and a chromic salt of the free acid is formed ($6FeSO_4 + K_2Cr_2O_7 + 7H_2SO_4 = 3Fe_2(SO_4)_3 + K_2SO_4 + Cr_2(SO_4)_3 + 7H_2O$).

Now, with 29·522 gr. potassium dichromate dissolved to 2 litres of fluid, 33·6 gr. iron may be changed from a ferrous to a ferric salt, (295·22 being the mol. weight of $K_2Cr_2O_7$, and 336 being 6 times the at. weight of iron;) 50 c.c. of the above solution correspond accordingly to ·84 grm. iron.

Care must be taken to use perfectly pure potassium dichromate; the salt is heated in a porcelain crucible until it is just fused; it is then allowed to cool under the desiccator, and the required quantity weighed off when cold. Besides the above solution, another should also be prepared, ten times more dilute.

It is always advisable to test the correctness of the standard solution of potassium dichromate, by oxidizing with it a known amount of pure iron dissolved to a ferrous salt (see p. 268, *aa*).

The ferrous solution is sufficiently diluted, mixed with a sufficient quantity of dilute sulphuric acid, and the standard solution of potassium dichromate slowly added from the burette, the liquid being stirred all the while with a thin glass rod. The fluid, which is at first nearly colorless, speedily acquires a pale green tint, which changes gradually to a darker chrome-green. A very small drop of the mixture is now from time to time taken out by means of the stirring-rod, and brought into contact with a drop of a solution of potassium ferricyanide (free from ferrocyanide) on a porcelain plate, which has been spotted with several of such drops. When the blue color thereby produced begins to lose the intensity which it exhibited on the first trials, and to assume a paler tint, the addition of the solution of potassium dichromate must be more carefully regulated than at first, and towards the end of the process, a fresh essay must be made, and with larger drops than at first, after each new addition of two drops, and finally, even of a single drop; drops must also be left for some time in contact before the observation is taken. When no further blue coloration ensues, the oxidation is terminated. From the remarkable sensitiveness of the reaction, the exact point may be easily hit to a drop. To heighten the accuracy of the results, the dilute (ten times weaker) standard fluid should, just at the end of the process, be substituted for the concentrated solution of potassium dichromate.

For the manner of proceeding in presence of ferric salts,

I refer to § 113. If there is a deficiency of free acid in the solution, brown chromic chromate may form, upon which the solution of ferrous salt exercises no longer a deoxidizing action.

This method is usually preferred to the preceding when hydrochloric acid is unavoidably present.

§ 113.

6. Ferric Iron.

a. Solution.

Many ferric compounds are soluble in water. Ferric oxide and most ferric compounds which are insoluble in water, dissolve in hydrochloric acid, but many of them only slowly and with difficulty; compounds of this nature are best dissolved in concentrated hydrochloric acid, in a flask, with the aid of heat; which, however, should not be allowed to reach the boiling-point; the compound must, moreover, be finely powdered, and even then it will often take many hours to effect complete solution. Fusion with sodium carbonate or potassium disulphate must sometimes be resorted to in case of native ferric compounds.

b. Determination.

The iron of ferric compounds is usually weighed as ferric oxide, but sometimes as ferrous sulphide (§ 81). It may, however, be estimated also indirectly, and also by volumetric analysis, both directly and after reduction to ferrous iron. The conversion of compounds of iron into ferric oxide is effected either by precipitation as ferric hydroxide, preceded in some cases by precipitation as ferrous sulphide, or as basic ferric acetate, succinate, or formate; or by ignition. While the volumetric and the now seldom-used indirect methods are applicable in almost all cases, we may convert into

1. Ferric Oxide.

a. By Precipitation as Ferric Hydroxide.

All salts soluble in water of inorganic or volatile organic acids, and likewise those which, insoluble in water, dissolve in hydrochloric acid, with separation of their acid.

b. By Precipitation as Ferrous Sulphide.

All compounds of iron without exception.

c. By ignition.

All ferric salts of volatile oxygen acids.

2. Ferrous Sulphide.

All compounds of iron without exception.

The method 1, *c*, is the most expeditious and accurate, and is therefore preferred in all cases where its application is admissible. The method 1, *a*, is the most generally used. The methods 1, *b*, and 2, serve principally to effect the separation of the iron from other bases; they are resorted to also in certain instances where *a* is inapplicable, especially in cases where sugar or other non-volatile organic substances are present; and also to determine iron in ferric phosphates and borates. For the manner of determining iron in ferric chromate and silicate, I refer to §§ 130 and 140. The volumetric methods for estimating the iron of ferric compounds are used in technical work almost to the exclusion of all others, and are very frequently employed in scientific analyses.

1. *Determination as Ferric Oxide.*

a. By Precipitation as Ferric Hydroxide.

Mix the solution in a dish or beaker with ammonia in excess, heat nearly to boiling, decant repeatedly on to a filter, wash the precipitate *carefully* with hot water, dry *thoroughly* (which very greatly reduces the bulk of the precipitate), and ignite in the manner directed in § 53.

For the properties of the precipitate and residue, see § 81. The method is free from sources of error. The precipitate, under all circumstances, even if there are no fixed bodies to be washed out, must be most *carefully* and *thoroughly* washed, since, should it retain any traces of ammonium chloride, a portion of the iron would volatilize in the form of ferric chloride. It is also highly advisable to dissolve the weighed residue, or a portion of it, in strong hydrochloric acid, to see whether it is quite free from silicic acid. The solution is most readily effected in hydrochloric acid if the oxide is previously reduced to metallic iron by ignition in hydrogen.

b. By Precipitation as Ferrous Sulphide.

The solution, in a not too large flask, is mixed with ammonia, till all the free acid is neutralized. (In the absence of organic, non-volatile substances, this leads to the precipitation of a little ferric hydroxide, which, however, is of no consequence.) Add ammonium chloride, if not already present in sufficient quantity, then colorless or yellowish ammonium sulphide in moderate excess, lastly water, till the fluid reaches to the neck of the flask. Cork it up and stand in a warm place till the precipitate has subsided, and the supernatant fluid has a clear yellowish appearance (without a tinge of green). Wash as directed in the case of manganese sulphide (§ 109, 1, *c*). Neglect of any of these precautions will occasion some loss of substance, the ferrous sulphide gradually combining with the oxygen of the air, and passing thus into the filtrate as ferrous sulphate. As this sulphate is reprecipitated by the ammonium sulphide present, the filtrate assumes, in such cases, a greenish color, and gradually deposits a black precipitate, the separation of which is highly promoted by addition of ammonium chloride. [See remarks in [ ] § 81, 5, c. p. 165.]

When the operation of washing is completed, the moist precipitate (if it is not dried and determined according to 2) is put, together with the filter, into a beaker, some water added, and then hydrochloric acid, until the whole is redissolved. Heat is now applied, until the solution smells no longer of hydrogen sulphide; the fluid is then filtered into a flask, the residual paper carefully washed, incinerated, the ash treated with warm strong hydrochloric acid. The solution thus obtained (if yellowish) is added to the main filtrate, which is next heated with nitric acid (see § 112, 1); the solution (now ferric) is finally precipitated with ammonia, as in *a*.

If a solution of potassium ferric, ammonium ferric, or sodium ferric tartrate contains a considerable excess of alkali carbonate, the precipitation of the iron as sulphide is prevented to a greater or less extent (BLUMENAU). In such cases the fluid must therefore be nearly neutralized with an acid, before the precipitation with the ammonium sulphide can be effected.

c. By Ignition.

Expose the compound, in a covered crucible, to a gentle heat at first, and gradually to the highest degree of intensity; continue the operation until the weight of the residuary ferric oxide remains constant.

2. *Determination as Anhydrous Ferrous Sulphide.*

The hydrated ferrous sulphide obtained, as in 1, *b*, may be very conveniently determined by conversion into the anhydrous sulphide. The process is the same as for zinc (§ 108, 2). The heat to which it is finally exposed in the current of hydrogen must be strong, as an excess of sulphur is retained with some obstinacy. In fact, it is advisable after weighing to re-ignite in hydrogen and weigh a second time. It is of no importance if the hydrated sulphide has oxidized on drying.

Ferrous sulphate and ferric hydroxide can be transformed into sulphide in the same manner, after having been dehydrated by ignition in a porcelain crucible (H. ROSE*).

The results obtained by OESTEN, and adduced by ROSE, as well as those obtained in my own laboratory, are exceedingly satisfactory. (Expt. No. 75.)

3. *Volumetric Determination.*

a. Preceded by Reduction of Ferric to Ferrous Iron.

We have to occupy ourselves simply with the reduction of ferric to ferrous solutions, the other part of the process having been fully discussed in § 112 (Ferrous Iron). This reduction can be effected by many substances (zinc, stannous chloride, hydrogen sulphide, sulphurous acid, &c.), but only those can be used with advantage, an excess of which may be added with impunity. If an excess must be very carefully avoided, or, being added, must be carefully removed, the method becomes troublesome, and a ready source of inaccuracy is introduced.

Reduction by Zinc.—Heat the hydrochloric or sulphuric acid solution, which must contain a moderate excess of acid, but be free from nitric acid, in a small long-necked flask, placed in a slanting position; drop in small pieces of iron-free zinc (§ 60), and conduct a slow current of carbon dioxide through the flask (fig. 52, p. 272). Evolution of hydrogen gas begins at once, and the color of the solution becomes paler in proportion as the ferric sulphate (or chloride) changes to ferrous sulphate (or chloride). Apply a moderate heat, to promote the action; and add also, if necessary, a little more zinc. As soon as the hot solution is completely decolorized (one cannot judge of the perfect reduction of a cold solution so well, as the color of a ferric salt is deeper when hot), allow to cool completely in the stream of carbon dioxiode; to hasten the cooling the flask may be immersed in cold water; then dilute the contents with water, pour off and wash carefully into a beaker, leaving behind any undissolved zinc, and also (as far as possible) any flocks of lead that may have separated from the zinc, and proceed as

* Pogg. Annal. 110, 126.

directed in § 112, 2. If the solution contain metals precipitable by zinc, these will separate, and may render filtration necessary. In this case the filtrate must be again heated with zinc before using the standard solution. If iron-free zinc cannot be procured, the percentage of iron in the metal used must be determined, and weighed portions of it employed in the process of reduction; the known amount of iron contained in the zinc consumed is then subtracted from the total amount of iron found.

[*Reduction by Hydrogen Sulphide.*—Pass hydrogen sulphide through the cold ferric solution in a flask. The solution should occupy about two-thirds of the capacity of the flask, and should not contain much more than ·2 gr. iron per 100 c.c., but may be more dilute when but little iron is present. Continue the treatment with hydrogen sulphide at least 10 minutes after the color due to the ferric salt has disappeared, or until the solution appears to be well saturated with that gas. Heat, at first cautiously, to boiling. Escape of hydrogen sulphide at this period indicates that enough of that reagent has been applied. Continue boiling so rapidly that air cannot enter the flask, the mouth of which may be partially closed by a loose roll of filter paper, or other means, until the solution is reduced to one half its first volume. This will insure the removal of excess of hydrogen sulphide. (The escaping vapor will cease to blacken paper dipped in an alkaline lead solution somewhat before this point is reached.) During the boiling, let the flask be inclined so as to prevent mechanical loss. When the boiling is discontinued fill the flask immediately with cold water to within an inch of the mouth, close with a stopper, and cool in a stream of water.

Before reducing the ferric solution by either of the above processes, it is desirable to remove hydrochloric acid, if it is present, so that the iron after reduction can be satisfactorily determined by means of potassium permanganate.

Chlorides can be decomposed and hydrochloric acid removed by evaporating the solution with excess of sulphuric acid so long as hydrochloric acid vapors are given off at a temperature slightly exceeding 100° C. A liberal excess of sulphuric acid is advantageous. After cooling add water and digest till the ferric sulphate is dissolved. This treatment is simple and safe when nothing is present which is thereby converted into a compound insoluble in dilute sulphuric acid (silicic acid, barium, strontium, much calcium, &c.). Such insoluble compounds may persistently retain iron.

When, therefore, by evaporation with sulphuric acid and subsequent treatment with water an insoluble residue remains, it should not be thrown away before testing it to ascertain whether it retains iron.]

b. Without previous Reduction to Ferrous Iron, after OUDEMANS.*

If an acid solution of ferric chloride is mixed with a little cupric sulphate and some potassium sulphocyanate and then sodium thiosulphate is added, the red color of the ferric sulphocyanate gets paler and paler, and finally when the ferric salts are reduced to ferrous, disappears altogether. Warming is unnecessary. To hit the point is not easy, so we add a slight excess of sodium thiosulphate, and then titrate back with standard iodine. The reaction is as follows: $Fe_2Cl_6 + 2Na_2S_2O_3 = 2FeCl_2 + 2NaCl + Na_2S_4O_6$; it is promoted by the addition of a small quantity of cupric sulphate, which is alternately reduced by the thiosulphate and oxidized by the ferric chloride. If a small quantity of cuprous salt is produced by the excess of thiosulphate this does not matter, as its action on the iodine solution is the same in extent as the action of the thiosulphate which produced it. The method is not accurate unless the fluid remains clear; neither cuprous sulphocyanate nor cuprous iodide nor sulphur must be thrown down. Hence care must be taken to maintain the proper amounts of the reagents and to dilute the fluid sufficiently.

We require—1. A solution of sodium thiosulphate containing about 24 gr. (of the crystallized salt) per litre. 2. A solution of ferric chloride of known strength, prepared by dissolving 10·04 grm. of clean, fine, and soft iron wire (= 10 grm. pure iron), in hydrochloric acid in a slanting long-necked flask, oxidizing the solution with potassium chlorate, *completely* removing the excess of chlorine by protracted gentle boiling, and finally diluting the solu-

* Sodium thiosulphate was first employed by SCHERER (Gel. Anz. der K. Bayerischen Akademie, vom 31 Aug. 1859), afterwards by KREMER and LANDOLT (Zeitschr. f. anal. Chem. 1, 214). The method of OUDEMANS is to be found in Zeitschr. f. anal. Chem. 6, 129; it was criticised and rejected in MOHR's Lehrb. d. Titrirmethode, 3 Aufl. 291. OUDEMANS replied to MOHR in Zeitschr. f. anal. Chem. 9, 342, and an examination of the method by C. BALLING, appeared in the same journal, 9, 99.

tion to 1 litre. 3. A solution of cupric sulphate, 1 in 100. 4. A solution of potassium sulphocyanate, 1 in 100. 5. A solution of iodine in potassium iodide, containing 5 or 6 grm. iodine in the litre (compare § 146, *a*). 6. Thin starch paste.

Measure off some of the sodium thiosulphate, add starch paste (§ 146, *a*), and then titrate with iodine solution, in order to determine the relation between the two solutions. Now transfer 10 or 20 c.c. of the ferric chloride to a beaker, add 2 c.c. concentrated hydrochloric acid, 100 or 150 c.c. water, 3 c.c. copper solution, and 1 c.c. potassium sulphocyanate, titrate with sodium thiosulphate till the fluid just loses its color, add at once some starch paste, and titrate back with iodine solution till the blue color appears. Deduct the thiosulphate equivalent to the iodine solution from the total quantity of thiosulphate used, and the remainder will represent the amount required to reduce the iron present. In the analysis the conditions should be similar to those in the standardizing of the thiosulphate.

This method is very rapid, and the results, though not so accurate as those by method *a*, are quite good enough for many technical purposes.

Supplement to the Fourth Group.

§ 114.

7. Uranium and Uranyl.

If the compound in which the uranium is to be determined contains no other fixed substances, it may often be converted into *uranous uranate* $U(UO_4)_2$—(called also uranoso-uranic oxide UO_2-$2UO_3$)—by simple ignition. If sulphuric acid is present, small portions of ammonium carbonate must be thrown into the crucible towards the end of the operation.

In cases where the application of this method is inadmissible, the solution of uranium (which, if it contains uranous salts, must first be warmed with nitric acid, until they are converted into uranyl salts) is nearly boiled in a platinum or porcelain dish, and precipitated with ammonia in slight excess. The yellow precipitate formed, which consists of hydrated *ammonium uranate*, is filtered off hot and washed with a dilute solution of ammonium chloride, to prevent the fluid passing milky through the filter. The precipitate is dried and ignited (§ 53). To make quite sure of obtaining the uranous uranate in the pure state, the crucible is ignited for some time in a slanting position and uncovered; the lid is then put on, while the ignition is still continuing; the crucible is allowed to cool under the desiccator, and weighed (RAMMELSBERG).

If the solution from which the uranyl is to be precipitated contains other basic radicals (alkali-earth metals, or even alkali metals), portions of these will precipitate along with the ammonium uranate. For the measures to be resorted to in such cases, I refer to Section V.

The reduction of the uranous uranate to the state of uranous oxide (UO_2) is an excellent means of ascertaining its purity for the purpose of control. This reduction should never be omitted, since PÉLIGOT has found the uranous uranate to be variable in composition. It is effected by ignition in a current of hydrogen gas, in the way described § 111, 1 (Cobalt). In the case of large quantities the ignition must be several times repeated, and the residue must be occasionally stirred with a platinum wire. While cooling increase the current of gas to prevent reabsorption of oxygen. By intense heating the property of spontaneous ignition in the air is destroyed. If after evaporating a solution of uranyl chloride, the residue is to be ignited in hydrogen, heat gently at first in the gas to avoid loss by volatilization. The separation of uranyl from phosphoric acid is effected by fusing the compound with potassium cyanide and sodium carbonate. Upon extracting the fused mass with water, the phosphoric acid is obtained in solution, whilst uranium is left as uranous oxide. KNOP and ARENDT* have employed this method.

Taking 237·6 as the atomic weight of uranium, $U(UO_4)_2$ uranous uranate contains 84·77 per cent. of uranium and 15·23 per cent. of oxygen. UO_2 uranous oxide contains 88·13 per cent. uranium and 11·87 per cent. of oxygen.

According to BELOHOUBECK,† uranium may be also determined volumetrically by reducing the solution of uranyl acetate or sulphate to uranous salts with zinc, as in the case of iron (§ 113, 3, *a*). As the color of the solution is no safe criterion of the end of the reduction, you must allow the action of the zinc to continue for a considerable time. BELOHOUBECK says, a quarter of an hour is sufficient for small quantities, half an hour for large quantities.

---

* Chem. Centralblatt, 1856, 773. † Zeitschr. f. anal. Chem. 6, 120.

The solution of the uranous salt is diluted, mixed with dilute sulphuric acid, and then titrated with permanganate to incipient reddening. The permanganate is standardized by § 112, 2, 1 at. uranium = 2 at. iron.

BELOHOUBECK obtained good results also in hydrochloric solutions, but experiments made in this laboratory have shown that these are liable to the error pointed out in the case of iron (Comp. p. 273, γ), at least in the presence of considerable quantities of hydrochloric acid.

Fifth Group.

SILVER—LEAD—MERCURY IN MERCUROUS COMPOUNDS—MERCURY IN MERCURIC COMPOUNDS—COPPER—BISMUTH—CADMIUM—(PALLADIUM).

§ 115.

1. SILVER.

a. Solution.

Metallic silver, and those of its compounds which are insoluble in water, are best dissolved in nitric acid (if soluble in that acid). Dilute nitric acid suffices for most compounds; silver sulphide, however, requires concentrated acid. The solution is effected best in a flask, which should be heated if necessary, and placed in a slanting position if gas is evolved. In the case of metallic silver, or silver sulphide, the solution is heated finally to gentle boiling to drive off nitrous acid. Silver chloride, bromide, and iodide are insoluble in water and in nitric acid. To get the silver contained in chloride and bromide in solution, proceed as follows:—Fuse the salt in a porcelain crucible (this operation, though not absolutely indispensable, had better not be omitted), pour water over it, put a piece of clean cadmium, zinc, or iron upon it, and add some dilute sulphuric acid. Wash the reduced spongy silver, first with dilute sulphuric acid, then with water, and finally dissolve it in nitric acid. However, as we shall see below, the quantitative analysis of these salts does not necessarily involve their solution.

b. Determination.

Silver may be weighed as *chloride*, *sulphide*, or *cyanide*, or in the *metallic* state (§ 82). It is also frequently determined by volumetric analysis.

We may convert into

1. SILVER CHLORIDE: All compounds of silver without exception.

2. SILVER SULPHIDE: 3. SILVER CYANIDE: All compounds soluble in water or nitric acid.

4. METALLIC SILVER: Silver oxide and some silver salts of readily volatile acids; silver salts of organic acids; silver chloride, bromide, iodide, sulphide, and sulphate.

The method 4 is the most convenient, especially when conducted in the dry way, and is preferred to the others in all cases where its application is admissible. The method 1 is that most generally resorted to. 2 and 3 serve mostly only to effect the separation of silver from other metals.

In assays for the Mint, silver is usually determined volumetrically by GAY-LUSSAC's method. PISANI's volumetric method is especially suited to the determination of very small quantities of silver. H. VOGEL's method is specially useful to photographers.

1. *Determination of Silver as Chloride.*

a. In the Wet Way.

Mix the moderately dilute solution in a beaker with nitric acid, heat to about 70°, and add hydrochloric acid with constant stirring till it ceases to produce a precipitate. A large excess of hydrochloric acid must be avoided, as the precipitate is not absolutely insoluble therein. While protecting the contents of the beaker from the action of direct sunlight continue the heat till the precipitate has fully settled, pour off the clear fluid through a small filter, rinse the precipitate on to the latter by means of hot water mixed with some nitric acid, wash with hot water containing nitric acid, then with pure hot water, dry thoroughly, transfer the precipitate to a watch-glass as nearly as possible, incinerate the filter in a weighed porcelain crucible, treat the ash (which always contains some metallic silver) with a few drops of nitric acid in the heat; add two or three drops of hydrochloric acid, evaporate cautiously to dryness, add the main bulk of the precipitate, using a camel's-hair brush to transfer the last portions, heat cautiously till it begins to fuse at the edge, allow to cool, and weigh.

To remove the fused mass without breaking the crucible, lay a small piece of iron or zinc upon it, and then add very dilute hydrochloric or sulphuric acid. The chloride will be reduced, and the silver may now be detached from the crucible with the greatest ease.

For the properties of the precipitate see § 82. The method gives very exact results, at all events in the absence of any considerable quantities of those salts in which silver chloride is somewhat soluble; compare § 82. To avoid error in this respect, it is well to test the clear filtrate with hydrogen sulphide.

b. In the Dry Way.

This method serves more exclusively for the analysis of silver bromide and iodide, although it can be applied in the case of other compounds.

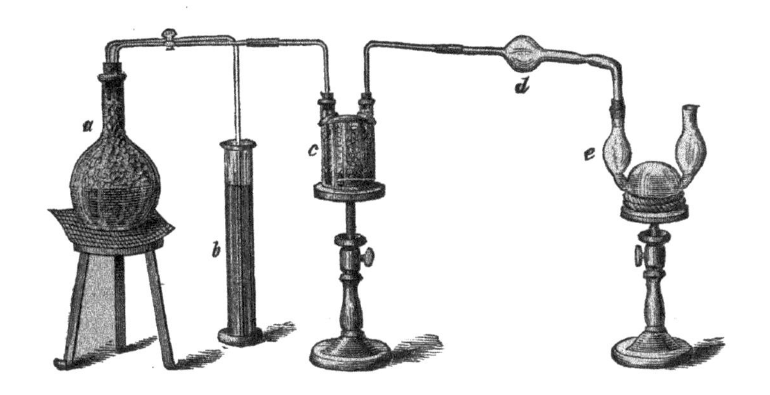

Fig. 53.

The process is conducted in the apparatus illustrated by fig. 53, leaving off the U tube *e*, and employing a straight bulb-tube or a plain tube with porcelain tray instead of the bent tube *d*.

a is a flask for disengaging chlorine, it is completely filled with pieces of pyrolusite (native manganese dioxide) of the size of hazel-nuts, and half filled with strong hydrochloric acid. The chlorine is conducted to the bottom of *c*, which contains a layer of sulphuric acid and is filled above with pumice-stone moistened with strong sulphuric acid. The flow of chlorine may be regulated by the stop-cock, while any excess accidentally produced is conducted by a second tube to the bottom of the cylinder *b*, in which it is absorbed by a soda solution; *d* is a bulb-tube intended for the reception of the silver iodide or bromide. The operation is commenced by introducing the compound to be analyzed into the bulb, and applying heat to the latter until its contents are fused; when cold, the tube is weighed and connected with the apparatus. Chlorine gas is then evolved from *a*; when the evolution of the gas has proceeded for some time, the contents of the bulb are heated to fusion, and kept in this state for about fifteen minutes, agitating now and then the fused mass. The bulb-tube is then removed from the apparatus, allowed to cool, and held in a slanting position to replace the chlorine by atmospheric air; it is subsequently weighed, then again connected with the apparatus, and the former process repeated, keeping the contents of *d* in a state of fusion for a few minutes. By means of a light glass tube attached by a piece of rubber tube to the end of *d* the chlorine escaping during the operation may be conducted into the open air or into a flue. The operation may, in ordinary cases, be considered concluded if the weight of the tube suffers no variation by the repetition of the process. If the highest degree of accuracy is to be attained, heat the silver chloride again to fusion, passing at the same time a slow stream of pure, dry carbon dioxide through the tube, in order to drive out the traces of chlorine absorbed by the fused chloride. Allow to cool, hold obliquely for a short time, so as to replace the carbon dioxide by air, and finally weigh. See § 82.

2. *Determination as Silver Sulphide.*

Hydrogen sulphide precipitates silver completely from acid, neutral, and alkaline solutions; ammonium sulphide precipitates it from neutral and alkaline solutions. The precipitate does not settle clearly and rapidly except a free acid or salt be present (such as nitric acid or an alkali nitrate). Recently prepared perfectly clear solution of hydrogen sulphide may be employed to precipitate small quantities of silver; to precipitate larger quantities, the solution of the salt of silver (which must not be too acid) is moderately diluted, and washed hydrogen sulphide gas conducted into it. After complete precipitation has been effected, and the silver sulphide has perfectly subsided (with exclusion of air), it is collected on a weighed filter, washed, dried at 100°, and weighed. For the properties of the precipitate, see § 82. This method, if properly executed, gives accurate results. The operator must take care to filter quickly, and to prevent the access of air as much as possible during the filtration, since, if this precaution be neglected, sulphur is likely to separate from the hydrogen sulphide water, which, of course, would add falsely to the weight of the silver sulphide. If the presence of a minute quantity of sulphur in the precipitate is suspected, treat it after drying with pure carbon disulphide on the filter repeatedly, till the fluid running through gives no residue on evaporation in a watch-glass; dry and weigh.

The sulphide must, however, never be weighed as just described, unless the analyst is satisfied that no considerable amount of sulphur has fallen down with it, as would occur if the fluid contained hyponitric acid, a ferric salt, or any other substance which decomposes hydrogen sulphide. In case the precipitate does contain much admixed sulphur, the simplest process is to convert it into metallic silver (H. Rose*). For this purpose it is transferred to a weighed porcelain crucible, the filter ash is added, and the whole is heated to redness in a stream of hydrogen, the apparatus described in § 108 being employed. Results accurate.

Should the apparatus in question not be at the operator's disposal, he may, after complete washing of the precipitate, carefully rinse it into a porcelain dish (without injuring the weighed filter), heat it once or twice with a moderately strong solution of pure sodium sulphite, retransfer the precipitate (now freed from admixed sulphur) to the old filter, wash well, dry and weigh (J. Löwe†); or he may treat the dried precipitate, together with the filter-ash, with moderately dilute chlorine-free nitric acid at a gentle heat, till complete decomposition has been effected (till the undissolved sulphur has a clean yellow appearance), filter, wash well, and proceed according to 1, *a*.

3. *Determination as Silver Cyanide.*

Mix the neutral solution of silver with potassium cyanide, until the precipitate of silver cyanide which forms at first is redissolved; add nitric acid in slight excess, and apply a gentle heat. If the solution contains free acid, this must be first neutralized with potash or sodium carbonate. After some time, collect the precipitated silver cyanide on a weighed filter, wash, dry at 100°, and weigh. For the properties of the precipitate, see § 82. The results are accurate.

* Pogg. Annal. 110, 139.

† Journ. f. prakt. Chem. 77, 73.

4. *Determination as Metallic Silver.*

Silver oxide, silver carbonate, &c., are easily reduced by simple ignition in a porcelain crucible. In the reduction of salts of organic acids, the crucible is kept covered at first, and a moderate heat applied; after a time the lid is removed, and the heat increased, until the whole of the carbon is consumed. For the properties of the residue, see § 82. The results are absolutely accurate, except as regards silver salts of organic acids; in the analysis of the latter, it not unfrequently happens that the reduced silver contains a minute portion of carbon, which increases the weight of the residue to a trifling extent.

If it is desired to transform silver chloride, bromide, or sulphide into metallic silver, for the purpose of analysis, they are heated in a current of pure hydrogen to redness, till the weight remains constant. The process may be conducted in a porcelain crucible or a bulb-tube. In the former case, the apparatus described p. 251, § 108, is used; in the latter the apparatus represented p. 285, with the substitution, of course, of hydrogen for chlorine. If the bulb-tube is used, it must, after cooling and before being weighed, be held in an inclined position, so that the hydrogen may be replaced by air. The results are perfectly accurate. Silver iodide cannot be reduced in this way.

5. *Volumetric Methods.*

I. GAY-LUSSAC'S.

This, the most exact of all known volumetric processes, was introduced by GAY-LUSSAC as a substitute for the assay of silver by cupellation, was thoroughly investigated by him, and will be found fully described in his work on the subject. This method has been rendered still more precise by the researches of G. J. MULDER, to whose exhaustive monograph* I refer the special student of this branch. I shall here confine myself to giving the process so far as to suit the requirements of the chemical laboratory, taking only for granted that the analyst has the ordinary measuring apparatus, &c., at his disposal. MULDER'S results will be made use of to the full extent possible under these circumstances.

a. REQUISITES.

α. SOLUTION OF SODIUM CHLORIDE. Take chemically pure

* Die Silberprobirmethode (see note p. 167).

sodium chloride—either artificially prepared or pure rock-salt—powder it roughly and ignite moderately (not to fusion*). Now dissolve 5·4202 grm. in distilled water to 1 litre, measured at 16°. 100 c.c. of this solution contains a quantity of sodium chloride equivalent to 1 grm. of silver. The solution is kept in a stoppered bottle and shaken before use.

β. DECIMAL SOLUTION OF SODIUM CHLORIDE. Transfer 50 c.c. of the solution described in *α* to a 500 c.c. measuring flask, fill up to the mark with distilled water and shake. Each c.c. of this decimal solution corresponds to ·001 grm. silver. The measuring must be performed at 16°. The solution is kept as the other.

γ. DECIMAL SILVER SOLUTION. Dissolve ·5 grm. chemically pure silver † in 2 to 3 c.c. pure nitric acid of 1·2 sp. gr., and dilute the solution with water exactly to 500 c.c. measured at 16°. Each c.c. contains ·001 grm. silver. The solution is kept in a stoppered bottle and protected against the influence of light.

δ. TEST-BOTTLES. These should be of colorless glass, holding easily 200 c.c., closed with well-ground glass-stoppers, running to a point below. The bottles fit into cases blackened on the inside, and reaching up to their necks. In order to protect the latter also from the action of light, a black-cloth cover is employed.

b. PRINCIPLE.

Suppose we know the value of a solution of sodium chloride, *i.e.*, the quantity that is necessary to precipitate a given amount of silver, say 1 grm., we are in the position, with the aid of this solu-

---

* On fusion, if the flame can in the least way act upon it, it takes an alkaline reaction, since under the influence of vapor of water and carbon dioxide, a little hydrochloric acid is formed and escapes, while a corresponding quantity of sodium carbonate remains.

† For the preparation of pure silver STAS recommends the following method: Take crude silver nitrate containing copper, fuse in order to decompose any platinum nitrate which may be present, dissolve in dilute ammonia, allow to stand 48 hours, filter and dilute till the fluid does not contain more than 2 per cent. silver. Add ammonium sulphite in excess. To ascertain how much sulphite will be required make a small preliminary test; as soon as after heating the blue solution loses all color, you may be sure that enough of the sulphide has been added. Warm on a water-bath to 60° or 70°, when all the silver will be thrown down as a metallic powder, allow to cool and wash by decantation with diluted ammonia till the washings are free from copper and sulphuric acid. Now digest the metal for several days with strong ammonia, wash, dry, and fuse with a flux of borax and sodium nitrate.

tion, to determine an unknown amount of silver, for if we put x for the unknown amount of silver, then

c.c. of solution used for 1 grm. : c.c. used for x :: 1 grm. : x.

But if we examine whether 1 mol. sodium chloride dissolved in water actually precipitates 1 at. of silver dissolved in nitric acid exactly, we find that this is not the case.* On the contrary, the clear supernatant fluid gives a small precipitate both on the addition of a little solution of sodium chloride, and on the addition of a little silver solution, as MULDER has most accurately determined. The value of a solution of sodium chloride in the sense explained above cannot, therefore, be reckoned from the amount of salt it contains, by calculating 1 at. silver for 1 mol. sodium chloride, but it can only be obtained by experiment. MULDER has shown that the temperature and the degree of dilution have some influence, and also that this fact is to be explained on the ground of the solvent power of the sodium nitrate produced on the silver chloride. In the solution thus formed we have to imagine $NaNO_3$ and NaCl with $AgNO_3$ in a certain state of equilibrium, which on the addition of either NaCl or $AgNO_3$ is destroyed, silver chloride being precipitated.

From this interesting observation it follows, that if to a silver-solution we add at first concentrated solution of sodium chloride, then decimal solution drop by drop, till the exact point is reached when no more precipitate appears, now, on addition of decimal silver-solution, a small precipitate will be again produced; and if we add the latter drop by drop, till the last drop occasions no turbidity, then again decimal solution of sodium chloride will give a small precipitate. On noticing the number of drops of both decimal solutions which are required to pass from one limit to the other, we find that the same number of each are used. Let us suppose that we had added decimal solution of sodium chloride till it ceased to react, and had then used 20 drops† of decimal silver-solution, till this ceased to produce a further turbidity, we must now again add 20 drops of decimal solution of sodium chloride, in

* If sodium bromide or potassium bromide is used, complete precipitation would ensue on addition of an equivalent quantity of silver solution, since bromide of silver is not at all soluble in the supernatant fluid (STAS, Compt. rend. 67, 1107).

† Twenty drops from MULDER's dropping apparatus are equal to 1 c.c.

order to reach the point at which this ceases to react. Were we to add only 10 instead of these 20 drops, we have the neutral point, as MULDER calls it, *i.e.*, the point at which both silver and sodium chloride produce equal precipitates.

We have, therefore, 3 different points to choose from for our final reaction: *a*, the point at which sodium chloride has just ceased to precipitate the silver; *b*, the neutral point; *c*, the point at which silver-solution has just ceased to precipitate sodium chloride. Whichever we may choose, we must keep to it, *i.e.*, we must not use a different point in standardizing the sodium chloride solution and in performing an analysis. The difference obtained, by using first *a* and then *b* is, according to MULDER, for 1 grm. silver, at 16°, about ·5 mgrm. silver; by employing first *a* and then *c*, as was permitted in the original process of GAY-LUSSAC, the difference is increased to 1 mgrm.

For our object, it appears most convenient to consider, once for all, the point *a* as the end, and never to finish with the silver-solution. If the point has been overstepped by the addition of too large an amount of decimal solution of sodium chloride, 2 or 3 c.c. of decimal silver-solution should be added all at once. The end-point is then found by carefully adding decimal solution of sodium chloride again, and the quantity of silver in the silver-solution added is added to the original amount of silver weighed off.

c. PERFORMANCE OF THE PROCESS.

This is divided into two operations—*α*, the titration of the sodium chloride solution; *β*, the assay of the silver alloy to be examined.

α. TITRATION OF THE SODIUM CHLORIDE SOLUTION.

Weigh off exactly from 1·001 to 1·003 grm. chemically pure silver,* put it into a test-bottle, add 5 c.c. perfectly pure nitric acid, of 1·2 sp. gr., and heat the bottle in an inclined position in a water- or sand-bath till complete solution is effected. Now blow out the nitrous fumes from the upper part of the bottle, and after it has cooled a little, place it in a stream of water, the temperature of which is about 16°, and let it remain there till its contents are cooled to this degree, wipe it dry, and place it in its case.

Now fill the 100 c.c. pipette with the concentrated solution of sodium chloride, which is then allowed to flow into the test-bottle

* See note, p. 289.

containing the silver-solution*. Insert the glass-stopper firmly (after moistening it with water), cover the neck of the bottle with the cap of black stuff belonging to it, and shake violently without delay, till the silver chloride settles, leaving the fluid perfectly clear. Then take the stopper out, rub it on the neck, so as to remove all silver chloride, replace it firmly, and by giving the bottle a few dexterous turns, rinse the chloride down from the upper part. After allowing to rest a little, again remove the stopper, and add, from a burette divided into $\frac{1}{10}$ c.c., decimal sodium chloride solution, allowing the drops to fall against the lower part of the neck, the bottle being held in an inclined position. If, as above directed, 1·001 to 1·003 grm. silver have been employed, the portions of sodium chloride solution at first added may be $\frac{1}{2}$ c.c. After each addition, raise the bottle a little out of its case, observe the amount of precipitate produced, shake till the fluid has become clear again, and proceed as above, before adding each fresh quantity of sodium chloride solution. The smaller the precipitate produced, the smaller should be the quantity of sodium chloride next added; towards the end only two drops should be added each time; and quite at the end read off the height of the fluid in the burette before each further addition. When the last two drops give no more precipitate, the previous reading is the correct one.

If by chance the point has been overstepped, and the time has been missed for the proper reading off of the burette, add 2 to 3 c.c. of the decimal silver solution (the silver in which is to be added to the quantity first weighed), and try again to hit the point exactly by careful addition of decimal sodium chloride solution.

The value of the sodium chloride solution is now known. Reckon it to 1 grm. silver.

Suppose we had used for 1·002 grm. silver, 100 c.c. of concentrated and 3 c.c. of decimal sodium chloride solution; this makes altogether 100·3 of concentrated; then

$$1{\cdot}002 : 1{\cdot}000 :: 100{\cdot}3 : x$$
$$x = 100{\cdot}0998$$

We may without scruple put 100·1 for this number. We now

* The pipette, having been filled above the mark, should be fixed in a support, before the excess is allowed to run out, otherwise the measurings will not be sufficiently accurate.

know that 100·1 c.c. of the concentrated solution of sodium chloride, measured at 16°, exactly precipitates 1 grm. of silver. This relationship serves as the foundation of the calculation in actual assaying, and must be re-examined whenever there is reason to imagine that the strength of the sodium chloride solution may have altered.

β. The actual assay of the Silver-Alloy.

Weigh off so much as contains about 1 grm. of silver, or better, a few mgrm. more;* dissolve in a test-bottle in 5 to 7 c.c. nitric acid, and proceed in all respects exactly as in α.

Suppose we had taken 1·116 grm. of the alloy, and in addition to the 100 c.c. of concentrated sodium chloride solution, had used 5 c.c. of the dilute (= ·5 concentrated), how much silver would the alloy contain?

Presuming that we use the same sodium chloride solution which served as our example in α, 100·1 c.c. of which = 1 grm. silver, then

$$100{\cdot}1 : 100{\cdot}5 :: 1{\cdot}000 : x$$
$$x = 1{\cdot}003996 \text{ (say } (1{\cdot}004).$$

We may also arrive at the same result in the following manner:—

	NaCl Solution.
For the precipitation of the silver in the alloy were used	100·5 c.c.
For 1 grm. silver are necessary	100·1 c.c.
Difference	·4 c.c.

There are, therefore, 4 mgrm. of silver present more than a grm., on the presumption that ·1 of the concentrated sodium chloride solution (= 1 c.c. of the decimal solution) corresponds to 1 mgrm.

* In coins containing 9 parts of silver and 1 part of copper, therefore take about 1·115 or 1·120. In weighing off alloys of silver and copper, which do not correspond to the formula Ag_3Cu_2 (standard $\frac{718{\cdot}67}{1000}$,) = we must remember that they are never homogeneous in the mass; thus, for instance, the pieces of metal, from which coins are stamped, often show 1·5 to 1·7 in a thousand more silver in the middle than at the edges. In assaying alloys, then, portions from various parts of the mass must be taken, in order to get a correct result. The inaccuracy, however, proceeding from the cause above-mentioned, can only be completely overcome by fusing the alloy and taking out a portion from the well-stirred mass for the assay.

silver. This supposition, although not absolutely correct, may be safely made, for the inexactness it involves is too minute, as is evident from the previous calculation.

Before we can execute this process exactly, we must know the quantity of silver the alloy contains very approximately. In assaying coins of known value this is the case, but with other silver alloys it is usually not so. Under the latter circumstances an approximate estimation must precede the regular assay. This is performed by weighing off ½ grm. (or in the case of alloys that are poor in silver, 1 grm.), dissolving in 3 to 6 c.c. nitric acid, and adding from the burette sodium chloride solution,—first in larger, then in smaller quantities—till the last drops produce no further turbidity. The last drops are not reckoned with the rest. The operation is conducted, as regards shaking, &c., as previously given. Suppose we had weighed off ·5 grm. of the alloy, and employed 25 c.c. of the sodium chloride solution—taking the above supposed value of the latter—

We have $$100{\cdot}1 : 25 :: 1{\cdot}000 : x$$
$$x = {\cdot}2497$$

that is, the silver in ·5 grm. of the alloy; and as to the quantity of alloy we have to weigh off for the assay proper,

We have $${\cdot}2497 : 1{\cdot}003 :: {\cdot}5 : x$$
$$x = 2{\cdot}008.$$

This quantity will, of course, require more nitric acid for solution than was previously used (use 10 c.c.). In cases where the highest degree of accuracy is not required, the results afforded by this rough preliminary estimation will be accurate enough, if the experiment is carefully conducted, since they give the quantity of silver present to within $\frac{1}{1000}$ or $\frac{1}{500}$.

With alloys which contain sulphur, and with such as consist of gold and silver, and contain a little tin, LEVOL* employs concentrated sulphuric acid (about 25 grm.) as solvent. The portion of the alloy is boiled with it till dissolved; after cooling, the fluid is treated in the usual manner. As, however, concentrated sulphuric acid fails to dissolve all the silver when there is much copper present, MASCAZZINI† digests the weighed portion of alloy (which

* Annal. de Chim. et de Phys. (3) 44, 347. † Chem. Centralbl. 1857, 300.

may contain small quantities of lead, tin, and antimony, besides gold) first with the least possible amount of nitric acid, as long as red vapors are formed; he then adds concentrated sulphuric acid, boils till the gold has settled well together, adds water after cooling, and then titrates. In the presence of mercury, the chloride of that metal is carried down with the silver, rendering the method inaccurate. If the quantity of mercury is but small, you may get over the difficulty by adding 25 c.c. ammonia and 20 c.c. acetic acid (LEVOL). The ammonium acetate acts by decomposing the mercuric chloride, and thus preventing its precipitation (DEBRAY*). If the quantity of mercury is large the addition of an alkali acetate is not effective, and DEBRAY recommends to drive off the mercury by igniting for four hours in a small crucible of gas carbon in a muffle. The presence of other volatile metals, such as zinc, does not interfere with this operation.

II. PISANI'S METHOD.†

This process depends on the following reaction: a solution of iodide of starch added to a very dilute neutral solution of silver nitrate, forms silver iodide and silver hypoiodite. The blue color consequently vanishes, and on continued addition of the iodide of starch, the fluid does not become permanently blue till all the silver nitrate present is decomposed in the above manner. The iodide of starch solution used is therefore proportional to the quantity of silver nitrate. Hence, if the value of the iodide of starch solution be determined, by allowing it to act on a certain amount of silver solution of known strength, we shall be able to estimate unknown quantities of silver with the greatest ease, provided that the silver solution is free from all other substances which exert a decomposing action on the iodide of starch. Besides the ordinary reducing agents, the following salts must be especially mentioned as possessing this power: mercurous and mercuric salts, stannous salts, manganous, ferrous, and antimonious salts, also auric chloride and arsenites; lead and copper salts, on the other hand, do not affect iodide of starch.

The iodide of starch is prepared as follows: make an intimate

* Compt. rend. 70, 849.

† Annal. d. Min. 10, 83.

mixture in a mortar of 2 grm. iodine and 15 grm. starch with the addition of 6 to 8 drops of water, and heat the slightly-moist mixture in a closed flask in a water-bath, till the original violet-blue color has passed into dark grayish-blue—it takes about an hour. The iodide of starch thus prepared is then digested with water; it dissolves completely to a deep bluish-black fluid.

The value of this fluid is determined by allowing it to act on 10 c.c. of a neutral solution of silver nitrate, containing 1 grm. of pure silver in 1 litre—the silver solution is mixed with a little pure precipitated calcium carbonate before adding the iodide of starch. The strength of this latter is right, if 50 to 60 c.c. are used in this experiment. On adding it, at first the blue color disappears rapidly, and the fluid becomes yellowish from the silver iodide. The end of the operation is attained as soon as the fluid is bluish-green. The point is pretty easy to hit, and an error of ·5 c.c. is of no importance, as it only corresponds to about ·0001 grm. silver. The calcium carbonate, besides neutralizing the free acid, has the effect of rendering the final change of the color more distinctly observable. To analyze an alloy of silver and copper, dissolve about ·5 grm. in nitric acid, dilute to 100 c.c. to lower the color of the copper, saturate 5 c.c. with calcium carbonate, and add iodide of starch till the coloration appears. Or you may determine very approximately the amount of silver in 2 c.c. of the solution, then precipitate the greater part (about 99%) of the silver from 50 c.c. of the solution with standard solution of potassium iodide, and without filtering estimate the remainder of the silver by means of iodide of starch. If the amount of silver to be determined is more than ·020 grm., it is always better to employ the latter method. In the case of a nitric acid solution containing silver with lead. the latter metal is first precipitated with sulphuric acid and filtered off, calcium carbonate is added to the filtrate till all free acid is neutralized, the fluid is filtered again (if necessary), and lastly, more calcium carbonate is added, and then the iodide of starch. Very dilute solutions must be concentrated, so that one may have no more than from 50 to 100 c.c. to deal with. The method is worthy of notice and specially suited for the estimation of small quantities of silver. With such it has afforded me perfectly satisfactory results. Instead of the standard iodide of starch, a dilute standard solution of iodine in potassium iodide may be equally well employed—with addition of starch sclution (FIELD*). If this is used you must bear in mind that any substance which decomposes potassium iodide with separation of iodine will interfere.

III. METHOD DEPENDING ON THE ACTION OF SILVER NITRATE ON SODIUM CHLORIDE IN THE PRESENCE OF POTASSIUM CHROMATE.

This is the reverse of the method for the estimation of chlorine § 141 *b*, *α*, and will be described in that place.

§ 116.

2. LEAD.

a. Solution.

Few of the lead salts are soluble in water. Metallic lead, lead oxide, and most of the lead salts that are insoluble in water dissolve in *dilute* nitric acid. Concentrated nitric acid effects neither complete decomposition nor complete solution, since, owing to the insolubility of lead nitrate in concentrated nitric acid, the first portions of nitrate formed protect the yet undecomposed parts of the salt from the action of the acid. For the solubility of lead chloride and sulphate, see § 83. As we shall see below, the analysis of these compounds may be effected without dissolving them, lead iodide dissolves readily in moderately dilute nitric acid upon application of heat, with separation of iodine. Solution of potassa is the only menstruum in which lead chromate dissolves without decomposition.

b. Determination.

Lead may be determined as *oxide*, *sulphate*, *chromate*, or *sulphide;* also by volumetric analysis.

We may convert into

1. LEAD OXIDE:

a. By Precipitation.

All lead salts soluble in water, and those of its salts which, insoluble in that menstruum, dissolve in nitric acid, with separation of their acid.

b. By Ignition.

α. Lead salts of readily volatile or decomposable inorganic acids.

β. Lead salts of organic acids.

---

* Chem. News, 2, 17.

2. LEAD SULPHIDE:

All lead salts in solution.

3. LEAD SULPHATE:

a. By Precipitation.

The salts that are insoluble in water, but soluble in nitric acid, whose acid cannot be separated from the solution.

b. By Evaporation.

α. All the oxides of lead, and also the lead salts of volatile acids.

β. Many of the organic compounds of lead.

4. LEAD CHROMATE:

The compounds of lead soluble in water or nitric acid.

The application of these several methods must not be understood to be rigorously confined to the compounds specially enumerated under their respective heads; thus, for instance, all the compounds enumerated sub 1, may likewise be determined as lead sulphate; and, as above mentioned, all soluble compounds of lead may be converted into lead sulphide; also, in lead sulphate the lead may be without difficulty determined as sulphide, Lead chloride, bromide, and iodide are most conveniently reduced to the metallic state in a current of hydrogen gas, in the manner described § 115 (Reduction of silver chloride), if it is not deemed preferable to dissolve them in water, or to decompose them by a boiling solution of sodium carbonate. If the reduction method is resorted to, the heat applied should not be too intense, since this might cause some lead chloride to volatilize.

The higher oxides of lead are reduced by ignition to the state of lead monoxide, and may thus be readily analyzed and dissolved. Should the operator wish to avoid having recourse to ignition, the most simple mode of dissolving the higher oxides of lead is to act upon them with dilute nitric acid, with the addition of alcohol. For the methods of analyzing lead sulphate, chromate, iodide, and bromide, I refer to the paragraphs treating of the corresponding acids, in the second part of this section. To effect the estimation of lead in the oxide and in many lead salts, especially also in the sulphate, the compound under examination may be fused with potassium cyanide, and the metallic lead obtained well washed, and weighed. From the sulphide also the greater portion of the lead may be separated by this method, but never the whole (H. Rose*).

1. *Determination as Oxide.*

a. By Precipitation.

Mix the moderately dilute solution with ammonium carbonate† slightly in excess, add some caustic ammonia, apply a gentle heat, allow to cool and filter through a small thin filter. Wash with pure water, dry, and transfer the precipitate to a watch-glass, removing it as completely as possible from the filter; burn the latter in a weighed porcelain crucible. After the crucible is cold, moisten the ash with nitric acid, allow it to evaporate, ignite gently, allow to cool, add the precipitate and ignite gently till all the carbonic acid is driven off. For the properties of the precipitate and residue, see § 83. The results are very satisfactory, although generally a trifle too low, owing to lead carbonate not being absolutely insoluble, particularly in fluids rich in ammonium salts (Expt. No. 47, *b*).

b. By Ignition.

Compounds like lead carbonate or nitrate are cautiously ignited in a porcelain crucible, until the weight remains constant. In case of lead salts of organic acids, the substance is very gently heated in a small covered porcelain crucible, which is included within a large one, also covered, until the organic matter is completely carbonized; the lids are then removed, when the mass begins to ignite, and a mixture of lead oxide with metallic lead results, which may still contain unconsumed carbon. A few pieces of recently fused ammonium nitrate are now thrown into the inner crucible, which has previously been removed from the flame, and both are again covered. The salt fuses, oxidizes the lead, and converts it partly into nitrate. The whole is now very gradually raised to a red heat, until no more fumes of hyponitric acid escape. The residuary oxide is then weighed.

The results are satisfactory.

2. *Determination as Sulphide.*

Lead may be completely precipitated from acid, neutral and

* Pogg. Annal. 91, 144.

† Ammonium oxalate, which has been so highly recommended as a precipitant for lead, is not so delicate as the carbonate. My experience in this respect coincides with F. Mohr's (Expt. No. 48).

alkaline solutions by hydrogen sulphide, and also from neutral and alkaline solutions by ammonium sulphide. Precipitation from acid solution is usually employed, especially in separations. A large excess of acid and also warming should both be avoided. The former is prejudicial to complete precipitation (§ 83, *f*), the latter may readily occasion the re-solution of the sulphide that has already been precipitated. In order to guard against incomplete precipitation, before filtering, test a portion of the supernatant fluid by mixing with a relatively large quantity of strong hydrogen sulphide water.

If the fluid contained no hydrochloric acid or metallic chloride, the lead sulphide is pure. After it has been filtered off, washed with cold water and dried, it is transferred, together with the filter-ash, to a porcelain crucible, a little sulphur added, and ignited in hydrogen at gentle redness till its weight is constant. It should always be allowed to cool in a current of the gas, before being weighed. As regards the apparatus, see § 108, 2, fig. 50. For the properties of the residue, see § 83, *f*. The results are satisfactory (H. ROSE). The heat of the ignition must not be too low, or the residue will contain too much sulphur, nor too high, or the lead sulphide will begin to volatilize, and lead disulphide will also be formed with loss of hydrogen sulphide. Drying the precipitate at 100° cannot be recommended (§ 83, *f*). If the fluid, on the contrary, contained hydrochloric acid or a metallic chloride, the lead sulphide contains chloride which cannot be removed even by boiling the precipitate with ammonium sulphide. If the precipitate were treated as above, we should obtain a tolerably pure sulphide, but not without loss from volatilization of chloride. A precipitate of this kind must therefore be decomposed with strong hydrochloric acid, the solution evaporated to dryness, the residue dissolved by heating with a concentrated solution of sodium acetate, and this solution diluted and poured with stirring into excess of strong hydrogen sulphide water. Or the lead chloride obtained may be evaporated, heated to 200°, and weighed as such (FINKENER*).

3. *Determination as Sulphate.*

a. By Precipitation.

α. Mix the solution (which should not be over dilute) with

* Handb. der anal. Chem. von H. ROSE, 6 Aufl. von FINKENER, 932.

moderately dilute pure sulphuric acid slightly in excess, and add to the mixture double its volume of common alcohol; wait a few hours, to allow the precipitate to subside; filter, wash the precipitate with common alcohol, dry, and ignite after the method described in § 53. Though a careful operator may use a platinum crucible, still a thin porcelain crucible is preferable. See also the remarks, 1, *a*.

β. In cases where the addition of alcohol is inadmissible, a greater excess of sulphuric acid must be used, and the precipitate, which is allowed some time to subside, filtered, and washed first with water acidulated with a few drops of sulphuric acid, then repeatedly with alcohol. The remainder of the process is conducted as in *α*.

If the fluid contained nitric acid, whether alcohol is used or not, it is advisable to evaporate on the water-bath after the addition of the sulphuric acid, till the nitric acid has escaped, otherwise the precipitation will not be complete. If the fluid contained hydrochloric acid or a metallic chloride, lead chloride is thrown down with the sulphate. In this case you must either evaporate the fluid with excess of sulphuric acid and heat the residue till sulphuric acid fumes escape to drive off the hydrochloric acid, or you must treat the precipitate and filter-ash in the crucible with concentrated sulphuric acid, evaporate and ignite to convert it into pure lead sulphate (FINKENER*).

For the properties of the precipitate, see § 83. The method *α* gives accurate results; those obtained by *β* are less exact (a little too low), but still however satisfactory, if the directions given are adhered to. If, on the contrary, a proper excess of sulphuric acid is not added, in the presence, for instance, of ammonium salts, the lead is not completely precipitated, and if pure water is used for washing, decided traces of the precipitate are dissolved.

b. *By Evaporation*.

α. Put the substance into a weighed dish, dissolve in dilute nitric acid, add moderately dilute pure sulphuric acid slightly in excess, and evaporate at a gentle heat; at last high over the lamp, until the excess of sulphuric acid is completely expelled. In the absence of organic substances, the evaporation may be effected

* Handb. der anal. Chem. von H. ROSE, 6 Aufl. von FINKENER, 932.

without fear in a platinum dish; but if organic substances are present, a light porcelain dish is preferable. With due care in the process of evaporation, the results are perfectly accurate.

β. Organic compounds of lead are converted into the sulphate by treating them in a porcelain crucible, with pure concentrated sulphuric acid in excess, evaporating cautiously in the well-covered crucible, until the excess of sulphuric acid is completely expelled, and igniting the residue. Should the latter not look perfectly white, it must be moistened once more with sulphuric acid, and the operation repeated. The method gives, when conducted with great care, accurate results; a trifling loss is, however, usually incurred, the escaping sulphur dioxide and carbon dioxide gases being liable to carry away traces of the salt.

4. *Determination as Lead Chromate.*

If the solution is not already distinctly acid render it so with acetic acid, then add potassium dichromate in excess, and, if free nitric acid is present, add sodium acetate in sufficient quantity to replace the free nitric acid by free acetic acid; let the precipitate subside at a gentle heat, and collect on a weighed filter dried at 100°, wash with water, dry at 100°, and weigh. The precipitate may also be ignited according to § 53, but in this case care must be taken that hardly any of the salt remains adhering to the paper, and that the heat is not too high. For the properties of the precipitate, see § 93, 2. The results are accurate. (Expt. No. 76.)

5. *Determination of Lead by Volumetric Analysis.*

Although there is no lack of proposed methods for the volumetric estimation of lead, we are still without a really good method for practical purposes, that is, a method which can be generally employed, and which is at the same time simple and exact. For the present, therefore, in almost all cases the gravimetric determination of lead is to be preferred to the volumetric. On my own part, at least, I cannot see that it is easier or any better, when one has the precipitate washed, to subject it to a volumetric process—whereby the accuracy is necessarily diminished—instead of igniting it gently and weighing. For this reason, the better volumetric methods will be but briefly described, the rest being altogether omitted.

a. The solution of the normal lead salt must be free from alkali salts, more especially from ammonium salts. It is precipitated with oxalic acid (not with ammonium oxalate), the well-washed precipitate is dissolved in nitric acid, sulphuric acid added, and the oxalic acid in the solution determined by potassium permanganate (§ 137) HEMPEL.

b. H. SCHWARZ's method.* To the nitric acid solution add ammonia or sodium carbonate, as long as the precipitate redissolves on shaking, mix with sodium acetate in not too small quantity, and then run in from a burette a solution of potassium dichromate (containing 14·761 grm. in the litre) till the precipitate begins to settle rapidly. Now place on a porcelain plate a number of drops of a neutral solution of silver nitrate, and proceed with the addition of the chromate, two or three drops at a time, stirring carefully after each addition. When the precipitate has settled tolerably clear, which takes only a few seconds, remove a drop of the supernatant liquid and mix it with one of the drops of silver solution on the plate. A small excess of chromate gives at once a distinct red coloration; the precipitated lead chromate does not act on the silver solution, but remains suspended in the drop. The number of c.c. of solution of chromate used (*minus* ·1 which SCHWARZ deducts for the excess) multiplied by ·0207 = the quantity of lead. If the fluid appear yellow before the reaction with the silver salt occurs, sodium acetate is wanting. In such a case first add more sodium acetate, then 1 c.c. of a solution containing ·0207 lead in 1 c.c., complete the process in the usual way, and deduct 1 c.c. from the quantity of chromate used on account of the extra lead added. Any iron present must be in the form of a ferric salt; metals whose chromates are insoluble, must be removed before the method can be employed.

c. The lead is precipitated according to 1, *a*, the carbonate (its composition is a matter of indifference in the present case) is washed, dissolved in a measured quantity of standard nitric acid, and a neutral solution of sodium sulphate added, whereby lead sulphate is precipitated and an equivalent quantity of sodium nitrate formed. If the nitric acid still free is now determined with standard alkali, we shall find the quantity of acid that has been neutralized by means of the lead, from which the amount of lead may be calculated. You may also determine the free nitric acid by adding standard sodium carbonate till, the vessel being on

* Dingl. polyt. Journ. 169, 284.

a black surface, a permanent turbidity is visible. Results good (F. MOHR*).

§ 117.

3. MERCURY IN MERCUROUS COMPOUNDS.

a. Solution.

Mercurous oxide and mercurous salts may generally be dissolved by means of dilute nitric acid, but without application of heat if conversion into mercuric compounds is to be avoided. If all that is required is to dissolve the mercury, the easiest way is to warm the substance for some time with nitric acid, then add hydrochloric acid, drop by drop, and continue the application of a moderate heat until a perfectly clear solution is produced, which now contains all the mercury in form of mercuric salts. Heating the solution to boiling, or evaporating, must be carefully avoided, as otherwise mercuric chloride may escape with the steam.

b. Determination.

If it is impracticable to produce a solution of the mercurous compound without formation of mercuric salts, it becomes necessary to convert the mercury completely into mercuric salts, when it may be determined as directed § 118. But if a solution of a mercurous compound has been obtained, quite free from mercuric salts, the determination of the mercury may be based upon the insolubility of mercurous chloride, and effected either gravimetrically or volumetrically. The process of determining mercury, described § 118, 1, *a*, may, of course, be applied equally well in the case of mercurous compounds.

1. *Determination as Mercurous Chloride.*

Mix the cold highly dilute solution with solution of sodium chloride, as long as a precipitate forms; let the precipitate subside, collect on a weighed filter, dry at 100°, and weigh. For the properties of the precipitate, see § 84. Results accurate. If the mercurous solution contains much free nitric acid, the greater part of this should be neutralized with sodium carbonate before adding the sodium chloride.

---

* His Lehrbuch der Titrirmethode, 3 Aufl. 115.

2. *Volumetric Methods.*

Several methods have been proposed under this head: the following are those which are most worthy of recommendation :—

a. Mix the cold solution with decinormal solution of sodium chloride (§ 141, *b*, *α*), until this no longer produces a precipitate, and is accordingly present in excess; filter and wash thoroughly, taking care, however, to limit the quantity of water used; add a few drops of solution of potassium chromate, then pure sodium carbonate, sufficient to impart a light yellow tint to the fluid, and determine by means of solution of silver nitrate (§ 141, *b*, *α*) the quantity of sodium chloride in solution, consequently the quantity which has been added in excess; this shows, of course, also the amount of sodium chloride consumed in effecting the precipitation. One mol. of Hg_2O is reckoned for 2 mols. of NaCl, consequently for every c.c. of the decinormal solution of sodium chloride, ·0208 grm. of mercurous oxide. As filtering and washing form indispensable parts of the process, this method affords no great advantage over the gravimetric; however, the results are accurate (Fr. Mohr*). The two methods, 1 and 2, *a*, may also be advantageously combined.

b. Precipitate the mercurous solution,† according to 1, with sodium chloride in a stoppered bottle, allow to subside, filter, wash, push a hole through the bottom of the filter, and rinse the precipitate into the bottle, which usually has some of the washed mercurous chloride adhering to its inside. Add a sufficient quantity of solution of potassium iodide, together with standard iodine solution (to 1 grm. Hg_2Cl_2 about 2·5 grm. KI and 100 c.c. $\frac{1}{10}$ normal iodine solution‡), insert the stopper, and shake till the precipitate has entirely dissolved ($Hg_2Cl_2 + 6KI + 2I = 2[HgI_2(KI)_2] + 2KCl$). As iodine is in excess, the solution appears brown. If any mercuric iodide separates, add potassium iodide to redissolve it. Now add from a burette solution of sodium thiosulphate—corresponding to decinormal iodine solution—till the fluid is decolorized and appears like water, transfer to a measuring flask, rinse and fill up to the mark, shake, take out an aliquot part, add starch paste to it, and determine the excess of sodium thiosulphate with $\frac{1}{10}$ iodine solution. After multiplying by the proper number, add the c.c. originally employed, subtract the c.c. of thiosulphate used, and

* Lehrbuch der Titrirmethode, 3 Aufl. 395.

† If oxide of mercury is also present, see § 118, 2.

‡ See § 146, 2.

reckon the quantity of mercury from the remainder. 2 at. iodine = 1 mol. Hg_2Cl_2. Results good (HEMPEL *).

§ 118.

4. MERCURY IN MERCURIC COMPOUNDS.

a. Solution.

Mercuric oxide, and those mercuric compounds which are insoluble in water, are dissolved, according to circumstances, in hydrochloric acid or in nitric acid. Mercuric sulphide is heated with hydrochloric acid, and nitric acid or potassium chlorate added until complete solution ensues; it is, however, most readily dissolved by suspending it in dilute potassa and transmitting chlorine, at the same time gently warming (H. ROSE). When a solution of mercuric chloride is evaporated on the water-bath, mercuric chloride escapes with the aqueous vapor.

b. Determination.

Mercury may be weighed in the *metallic state*, or as *mercurous chloride*, *mercuric sulphide*, or *mercuric oxide* (84); in separations it is sometimes determined as loss on ignition. It may also be estimated volumetrically.

The three first methods may be used in almost all cases; the determination as mercuric oxide, on the contrary, is possible only in mercurous or mercuric nitrates. The methods by which the mercury is determined as mercurous chloride or mercuric sulphide are to be preferred before those in which it is separated in the metallic form. The volumetric method 5, is of very limited application. The mercurous chloride obtained by method 2, instead of being weighed, may be determined volumetrically as in § 117, 2, *b*.

1. *Determination as Metallic Mercury.*

a. In the Dry Way.

The process is conducted in the apparatus illustrated by fig. 54.

Take a tube 18 inches long, and about 4 lines wide, made of difficultly fusible glass, and sealed at one end. First put into the tube a mixture of sodium hydrogen carbonate and powdered chalk, then a layer of quick-lime; these two will occupy the space from *a* to *b*. (Let the mixture for generating carbon dioxide take up about two inches.) Then add the intimate mixture of the sub-

* Annal. d. Chem. u. Pharm. 110, 176.

stance with an excess of quick-lime (*b–c*), then the lime-rinsings of the mortar (*c-d*), then a layer of quick-lime (*d–e*), and lastly, a loose stopper of asbestus (*e–f*). The anterior end of the tube is then drawn out, and bent at a somewhat obtuse angle. The manipulations in the processes of mixing and filling being the same as in organic analysis, they will be found in detail in the chapter on that subject.

A few gentle taps upon the table are sufficient to shake the contents of the tube down so as to leave a free passage through the whole length of the tube. The tube, so prepared and arranged, is now placed in a combustion furnace, the point being inserted into a flask containing water, the surface of which it should just touch, so that the opening may be just closed.

The tube is now surrounded with red-hot charcoal, in the same way as in organic analysis, proceeding slowly from *e* to *a*, the last traces of mercurial vapor being expelled by heating the mixture at

Fig. 54.

the sealed end of the tube. Whilst the tube still remains in a state of intense ignition, the neck is cut off at *f*, and carefully and completely rinsed into the receiving flask, by means of a washing-bottle. The small globules of mercury which have distilled over are united into a large one, by agitating the flask, and, after the lapse of some time, the perfectly clear water is decanted, and the mercury poured into a weighed porcelain crucible, where the greater portion of the water still adhering to it is removed with blotting-paper. The mercury is then finally dried under a bell-jar, over concentrated sulphuric acid, until the weight remains constant. Heat must not be applied. For the properties of the metal, see § 84. In the case of sulphides, in order to avoid the presence of vapor of water in the tube, which would give rise to the formation of sulphuretted hydrogen, the mixture of sodium hydrogen carbonate and chalk is replaced by magnesite. Mercuric iodide cannot be completely decomposed by lime. To analyze this in the dry way, substitute finely divided metallic copper for the lime (H. Rose*). The accuracy of the results is entirely dependent upon the care bestowed. The most highly accurate results are, however, obtained by the application of the somewhat more complicated modification adopted by Erdmann and Marchand for the determination of the atomic weight of mercury and of sulphur. For the details of this modified process, I refer to the original essay,† simply remarking here, that the distillation is conducted, in a combustion-tube, in a current of carbon dioxide gas, and that the distillate is received in a weighed bulb apparatus with the outer end filled with gold-leaf, to insure the condensation of every trace of mercury vapor. This way of receiving and condensing may be employed also in the analysis of amalgams (König‡).

b. In the Wet Way.

The solution, free from nitric acid, and mixed with free hydrochloric acid, is precipitated, in a flask, with an excess of a clear solution of stannous chloride, containing free hydrochloric acid; the mixture is boiled for a short time, and then allowed to cool. After some time, the perfectly clear supernatant fluid is decanted from the metallic mercury, which, under favorable circumstances, will be found united into one globule; if this is the case, the globule of mercury may be washed at once by decantation, first with water acidulated with hydrochloric acid, and finally with pure water; it is dried as in *a*.

If, on the other hand, the particles of the mercury have not united, their union in one globule may as a rule be readily effected by boiling a short time with some moderately dilute hydrochloric acid mixed with a few drops of stannous chloride (having, of course, previously removed by decantation the supernatant clear fluid). For the properties of metallic mercury, see § 84.

Instead of stannous chloride, other reducing agents may be used, especially phosphorous acid at a boiling temperature. This method gives accurate results only when conducted with the greatest care. In general, a little mercury is lost (Comp. Expt. No. 77).

---

* Pogg. Annal. 110, 546.

† Journ. f. prakt. Chem. 31, 385; also Pharm. Centralbl. 1844, 354.

‡ Journ. f. prakt. Chem. 70, 64.

2. *Determination as Mercurous Chloride.*

a. After H. ROSE.* Mix the mercuric solution (which may contain nitric acid) with hydrochloric acid and excess of phosphorous acid (obtained by the deliquescence of phosphorus in moist air), allow to stand for 12 hours in the cold or at a very gentle heat (at all events under 60°), collect the mercury, now completely separated as mercurous chloride, on a weighed filter, wash with hot water, dry at 100°, and weigh. Results perfectly satisfactory.

3. *Determination as Mercuric Sulphide.*

The solution is sufficiently diluted, acidulated with hydrochloric acid, and precipitated with clear saturated hydrogen sulphide water (or in the case of large quantities, by passing the gas); filter after allowing the precipitate a short time to deposit, wash quickly with cold water, dry at 100°, and weigh. Results very satisfactory.

If from any cause (*e.g.* presence of ferric salts, free chlorine, or the like) the precipitate should contain free sulphur, the filter is spread out on a glass plate, the precipitate removed to a porcelain dish by the aid of a jet from the wash-bottle, and warmed for some time with a moderately strong solution of sodium sulphite. The filter, having been in the mean while somewhat dried on the glass plate, is replaced in the funnel, the supernatant fluid is poured on to it, the treatment with sodium sulphite is repeated, and the precipitate (now free from sulphur) is finally collected on the filter, washed, dried, and weighed. Results very good (J. LÖWE†).

Should the quantity of sulphur mixed with the precipitate be not very large, it may be removed also as follows: the precipitate is first washed with water, then twice with strong alcohol, then repeatedly with carbon disulphide, till a few drops of the washings evaporate on a watch-glass without leaving a residue. (The precipitate is retained on the filter throughout this operation.)

Properties of mercuric sulphide, § 84

4. *Determination as Oxide.*

In the mercurous and mercuric salts of the nitrogen acids, the metal may be very conveniently determined in the form of mercuric oxide (MARIGNAC*). For this purpose, the salt is heated in

* Pogg. Annal. 110, 529.

† Journ. f. prakt. Chem. 77, 73.

‡ Jahresber. von Liebig u. Kopp, 1849, 594.

a bulb-tube, of which the one end, drawn out to a point, dips under water, the other end being connected with a gasometer, by means of which dry air is transmitted through the tube, as long as the application of heat is continued. In this way complete decomposition of the salt is readily effected, without reaching the temperature at which the oxide itself would be decomposed.

5. *Volumetric Methods.*

After J. J. SCHERER.* Mercuric nitrate or chloride may be directly determined with sodium thiosulphate. The reactions are as follows: $2H_2O + 3Hg(NO_3)_2 + 2Na_2S_2O_3 = (HgS)_2 \cdot Hg(NO_3)_2 + 2Na_2SO_4 + 4HNO_3$; or, $2H_2O + 3HgCl_2 + 2Na_2S_2O_3 = (HgS)_2 \cdot HgCl_2 + 2Na_2SO_4 + 4HCl$. The process is conducted as follows in the case of mercuric nitrate: Mix the highly dilute solution with a little free nitric acid in a tall glass, and add drop by drop solution of sodium thiosulphate—12·4 grm. in a litre. Each drop produces an intense yellow cloud, which on shaking quickly subsides in the form of a heavy flocculent precipitate $(HgS)_2 \cdot Hg(NO_3)_2$. In order to distinguish clearly the exact end of the reaction, SCHERER recommends to transfer the fluid towards the end to a measuring flask, to take out $\frac{1}{3}$ or $\frac{1}{2}$ of the clear fluid and to finish with this. The portion of thiosulphate last used is multiplied by 3 or 2, as the case may be, and added to the quantity first used. 1 c.c. of the solution corresponds to ·015 mercury, or ·0162 mercuric oxide. The relation is not changed even when the fluid contains another acid (sulphuric, phosphoric).

In the case of mercuric chloride, the highly dilute solution is mixed with a little hydrochloric acid and warmed, nearly to boiling, before beginning to add the sodium thiosulphate. At first a white turbidity is formed, then the precipitate separates in thick flocks. When the solution begins to appear transparent, the precipitant is added more slowly. In order to hit the end of the reaction exactly, small portions must be filtered off towards the close. The precipitate must be completely white; if too much thiosulphate has been added, it is gray or blackish, and the experiment must be repeated. SCHERER obtained very accurate results. Of course no other metals must be present that exert a decomposing action on sodium thiosulphate.

* His Lehrbuch der Chemie, i. 513.

§ 119.

5. COPPER.

a. Solution

Many cupric salts dissolve in water. Metallic copper is best dissolved in nitric acid. Cupric oxide, and those cupric salts which are insoluble in water, may be dissolved in nitric, hydrochloric, or sulphuric acid. Cupric sulphide is treated with fuming nitric acid, or it is heated with moderately dilute nitric acid, until the separated sulphur exhibits a pure yellow tint; addition of a little hydrochloric acid or potassium chlorate greatly promotes the action of the dilute acid.

b. Determination.

Copper may be weighed in the form of *cupric oxide*, or in the *metallic state*, or as *cuprous sulphide* (§ 85). Into the form of cupric oxide it is converted by precipitation, or ignition, sometimes with previous precipitation as sulphide. The determination as cuprous sulphide is preceded usually by precipitation either as cupric sulphide or as cuprous sulphocyanate. Copper may be determined also by various volumetric and indirect methods.

We may convert into

1. CUPRIC OXIDE:

a. By Precipitation as hydrated cupric oxide and subsequent ignition: All cupric salts soluble in water, and also those insoluble salts, the acids of which may be removed upon solution in nitric acid, provided no non-volatile organic substances be present.

b. By Precipitation, preceded by Ignition of the compound: Such of the salts enumerated under *a* as contain a non-volatile organic substance, thus more particularly cupric salts of non-volatile organic acids.

c. By Ignition: Cupric salts of oxygen acids that are readily volatile or decomposable at a high temperature (cupric carbonate, cupric nitrate).

2. METALLIC COPPER: Copper in all solutions free from other metals precipitable by zinc or the galvanic current, also the oxides of copper.

3. CUPROUS SULPHIDE: Copper in all cases in which no other metals are present that are precipitable by hydrogen sulphide or potassium sulphocyanate.

Of the several methods of effecting the estimation of copper, No. 3 is particularly to be recommended for use in laboratories; method 2 is also very convenient, and well adapted for assaying. Of the volumetric methods, one is suited for technical purposes, the other for the estimation of small quantities of copper. For technical purposes there are, besides, also several colorimetric methods, proposed by HEINE, VON HUBERT, JACQUELAIN, A. MÜLLER, and others, which are, all of them, based upon the comparison of an ammoniacal solution of copper, of unknown strength, with others of known strength.*

LEVOL'S indirect method of estimating copper, which is based upon the diminution of weight suffered by a strip of copper when digested in a close-stoppered flask with ammoniacal solution of copper till decolorization is effected, takes too much time, and is apt to give false results (PHILLIPS,† ERDMANN‡). The latter remark applies also to the indirect method proposed by RUNGE, which consists in boiling the solution of copper, free from nitric acid and ferric salts, in presence of some free hydrochloric acid, in a flask, with a weighed strip of copper, and, after decolorization of the fluid, determining the loss of weight suffered by the copper.

1. *Determination as Cupric Oxide.*

a. By direct Precipitation as Oxide.

Heat the rather dilute *neutral* or *acid* solution in a platinum or porcelain dish, to incipient ebullition, add a somewhat dilute solution of pure soda or potassa until the formation of a precipitate ceases, and keep the mixture a few minutes longer at a temperature near boiling. Allow to subside, filter, wash by decantation twice or thrice, boiling up each time, then collect it on the filter, wash thoroughly with hot water, dry, and ignite in a porcelain or platinum crucible, as directed § 53. Do not use the blow-pipe. After ignition, and having added the ash of the filter, let the crucible cool in the desiccator, and weigh. The action of reducing gases must be carefully guarded against in the process of ignition.

It will sometimes happen, though mostly from want of proper attention to the directions here given, that particles of the precipi-

* This subject hardly comes within the scope of the present work. I therefore refer to AL. MÜLLER, das Complementärcolorimeter; Chemnitz, 1854; BODEMANN'S Probirkunst von KERL, 222; also to DEHMS, Zeitschr. f. anal. Chem. 3, 218, and GUSTAV BISCHOF, jun., *Ib.* 6, 459.

† Annal. d. Chem. u. Pharm. 81, 208. ‡ Jour. f. prakt. Chem. 75, 211.

tate adhere so tenaciously to the dish as to be mechanically irremovable. In a case of this kind, after washing the dish thoroughly, dissolve the adhering particles with a few drops of nitric acid, and evaporate the solution over the principal mass of the precipitated oxide, before you proceed to ignite the latter. Should the solution be rather copious, it must first be concentrated by evaporation, until only *very little* of it is left. For the properties of the precipitate, see § 85.

With proper attention to the directions here given, the results obtained by this method are quite accurate, otherwise they may be either too high or too low. Thus, if the solution be not sufficiently dilute, the precipitant will fail to throw down the whole of the copper; or if the precipitate be not thoroughly washed with hot water, it will retain a portion of the alkali; or if the ignited precipitate be allowed to stand exposed to the air before it is weighed, an increase of weight will be the result; and so, on the other hand, a diminution of weight, if the oxide be ignited with the filter or under the influence of reducing gases, as thereby cuprous oxide would be formed. Should a portion of the oxide have suffered reduction, it must be reoxidized by moistening with nitric acid, evaporating cautiously to dryness, and exposing the residue to a gentle heat, increasing this gradually to a high degree of intensity.

Let it be an invariable rule to test the filtrate for copper with hydrogen sulphide water. If, notwithstanding the strictest compliance with the directions here given, the addition of this reagent produces a precipitate, or imparts a brown tint to the fluid, this is to be attributed to the presence of organic matter; in that case, concentrate the filtrate and wash-water by evaporation, acidify, precipitate with hydrogen sulphide water, filter, incinerate the filter, heat with nitric acid, dilute, filter, concentrate, precipitate with soda, and add the oxide obtained to the main quantity.

Never neglect to test the cupric oxide after weighing for alkali or alkali salt by boiling it with water. If either is present, the oxide must be exhausted with hot water, and then reignited and reweighed. Finally, dissolve the oxide in hydrochloric acid to detect and if necessary to estimate any silicic acid it may contain.

In default of sufficiently pure potash or soda, the carbonate may be used, but the solution must not contain more than 1 grm. copper in the litre; the alkali carbonate must only be added slightly in excess, and the mixture must be boiled for half an hour.

The bluish-green precipitate will then turn dark brown and granular, and may be easily washed (GIBBS*).

From *ammoniacal* solutions, also, copper may be precipitated by soda or potassa. In the main, the process is conducted as above. After precipitation the mixture is heated, until the supernatant fluid has become perfectly colorless; the fluid is then filtered off with the greatest possible expedition. If allowed to cool with the precipitate in it, a small portion of the latter would redissolve.

b. By Precipitation as Oxide, preceded by Ignition of the Substance.

Heat the substance in a porcelain crucible, until the organic matter present is totally destroyed; dissolve the residue in dilute nitric acid, filter if necessary, and treat the clear solution as directed in *a*.

c. By Ignition.

The salt is put into a platinum or porcelain crucible, and exposed to a very gentle heat, which is gradually increased to intense redness; the residue is then weighed. As cupric nitrate spirts strongly when ignited, it is always advisable to put it into a small covered platinum crucible, and to place the latter in a large one, also covered. With proper care, the results are accurate. Cupric salts of organic acids may also be converted into cupric oxide by simple ignition. To this end, the residue first obtained, which contains cuprous oxide, is completely oxidized by ignition with mercuric oxide (which leaves no residue on ignition), or, with less advantage, by repeated moistening with nitric acid, and ignition. A loss of substance is generally incurred by the use of nitric acid from the difficulty of avoiding spirting.

2. *Determination as Metallic Copper.*

a. By Precipitation with Zinc or Cadmium.†

Introduce the solution of copper, after having, if required, first freed it from nitric acid, by evaporation with hydrochloric acid or

---

* Zeitschr. f. anal. Chem. 7, 258.

† The method of precipitating copper by iron or zinc, and weighing it in the metallic form, was proposed long ago; see PFAFF's Handbuch der analytischen Chemie, Altona, 1822, 2, 269; where the reasons are given for preferring zinc as a precipitant, and hydrogen sulphide is recommended as a test for ascertaining whether the precipitation is complete. I mention this with reference to F. MOHR's paper in the Annal. d. Chem. u. Pharm. 96, 215, and BODEMANN's Probirkunst von KERL, 220.

sulphuric acid, into a weighed platinum dish, dilute, if necessary with some water, throw in a piece of zinc (soluble in hydrochloric acid without residue), and add, if necessary, hydrochloric acid in sufficient quantity to produce a moderate evolution of hydrogen. If, on the other hand, this evolution should be too brisk, owing to too large excess of acid, add a little water. Cover the dish with a watch-glass, which is afterwards rinsed into the dish with the aid of a washing-bottle. The separation of the copper begins immediately; a large proportion of it is deposited on the platinum in form of a solid coating; another portion separates, more particularly from concentrated solutions, in the form of red spongy masses. Application of heat, though it promotes the reaction, is not absolutely necessary; but there must always be sufficient free acid present to keep up the evolution of hydrogen. After the lapse of about an hour or two, the whole of the copper has separated. To make sure of this, test a small portion of the supernatant fluid with hydrogen sulphide water; if this fails to impart a brown tint to it, you may safely assume that the precipitation of the copper is complete. Ascertain now, also, whether the zinc is entirely dissolved, by feeling about for any hard lumps with a glass rod, and observing whether renewed evolution of hydrogen will take place upon addition of some hydrochloric acid. If the results are satisfactory in this respect also, press the copper together with the glass rod, decant the clear fluid, which is an easy operation, pour, without loss of time, boiling water into the dish, decant again, and repeat this operation until the washings are quite free from hydrochloric acid. Decant the water now as far as practicable, rinse the dish with strong alcohol, dry at 100°, let it cool, and weigh. If you have no platinum dish, the precipitation may be effected also in a porcelain crucible or glass dish; but it will, in that case, take a longer time; and the whole of the copper will be obtained in loose masses, and not firmly adhering to the sides of the crucible or dish, as in the case of precipitation in platinum vessels.

The results are very accurate. The direct experiment, No. 78, gave 100·0 and 100·06, instead of 100. Fr. Mohr (*loc. cit.*) obtained equally satisfactory results by precipitating in a porcelain crucible.*

* Storer (On the alloys of copper and zinc, Cambridge, 1860, p. 47) says that the precipitated copper retains water, but I have not found this to be the case.

Zinc being sometimes difficult to obtain of sufficient purity, cadmium may be used instead; it dissolves with less violence in strongly acid copper solutions. It may be used in the form of rod in which it usually occurs in commerce (CLASSEN*).

b. By Precipitation with the Galvanic Current.

This method makes us independent of pure zinc or cadmium, and yields the copper in a compact form, readily washed and determined. It is now largely used in copper works, constant batteries have been employed for it, and the whole process has been organized for use on a large scale by LUCKOW, and adopted by the Mansfeld Ober-Berg-und Hütten-Direction in Eisleben.† A small electrolytic apparatus without separate battery, for single precipitations, has been described by ULLGREN.‡

c. By Ignition in Hydrogen.

The oxides of copper when ignited in a current of pure hydrogen are converted into metallic copper, and may thus be conveniently analyzed. Occasionally the cupric oxide obtained by 1, *a* or *b*, is reduced either at once, or after weighing; in the latter case the reduction serves as a control.

3. *Determination as Cuprous Sulphide.*

a. By Precipitation as Cupric Sulphide.

Precipitate the solution—which is best moderately acid, but should not contain a great excess of nitric acid—according to the quantity of copper present, either by the addition of strong hydrogen sulphide water, or by passing the gas. In the absence of nitric acid it is well to heat nearly to boiling while the gas is passing, as this makes the precipitate denser, and it is more easily washed. When the precipitate has fully subsided, and you have made sure that the supernatant fluid is no longer colored or precipitated by strong *hydrogen sulphide water*, filter quickly, wash the precipitate without intermission with water containing hydrogen sulphide, and dry on the filter with some expedition. Transfer to a weighed porcelain crucible, add the filter-ash and some pure powdered sulphur, and ignite strongly in a stream of hydrogen (§ 108, fig. 50). It is advisable to use a glass blow-pipe. The results are very accurate (H. ROSE§).

---

* Journ. f. prakt. Chem. 96, 259.

† Zeitschr. f. anal. Chem. 8, 23 and 11, 1. Compare also GIBBS, *Ib.* 3, 334, and LECOQ DE BOISBAUDAN, *Ib.* 7, 253. ‡ *Ib.* 7, 442. § Pogg. Annal. 110, 138.

This method, which was recommended by BERZELIUS, and afterwards by BRUNNER, has only lately received a very practical form, from the apparatus introduced by H. ROSE. I feel great pleasure in recommending it. In my own laboratory it is in frequent use.

b. By Precipitation as Cuprous Sulphocyanate, after RIVOT.*

The solution should be as free as possible from nitric acid and free chlorine, and should contain little or no free acid. Add sulphurous or hypophosphorous acid in sufficient quantity, and then solution of potassium sulphocyanate in the least possible excess. The copper precipitates as white cuprous sulphocyanate. It is filtered after standing some time, washed and dried, mixed with sulphur, ignited in hydrogen in the apparatus mentioned in *a*, and this ignition with sulphur is repeated till the weight is constant. The precipitate may also be collected on a weighed filter, dried at 100°, and then weighed. The experiment, No. 80, conducted in the latter way, gave 99·66 instead of 100. The process yields satisfactory results, but they are always inclined to be a little too low, as the cuprous sulphocyanate is not absolutely insoluble. The loss is larger in the presence of much free acid.

c. Cuprous and cupric oxide, cupric sulphate, and many other salts of copper (but not chloride, bromide, or iodide) may be directly converted into cuprous sulphide, by mixing with sulphur and igniting in hydrogen as in *a* (H. ROSE, *loc. cit.*). The results are thoroughly satisfactory.

4. *Volumetric Methods.*

a. DE HAEN'S METHOD.†

I recommend this method, which was devised in my own laboratory, as more especially applicable in cases where small quantities of copper are to be estimated in an expeditious way. The method is based upon the fact that, when a cupric salt in solution is mixed with potassium iodide in excess, cuprous iodide and free iodine are formed, the latter remaining dissolved in the solution of potassium iodide: $CuSO_4 + 2KI = CuI + K_2SO_4 + I$. Now, by estimating the iodine by BUNSEN'S method, or with sodium thiosulphate (§ 146), we learn the quantity of copper, as 1 at. iodine (126·85) corresponds to 1 at. copper (63·4). The following is the most convenient way of proceeding: Dissolve the compound

* Compt. Rend. 38, 868; Journ. f. prakt. Chem. 62, 252.

† Annal. d. Chem. u. Pharm. 91, 237.

of copper in sulphuric acid, best to a neutral solution; a moderate excess of free sulphuric acid, however, does not injuriously affect the process. Dilute the solution, in a measuring flask, to a definite volume; 100 c.c. should contain from 1 to 2 grm. of copper. Introduce now about 10 c.c. of potassium iodide solution (1 in 10) into a stoppered bottle, add 10 c.c. of the copper solution, mix, allow to stand 10 minutes, and then determine the separated iodine, either with sulphurous acid and iodine (§ 146, 1), or with sodium thiosulphate (§ 146, 2). The copper solution must be free from ferric salts and other bodies which decompose potassium iodide, also free nitric acid, and free hydrochloric acid; and the solution must not be allowed to stand too long before titration. With strict attention to these rules, the results are accurate. De Haen obtained, for instance, ·3567 instead of ·3566 of cupric sulphate, 99·89 and 100·1 instead of 100 of metallic copper. Further experiments (No. 81) have convinced me, however, that, though the results attainable by this method are satisfactory, they are not always quite so accurate as would be supposed from the above figures given by De Haen. Acting upon Fr. Mohr's suggestion, I tried to counteract the injurious influence of the presence of nitric acid, by adding to the solution containing nitric acid, first, ammonia in excess, then hydrochloric acid to slight excess; the result was by no means satisfactory. The reason of this is that a solution of ammonium nitrate, mixed with some hydrochloric acid, will, even after a short time, begin to liberate iodine from solution of potassium iodide.

§ 120.

6. Bismuth.

a. Solution.

Metallic bismuth, bismuth trioxide, and all other compounds of that metal, are dissolved best in nitric acid, more or less diluted. It must be borne in mind that hydrochloric acid solutions of bismuth, if concentrated, cannot be evaporated without loss of bismuth chloride.

b. Determination.

Bismuth is weighed in the form of *trioxide*, of *chromate*, of *sulphide*, or in the *metallic state*. The compounds of bismuth are converted into trioxide by ignition, by precipitation as basic carbonate, or by repeated evaporation of the nitric solution. These are sometimes preceded by separation as sulphide. The determination as metallic bismuth is frequently preceded by precipitation as sulphide or as basic chloride.

We may convert into

1. Bismuth trioxide:

a. By Precipitation as basic Bismuth Carbonate. All compounds of bismuth which dissolve in nitric acid to nitrate, *no* other acid remaining in the solution.

b. By Ignition.

α. Bismuth salts of readily volatile oxygen acids.

β. Bismuth salts of organic acids.

c. By Evaporation. Bismuth in nitric acid solution.

d. By Precipitation as Bismuth Trisulphide. All compounds of bismuth without exception.

2. Bismuth chromate. All compounds named in 1, *a*.

3. Bismuth trisulphide. The compounds of bismuth without exception.

4. Metallic bismuth: The trioxide and oxygen salts, the sulphide, the basic chloride, in which latter form the bismuth may be precipitated out of all its solutions.

1. *Determination of Bismuth as Trioxide.*

a. By Precipitation as Bismuth Carbonate.

If the solution is concentrated add water, taking no notice of any precipitate of basic nitrate that may be formed. Mix with ammonium carbonate in *very slight* excess, and heat for some time nearly to boiling; filter, dry the precipitate, and ignite in the manner directed § 116, 1 (Ignition of lead carbonate); the process of ignition serves to convert the carbonate into bismuth trioxide. For the properties of the precipitate and residue, see § 86. The method gives accurate results, though generally a trifle too low, owing to the circumstance that bismuth carbonate is not absolutely insoluble in ammonium carbonate. Were you to attempt to precipitate bismuth, by means of ammonium carbonate, from solutions containing sulphuric acid or hydrochloric acid, you would obtain incorrect results, since with the basic carbonate, basic sulphate or basic chloride would be precipitated, which are not decomposed by excess of ammonium carbonate. Were you to filter off the precipitate without warming, a considerable loss would be sustained, as the whole of the basic carbonate would not have been separated (Expt. No. 83).

b. By Ignition.

α. Compounds like bismuth carbonate or nitrate are ignited in a porcelain crucible until their weight remains constant.

β. Salts of organic acids are treated like the corresponding compounds of copper (§ 119, 1, *c*).

c. By Evaporation.

The solution of the nitrate is evaporated, in a porcelain dish on the water-bath, till the neutral salt remains in syrupy solution; add water, loosen the white crust that is formed with a glass rod from the sides, evaporate again on a water-bath, reprecipitate with water, and repeat the whole operation three or four times. After the dry mass on the water-bath has ceased to smell of nitric acid, it is allowed to cool thoroughly, and then treated with cold water containing a little ammonium nitrate (1 in 500); after the residue and fluid have been a short time together, filter, wash with the weak solution of ammonium nitrate, dry and ignite (§ 53). Results very satisfactory (J. Löwe*).

d. By Precipitation as Bismuth Trisulphide.

Dilute the solution with water slightly acidulated with acetic acid (to prevent the precipitation of a basic salt), and precipitate with hydrogen sulphide water or gas; allow the precipitate to subside, and test a portion of the supernatant fluid with *hydrogen sulphide water:* if it remains clear, which is a sign that the bismuth is completely precipitated, filter (the filtrate should smell strongly of H_2S), and wash the precipitate with water containing hydrogen sulphide. Or mix with ammonia until the free acid is neutralized, then add ammonium sulphide in excess, and allow to digest for some time.

The washed precipitate may now be weighed in three different forms, viz., as trisulphide, as metal, or as trioxide. The treatment in the two former cases will be described in 3 and 4: in the latter case proceed as follows:

Spread the filter out on a glass plate and remove the precipitate to a vessel by means of a jet of water from the wash-bottle—or, if this is not practicable, put the precipitate and filter together into the vessel—and heat gently with moderately strong nitric acid

* Journ. f. prakt. Chem. 74, 344.

until complete decomposition is effected; the solution is then diluted with water slightly acidulated with acetic or nitric acid, and filtered, the filter being washed with the acidulated water; the filtrate is then finally precipitated as directed in *a*.

2. *Determination of Bismuth as Chromate* (J. Löwe*).

Pour the solution of bismuth, which must be as neutral as possible, and must, if necessary, be first freed from the excess of nitric acid by evaporation on the water-bath, into a warm solution of pure potassium dichromate in a porcelain dish, with stirring, and take care to leave the alkali chromate slightly in excess. Rinse the vessel which contained the solution of bismuth with water containing nitric acid into the porcelain dish. The precipitate formed must be orange-yellow, and dense throughout; if it is flocculent, and has the color of the yolk of an egg, this is a sign that there is a deficiency of potassium dichromate; in which case add a fresh quantity of this salt, taking care, however, to guard against too great an excess, and boil until the precipitate presents the proper appearance. Boil the contents of the dish for ten minutes, with stirring; then wash the precipitate, first by repeated boiling with water and decantation on to a weighed filter, at last thoroughly on the latter with boiling water; dry at about 120°, and weigh. For the properties and composition of the precipitate, see § 86. Results very satisfactory.

3. *Determination of Bismuth as Trisulphide.*

Precipitate the bismuth as trisulphide according to 1, *d*. If the precipitate contains free sulphur, extract the latter by boiling with solution of sodium sulphite, or by treatment with carbon disulphide (compare the determination of mercury as sulphide, § 118, 3), collect on a weighed filter, dry at 100°, and weigh.

The drying must be conducted with caution. At first the precipitate loses weight, by the evaporation of water, then it gains weight, from the absorption of oxygen. Hence you should weigh every half hour, and take the lowest weight as the correct one. Compare Expt. No. 58. Properties and composition, § 86, *g*.

The bismuth sulphide cannot be conveniently converted into the metallic state by ignition in hydrogen, as its complete decomposition is a work of considerable time. As regards reduction with potassium cyanide, see 4.

* Journ. f. prakt. Chem. 67, 464.

4. *Determination of Bismuth as Metal.*

The oxide, sulphide, or basic chloride that are to be reduced are fused in a porcelain crucible with five times their quantity of ordinary potassium cyanide. The crucible must be large enough. In the case of oxide and basic chloride, the reduction is completed in a short time at a gentle heat; sulphide, on the other hand, requires longer fusion and a higher temperature. The operation has been successful if on treatment with water metallic grains are obtained. These grains are first washed completely and rapidly with water, then with weak and lastly with strong alcohol, dried and weighed. If you have been reducing the sulphide, and on treating the fused mass with water a black powder (a mixture of bismuth with bismuth sulphide) is visible, besides the metallic grains, it is necessary to fuse the former again with potassium cyanide.

It sometimes happens that the crucible is attacked, and particles of porcelain are found mixed with the metallic bismuth; to prevent this from spoiling the analysis, weigh the crucible together with a small dried filter before the experiment, collect the metal on the filter, dry and weigh the crucible with the filter and bismuth again. Results good (H. Rose*).

The precipitation of bismuth as basic chloride, and the reduction of the latter with potassium cyanide, has been recommended by H. Rose.† The process is conducted as follows: Nearly neutralize any large excess of acid that may be present with potassa, soda, or ammonia, add ammonium chloride in sufficient quantity (if hydrochloric acid is not already present), and then a rather large quantity of water. After allowing to stand some time, test whether a portion of the clear supernatant fluid is rendered turbid by a further addition of water; and then, if required, add water to the whole till the precipitation is complete. Finally filter, wash completely with cold water, dry and fuse according to the directions just given with potassium cyanide. It is less advisable to dry the precipitate at 100°, weigh and calculate the metal present from the formula BiOCl, as washing causes a slight alteration in its composition (unless a little hydrochloric acid is added to the wash-water, which is inconvenient when the precipitate is collected on a weighed filter), and if precipitated in the presence of sulphuric, phosphoric acids, &c., it is liable to contain small quantities of these acids. Results accurate.

* Pogg. Annal. 91, 104, and 110, 136. † *Ib.* 110, 425.

§ 121.

7. Cadmium.

a. Solution.

Cadmium, its oxide, and all the other compounds insoluble in water, are dissolved in hydrochloric acid or in nitric acid.

b. Determination.

Cadmium is weighed either in the form of *oxide*, or in that of *sulphide* (§ 87). It may also be weighed as *sulphate*, and in the absence of other bases precipitable by oxalic acid, it may be estimated volumetrically.

We may convert into

1. Cadmium oxide:

a. By Precipitation. The compounds of cadmium which are soluble in water; the insoluble compounds, the acid of which is removed upon solution in hydrochloric acid; cadmium salts of organic acids.

b. By Ignition. Cadmium salts of readily volatile or easily decomposable inorganic oxygen acids.

2. Cadmium sulphide: All compounds of cadmium without exception.

3. Cadmium sulphate: All compounds of cadmium, in the absence of other non-volatile substances.

1. *Determination as Cadmium Oxide.*

a. By Precipitation.

Precipitate with potassium carbonate, wash the precipitated cadmium carbonate, and convert it, by ignition, into oxide. The precipitation is conducted as in the case of zinc, § 108, 1, *a*. The cadmium oxide which adheres to the filter may easily be reduced and volatilized; it is therefore necessary to be cautious. In the first place choose a thin filter, transfer the dried precipitate as completely as possible to the crucible, replace the filter in the funnel, and moisten it with ammonium nitrate solution, allow to dry, and then burn carefully in a coil of platinum wire. Let the ash fall into the crucible containing the mass of the precipitate, ignite carefully, avoiding the action of reducing gases, and finally weigh. It is difficult to remove the last portions of carbonic acid; you must therefore repeat the ignition till the weight remains constant.

Properties of precipitate and residue, § 87. Results generally a little too low.

b. By Ignition.

Same process as for zinc, § 108, 1, *c*.

2. *Determination as Cadmium Sulphide.*

It is best to precipitate the moderately acid solution with hydrogen sulphide water or gas, which must be used in sufficient excess. The presence of a considerable quantity of free hydrochloric or nitric acid may—especially if the solution is not enough diluted—prevent complete precipitation, hence such an excess should be avoided, and the clear supernatant fluid should in all cases be tested, by the addition of a relatively large amount of hydrogen sulphide water to a portion, before being filtered. Alkaline solutions of cadmium may be precipitated with ammonium sulphide. If the cadmium sulphide is free from admixed sulphur, it may be at once collected on a weighed filter, washed first with diluted hydrogen sulphide water mixed with a little hydrochloric acid, then with pure water, dried at 100°, and weighed; if, on the contrary, it contains free sulphur, it may be purified by boiling with a solution of sodium sulphite, or by treatment with carbon disulphide (see Mercuric Sulphide, § 118, 3). Results accurate. The precipitation of sulphur may occasionally be obviated by adding to the cadmium solution potassium cyanide till the precipitate first formed is redissolved, and then precipitating this solution with hydrogen sulphide.

If the cadmium sulphide is not to be weighed as such, warm it, together with the filter, with moderately strong hydrochloric acid, till the precipitate has dissolved and the odor of hydrogen sulphide is no longer perceptible, filter and precipitate the solution as in 1, *a*, after having removed the excess of free acid for the most part by evaporation.

3. *Determination as Cadmium Sulphate.*

Same process as for magnesium (§ 104, 1). The $CdSO_4$ may be rather strongly ignited without decomposition.

4. W. GIBBS* determines cadmium *volumetrically* by mixing the concentrated solution of the sulphate, nitrate, or chloride with excess of oxalic acid and a quantity of strong alcohol, filtering, washing with alcohol, dissolving in hot hydrochloric acid and

* Zeitschr. f. anal. Chem. 7, 259.

determining the oxalic acid with permanganate (§ 137). W. G. LEISON* obtained satisfactory results by this process.

Supplement to the Fifth Group.

§ 122.

8. PALLADIUM.

Palladium is converted, for the purpose of estimation, into the *metallic state*; or—in many separations—into *potassium palladic chloride.*

1. *Determination as Palladium.*

a. Neutralize the solution of palladious chloride almost completely with sodium carbonate, mix with solution of mercuric cyanide; and heat gently for some time, until the odor of hydrocyanic acid has gone off. A yellowish-white precipitate of palladious cyanide will subside; from dilute solutions, only after the lapse of some time. Wash first by decantation, then on the filter, dry thoroughly, ignite cautiously, finally over the gas blowpipe till the palladium paracyanide first formed is decomposed, then ignite in hydrogen, since the palladium has been slightly oxidized. As soon as the lamp is removed, stop the hydrogen to prevent absorption, and weigh the metal. If the solution contains palladious nitrate, evaporate it first with hydrochloric acid to dryness; as otherwise the precipitate obtained deflagrates upon ignition (WOLLASTON). Results exact.

b. Mix the solution of palladious chloride or nitrate with sodium or potassium formate, and warm until no more carbonic acid escapes. The palladium precipitates in brilliant scales (DÖBEREINER).

c. Precipitate the acid solution of palladium with hydrogen sulphide, filter, wash with boiling water, roast, dissolve in hydrochloric acid and nitric acid, and precipitate as in *a*.

Exposed to a moderate red heat *metallic palladium* becomes covered with a film varying from violet to blue, but at a higher temperature it recovers its lustre, which it keeps after being suddenly cooled, for instance, with cold water. This tarnishing and recovery of the metallic lustre is not attended with any percepti-

* Zeitschr. f. anal. Chem. 10, 343.

ble difference of weight. Palladium which has taken up oxygen is immediately reduced in hydrogen; when cooled in the current of gas, it retains some absorbed hydrogen. Palladium requires the very highest degree of heat for its fusion. It dissolves readily in nitrohydrochloric acid, with difficulty in pure nitric acid, more easily in nitric acid containing nitrous acid, with difficulty in boiling concentrated sulphuric acid.

2. *Determination as Potassium Palladic Chloride.*

Evaporate the solution of palladic chloride with potassium chloride and nitric acid to dryness, and treat the mass when cold with alcohol of ·833 sp. gr., in which the double salt is insoluble. Collect on a weighed filter, dry at 100°, and weigh. Results a little too low, as traces of the double salt pass away with the alcohol washings (BERZELIUS). Instead of weighing the double salt you may ignite in hydrogen, remove the potassium chloride with water and weigh the metal obtained. This method is indeed to be preferred, as it prevents any potassium chloride in the precipitate from affecting the result.

POTASSIUM PALLADIC CHLORIDE consists of microscopic octahedra; it presents the appearance of a vermilion or, if the crystals are somewhat large, of a brown powder. It is very slightly soluble in cold water; it is almost insoluble in cold alcohol of the above strength. It contains 26·806% palladium.

Sixth Group.

GOLD—PLATINUM—ANTIMONY—TIN IN STANNIC COMPOUNDS—TIN IN STANNOUS COMPOUNDS—ARSENIOUS AND ARSENIC ACIDS—(MOLYBDIC ACID).

§ 123.

1. GOLD.

a. Solution.

Metallic gold, and all compounds of gold insoluble in water, are warmed with hydrochloric acid, and nitric acid is gradually added until complete solution is effected; or they are repeatedly digested with strong chlorine water. The latter method is resorted to more especially in cases where the quantity of gold to be dissolved is small, and mixed with foreign oxides which it is wished to leave undissolved. According to W. SKEY* tincture of iodine, or, for larger quantities of gold, bromine water, is better than chlorine water. They give solutions freer from other metals than the chlorine water gives.

b. Determination.

Gold is always weighed in the *metallic state.* The compounds are brought into this form, either by ignition or by precipitation, as gold, or auric sulphide.

We convert into

METALLIC GOLD:

a. By Ignition. All compounds of gold which contain no fixed acid, or other body.

b. By Precipitation as metallic gold. All compounds of gold without exception in cases where *a* is inapplicable.

c. By Precipitation as auric sulphide. This method serves to effect the separation of gold from certain other metals which may be mixed with it in a solution.

Determination as Metallic Gold.

a. By Ignition.

Heat the compound, in a covered porcelain crucible, very gently at first, but finally to redness, and weigh the residuary pure gold. For properties of the residue, see § 88. The results are most accurate.

b. By Precipitation as Metallic Gold.

α. The solution is free from Nitric Acid. Mix the solution with a little hydrochloric acid, if it does not already contain some of that acid in the free state, and add a clear solution of ferrous sulphate in excess; heat gently for a few hours until the precipitated fine gold powder has completely subsided; filter, wash, dry, and ignite according to § 52. A porcelain dish is a more appropriate vessel to effect the precipitation in than a beaker, as the heavy fine gold powder is more readily rinsed out of the former than out of the latter. There are no sources of error inherent in the method.

β. The solution of Gold contains Nitric Acid. Evaporate the solution, on a water-bath, to the consistence of syrup, adding from time to time hydrochloric acid; dissolve the residue in water con-

---

* Zeitschr. f. anal. Chem. 10, 221.

taining hydrochloric acid, and treat the solution as directed in *α*. It will sometimes happen that the residue does not dissolve to a clear fluid, in consequence of a partial decomposition of auric chloride into aurous chloride and metallic gold; however, this is a matter of perfect indifference.

γ. In cases where it is wished to avoid the presence of iron in the filtrate, the gold may be reduced by means of oxalic acid. To this end, the dilute solution—freed previously, if necessary, from nitric acid, in the manner directed in *β*—is mixed, in a beaker, with oxalic acid, or with ammonium oxalate in excess, some sulphuric acid added (if that acid is not already present in the free state), and the vessel, covered with a glass plate, is kept standing for two days in a moderately warm place. At the end of that time, the whole of the gold will be found to have separated in small yellow scales, which are collected on a filter, washed first with dilute hydrochloric acid, then with water, dried, and ignited. If the gold solution contains a large excess of hydrochloric acid, the latter should be for the most part evaporated, before the solution is diluted and the oxalic acid added. If the gold solution contains chlorides of alkali metals, it is necessary to dilute largely, and allow to stand for a long time, in order to effect complete precipitation (H. Rose).

δ. The gold may also be thrown down in the metallic form by hydrate of chloral* in the presence of potash. Warm the solution, add the chloral, then pure potash in excess, and boil for a minute or so. The gold is precipitated with evolution of chloroform.

ε. Finally, gold may be thrown down by many metals, such as zinc, cadmium, magnesium, &c. The latter has been recommended by Scheibler† for the analysis of the gold salts of organic bases. The precipitate is first washed with hydrochloric acid, then with water.

c. By Precipitation as Auric Sulphide.

Hydrogen sulphide gas is transmitted in excess through the dilute solution containing some free acid; the precipitate formed is speedily filtered off, without heating, washed, dried, and ignited in a porcelain crucible. For the properties of the precipitate, see § 88. No sources of error.

---

* Hager's pharmac. Centralhalle, 11, 393.

† Ber. der deutsch. chem. Gesellsch. 1869, 295.

§ 124.

2. Platinum.

a. Solution.

Metallic platinum, and the compounds of platinum which are insoluble in water, are dissolved by digestion, at a gentle heat, with nitrohydrochloric acid.

b. Determination.

Platinum is invariably weighed in the *metallic state*, to which condition its compounds are brought, either by precipitation as ammonium platinic chloride, potassium platinic chloride, or platinic sulphide, or by ignition, or by precipitation with reducing agents. All compounds of platinum, without exception, may, in most cases, be converted into platinum by either of these methods. Which is the most advantageous process to be pursued in special instances, depends entirely upon the circumstances. The reduction to the metallic state by simple ignition is preferable to the other methods, in all cases where admissible. The precipitation as platinic sulphide is resorted to exclusively to effect the separation of platinum from other metals.

Determination as Metallic Platinum.

a. By Precipitation as Ammonium Platinic Chloride.

The solution must be concentrated if necessary by evaporation on a water-bath. Mix, in a beaker, with ammonia until the excess of acid (that is, supposing an excess of acid to be present) is *nearly* saturated; add ammonium chloride in excess, and mix the fluid with a pretty large quantity of strong alcohol. Cover the beaker now with a glass plate, and let it stand for twenty-four hours, after which filter, wash the precipitate with alcohol of about 80 per cent., till the substances to be separated are removed, dry carefully, ignite according to § 99, 2, and weigh. In the case of large quantities the final ignition is advantageously conducted in a stream of hydrogen (§ 108, fig. 50), in order to be quite sure of effecting complete decomposition. For the properties of the precipitate and residue, see § 89. The results are satisfactory, though generally a little too low, as the ammonium platinic chloride is not altogether insoluble in alcohol of the above strength (Expt. No. 16), and as the fumes of ammonium chloride are liable to carry away traces of the yet undecomposed double chloride, if the application of heat is not conducted with the greatest care.

If the precipitated ammonium platinic chloride were weighed in that form, the results would be inaccurate, since, as I have convinced myself by direct experiments, it is impossible to completely free the double chloride, by washing with alcohol, from all traces of the ammonium chloride thrown down with it, without dissolving at the same time a notable portion of the double chloride. As a general rule, the results obtained by weighing the ammonium platinic chloride in that form are one or two per cent. too high.

b. By Precipitation as Potassium Platinic Chloride.

Mix the solution, in a beaker, with potassa, until the greater part of the excess of acid (if there be any) is neutralized; add potassium chloride slightly in excess, and finally a pretty large quantity of strong alcohol; should your solution of platinum be very dilute, you must concentrate it previously to the addition of the alcohol. After twenty-four hours, collect the precipitate upon a rather small unweighed filter, wash with alcohol of 80 per cent., dry thoroughly at 100°, and transfer to a porcelain crucible, dissolving the portion which adheres to the filter, and evaporating the solution in the crucible. See § 97, 3. Next, by igniting with hydrogen by means of apparatus described in § 108, page 251, convert the compound into metallic platinum and potassium chloride. Reduction is best effected if the heat is very gradually applied, and does not at all quite reach the point at which potassium chloride fuses. After reduction, wash out the potassium chloride, ignite and weigh the platinum. For the properties of the precipitate and residue, see § 89.

The results are more accurate than those obtained by method *a*, since, on the one hand, the potassium platinic chloride is more insoluble in alcohol than the corresponding ammonium salt; and, on the other hand, loss of substance is less likely to occur during ignition. To weigh the potassium platinic chloride in that form would not be practicable, as it is impossible to remove, by washing with alcohol, all traces of the potassium chloride thrown down with it, without, at the same time, dissolving a portion of the double chloride.

c. By Precipitation as Platinic Sulphide.

Precipitate the solution with hydrogen sulphide water or gas, according to circumstances, heat the mixture to incipient ebullition, filter, wash the precipitate, dry, and ignite according to § 52. For the properties of the precipitate and residue, see § 89. The results are accurate.

d. By Ignition.

Same process as for gold, § 123. For the properties of the residue, see § 89. The results are most accurate.

e. By Precipitation with Reducing Agents.

Various reducing agents may be employed to precipitate platinum from its solutions in the metallic state. The reduction is very promptly effected by ferrous sulphate and potassa or soda (the protosesquioxide of iron being removed by subsequent addition of hydrochloric acid, HEMPEL), or by pure zinc or magnesium (the excess of which is removed by hydrochloric acid); somewhat more slowly, and only with application of heat, by alkali formiates. Mercurous nitrate also precipitates the whole of the platinum from solution of platinic chloride; upon igniting the brown precipitate obtained, fumes of mercurous chloride escape, and metallic platinum remains.

§ 125.

3. ANTIMONY.

a. Solution.

Antimonious oxide, and the compounds of antimony which are insoluble in water, or are decomposed by that agent, are dissolved in more or less concentrated hydrochloric acid. Metallic antimony is dissolved best in nitrohydrochloric acid. The ebullition of a hydrochloric acid solution of antimonious chloride is attended with volatilization of traces of the latter; the concentration of a solution of the kind by evaporation involves accordingly loss of substance. Solutions so highly dilute as to necessitate a recourse to evaporation must therefore previously be supersaturated with potassa. Solutions of antimonious chloride, which it is intended to dilute with water, must previously be mixed with tartaric acid, to prevent the separation of basic salt. In diluting an acid solution of antimonic acid in hydrochloric acid, the water must not be added gradually and in small quantities at a time, which would make the fluid turbid, but in sufficient quantity at once, which will leave the fluid clear.

b. Determination.

Antimony may be weighed as *antimonious sulphide* or *antimony tetroxide*, in separations it is sometimes weighed as *metallic antimony;* or it is estimated volumetrically.

Antimony in solution is almost invariably first precipitated as sulphide, which is then, with the view of estimation, converted into anhydrous sulphide, or determined volumetrically.

1. *Precipitation as Antimonious Sulphide.*

Add to the antimony solution hydrochloric acid, if not already present, then tartaric acid, and dilute with water, if necessary. Introduce the clear fluid into a flask, closed with a doubly perforated cork; through one of the perforations passes a tube, bent outside at a right angle, which nearly extends to the bottom of the flask; through the other perforation passes another tube, bent outside twice at right angles, which reaches only a short way into the flask; the outer end of this tube dips slightly under water. Conduct through the first tube hydrogen sulphide gas, until it predominates strongly; put the flask in a moderately warm place, and after some time conduct carbon dioxide into the fluid, until the excess of the other gas is almost completely removed. If there is no reason against it, from the presence of a large quantity of hydrochloric acid, or from the presence of nitric acid, it is well to heat the solution during the passing of the gas, finally even boiling. The precipitate is then denser, and may be very easily washed (SHARPLES*).

If the amount of the precipitate is at all considerable, filter without intermission through a weighed filter, wash rapidly and thoroughly with water mixed with a few drops of hydrogen sulphide water, dry at 100°, and weigh. The precipitate so weighed always retains some water, and may, besides, contain free sulphur; in fact, it always contains the latter in cases where the antimony solution, besides antimonious salts, contains antimonic acid or pentachloride of antimony, since the precipitation under these circumstances is preceded by a reduction of antimonic to antimonious compounds, accompanied by separation of sulphur (H. ROSE). A further examination of the precipitate is accordingly indispensable. To this end, treat a sample of the weighed precipitate with strong hydrochloric acid. If

* Zeitschr. f. anal. Chem. 10, 343.

a. The sample dissolves to a clear fluid, this is a proof that the precipitate only contains Sb_2S_3; but if

b. Sulphur separates, this shows that free sulphur is present.

In *case a* (in order to remove the water retained at 100°) the greater portion of the dried precipitate is weighed in a porcelain boat, which is then inserted into a glass tube, about 2 decimetres long; a slow current of dry carbon dioxide is transmitted through the latter, and the boat cautiously heated by means of a lamp, moved to and fro under it, until the orange precipitate becomes black. The precipitate is then allowed to cool in the current of carbon dioxide, and weighed; from the amount found, the total quantity of anhydrous antimonious sulphide contained in the entire precipitate is ascertained by a simple calculation. The results are accurate. Expt. No. 84 gave 99·22 instead of 100. But if the precipitate is simply dried at 100°, the results are about 2 per cent. too high—see the same experiment. For the properties of the precipitate, see § 90.

In *case b*, the precipitate is subjected to the same treatment as in *a*, with this difference only, that the contents of the boat are heated much more intensely, and the process is continued until no more sulphur is expelled. This removes the whole of the admixed sulphur; the residue consists of pure antimonious sulphide. It must be completely soluble in fuming hydrochloric acid on heating.

If the amount of the precipitate is small, collect it in a weighed asbestos filtering tube, dry in a slow current of carbon dioxide at a gentle heat, heat finally rather more strongly till the sulphide has turned black and any free sulphur present has volatilized, allow to cool, replace the gas in the tube by air, and weigh. Results quite satisfactory.*

For the method of estimating the antimony in the sulphide volumetrically and indirectly, see 3.

2. *Determination as Antimony Tetroxide.*

a. In the case of antimonious oxide or a compound of the same with an easily volatile or decomposable oxygen acid, evaporate carefully with nitric acid, and ignite finally for some time till the weight is constant. The experiment may be safely made in a platinum crucible. With antimonic acid, the evaporation with nitric acid is unnecessary.

* Zeitschr. f. anal. Chem. 8, 155.

b. If antimonious sulphide is to be converted into antimony tetroxide, one of the two following methods given by BUNSEN* is employed:

α. Moisten the dry sulphide of antimony with a few drops of nitric acid of 1·42 sp. gr., then treat, in a weighed porcelain crucible, with concave lid, with 8—10 times the quantity of fuming nitric acid,† and let the acid gradually evaporate on the water-bath. The sulphur separates at first as a fine powder, which, however, is readily and completely oxidized during the process of evaporation. The white residual mass in the crucible consists of antimonic acid and sulphuric acid, and may by ignition be converted, without loss, into antimony tetroxide. If the sulphide of antimony contains a large excess of free sulphur, this must be removed by washing with bisulphide of carbon.

β. Mix the sulphide of antimony with 30—50 times its quantity of pure mercuric oxide,‡ and heat the mixture gradually in an open porcelain crucible. As soon as oxidation begins, which may be known by the sudden evolution of gray mercurial fumes, moderate the heat. When the evolution of mercurial fumes diminishes raise the temperature again, always taking care, however, that no reducing gases come in contact with the contents of the crucible. Remove the last traces of mercuric oxide over the blast gas-lamp, then weigh the residual fine white powder of antimony tetroxide. As mercuric oxide generally leaves a trifling fixed residue upon ignition, the amount of this should be determined once for all, the mercuric oxide added approximately weighed, and the corresponding amount of fixed residue deducted from the antimony tetroxide. The volatilization of the oxide of mercury proceeds much more rapidly when effected in a platinum crucible instead of a porcelain one. But, if a platinum crucible is employed, it must be effectively protected from the action of antimony upon it, by a *good lining* of mercuric oxide.§ If the sulphide of anti-

---

* Annal. d. Chem. u. Pharm. 106, 3.

† Nitric acid of 1·42 sp. gr. is not suitable for this purpose, as its boiling point is almost 10° above the fusing point of sulphur, whereas fuming nitric acid boils at 86°, consequently below the fusing point of sulphur. With nitric acid of 1·42 sp. gr., therefore, the separated sulphur fuses and forms drops, which obstinately resist oxidation.

‡ Prepared by precipitation from mercuric chloride by excess of soda solution and thorough washing.

§ This is effected best, according to BUNSEN, in the following way: Soften the

mony contains free sulphur, this must first be removed by washing with bisulphide of carbon, before the oxidation can be proceeded with, since otherwise a slight deflagration is unavoidable.

According to later experiments made by BUNSEN,* it is somewhat difficult to obtain good results by this method, because a temperature a little above that required to reduce Sb_2O_5 to Sb_2O_4 will reduce the latter Sb_2O_3. Ignition over a blast-lamp in a very large covered platinum, or rather large open porcelain crucible, keeping only the bottom at a full red heat, is recommended as a method by which it is possible to drive off just one atom O from Sb_2O_5.

3. *Volumetric Methods.*

a. Conversion of Antimonious Chloride to Antimonic Chloride by Hydrochloric Acid and Potassium Chromate or Permanganate.

F. KESSLER'S† first description of this method was so wanting in precision, that it could not be depended upon. However, he has since‡ determined most accurately the conditions under which antimony in acid solution may be satisfactorily titrated either with potassium chromate (the excess of the standard solution being determined with ferrous sulphate) or with potassium permanganate.

I. *Titration with Potassium Dichromate.*

1. REQUISITES.

α. Standard Solution of Arsenious Acid. Dissolve exactly 5 grm. pure arsenious oxide by the aid of some soda solution, add hydrochloric acid till slightly acid, then 100 c.c. more of hydrochloric acid of 1·12 sp. gr., and dilute to 1000 c.c. Each c.c. contains ·005 grm. arsenious oxide and corresponds to ·007374 antimonious oxide.

---

sealed end of a common test-tube before the glass-blower's lamp; place the softened end in the centre of the platinum crucible, and blow into it, which will cause it to expand and assume the exact form of the interior of the crucible. Crack off the bottom of the little flask so formed, and smooth the sharp edges cautiously by fusion. A glass is thus obtained, open at both ends, which exactly fits the crucible. To effect the lining by means of this instrument, fill the crucible loosely with mercuric oxide up to the brim, then force the glass gradually and slowly down to the bottom of the crucible, occasionally shaking out the oxide of mercury from the interior of the glass. The inside of the crucible is thus covered with a layer of oxide of mercury ½—1 line thick, which, after the removal of the glass, adheres with sufficient firmness, even upon ignition.

* Zeitschr. f. anal. Chem. 18, 268. † Pogg. Annal. 95, 204.

‡ *Ib.* 118, 17; and Zeitschr. f. anal. Chem. 2, 383.

β. Solution of Potassium Dichromate. Dissolve about 2·5 grm. to 1 litre.

γ. Solution of Ferrous Sulphate. Dissolve about 1·1 grm. iron wire in 20 c.c. dilute sulphuric acid (1 to 4), filter, and dilute to 1 litre.

δ. Solution of Potassium Ferricyanide. Should be tolerably dilute and freshly prepared.

2. DETERMINATION OF THE SOLUTIONS.

α. Relation between the Solution of Chromate and the Solution of Ferrous Sulphate. Run into a beaker 10 c.c. of the chromate solution from the burette, add 5 c.c. of hydrochloric acid and 50 c.c. water, and then add iron solution from a burette till the fluid is green. Continue adding the iron solution, a c.c. at a time, testing after each addition whether a drop of the fluid, when brought in contact with a drop of the potassium ferricyanide, on a porcelain plate, manifests a distinct reaction for ferrous iron. As soon as this point is attained, add ·5 c.c. of chromate solution and then iron solution two drops at a time, till the blue reaction just occurs. Now read off both burettes, and calculate how much chromate solution corresponds to 10 c.c. of iron solution. This experiment is to be repeated before every fresh series of analyses, as the iron solution gradually oxidizes.

β. Relation between the Chromate Solution and the Solution of Arsenious Acid. Transfer 10 c.c. of the arsenious solution to a beaker, add 20 c.c. hydrochloric acid of 1·2 sp. gr., and 80—100 c.c.* water, run in chromate solution till the yellow color of the fluid shows an excess, wait a few minutes, add excess of iron solution, then again ·5 chromate solution, and finally again iron solution till the end-reaction appears (see above). Deduct from the total quantity of chromate solution employed, the amount corresponding to the iron used, and from the datum thus afforded calculate how much antimony corresponds to 100 c.c. of chromate solution; in other words, how much antimony is converted by the quantity of chromate mentioned from $SbCl_3$ into $SbCl_5$.

3. THE ACTUAL ANALYSIS.

In the absence of organic matter, heavy metallic oxides, and other

* The water must be measured, for the action of chromic acid on arsenious acid (and also on antimonious chloride) is normal only if the fluid contains at least one sixth of its volume of hydrochloric acid of 1·12 sp. gr.

bodies which are detrimental to the reaction, dissolve the antimonious compound at once in hydrochloric acid. The solution should contain not less than $\frac{1}{6}$ of its volume of hydrochloric acid of 1·12 sp. gr. It is not advisable, on the other hand, that it should contain more than $\frac{1}{2}$, otherwise the end-reaction with potassium ferricyanide is slower in making its appearance and loses its nicety. Tartaric acid cannot be employed as a solvent, since it interferes with the action of chromic acid on ferrous salts. Now proceed as directed in 2. If the direct determination of antimony in the hydrochloric acid solution is not practicable, precipitate it with hydrogen sulphide. Wash the precipitate, transfer it, together with the filter, to a small flask; treat it with a sufficiency of hydrochloric acid, dissolve by digestion on the water-bath, add a sufficient quantity of a nearly saturated solution of mercuric chloride in hydrochloric acid of 1·12 sp. gr. to remove the hydrogen sulphide, and then proceed as directed.

II. *Titration with Potassium Permanganate.*

Here also the fluid must contain at least $\frac{1}{6}$ of its volume of hydrochloric acid of 1.12 sp. gr. The permanganate solution, which may contain about 1·5 grm. of the crystallized salt in a litre, is added to permanent reddening. The end-reaction is exact, and the conversion of antimonious to antimonic chloride goes on uniformly, although the degree of dilution may vary, provided the above relation between hydrochloric acid and water is kept up. It is not well that the hydrochloric acid should exceed $\frac{1}{3}$ of the volume of the fluid, as in that case the end-reaction would be too transitory. Tartaric acid, at least in the proportion to antimony in which it exists in tartar emetic, does not interfere with the reaction. Hence the permanganate may be standardized by the aid of solution of tartar emetic of known strength.

If you have to analyze antimonious sulphide, proceed as directed I. 3; make the fluid mixed with mercuric chloride up to a certain volume, allow to settle, and use a measured portion of the *perfectly clear* solution for the experiment.

My own experiments* have shown that KESSLER's methods are also suitable for the estimation of very small quantities of antimony.

---

* Zeitschr. f. anal. Chem. 8, 155.

b. Volumetric Estimation by determining the Hydrogen Sulphide given up by the Sulphide (R. SCHNEIDER*).

Both antimonious and antimonic sulphides yield under the action of boiling hydrochloric acid 3 mol. hydrogen sulphide for every 2 atoms of antimony. Hence, if the amount of the gas evolved under such circumstances is estimated, the amount of antimony is known.

For decomposing the sulphide and absorbing the gas, the same apparatus serves as BUNSEN employs for his iodimetric analyses (§ 130). The size of the boiling-flask should depend on the quantity of sulphide; for quantities up to ·4 grm. Sb_2S_3, a flask of 100 c.c. is large enough; for ·4—1 grm., use a 200 c.c. flask. The body of the flask should be spherical, the neck rather narrow, long, and cylindrical. If the sulphide of antimony is on a filter, put both together into the flask. The hydrochloric acid should not be too concentrated.

The determination of the hydrogen sulphide is best conducted according to the method given in § 148, *b*. The results obtained by SCHNEIDER are satisfactory. If the precipitate contains antimonious chloride, the results are of course false, and this would actually be the case if on precipitation with hydrogen sulphide the addition of the tartaric acid were omitted.

§ 126.

4. TIN IN STANNOUS COMPOUNDS, and 5. TIN IN STANNIC COMPOUNDS.

a. Solution.

In dissolving compounds of tin soluble in water, a little hydrochloric acid is added to insure a clear solution. Nearly all the compounds of tin insoluble in water dissolve in hydrochloric acid, or in aqua regia. The hydrate of metastannic acid may be dissolved by boiling with hydrochloric acid, decanting the fluid, and treating the residue with a large proportion of water. Ignited stannic oxide, and stannic compounds insoluble in acids, are prepared for solution in hydrochloric acid, by reducing them to the state of a fine powder, and fusing in a silver crucible with potassium or sodium hydroxide, in excess. Metallic tin is dissolved best in aqua regia; the solution frequently contains metastannic chloride mixed with the stannic chloride (TH. SCHEERER†). It is generally determined,

* Pogg. Annal. 110, 634. † Journ. f. prakt. Chem. N. F. 3, 472.

however, by converting it into stannic oxide, without previous solution. Acid solutions of stannic salts, which contain hydrochloric acid, or a chloride, cannot be concentrated by evaporation, not even after addition of nitric acid or sulphuric acid, without volatilization of stannic chloride taking place.

b. Determination.

Tin is weighed in the form of *stannic oxide*, into which it is converted, either by the agency of nitric acid, or by precipitation as stannic (or metastannic) acid, or by precipitation as sulphide. A great many volumetric methods of estimating tin have been proposed. They all depend on obtaining the tin in solution in the condition of stannous chloride, and converting this into stannic chloride either in alkaline or acid solution. A few only yield satisfactory results.

We may convert into

STANNIC OXIDE:

a. By the Agency of Nitric Acid. Metallic tin, and those compounds of tin which contain no fixed acid, provided no compounds of chlorine be present.

b. By Precipitation as Stannic (or Metastannic) Acid. All tin salts of volatile acids, provided no non-volatile organic substances nor ferric salts be present.

c. By Precipitation as Sulphide. All compounds of tin without exception.

In methods *a* and *c*, it is quite indifferent whether the tin is present as a stannous or a stannic compound. The method *b* requires the tin to be present as a stannic salt. The volumetric methods may be employed in all cases; but the estimation is simple and direct only where the tin is in solution as stannous chloride and free from other oxidizable bodies, or can readily be brought into this state. For the methods of determining stannous and stannic tin in presence of each other, I refer to Section V.

1. *Determination of Tin as Stannic Oxide.*

a. By Treating with Nitric Acid.

This method is resorted to principally to convert the metallic tin into stannic oxide. For this purpose the finely-divided metal is put into a capacious flask, and moderately concentrated pure nitric acid (about 1·3 sp. gr.) gradually poured over it; the flask is covered with a watch glass. When the first tumultuous action of the acid has somewhat abated, a gentle heat is applied until the metastannic acid formed appears of a pure white color, and further action of the acid is no longer perceptible. The contents of the flask are then transferred to a porcelain dish and evaporated on a water-bath nearly to dryness, water is then added, and the precipitate is collected on a filter, washed, till the washings scarcely redden litmus paper, dried, ignited, and weighed. The ignition is effected best in a small porcelain crucible, according to § 53; still a platinum crucible may also be used. A simple red heat is not sufficient to drive off all the water; the ignition must therefore be finished over a gas blowpipe. Compounds of tin which contain no fixed substances may be converted into stannic oxide by treating them in a porcelain crucible with nitric acid, evaporating to dryness, and igniting the residue. If sulphuric acid be present, the expulsion of that acid may be promoted, in the last stages of the process, by ammonium carbonate, as in the case of acid potassium sulphate (§ 97); here also the heat must be increased as much as possible at the end. For the properties of the residue, see § 91. There are no inherent sources of error.

b. By Precipitation as Stannic (or Metastannic) Acid.

The application of this method presupposes the whole of the tin to be present in the state of stannic salts. Therefore, if a solution contains stannous salts, either mix with chlorine water, or conduct chlorine gas into it, or heat gently with chlorate of potassa, until the conversion of the stannous into stannic salts is effected. When this has been done, add ammonia until a permanent precipitate just begins to form, and then hydrochloric acid, drop by drop, until this precipitate is completely redissolved; by this means a large excess of hydrochloric acid in the solution will be avoided. Add to the fluid so prepared a concentrated solution of ammonium nitrate (or sodium sulphate), and apply heat for some time, whereupon the whole of the tin will precipitate as stannic acid. Decant three times on to a filter, then collect the precipitate on the latter, wash thoroughly, dry, and ignite. To make quite sure that the whole of the tin has separated, you need simply, before proceeding to filter, add a few drops of the clear supernatant fluid to a hot solution of ammonium nitrate, or sodium sulphate, when the formation or non-formation of a precipitate will at once decide the question. The tin is also precipitated from metastannic chloride by the above reagents.

This method, which we owe to J. LÖWENTHAL, has been repeatedly tested by him in my own laboratory,* is easy and convenient, and gives very accurate results. The decomposition is expressed by the equation, $SnCl_4 + 4Na_2SO_4 + 3H_2O = H_2SnO_3 + 4NaCl + 4NaHSO_4$, or in precipitating with ammonium nitrate: $SnCl_4 + 4NH_4NO_3 + 3H_2O = H_2SnO_3 + 4NH_4Cl + 4HNO_3$.

Tin may also, according to H. ROSE,† be completely precipitated from stannic solutions by sulphuric acid. If the solution contains metastannic acid or metastannic chloride, the precipitation is effected without extraordinary dilution; the other stannic compounds, however, require very considerable dilution. If free hydrochloric acid is absent, the precipitation is rapid; in other cases 12 or 24 hours at least are required for perfect precipitation. Allow to settle thoroughly, before filtering, wash well (if hydrochloric acid was present, till the washings give no turbidity with silver nitrate), dry and ignite, at last intensely with addition of some ammonium carbonate. The results obtained by OESTEN, and communicated by H. ROSE, are exact.

c. By Precipitation as Stannous or Stannic Sulphide.

Precipitate the dilute moderately acid solution with hydrogen sulphide water or gas. If the tin was present in the solution as a stannous salt, and the precipitate consists accordingly of the brown stannous sulphide, keep the solution, supersaturated with hydrogen sulphide, standing for half an hour in a moderately warm place, and then filter. If, on the other hand, the solution contain a stannic salt, or metastannic acid, and the precipitate is yellow and consists of stannic sulphide mixed with stannic oxide, or yellowish brown and consists of hydrated metastannic sulphide mixed with metastannic acid (BARFOED, p. 189, TH. SCHEERER‡), put the fluid, loosely covered, in a warm place, until the odor of hydrogen sulphide has nearly gone off, and then filter. The washing of the stannic sulphide precipitate, which has a great inclination to pass through the filter, is best effected with a concentrated solution of sodium chloride, the remains of the latter being got rid of by a solution of ammonium acetate containing a small excess of acetic acid. If there is no objection to having the latter salt in the filtrate, the washing may be entirely effected by its means (BUNSEN§).

* Journ. f. prakt. Chem. 56, 366.
† Pogg. Annal. 112, 164.
‡ Journ. f. prakt. Chem. N. F. 3, 472.
§ Annal. d. Chem. u. Pharm. 106, 13.

Transfer the dry precipitate as completely as possible to a watch glass, burn the filter carefully in a weighed porcelain crucible, moisten the ash with nitric acid, ignite, allow to cool, add the precipitate, cover the crucible, heat gently for some time (slight decrepitation often occurs), remove the lid and heat gently with access of air, till sulphur dioxide has almost ceased to be formed. (If too much heat is applied at first, stannic sulphide volatilizes, the fumes of which give stannic oxide.) Now heat strongly, allow to cool, and heat repeatedly with pieces of ammonium carbonate to a high degree, to drive out the last portions of sulphuric acid. When the weight remains constant the experiment is ended (H. ROSE). For the properties of the precipitates, see § 91. The results are accurate.

2. *Volumetric Methods.*

The determination of tin by the conversion of stannous into stannic chloride with the aid of oxidizing agents (potassium dichromate iodine, potassium permanganate, etc.) offers peculiar difficulties, inasmuch as on the one hand the stannous chloride takes up oxygen from the air and from the water used for dilution, with more or less rapidity, according to circumstances; and on the other hand, the energy of the oxidizing agent is not always the same, being influenced by the state of dilution and the presence of a larger or smaller excess of acid.

In the following methods, these sources of error are avoided or limited in such a manner as to render the results satisfactory.

1. *Determination of Stannous Chloride by Iodine in Alkaline Solution* (*after* LESSEN*).

Dissolve the stannous salt or the metallic tin† in hydrochloric acid (preferably in a stream of carbon dioxide), add Rochelle salt, then sodium hydrogen carbonate in excess. To the clear slightly alkaline solution thus formed add some starch-solution, and afterwards the iodine solution of § 146, till a permanent blue coloration appears. 2 at. free iodine used corresponds to 1 at. tin.

LENSSEN's results are entirely satisfactory.

---

* Journ. f. prakt. Chem. 78, 200; Annal. d. Chem. u. Pharm. 114, 113.

† The solution of metallic tin is much assisted by the presence of platinum foil, which is accordingly added. LENSSEN found this addition of platinum to be objectionable; but no other experimenter has observed that it interferes with the accuracy of the results.

2. *Determination of Stannous Chloride after addition of Ferric Chloride.*

The fact that stannous chloride in acid solution can be far more accurately converted into stannic by oxidizing agents after being mixed with ferric chloride (or even with cupric chloride) than without this addition, was first settled by LÖWENTHAL.* Subsequently STROMEYER† published some experiments leading to the same results, together with practical remarks on the best way of carrying out the method in different cases. The processes thus originated, and which have been well tested, are as follows:

a. The given substance is a stannous salt. Dissolve in pure ferric chloride (free from ferrous chloride) with addition of hydrochloric acid, dilute and add standard permanganate from the burette. Now make another experiment with the same quantity of water similarly colored with ferric chloride to ascertain how much permanganate is required to tinge the liquid, and subtract the quantity so used from the amount employed in the actual analysis, and from the remainder calculate the tin.

The reaction between the tin salt and the iron solution is $SnCl_2 + Fe_2Cl_6 = SnCl_4 + 2FeCl_2$. The solution thus contains ferrous chloride in the place of stannous salt, the former being, as is well known, far less susceptible of alteration from the action of free oxygen than the latter. 2 at. iron found correspond to 1 at. tin. It must not be forgotten that the titration takes place in presence of hydrochloric acid. The results cannot be considered accurate unless the standardizing of the permanganate and the analysis take place under similar conditions as regards dilution and amount of hydrochloric acid.

b. The given substance is metallic tin. Either dissolve in hydrochloric acid—preferably with addition of platinum and in an atmosphere of carbon dioxide—and treat the solution according to *a*, or place the substance at once in a concentrated solution of ferric chloride mixed with a little hydrochloric acid; under these circumstances it will, if finely divided, dissolve quickly even in the cold and without evolution of hydrogen. Gentle warming is unobjectionable. Now add the permanganate. The reaction is $Sn + 2Fe_2Cl_6 = SnCl_4 + 4FeCl_2$, therefore every 4 at. iron found reduced correspond to 1 at. tin. The results are of course only

* Journ. f. prakt. Chem. 76, 484. † Annal. d. Chem. u. Pharm. 117, 261.

correct when iron is not present. Where this is the case, proceed with the impure tin solution according to *c*.

c. The given substance is stannic chloride or stannic oxide, or a compound of tin containing iron. Dissolve in water with addition of hydrochloric acid, place a plate of zinc in the solution and allow to stand twelve hours, then remove the precipitated tin with a brush, wash it, dissolve in ferric chloride, and proceed as in *b*.

d. The given substance is pure stannic sulphide, precipitated out of an acid stannic solution containing no stannous salt. Mix with ferric chloride, heat gently, filter off the sulphur, and then add the permanganate. 4 at. iron correspond to 1 at. tin, for $SnS_2 + 2Fe_2Cl_6 = SnCl_4 + 4FeCl_2 + 2S$. The results obtained by STROMEYER are quite satisfactory. As regards the precipitated stannic sulphide, see BARFOED, p. 189.

§ 127.

6. ARSENIOUS ACID, and 7. ARSENIC ACID.

a. Solution.

The compounds of arsenious and arsenic acids which are not soluble in water are dissolved in hydrochloric acid or in nitrohydrochloric acid. Some native arsenates require fusing with sodium carbonate. Metallic arsenic, arsenious sulphide, and metallic arsenides are dissolved in fuming nitric acid or nitrohydrochloric acid, or a solution of bromine in hydrochloric acid; those metallic arsenides which are insoluble in these menstrua are fused with sodium carbonate and potassium nitrate, by which means they are converted into soluble alkali arsenates and insoluble metallic oxides, or they may be suspended in potassa solution and treated with chlorine (§ 164, **137** and **138**). In this last manner, too, arsenious sulphide, dissolved in concentrated potassa, may be very easily rendered soluble. All solutions of compounds of arsenic which have been effected by long heating with fuming nitric acid, or by warming with excess of nitrohydrochloric acid, or chlorine, contain arsenic acid. A solution of arsenious acid in hydrochloric acid cannot be concentrated by evaporation, since arsenious chloride would escape with the hydrochloric acid fumes. This, however, less readily takes place if the solution contains arsenic acid; in fact, it only occurs in the presence of a large proportion of hydrochloric acid (for instance, half the volume of hydrochloric acid of

1·12 sp. gr.*). It is therefore advisable in most cases where a hydrochloric acid solution containing arsenic is to be concentrated, previously to render the same alkaline.

b. Determination.

Arsenic is weighed as *lead arsenate*, *as ammonium magnesium arsenate*, as *magnesium pyroarsenate*, as *uranyl pyroarsenate*, or as *arsenious sulphide*. The determination as ammonium magnesium arsenate is sometimes preceded by precipitation as ammonium arsenio-molybdate. The method recommended by BERTHIER and modified by v. KOBELL of separating the arsenic as basic ferric arsenate is only used in separations. Arsenic may be estimated also in an *indirect way*, and by *volumetric methods*.

We may convert into

1. LEAD ARSENATE: Arsenious and arsenic acids in aqueous or nitric acid solution. (Acids or halogens forming fixed salts with lead, and also ammonium salts, must not be present.)

2. AMMONIUM MAGNESIUM ARSENATE, or MAGNESIUM PYROARSENATE:

a. By direct Precipitation. Arsenic acid in all solutions free from bases or acids precipitable by magnesia or ammonia.

b. Preceded by Precipitation as Ammonium Arsenio-molybdate. Arsenic acids in all cases where no phosphoric acid is present, little or no hydrochloric acid, nor any substance which decomposes molybdic acid.

3. URANYL PYROARSENATE: Arsenic acid in all combinations soluble in water and acetic acid.

4. ARSENIOUS SULPHIDE: All compounds of arsenic without exception.

Arsenic may be determined volumetrically in a simple and exact manner, whether present in the form of arsenious acid or an alkali arsenite, or as arsenic acid or an alkali arsenate. The volumetric methods have now almost entirely superseded the indirect gravimetric methods formerly employed to effect the determination of arsenious acid.

1. *Determination as Lead Arsenate.*

a. Arsenic Acid in Aqueous Solution.

A weighed portion of the solution is put into a platinum or porcelain dish, and a weighed amount of recently ignited pure lead

---

* Zeitschr. f. Chem. 1, 448.

oxide added (about five or six times the supposed quantity of arsenic acid present); the mixture is cautiously evaporated to dryness, and the residue heated to gentle redness, and maintained some time at this temperature. The residue is lead arsenate + lead oxide. The quantity of arsenic acid is now readily found by subtracting from the weight of the residue that of the oxide of lead added. For the properties of lead arsenate, see § 92. The results are accurate, provided the residue be not heated beyond gentle redness.

b. Arsenious Acid in Solution.

Mix the solution with nitric acid, evaporate to a small bulk, add a weighed quantity of lead oxide in excess, evaporate to dryness, and ignite the residue most cautiously in a covered crucible, until the whole of the lead nitrate is decomposed. The residue consists here also of arsenic acid + lead oxide. This method requires considerable care to guard against loss by decrepitation upon ignition of the lead nitrate.

2. *Estimation as Ammonium Magnesium Arsenate, or Magnesium Pyroarsenate.*

a. By direct Precipitation.

This method, which was first recommended by LEVOL, presupposes the whole of the arsenic in the form of arsenic acid. Where this is not the case, the solution is gently heated, in a capacious flask, with hydrochloric acid, and potassium chlorate added in small portions, until the fluid emits a strong smell of chlorous acid; it is then allowed to stand at a gentle heat until the odor of this gas is nearly gone off.

The arsenic acid solution is now mixed with ammonia in excess, which must not produce turbidity, even after standing some time; magnesia mixture is then added (p. 113, § 62, 6). The fluid, which smells strongly of ammonia, is allowed to stand 24 or 48 hours in the cold, well covered, and then filtered through a weighed filter. The precipitate is then transferred to the filter, with the aid of portions of the filtrate, so as to use no more washing water than necessary, and washed with small quantities of a mixture of three parts water and one part ammonia, till the washings, on being mixed with nitric acid and silver nitrate, show no opalescence. The precipitate is dried at 102° to 103°, and weighed. It has the formula $(MgNH_4AsO_4)_2+H_2O$.* As the drying of ammonium magnesium arsenate till its weight is constant, requires much time and repeated weighings, it is a great advantage that we can now convert it without loss of arsenic into magnesium pyroarsenate ($Mg_2As_2O_7$), thanks to the researches of H. ROSE,† WITTSTEIN‡ and PULLER.§ For this purpose first transfer the dried precipitate as completely as possible to a watch-glass, saturate the filter with a solution of ammonium nitrate, dry and burn it cautiously in a porcelain crucible. After cooling, transfer the precipitate to the crucible, heat in an air-bath to about 130°, continue heating for 2 hours on a sand-bath, then heat for an hour or two on an iron plate a little more strongly, and when the ammonia has been thus entirely expelled ignite strongly for some time over the lamp. The process may be shortened by conducting the heating in a ROSE's crucible in a slow current of oxygen. The ammonia may then be driven off in 10 minutes, and after the precipitate has been at last strongly heated it will be ready to weigh. For the properties of the ammonium magnesium arsenate and magnesium pyroarsenate, see § 92. The method yields satisfactory results, since the small loss of precipitate dissolved in the filtrate and washings is counterbalanced by the presence of a trace of basic magnesium sulphate (PULLER). PULLER with a quantity of ·37 grm. ammonium magnesium arsenate lost only a fraction of a milligramme ; on the addition of a large proportion of ammonium chloride the loss rose to about ·002 grm. The correction for the solubility of the precipitate in the ammoniacal filtrate containining excess of magnesia mixture is ·001 grm. of $(MgNH_4AsO_4)_2+H_2O$ for 30 c.c.

b. Preceded by Precipitation as Ammonium Arsenio-molybdate.

Mix the acid solution, which must be free from phosphoric and silicic acids, with an excess of solution of ammonium molybdate. The ammonium molybdate solution should have been previously mixed with nitric acid in excess, and the whole process is conducted exactly as in the case of phosphoric acid—see § 134, *b*, *β*.

* If it is dried in a water-bath, the drying must be extremely prolonged, or otherwise more than 1 eq. will be left. After brief drying in the water-bath the compound contains between 1 and 3 eq. water. If it is dried between 105° and 110°, part of the 1 eq. water is lost.

† His Handbuch der anal. Chem. 6 Aufl. 2, 390.

‡ Zeitschr. f. anal. Chem. 2, 19.

§ *Ib.* 10, 63.

After dissolving the ammonium arsenio-molybdate in ammonia, neutralize the latter partially with hydrochloric acid. Treat the ammonium magnesium arsenate as in *a*. Results satisfactory.

3. *Estimation as Uranyl Pyroarsenate.*

This method was first proposed by WERTHER.* It has been carefully studied by PULLER† in my laboratory, and gives thoroughly satisfactory results. Mix the arsenic acid solution with potash or ammonia in excess, and then a good excess of acetic acid. (If a precipitate of ferric or aluminium arsenate here remains insoluble, the method would be inapplicable.) Add uranyl acetate in excess, and boil. Wash the slimy precipitate of uranyl arsenate or of ammonium uranyl arsenate by decantation with boiling water, and then transfer to a filter. The addition of a few drops of chloroform to the partly cool fluid will hasten the deposition of the precipitate. Dry, transfer the precipitate to a watch-glass, cleaning the filter as much as possible; saturate the latter with ammonium nitrate, dry it, incinerate in a porcelain crucible, and add the precipitate. If the precipitate contains ammonium, heat very cautiously, finally adding nitric acid, or ignite in oxygen. (See 2, *a*.) If the precipitate is free from ammonium, ignite in the ordinary way. Ammonium salts do not interfere. Properties of the precipitate and residue, § 92, *e*.

4. *Estimation as Arsenious Sulphide.*

a. In solutions of Arsenious Acid or Arsenites free from Arsenic Acid.

The solution should be strongly acid with hydrochloric acid. Precipitate with hydrogen sulphide and expel the excess with carbon dioxide. Pass the latter through the solution for an hour, a longer time is useless. (See § 125, 1.) Wash the precipitate thoroughly and dry at 100° till the weight is constant. Particles of the precipitate which adhere so firmly to the glass that they cannot be removed mechanically are dissolved in ammonia and reprecipitated with hydrocholric acid. Properties of the precipitate, § 92. Do not omit to test a weighed portion to see whether it completely volatilizes on heating. If a residue remains it is to be weighed and the proportional quantity deducted from the total weight of the precipitate. Results accurate.

---

* Journ. f. prakt. Chem. 43, 346. † Zeitschr. f. analyt. Chem. 10, 72.

If the solution contains any substance which decomposes hydrogen sulphide, such as ferric chloride, chromic acid, etc., the precipitate produced in the cold contains an admixture of finely divided sulphur. It should be collected in the same manner on a filter dried at 100°, and weighed, washed and dried. Extract the admixed sulphur with purified carbon disulphide (which should leave no residue on evaporation), continuing till the fluid which runs through leaves no residue. Dry at 100° till the weight is constant. From experiments made in my laboratory it appears that the results thus obtained are quite accurate, even when the amount of admixed sulphur is large; but the precipitation must have been effected in the cold. If, on the contrary, heat is used, the sulphur is in the form of small agglutinated grains and cannot be completely extracted by cold carbon disulphide on the filter. However, it may be extracted by removing the precipitate from the filter and repeatedly digesting it with the disulphide on a water-bath (PULLER*).

Instead of purifying the arsenious sulphide you may estimate the arsenic in the mixture of the sulphide with sulphur as follows: Dissolve the precipitate in strong potash, and pass chlorine into the solution (§ 148, II. 2, *b*). The arsenic and the sulphur are converted into arsenic and sulphuric acid respectively; the former may be estimated according to 2, *a*, or the latter according to § 132. In the latter case, deduct the sulphur found from the weight of the arsenical precipitate. There is no loss of arsenic in this process from volatilization of the chloride, as the solution remains alkaline. The object may also be conveniently attained by the use of nitric acid. A very strong fuming acid, of 86° boiling point, is employed; an acid of 1·42 sp. gr. which boils at a higher temperature does not answer the purpose, as the separated sulphur would fuse, and its oxidation would be much retarded. The well dried precipitate is shaken into a small porcelain dish, treated with a tolerably large excess of the fuming nitric acid, the dish immediately covered with a watch-glass, and as soon as the turbulence of the first action has somewhat abated, heated on a water-bath till all the sulphur has disappeared, and the nitric acid has evaporated to a small volume. The filter to which the unremovable traces of arsenious sulphide adhere is treated separately in the same manner,

* Zeitschr. f. anal. Chem. 10, 46 et seq.

the complete destruction of the organic matter being finally effected by *gently* warming the somewhat dilute solution with potassium chlorate (BUNSEN*). Or the filter may instead be extracted with ammonia, the solution evaporated in a separate dish, and the residual sulphide treated as above. In the mixed solution the arsenic acid is finally precipitated as ammonious magnesium arsenate. (§ 127, 2, *a*). Treatment of the impure precipitate with ammonia, whereby the sulphide is dissolved, and the sulphur is supposed to remain behind, only gives approximate results, as the ammoniacal solution of arsenious sulphide takes up a little sulphur.

b. In solutions of Arsenic Acid, or of a mixture of the two Oxides of Arsenic.

Heat the solution in a flask (preferably on an iron plate) to about 70°, and conduct hydrogen sulphide at the same time into the fluid, as long as precipitation takes place. The precipitate formed is always a mixture of sulphur and arsenious sulphide, since the arsenic acid is first reduced to arsenious acid with separation of sulphur, and then the latter is decomposed (H. ROSE†). Only in the case when a sulphosalt containing pentasulphide of arsenic is decomposed with an acid, is the precipitate actually pentasulphide, and not merely a mixture of sulphur with arsenious sulphide (A. FUCHS‡). To convert this mixture of arsenious sulphide and granular sulphur into pure arsenious sulphide, suitable for weighing, treat it as follows: Extract the washed and still moist precipitate on the filter with ammonia, wash the residual sulphur, precipitate the solution with hydrochloric acid without heat, filter, dry, extract with carbon disulphide, dry at 100°, and weigh. Results accurate. The mixture of arsenious sulphide and sulphur obtained by hot precipitation may, of course, also be estimated directly or indirectly after one of the other methods in 4, *a*.

5. *Volumetric Methods.*

a. Method which presupposes the presence of Arsenious Acid.

BUNSEN'S method.§ This method is based upon the following facts:

aa. If potassium dichromate is boiled with concentrated hydrochloric acid, 6 at. chlorine are disengaged to every 2 mol. chromic acid $2CrO_3 + 12HCl = Cr_2Cl_6 + 6H_2O + 6Cl$.

* Annal. d. Chem. u. Pharm. 106, 10. † Pogg. Annal. 107, 186.
‡ Zeitschr. f. anal. Chem. 1, 189. § Annal. d. Chem. u. Pharm. 86, 290.

bb. But if arsenious acid is present (not in excess) there is not the quantity of chlorine disengaged corresponding to the chromic acid, but so much less of that element as is required to convert the arsenious into arsenic acid ($H_3AsO_3 + 2Cl + H_2O = H_3AsO_4 + 2 HCl$). Consequently, for every 2 at. chlorine wanting is to be reckoned 1 mol. arsenious acid.

cc. The quantity of chlorine is estimated by determining the quantity of iodine liberated by it from potassium iodide.

These are the principles of BUNSEN's method. For the manner of execution I refer to the Estimation of Chromic Acid.

b. Method, which presupposes the presence of Arsenic Acid.

This method depends on the precipitation of the arsenic acid by uranium solution and the recognition of the end of the reaction by means of potassium ferrocyanide. It is therefore the same as was suggested for phosphoric acid by LECOMTE, and brought into use by NEUBAUER,* and afterwards by PINCUS.†

BÖDEKER,‡ who first employed the process for arsenic acid, recommends the employment of a solution of uranyl nitrate, as this is more permanent than the hitherto used acetate, which is gradually decomposed by the action of light.

The uranium solution has the correct degree of dilution, if it contains about 17 grm. of uranium in 1 litre. It should contain as little free acid as possible. The determination of its value may be effected with the aid of pure sodium arsenate or by means of arsenious acid—the latter is converted into arsenic acid by boiling with fuming nitric acid. The solution is rendered strongly alkaline with ammonia, and then distinctly acid with acetic acid. The uranium solution is now run in from the burette slowly, the liquid being well stirred all the while, till a drop of the mixture spread out on a porcelain plate, gives with a drop of potassium ferrocyanide placed in its centre, a distinct reddish-brown line where the two fluids meet. The height of the fluid in the burette is now read off, the level of the mixture in the beaker is marked with a strip of gummed paper, and the beaker is emptied and washed, filled with water with addition of about as much ammonia and acetic acid as was before employed, and the uranium solution is cautiously dropped in from the burette, till a drop taken out of the beaker and tested as above, gives an equally distinct reaction. The

* Archiv für wissenschaftliche Heilkunde, 4, 228.

† Journ. f. prakt. Chem. 76, 104. ‡ Annal. de Chem. u. Pharm. 117, 195.

quantity of uranium solution used in this last experiment is the excess, which must be added to make the end-reaction plain for the dilution adopted. This amount is subtracted from that used in the first experiment, and we then know the exact value of the uranium solution with reference to arsenic acid.

In an actual analysis, the arsenic is first brought into the form of arsenic acid, a clear solution is obtained containing ammonium acetate and some free acetic acid,* and the process is conducted exactly as in determining the value of the standard solution. The experiment to ascertain the correction must not be omitted here, otherwise errors are sure to arise from the different degrees of dilution of the arsenic acid solutions used in the determination of the value of the standard solution and in the actual analyses. The results of two determinations of arsenic given by Bödeker are satisfactory. To execute the method well requires practice. The results are not exact enough unless the conditions as regards amount and quality of alkali salts are nearly similar in the standardizing of the uranium solution and in its use. Compare Waitz.†

6. *Estimation of Arsenious Acid by Indirect Gravimetric Analysis.*

a. Rose's method. Add to the hydrochloric acid solution, in the preparation of which care must be taken to exclude oxidizing substances, a solution of sodium- or ammonium-auric chloride in excess, and digest the mixture for several days, in the cold, or, in the case of dilute solutions, at a gentle warmth; then weigh the separated gold as directed in § 123. Keep the filtrate to make quite sure that no more gold will separate. 2 at. gold correspond to 3 mol. arsenious acid.

b. Vohl's‡ method. Mix the solution with a weighed quantity of potassium dichromate, and free sulphuric acid; estimate the chromic acid still present by the method given in § 130, *c*, and deduce from the quantity of that acid consumed in the process, *i.e.*, reduced by the arsenious acid, the quantity of the latter, after the formula $3H_3AsO_3 + 2CrO_3 = 3H_3AsO_4 + Cr_2O_3$.

* Alkalies, alkali earths, and zinc oxide may be present, but not such metals as yield colored precipitates with ferrocyanide of potassium, as, for instance, copper.

† Zeitschr. f. anal. Chem. 10, 182.

‡ Annal. de Chem. u. Pharm. 94, 219.

Supplement to the Sixth Group.

§ 128.

8. Molybdic Acid.

Molybdic acid is converted, for the purpose of its determination, either into molybdenum dioxide, or into lead molybdate, or into molybdenum disulphide.

a. Molybdic anhydride (MoO_3), and also ammonium molybdate, may be reduced to dioxide by heating in a current of hydrogen gas. This may be done either in a porcelain boat, placed in a wide glass tube, or in a platinum or porcelain crucible with perforated cover (§ 108, fig. 50). The operation is continued till the weight remains constant. The temperature must not exceed a gentle redness, otherwise the dioxide itself might lose oxygen and become partially converted into metal. In the case of ammonium molybdate the heat must be very low at first on account of the frothing. If you have a platinum tube it is safer to ignite the molybdic acid in this for 2 or 3 hours in a slow current of hydrogen, thus reducing it to the metallic state. When reducing to dioxide the contents of the crucible are frequently gray below, and brown above (Rammelsberg*).

b. The following is the best method of precipitating molybdic acid from an alkaline solution: Dilute the solution, if necessary, neutralize the free alkali with nitric acid, and allow the carbonic acid, which may be liberated in the process, to escape, then add neutral mercurous nitrate. The yellow precipitate formed appears at first bulky, but after several hours' standing it shrinks; it is insoluble in the fluid, which contains an excess of mercurous nitrate. Collect on a filter, and wash with a dilute solution of mercurous nitrate, as it is slightly soluble in pure water. Dry, remove the precipitate as completely as practicable from the filter, and determine the molybdenum in it as directed in *a* (H. Rose); or mix the precipitate, together with the filter-ash, with a weighed quantity of ignited lead oxide, and ignite until all the mercury is expelled; then add some ammonium nitrate, ignite again and weigh. The excess obtained, over and above the weight of the lead oxide used, is molybdenum trioxide (Seligsohn†).

---

* Pogg. Annal. 127, 281; Zeitschr. f. anal. Chem. 5, 203.

† Journ. f. prakt. Chem. 67, 472.

c. CHATARD* recommends estimating molybdic acid in the solution of its alkali salts by adding lead acetate in slight excess to the boiling solution and boiling for a few minutes. The precipitate which is at first milky becomes granular, deposits well, and may be easily washed with hot water. It is dried, removed from the filter as much as possible, ignited and weighed as $PbMoO_4$. The method is only applicable for solutions of *pure* alkali molybdates.

d. The precipitation of molybdenum as sulphide is always a difficult operation. If the acid solution is supersaturated with hydrogen sulphide, warmed, and filtered, the filtrate and washings are generally still colored. They must, accordingly, be warmed, and hydrogen sulphide again added, and the operation must afterwards, if necessary, be repeated until the washings appear almost colorless. The precipitation succeeds better when the molybdenum sulphide is dissolved in a relatively large excess of ammonium sulphide, and, after the fluid has acquired a reddish-yellow tint, precipitated with hydrochloric acid. ZENKER† advises then to boil, until the hydrogen sulphide is expelled, and to wash with hot water, at first slightly acidified. To make quite sure that all the molybdenum is precipitated, treat the filtrate and washings again with hydrogen sulphide and allow to stand for some time. The brown molybdenum sulphide is collected on a weighed filter, and the molybdenum determined in an aliquot part of it, by gentle ignition in a current of hydrogen gas, as in *a.* The brown molybdenum sulphide changes in this process to the gray disulphide (H. ROSE).

e. F. PISANI‡ gives the following method for estimating molybdic acid volumetrically. Digest the molybdic acid with hydrochloric acid and zinc, dissolving any precipitate which may form from want of acid and also the excess of zinc. The molybdic acid is thus reduced to a molybdenum salt corresponding to molybdenum sesquioxide. Convert the molybdenum in this solution again into molybdic acid by standard permanganate of potash. The brown color of the solution turns first green, and then disappears. RAMMELSBERG§ confirms the statements of PISANI.

* Sill. Amer. Journ. (3), 1, 416.
† Journ. f. prakt. Chem. 58, 259.
‡ Compt. rend. 59, 301.
§ Pogg. Annal. 127, 281; Zeitschr. f. anal. Chem. 5, 203.

II. DETERMINATION OF ACIDS IN COMPOUNDS CONTAINING ONLY ONE ACID, FREE OR COMBINED;—AND SEPARATION OF ACID FROM BASIC RADICALS.

First Group.

FIRST DIVISION.

Arsenious Acid—Arsenic Acid—Chromic Acid—(Selenious Acid, Sulphurous and Hyposulphurous Acids, Iodic Acid).

§ 129.

1. Arsenious and Arsenic Acids.

These have been already treated of among the bases (§ 127) on account of their behavior with hydrogen sulphide; they are merely mentioned here to indicate the place to which they properly belong. The methods of separating them from the bases will be found in Section V.

§ 130.

2. Chromic Acid.

I. Determination.

Chromic acid is determined either as *chromic oxide* or *lead chromate*. But it may be estimated also from the quantity of carbon dioxide disengaged by its action upon oxalic acid in excess, and also by volumetric analysis. In employing the first method it must be borne in mind that 1 mol. chromic oxide corresponds to 2 mol. chromic acid.

a. Determination as Chromic Oxide.

α. The chromic acid is reduced to the state of a chromic salt and the amount of chromium in the latter determined (§ 106). The reduction is effected either by heating the solution with hydrochloric acid and alcohol; or by mixing hydrochloric acid with the solution, and conducting hydrogen sulphide into the mixture; or by adding a strong solution of sulphurous acid, and applying a gentle heat. With concentrated solutions the first method is generally resorted to, with dilute solutions one of the two latter. With respect to the first method, I have to remark that the alcohol must be expelled before the chromium can be precipitated as hydroxide by ammonia; and with respect to the second, that the solution supersaturated with hydrogen sulphide must be allowed to stand in a moderately warm place, until the separated sulphur has completely subsided. The results are accurate, unless the weighed precipitate contains silica and lime, which is always the case if the precipitation is effected in glass vessels.

β. The neutral or slightly acid (nitric acid) solution is precipitated with mercurous nitrate, after long standing the red precipitate of mercurous chromate is filtered off, washed with a dilute solution of mercurous nitrate, dried, ignited, and the residuary chromic oxide weighed (H. Rose). Results accurate.

b. Determination as Lead Chromate.

The solution is mixed with sodium acetate in excess, and acetic acid added until the reaction is strongly acid; the solution is then precipitated with neutral lead acetate. The washed precipitate is either collected on a weighed filter, dried in the water-bath, and weighed; or it is gently ignited as directed § 53, and then weighed. For the properties of the precipitate, see § 93, 2. Results accurate.

c. Determination by means of Oxalic Acid (after Vohl).

When chromic acid and oxalic acid are brought together in the presence of water and excess of sulphuric acid, chromic sulphate and carbon dioxide are formed, $3H_2C_2O_4 + 2H_2CrO_4 + 3H_2SO_4 = 6CO_2 + Cr_2(SO_4)_3 + 8H_2O$. Accordingly the amount of chromic acid can be calculated from the weight of carbon dioxide evolved. The process is the same as in the analysis of manganese ores (§ 203). 1 part of chromic acid requires $2\frac{1}{4}$ parts of sodium oxalate. If it is intended to determine potassium or sodium in the residue, ammonium oxalate is used.

d. Determination by Volumetric Analysis.

α. Schwarz's *method.*

The principle of this very accurate method is identical with that upon which Penny's method of determining iron is based (§ 112, 2, *b*). The execution is simple: acidify the not too dilute solution of the chromate with sulphuric acid, add in excess a measured quantity of solution of a ferrous salt, the strength of which you have previously ascertained, according to the directions of § 112, 2, *a*, or *b*, or the solution of a weighed quantity of ammonium ferrous sulphate, free from ferric salt, and then determine in the manner directed § 112, 2, *a*, or *b*, the quantity of ferrous iron remaining. The difference shows the amount of iron that has been converted by the chromic acid from a ferrous to a ferric salt. 1 grm. of iron corresponds to 0·5981 of chromic anhydride (CrO_3). To determine the chromic acid in lead chromate, the latter is, after addition of the ammonium ferrous sulphate, most thoroughly triturated with hydrochloric acid, water added, and the analysis then proceeded with.

β. BUNSEN'S *method.**

If a chromate is boiled with an excess of fuming hydrochloric acid, there are disengaged for every atom of chromium 3 at. chlorine; for instance, $K_2Cr_2O_7 + (HCl)_{14} = (KCl)_2 + Cr_2Cl_6 + 6Cl + 7H_2O$. If the escaping gas is conducted into solution of potassium iodide in exces, the 3 at. chlorine set free 3 at. iodine. The liberated iodine may next be determined as described in § 146. 380·55 of iodine correspond to 100·48 of chromic anhydride (CrO_3).

The analytical process is conducted as follows: Put the weighed sample of the chromate (say ·3 to ·4 grm.) into the little flask *d*, fig. 55 (blown before the lamp. and holding only from 36 to 40 c.c.), and fill the flask two thirds with pure fuming hydrochloric acid (free from Cl and SO_2), add a compact lump of magnesite, to keep up a constant current of gas and prevent the fluid from receding. Connect the bulbed evolution tube *a*

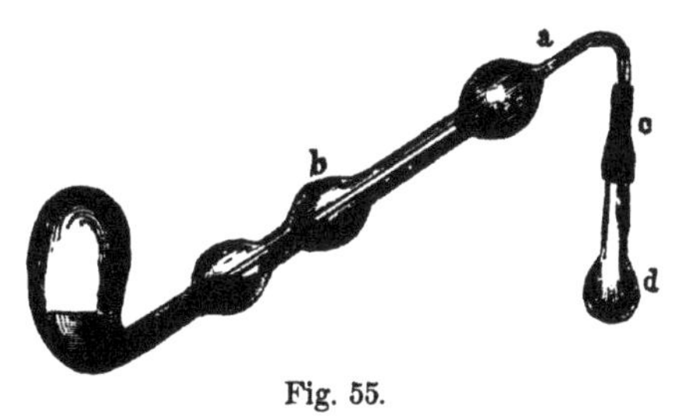

Fig. 55.

with the neck of the flask by means of a stout india-rubber tube *c*. As shown in the engraving, *a* is a bent pipette, drawn out at the lower end into an upturned point. A loss of chlorine need not be apprehended on adding the hydrochloric acid, as the disengagement of that gas begins only upon the application of heat. Insert the evolution tube into the neck of the retort, which is one-third filled with solution of potassium iodide.† This retort holds about

* Annal. d. Chem. u. Pharm. 86, 279.

† 1 part of pure potassium iodide, free from iodic acid, dissolved in 10 parts of water. The fluid must show no brown tint immediately after addition of dilute sulphuric acid.

160 c.c. The neck presents two small expansions, blown before the lamp, and intended, the lower one, to receive the liquid which is forced up during the operation, the upper one to serve as an additional guard against spirting. Apply heat now, cautiously, to the little flask. After two or three minutes ebullition the whole of the chlorine has passed over, and liberated its equivalent quantity of iodine in the potassium iodide solution. When the ebullition is at an end, take hold of the caoutchouc tube *c* with the left hand, and, whilst steadily holding the lamp under the flask with the right, lift *a* so far out of the retort that the curved point is in the bulb *b*. Now remove first the lamp, then the flask, dip the retort in cold water to cool it, and shake the fluid in it about to effect the complete solution of the separated iodine in the excess of potassium iodide solution. When the fluid is quite cold, transfer it to a beaker, rinsing the retort into the beaker, and proceed as directed § 146. The method gives very satisfactory results. The apparatus here recommended differs slightly from that used by BUNSEN, the retort of the latter having only one bulb in the neck, and the evolution tube no bulb, being closed instead, at the lower end, by a glass or caoutchouc valve, which permits the exit of the gas from the tube, but opposes the entrance of the fluid into it. I think the modifications which I have made in BUNSEN's apparatus are calculated to facilitate the success of the operation. Instead of this apparatus, that described § 142 may also be very conveniently used.

II. SEPARATION OF CHROMIC ACID FROM THE BASIC RADICALS.

a. Of the First Group.

α. Reduce the chromic acid to a chromic salt, as directed in I., and separate the chromium from the alkalies as directed in § 155.

β. Mix the potassium or sodium chromate with about 5 parts of dry pulverized ammonium chloride, and heat the mixture cautiously. The residue contains the chlorides of the alkali metals and chromic oxide, which may be separated by means of water.

γ. Precipitate the chromic acid according to I., *a*, *β*, and separate the mercury and alkali metals in the filtrate by § 162.

b. Of the Second Group.

α. Fuse the compound with 4 parts of sodium and potassium carbonates, and treat the fused mass with hot water, which dissolves the chromic acid in the form of an alkali chromate. The residue contains the alkali earth metals in the form of carbonates; but as they contain alkali, they cannot be weighed directly. The chromic acid in the solution is determined as in I. Strontium and calcium chromates may be decomposed by boiling with potassium or sodium carbonate. Barium chromate may also be decomposed in the same way, but the boiling must be repeated a second time with fresh solution of alkali carbonate (H. ROSE).

β. Dissolve in hydrochloric acid, reduce the chromic acid according to I., *a*, and separate the chromium from the alkali earth metals according to § 156.

γ. Magnesium chromate, as well as other chromates of the alkali earth metals soluble in water, may be easily decomposed also, by determining the chromic acid according to I., *a*, *β*, or I., *b*, and separating the magnesium, etc., in the filtrate from the excess of the salt of mercury or lead as directed § 162.

δ. Barium strontium and calcium chromates may also be decomposed by the method described II., *a*, *β*. Compare BAHR, Analysis of barium and calcium dichromates, etc.*

c. Of the Third Group.

α. From Aluminium.

If you have chromic acid to separate from aluminium in acid solution, precipitate the aluminium with ammonia or ammonium carbonate (§ 105, *a*), and determine the chromic acid in the filtrate according to I. If the washed aluminium hydroxide has a yellow color, treat on the filter with ammonia, and wash with boiling water; this will remove the last traces of chromic acid. However, a little aluminium hydroxide dissolves in the ammonia, therefore heat the ammoniacal fluid in a platinum dish till it has almost lost its alkaline reaction, and collect on a filter the flocks of aluminium hydroxide which separate, and add them to the principal precipitate.

β. From Chromium.

aa. Determine in one portion the quantity of the chromic acid according to I., *c*, or I., *d*, *α*, or *β*, and in another portion the total amount of the chromium, by converting it into sesquioxide by cautious ignition with ammonium chloride, or by I., *a*, or by converting it entirely into chromic acid by § 106, 2.

bb. In many cases the chromic acid may be precipitated accord-

* Journ. f. prakt. Chem. 60, 60.

ing to I., *a*, *β*, or I., *b*. The chromium and mercury, or lead, in the filtrate, are separated as directed § 162.

cc. The hydrated compounds of sesquioxide of chromium with chromium trioxide, or chromic chromates, such as are obtained by precipitating a solution of chromic salt with potassium chromate, etc., may also be analyzed by ignition in a stream of dry air, in a bulb tube, to which a calcium chloride tube is attached (fig. 25, § 36). The loss of weight represents the joint amount of oxygen and water that have escaped. If the increment of the $CaCl_2$ tube is deducted, we shall have the oxygen. Now every 3 at. oxygen correspond to 2 mol. CrO_3. The amount of the latter being thus calculated, we have only to subtract its equivalent quantity of sesquioxide from the weight of residue after the ignition, and the remainder is the quantity of sesquioxide originally present. Vogel* and also Storer and Elliot† have employed this method.

d. Of the Fourth Group.

α. Proceed as directed in *b*, *α*. Upon treating the fused mass with hot water, oxides of the basic metals are left. In the case of manganese the fusion must be effected in an atmosphere of carbon dioxide. Apparatus, fig. 50 in § 108.

β. Reduce the chromic acid as directed in I., *a*, and separate the chromium from the metals in question, as directed in § 160.

e. Of the Fifth and Sixth Groups.

α. Acidify the solution, and precipitate, either at once or after reduction of the chromic acid by sulphurous acid, with hydrogen sulphide. The metals of the fifth and sixth groups precipitate in conjunction with free sulphur (§§ 115 to 127), the chromic acid is reduced. Filter and determine the chromium in the filtrate, as directed in I., *a*.

β. Lead chromate may be conveniently decomposed by heating with hydrochloric acid and some alcohol; the lead chloride and chromic chloride formed are subsequently separated by means of alcohol (compare § 162). The alcoholic solution ought always to be tested with sulphuric acid; should a precipitate of lead sulphate form, this must be filtered off, weighed, and taken into account.

---

* Journ. f. prakt. Chem. 77, 484.

† Proceedings of the American Academy, 5, 198.

Supplement to the First Division.

§ 131.

1. Selenious Acid.

From aqueous or hydrochloric acid solutions of selenious acid, the selenium is precipitated by sulphurous acid gas, or, in presence of an excess of acid, by sodium sulphite, or ammonium sulphite. The liquid containing the precipitate is heated to boiling for ¼ hour, which changes the precipitate from its original red color to black, and makes it dense and heavy. The liquid is tested by a further addition of the reagent to see whether any more selenium will separate; the precipitate is finally collected on a weighed filter, dried at a temperature somewhat below 100°, and weighed. Since H. Rose* has shown that the presence of hydrochloric acid is an essential condition to the complete reduction of selenious acid, the former acid must be added, if not already present. To make quite sure that all the selenium has been removed, the filtrate is evaporated to a small volume, with addition of potassium or sodium chloride, boiled with strong hydrochloric acid, so as to reduce any selenic acid to selenious acid, and tested once more with sulphurous acid. If the solution contains nitric acid it must be evaporated repeatedly with hydrochloric acid, with addition of sodium or potassium chloride. If the latter were omitted there would be considerable loss of selenious acid (Rathke†).

As regards the separation of selenious acid from basic radicals, the following brief directions will suffice:

a. If the basic radicals are not liable to be altered by the action of sulphurous acid and hydrochloric acid, the selenium may be at once precipitated in the way just given; the filtrate, when evaporated with sulphuric acid, yields the base as sulphate.

b. From basic metals which are not thrown down from acid solution by hydrogen sulphide, the selenious acid may be separated by precipitation with that reagent. The precipitate (according to Rathke‡, a mixture of SeS_2, Se_2S and S) contains 2 at. sulphur to 1 at. selenium. If it is dried at or a little below 100°, the weight

* Zeitschr. f. anal. Chem. 1, 73.

† Journ. f. prakt. Chem. 108, 249; Zeitschr. f. anal. Chem. 9, 484.

‡ Journ. f. prakt. Chem. 108, 252.

of the selenium may be accurately ascertained. Should, however, extra sulphur be mixed with the precipitate, the latter is oxidized while still moist with hydrochloric acid and potassium chlorate, or by treatment with potassa solution with simultaneous heating and transmission of chlorine. It is necessary here to oxidize the sulphur completely, as it may enclose selenium, The solution now containing selenic acid is heated till it smells no longer of chlorine, hydrochloric acid is added, and the mixture is reheated. The selenic acid is hereby reduced to selenious acid, and when the solution has again ceased to smell of chlorine, the selenium is precipitated with sulphurous acid. Instead of this process you may digest the precipitate of sulphur and selenium for some hours with concentrated potassium cyanide, which will completely dissolve it, and then throw down the selenium from the dilute solution with hydrochloric acid as in *c* (RATHKE, *loc. cit.*).

c. In many selenites or selenates the selenium may also be determined by converting first into potassium selenocyanate, and precipitating the aqueous solution of the latter with hydrochloric acid (OPPENHEIM*). To this end the substance is mixed with 7 or 8 times its quantity of ordinary potassium cyanide (containing cyanic acid), the mixture is put into a long-necked flask, or a porcelain crucible, covered with a layer of potassium cyanide, and fused in a stream of hydrogen. The temperature is kept so low that the glass or porcelain is not attacked, and while cooling care must be taken to exclude atmospheric air. When cold, the brown mass is treated with water, and the colorless solution filtered, if necessary. The liquid should be somewhat but not immoderately diluted. Now boil some time (in order to convert the small quantity of potassium selenide that may be present into potassium selenocyanate, by the excess of potassium cyanide, allow to cool, supersaturate with hydrochloric acid, and heat again for some time. At the end of 12 or 24 hours all selenium will have separated, filter, dry at 100°, and weigh. The results obtained by this process are accurate (H. ROSE†). If the selenium agglomerates together on heating, it may enclose salts. In such cases, by way of control, it should be redissolved in nitric acid, and, after addition of hydrochloric acid, precipitated with sulphurous acid. The fluid filtered from the selenium precipitate is, as a rule, free from selenium ; it

---

* Journ. f. prakt. Chem. 71, 280. † Zeitschr. f. anal. Chem. 1, 73.

is, however, always well to satisfy one's self on this point by the addition of sulphurous acid.

d. From many basic radicals selenious acid (and also selenic acid) may be separated by fusing the compound with 2 parts of sodium carbonate and one part of potassium nitrate, extracting the fused mass thoroughly by boiling with water, saturating the filtrate, if necessary, with carbonic acid, to free it from lead which it might contain, then boiling down with hydrochloric acid in excess (to reduce the selenic acid and drive off the nitric acid), and precipitating finally with sulphurous acid.

Selenium, if pure, must volatilize without residue when heated in a tube.

2. Sulphurous Acid.

To estimate free sulphurous acid in a fluid which may contain also other acids (sulphuric acid, hydrochloric acid, acetic acid), a weighed quantity of the fluid is diluted with water, absolutely free from air,* until the diluted liquid contains not more than ·05 per cent. by weight of sulphurous acid, the solution is poured with stirring into an excess of standard solution of iodine, the free iodine remaining is titrated with sodium thiosulphate, and the iodine used for the conversion of sulphurous into sulphuric acid is thus found. The reaction is expressed by the equation, $SO_2 + 2H_2O + 2I = H_2SO_4 + 2HI$. According to Finkener, if the iodine is added to the sulphurous acid the reaction is not quite normal. Anyhow this method of operating prevents any loss of sulphurous acid. For the details, see § 146. In case of sulphites soluble in water or acids, water perfectly free from air is poured over the substance, in sufficient quantity to attain the degree of dilution stated above, sulphuric or hydrochloric acid is added in excess, and then the titration is effected as above. The greatest care must be taken in this method, to use, for the purpose of dilution, water absolutely free from air.

Sulphurous acid may also be determined in the gravimetric way, by conversion into sulphuric acid, and precipitation of the latter with barium chloride, according to § 132. This method is especially applicable in the case of sulphites quite free from sulphuric acid. The conversion of the sulphurous into sulphuric acid is

* Prepared by long-continued boiling and subsequent cooling with exclusion of air.

effected in the wet way, best by pouring the dilute solution with stirring into excess of chlorine or bromine water. Sulphites insoluble in water are decomposed by boiling with sodium carbonate, and the solution of sodium sulphite is treated as directed. After driving off the excess of chlorine or bromine by heating, the moderately acid solution is precipitated with barium chloride. Sulphites may be oxidized in the dry way by heating in a platinum crucible, with 4 parts of a mixture of equal parts sodium carbonate and potassium nitrate.

3. Thiosulphuric Acid.

Thiosulphuric acid, in form of soluble thiosulphates, may be determined by means of iodine, in a similar way to sulphurous acid. The reaction is represented by the equation, $2Na_2S_2O_3 + 2I = 2NaI + Na_2S_4O_6$. The salt under examination is dissolved in a large amount of water, starch-paste added, and then the neutral solution is titrated with iodine. That this method can give correct results only in cases where no other substances acting upon iodine are present, need hardly be mentioned. Thiosulphuric may like sulphurous acid be converted into sulphuric acid by means of chlorine or bromine water, and then determined.

4. Iodic Acid.

Iodic acid may be determined by the following easy method:—Distil the free acid or iodate with an excess of pure fuming hydrochloric acid, in the apparatus described in § 130, *d*, *β* (chromic acid), receive the disengaged chlorine in solution of potassium iodide, and determine the separated iodine as directed in § 130, I, *d*, *β*. The decomposition of iodic acid by hydrochloric acid is represented by the equation, $HIO_3 + 5HCl = ICl + 4Cl + 3H_2O$. Since the 4 at. Cl set free 4 at. I, the amount of iodic acid or iodic anhydride can be calculated from the weight of the latter; 1014·8 iodine correspond to 333·7 iodic anhydride (I_2O_5) (Bunsen*). The following method also yields good results. Mix the solution with dilute sulphuric acid, add potassium iodide in excess, and determine the amount of liberated iodine, after § 146. One sixth of the iodine thus formed is derived from the iodic acid ($HIO_3 + 5HI = 3H_2O + I_6$). See Rammelsberg.†

* Annal. d. Chem. u. Pharm. 86, 285.

† Pogg. Annal. 135, 493; Zeitschr. f. anal. Chem. 8, 456.

5. Nitrous Acid.

The nitrous acid in nitrites which are free from nitrates may be estimated by converting the nitrogen into ammonia and determining the latter, or by determining the oxidizing action on ferrous salt. This method is conducted exactly as described under nitric acid (§ 149). When nitric acid is also present, nitrous acid may be determined very satisfactorily with a solution of pure potassium permanganate, provided the fluid be sufficiently diluted to prevent the nitrous acid, which is liberated by the addition of a stronger acid, being decomposed by water with formation of nitric acid and nitric oxide. For 1 part of nitrous anhydride at least 5000 parts of water should be present. The decomposition is represented by the following equation, $5HNO_2 + K_2Mn_2O_8 + 3H_2SO_4 = 5HNO_3 + K_2SO_4 + 2MnSO_4 + 3H_2O$. If the permanganate be standardized with iron, 4 at. iron correspond to 1 mol. N_2O_3, since both of these require 2 at. oxygen. Nitrites are dissolved in *very slightly* acidulated water, the permanganate is added till the oxidation of the nitrous acid is nearly completed, the solution is then made strongly acid, and finally permanganate is added to light-red coloration.

To determine nitrogen tetroxide N_2O_4 in red fuming nitric acid, transfer a few c.c. to about 500 c.c. cold pure distilled water with stirring, and determine the nitrous acid produced. 1 mol. nitrous anhydride found corresponds to 2 mol. nitrogen tetroxide, for the latter—when mixed with such a large quantity of water as is indicated above—is decomposed in accordance with the following equation :—$N_2O_4 + H_2O = HNO_3 + HNO_2$ (Sig. Feldhaus*).

Nitrous acid and nitrogen tetroxide in presence of nitric acid may also be estimated by the reduction of chromic acid. An excess of standard potassium dichromate is added, and the undecomposed residue of chromic acid is estimated with standard solution of ferrous salt (F. Mohr†).

As regards the estimation of nitrous acid with lead dioxide, comp. Feldhaus, *loc. cit.* p. 431, also Lang‡ and J. Löwenthal.§

* Zeitschr. f. anal. Chem. 1, 426.
† His Lehrbuch der Titrirmethode, 3 Aufl. 236.
‡ Zeitschr. f. anal. Chem. 1, 485. § *Ib.* 3, 176.

Second Division of the First Group of the Acids.

SULPHURIC ACID; (Hydrofluosilicic Acid).

§ 132.

SULPHURIC ACID.

I. DETERMINATION.

Sulphuric acid is usually determined in the gravimetric way as *barium sulphate.* The acid may, however, be estimated also by the acidimetric method (§ 192), and by certain volumetric methods, based upon the insolubility of the barium sulphate (and lead sulphate).

1. *Gravimetric Method.*

The exact estimation of sulphuric acid as barium sulphate is by no means so simple and easy as it was formerly supposed to be, but requires, on the contrary, great care and attention. This arises from three causes: first, the barium sulphate is found to be far more soluble than was imagined in solutions of free acids and of many salts; secondly, it is extremely liable to carry down with it foreign salts, which are of themselves soluble in water; thirdly, when the precipitate has once separated in an impure state, it is often very difficult to purify it completely.

The solution should contain but little free hydrochloric acid, and no nitric or chloric acid. If either of the two last are present, evaporate repeatedly, on the water-bath with pure hydrochloric acid. Dilute considerably, heat nearly to boiling, add barium chloride in moderate excess, and allow to settle for a long time at a gentle heat. Decant the clear fluid through a filter, treat the precipitate with boiling water, allow to settle, decant again, and so on, till the washings are free from chlorine. Finally transfer the precipitate to the filter, dry and treat according to § 53, using only a moderate red heat.

After the precipitate has been weighed it is well to warm it for some time with dilute hydrochloric acid on the water-bath. Then pour off the hydrochloric acid through a small filter, wash the precipitate by decantation with boiling water without removing it to the filter, evaporate the filtrate and washings nearly to dryness in a platinum or porcelain dish, add water, collect the minute amount of barium sulphate here left undissolved upon the small filter, wash, dry, incinerate, add the ash to the bulk of the precipitate, ignite again, and weigh. If the precipitate has lost weight, this shows that it at first contained foreign salts.

This method of purification sometimes fails when the precipitate contains ferric oxide or platinum (CLAUS*), and it invariably fails when the solution contained any notable quantity of nitric acid.† In such cases there is only one resource, namely, to fuse with about four parts of sodium carbonate, warm with water, filter, wash with boiling water, acidify the filtrate slightly with hydrochloric acid, and determine the sulphuric acid again.

The results are thoroughly satisfactory if these directions are attended to; if not, the result may be two or three per cent. too high or too low.

2. *Volumetric Methods.*

a. After CARL MOHR.‡ We require a normal solution of barium chloride, containing 121·96 grm. of the pure crystallized salt in 1 litre, and also normal nitric or hydrochloric acid and normal soda (§ 192, *c.* *δ*). Add to the fluid to be examined for sulphuric acid—which, should it contain much free acid, is previously to be nearly neutralized with pure sodium carbonate—a measured quantity of barium chloride solution, best a round number of cubic centimetres, in more than sufficient proportion to precipitate the sulphuric acid, but not in too great excess. Digest the mixture for some time in a warm place, then precipitate, without previous filtration, the excess of barium chloride with ammonium carbonate and a little ammonia, filter off the barium sulphate and carbonate, wash until the water running off acts no longer upon red litmus paper, and then determine the barium carbonate by the alkalimetric method given in § 198. Deduct the c.c. of normal acid used from the c.c. of barium chloride, and the remainder will be the c.c. of barium chloride corresponding to the sulphuric acid present. The results of this method are quite satisfactory, if the solution does not contain too much free acid; but in presence of a large excess of free acid, the action of the salt of ammonia will retain barium carbonate in solution, which, of course, will make

* Jahresber. von KOPP und WILL. 1861, 323, note.

† Compare my paper in Zeitschr. f. anal. Chem. 9, 52.

‡ Ann. der Chem. u. Pharm. 90, 165.

the amount of sulphuric acid appear higher than is really the case. It need hardly be mentioned that this method is altogether inapplicable in presence of phosphoric acid, oxalic acid, or any other acid precipitating barium salt from neutral solutions, and that no basic radicals except the alkalies may be present.

b. After R. WILDENSTEIN.* Of all the methods for the volumetric estimation of sulphuric acid, the simplest and that which is capable of the most general application, is to drop into the solution containing excess of hydrochloric acid, standard barium chloride solution, till the exact point is reached when no more precipitation takes place. This point is difficult to hit, and hence the method has only found a very limited use.

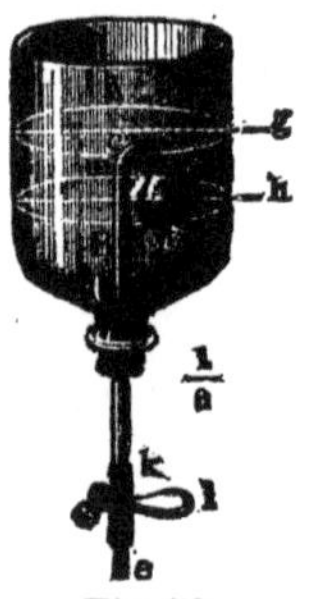

Fig. 56.

WILDENSTEIN has given this method a practical form, which renders it possible to complete an analysis in about half an hour, and at the same time to obtain satisfactory results. He employs the apparatus, fig. 56. *A* is a bottle of white glass, whose bottom has been removed, it contains 900—950 c.c. *B* is a strong funnel-tube, with bell-shaped funnel, and bent as shown, provided below with a piece of india-rubber tube, a screw compression-cock, and a small piece of tubing not drawn out. The length from *c* to *d* is about 7½-8, from *d* to *e* about 12 cm. The opening of the funnel-tube *f*, which should have a diameter of 2·5 to 3 cm., is covered as follows: Take a piece of fine new calico or muslin, free from sulphuric acid, and about 6 cm. square, lay on it two pieces of Swedish paper of the same size, and then another piece of stuff like the first, now bind these altogether over the opening *f*, carefully and without injuring the paper, by means of a strong linen thread which has been drawn a few times over wax, and cut it off even all round. We have now a small syphon-filter, which enables us to filter off a portion of fluid contained in *A*, and turbid from barium sulphate, clear and with comparative rapidity.

On gradually adding barium chloride to the dilute acid solution of a sulphate a point occurs which may be compared to the neutral point in precipitating silver with sodium chloride (see § 115, 5, *b*.); *i.e.*, there is a certain moment, when a portion filtered off will give

---

* Zeitschr. f. anal. Chem. 1, 432.

a turbidity both with sulphuric acid and barium chloride after the lapse of a few minutes. On this account we must either proceed on the principle recommended for the estimation of silver, *i.e.*, disregarding the quantity of barium chloride in the solution, to standardize it by adding it to a known amount of sulphate, till a precipitate ceases to be formed; or else we must—and WILDENSTEIN recommends this latter course—consider as the end-point of the reaction the point at which barium chloride ceases to produce a distinctly visible precipitation in the clear filtrate after a lapse of two minutes.

The barium chloride solution is prepared so that 1 c.c. corresponds to ·02 sulphuric anhydride by making a solution containing the requisite calculated and carefully weighed amount of the pure salt per litre.—A solution of sulphuric acid containing ·02 gr. SO_3 per c.c. may also be required. The process is as follows:

First prepare the solution of the sulphate to be analyzed (using about 3 or 4 grm.), then fill *A* with hot water, open the cock with the screw or by the aid of a glass rod, and wait till the syphon *B* is quite full of water. If the water runs down the tube *c e* without filling it entirely, close and open the cock a few times, and this inconvenience will be removed. (It is not allowable to suck at *e*, or to fill the syphon with the wash-bottle at *e*, as either proceeding would inevitably lead to injuring the filter.) Now close the cock and pour out the hot water, replace it by 400 c.c. of boiling water, add the ready-prepared solution of the sulphate, and a small quantity of hydrochloric acid, if necessary, and run in the barium chloride solution, at first in rather large portions, at last in $\frac{1}{5}$ c.c. Before each fresh addition of barium chloride open the cock and allow rather more liquid to flow into a beaker than corresponds to the contents of the syphon. This quantity should be previously ascertained, and a mark indicating it made on the beaker. Now close the cock and pour the filtrate without loss back into *A*. (As the beaker is used over and over again for the same purpose, it need not be rinsed out.) Now run some of the fluid into a test-tube, so as to one third fill it, add to the clear fluid 2 drops of barium chloride from the burette and shake. If a precipitate or turbidity is produced, return the portion to the main quantity. The experiment is finished when the last portion tested shows after the lapse of exactly two minutes no distinctly visible turbidity. The drops of barium chloride used for the last testing are of course not reckoned. The slight error involved from the fact that the small quantity of fluid in the syphon is finally unacted on, is too small to be noticed. During the experiment the filter must not be injured by the stirring. In case the end reaction has been overstepped, add 1 c.c. of dilute sulphuric acid (equivalent to the barium chloride) to *A*, and endeavor to hit it again. Here 1 c.c. will have to be subtracted from the c.c. of barium chloride used.

The results obtained by WILDENSTEIN are of sufficient accuracy for technical purposes. Some experiments made in my own laboratory were also quite satisfactory.

II. SEPARATION OF SULPHURIC ACID FROM THE BASIC RADICALS.

a. In Sulphates which are soluble in Water or Hydrochloric Acid.

The solution should be free from nitric acid. Precipitate the sulphuric acid according to I. by barium chloride (or barium acetate). The filtrate contains the excess of barium chloride, together with the chlorides of the metals present; separate barium from the latter by methods given in the fifth section. The fluid obtained by treating the ignited barium sulphate with hydrochloric acid, evaporating and filtering from the small amount of barium sulphate, must be added to the first solution before separating barium from it.

b. In Sulphates which are insoluble or difficultly soluble in Water or in Hydrochloric Acid.

α. From barium, strontium and calcium: Fuse the finely pulverized substance in a platinum crucible, with 5 parts of mixed sodium and potassium carbonates. Put the crucible, with its contents, into a beaker, or into a platinum or porcelain dish, pour water over it, and apply heat until the alkali sulphates and carbonates are completely dissolved; filter the hot solution from the residuary alkali-earth carbonates, wash the latter thoroughly with water, to which a little ammonia and ammonium carbonate has been added, and determine according to §§ 101 to 103. If the precipitates have been well washed, it is perfectly admissible to ignite and weigh at once. Precipitate the sulphuric acid from the filtrate, as in I., after acidifying with hydrochloric acid. Finely pulverized calcium and strontium sulphates may be completely decomposed also by boiling with a solution of potassium carbonate.*

* Sodium carbonate does not answer as well.

β. From lead: The simplest way of effecting the decomposition of lead sulphate consists in digesting it, at the common temperature, with a solution of hydrogen sodium or hydrogen potassium carbonate, filtering, washing the precipitate, determining the sulphuric acid in the filtrate as in I., dissolving the precipitate, which contains alkali, in nitric or acetic acid, and determining the lead in the solution, by one of the methods given in § 162.

Presence of strontium and calcium necessitates no alteration in this method; but if barium also is present, and it is accordingly necessary to ignite* the mixture with alkali carbonates, a small portion of lead always remains in solution in the alkaline fluid; this must be precipitated by passing through it carbon dioxide before filtering.

γ. From mercury in mercurous sulphate: Mercurous sulphate is best dissolved by warming with dilute hydrochloric acid with addition of potassium chlorate or bromine, and the solution is treated according to *a*. If the salt is boiled with solution of potassium carbonate, the mercurous carbonate first formed is decomposed, and the residue contains metallic mercury and mercuric oxide; a small part of the latter passes into the filtrate.

III. Estimation of Free Sulphuric Acid in presence of Sulphates.

We have occasionally to estimate the free acid in presence of sulphates, as, for instance, in vinegar, wine, etc. According to A. Girard† the following is the only direct method which can be relied on. Evaporate on the water-bath to dryness and exhaust the residue with absolute alcohol; determine the combined acid in the residue, and the free acid in the alcoholic extract, after mixing with water and evaporating off the alcohol. It has been said that the object may be obtained by the use of barium carbonate, which is supposed to throw down the free acid only, but this is erroneous, since alkali sulphates in aqueous solution are partially decomposed at the ordinary temperature by barium carbonate. In some cases the amount of free sulphuric acid present may be calculated after having determined the total amount of basic and acid radicals present. When no other free acid is present, free sulphuric acid may be determined by the acidimetric process.

---

* This is best done in a porcelain crucible.

† Compt. rend. 58, 515; Zeitschr. f. anal. Chem. **4**, **219**.

Supplement to the Second Division.

§ 133.

Hydrofluosilicic Acid.

If you have hydrofluosilicic acid in solution, add solution of potassium chloride, then a volume of strong alcohol equal to the fluid present, collect the precipitated potassium silicofluoride on a weighed filter, and wash with a mixture of equal volumes of alcohol and water. Dry the washed precipitate at 100°, and weigh. Mix the alcoholic filtrate with hydrochloric acid, evaporate to dryness, and treat the residue with hydrochloric acid and water. If this leaves an undissolved residue of silicic acid, this is a sign that the examined acid contained an excess of silicic acid; the weight of the residue shows the amount of excess. Potassium silicofluoride dried at 100° has the formula $(KF)_2 SiF_4$; for its properties, see § 68. Instead of weighing it, it may be estimated volumetrically according to § 97, 4. The analysis of metallic silicofluorides is best effected by heating in platinum vessels, with concentrated sulphuric acid; silicon fluoride and hydrofluoric acid volatilize, the basic metals are left behind in the form of sulphates, and may, in many cases, after volatilization of the excess of sulphuric acid, be weighed as such. If the metallic silicofluorides to be analyzed contain water, the latter cannot be estimated by mere ignition, since silicon fluoride would escape with the water. H. Rose recommends the following method: Mix them most intimately with 6 parts of recently ignited lead oxide, cover the mixture in a small retort, with a layer of pure lead oxide, weigh the retort, heat cautiously until the contents begin to fuse together, remove the aqueous vapor still remaining in the vessel by suction, and weigh the retort again when cold. The diminution of weight shows the quantity of water expelled. Do not neglect testing the drops of the escaping water with litmus paper; the result is accurate only if they have no acid reaction.

F. Stolba* proposes the following process, at least for compounds soluble in water: Put into a crucible double as much magnesia as is necessary to decompose the silicofluoride to be analyzed, ignite it as strongly as possible, allow to cool, and weigh. Add water

* Zeitschr. f. anal. Chem. 7, 93.

to form a thick paste, and then the weighed silicofluoride; if the amount of water present is not enough to dissolve the compound, add some more, mix with a platinum wire which must afterwards be wiped off clean, dry, ignite, and weigh. The increase in weight shows the amount of anhydrous silicofluoride, provided no oxide is present which takes up oxygen.

Third Division of the First Group of the Acids.

PHOSPHORIC ACID—BORACIC ACID—OXALIC ACID—HYDROFLUORIC ACID.

§ 134.

1. Phosphoric Acid.

I. Determination.

Orthophosphoric acid may be determined in a great variety of ways. The forms in which this determination may be effected have been given already in § 93, 4. The most appropriate forms for the purpose, however, are *magnesium pyrophosphate* and *uranyl pyrophosphate*. The determination as magnesium pyrophosphate is frequently preceded by precipitation in another way, especially as ammonium phospho-molybdate, occasionally as stannic phosphate or mercurous phosphate. The other forms in which phosphoric acid may be determined give also, in part, very good results, but admit only of a more limited application. With respect to volumetric methods, those which depend upon the use of standard solution of uranium are the best.

With regard to meta- and pyrophosphoric acids, I have simply to remark here that these acids cannot be determined by any of the methods given below. The best way to effect their determination is to convert them into orthophosporic acid, as follows:

α. In the dry way. By protracted fusion with from 4 to 6 parts of mixed sodium and potassium carbonates. This method is, however, applicable only in the case of alkali meta- and pyrophosphates, and of those metallic meta- or pyrophosphates which are completely decomposed by fusion with alkali carbonates; it fails, accordingly, for instance, with the salts of the alkali-earth metals, magnesium excepted.

β. In the wet way. The salt is heated for some time with a strong acid, best with concentrated sulphuric acid (WEBER*). This method leads only to the attainment of approximate results, in the case of all salts whose basic radicals form soluble salts of the acid added, since in these cases the meta- or pyrophosphoric acid is never completely liberated; but the desired result may be fully attained by the use of any acid which forms insoluble salts compounds with the basic radicals present. Respecting the partial conversion in the former case, I have found that it approaches the nearer to completeness the greater the quantity of free acid added,† and that the ebullition must be long continued.

BUNCE's statement,‡ that phosphoric acid volatilizes when a phosphate is evaporated to dryness with hydrochloric or nitric acid and the residue heated a little, is quite erroneous (compare my paper on the subject, in Annal. der Chem. und Pharm., 86, 216). But, on the other hand, it must be borne in mind that orthophosphoric acid under these circumstances changes, not indeed at 100°, but at a temperature still below 150°, to pyrophosphoric acid; thus, for instance, upon evaporating common hydrogen sodium phosphate with hydrochloric acid in excess, and drying the residue at 150°, we obtain $2\,NaCl + Na_2H_2P_2O_7$.

a. Determination as Lead Phosphate.

Proceed as with arsenic acid, § 127, 1, *a*—*i.e.*, evaporate with a weighed quantity of oxide of lead, and ignite. This method presupposes that no other acid is present in the aqueous or nitric acid solution; it has this great advantage, that it gives correct results, no matter whether ortho-, meta-, or pyrophosphoric acid is present.

b. Determination as Magnesium Pyrophosphate.

α. Direct determination. Suitable in all cases in which it is quite certain that the acid present is orthophosphoric, either free or combined as an alkali phosphate.

The solution should be neutral, or only moderately ammoniacal. Add ammonium chloride, and then the usual magnesia mixture (§ 62, 6), in sufficient but not too excessive quantity (see § 62, 6). The precipitate being under these conditions somewhat slowly formed, appears distinctly crystalline. After some time add ammonia gradually to the amount of one third of the fluid. Allow

---

* Pogg. Annal. 73, 137.

† There are, however, other considerations which forbid going too far in this respect.

‡ Sillim. Journ. May, 1851. 405.

to stand 12 hours in a well-covered vessel in the cold, filter, test the filtrate with magnesia mixture and ammonia, and wash the precipitate with ammonia diluted with 3 volumes of water till the washings, when acidified with nitric acid and tested with silver nitrate, are no longer rendered turbid; proceed according to § 104, 2. The precipitate is not absolutely insoluble in ammoniated water, therefore it is well to wash by suction, as this reduces the necessary amount of wash water to a minimum. The results are accurate (Expt. No. 89, also KISSEL*). If there is reason to suspect the purity of the precipitate, dissolve it in hydrochloric acid, and throw down again with ammonia, adding some magnesia mixture. If the magnesia mixture is omitted, the solution being free from magnesia will dissolve some of the precipitate. Compare KISSEL, *loc. cit.* Properties of the precipitate and residue, § 74. If the solution contains pyrophosphoric acid, the precipitate is flocculent and dissolves to a notable degree in ammoniated water (WEBER).

β. Indirect determination, with previous precipitation as *ammonium phosphomolybdate*, after SONNENSCHEIN.†

Applicable in all cases in which the phosphoric acid present is *orthophosphoric*, even in presence of salts of the alkali-earth metals, aluminium, ferric iron, &c. Tartaric acid, however, and similarly acting organic substances must be absent. No considerable quantity of free hydrochloric acid may be present. Large quantities of ammonium chloride, and of metallic chlorides generally, also of certain ammonium salts, especially the oxalate and citrate (KÖNIG)‡, are to be avoided. Ammonium nitrate assists the precipitation and neutralizes the injurious action of very large quantities of nitrates and sulphates (E. RICHTERS)§. The molybdenum solution described "Qual. Anal.," § 55, is employed as the precipitant. It contains 5 per cent. of molybdic acid. The fluid to be examined for phosphoric acid should be concentrated, it may contain free nitric or free sulphuric acid. Transfer to a beaker and add a considerable quantity of the molybdenum solution. About 40 parts molybdic acid must be added for every 1 part phosphoric anhydride, therefore 80 c.c. of the molybdic solution for ·1 grm. Stir, without touching the sides, and keep covered 12 hours at about 40°. Then remove a portion of the clear supernatant fluid with a pipette, mix

* Zeitschr. f. anal. Chem. 8, 170.
† Journ. f. prakt. Chem. 53, 343.
‡ Zeitschr. f. anal. Chem. 10, 305.
§ *Ib.* 10, 469.

it with an equal volume of molybdenum solution, and allow it to stand some time at 40°. If a further precipitation takes place, return the portion to the main quantity, add more molybdenum solution, allow to stand again 12 hours, and test again. When complete precipitation has been effected pour the fluid off through a small filter and wash the precipitate entirely by decantation, using a mixture of 100 parts molybdate solution, 20 parts nitric acid of 1·2 sp. gr., and 80 parts water.* The washing must be thorough, and the last runnings must not be precipitated by excess of ammonia, even if lime, iron, &c., was present in the solution. Now dissolve the precipitate in the least quantity of ammonia, pour the fluid through the small filter, when the minute amount of precipitate thereon will be dissolved, wash the filter with ammonia diluted with three volumes of water, mix the filtrate and washings, and add hydrochloric acid carefully till the precipitate produced, instead of redissolving instantly, takes a little time to disappear; finally throw down with magnesia mixture (compare *α*). If the ammonia leaves a small amount of the precipitate undissolved, treat the residue with nitric acid and test the filtrate with molybdic solution in order to save any phosphoric acid. Results accurate.†

As this method requires so large a quantity of molybdic acid, it is usually resorted to only in cases where methods *b*, *α*, and *c* are inapplicable; and the phosphoric acid in the quantity of substance taken is not allowed to exceed ·3 grm. Arsenic acid and silicic acid,‡ if present, must first be removed. Of all the methods for determining phosphoric acid which are admissible in the presence of ferric and aluminium salts, this is the best in my opinion, especially for the estimation of small quantities of the acid in presence of large quantities of these salts.

---

* According to E. RICHTERS (Zeitschr. f. anal. Chem. 10, 471) you may also wash with a solution of ammonium nitrate containing 15 grm. in 100 c.c. slightly acidified with nitric acid and containing a few per-cents of molybdic acid solution.

† Zeitschr. f. anal. Chem. 3, 446, and 6, 403.

‡ Silicic acid may also be thrown down, in form of a yellow precipitate, by acid solution of ammonium molybdate, especially in presence of much ammonium chloride (W. KNOP, Chem. Centralb. 1857, 691). Mr. GRUNDMANN, who repeated KNOP'S experiments in my laboratory, obtained the same results. The precipitate dissolves in ammonia. If the solution, after addition of some ammonium chloride, is allowed to stand for some time, the silicic acid separates, and the phosphoric acid may then be precipitated from the filtrate with magnesia-mixture; it is, however, always the safer way to remove silicic acid first.

γ. *Indirect determination*, with previous precipitation as *mercurous phosphate*, after H. Rose.*

Applicable for the separation of phosphoric acid (also of pyro- and metaphosphoric acid) from all basic radicals, except aluminium. Comp. § 135, *k*.

Dissolve the phosphate in neither too large nor too small a quantity of nitric acid, in a porcelain dish, add pure metallic mercury in sufficient quantity to leave a portion, even though only a small one, undissolved by the free acid. Evaporate on the water-bath to dryness. If the warm mass still evolves an odor of nitric acid, moisten it with water, and heat again on the water-bath, until it smells no longer of nitric acid. Add now hot water, pass through a small filter, and wash until the washings leave no longer a fixed residue upon platinum. Dry the filter, which, besides mercurous phosphate, contains also basic mercurous nitrate and free mercury, mix its contents, in a platinum crucible, with mixed sodium and potassium carbonates in excess, roll the filter into the shape of a ball, place it in a hollow made in the mixture, and cover the whole with a layer of the mixed carbonates. Expose the crucible, under a chimney with good draught, for about half an hour to a moderate heat, so that it does not get red-hot. At this temperature, the mercurous nitrate and the metallic mercury volatilize. Heat now over the lamp to bright redness, and treat the residue with hot water, which will dissolve it completely, if no ferric oxide be present, and if no oxide of platinum has been formed. The latter may occur on account of too rapid heating, which might produce sodium nitrate, which would act upon the platinum. Supersaturate the clear (if necessary, filtered) solution with hydrochloric acid, add ammonia and magnesia-mixture, and proceed as in *α*.

δ. *Indirect determination*, with previous precipitation as *stannic phosphate*.

After Girard.† Dissolve the substance in highly concentrated nitric acid, remove all chlorine either by precipitation with silver nitrate or by repeated evaporation with nitric acid, add 8 times as much tinfoil as there is phosphoric acid present, and warm the mixture 5 or 6 hours, until the precipitate has completely subsided, leaving the supernatant fluid clear. Wash with hot water by decantation and finally by filtration. The precipitate consists of

* Pogg. Annal. 76, 218.

† Compt. rend. 54, 468; Zeitschrift f analyt. Chem. 1, 366.

metastannic acid and stannic phosphate, together with a little ferric and aluminium phosphate. Heat it either at first with a small quantity of aqua regia, and then with ammonia and ammonium sulphide, or immediately with ammonium sulphide in excess. The last process is recommended by O. BÄBER,* on the ground that the former leaves a little phosphoric acid in the precipitate. The whole is digested about two hours, and then filtered; the precipitate, consisting of ferrous sulphide and aluminium hydroxide, is washed with warm ammonium sulphide, then with water containing a little ammonium sulphide, dissolved in nitric acid, and the solution thus formed mixed with the filtrate from the tin precipitate which contains the principal quantity of the basic metals. From the ammonium sulphide filtrate, which contains stannic sulphide and ammonium phosphate, the phosphoric acid is at once precipitated by magnesia-mixture. I may add that GIRARD considers 4 to 5 parts tin sufficient for 1 part P_2O_5. The results afforded by his test-analyses are unexceptionable. According to JANOVSKY,† at least six parts of tin must be used. The tin should be free from arsenic.

c. Determination as Uranyl Pyrophosphate.

After LECONTE, A. ARENDT, and W. KNOP.‡ (Very suitable in presence of alkali and alkali-earth metals, but not in presence of any notable amount of aluminium; in presence of ferric iron, the method can be applied only with certain modifications.)§ Where it is possible, prepare an acetic acid solution of the compound. If you have a nitric or hydrochloric acid solution, remove the greater portion of the free acid by evaporation, add ammonia until red litmus paper dipped into it turns very distinctly blue, and then redissolve the precipitate formed in acetic acid. If mineral acids were present, add also some ammonium acetate; this addition is beneficial under any circumstances. Mix the fluid now with solution of uranyl acetate, and heat the mixture to boiling, which will cause the phosphoric acid to separate, in form of pale greenish-yellow ammonium uranyl phosphate.

* Zeitschr. f. die gesammten Naturwissensch. 1864, 293.

† Zeitschr. f. anal. Chem. 11, 157.

‡ LECONTE was the first to recommend the method of precipitating phosphoric acid from acetic acid solutions by means of a salt of uranium (Jahresb. von LIEBIG und KOPP, für 1853, 642); A. ARENDT and W. KNOP have subsequently subjected it to a careful and searching examination (Chem. Centralbl. 1856, 769, 803; and 1857, 177).

§ Chem. Centralbl. 1857, 182.

Wash the precipitate, first by decantation, boiling up each time, then by filtration; the operation may be materially facilitated by adding a few per-cents of ammonium nitrate to the water. Dry the precipitate, and ignite as directed § 53. It is advisable to evaporate small quantities of nitric acid on the ignited precipitate repeatedly, and to reignite. The residue must have the color of the yolk of an egg. For the properties of the precipitate and residue, see § 93, 4, *e*. Should it be necessary to dissolve the ignited residue again, for the purpose of reprecipitating it, this can be done only after fusing it with a large excess of mixed sodium and potassium carbonates, and thereby converting the pyrophosphoric into orthophosphoric acid. Results accurate; compare the test-analyses given by the authors, Expt. No. 90, and KISSEL's experiments.*

d. Determination as Basic Ferric Phosphate.

α. Mix the acid fluid containing the phosphoric acid with an excess of solution of ferric chloride of known strength, add, if necessary, sufficient ammonia to neutralize the greater portion of the free acid, mix with ammonium acetate in not too large excess, and boil. If the quantity of solution of ferric chloride added was sufficient, the precipitate must be brownish-red. This precipitate consists of basic ferric phosphate and basic ferric acetate, and contains the whole of the phosphoric acid and of the ferric iron. Filter off boiling, wash with boiling water mixed with some ammonium acetate, dry carefully. [Detach the greater part of the precipitate from the filter, incinerate the filter, transfer to the crucible the main part of the precipitate, moisten with strong nitric acid, dry, moisten again with nitric acid and dry and ignite —without these precautions reduction of ferric oxide to magnetic oxide is liable to occur.] Deduct from the weight of the residue that ferric oxide produced from the solution added; the difference is the P_2O_5.

[This modification of SCHULZE's method was first recommended by A. MÜLLER;† it has been adopted also by WAY and OGSTON, in their analyses of ashes.‡ MÜLLER's improvement consists in the use of a solution of ferric chloride of known strength, whereby the determination of iron in the residue is dispensed with.]

β. J. WEEREN's method, suitable for the estimation of the phos-

* Zeitschr. f. anal. Chem. 8, 167. † Journ. f. prakt. Chem. 47, 341.
‡ Journal of the Royal Agricultural Society, viii. part i.

phoric acid in phosphates of the alkali and alkali-earth metals.* Mix the nitric acid solution of the phosphate under examination, which must contain no other strong acid, with a solution of ferric nitrate, of known strength, in sufficient proportion to insure the formation of a basic salt (2 or 3 parts of iron should be present for 1 part P_2O_5); evaporate to dryness, heat the residue to 160°, until no more nitric acid fumes escape, treat with hot water containing ammonium nitrate until all nitrates of the alkali and alkali-earth metals are removed, collect the yellow-ochreous precipitate on a filter, dry, ignite (see § 53), weigh, and deduct from the weight the quantity of iron added reckoned as ferric oxide. LATSCHINOW† recommends heating the residue to 200°, warming with water and a few drops of sulphuric acid, adding ammonia and then treating with hot solution of ammonium nitrate. He says that the phosphoric acid is thus more completely separated, and the precipitate may be more readily filtered off.

e. Determination as Normal Magnesium Phosphate $Mg_3(PO_4)_2$.

(FR. SCHULZE's method, suitable more particularly to effect the separation of phosphoric acid from the alkalies.‡)

Mix the solution of the alkali phosphate, which contains ammonium chloride, with a weighed excess of pure magnesium oxide, evaporate to dryness, ignite the residue until the ammonium chloride is expelled, and separate the magnesium, which is still present in form of magnesium chloride, by means of mercuric oxide (§ 153, 4, γ). Treat the ignited residue with water, filter the solution of the chlorides of the alkali metals, wash the precipitate, dry, ignite, and weigh. The excess of weight over that of the magnesium oxide used shows the quantity of the P_2O_5. Results satisfactory.

f. SCHLÖSING's method§ does not appear to offer any advantages. The phosphate is mixed with silica and ignited in carbon monoxide, the expelled phosphorus being taken up by copper or by silver nitrate.

g. Determination by Volumetric Analysis (*With Uranium Solution*).

This method was recommended originally by LECONTE.‖ It

* Journ. f. prakt. Chem. 67, 8. † Zeitschr. f. anal. Chem. 7, 213.
‡ Journ. f. prakt. Chem. 63, 440.
§ Zeitschr. f. anal. Chem. 4, 118, and 7, 473.
‖ Jahresber. von LIEBIG u. KOPP, für 1853, 642.

was improved and described in detail by NEUBAUER,* and was afterwards recommended by PINCUS,† and subsequently by BÖDEKER.‡ The *principle* of the method is as follows: uranyl acetate precipitates from solutions rendered acid by acetic acid, hydrogen uranyl phosphate, or—in the presence of considerable quantities of ammonium salts—ammonium uranyl phosphate. The proportion between the uranium and the phosphoric acid is the same in both compounds. Both compounds when freshly precipitated and suspended in water are left unchanged by potassium ferrocyanide; uranyl acetate, on the other hand, is indicated by this reagent with great delicacy by the formation of an insoluble reddish-brown precipitate.

According to NEUBAUER§ the following *solutions* are employed:

a. A solution of phosphoric acid of known strength. Prepared by dissolving 10·085 grm. pure, crystallized, uneffloresced, powdered, and pressed hydrogen sodium phosphate in water to 1 litre. 50 c.c. contain ·1 grm. P_2O_5. It is well to control this solution by evaporating 50 c.c. in a weighed platinum dish to dryness, igniting strongly, and weighing. The weight should be ·1874 grm.

b. An acid solution of sodium acetate. Prepared by dissolving 100 grm. sodium acetate in 900 water, and adding acetic acid of 1·04 sp. gr. to 1 litre.

c. A solution of uranyl acetate (§ 63, 3). This is standardized by means of the hydrogen sodium phosphate solution. 1 c.c. indicates ·005 grm. P_2O_5. The solution is made at first a little stronger than necessary, so that it may contain in the litre, say, 32·5 grm. $UO_2(C_2H_3O_2)_2 + 2H_2O$ or 34 grm. $UO_2(C_2H_3O_2)_2 + 3H_2O$ (corresponding to 22 grm. UO_2O), its value is determined, and it is diluted accordingly. To determine its value proceed as follows: Transfer 50 c.c. of the *a* solution to a beaker, add 5 c.c. of the *b* solution, and heat in a water-bath to 90—100°. Now run in uranium solution, at first a large quantity, at last in ½ c.c., testing after each addition whether the precipitation is finished or not. For this purpose spread out one or two drops of the mixture on a white porcelain surface and introduce into the middle, by means of a thin glass rod, a small drop of freshly prepared potassium ferrocyanide solution or a little of the powdered salt. As soon as a trace of

* Archiv. für wissenschaftliche Heilkunde, 4, 228.

† Journ. f. prakt. Chem. 76, 104. ‡ Anal. d. Chem. u. Pharm. 117, 195.

§ His Anleitung zur Harnanalyse, 6 Aufl. 171.

excess of uranyl acetate is present, a reddish-brown spot forms in the drop, which, surrounded as it is by the colorless or almost colorless fluid, may be very distinctly perceived. When the final reaction has just appeared, heat a few minutes in the water-bath and repeat the testing on the porcelain. If now the reaction is still plain the experiment is concluded. If the uranium solution had been exactly of the required strength, 20 c.c. would have been used; but it is actually too concentrated, hence less than 20 c.c. must have been used. Suppose it was 18 c.c., then the solution will be right, if for every 18 c.c. we add 2 c.c. of water. If in this first experiment we find that the solution is much too strong, the solution is diluted with somewhat less water than is properly speaking required, another experiment is made, and it is then diluted exactly.

The *actual analysis* must be made under as nearly as possible similar circumstances to those under which the standardizing of the uranium solution was performed, especially as regards the sodium acetate. This salt retards the precipitation of uranium by potassium ferrocyanide, hence the test-drop on the porcelain plate becomes darker and darker. The analyst should accustom himself to observing the first appearance of the slightest brownish coloration in the middle of the drop, and should take this as the end-reaction. It need hardly be added that the same person must make the analysis who has standardized the solution (NEUBAUER).

The method is applicable to free phosphoric acid, alkali phosphates, and magnesium phosphate, also in the presence of small quantities of the phosphates of other alkali-earth metals, but cannot be employed in presence of ferric and aluminium salts. Dissolve the substance in water or the least possible quantity of acetic acid, add 5 c.c. of the *b* solution, dilute to 50 c.c., and proceed with the addition of uranium as above. The results are very satisfactory. Compare KISSEL's experiments.* If the above process is followed in the presence of much calcium, for instance with a solution of calcium phosphate in dilute acetic acid, the results are almost always too low, as little calcium phosphate is precipitated along with uranyl phosphate. [The best means of obviating that error is, according to ABESSER, JANI, and MÄRCKER,† to standardize the uranium solution under the same conditions as near as possible

* Zeitschr. f. anal. Chem. 8, 167. † *Ib.* 12, 262.

as exist when the solution is used for the actual determination of phosphoric acid. It must therefore be standardized with calcium phosphate. Prepare a solution of suitable strength by dissolving pure $Ca_3(PO_4)_2$ in the smallest possible quantity of nitric acid and diluting to the desired volume. Determine accurately the amount of $Ca_3(PO)_4$ in this solution by evaporating to dryness in a platinum vessel 50 c.c., moistening the residue with ammonia and igniting. The residual somewhat hygroscopic calcium phosphate is quickly weighed in the covered platinum vessel].

II. Separation of Phosphoric Acid from the Basic Radicals.

§ 135.

a. From the Alkalies (see also *d*, *k*, and *l*).

α. Add ammonium chloride, or hydrochloric acid, then lead acetate, exactly, till no more precipitate is produced, and lastly some pure lead carbonate (prepared by precipitating lead acetate with ammonium carbonate, Bäber*), allow to digest for some time, filter off the precipitate consisting of lead phosphate, chloride, and carbonate, wash, precipitate from the filtrate the slight excess of lead by hydrogen sulphide, filter and evaporate with hydrochloric acid (in the case of lithium, sulphuric acid). If the phosphoric acid is to be estimated in the same portion, proceed with the first precipitate (after washing to remove the larger quantity of chloride), according to *b*.

β. (Only applicable in the case of fixed alkalies.) Separate the phosphoric acid as ferric phosphate, according to one of the methods given § 134, *d*. Or if you do not wish to determine the phosphoric acid it is very convenient to acidify with hydrochloric acid, add ferric chloride, dilute rather considerably, add ammonia till the fluid is neutral, and boil; all the phosphoric acid will then separate with ferric oxychloride as ferric phosphate. The separation of phosphoric acid may also be effected as magnesium phosphate (§ 134, *e*). The alkalies are contained in the filtrate as nitrates or chlorides.

b. From Barium, Strontium, Calcium, and Lead.

The compound under examination is dissolved in hydrochloric or nitric acid, and the solution precipitated with sulphuric acid in

* Zeitschr. f. die ges. Naturwiss. 1864, 298; Zeitschr. f. anal. Chem. 4, 120.

slight excess. In the separation of phosphoric acid from strontium, calcium, and lead, alcohol is added with the sulphuric acid. The phosphoric acid in the filtrate is determined according to § 134, *b*, *α*, after removal of the alcohol by evaporation. The determination of the phosphoric acid is effected most accurately by saturating the fluid with sodium carbonate, evaporating to dryness, and fusing the residue with sodium and potassium carbonates. The fused mass is then dissolved in water, and the further process conducted as in § 134, *b*, *α*.

c. From Magnesium (see also *d*, *h*, *k*, *l*).

Add ferric chloride in sufficient excess, dilute, add excess of barium carbonate, allow to remain for several hours with frequent stirring, filter and separate magnesium and barium in the filtrate after § 154.

d. From the whole of the Alkali-earth Metals and fixed Alkalies (comp. *h*, *k*, *l*).

α. Dissolve in the least possible quantity of nitric acid, add a little ammonium chloride, precipitate exactly with lead acetate, add a little lead carbonate (precipitated), digest, filter, precipitate the excess of lead rapidly from the filtrate by hydrogen sulphide, filter and determine the basic metals in the filtrate. Results good.

β. Dissolve in water, and—in case of phosphates of the alkali-earth metals—the least possible nitric acid, add neutral silver nitrate and then silver carbonate, till the fluid reacts neutral. All phosphoric acid now separates as Ag_3PO_4. Warming is unnecessary. Filter, wash the precipitate, dissolve it in dilute nitric acid, precipitate the silver with hydrochloric acid, and determine the phosphoric acid in the filtrate according to § 134, *b*, *α*. The filtrate from the silver phosphate is freed from silver by hydrochloric acid, and the basic metals are then determined according to the methods already given (G. CHANCEL*). A good and convenient method unless the proportion of alkali is very large. (If the substance contains aluminium or ferric iron, they are completely precipitated by the silver carbonate, and are found with the silver phosphate.)

γ. Separate the phosphoric acid as uranyl phosphate (§ 134, *c*), and the excess of uranium from the alkali-earth metals, &c., in the filtrate, according to §§ 160 and 161, Supplement. Results good.

δ. Separate the phosphoric acid according to §134, *d*, *α* or *β*.

---

* Compt. rend. **49**, **997**

The alkali-earth metals are obtained in solution in the first case, as chlorides, together with alkali acetate and chloride; in the second case as nitrates. Results good.

e. From Aluminium.

The best method of separating phosphoric acid from aluminium is that depending on precipitation by ammonium molybdate (*l*). The separation of the acid as stannic phosphate (*h*, *α*) is also satisfactory.

Of several other methods which have been used, the following (by WACKENRODER and FRESENIUS) is one of easiest to carry out: Precipitate the not too acid solution with ammonia, taking care not to use a great excess of that reagent, and add barium chloride as long as a precipitate continues to form. Digest for some time, and then filter. The precipitate contains the whole of the aluminium and the whole of the phosphoric acid; the latter combined partly with aluminium, partly with barium. Filter it off, wash it a little, and dissolve in the least possible quantity of hydrochloric acid. Warm, saturate the solution with barium carbonate, add pure solution of potassa in excess, apply heat, precipitate the barium which the solution may contain with sodium carbonate, and filter. You have now the whole of the aluminium in the solution, the whole of the phosphoric acid in the precipitate. Acidify the solution with hydrochloric acid, boil with some potassium chlorate, and precipitate as directed § 105. Dissolve the precipitate in hydrochloric acid, precipitate the barium with dilute sulphuric acid, filter, and determine the phosphoric acid in the filtrate by precipitation with solution of magnesium in the manner described in § 134, *b*, *α*. (HERMANN has applied a perfectly similar method in his analysis of [impure] gibbsite.)

f. From Chromium (see also *h*, *k*, *l*).

Fuse with sodium carbonate and nitrate, and separate the chromic acid and phosphoric acid in the manner described § 166.

g. From the Metals of the Fourth Group (see also *h*, *k*, *l*).

α. The method so often used of fusing with sodium carbonate does not give accurate results on account of the constant presence of some phosphoric acid in the washed residue. Compare W. SCHWEIKERT* and G. SCHWEITZER.† The former has studied the

* Annal. d. Chem. u. Pharm. 145, 57; Zeitschr. f. anal. Chem. 7, 246.
† Zeitschr. f anal. Chem. 9, 84.

separation of zinc from phosphoric acid by this method, the latter the separation of iron.

β. Dissolve in hydrochloric acid, add tartaric acid, ammonium chloride and ammonia, and finally, in a flask which is to be closed afterwards, ammonium sulphide, put the flask in a moderately warm place, allowing the mixture to deposit until the fluid appears of a yellow color, without the least tint of green; filter, and determine the metals as directed in §§ 108 to 114. The phosphoric acid is found from the loss or determined according to § 134, *b*, *α*. The magnesia-mixture may immediately be added to the filtrate, which contains ammonium sulphide. The washed precipitate is redissolved in just sufficient hydrochloric acid, and the solution reprecipitated by ammonia with addition of magnesia-mixture. This method is not well adapted for nickelous phosphate.

h. From Metals of the Second, Third, and Fourth Groups.

α. More especially from the second group, aluminium, manganese, nickel, cobalt, zinc; and also from ferric iron, if the quantity of the latter is not too considerable.

The phosphoric acid is precipitated as stannic phosphate, according to § 134, b, *δ*. The filtrate contains the bases free from any foreign body requiring removal, which, of course, greatly facilitates their estimation.*

i. From the Metals of the Fifth and Sixth Groups.

Dissolve in hydrochloric or nitric acid, precipitate with hydrogen sulphide, filter, determine the bases by the methods given in §§ 115 to 127, and the phosphoric acid in the filtrate by the method described § 134, *b*, *α*. From silver the phosphoric acid is separated in a more simple way still, by adding hydrochloric acid to the nitric acid solution; from lead it is separated most readily by the method described in *b*.

k. From all Basic Metals, except Mercury (H. ROSE).

The phosphoric acid is separated as mercurous phosphate by ROSE'S method (§ 134, *b*, *γ*).

α. *If the substance is free from iron and aluminium*, the filtrate from the mercurous phosphate contains all the metals as nitrates, together with much mercurous nitrate, and occasionally

---

* If the nitric acid is not concentrated, a little nitrate of protoxide of tin is formed, which dissolves and must afterwards be precipitated from the acid fluid by sulphuretted hydrogen. BÄBER, Zeitschr. f. d. ges. Naturwiss. 1864, 324.

also some mercuric salt. The former is removed by the addition of hydrochloric acid. The precipitated mercurous chloride is free from other metals: if large in quantity, it should be separated by filtering; if slight, filtering may be omitted. Add next ammonia to slight alkaline reaction (with previous addition of ammonium chloride if magnesium is present). Filter rapidly from the mercury compound which will be precipitated so as to avoid formation of calcium carbonate by contact with air. The filtrate contains the basic radicals from which phosphoric acid has been separated. The mercury compound which has been separated by ammonia is dried and ignited (under a chimney with good draught). Should a residue remain, this must be examined. If it consists of phosphates of the alkali-earth metals, the treatment with mercury and nitric acid must be repeated; if, on the contrary, it consists of magnesium oxide or of carbonates of the alkali-earth metals, it is dissolved in hydrochloric acid, and the solution added to the fluid containing the chief portion of the basic metals, which may then be separated and determined in the usual manner. The following method is often advantageously resorted to instead of the one described: The filtrate from the mercurous phosphate is evaporated to dryness, in a platinum dish, and the residue ignited, in a platinum crucible, under a chimney with good draught. If alkali nitrates are present, some ammonium carbonate must be added from time to time during the process of ignition, to guard against injury to the crucible from the formation of caustic alkali. The ignited residue is treated, according to circumstances, first with water and then with nitric acid, or at once with nitric acid.

β. If the substance contains iron but not aluminium, the greater part of the iron is left undissolved with the mercurous phosphate. The dissolved part is separatd from the other bases by the methods given in Section V.; the iron in the undissolved part is obtained, after ignition of the residue with sodium carbonate and treating the ignited mass with water, as ferric oxide containing alkali (and generally also some phosphoric acid). This is dissolved in hydrochloric acid, and precipitated with ammonia.

γ. If the substance contains aluminium, the process just given cannot be used, as aluminium phosphate is not decomposed by fusion with alkali carbonates, while aluminium nitrate, like ferric nitrate, is decomposed by simple evaporation. In this case proceed as follows: Dissolve the substance in the least quantity of nitric acid, precipitate hot with mercurous nitrate, add a little mercuric nitrate, and then pure potash or soda, till a permanent red precipitate appears. The precipitate contains no aluminium, it is to be treated according to α or β (H. ROSE, E. E. MUNROE*).

l. From all Bases without exception.

Apply SONNENSCHEIN'S method (§ 134, *b*, β), and in the filtrate from the ammonium phospho-molybdate separate the bases from the molybdic acid. As molybdic acid comports itself with hydrogen sulphide and ammonium sulphide like a metal of the sixth group, it is best to precipitate metals of the sixth and also of the fifth group from acid solution with hydrogen sulphide, before proceeding to precipitate the phosphoric acid with molybdic acid; the latter will then have to be separated only from the metals of the first four groups. This is done in the following manner: Mix the acid fluid, in a flask, with ammonia till it acquires an alkaline reaction, add ammonium sulphide in sufficient excess, close the mouth of the flask, and digest the mixture. As soon as the solution appears of a reddish-yellow color, without the least tint of green, filter off the fluid, which contains molybdenum and ammonium sulphide, wash the residue with water mixed with some ammonium sulphide, and separate the remaining metallic sulphides and hydroxides of the fourth and third groups by the methods which will be found in Section V. Mix the filtrate cautiously with hydrochloric acid in moderate excess, remove the molybdenum sulphide according to § 128, *d*, and determine the metals of the first and second groups in the filtrate.

This method of separating the phosphoric acid from basic radicals is highly to be recommended; especially in cases where a small quantity of phosphoric acid has to be determined in presence of a very large quantity of ferric and aluminium salts, as, for example, in iron ores, soils, &c. As arsenic acid and silicic acid give, with molybdic acid and ammonia, similar yellow precipitates, it is necessary, if these acids are present, to remove them first.

As the separation of the basic metals from the large excess of molybdic acid used is somewhat tedious, the best way is to arrange matters so that this process may be altogether dispensed with. Supposing, for instance, you have a fluid containing ferric iron, aluminium, and phosphoric acid, estimate, in one portion, by cau-

* Amer. Journ. of Sci. and Arts, May, 1871; Zeitschr. f. anal. Chem. 10, 467.

tious precipitation with ammonia, the total amount of the three bodies; in another portion the phosphoric acid, by means of molybdic acid; and in a third, the iron, in the volumetric way. The aluminium can then be calculated by difference.

§ 136.

Boric Acid (H_3BO_3) and Boric Anhydride (B_2O_3).

I. *Determination.*

Boric acid is estimated either *indirectly* or in the form of *potassium borofluoride.*

1. The determination of the boric acid in an aqueous or alcoholic solution cannot be effected by simply evaporating the fluid and weighing the residue, as a notable portion of the acid volatilizes and is carried off with the aqueous or alcoholic vapor. This is the case also when the solution is evaporated with lead oxide in excess.

a. Mix the solution of the boric acid with a weighed quantity of perfectly anhydrous pure sodium carbonate, in amount about 1½ times the supposed quantity of B_2O_3 present. Evaporate the mixture to dryness, heat the residue to fusion, and weigh. The residue contains a known amount of Na_2O, and unknown quantities of CO_2 and B_2O_3 combined as sodium borate and carbonate. Determine the CO_2 by one of the methods given in § 139, and find the B_2O_3 from the difference (H. Rose).

b. In the method *a*, if between 1 and 2 mol. sodium carbonate (Na_2CO_3) are used to 1 mol. B_2O_3—and this can easily be done if one knows approximately the amount of the latter present—all the carbonic acid is expelled by the boracic acid. Hence we have only to deduct the Na_2O from the residue to find the B_2O_3. As the tumultuous escape of carbonic acid may lead to loss, it is well, after having thoroughly dried the residual saline mass, to project it in small portions cautiously into the red-hot crucible. Results good (F. G. Schaffgotsch).*

c. When the amount of acid is quite unknown, and an estimation of carbonic acid in the residue is objected to, you may proceed thus: Evaporate the solution of the acid with addition of a weighed quantity of anhydrous neutral borax (sodium metaborate $NaBO_2$)

* Pogg. Ann. 107, 427.

free from carbonic acid to dryness, and heat the residue to redness with great caution (on account of the intumescence) till the weight is constant. The amount of neutral borax must be so adjusted that it may not be entirely converted into common borax ($2NaBO_2$ B_2O_3) (H. ROSE).

d. If a solution contains, besides boric acid, only alkalies or magnesium, the acid may be determined, according to C. MARIGNAC,* in the following manner: Neutralize the solution with hydrochloric acid, add double magnesium and ammonium chloride in sufficient quantity to give at least 2 parts of MgO to 1 part of B_2O_3, then add ammonia and evaporate to dryness. If a precipitate is formed on adding the ammonia which does not redissolve readily on warming, add more ammonium chloride. The evaporation is conducted, at least towards the end, in a platinum dish, a few drops of ammonia being added from time to time. Ignite the dry mass, treat with boiling water, collect the insoluble precipitate (consisting of magnesium borate mixed with excess of magnesium oxide) on a filter, and wash with boiling water till the washings remain clear with nitrate of silver. The filtrate and washings are mixed with ammonia, evaporated to dryness, ignited, and washed with boiling water as before.

The two insoluble residues are ignited together in the platinum dish before used, as strongly as possible, and for a sufficiently long time, in order to decompose the slight traces of magnesium chloride that might still be present. After weighing determine the magnesium oxide, and find the boric acid from the difference. The determination of the magnesium may be made by dissolving the residue in hydrochloric acid and precipitating as ammonium magnesium phosphate, or more quickly, and almost as accurately, by dissolving in a known quantity of standard sulphuric acid at a boiling temperature and determining the excess of acid with standard soda (comp. Alkalimetry).

Should a little platinum remain behind on dissolving the residue, it must be weighed and subtracted from the weight of the whole (unless the dish was weighed first). Results satisfactory. MARIGNAC obtained in two experiments ·276 instead of ·280.

2. If boric acid is to be determined as *potassium borofluoride*, alkalies only (preferably only potash) may be present. The process

* Zeitschr. f. anal. Chem. 1, 405.

is conducted as follows: Mix the fluid with pure solution of potassa, adding for each mol. boric acid supposed to be present, at least 1 mol. potassa; add pure hydrofluoric acid (free from silicic acid) in excess, and evaporate, in a platinum dish, on the water-bath, to dryness. The fumes from the evaporating fluid should redden litmus paper, otherwise there is a deficiency of hydrofluoric acid. The residue consists now of KF,BF_3 and KF,HF. Treat the dry saline mass, at the common temperature, with a solution of 1 part of potassium acetate in 4 parts of water, let it stand a few hours, with stirring, then decant the fluid portion on to a weighed filter, and wash the precipitate repeatedly in the same way, finally on the filter, with solution of potassium acetate, until the last rinsings are no longer precipitated by calcium chloride. By this course of proceeding, the hydrogen potassium fluoride is removed, without a particle of the potassium borofluoride being dissolved. To remove the potassium acetate, wash the precipitate now with alcohol of 78 per cent., dry at 100°, and weigh. As potassium chloride, nitrate, and phosphate, sodium salts, and even, though with some difficulty, potassium sulphate, dissolve in solution of potassium acetate, the presence of these salts does not interfere with the estimation of the boric acid; however, sodium salts must not be present in considerable proportion, as sodium fluoride dissolves with very great difficulty. The results obtained by this method are satisfactory. STROMEYER's experiments gave from 97·5 to 100·7 instead of 100. When the amount of alkali salt to be removed is very large, the saline mass left on evaporation should be warmed with the solution of potassium acetate, allowed to stand 12 hours in the cold and then filtered. In this way the quantity of potassium acetate required will be much reduced. For the composition and properties of potassium borofluoride, see § 93, 5. As the salt is very likely to contain potassium silicofluoride it is indispensable to test it for that substance; this is done by placing a small sample of it on moist blue litmus paper, and putting another sample into cold concentrated sulphuric acid. If the blue paper turns red, and effervescence ensues in the sulphuric acid, the salt is impure, and contains potassium silicofluoride. To remove this impurity, dissolve the remainder of the salt, after weighing it, in boiling water, add ammonia, and evaporate, redissolve in boiling water, add ammonia, &c., repeating the same operation at least six times. Finally, after warming once more with ammonia, filter off the silicic acid, evaporate to dryness, and treat again with solution of potassium acetate and alcohol (A. STROMEYER).* I was obliged to modify STROMEYER's method for effecting the separation of the silicic acid, the results of my experiments having convinced me that treating the salt only once with ammonia, as recommended by that chemist, is not sufficient to effect the object in view.

II. *Separation of Boric Acid from the Basic Radicals.*

a. From the Alkalies.

Dissolve a weighed quantity of the borate in water, add an excess of hydrochloric acid, and evaporate the solution on the water-bath. Towards the end of the operation add a few more drops of hydrochloric acid, and keep the residue on the water-bath, until no more hydrochloric acid vapors escape. Determine now the chlorine in the residue (§ 141), calculate from this the alkali, and you will find the boric acid from the difference.

E. SCHWEIZER, with whom this method originated, states that it gave him very satisfactory results in the analysis of borax. It will answer also for the estimation of the basic metals in the case of some other borates. It is self-evident that the boric acid may be estimated, in another portion of the salt, by I., 1, *c*, or 2. If you have to estimate boric acid in presence of large proportions of alkali salts, make the fluid alkaline with potassa, evaporate to dryness, extract the residue with alcohol and some hydrochloric acid, add solution of potassa to strongly alkaline reaction, distil off the alcohol, and then proceed as in I., 1, *c*, or 2 (AUG. STROMEYER, *loc. cit.*).

LUNGE† determined the soda in boronatrocalcite alkalimetrically, by dissolving the mineral in normal nitric acid and titrating back with normal soda, till the tint of the litmus added becomes violet.

b. From Calcium.

Dissolve in hydrochloric acid in the heat, avoiding too large an excess, neutralize with ammonia and precipitate with ammonium oxalate (LUNGE, *loc. cit.*).

c. From almost all other Bases except Alkalies.

The compounds are decomposed by boiling or fusing with potassium carbonate or hydroxide; the precipitated base is filtered off, and the boric acid determined in the filtrate, according to I., 1,

* Annal. d. Chem. u. Pharm. 100, 82.

† *Ib.* 138, 53.

d, or 2. If magnesium was present, a little of this is very likely to get into the filtrate, and—if process I., 2, is employed—upon neutralizing with hydrofluoric acid, this separates an insoluble magnesium fluoride, which may either be filtered off at once, or removed subsequently, by treating the potassium borofluoride with boiling water, in which that salt is soluble, and the magnesium fluoride insoluble.

d. From the Metallic Oxides of the Fourth, Fifth, and Sixth Groups.

The metallic oxides are precipitated by hydrogen sulphide, or, as the case may be, ammonium sulphide,* and determined by the appropriate methods. The quantity of boric acid may often be inferred from the loss. If it has to be estimated in the direct way, the filtrate, after addition of solution of potassa and some potassium nitrate, is evaporated to dryness, the residue ignited, and the boric acid estimated by I., 1, *d*, or 2. In cases where the metal has been precipitated by hydrogen sulphide from acid or neutral solutions, the boric acid may also be determined in the filtrate—in the absence of other acids—by I., 1, *a* or *b* or *c*, after the complete removal of the hydrogen sulphide by transmitting carbon dioxide through the fluid.

e. From the whole of the Fixed Basic Radicals.

A portion of the very finely pulverized substance is weighed, put into a capacious platinum dish, and digested with a sufficient quantity of hydrofluoric acid (which leaves no residue when evaporated in a platinum dish); pure concentrated sulphuric acid is then gradually added, drop by drop, and the mixture heated, gently at first, then more strongly, until the excess of the sulphuric acid is completely expelled. In this operation the boric acid goes off in the form of fluoride of boron ($B_2O_3 + 6HF = 2BF_3 + 3H_2O$). The basic metals contained in the residue in the form of sulphates are determined by the appropriate methods, and the quantity of the boric acid is found by difference. It is of course taken for granted that the substance is decomposable by sulphuric acid.

* Boric acid cannot be separated completely from aluminium by precipitation of the hydrochloric acid solution with ammonium sulphide or with ammonium carbonate (WÖHLER, Ann. d. Chem. u. Pharm. 141, 268).

§ 137.

3. Oxalic Acid.

I. *Determination.*

Oxalic acid is either precipitated as *calcium oxalate*, and estimated after determination of the calcium in the latter as *oxide*, *carbonate*, or *sulphate;* or the amount contained in a compound is inferred from the quantity of solution of potassium permanganate required to effect its conversion into carbonic acid; or from the quantity of gold which it reduces; or from the amount of carbonic acid which it affords by oxidization.

a. Determination as Calcium Carbonate, &c.

Precipitate with solution of calcium acetate, added in moderate excess, and treat the precipitated calcium oxalate as directed in § 103. If this method is to yield accurate results, the solution must be neutral or slightly acid with *acetic acid;* it must not contain salts of aluminium, chromium, or of the heavy metals, more especially cupric or ferric salts; therefore, where these conditions do not exist, they must first be supplied.

b. Determination by means of Solution of Potassium Permanganate.

Standardize the solution of potassium permanganate, as directed § 112, 2, *a*, *cc*, by means of oxalic acid; then dissolve the substance in about 150 c.c. water, or acid and water (sulphuric acid is the best acid to use); add, if necessary, a further quantity of sulphuric acid (about 6 or 8 c.c. strong sulphuric acid should be present), heat to about 60°, and then run in the permanganate, with constant stirring, until the fluid just shows a red tint. Knowing the quantity of oxalic acid which 100 c.c. of the standard permanganate will oxidize, a simple calculation will give the quantity of oxalic acid corresponding to the c.c. of permanganate used in the experiment. The results are very accurate.

c. Determination from the reduced Gold (H. Rose).

α. In compounds soluble in water. Add to the solution of the oxalic acid or the oxalate a solution of sodium auric chloride, or ammonium auric chloride, and digest for some time at a temperature near ebullition, with exclusion of direct sunlight. Collect the precipitated gold on a filter, wash, dry, ignite, and weigh. 2 at.

Au. ($196 \cdot 71 \times 2 = 393 \cdot 42$) correspond to 3 mol. C_2O_3 ($72 \times 3 = 216$).

β. In compounds insoluble in water. Dissolve in the least possible amount of hydrochloric acid, dilute with a very large quantity of water, in a capacious flask, cleaned previously with solution of soda; add solution of gold in excess, boil the mixture some time, let the gold subside, taking care to exclude sunlight, and proceed as in *α*.

d. Determination as Carbonic Acid.

This may be effected either,

α. By the method of organic analysis; or

β. By mixing the oxalic acid or oxalate with finely pulverized manganese dioxide in excess, and adding sulphuric acid to the mixture, in an apparatus so constructed that the disengaged CO_2 passes off perfectly dry. The theory of this method may be illustrated by the following equation: $H_2C_2O_4 + MnO_2 + H_2SO_4 = MnSO_4 + 2H_2O + 2CO_2$. For the apparatus and process, I refer to the chapter on the examination of manganese ores, in the Special Part of this work. Here I may remark that free oxalic acid must first be prepared for the process by slight supersaturation with alkali free from carbonic acid, and also that 9 parts of oxalic anhydride (C_2O_3) require theoretically 11 parts of (pure) manganese dioxide. Since an excess of the latter substance does not interfere with the accuracy of the results, it is easy to find the amount to be added. The manganese dioxide need not be pure, but it must contain no carbonate. This method is expeditious, and gives very accurate results, if the process is conducted in an apparatus sufficiently light to admit of the use of a delicate balance. Instead of manganese dioxide, potassium chromate may be used (compare § 130, 1, *c*), and instead of estimating the carbonic acid by loss it may be collected by an absorbent and weighed (§ 139, II., *e*); the latter method is always to be preferred in the case of small quantities.

II. *Separation of Oxalic Acid from the Basic Radicals.*

The most convenient way of analyzing oxalates is, in all cases, to determine in one portion the acid, by one of the methods given in I., in another portion the basic radical, particularly as the latter object may be generally effected by simple ignition in the air, which reduces the salt either to the metallic state (*e.g.*, silver oxalate), or to pure oxide (*e.g.*, lead oxalate), or to carbonate (*e.g.*, the oxalates of the alkalies and alkali-earth metals).

If the acid and basic radical have to be determined in one and the same portion of the oxalate, the following methods may be resorted to:

a. The oxalic acid is determined by I., *c*, and the gold separated from the basic metals in the filtrate by the methods given in Section V.

b. In many soluble salts the oxalic acid may be determined by the method I., *a*; separating the basic metals afterwards from the excess of the calcium salt by the methods given in Section V.

c. Many oxalates of metals which are completely precipitated as carbonates or oxides by excess of sodium or potassium carbonate, may be decomposed by boiling with excess of these reagents, metallic oxide or carbonate being formed on the one, and alkali oxalate on the other side.

d. All oxalates of the metals of the fourth, fifth, and sixth groups may be decomposed with hydrogen sulphide or ammonium sulphide.

§ 138.

4. Hydrofluoric Acid.

I. Determination.

Free hydrofluoric acid in aqueous solution* is determined either with standard alkali or as *calcium fluoride.* In the latter case sodium carbonate is added in moderate excess, then the solution being boiled, calcium chloride is added as long as a precipitate continues to form; when the precipitate, which consists of calcium fluoride and carbonate, has subsided, it is washed, first by decantation, afterwards on the filter, and dried; when dry, it is ignited in a platinum crucible (§ 53); water is then poured over it in a platinum or porcelain dish, acetic acid added in slight excess, the mixture evaporated to dryness on the water-bath, and heated on the latter until all odor of acetic acid disappears. The residue, which consists of calcium fluoride and acetate, is heated with water, the

* In analyzing fluorides you must always avoid bringing acid solutions in contact with glass or porcelain. If platinum or silver dishes of sufficient size are not at hand you may sometimes use gutta-percha vessels, or glass vessels coated with wax or paraffin.

calcium fluoride filtered off, washed, dried, ignited (§ 53), and weighed. As a control of the purity of the calcium fluoride, it is well to convert it after weighing into sulphate. If the precipitate of calcium fluoride and carbonate were treated with acetic acid, without previous ignition, the washing of the fluoride would prove a difficult operation. Presence of nitric or hydrochloric acid in the aqueous solution of the hydrofluoric acid does not interfere with the process (H. Rose).

II. Separation of Fluorine from the Metals.

1. *Fluorides Soluble in Water.*

If the solutions have an acid reaction, sodium carbonate is added in excess. If there is an odor of ammonia now, heat till the latter is expelled. If the sodium carbonate produces no precipitate, the fluorine is determined by the method given in I., and the metals in the filtrate are separated from calcium and sodium by the methods given in Section V. But if the sodium carbonate produces a precipitate, the mixture is heated to boiling, then filtered, and the fluorine determined in the filtrate by the method given in I.; the metals are in the precipitate, which must, however, first be tested, to make sure that it contains no fluorine. Neutral solutions are mixed with a sufficient quantity of calcium chloride, and the mixture heated to boiling in a platinum dish or, but less appropriately, in a porcelain dish; the precipitate of calcium fluoride is allowed to subside, thoroughly washed with hot water by decantation, transferred to the filter, dried, ignited, and weighed. The basic metals in the filtrate are then separated from the excess of the calcium salt by the usual methods. That the basic metals may be determined also in separate portions by the methods given in 2 *a*, need hardly be stated.

2. *Insoluble Fluorides.*

a. Decomposition by Sulphuric Acid (Indirect Estimation of the Fluorine).

α. Anhydrous Compounds.

The finely pulverized and weighed substance is heated for some time with pure concentrated sulphuric acid, and finally ignited until the free sulphuric acid is completely expelled. In the presence of alkalies, ammonium carbonate must be added during the ignition. The residuary sulphate is weighed, and the metal contained in it calculated; the fluorine is estimated by loss. In cases where we have to deal with a metal whose sulphate gives off part of the sulphuric acid upon ignition, or where the residue contains several metals, it is necessary to subject the residue to analysis before this calculation can be made. In the case of many compounds, for instance of aluminium fluoride (which after ignition requires prolonged heating with sulphuric acid for its decomposition), long continued strong ignition does not leave the sulphate, but the oxide in a pure state. Topaz (a silicate of aluminium in isomorphous mixture with aluminium silicofluoride) is not decomposed by boiling sulphuric acid, but it is decomposed by fusion with potassium disulphate.

β. Hydrated Fluorides.

A sample of the substance is heated in a tube.

aa. The Water expelled does not redden Litmus Paper. The water is determined by ignition; the fluorine and metal as directed in *a*, *α*.

bb. The Water expelled has an acid reaction. The substance is treated with sulphuric acid as directed in *a*, *α*, to determine the metal on the one hand, and the water + fluorine on the other. Another weighed portion is then mixed, in a small retort, with about 6 parts of recently ignited lead oxide; the mixture is covered with a layer of lead oxide, the retort weighed, and the water expelled by the application of heat, increased gradually to redness. No hydrofluoric acid escapes in this process. The weight of the expelled water is inferred from the loss. The first operation having given us the water + fluorine, and the second the water alone, the difference is consequently the fluorine.

b. Decomposition by Fusion with Alkali Carbonates.

Many insoluble fluorides, aluminium fluoride for instance, may be completely decomposed by fusion with alkali carbonate alone; others, such as calcium fluoride, require the addition of silicic acid. In the first case the fluorine is estimated in the aqueous solution of the fusion according to I., in the latter according to § 166, 5. The temperature must not be too high, or some alkali fluoride may be lost.

3. *Fluorides completely Decomposable by Sulphuric Acid.*

As might be inferred from 2, almost all fluorides are decomposed by heating with sulphuric acid with evolution of hydrofluoric acid. If silica or silicate is added to the fluoride in sufficient quantity, silicon fluoride and water escape instead of hydrofluoric acid: $SiO_2 + 4HF = SiF_4 + 2H_2O$.

On this reaction methods of determining fluorine have been based. In the first, which I published some years ago,* the fluoride of silicon is determined by increase of weight of absorption tubes; this I believe to be in many cases the only method which is applicable, and when carefully carried out yields the most accurate results.

a. Estimation by Absorption of the evolved Fluoride Silicon.

The method as here given is the result of a long series of experiments; the conditions laid down must be most carefully attended to. The fluoride must be in the finest powder. As silicic acid we use finely powdered quartz, which has been ignited in the air to destroy any organic admixture. The sulphuric acid should have a sp. gr. of 1·848, it must be colorless and free from oxides of nitrogen and sulphurous acid. The gasometer must be filled with clean air, and not with air from the laboratory, for any dust of organic matter, traces of coal gas, &c., would interfere with the accuracy of the result. The apparatus required is shown fig. 57. *A* contains atmospheric air, *b* is half filled with sulphuric acid, *c* contains soda-lime with plugs of cotton, *d* pieces of glass moistened with sulphuric acid. The air is thus freed from carbonic acid and suspended matter, and dried by sulphuric acid (p. 61). *e* is the decomposing flask; it has a capacity of about 250 c.c. *f* is half filled with sulphuric acid; its cork, which should not fit air-tight, bears a thermometer whose bulb dips into the acid. *e* and *f* should be so placed on the iron plate that the temperature in both may be equal. *g* is empty; *h* contains fused calcium chloride in the first limb, and pumice impregnated with anhydrous cupric sulphate in the second. These U-tubes serve to retain the small amount of sulphuric acid and the hydrochloric acid which may accompany it. The calcium chloride and the cupric sulphate must both be anhydrous, or they will decompose and retain silicon fluoride. *i*, *k*, and *l* are the weighed absorption tubes; they are 10 or 12 cm. high, and about 12 mm. wide. *i* contains in the first limb pumice moistened with water between plugs of cotton, in the bend and half of the second limb soda-lime, in the upper half of the second limb fused calcium chloride

* Zeitschr. f. anal. Chem. 5, 190.

between plugs of cotton. The tube after being charged weighs about 40 or 50 grm. *k* completes the absorption; it is filled half with

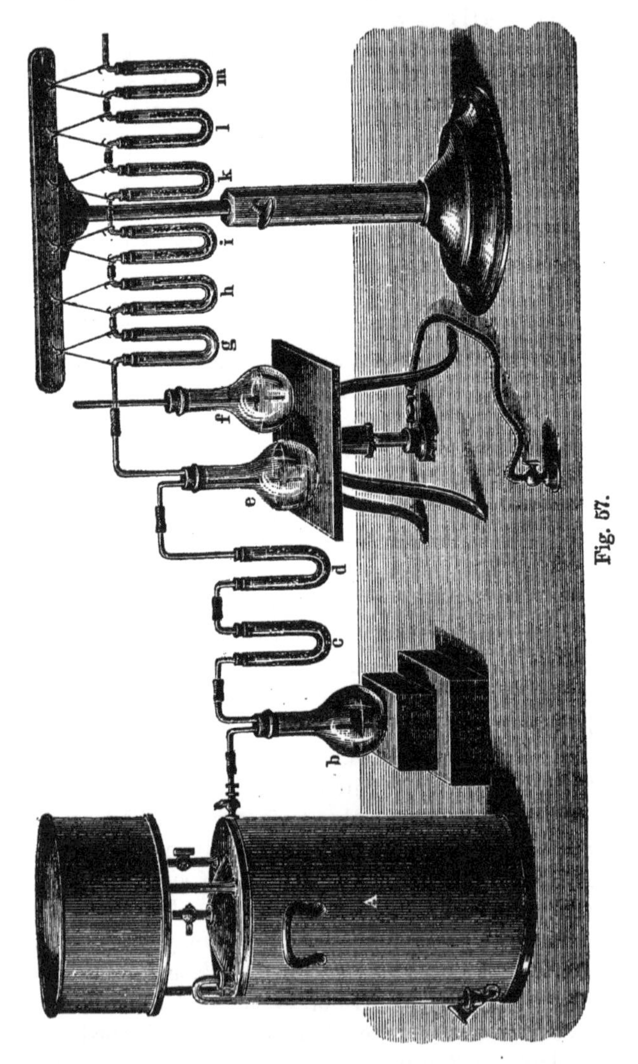

Fig. 57.

soda-lime and half with fused calcium chloride. *l* takes up again the small amount of water carried away from *i* and *k*; the bend is filled with pieces of glass moistened with sulphuric acid. These absorption tubes retain the silicon fluoride, the carbonic acid which may be possibly evolved from the soda-lime by hydrofluosilicic acid, and the aqueous vapor; and the air escapes through the unweighed guard tube *m* into the atmosphere. The latter contains in the first limb calcium chloride, in the second soda-lime. The flexible connections should not be long, and should be washed and dried before use.

When the apparatus has been tested and found air-tight, place the weighed and very finely divided substance in *e*. The substance should be free from carbonic acid, and the quantity taken should give not less than ·1 grm. silicon fluoride if possible. Add for every part of fluoride supposed to be present 10 or 15 parts of finely powdered quartz (previously strongly ignited in the air), and then 40 or 50 c.c. pure concentrated sulphuric acid. Connect *e*, on the one hand, with *d*, and, on the other, with *g*, and pass a moderate current of air, which should enter the fluid in the decomposing flask from the bottom. Heat the iron plate, shake *e* frequently and raise the temperature very gradually, till the thermometer in *f* indicates 150° to 160°. The commencement of the decomposition shows itself not only by the appearance of bubbles of gas in the fluid, more particularly at the edge, but also by the separation of hydrated silica in *i*. The bubbles of gas will disappear on shaking the fluid; as soon as they cease to form again remove the lamp; the time usually occupied in the decomposition is one hour for small quantities of fluoride (·1 grm.), two or three hours for large quantities (1 grm.). After a while shut off the current of air, remove the weighed tubes *i*, *k*, and *l*, and during the weighing of these connect *h* with *m* by means of a glass tube. After weighing replace *i*, *k*, and *l*, heat again to 150° or 160°, and pass the air again for half an hour or an hour, weighing *i*, *k*, and *l* again. If any alteration of weight has occurred, the process must be continued.

The increase in weight of the absorption tubes after deducting ·001 grm. for every hour during which the air has been passing (*i.e.*, for every 6 litres of air) represents the amount of silicon fluoride. The small correction is necessary because air, even when it comes in contact only with short washed pieces of india-rubber, always gives traces of sulphurous and carbonic acid when passed through hot concentrated sulphuric acid. The results thus obtained are very satisfactory, and differ from the truth at the most by a few milligrammes.

b. Other methods of Estimating the Silicon Fluoride expelled.

α. Method of WÖHLER. Only applicable when the substance is readily decomposed by sulphuric acid, and the amount of fluorine is large. Transfer the very finely divided substance, if necessary, intimately mixed with 10 or 15 parts of ignited quartz powder, to a small flask, add pure sulphuric acid, close quickly with a cork fitted with a small tube filled with fused calcium chloride (or better still, half with fused calcium chloride and half with anhydrous cupric sulphate on pumice), weigh the whole apparatus as quickly as possible, warm it till no more fumes of silicon fluoride escape, remove the last particles of gas in the apparatus by an air pump, allow to cool, and weigh. The loss of weight indicates the amount of silicon fluoride.

β. [S. L. PENFIELD* determines the amount of expelled silicon fluoride by an indirect volumetric method; viz.: by passing it into a solution of potassium chloride, and titrating the hydrochloric acid which is set free with standard ammonia solution. $3SiF_4 + 2H_2O = 2H_2F_2SiF_4 + SiO_2$ and $H_2F_2SiF_4 + 2KCl = (KF)_2SiF_4 + 2HCl$. Two mol. HCl thus liberated correspond to six at. F.

The process of decomposing the fluorine compound is conducted as in *a*, and the same apparatus may be used except that the four last U-tubes *i*, *k*, *l*, *m*, are replaced by two larger U-tubes for holding the solution of potassium chloride.

The aqueous solution of KCl is mixed with an equal volume of alcohol to effect complete precipitation of the hydrofluosilicic acid. The titration may be either effected directly in U-tubes (the second of which will contain but a very small quantity of acid) or after transferring to a beaker and rinsing the tubes with alcohol and water. Care must be taken to loosen and break up the silicic acid and to have at least half of the final volume at the end of the titration consist of alcohol. Results given by the author (loc. cit.) very satisfactory.]

* American Chem. Journ. i. p. 27.

Fourth Division of the First Group of the Acids.

CARBONIC ACID—SILICIC ACID.

§ 139.

1. Carbonic Acid.

I. *Determination.*

a. In a mixture of Gases.

After thoroughly drying the gases with a ball of calcium chloride, or saturating with moisture (§ 16), measure them accurately in a graduated tube over mercury, insert a ball of hydrate of potassa,* cast on a platinum wire in a pistol bullet-mould, take care that the end of the platinum wire remains under the surface of the mercury, leave in the tube for 24 hours, or until the volume of the gas ceases to show further diminution; withdraw the ball, and measure the gas remaining, reinsert the same or a fresh ball of potassa, and repeat till no further absorption takes place. The carbonic acid gas is inferred from the difference, provided the gaseous mixture contained no other gas liable to absorption by potassa (compare §§ 12–16). In very accurate analyses you must bear in mind that carbonic acid does not exactly follow the law of Mariotte.

If the amount of carbonic acid is very small, this process does not yield sufficiently accurate results. In such cases one of the methods recommended in "The Analysis of Atmospheric Air" should be employed. Several kinds of special apparatus are in use for the estimation of carbonic acid in coal gas and for the purposes of sugar works. I may mention those proposed by F. Rüdorff† and Lehmann and H. Wählert‡ for the first purpose, and by C. Scheibler§ and C. Stammer‖ for the second. Besides these volumetric methods the gravimetric processes given by myself for the analysis of gaseous mixtures¶ may often be used with great advantage.

---

* The ordinary hydrate is not adapted for the purpose. It should be fused with a quarter of its weight of water in a platinum crucible.

† Pogg. Annal. 125, 71.

‡ Zeitschr. f. anal. Chem. 7, 58.

§ Dingler's polyt. Journ. 183, 306.

‖ *Ib.* 102, 368.

¶ Zeitschr. f. anal. Chem. 3, 343.

b. In Aqueous Solution.

α. WITH CALCIUM HYDROXIDE.

Into a flask, holding about 300 c.c., put 2·5 to 3 grm. calcium hydroxide perfectly free from carbonate.* Provide the flask with a good india-rubber stopper, tare or weigh exactly, add the carbonic acid water with gentle agitation till the flask is two thirds or three quarters full, and close at once.

In adding the carbonic acid water every care must of course be taken to guard against loss of carbonic acid. If the water flows from a pipe, it is allowed simply to run in. If it is in a jug or bottle, cool it to 4°, and transfer the quantity required with a syphon.† If the water is in a basin or well, provide the flask with a stopper in which two glass tubes are inserted, one a few inches long, pushed down only to the lower surface of the stopper, the other extending through the stopper a short distance into the flask, but only to the *upper* surface of the stopper. Sink the flask into the water, and water will enter one tube and air escape through the other. Water which is not very rich in free carbonic acid may be removed from the basin or well by a plunging-syphon.

Now weigh the flask with its stopper again, and you will find the quantity of water taken. No way of measuring the water is so accurate in retaining all the carbonic acid and in giving the quantity of water taken.

If there is much interval between the mixing of the water and the lime and the estimation of the carbonic acid in the precipitate, the calcium carbonate, which is at first amorphous, passes spontaneously into the crystalline condition; but if the carbonic acid is to be determined soon after the mixing, heat for some time on the water-bath, raising the stopper occasionally, in order to hasten the change of the calcium carbonate. Now, without disturbing the precipitate, filter the clear fluid through a small plaited filter, which will take a very short time, throw the filter at once into the flask containing the precipitate and the rest of the fluid, and proceed according to II., *e*. This process has been in use for 10 years in my laboratory for all mineral water analyses; it is extremely

* This is prepared by slaking freshly burnt lime with water in such a manner that the hydrate obtained appears dry and pulverulent. It is preserved in small bottles, the corks or stoppers of which are covered with sealing wax.

† If the water is poured directly from the jug into the flask, carbonic acid gas is very likely to get into the latter as well as the water.

simple, and gives excellent results.* If the water contains alkali carbonate, put a quantity of calcium chloride sufficient to decompose the alkali carbonate with the lime in the flask before adding the water.

β. After Pettenkofer.†

The principle of this simple and expeditious process consists in mixing the carbonic acid water with a measured quantity of standard lime water (or, under certain circumstances, baryta water) in excess. After complete separation of the calcium or barium carbonate, the excess of calcium or barium in the fluid is determined in an aliquot part by means of standard solution of oxalic acid; the difference gives the calcium or barium precipitated by the carbonic acid, and consequently the amount of the latter present.

If a water contains only free carbonic acid, the analyst has only to bear in mind—if lime water is employed—that the calcium carbonate formed is at first, as long as it remains amorphous, very perceptibly soluble in water, to which it communicates an alkaline reaction. Hence the unprecipitated lime in the fluid cannot be estimated till the calcium carbonate has separated in the crystalline form, which takes 8 or 10 hours, unless the mixture is warmed to 70° or 80°. On this account it is generally best to use baryta water (see "Analysis of Atmospheric Air").

If, on the contrary, a water contains an alkali carbonate or any other alkali salt whose acid would be precipitated by lime or baryta, a neutral solution of calcium or barium chloride must first be added to decompose the same. This addition, too, prevents any inconvenience arising from the presence of free alkali in the lime or baryta water, or of magnesium carbonate in the carbonic acid water; this inconvenience consists in the fact that oxalate of an alkali or of magnesium enters into double decomposition with calcium carbonate (which is seldom entirely absent from the fluid to be analyzed), forming calcium oxalate and carbonate of the alkali or of magnesium, which latter will of course again take up oxalic acid.

In the presence of magnesium salts in the carbonic acid water, in order to avoid the precipitation of the magnesium, a little ammonium chloride must also be added, but in this case heat must

* Zeitschr. f. anal. Chem. 2, 49 and 341.

† Buchner's neues Repert. 10, 1; Journ. f. prakt. Chem. 82, 32; Annal. d. Chem. u. Pharm. ii., Supplementb. 1; Zeitschr. f. anal. Chem. 1, 92.

not be applied to induce the calcium carbonate to become more quickly crystalline, as ammonia would be thereby expelled.

In making the determination the first thing to be done is to ascertain the relation between the lime or baryta water and a standard solution of oxalic acid. PETTENKOFER makes the latter solution by dissolving 2·8636 grm. pure uneffloresced dry crystallized oxalic acid to 1 litre; 1 c.c. of this is equivalent to 1 mgrm. carbonic acid. The lime water is standardized as follows: Measure 45 c.c. into a little flask which can be closed by the thumb, and then run in from the burette the solution of oxalic acid till the alkaline reaction has just vanished. During the operation the flask is closed with the thumb and gently shaken. The end is attained as soon as a drop taken out with a glass rod and applied to delicate turmeric paper* produces no brown ring. The first experiment is a rough one, the second should be exact.

The analysis of a carbonic acid water (a spring water, for instance) is performed by transferring 100 c.c. to a dry flask, adding 3 c.c. of a neutral and nearly saturated solution of calcium or barium chloride, and 2 c.c. of a saturated solution of ammonium chloride, then 45 c.c. of the standard lime or baryta water; close the flask with an india-rubber stopper, shake and allow to stand 12 hours. The fluid contents of the flask measure consequently 150 c.c. From the clear fluid† take out by means of a pipette two portions of 50 c.c. each, and determine the free lime or baryta by means of oxalic acid, in the first portion approximately, in the second exactly. Multiply the c.c. used in the last experiment by 3 and deduct the product from the c.c. of oxalic acid which correspond to 45 c.c. of lime or baryta water. The difference shows the lime or baryta precipitated by carbonic acid, each c.c. corresponds to 1 mgrm. carbonic acid.

---

* For the preparation of this bibulous paper should be used, the ash of which is free from carbonate of lime. Swedish filtering-paper answers best. J. GOTTLIEB (Journ. f. prakt. Chem. 107, 488; Zeitschr. f. anal. Chem. 9, 251) prefers aqueous tincture of litmus, prepared from litmus first exhausted with spirit and used in a very dilute state. E. SCHULZE and M. MÄRCKER (Zeitschr. f. anal. Chem. 9, 334) employ corallin or rosolic acid, which they say is specially adapted for the purpose. The alcoholic solution is cautiously neutralized with potash, and a drop or two of this tincture is added. F. SCHULZE (Zeitschr. f. anal. Chem. 9, 292) recommends spirituous tincture of turmeric.

† It is not admissible to use a filter (A. MÜLLER, Zeitschr. f. anal. Chem. 1, 84).

The method is convenient and good; it is especially to be recommended for dilute carbonic acid water. When calcium sulphate or carbonate is present, as is almost always the case in spring water, you must always before titrating await the conversion of the amorphous calcium carbonate to the crystalline state, even if baryta water is used (K. KNAPP*). Baryta water therefore possesses no advantages over lime water for the analysis of spring waters.

II. *Separation of Carbonic Acid from the Basic Radicals, and its Estimation in Carbonates.*

a. Estimation in Normal Alkali Carbonates and Alkali-earth Carbonates.

If the salts are unquestionably normal carbonates, and there is no other salt with power to neutralize an acid present, we may determine the quantity of the basic radical by the alkalimetric method (§§ 196, 198), and calculate the amount of CO_2 necessary to form with it normal carbonate.

b. Separation from Basic Metals in Salts which upon ignition readily and completely yield their Carbonic Acid.

Such are, for instance, the carbonates of zinc, cadmium, lead, copper, magnesium, &c.

α. Anhydrous Carbonates.—Ignite the weighed substance, in a platinum crucible (cadmium and lead carbonates in a porcelain crucible), until the weight of the residue remains constant. The results are, of course, very accurate. Substances liable to absorb oxygen upon ignition in the air are ignited in a bulb-tube, through which a stream of dry carbon dioxide gas is conducted. The carbonic acid is inferred from the loss.

β. Hydrated Carbonates.—The substance is ignited in a bulb-tube through which dried air or, in presence of oxidizable substances, carbon dioxide is transmitted, and which is connected with a calcium chloride tube, by means of a dry, close-fitting cork. During the ignition, the posterior end of the bulb-tube is, by means of a small lamp, kept sufficiently hot to prevent the condensation of water in it, care being taken, however, to guard against burning the cork. The loss of weight of the tube gives the amount of the water + the carbonic acid; the increase of weight gained by the calcium chloride tube gives the amount of the water, and the difference accordingly that of the carbonic acid. A somewhat

* Annal. d. Chem. u. Pharm. 158, 112; Zeitschr. f. anal. Chem. 10, 361.

wide glass tube may also be put in the place of the bulb-tube, and the substance introduced into it in a little boat, which is weighed before and after the operation.

c. Separation from all fixed Basic Radicals, without exception, in Anhydrous Carbonates.

Fuse vitrified borax in a weighed platinum crucible, allow to cool in the desiccator, weigh, then transfer the well-dried substance to the crucible and weigh again. The weights of both carbonate and borax are thus ascertained. They should be in about the proportion of 1 : 4. Heat is then applied, which is gradually increased to redness, and maintained at this temperature until the contents of the crucible are in a state of calm fusion. The crucible is now allowed to cool, and weighed. The loss of weight is carbonic acid. The results are very accurate (SCHAFFGOTSCH).

I must add that borax-glass may be kept in a state of fusion at a red heat for ¼ to ½ an hour without the occurrence of any volatilization, but that at a white heat (by igniting over the gas-bellows), even in a few minutes, it suffers a decided loss.* A few bubbles of carbonic acid remaining in the fusing mass are without any influence on the result.

Instead of vitrified borax fused potassium dichromate may be used, in the proportion of 5 to 1 of the carbonate (H. ROSE†). The heat applied in this case must be low, and great caution must be used, or the dichromate will lose weight of itself.‡ The carbonic acid may be expelled from alkali carbonates, by strong ignition with ignited silica (H. ROSE§).

d. Separation by decomposition with Acids. (Estimation from the loss of weight.)

α. Carbonates of metals which form Soluble Salts with Sulphuric Acid.

The process is conducted in the apparatus illustrated by fig. 58.

The size of the flask depends upon the capacity of the balance. *B* may be smaller than *A*. The tube *a* is closed at *b* with a little wax ball, or a small piece of india-rubber tube, stopped with half an inch of rod; the other end of the tube *a* is open, as are also both ends of *c* and *d*. The flask *B* is nearly half filled with concentrated sulphuric acid, free from oxides of nitrogen and sulphurous acid.

---

* Zeitschr. f. anal. Chem. 1, 65.

† Pogg. Annal. 116, 131.

‡ Zeitschr. f. anal. Chem. 1, 183.

§ Pogg. Annal. 116, 686.

The tubes must fit air-tight in the corks, and the latter equally so in the flasks. The weighed substance is put into *A*; this flask is then filled about one third with water, the cork properly inserted, and the apparatus tared on the balance. A few bubbles of air are now sucked out of *d*, by means of an india-rubber tube. This serves to rarefy the air in *A* also, and causes the sulphuric acid in *B* to ascend in the tube *c*. The latter is watched for some time, to ascertain whether the column of sulphuric acid in it remains stationary, which is a proof that the apparatus is air-tight. Air is then again sucked out of *d*, which causes a portion of the sulphuric acid to flow over into *A*. The carbonate in the latter flask is decomposed by the sulphuric acid, and the liberated carbonic acid, completely dried in its passage through the sulphuric acid in *B*, escapes through *d*. When the evolution of the gas slackens a fresh portion of sulphuric acid is made to pass over into *A*, by renewed suction through *d*; the operation being repeated until the whole of the carbonate is decomposed. A more vigorous suction is now applied, to make a large amount of sulphuric acid pass over into *A*, whereby the contents of that flask are considerably heated; when the evolution of gas bubbles has completely ceased, the stopper on *a* is opened, and suction applied to *d*, until the air sucked out tastes no longer of carbonic acid.* When the apparatus is quite cold it is replaced upon the balance, and the equilibrium restored by additional weights. The sum of the weights so added indicates the amount of carbonic acid originally present in the substance.

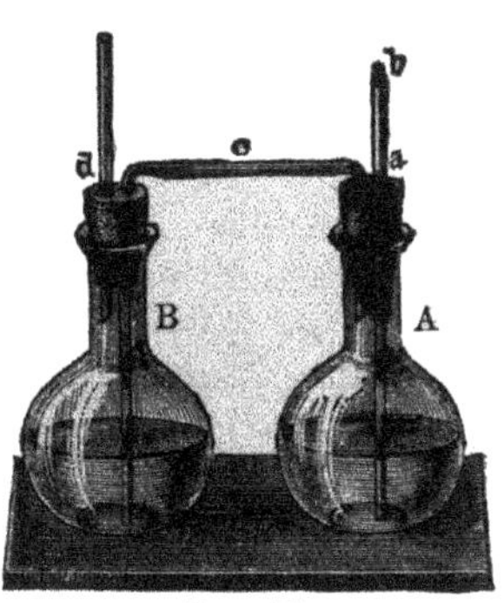

FIG. 58.

If the flasks *A* and *B* are selected of small size, the apparatus may be so constructed that, together with the contents, it need not weigh above 70 grammes, admitting thus of being weighed on a delicate balance. The results obtained by the use of this apparatus, first suggested by WILL and myself, are very accurate, provided the quantity of the carbonic acid be not too trifling. Various

* In accurate experiments, it is advisable to connect the end *b* of the tube *a* with a calcium chloride tube during the process of suction, and to use an aspirator or hydraulic air-pump instead of the mouth.

modifications of the apparatus have been proposed, principally in order to make it lighter.

If sulphites or sulphides are present, together with the carbonates, their injurious influence is best obviated by adding to the carbonate solution of normal potassium chromate in more than sufficient quantity to effect their oxidation. If chlorides are present, in order to prevent the evolution of hydrochloric acid, add to the evolution flask a sufficient quantity of silver sulphate in solution, or connect the exit tube *d* with a small prepared U-tube, which is, of course, first tared with the apparatus, and afterwards weighed with it. This U-tube is prepared—in accordance with the happy proposal of STOLBA—by filling with fragments of pumice which have been boiled with an excess of concentrated solution of cupric sulphate, till the air has been expelled, and then dried and heated to complete dehydration of the copper salt. If the U-tube is only 8 cm. high and has a bore of 1 cm., it answers the purpose very well. The outer end is provided with a perforated cork and short glass tube. We apply suction to this by means of a flexible tube, instead of to *d*.

β. After S. W. JOHNSON.* *All Carbonates which dissolve freely in cold dilute acid.*

The apparatus may consist of a light flask or bottle with wide mouth which is closed by a soft rubber stopper, through which there passes, on the one hand, a calcium chloride tube, the lower bulb of which contains cotton, and, on the other, the neck of a vessel which contains the dilute acid. This acid reservoir is so constructed that on suitably inclining it, its contents will flow freely into the flask. For this purpose the tube connecting with the latter has an internal diameter of seven millimetres, and its extremity is cut off obliquely; at its other end, the acid reservoir terminates in an upturned narrow tube. This and the upper termination of the calcium chloride tube are chosen of such diameter that they fit quite snugly into short, narrow, and thick-walled rubber connectors which are again provided with glass-rod stoppers; all these joints must be gas-tight. In figure 59 the apparatus is represented in one third its proper dimensions.

The weighed substance, in case of calcium carbonate, *e.g.*, is placed at the bottom of the flask, most conveniently in the

* American Journal of Science and Arts, vol. xlv. iii., July, 1869.

form of small fragments. The acid vessel is nearly filled with hydrochloric acid of sp. gr. 1·1. It and the calcium chloride tube are tightly adjusted to the neck of the flask, and the glass-rod stoppers being removed, the apparatus is connected at *c* with a self-regulating generator of washed carbonic acid, and a rather rapid stream of the gas is transmitted through the apparatus for 15 minutes, or until the liquid is saturated and the air is thoroughly displaced. Then the opening at *d* is stopped and afterward the apparatus is disconnected with the carbonic acid generator and stopped at *c*. During these as well as the subsequent operations, the apparatus must be so handled that its temperature shall not change. It is immediately weighed. When removed from the balance, loosen the stopper at *d*, and, holding the flask by a wooden clamp, incline it so that the acid may flow over upon the carbonate. The decomposition should proceed slowly, so that the escaping gas may be thoroughly dried. As soon as solution of the carbonate is complete, replace the stopper at *d*, and weigh again. Should there be any leak in the apparatus the fact is made evident by a slow but steady loss of weight, when it is brought upon the balance. If all the joints are sufficiently tight, the weight remains the same for at least fifteen minutes.

Fig. 59.

When properly executed the process gives extremely accurate results; a slight change of temperature or of atmospheric pressure between the two weighings of course greatly impairs the results or renders them worthless. Since the apparatus usually rises a little in temperature during the solution of the carbonate, it is better, as soon as the substance is decomposed, to stopper the $CaCl_2$ tube and let the whole stand fifteen minutes, then to connect as before with the gas-generator and pass *dried* CO_2 for a minute, and finally to stopper again and bring upon the balance. In seven analyses of pure calcite in quantities ranging from 0·5 to 0·9 grm., the following percentages of carbonic acid were obtained, viz.: 44·07, 44·07, 43·98, 44·01, 44·04, 44·11, 44·16; calculation requires 44·00.

In case of alkali-carbonates which absorb carbonic acid gas, it is necessary to modify the apparatus. Instead of the light flask, we may employ a small bottle of thick glass and wider mouth, and a thrice perforated rubber stopper. Through the third orifice pass a narrow tube 3 to 4 inches long enlarged below to a small bulb to contain the carbonate. This bulb must be so thin that on pushing down the tube within the bottle it shall be easily crushed to pieces against the bottom of the latter. The carbonate is weighed into the bulb-tube, the latter is wiped clean down to the bulb, corked and fixed in the stopper. The apparatus is filled as before with CO_2 and weighed. Then the bulb is broken and the process finished as before described. In three estimations on sodium carbonate, 41·54, 41·64, and 41·58 per cent. of CO_2 were obtained. Calculation requires 41·51 per cent.]

e. All Carbonates without exception (Determination by absorption and weighing of CO_2), H. ROSE.

The flask for decomposing the carbonate should be small (150 c.c.), in order to facilitate subsequent removal of carbonic acid by aspiration, unless the substance froths strongly during its decomposition, in which case a larger flask must be used. The end of the funnel tube, after it is inserted in the rubber stopper which is fitted to the flask, is drawn to a less diameter and bent upwards in the form of a hook, to prevent the entrance of gas-bubbles. Above the stop-cock its internal diameter should not be so small as to prevent water when poured in from filling it, and this portion should be so long that the pressure of the liquid filling it will suffice to force gas through the apparatus. A piece of glass tube bent at a right angle is fitted to the funnel by means of a piece of rubber tube slipped over it.

The nearly horizontal glass tube (about 0·7 metre long) is of thin glass, and of a diameter not less than 12 millimetres. It is inclined to such extent that water condensing in it may flow back. The upper half is filled with granulated dried calcium chloride, secured in place by a little cotton or asbestos at each end. In the end of the large tube a small tube is fitted by means of a rubber stopper, and to this is joined by a rubber tube the potash apparatus and soda-lime tube (weighable either jointly or separately) charged with absorbents, as described §§ 174, 175. The flask is removed to receive the weighed substance, and replaced without disturbing the position of the rest of the apparatus. It can now be ascertained whether the apparatus will leak gas by forcing a little air (free from carbonic acid) through the funnel tube, closing the stop-cock, and observing whether the unequal height of liquid in the two limbs of the potash apparatus remains for a few minutes. Introduce a little water through the funnel tube, and next acid slowly by turning the stop-cock until evolution of CO_2 ceases. The small right-angled tube, to which is attached a large tube filled with fragments of potash (see § 175), is now inserted in the glass funnel, and a slow current of air (1 bubble per second) is drawn through the apparatus by means of an aspirator (fig. 62) connected with the soda-lime tube. The aspirator should not be connected directly to the soda-lime tube, but to a calcium-chloride tube, which ought to be connected with the latter during the whole operation. As soon as the current of air is established,

Fig. 60.

apply the smallest possible flame of a Bunsen lamp, best maintained constant by capping the burner with wire gauze until the fluid just boils. Keep up the gentle boiling a few minutes until water condenses in the tube, but not until condensed drops appear quite up to the calcium chloride. Remove then the lamp, and aspirate a while longer somewhat faster. The volume of air necessary to remove the carbonic acid depends upon the size of the decomposing flask. When the operation is completed, disconnect the absorbing apparatus, close the ends with caps of rubber tubing, and weigh after lapse of half an hour.

For liberating the carbonic acid, sulphuric acid (the concentrated diluted with 4 or 5 times its volume of water) is best adapted, provided it readily decomposes the substance without formation of insoluble sulphates.

When there are objections to using sulphuric acid, dilute hydrochloric acid (containing about 10 per cent) may be used, or more rarely nitric acid. Nitric acid cannot be used when substances are present which cause its decomposition; *e.g.*, ferrous salts and sulphides.

When sulphuric acid is used, the evolution of H_2S from sul-

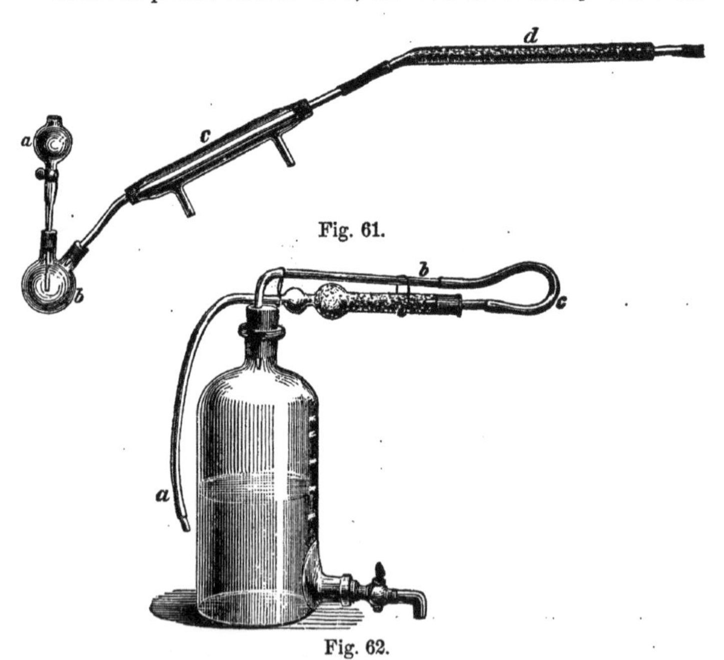

Fig. 61.

Fig. 62.

phides, if present, may be prevented by adding first a solution of chromic acid or mercuric chloride. If sulphites are present, use chromic acid or potassium chromate. When hydrochloric acid is employed, the disturbing influence of compounds which cause evolution of chlorine may be prevented by allowing some concentrated solution of stannous chloride to run into the flask before addition of the acid. When hydrochloric acid is used, or even sulphuric in the presence of chlorides, it is best to guard against the possibility of carrying HCl gas into the potash apparatus by substituting STOLBA's preparation of anhydrous copper sulphate and pumice-stone (see page 410) for that portion of the calcium chloride which fills 10–15 cm. of the end of the tube.

A modification* of the above-described apparatus, possessing some obvious advantages, is shown by fig. 61. In place of the empty part of the long glass tube shown in fig. 60 is substituted a smaller strong tube, provided with a cooling apparatus through which water circulates. This is connected by a piece of close-fitting rubber tube with the remaining part *d*. Some suitable form of apparatus for absorbing CO_2 must, of course, be attached to *d* in the manner shown by fig. 60. The calcium-chloride tube, used to prevent moist air from entering the absorbing apparatus, is conveniently supported by attaching it to the aspirator (fig. 62). The aspirator may be connected with the apparatus from the beginning to the end of the operation, with its stop-cock so adjusted that water flows from it drop by drop. In conducting the operation, a little variation from the before described manipulation is admissible on account of the presence of the condensing apparatus. After enough acid has been admitted to effect decomposition, the stop-cock of *a* is closed, a little liquid still being allowed to remain above it. Heat is then applied as before directed, but continued longer until the CO_2 is almost or quite expelled from the flask by steam. This point is indicated by almost, or nearly, entire cessation of dropping of water from the aspirator. Diminish now the heat, and immediately after open the stop-cock of *a* and let air (free from CO_2) enter and replace the condensing steam. Boil again to expel the air which has entered, after which a small volume of air drawn through the apparatus by the aspirator will ensure the bringing of all the CO_2 into the absorbing apparatus.

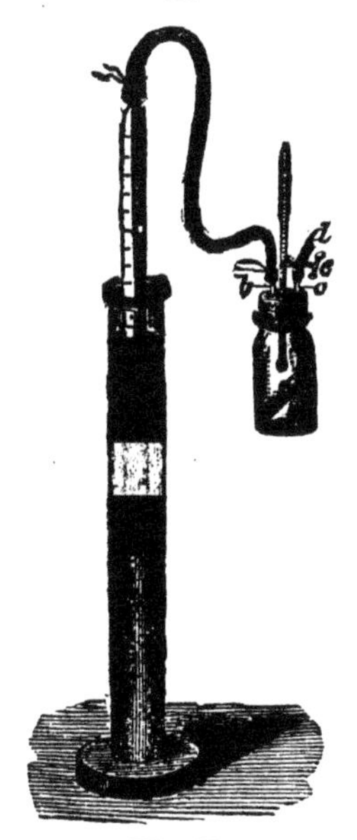

Fig. 63.

f. Estimation by Measuring the Gas.

This process is applicable in the case of all salts which are

* Devised by H. L. WELLS, of the Sheffield Laboratory.

TABLE OF THE WEIGHT OF A CUBIC

In Milligrammes from 720 to 770 mm. of press-

MILLIMETRES.

TEMPERATURE CENT.	720	722	724	726	728	730	732	734	736	738	740	742	744
10°	1.77446	1.77945	1.78445	1.78944	1.79443	1.79942	1.80441	1.80941	1.81440	1.81940	1.82438	1.82937	1.83437
11°	1.76668	1.77165	1.77662	1.78160	1.78657	1.79155	1.79652	1.80149	1.80647	1.81144	1.81642	1.82139	1.82636
12°	1.75881	1.76377	1.76873	1.77368	1.77864	1.78359	1.78855	1.79351	1.79846	1.80342	1.80838	1.81333	1.81829
13°	1.75092	1.75587	1.76081	1.76576	1.77070	1.77565	1.78059	1.78554	1.79048	1.79543	1.80037	1.80532	1.81026
14°	1.74301	1.74795	1.75288	1.75781	1.76275	1.76768	1.77261	1.77754	1.78248	1.78741	1.79234	1.79728	1.80221
15°	1.73502	1.73993	1.74484	1.74974	1.75465	1.75955	1.76446	1.76937	1.77427	1.77918	1.78408	1.78899	1.79390
16°	1.72699	1.73188	1.73677	1.74166	1.74655	1.75144	1.75633	1.76122	1.76611	1.77100	1.77590	1.78078	1.78567
17°	1.71888	1.72376	1.72862	1.73349	1.73836	1.74322	1.74809	1.75296	1.75783	1.76269	1.76756	1.77243	1.77729
18°	1.71069	1.71554	1.72040	1.72525	1.73011	1.73497	1.73982	1.74468	1.74953	1.75439	1.75925	1.76410	1.76896
19°	1.70239	1.70723	1.71207	1.71691	1.72175	1.72659	1.73143	1.73627	1.74111	1.74595	1.75078	1.75562	1.76046
20°	1.69412	1.69894	1.70377	1.70859	1.71341	1.71823	1.72305	1.72788	1.73270	1.73725	1.74234	1.74716	1.75199
21°	1.68571	1.69051	1.69532	1.70012	1.70493	1.70974	1.71454	1.71935	1.72415	1.72896	1.73377	1.73857	1.74338
22°	1.67722	1.68201	1.68680	1.69151	1.69638	1.70117	1.70596	1.71075	1.71554	1.72033	1.72512	1.72991	1.73470
23°	1.66862	1.67340	1.67817	1.68294	1.68772	1.69249	1.69727	1.70204	1.70681	1.71159	1.71636	1.72114	1.72591
24°	1.65994	1.66470	1.66945	1.67421	1.67897	1.68372	1.68848	1.69324	1.69799	1.70275	1.70751	1.71227	1.71702
25°	1.65113	1.65587	1.66061	1.66535	1.67009	1.67484	1.67958	1.68432	1.68906	1.69380	1.69854	1.70329	1.70803
	720	722	724	726	728	730	732	734	736	738	740	742	744

MILLIMETRES.

CENTIMETRE OF CARBONIC ACID.

ure of mercury, and from 10° *to* 25° *Cent.*

MILLIMETRES.

746	748	750	752	754	756	758	760	762	764	766	768	770	Temperature Cent.
1.83936	1.84435	1.84934	1.85433	1.85933	1.86432	1.86931	1.87430	1.87930	1.88429	1.88928	1.89427	1.89926	10°
1.83134	1.83631	1.84129	1.84626	1.85123	1.85621	1.86118	1.86616	1.87113	1.87610	1.88108	1.88605	1.89103	11°
1.82324	1.82820	1.83315	1.83811	1.84307	1.84802	1.85298	1.85793	1.86289	1.86785	1.87280	1.87776	1.88271	12°
1.81521	1.82015	1.82510	1.83004	1.83499	1.83993	1.84488	1.84982	1.85477	1.85971	1.86466	1.86960	1.87455	13°
1.80714	1.81208	1.81701	1.82194	1.82687	1.83181	1.83674	1.84167	1.84661	1.85154	1.85647	1.86141	1.86634	14°
1.79880	1.80371	1.80861	1.81352	1.81843	1.82333	1.82824	1.83314	1.83805	1.84296	1.84786	1.85277	1.85767	15°
1.79056	1.79545	1.80034	1.80523	1.81012	1.81501	1.81990	1.82479	1.82968	1.83457	1.83946	1.84435	1.84924	16°
1.78216	1.78703	1.79189	1.79676	1.80163	1.80650	1.81136	1.81623	1.82110	1.82596	1.83083	1.83570	1.84056	17°
1.77381	1.77867	1.78353	1.78838	1.79324	1.79809	1.80295	1.80781	1.81266	1.81752	1.82337	1.82723	1.83209	18°
1.76530	1.77014	1.77498	1.77982	1.78466	1.78950	1.79434	1.79917	1.80401	1.80885	1.81369	1.81853	1.82337	19°
1.75681	1.76113	1.76645	1.77127	1.77610	1.78092	1.78574	1.79056	1.79538	1.80021	1.80503	1.80985	1.81467	20°
1.74818	1.75299	1.75780	1.76260	1.76741	1.77221	1.77702	1.78183	1.78663	1.79144	1.79624	1.80105	1.80586	21°
1.73949	1.74428	1.74907	1.75386	1.75865	1.76344	1.76823	1.77302	1.77781	1.78260	1.78739	1.79218	1.79697	22°
1.73068	1.73546	1.74023	1.74501	1.74978	1.75455	1.75933	1.76410	1.76888	1.77365	1.77842	1.78320	1.78797	23°
1.72178	1.72654	1.73129	1.73605	1.74081	1.74556	1.75032	1.75508	1.75984	1.76459	1.76935	1.77411	1.77886	24°
1.71277	1.71751	1.72225	1.72699	1.73173	1.73648	1.74122	1.74596	1.75070	1.75544	1.76018	1.76492	1.76967	25°
746	748	750	752	754	756	758	760	762	764	766	768	770	

MILLIMETRES.

TABLE OF THE ABSORPTION OF CARBONIC ACID

in 5 c. c. of H Cl., sp. gr. 1·125, for an evolution of 1 to 100 c. c.

Evolved........	1	2	3	4	5	6	7	8	9	10	11	12	13	14	15	16	17	18	19	20
Absorbed.......	1.85	2.00	2.16	2.31	2.47	2.62	2.78	2.93	3.09	3.24	3.40	3.55	3.71	3.86	4.02	4.17	4.33	4.48	4.64	4.79
Evolved........	21	22	23	24	25	26	27	28	29	30	31	32	33	34	35	36	37	38	39	40
Absorbed.......	4.95	4.96	4.97	4.98	5.00	5.03	5.04	5.06	5.07	5.09	5.10	5.11	5.13	5.14	5.16	5.17	5.18	5.20	5.21	5.23
Evolved........	41	42	43	44	45	46	47	48	49	50	51	52	53	54	55	56	57	58	59	60
Absorbed.......	5.24	5.25	5.26	5.27	5.28	5.30	5.31	5.32	5.34	5.35	5.36	5.37	5.38	5.40	5.41	5.43	5.44	5.45	5.47	5.48
Evolved........	61	62	63	64	65	66	67	68	69	70	71	72	73	74	75	76	77	78	79	80
Absorbed.......	5.50	5.51	5.52	5.54	5.55	5.57	5.58	5.59	5.61	5.62	5.64	5.65	5.66	5.68	5.69	5.71	5.72	5.73	5.75	5.76
Evolved........	81	82	83	84	85	86	87	88	89	90	91	92	93	94	95	96	97	98	99	100
Absorbed.......	5.78	5.79	5.80	5.82	5.83	5.85	5.86	5.87	5.89	5.90	5.92	5.93	5.94	5.96	5.97	5.99	6.00	6.01	6.03	6.04

decomposed by hydrochloric acid in the cold. It is distinguished for rapid and convenient execution and very satisfactory results. [The azotometer, fig. 63, is employed, and the details of the process are for the most part similar to those followed in the estimation of ammonia as described on page 222. The weighed carbonate is put in the bottle *a*, and the tube *f* is charged with 5 c.c. of H. Cl., sp. gr. 1·125. When the burette is adjusted to zero, the acid is poured *at once* upon the carbonate. The precautions to be observed in the measurement of the gas are as detailed on page 222. It is not needful to wait so long for the gas to cool. The necessary corrections are applied by aid of the tables given by Dietrich, pages 416–418. Their use is perfectly similar to that of the tables given on pages 223–225.]

§ 140.

2. Silicic Acid.

I. Determination.

The direct estimation of silicic acid is almost invariably effected by converting the soluble modification of the acid into the insoluble modification, by evaporating and completely drying; the insoluble modification is then, after removal of all foreign matter, ignited strongly (over the bellows blowpipe) and weighed.

For the guidance of the student I would observe here that, to guard against mistakes, he should always *test the purity of the weighed silicic acid.* The methods of testing will be found below.

If you have free silicic acid in the state of hydrate, in an aqueous or acid solution free from other fixed bodies, simply evaporate the solution in a platinum dish, ignite and weigh the residue.

Respecting a volumetric estimation of silicic acid (conversion into potassium silicofluoride and acidimetric determination of the same, see § 97, 4), I must refer to Stolba.*

II. Separation of Silicic Acid from the Basic Radicals.

a. In all compounds which are decomposed by Hydrochloric or Nitric Acid, on digestion in open vessels.

* Zeitschr. f. anal. Chem. 4, 163.

To this class belong the silicates soluble in water, as well as many of the insoluble silicates, as, for instance, nearly all zeolites. Several minerals not decomposable of themselves by acids, become so by persistent ignition in a state of fine powder (F. MOHR*). If the ignition is too strong, particles of alkali may be lost.

The substance is very finely powdered,† dried at 100°, and put into a platinum or porcelain dish (in the case of silicates whose solution might be attended with disengagement of chlorine, platinum cannot be used); a little water is then added, and the powder mixed to a uniform paste. Moderately concentrated hydrochloric acid, or—if the substance contains lead or silver—nitric acid, is now added, and the mixture digested at a very gentle heat, with constant stirring, until the substance is completely decomposed, in other terms, until the glass rod, which is rounded at the end, encounters no more gritty powder, and the stirring proceeds smoothly without the least grating.

The silicates of this class do not all comport themselves in the same manner in this process, but show some differences; thus most of them form a bulky gelatinous mass, whilst in the case of others the silicic acid separates as a light pulverulent precipitate; again, many of them are decomposed readily and rapidly, whilst others require protracted digestion.

When the decomposition is effected, the mixture is evaporated to dryness on the water-bath, and the residue heated, with frequent stirring, until all the small lumps have crumbled to pieces, and the whole mass is thoroughly dry, and until no more acid fumes escape. It is always the *safest* way to conduct the drying on the water-bath. Occasionally it is well to moisten the dry mass with water and evaporate again. In cases where it appears desirable to accelerate the desiccation by the application of a stronger heat, an air-bath may be had recourse to; which may be constructed in a simple way, by suspending the dish containing the substance, with the aid of wire, in a somewhat larger dish of silver or iron, in a manner to leave everywhere between the two dishes a small space of uniform width. Direct heating over the lamp is not advisable, as in the most strongly heated parts the silicic acid is liable to unite again with

* Zeitschr. f. anal. Chem. 7, 293.

† Very hard silicates cannot be powdered in an agate mortar without taking up silica; these must, therefore, be powdered in a steel mortar, sifted, and freed from particles of steel with the magnet.

the separated bases to compounds which are not decomposed, or only imperfectly, by hydrochloric acid.

When the mass is cold, it is brought to a state of semi-fluidity by thoroughly moistening it with hydrochloric acid; after which it is allowed to stand for half an hour, then warmed on a water-bath, diluted with hot water, stirred, allowed to deposit, and the fluid decanted on to a filter; the residuary silicic acid is again stirred with hydrochloric acid, warmed, diluted, and the fluid once more decanted; after a third repetition of the same operation, the precipitate also is transferred to the filter, thoroughly washed with hot water, well dried, and ignited at last as strongly as possible, as directed in § 52. For the properties of the residue, see § 93, 9. The results are accurate. The basic metals, which are in the filtrate as chlorides, are determined by the methods given above. Deviations from the instructions here given are likely to entail loss of substance; thus, for instance, if the mass is not thoroughly dried, a not inconsiderable portion of the silicic acid passes into the solution, whereas, if the instructions are strictly complied with, only traces of the acid are dissolved; in accurate analyses, however, even such minute traces must not be neglected, but should be separated from the metals precipitated from the solution. The separation may, as a rule, be readily effected by dissolving them, after ignition and weighing, in hydrochloric or sulphuric acid, by long digestion in the heat, the traces of silicic acid being left undissolved. Sometimes it is better to fuse the metallic oxides with potassium disulphate, or to reduce them to the metallic state by ignition in hydrogen, and then to treat with hydrochloric acid. Again, if the silicic acid is not thoroughly dried previous to ignition, the aqueous vapor disengaged upon the rapid application of a strong heat may carry away particles of the light and loose silica.

The silicic acid may be tested as follows: This testing must on no account be omitted if the silica has been separated in a pulverulent and not in a gelatinous form. Heat a portion on a water-bath with moderately concentrated solution of sodium carbonate for an hour in a platinum or silver dish; with less advantage in a porcelain dish. EGGERTZ* recommends, for ·1 grm. silicic acid, 6 c.c. of a saturated solution of sodium carbonate and 12 c.c. of water. Pure silica would dissolve. If a residue remains, pour off the clear

* Zeitschr. f. anal. Chem. 7, 502.

fluid and heat again with a small quantity of sodium carbonate. If a residue still remains, weigh the rest of the impure silica and treat it according to *b*, to estimate the amount of impurity.

If you have *pure* hydrofluoric acid, you may also test the silicic acid in a very easy manner, by treating it with this acid and a few drops of sulphuric acid in a platinum dish; upon the evaporation of the solution, the silicic acid, if pure, will volatilize completely (as fluoride of silicon). If a residue remains, moisten this once more with hydrofluoric acid, add a few drops of sulphuric acid, evaporate, and ignite; the residue consists of the sulphates of the metals retained by the silicic acid, as well as any titanic acid that was present (BERZELIUS). Ammonium fluoride may be used instead of hydrofluoric acid.

b. Compounds which are not decomposed by Hydrochloric or Nitric Acid, on digestion in open vessels.

α. Decomposition by fusion with Alkali Carbonate.

Reduce the substance to an impalpable powder, by trituration and, if necessary, sifting (§ 25); transfer to a platinum crucible, and mix with about 4 times the weight of pure anhydrous sodium carbonate or sodium and potassium carbonate, with the aid of a rounded glass rod; wipe the rod against a small portion of sodium carbonate on a card, and transfer this also from the card to the crucible. Cover the latter well, and heat, according to size, over a gas or spirit-lamp with double draught, or a blast gas-lamp; or insert in a Hessian crucible, compactly filled up with calcined magnesia, and heat in a charcoal fire.

Apply at first a moderate heat for some time to make the mass simply agglutinate; the carbonic acid will, in that case, escape from the porous mass with ease and unattended with spirting. Increase the heat afterwards, finally to a very high degree, and terminate the operation only when the mass appears in a state of calm fusion, and gives no more bubbles.

The platinum crucible in which the fusion is conducted must not be too small; in fact, the mixture should only half fill it. The larger the crucible, the less risk of loss of substance. As it is of importance to watch the progress of the operation, the lid must be easily removable; a concave cover, simply lying on the top, is therefore preferable to an overlapping lid. If the process is conducted over the spirit or simple gas-lamp, the mixed sodium and potassium carbonates are preferable to sodium carbonate, as they fuse much more readily than the latter. In heating over a lamp, the crucible should always be supported on a triangle of platinum wire, with the opening just sufficiently wide to allow the crucible to drop into it fully one third, yet to retain it firmly, even with the wire at an intense red heat. When conducting the process over a spirit-lamp with double draught, or over a simple gas-lamp, it is also advisable, towards the end of the operation, when the heat is to be raised to the highest degree, to put a chimney over the crucible, with the lower border resting on the ends of the iron triangle which supports the platinum triangle; this chimney should be about 12 or 14 cm. high, and the upper opening measure about 4 cm. in diameter. The little clay chimneys recommended by O. L. ERDMANN are still more serviceable (fig. 21, p. 24, "Qual. Anal."). When the fusion is ended, the red-hot crucible is removed with tongs, and placed on a cold, thick, clean iron plate, on which it will rapidly cool; it is then generally easy to detach the fused cake in one piece.

The cake (or the crucible with its contents) is put into a beaker, from 10 to 15 times the quantity of water poured over it, and heat applied for half an hour, then hydrochloric acid is gradually added, or, under certain circumstances, nitric acid; the beaker is kept covered with a glass plate, or, which is much better, with a large watch-glass or porcelain dish, perfectly clean outside, to prevent the loss of the drops of fluid which the escaping carbonic acid carries along with it; the drops thus intercepted by the cover are afterwards rinsed into the beaker. The crucible is also rinsed with water mixed with dilute acid, and the solution obtained added to the fluid in the beaker.

The solution is promoted by the application of a gentle heat, which is continued for some time after this is effected to insure the complete expulsion of the carbonic acid; since otherwise some loss of substance might be incurred, in the subsequent process of evaporation, by spirting caused by the escape of that gas. If in the process of treating the fused mass with hydrochloric acid, a saline powder subsides (sodium or potassium chloride), this is a sign that more water is required.

If the decomposition of the mineral has succeeded to the full extent, the hydrochloric acid solution is either perfectly clear, or light flakes of silicic acid only float in it. But if a heavy powder subsides, which feels gritty under the glass rod, this consists of undecomposed mineral. The cause of such imperfect decomposition is generally to be ascribed to imperfect pulverization. In such cases the undecomposed portion may be fused once more with alkali carbonate; the better way, however, is to repeat the process with a fresh portion of mineral more finely pulverized.

The hydrochloric or nitric acid solution obtained is poured, together with the precipitate of silicic acid, which is usually floating in it, into a porcelain or, better, into a platinum dish, and treated as directed in II., *a*. That the fluid may not be too much diluted, the beaker should be rinsed only once, or not at all, and the few remaining drops of solution dried in it; the trifling residue thus obtained is treated in the same way as the residue left in the evaporating basin. This is the method most commonly employed to effect the decomposition of silicates that are undecomposable by acids; that it cannot be used to determine alkalies in silicates is self-evident.

β. Decomposition by means of Hydrofluoric Acid.

aa. By Aqueous Hydrofluoric Acid.

The silicate should be finely pulverized, dried at 100° (in some cases ignition is advisable*). It is mixed, in a platinum dish, with rather concentrated, slightly fuming hydrofluoric acid, the acid being added gradually, and the mixture stirred with a thick platinum wire. The mixture, which has the consistence of a thin paste, is digested some time on a water-bath at a gentle heat, and pure concentrated sulphuric acid, diluted with an equal quantity of water, is then added, drop by drop, in more than sufficent quantity to convert all the basic metals present into sulphates. The mixture is now evaporated on the water-bath, during which operation silicon fluoride gas and hydrofluoric acid gas are continually volatilizing; then it is finally exposed to a stronger heat at some height above the lamp, until the excess of sulphuric acid is almost completely expelled. The mass, when cold, is thoroughly moistened with concentrated hydrochloric acid, and allowed to stand at rest for one hour; water is then added, and a gentle heat applied. If the decomposition has fully succeeded, the whole must dissolve to a clear fluid. If an undissolved residue is left, the mixture is heated

* Many minerals are much more readily decomposed by hydrofluoric acid also, if they are previously ignited in a state of fine division (HERMANN, RAMMELSBERG, FR. MOHR, Zeitschr. f. anal. Chem. 7, 291).

for some time on the water-bath, then allowed to deposit, the clear supernatant fluid decanted as far as practicable, the residue dried, and then treated again with hydrofluoric acid and sulphuric acid, and, lastly, with hydrochloric acid, which will now effect complete solution, provided the analyzed substance was very finely pulverized, and free from barium, strontium (and lead). The solution is added to the first. The basic metals in the solution (which contains them as sulphates, and contains also free hydrochloric acid) are determined by the methods which will be found in Section V.

This method, which is certainly one of the best to effect the decomposition of silicates, was proposed by BERZELIUS. It has been but little used hitherto, because we did not know how to prepare hydrofluoric acid, except with the aid of a distilling apparatus of platinum, or, at least, with a platinum head; nor to keep it, except in platinum vessels. These difficulties can now be considered as overcome, comp. § 58, 2. Never omit testing the acid before using it.

The hydrofluoric acid may also be employed in combination with hydrochloric acid; thus 1 grm. of finely elutriated felspar, mixed with 40 c.c. water, 7 c.c. hydrochloric acid of 25$\frac{8}{9}$ and 3$\frac{1}{2}$ c.c. hydrofluoric acid, and heated to near the boiling point, dissolves completely in three minutes. 4 c.c. sulphuric acid are then added, the barium sulphate which may separate is filtered off, and the filtrate evaporated till no more hydrofluoric acid escapes (AL. MITSCHERLICH*).

The execution of the method requires the greatest possible care, both the liquid and the gaseous hydrofluoric acid being most injurious substances. The treatment of the silicate with the acid and the evaporation must be conducted in the open air, otherwise the windows and all glass apparatus will be attacked. As the silicic acid is in this method simply inferred from the loss,† a combination with method α is often resorted to.

bb. By Ammonium Fluoride.

Mix the very finely powdered substance in a platinum dish with four times its weight of ammonium fluoride, moisten well with concentrated sulphuric acid, heat on the water-bath till the

* Journ. f. prakt. Chem. 81, 108.

† The silicon escaping in the form of fluoride may sometimes be determined directly, by the method of STORY MASKELYNE (Zeitschr. f. anal. Chem. 9, 380), which, however, requires a platinum retort of peculiar construction.

evolution of silicon fluoride and hydrofluoric acid slackens, add more sulphuric acid, heat again, finally somewhat more strongly till the greater part of the sulphuric acid has escaped, and treat the residue according to *aa* (L. v. BABO, J. POTYKA, R. HOFFMANN*). H. ROSE† first warms the silicate gently with seven times its amount of the fluoride and some water, then heats gradually to redness till no more fumes escape, and finally treats with sulphuric acid.

cc. By Fluoride of Hydrogen and Potassium, &c.

In silicates, which more or less resist the action of hydrofluoric acid, such as zircon and beryl, the basic metals with the exception of the alkalies may be determined by fusing with fluoride of hydrogen and potassium (MARIGNAC, GIBBS‡), or by mixing with 3 parts of sodium fluoride, adding 12 parts of potassium disulphate to the crucible, and then heating at first very gently, afterwards more strongly till the mass fuses calmly. The residue is dissolved in water or hydrochloric acid (CLARKE§).

[*γ. Decomposition by ignition with Calcium Carbonate and Ammonium Chloride.* PROF. J. L. SMITH'S METHOD for separating alkalies.

Mix 1 part of the pulverized silicate with 1 part of dry ammonium chloride, by gentle trituration in a smooth mortar, then add 8 parts of calcium carbonate ("Qual. Anal." p. 87) and mix intimately. Bring the mixture into a platinum crucible, rinsing the mortar with a little calcium carbonate. Warm the crucible gradually over a small Bunsen burner until fumes of ammonium salts no longer appear, then heat with the flame of a Bunsen burner until the lower *three-fourths only* of the crucible are brought to a red heat. Keep this temperature constant from 40 to 60 minutes. The temperature desired is that which suffices to keep in state of fusion the calcium chloride formed by the reaction of ammonium chloride with calcium carbonate. The mass, however, does not become liquid since the fused calcium chloride is absorbed by the large quantity of calcium carbonate present. If the silicate is fused by application of too strong heat, disintegration of the mass at the end of the operation with water cannot be effected. Moreover, too high a temperature causes volatilization of alkali chlo-

---

* Zeitschr. f. anal. Chem. 6, 366.
† Pogg. Annal. 108, 20.
‡ Zeitschr. f. anal. Chem. 3, 399.
§ *Ib.* 7, 463.

rides. Certain silicates—*e.g.*, those which contain much ferrous iron—may fuse when heated with the above mixture, even if no higher temperature is employed than is necessary to effect decomposition. If this occurs, it is better to repeat the ignition with a new portion of the silicate, using 8 to 10 parts of calcium carbonate. The mass contracts in volume during the ignition, and is usually easily detached from the crucible. Boil it in a covered porcelain dish, with 50–75 c.c. water, half an hour, replacing water lost by evaporation. Decant the solution from the residue upon a filter, boil the residue a few minutes with water, and decant again. If the residue is now all in a finely disintegrated state, it may be brought upon the filter and washed. But if, as is often the case, a portion remains coherent or in a coarsely granular state, it must be reduced to a fine state of division by trituration with a porcelain or agate pestle in the dish, and boiling with water again. By a few repetitions of the trituration, boiling and decanting, allowing the fine suspended portion to pass upon the filter each time, the whole can usually be transferred to the filter in properly disintegrated condition in course of an hour. Next wash until a few drops of the washings acidified with nitric acid give but a slight turbidity with silver nitrate. The filtrate now contains the alkalies of the silicate as chlorides together with calcium chloride and hydroxide. It is not advisable to concentrate this filtrate in a glass vessel, since it might take an appreciable quantity of sodium from the glass. Precipitate, therefore, the calcium at once with ammonium carbonate; allow the precipitate to settle, and concentrate the supernatant solution in a porcelain (or platinum) dish, decanting it into the latter, portionwise if necessary, rinsing finally the precipitate into the porcelain dish. When the whole is thus reduced to about 30 c.c., add a little more ammonium carbonate and ammonia, heat and filter into a platinum (or porcelain) dish, evaporate to dryness on a water-bath, expel ammonium chloride by ignition, dissolve the residual alkali chlorides in 3 to 5 c.c. of water. A little black or dark-brown flocculent matter usually remains undissolved, while the solution may still contain traces of calcium. Add two or three drops of ammonium carbonate and ammonia, warm gently, and filter through a very small filter into a weighable platinum vessel. Evaporate to dryness on a water-bath, heat to incipient fusion of the alkali chlorides, and after cooling weigh.

Prof. SMITH's method is the most convenient of all methods

for extracting alkalies from silicates, and is universally applicable, except perhaps in presence of boric acid. When carried out as here described, the results are sufficiently accurate in most cases. If, however, the silicate is rich in alkalies, a loss amounting to 0·1 or 0·2 per cent of the mineral is possible. If great accuracy is desired in such cases, a repetition of the whole process may be applied to the residue left by treatment of the ignited mass with water. It need hardly be mentioned that unless care be taken to use reagents perfectly free from soda and to avoid action of solutions on glass, an amount of soda may be introduced from these sources equal to 0·1 or 0·2 per cent of the silicate.]

Second Group.

CHLORINE—BROMINE—IODINE—CYANOGEN—SULPHUR.

§ 141.

1. Chlorine.

I. *Determination.*

Chlorine may be determined very accurately in the gravimetric as well as in the volumetric way.*

a. Gravimetric Method.—Determination as Silver Chloride.

Solution of silver nitrate, mixed with some nitric acid, is added in excess to the solution of the chloride, the precipitated chloride is made to unite by heating and agitating, washed by decantation and filtration, dried, and ignited. The details of the process have been given in § 115, 1, *a*. Care must be taken not to heat the solution mixed with nitric acid, before the nitrate of silver has been added in excess. As soon as the latter is present in excess, the silver chloride separates immediately and completely upon shaking or stirring, and the supernatant fluid becomes perfectly clear after standing a short time in a warm place. The determination of chlorine by means of silver is therefore more readily effected than that of silver by means of hydrochloric acid.

b. Volumetric Methods.

α. By Solution of Silver Nitrate.

In § 115, 5, we have seen how the silver in a fluid may be estimated by adding a standard solution of sodium chloride until no

* For the acidimetric estimation of free hydrochloric acid, see § 192.

further precipitation ensues; in the same way we may determine also, by means of a standard solution of silver, the amount of hydrochloric acid in a fluid, or of chlorine in combiuation with a metal. FELOUZE has used this method for the determination of several atomic weights. LEVOL* proposed a modification which serves to indicate more readily the exact point of complete precipitation. To the fluid, which must be *neutral*, he added one tenth volume of a saturated solution of sodium phosphate. When the whole of the chlorine has been precipitated by the silver, the further addition of the solution of silver produces a yellow precipitate which does not disappear upon shaking the vessel. FR. MOHR has since replaced, with the most complete success, the sodium phosphate by potassium chromate.

This convenient and accurate method requires a perfectly neutral solution of silver nitrate of known value. The strength most convenient is, 1 litre = 1 at. Cl. I recommend the following method of preparation: Dissolve 18·80 to 18·85 grm. pure fused silver nitrate in 1100 c.c. water, and filter the solution if required; the solution is purposely made too strong at first. Now weigh off exactly four portions of pure sodium chloride, each of ·10 to ·18 grm., one after another. The salt should be moderately ignited, not fused, powdered roughly while still warm, and introduced into a small dry tube, that can be well closed. The weighing off is performed by first weighing the filled tube, then shaking out into a dry beaker the quantity required, weighing again, dropping a second portion into beaker No. 2, weighing again, and so on. Each portion is dissolved in 20 to 30 c.c. water, and about 3 drops of a cold saturated solution of pure normal potassium chromate added.

Fill a MOHR's burette (in very accurate analysis an ERDMANN's float should be used) with the silver solution, and run it slowly, with constant stirring, into the light yellow solution contained in one of the beakers. Each drop produces, where it falls, a red spot, which on stirring disappears, owing to the instant decomposition of the silver chromate with the sodium chloride. At last, however, the slight red coloration remains. Now all chlorine has combined with silver, and a little silver chromate has been permanently formed. Read off the burette and reckon how much silver solu-

* Journ. f. prakt. Chem. 60, 384.

tion would have been required for ·1 mol. sodium chloride, *i.e.*, 5·85 grm. Suppose we have used to ·110 sodium chloride 18·7 c.c. silver solution.

$$\cdot110 : 5\cdot85 :: 18\cdot7 : x;\ x = 994\cdot5.$$

Now, without throwing away the contents of the first beaker, make a second and third experiment in the same manner, of course always taking notice to regard the same shade of red as the sign of the end. The results of these are reckoned out in the same way as the first. Suppose they gave for 5·85 NaCl 995·0 and 993·0 respectively, we take the mean of the three numbers, which is 994·2, and we now know that we have only to take this number of c.c. of silver solution, and make it up to 1000 c.c. with 5·8 water, in order to obtain a solution of the required strength, *i.e.*, 1000 c.c. = ·1 mol. NaCl. But if 994·2 requires 5·8 water, 1000 requires 5·83. Hence we fill a litre-flask (previously dried or rinsed with a small portion of the solution) up to the "holding" mark with the solution, add 5·83 c.c. water, insert a caoutchouc stopper, and shake.

The solution must now be correct; however, to make quite sure, we perform another experiment with it. To this end rinse the empty burette with the new solution, fill it with the same and test with the portion of salt in beaker No. 4. The c.c. used of silver solution must now, if multiplied by ·00585, give exactly the weight of the salt.

Being now in possession of a standard silver solution, and being practised in exactly hitting the transition from yellow to the shade of red, we are in the position to determine with precision chlorine in the form of hydrochloric acid or of a metallic chloride soluble in water. The fluid to be tested must be neutral—free acids dissolve the silver chromate. The solution of the substance is therefore, if necessary, rendered neutral by addition of nitric acid or sodium carbonate (it should be rather alkaline than acid), about 3 drops of the solution of chromate added, and then silver from the burette, till the reddish coloration is just perceptible. The number of c.c. used has only to be multiplied by the atomic weight of chlorine or the mol. weight of the metallic chloride and divided by 10,000 to give the amount of these respectively present.

If the operator fears he has added too much silver solution, *i.e.*, if the red color is too strongly marked, he may add 1 c.c. of a solution of sodium chloride containing 5·85 in a litre (and therefore corresponding to the silver solution), and then add the silver drop by drop again. Of course in this case 1 c.c. must be deducted from the amount of silver solution used.

The results are very satisfactory. The fluid to be analyzed should be about the same volume as the solutions employed in standardizing the silver solution, and also about the same strength, otherwise the small quantity of silver which produces the coloration will not stand in the same proportion to the chlorine present. This small quantity of silver solution is extremely small, varying between ·05 and ·1 c.c.: the inaccuracy hereby arising even in the case of quantities of chlorine differing widely from that originally used in standardizing the silver solution is therefore almost inconsiderable. If the amount of silver solution necessary to impart the coloration always remained the same, we should have simply to deduct the amount in question with all experiments, in order to avoid this small inaccuracy entirely, since, however, the greater the quantity of silver chloride the more silver chromate is required for visible coloration, this method of proceeding would not increase the exactness of the results.

β. By Solution of Silver Nitrate and Iodide of Starch (PISANI's method*).

Add to the solution of the chloride, acidified with nitric acid, a slight excess of standard solution of silver nitrate, warm, and filter. Determine the excess of silver in the filtrate by means of solution of iodide of starch (see p. 295), and deduct this from the amount of silver solution used. The difference shows the quantity of silver which has combined with the chlorine; calculate from this the amount of the latter. Results satisfactory.

Of these volumetric methods of estimating chlorine, the first deserves the preference in all ordinary cases. PISANI's method (*b*, *β*) is especially suited for the estimation of very minute quantities of chlorine, but is not applicable when—as in nitre analyses—large quantities of alkaline nitrate are present (p. 290).

II. *Separation of Chlorine from the Metals.*

a. In Soluble Chlorides.

The same method as in I., *a*. The metals in the filtrate are separated from the excess of the salt of silver by the methods

* Annal. d. Mines, 10, 83; LIEBIG and KOPP's Jahresbericht, 1856, 751.

which will be found in Section V. Chlorides soluble in water may also be completely decomposed by cold digestion with oxide or carbonate of silver. Silver chloride is obtained, while the metal combined with the chlorine is converted into oxide or carbonate and either remains in solution or falls down with the silver chloride. Take care that no traces of oxide or carbonate of silver pass into the filtrate.

Stannous chloride, *mercuric chloride*, *platinic chloride*, the *chlorides of antimony*, and the *green chloride of chromium*, form exceptions from the rule.

α. From *stannic chloride*, silver nitrate would precipitate, besides silver chloride, a compound of stannic oxide and silver oxide. To precipitate the tin, therefore, the solution is mixed with concentrated solution of ammonium nitrate, boiled, allowed to deposit, decanted, and filtered (compare § 126, 1, *b*), and the chlorine in the filtrate is precipitated with solution of silver. LÖWENTHAL, the inventor of this method, has proved its accuracy.*

β. When *mercuric chloride* is precipitated with solution of silver nitrate, the silver chloride thrown down contains an admixture of mercury. The mercury is, therefore, first precipitated by hydrogen sulphide, and the chlorine in the filtrate determined as directed in § 169.

γ. *The chlorides of antimony* are also decomposed in the manner described in β. The separation of basic salt upon the addition of water may be avoided by addition of tartaric acid. The antimonious sulphide should be tested for chlorine.

δ. Solution of silver fails to precipitate the whole of the chlorine from solution of the *green chloride of chromium* (PÉLIGOT). The chromium is, therefore, first precipitated with ammonia, the fluid filtered, and the chlorine in the filtrate precipitated as in I., *a*.

ε. From *platinic chloride* silver nitrate throws down a compound of platinous chloride and silver chloride (COMAILLE†). We may either ignite the platinic chloride in a current of hydrogen and pass the hydrochloric acid produced into solution of silver (BONSDORFF); or we may evaporate the solution with sodium carbonate, fuse the residue in a platinum crucible and determine the chloride in the aqueous solution of the fusion. Or, thirdly, we may (after TOPSÖE‡) digest the moderately dilute solution in the

* Journ. f. prakt. Chem. 66, 371. † Zeitschr. f. anal. Chem. 6, 121.
‡ Zeitschr. f. anal. Chem. 9, 30.

cold with zinc clippings till hydrogen ceases to escape, add ammonia in excess, heat on a water-bath till the fluid is fully decolorized, all the platinum being precipitated, and finally determine the chlorine in the filtrate.

b. In Insoluble Chlorides.

α. Chlorides soluble in Nitric Acid.

Dissolve the chloride in nitric acid, without applying heat, and proceed as in I., *a.*

β. Chlorides insoluble in Nitric Acid (lead chloride, silver chloride, mercurous chloride).

aa. Lead chloride is decomposed by digestion with alkali hydrogen carbonate and water. The process is exactly the same as for the decomposition of lead sulphate (§ 132, II., *b*, *β*).

bb. Silver chloride is ignited in a porcelain crucible, with 3 parts of sodium and potassium carbonate, until the mass commences to agglutinate. Upon treating with water, the metallic silver is left undissolved; the solution contains the alkali chloride, which is then treated as in I., *a.*

Silver chloride may also be readily decomposed by long digestion with pure iron (reduced by hydrogen) and dilute sulphuric acid. Zinc may be used instead of iron, but it does not answer so well. The separated metallic silver may be washed, heated with dilute sulphuric acid, washed again and weighed; it must afterwards be ascertained, however, whether it dissolves in nitric acid. The chlorine is determined in the chloride of iron or zinc as in I., *a.*

cc. Mercurous chloride is decomposed by digestion with solution of soda or potassa. The hydrochloric acid in the filtrate is determined as in I., *a.* The mercurous oxide is dissolved in nitric or nitrohydrochloric acid, and the mercury determined as directed in § 117 or § 118.

c. The soluble chlorides of the metals of the fourth, fifth, and sixth groups may generally be decomposed also by hydrogen sulphide or ammonium sulphide. The chlorine in the filtrate is determined as in § 169. It must not be omitted to test the precipitated sulphides for chlorine. Several chlorides, cadmium chloride for instance, give sulphides free from chlorine with ammonium sulphide, but not with hydrogen sulphide.

d. In many metallic chlorides, for instance in those of the first and second groups, the chlorine may be determined also by evaporating with sulphuric acid, converting the metal thus into a sulphate, which is then ignited and weighed as such; the chlorine being calculated from the loss. This method is not applicable in the case of silver chloride and lead chloride, which are only imperfectly and with difficulty decomposed by sulphuric acid; nor in the case of mercuric chloride and stannic chloride, which sulphuric acid fails almost or altogether to decompose.

Supplement.

§ 142.

DETERMINATION OF CHLORINE IN THE FREE STATE.

Chlorine in the free state may be determined both in the volumetric and in the gravimetric way. The volumetric methods, however, deserve the preference in most cases. They are very numerous.

I shall only here adduce that one which is undoubtedly the most accurate and at the same time the most convenient.*

1. *Volumetric Method.*

With Potassium Iodide (*after* BUNSEN).

Bring the chlorine, in the gaseous form or in aqueous solution, into contact with an excess of solution of potassium iodide in water. Each at. chlorine liberates 1 at. iodine, which remains dissolved in the excess of potassium iodide. By determining the liberated iodine by means of sodium thiosulphate as in § 146, you will accordingly learn the quantity of chlorine, and, in fact, with the greatest accuracy. If you have to determine the chlorine of chlorine water, measure a portion off with a pipette. So as to prevent any of the gas entering the mouth, connect the upper end of the pipette with a tube containing moist hydrate of potassa laid between cotton. When the pipette has been correctly filled allow its contents to flow, with stirring, into an excess of solution of potassium iodide (1 in 10). There is no difficulty about knowing whether the latter is sufficiently in excess, for if not, a black precipitate is formed. If the chlorine is evolved in the gaseous condition, you may employ either the apparatus given in § 130, I., *d*, *β*, or the following, which is especially suitable where the chlorine is not pure, but is mixed with other gases.

* Compare "Chlorimetry" in the Special Part.

a is a little flask, from which the chlorine is evolved by boiling the substance with hydrochloric acid, a small lump of magnesite being added; it is connected with the tube, *b*, by means of a flexible tube. The latter must be free from sulphur—should it contain sulphur it is well boiled with dilute potassa and then thoroughly washed. The thinner tube, *c*, which has been fused to the bulb of *b*, leads through the caoutchouc stopper (which has been deprived of sulphur) to the bulbed U-tube, *d*, which contains solution of potassium iodide, and which for safety is connected with the plain U-tube, *e*, also containing potassium iodide solution. Both tubes stand in a beaker

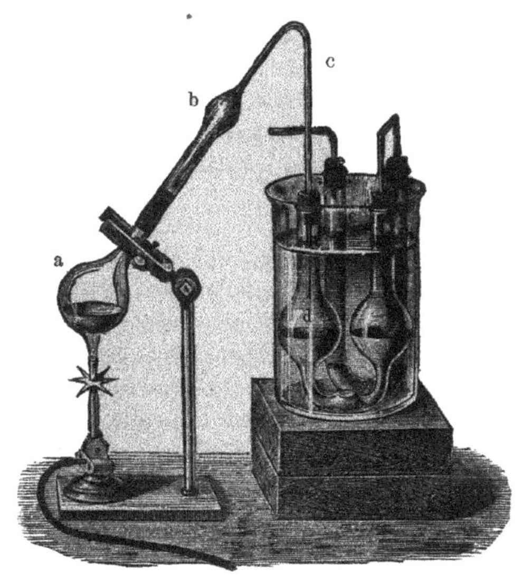

Fig. 64.

filled with water. The apparatus offers the advantages that the fluid cannot return, that the potassium iodide remains cold, and that the absorption is complete. After all the chlorine has been expelled by boiling long enough, rinse *d* and *e* out into a beaker and titrate with standard sodium thiosulphate (§ 146).

2. *Gravimetric Method.*

The fluid under examination, which must be free from sulphuric acid, say, for instance, 30 grm. chlorine water, is mixed in a stoppered bottle, with a slight excess of sodium thiosulphate, say ·5 grm., the stopper inserted, and the bottle kept for a short time in a warm place; after which the odor of chlorine is found to have gone off. The mixture is then heated to boiling with some hydrochloric acid in excess, to destroy the excess of sodium thiosulphate, filtered, and the sulphuric acid in the filtrate determined by barium chloride (§ 132). 1 mol. sulphuric acid corresponds to 4 at. chlorine (WICKE*).

In fluids containing, besides free chlorine, also hydrochloric acid, or a metallic chloride, the chlorine existing in a state of combination may be determined, in presence of the free chlorine, in the following way:

A weighed portion of the fluid is mixed with solution of sulphurous acid in excess, after some time nitric acid is added, and then potassium chromate to destroy the excess of sulphurous acid, and the whole of the chlorine is precipitated as silver chloride. The quantity of the free chlorine is then determined in another weighed portion, by means of potassium iodide; the difference gives the amount of combined chlorine.†

Having thus seen in how simple and accurate a manner the quantity of free chlorine may be determined by BUNSEN's method, it will be readily understood that all oxides and peroxides which yield chlorine when heated with hydrochloric acid, may be analyzed by heating them with concentrated hydrochloric acid, with addition of a small lump of magnesite, and determining the amount of chlorine evolved.

§ 143.

2. BROMINE.

I. *Determination.*

a. Gravimetric Methods.

Estimation as silver bromide. Free hydrobromic acid—in a solution free from hydrochloric acid or chlorides—is precipitated

---

* Annal. d, Chem. u. Pharm. 99, 99.

† If chlorine water is mixed at once with silver nitrate, $\frac{5}{6}$ only of the chlorine is obtained as silver chloride: $6Cl + 3Ag_2O = 5AgCl + AgClO_3$ (H. ROSE, WELTZIEN, Annal. d. Chem. u. Pharm. 91, 45). If chlorine water is mixed with ammonia in excess, there are formed at first ammonium chloride and ammonium hypochlorite, the latter then gradually decomposes into nitrogen and ammonium chloride; however, a little ammonium chlorate is also formed besides (SCHÖNBEIN, Journ. f. prakt. Chem. 84, 386; Zeitschrift f. analyt. Chem. 2, 59).

by silver solution, and the further process is conducted as in the case of hydrochloric acid (§ 141). For the properties of silver bromide, see § 94, 2. The results are perfectly accurate.

b. Volumetric Methods.

Like chlorine in hydrochloric acid and alkali chlorides, bromine may be estimated in the analogous compounds by *standard silver solution* (§ 141, I., *b*, *α*), by *solution of silver and iodide of starch* (§ 141, I., *b*, *β*). But these methods are seldom applicable, as they cannot be used in the presence of hydrochloric acid and metallic chlorides.

The following methods must therefore be detailed; they are especially useful for the estimation of small quantities of bromine in solutions containing chlorides, but in point of accuracy they leave much to be desired.*

α. With chlorine water and chloroform (*after* A. REIMANN†). This method depends on the facts that chlorine when added to bromides first liberates the bromine and then combines with it, and that bromine colors chloroform yellow or orange, while bromine chloride merely communicates a yellowish tinge to that fluid. The process is as follows: Mix the liquid containing a bromide of an alkali metal in neutral solution, in a stoppered bottle with a drop of pure chloroform about the size of a hazel-nut, then add standard chlorine water from a burette, protected from the light by being surrounded with black paper. On shaking, the chloroform becomes yellow, on further addition of chlorine water, orange, then yellow again, and lastly—at the moment when 2 at. chlorine have been used for 1 at. bromine—yellowish white ($KBr + 2Cl = KCl + BrCl$). Considerable practice and skill are required before the operator can tell the end-reaction. He will be assisted by placing the bottle on white paper and comparing the color of the chloroform with that of a dilute solution of yellow potassium chromate of the required color. The strength of the chlorine water should depend on the amount of the bromine to be determined. It should be so adjusted that about 100 c.c. may be used. The chlorine water is standardized with potassium iodide and sodium thiosulphate (§ 142, 1). The method is especially suited for the determination of small quantities of bromine in mother liquors, kelp, &c. The results are approximate: *e.g.*, ·0180 instead of ·0185—·055 instead

* Compare § 169. † Annal. d. Chem. u. Pharm. 115, 140.

of ·059—·0112 instead of ·0100, &c. If the fluid contains organic substances, it is—after being rendered alkaline with caustic soda—evaporated to dryness, the residue ignited in a silver dish, extracted with water, the solution neutralized exactly with hydrochloric acid, and then tested.

β. HEINE'S *colorimetric method.** The bromine is liberated by means of chlorine, and taken up with ether; the solution is compared, with respect to color, with an ethereal solution of bromine of known strength, and the quantity of bromine in it thus ascertained. FEHLING† obtained satisfactory results by this method. It will at once be seen that the amount of bromine contained in the fluid to be analyzed must be known in some measure before this method can be resorted to. As the brine examined by FEHLING could contain at the most ·02 grm. bromine in 60 grm., he prepared ten different test fluids, by adding to ten several portions of 60 grm. each of a saturated solution of common salt increasing quantities of potassium bromide, containing respectively from ·002 grm. to ·020 grm. bromine. He added an equal volume of ether to the test fluids, and then chlorine water, until there was no further darkening observed in the color of the ether. It being of the highest importance to hit this point exactly, since too little as well as too much chlorine makes the color appear lighter, FEHLING prepared three samples of each test fluid, and then chose the darkest of them for the comparison. 60 grm. are now taken‡ of the mother liquor to be examined, the same volume of ether added as was added to the test fluids, and then chlorine water. Every experiment is repeated several times. Direct sunlight must be avoided, and the operation conducted with proper expedition. In my opinion it is well to replace the ether by chloroform or carbon bisulphide. CAIGNET§ substituted sodium hypochlorite for the chlorine water, and removed the colored carbon bisulphide from time to time.

II. *Separation of Bromine from the Metals.*

The metallic bromides are analyzed exactly like the corresponding chlorides (§ 141, II., *a* to *d*), the whole of these methods being

---

* Journ. f. prakt. Chem. 36, 184. Proposed to effect the determination of bromine in mother liquors.

† Journ. f prakt. Chem. 45, 269.

‡ The best way is to take them by measure.

§ Zeitschr. f. anal. Chem. 9, 427.

applicable to bromides as well as chlorides. In the decomposition of bromides by sulphuric acid (§ 141, II., *d*), porcelain crucibles must be used instead of platinum ones, as the latter would be attacked by the liberated bromine. Some bromides, it must be remembered, are not completely decomposed by sulphuric acid; for instance, mercuric bromide is not. The soluble bromides may be converted into chlorides by evaporation with hydrochloric acid and excess of chlorine water; but this process cannot be applied where the chloride is liable to be carried away with the steam; for instance, in the case of mercuric bromide.

Supplement.

§ 144.

DETERMINATION OF FREE BROMINE.

Free bromine in aqueous solution, or evolved in the gaseous form, is caused to act on excess of solution of potassium iodide. Each at. bromine liberates 1 at. iodine, which is most conveniently determined by means of sodium thiosulphate (§ 146). As regards the best mode of bringing about the action of the bromine on the potassium iodide, compare § 142, 1.

The determination of free bromine in presence of hydrobromic acid or metallic bromides is effected in the same manner as that of free chlorine in presence of hydrochloric acid (see § 142).

§ 145.

3. IODINE.

I. *Determination.**

a. Gravimetric Methods.

α. Estimation as silver iodide. If you have hydriodic acid in solution, free from hydrochloric and hydrobromic acids, precipitate with silver nitrate, and proceed exactly as with hydrochloric acid (§ 141). If the solution is colored with free iodine, first add sulphurous acid cautiously till the color is removed. The particles of silver iodide adhering to the filter are not reduced on incineration, but a little of the iodide is liable to volatilize if the heat is

* For the methods to be adopted in the presence of bromine and chlorine, see § 169.

too high. Hence the filter should be got as clean as possible, and the heat during incineration should not be unduly raised. For the properties of silver iodide, see § 94, 3. The results are perfectly accurate.

β. Estimation as palladious iodide. The following method, recommended first by LASSAIGNE, is resorted to exclusively to effect the separation of iodine from chlorine and bromine, for which purpose it is extremely well adapted. The solution may not contain any alcohol. Acidify it slightly with hydrochloric acid, and add a solution of palladious chloride, as long as a precipitate forms; let the mixture stand from 24 to 48 hours in a warm place, filter the brownish-black precipitate off on a weighed filter, wash with warm water, and dry at 100°, until the weight remains constant. For the properties of the precipitate, see § 94, 3. This method gives very accurate results. Instead of simply drying the palladious iodide, and weighing it in that form, you may ignite it in a current of hydrogen in a crucible of porcelain or platinum,* and calculate the iodine from the residuary palladium (H. ROSE). Compare § 122, 1.

b. Volumetric Methods.

α. The methods given for hydrochloric acid by *precipitating with silver solution* (§ 141, I., *b*, *α*), and by *silver solution and iodide of starch* (§ 141, I., *b*, *β*), may be used for hydriodic acid and alkali iodides; the absence of chlorine and bromine being of course presupposed.

β. With nitrous acid and carbon disulphide. This excellent method has been in frequent use in my laboratory for a length of time; it may be used for small or large quantities of iodine. We require:

aa. Solution of potassium iodide of known strength. Made by drying the pure salt at 180° (see p. 121) and dissolving an exactly weighed quantity (about 5 grm.) to 1 litre.

bb. Solution of sodium thiosulphate containing about 13 or 13·5 grm. of the pure crystallized salt in 1 litre.

cc. Solution of nitrous acid in sulphuric acid. Prepared by passing nitrous acid gas into sulphuric acid to saturation.

dd. Pure carbon disulphide.

ee. Solution of sodium hydrogen carbonate. Made by dissolv-

---

* This substance is not injured by the operation.

ing 5 grm. in 1000 c.c. cold water and adding 1 c.c. of hydrochloric acid to the solution.

Begin by standardizing the thiosulphate as follows: Take a well-stoppered bottle of about 400 c.c. capacity, transfer to it 50 c.c. of the potassium iodide solution, add about 150 c.c. water, 20 c.c. carbon disulphide, some dilute sulphuric acid, and 10 drops of the solution of nitrous acid in sulphuric acid. Insert the stopper and shake the bottle violently for some time, allow to settle, and ascertain by adding a few more drops of the nitrous acid that the whole of the iodine has been liberated. Shake again, allow to settle, and pour the supernatant fluid as completely as possible into a flask, leaving the carbon disulphide in the bottle, add 200 c.c. water to the latter, shake well, pour off the water into the flask and repeat the washing till the last water has no acid reaction. To the contents of the flask add 10 c.c. carbon disulphide, shake well, pour off into a second flask, wash the disulphide a little, and finally shake the contents of the second flask again with some fresh disulphide, which should now be barely tinged. Collect the disulphide from both flasks on a filter moistened with water, wash it till the washings are no longer acid, place the funnel in the bottle and pierce the point of the filter so that the disulphide from all the operations may be mixed. Add 30 c.c. of the sodium hydrogen carbonate and then the thiosulphate from a burette, with continual shaking, till the disulphide has lost its color. The number of c.c. of thiosulphate used will correspond to the iodine in 50 c.c. of potassium iodide solution.

The analysis is performed exactly as above. The thiosulphate requires to be standardized before every fresh series of experiments, as it is liable to slight alteration. The presence of chlorides has no influence whatever on the results. In determining minute quantities of iodine let the solutions be ten times weaker, and use smaller quantities and smaller vessels.

The results are entirely concordant and exact.

γ. *By distillation with ferric chloride* (DUFLOS). When hydriodic acid or a metallic iodide is heated in a retort with solution of pure ferric chloride, the whole of the iodine escapes with the aqueous vapor, and ferrous chloride is formed ($Fe_2Cl_6 + 2HI = 2FeCl_2 + 2HCl + 2I$). The iodine passing over is received in solution of potassium iodide and determined by sodium thiosulphate, as directed § 146. In employing this method it must be borne in mind that the ferric chloride must be free from chlorine and nitric acid. It is best to prepare it from ferric oxide and hydrochloric acid. We must not forget too that the separated iodine is liable to act on cork and caoutchouc; the apparatus should therefore be constructed according to fig. 64, p. 435.

δ. H. STRUVE's *colorimetric method* may be used in many cases. In this method the amount of iodine is estimated by the depth of color which the separated iodine gives to a measured quantity of carbon disulphide.

II. *Separation of Iodine from the Metals.*

The metallic iodides are in general analyzed like the corresponding chlorides. From iodides of the alkali metals containing free alkali the iodine may be precipitated as silver iodide, by first saturating the free alkali almost completely with nitric acid, then adding solution of silver nitrate in excess, and finally nitric acid to strongly acid reaction. If an excess of acid were added at the beginning, free iodine might separate, which is not converted completely into silver iodide by solution of silver nitrate. In compounds soluble in water the iodine may generally be precipitated as palladious iodide; you may also determine the base in one portion (decomposing the compound with concentrated sulphuric acid) and the iodine in another portion according to I., *b*, γ.

Iodine cannot be separated from platinum directly with silver nitrate, as insoluble platinum salts would be thrown down with the silver iodide. For this purpose H. TOPSÖE* recommends the following process: Dissolve the substance in a good amount of water, add solution of sodium hydrogen sulphite and sulphurous acid, heat on a water-bath till the color has entirely disappeared, and the platinum is consequently converted into platinous sulphite. In this operation a white flocculent precipitate of sodium platinous sulphite which is difficultly soluble separates; it redissolves on addition of sulphurous acid. After heating on the water-bath for some time, allow to cool completely, precipitate with silver solution, which should not be added in large excess, add nitric acid, heat for about an hour to redissolve the silver sulphite first thrown down with the iodide, and then filter off the latter. Occasionally it is to be preferred to add sulphurous acid instead of the sulphite, and then, when the fluid has been heated and the color has gone,

* Zeitschr. f. anal. Chem. 9, 30.

to add an excess of ammonia. In this way the platinum compound is not thrown down, and the silver sulphite does not separate after the addition of silver solution till nitric acid is added, and is immediately redissolved by the excess of the same.

For the analysis of insoluble iodides, especially silver and lead iodides, mercurous and cuprous iodides, E. MEUSEL* strongly recommends sodium thiosulphate, in which these salts dissolve. Very little water should be used, and as small a quantity of the thiosulphate as possible. The metal is precipitated from the solution by ammonium sulphide in the form of sulphide. Evaporate the filtrate with soda, and heat the residue in a platinum dish to incipient redness to destroy sodium thiosulphate and tetrathionate. Dissolve the fusion in water by the aid of heat, and determine the iodine in it by I., *b*, γ. A large quantity of ferric chloride will be required to decompose the sodium sulphite; the residue in the retort should have a deep reddish-brown color.

Silver iodide may be decomposed also by fusing with sodium carbonate, but not by igniting in a current of hydrogen, and not completely by zinc or iron. Mercurous iodide may be easily decomposed by distilling with 8 or 10 parts of a mixture of 1 part potassium cyanide and 2 parts quicklime. For the apparatus, see fig. 54, p. 307; *ab* is filled with magnesite (H. ROSE†). Palladious iodide may be decomposed by igniting in hydrogen. Cuprous iodide and many other iodides may be decomposed by boiling with potassium or sodium carbonate. Portions of metal, which may pass into the alkaline solution, may be thrown down by ammonium sulphide, or by acidifying with acetic acid, and passing hydrogen sulphide.

Supplement.

§ 146.

DETERMINATION OF FREE IODINE.

The determination of free iodine is an operation of great importance in analytical chemistry, since, as BUNSEN‡ first pointed out, it is a means for the estimation of all those substances which, when brought in contact with potassium iodide, separate from the same a definite quantity of iodine (*e.g.*, chlorine, bromine, &c.), or, when

* Zeitschr. f. anal. Chem. 9, 208. † *Ib.* 2, 1.
‡ Annal. d. Chem. u. Pharm. 86, 265.

boiled with hydrochloric acid, yield a definite quantity of chlorine (*e.g.*, chromic acid, peroxide of manganese, &c.). By causing the chlorine produced to act on potassium iodide, we obtain the equivalent quantity of free iodine.

Of the various methods which have been proposed for the estimation of free iodine, the oldest is that of SCHWARZ.* It is based upon the following reaction: $2Na_2S_2O_3 + 2I = 2NaI + Na_2S_4O_6$. 24·8 grm. pure crystallized sodium thiosulphate are dissolved to 1 litre. 1000 c.c. of the solution correspond to 12·685, *i.e.*, to 1 at. iodine. This solution is added to the solution of the substance in potassium iodide until the fluid appears bright yellow, 3 or 4 c.c. thin and very clear starch-paste are then added, which must produce blue coloration, and finally again sodium thiosulphate, until the blue fluid is decolorized.

This method, though in itself excellent, is open to this objection, that it is difficult to obtain a solution of absolutely exact value by weighing off sodium thiosulphate, as the salt is not readily procurable in a perfectly pure and dry condition, and although the solution does not change rapidly or to any great extent, it is still liable to gradual alteration, especially under the influence of light.

BUNSEN'S researches on the volumetric estimation of iodine cited above produced a very important and beneficial effect on the whole domain of chemical analysis. His process depends on the fact that when iodine comes in contact with an aqueous solution of sulphurous acid, a decomposition takes place in accordance with the equation $H_2SO_3 + H_2O + 2I = H_2SO_4 + 2(HI)$, provided the solution does not contain more than ·04 to ·05 per cent. of anhydrous sulphurous acid. If the solution is more concentrated, another reaction also takes place to a greater or less extent—namely, $H_2SO_4 + 2HI = H_2SO_3 + H_2O + 2I$.

In this method, a solution of iodine in potassium iodide containing a known quantity of free iodine is employed, and we commence by determining the relation between it and a sufficiently dilute solution of sulphurous acid. In applying the method, the iodine to be estimated is dissolved in potassium iodide, the standard sulphurous acid is added to decoloration, then thin starch-paste, and finally standard iodine solution till the blue color of iodide of starch is just visible.

* Anleit. zu Maassanal. Nachträge, 1853, 22.

We calculate now the c.c. of iodine solution which correspond to the sulphurous acid employed, and deduct therefrom the c.c. of iodine added to destroy the excess of sulphurous acid. The remainder gives the number of c.c. of iodine solution which contain a quantity of iodine equal to that in the substance analyzed.

On account of the rapidity with which solution of sulphurous acid changes, this method is somewhat inconvenient, and has given place to the following, which is now universally employed. It retains the basis of BUNSEN's method, but substitutes sodium thiosulphate for sulphurous acid, employing the reaction of SCHWARZ's method. With F. MOHR* I give this "combined method" the preference, because, first, we are not bound to a definite strength of the thiosulphate; secondly, the solution of thiosulphate is far less affected by the oxygen of the air than sulphurous acid; and thirdly, it loses nothing by evaporation. FINKENER† even says, that the use of thiosulphate makes the method more accurate, his experiments having shown that in using BUNSEN's method the results differ; if, on one occasion, we add the sulphurous acid to the iodine, and, on another, the iodine to the sulphurous acid.

a. REQUISITES FOR THE COMBINED METHOD.

α. Iodine solution of known strength. Dissolve 6·2 to 6·3 grm. iodine with the aid of about 9 grm. potassium iodide (free from iodic acid) to about 1200 c.c.

β. Solution of sodium thiosulphate. Dissolve 12·2 to 12·3 grm. of the pure and dry salt to about 1200 c.c.

γ. Solution of potassium iodide. Dissolve 1 part of the salt (free from iodic acid) in about 10 parts of water. The solution must be colorless and must remain so immediately after the addition of dilute sulphuric or hydrochloric acid (either must be iron-free).

δ. Starch solution. Stir the purest starch powder gradually with about 100 parts cold water and heat to boiling with constant stirring. Allow to cool quietly, and pour off the fluid from any deposit. The solution should be almost clear and free from all lumps. The starch solution is best prepared fresh before each series of experiments.

---

* Lehrb. d. chem.-analyt. Titrirmethode, 3 Aufl. 256.

† H. ROSE, Handb. d. anal. Chem. 6 Aufl. von FINKENER, 2, 937.

b. PRELIMINARY DETERMINATIONS.

α. Determination of the relation between the Iodine Solution and Thiosulphate Solution.

Fill two burettes with the solutions. Run 20 c.c. of the thiosulphate into a beaker, add some water and 3 or 4 c.c. starch solution, then add the iodine till a blue coloration is just produced. If you have added a drop too much, run in one or two drops more of the thiosulphate, and then more cautiously the iodine solution. After a few minutes read off the height of the fluid in both burettes. Suppose we had used 20 c.c. thiosulphate to 20·2 c.c. iodine.

β. Exact Determination of the Iodine in the Solution.

This is done immediately before each series of analyses with the aid of an exactly weighed quantity of pure and dry iodine. Experience has convinced me that solution of iodine in potassium iodide, even when kept cool and in the dark, is much more liable to change than is usually supposed.*

The process is conducted in the following manner: Select three watch-glasses, *a*, *b*, and *c*, which fit each other; weigh *b* and *c* together accurately. Put about 0·5 grm. pure dry iodine (prepared according to § 65, 6) into *a*, place it on an iron plate, heat gently, till dense fumes of iodine escape. Now cover it with *b* and regulate the heat so that the iodine may sublime entirely or almost entirely into *b*. Next remove *b* while still hot, and give it a gentle swing in the air to remove the still uncondensed iodine fumes and any traces of aqueous vapor, cover it with *c*, allow to cool under the desiccator, weigh and transfer the two watch-glasses together with the weighed iodine to a capacious beaker, containing a sufficient quantity of potassium iodide solution to dissolve the whole of the iodine to a clear fluid. Add water and then thiosulphate from a burette till the color is gone; now add 3 or 4 c.c. of starch-paste and iodine solution (*a*, *α*) from a second burette till a blue tinge just appears. Having read off both burettes, the following simple calculation will give you the iodine in the solution *a*, *α*:

Suppose we had weighed off ·150 grm. iodine, and used 29·5 c.c. thiosulphate and ·3 c.c. iodine solution.

---

* I filled several small well-stoppered bottles with some solution of iodine in potassium iodide, whose standard had been accurately determined, and placed them in a cellar. Even in the course of a few weeks the standard had altered. I now never rely on the strength of a solution of iodine, unless I have determined it shortly before.

From *b*, *α*, we know that 20 c.c. thiosulphate correspond to 20·2 c.c. iodine solution; 29·5 c.c. therefore correspond to 29·8 c.c.

Now 29·5 c.c. thiosulphate correspond to ·150 grm. iodine + ·3 c.c. iodine solution.

But 29·5 c.c. thiosulphate also correspond to 29·8 c.c. iodine solution.

∴ ·150 grm. iodine + ·3 c.c. iodine solution = 29·8 c.c. iodine solution.

∴ ·150 grm. iodine ≐ 29·5 c.c. iodine solution.

∴ 1 c.c. iodine solution = ·0050847 grm. iodine.

The experiment just described is repeated and the mean of the two results taken, provided they exhibit sufficient uniformity.

γ. Dilution of the standard fluids to a convenient strength.

With the aid of the iodine solution the strength of which we now know exactly, and the solution of sodium thiosulphate which stands in a known relation to the same, we might make any determinations of iodine. The calculation, although in principle extremely simple, is yet somewhat hampered by reason of the long decimal which expresses the quantity of iodine in 1 c.c. of the solution. It is therefore convenient to dilute the iodine solution so that 1 c.c. may exactly contain ·005 grm. iodine. This is done by filling a litre flask therewith, and adding the necessary quantity of water; in our case 16·94 c.c., for 5 : 5·0847 :: 1000 : 1016·94. If the litre flask will hold above the mark this 16·94 c.c., it is simply added, otherwise it is put into the dry bottle destined to receive the iodine solution, the iodine solution added, the whole shaken together, a portion of the fluid returned to the flask, shaken, poured back into the bottle, and the whole shaken again.

The solution of thiosulphate may now be diluted in a corresponding manner. In our case we should have had to add 27·11 c.c. water to 1000 c.c. of the solution, as will be seen from the following consideration:

20·2 c.c. of the original iodine solution correspond to 20 c.c. of the thiosulphate solution.

∴ 1000 c.c. correspond to 990·1 c.c.

Now these 1000 c.c. were made up to 1016·94 by addition of water; if therefore we make up 990·1 c.c. of the sodium thiosulphate to the same bulk by addition of water we shall have equivalent solutions. Hence, to 990·1 c.c. we must add 26·84 c.c. water, or to 1000 c.c. 27·11 water.

In such cases of dilution I always prefer to take exactly 1 litre instead of an uneven number of c.c., as in measuring the latter errors and inaccuracies may readily occur; I have therefore above recommended the preparation of 1200 c.c. of the fluids, so that after their determination 1000 c.c. may be sure to remain.

c. THE ACTUAL ANALYSIS.

Weigh the iodine to be determined in a glass-stoppered tube, dissolve in potassium iodide solution as in *b*, *β*, add thiosulphate solution from the burette till decoloration is just produced, then 3 or 4 c.c. starch solution, then iodine solution from a second burette to incipient blueness. The substance contains the same amount of iodine as the c.c. of iodine solution corresponding to the thiosulphate used *minus* the c.c. of the former used to destroy the excess of the latter. Where the solutions are of equal value and 1 c.c. corresponds to ·005 grm. iodine, the calculation is in the highest degree simple; for suppose we had used 21 c.c. $Na_2S_2O_3$ and 1 c.c. iodine, the quantity of iodine present is ·100 grm.

$$21 - 1 = 20, \text{ and } 20 \times \cdot005 = \cdot100\cdot$$

Where you are analyzing chromic acid or manganese dioxide by boiling with hydrochloric acid, and passing the chlorine evolved into potassium iodide, you must allow the solution to cool before titrating with thiosulphate; for at a high temperature a portion of the sodium tetrathionate produced is converted into sodium sulphate by the iodine (WRIGHT*).

Free acid in the iodine solution to be estimated is not injurious; when such is present, however, the excess of the thiosulphate must be titrated without delay, or the free thiosulphuric acid may be decomposed before the iodine is added.

d. KEEPING OF THE SOLUTIONS.

The iodine solution and the thiosulphate solution are kept in glass-stoppered bottles in a cool, dark place. But the relation between the two solutions must be tested before each new series of experiments, and the iodine in the iodine solution must be redetermined.

If a fluid contains free iodine in presence of iodine in combination, determine the former in one portion by the combined method, and the total quantity in another portion. For this purpose you

* Zeitschr. f. anal. Chem. 9, 482.

may either (1) add sulphurous acid to decoloration, precipitate with silver nitrate (§ 145, I., *a*, *α*), digest the precipitate with nitric acid to remove any silver sulphite which it may contain, filter, &c.; or (2) distil with ferric chloride as directed, § 145, I., *b*, *γ*.

§ 147.

4. Cyanogen.*

I. *Determination.*

a. Gravimetric Estimation.—If you have free hydrocyanic acid in solution run it into an excess of solution of silver nitrate, add a little nitric acid, allow to settle without warming, and determine the precipitated silver cyanide either by collecting on a weighed filter, drying at 100° and weighing (§ 115, 3), or by collecting on an unweighed filter and converting into metallic silver. The latter operation is performed by igniting the precipitate in a porcelain crucible for ¼ hour, or till it ceases to lose weight (H. Rose). If you wish to determine in this way the hydrocyanic acid in bitter almond water or cherry laurel water, add ammonia after the addition of the solution of silver nitrate till the fluid is strongly alkaline (it is not necessary to dissolve all the silver cyanide), and *at once* acidify with nitric acid. When the precipitate has settled, filter. The whole of the cyanogen in the fluid will have been now converted into silver cyanide. (The cyanogen was originally present partly as hydrocyanic acid, partly as ammonium cyanide, but principally as hydrocyanate of benzaldehyd—S. Feldhaus.†)

Feldhaus recommends the following proportions: 100 grm. bitter almond water, about 1·2 grm. silver nitrate, dissolved in water and 2 to 3 c.c. ammonia sp. gr. ·96· A portion of the filtrate should be tested to make sure that it contains silver salt in excess, another portion should be tested by making it strongly alkaline with ammonia, and then acid again with nitric acid. If a precipitate is formed in the latter case it shows that the whole of the hydrocyanate of benzaldehyd was not decomposed, and the precipitation must be repeated. If you want to measure off a fluid containing hydrocyanic acid with a pipette, insert a little tube with

* With regard to Herapath's colorimetric method, which is founded on the intensity of the color of a solution of persulphocyanide of iron, compare Chem. Gaz. Aug. 1853, 294.

† Zeitschr. f. anal. Chem. 3, 34.

soda-lime between the pipette and the flexible tube which you put into your mouth.

b. LIEBIG'S *Volumetric Method**.—If hydrocyanic acid is mixed with potassa to strong alkaline reaction, and a dilute solution of silver nitrate is then added, a permanent turbidity of silver cyanide —or, if a few drops of solution of sodium chloride have been added, of silver chloride—forms only after the whole of the cyanogen is converted into double cyanide of silver and potassium. The first drop of solution of silver nitrate added in excess produces the permanent precipitate. 1 at. silver consumed in the process corresponds, therefore, exactly to 2 mol. hydrocyanic acid ($2KCy + AgNO_3 = AgCy.KCy + KNO_3$). A decinormal solution of silver nitrate, containing consequently 10·793 grm. silver in the litre, should be used; 1 c.c. of this solution corresponds to ·005408 of hydrocyanic acid. In examining medicinal hydrocyanic acid, 5 to 10 grm. ought to be used, but of bitter almond water about 50 grm.; if exactly 5·408 or 54·08 grm. are used, the number of c.c. of the silver solution, divided by 10, or by 100, expresses exactly the percentage of hydrocyanic acid. Medicinal hydrocyanic acid is suitably diluted first by adding from 5 to 8 volumes of water; bitter almond water also is slightly diluted; if the latter is turbid the end-reaction will not be sufficiently distinct, and the gravimetric method is to be preferred.

LIEBIG has examined hydrocyanic acid of various degrees of dilution, and has obtained results by this method corresponding exactly with those obtained by *a.* SOUCHAY,† too, obtained results almost identical; with pure dilute hydrocyanic acid, the gravimetric results were to the volumetric as 100 to 100·5—101; with clear or nearly clear bitter almond water as 100 to 102. FELDHAUS (*loc. cit.*) obtained very nearly similar results. The slightly higher results of the volumetric process are to be explained from the fact that a small excess of silver solution is necessary to produce the final reaction. The less the amount of the substance taken the greater importance does this error assume. We should also notice that in the bitter almond water, which contains ammonium cyanide, some ammonia is set free which has a solvent action on the silver cyanide. In this method it does not matter whether the hydrocyanic acid

* Annal. d. Chem. u. Pharm. 77, 102.
† Zeitschr. f. anal. Chem. 2, 180.

contains an admixture of hydrochloric acid or formic acid. A considerable excess of potassa must be avoided.

If it is intended to determine potassium cyanide by this method, a solution of that salt must be prepared of known strength, and a measured quantity used containing about ·1 grm. of the salt. Should it contain potassium sulphide, a small quantity of freshly precipitated lead carbonate must be first added, and the solution filtered before proceeding to the determination.

II. *Separation of Cyanogen from the Metals.*

a. In Cyanides of the Alkali Metals.

Mix the substance (if solid, without previous solution in water) with excess of silver nitrate solution, then add water, finally nitric acid in slight excess, allow to settle without warming, and determine the silver cyanide as in I., *a*. The basic metals are determined in the filtrate after separating the excess of silver.

b. In Cyanides and double Cyanides, which are completely decomposed by Silver Nitrate and Nitric Acid or Silver Nitrate and Ammonia.

Digest for some time with a dilute solution of silver nitrate, stirring frequently,* then add nitric acid in moderate excess, and digest at a gentle heat, till the foreign cyanide is fully dissolved and the silver cyanide has become pure and quite white. Then add water and filter. As a precautionary measure it is well to test the metal obtained by long ignition of the silver cyanide, whether it is free from those metals which were combined with the cyanogen. The filtrate is used for estimating the basic metals, the silver being first precipitated with hydrochloric acid. This method affords us an exact analysis of the double cyanides of potassium with nickel, copper, and zinc (H. Rose).

W. Weith† recommends a solution of silver nitrate in ammonia for the decomposition of many cyanogen compounds, such as potassium ferrocyanide, Prussian blue, and even potassium cobalticyanide. He digests them in sealed tubes at 100° (in the case of potassium cobalticyanide, 150°) for 4 or 5 hours. Warm the contents of the tube gently in a dish, until the crystals of ammonio-cyanide of silver are dissolved, filter off the separated metallic

* Double cyanide of nickel and potassium yields by this process a mixture of silver cyanide with nickel cyanide. Like double cyanides are similarly decomposed.

† Zeitschr. f. anal. Chem. 9, 379.

oxide, wash it with ammonia, dilute, and precipitate the silver cyanide by acidifying with nitric acid. In the filtrate separate the silver from the alkalies, &c. In respect to the undissolved oxides it should be noted that metallic silver is always mixed with the ferric oxide.

c. In Mercuric Cyanide.

Precipitate the aqueous solution with hydrogen sulphide; the mercuric sulphide may be filtered without difficulty if a little ammonia or hydrochloric acid be added; it is determined according to § 118, 3. If the compound is in the solid condition, the cyanogen may be determined in another portion by ignition with cupric oxide, the nitrogen and carbonic acid being collected and separated (comp. Organic Analysis).

H. ROSE and FINKENER* have, after much trouble, succeeded in finding out a method for determining cyanogen with precision also in solutions of mercuric cyanide. Mix the solution of the mercuric cyanide with zinc nitrate dissolved in ammonia. To 1 part of mercuric salt you may add about 2 parts of the zinc-salt. Add to the clear solution hydrogen sulphide water gradually till it produces a perfectly white precipitate of zinc sulphide. The precipitate, which is a mixture of the mercuric and zinc sulphides, settles well. After a quarter of an hour filter it off and wash with very dilute ammonia. The filtrate contains zinc cyanide dissolved in ammonia, together with ammonium nitrate. It does not smell of hydrocyanic acid, and consequently no escape of the latter takes place. Mix it with silver nitrate and then add dilute sulphuric acid in excess. The silver cyanide is next washed a little by decantation, then—to free it from any zinc cyanide simultaneously precipitated—heated with a solution of silver nitrate, finally filtered off, washed, and determined after I., *a*. The precipitated sulphides may be dissolved in aqua regia, and the mercury precipitated as mercurous chloride according to § 118, 2. The test-analyses communicated by ROSE yielded excellent results.

d. In compounds decomposable by Mercuric Oxide in the Wet Way.

Many simple cyanides, and also double cyanides—both of the character of the double cyanide of nickel and potassium, and of the ferro- or ferricyanides (not, however, cobalticyanides)—may, as

* Zeitschr. f. anal. Chem. 1, 288.

is well known, be completely decomposed by boiling with excess of mercuric oxide and water, all cyanogen being obtained as mercuric cyanide and the metals passing into oxides.

H. ROSE (*loc. cit.*) has shown that Prussian blue, potassium ferro- and ferricyanide, more particularly, may be readily analyzed in this manner.

Boil a few minutes with water and *excess* of mercuric oxide till complete decomposition is effected, add—in order to render the ferric hydroxide and mercuric oxide removable by filtration—nitric acid in small portions, till the alkaline reaction has nearly disappeared, filter, wash with hot water, dry the precipitate, ignite—very gradually raising the heat—under a hood (with a good draught), and weigh the ferric oxide remaining. In the filtrate the cyanogen is determined according to *c*, and any potassium that may be present is determined in the filtrate from the silver cyanide.

e. Determination of Metals contained in Cyanides with decomposition and volatilization of the Cyanogen.

Of the various means for completely decomposing compounds of cyanogen, especially also the double cyanides, according to H. ROSE (*loc. cit.*) three particularly are worthy of recommendation—viz., concentrated sulphuric acid, mercuric sulphate, and ammonium chloride. The nitrates seemed decidedly less suitable on account of their too violent action.

α. DECOMPOSITION BY SULPHURIC ACID. All cyanogen compounds, simple or double, are completely decomposed and converted into sulphates or oxides, as the case may be, if treated in a powdered condition in a platinum dish or a capacious platinum crucible with a mixture of about 3 parts concentrated sulphuric acid and 1 part water, and heated till almost all the sulphuric acid had been expelled. The residual mass is then free from cyanogen. It is dissolved in water, if necessary with addition of hydrochloric acid, and the metals determined by the usual methods. This way is not adapted for mercuric cyanide, as a little of the metal would escape with the fumes of the sulphuric acid.

β. DECOMPOSITION BY MERCURIC SULPHATE. Of the mercuric sulphates, those suitable to our present purpose are the normal and the basic (Turpeth mineral). The substance is mixed with 6 parts of the latter, heated in a platinum crucible gradually, and finally maintained for a long time at a red-heat, till all the mercury has volatilized, and the weight of the crucible remains constant. If alkalies are present, a little ammonium carbonate is added during the final ignition, from time to time, in order to convert the acid sulphates into normal. The residue may usually be analyzed by simple treatment with water; in the case of potassium ferrocyanide, for instance, the potassium sulphate dissolves, and pure (alkali-free) ferric oxide remains behind. The test-analyses that have been communicated yielded excellent results.

γ. Decomposition by Ammonium Chloride. Mix the substance with twice or thrice the amount of this salt, and ignite the mixture moderately in a stream of hydrogen (apparatus, p. 251, fig. 50). From the cooled mass water extracts alkaline chloride, while the reducible metals remain in the metallic state. The method is peculiarly adapted for the analysis of double cyanide of nickel and potassium and cobalticyanide of potassium, not so for iron compounds, since the iron obtained is not pure, but contains carbon.

If one of the methods described in *e* is employed, the nitrogen and carbon (the cyanogen) must be determined by a combustion, if an estimation by the loss is not sufficient.

f. Determination of the Alkalies, especially of Ammonia in Soluble Ferrocyanides.

Mix the boiling solution with a solution of cupric chloride, in moderate excess, filter off the precipitated cupric ferrocyanide, free the filtrate from copper by means of hydrogen sulphide, and then determine the alkalies (Reindel*).

g. Volumetric Determination of Ferro- and Ferricyanogen.

α. After E. de Haen. This method, devised in my laboratory, is founded upon the simple fact that a solution of potassium ferrocyanide acidified with sulphuric acid (and which may accordingly be assumed to contain free hydroferrocyanic acid) is by addition of potassium permanganate converted into the corresponding ferricyanide. If this conversion is effected in a very dilute fluid, containing about ·2 grm. potassium ferrocyanide in from 100 to 200 c.c., the termination of the reaction is clearly and unmistakably indicated by the change of the originally pure yellow color of the fluid to reddish-yellow.†

---

* Journ. f. prakt. Chem. 65, 452.

† Instead of the permanganate you may use chromate of potash. The solution is added till spots of sesquichloride of iron on a plate are no longer colored blue or green, but brownish. E. Meyer, Zeitschr. f. anal. Chem. 8, 508.

The process requires two test-fluids of known strength, viz.:

1. A solution of pure potassium ferrocyanide.
2. A solution of potassium permanganate.

The *former* is prepared by dissolving 20 grm. perfectly pure and dry crystallized potassium ferrocyanide in water to 1 litre; each c.c. therefore contains 20 mgrm. The *latter* is diluted so that somewhat less than a buretteful is required for 10 c.c. of the solution of potassium ferrocyanide.

To determine the strength of the potassium permanganate solution in its action upon the potassium ferrocyanide, measure off, by means of a pipette, 10 c.c. of the solution of potassium ferrocyanide (containing ·2 grm.), dilute with 100 to 200 c.c. water, acidify with sulphuric acid, place the glass on a sheet of white paper, and allow the permanganate to drop into the fluid, stirring it at the same time, until the change from yellow to *reddish*-yellow indicates that the conversion is complete.* Repetitions of the experiment always give very accurately corresponding results. If at any time you have reason to suspect that the permanganate has suffered alteration, recourse must be had again to this experiment. If after acidifying the potassium ferrocyanide with sulphuric acid you add a trace of ferric chloride to produce a bluish-green color, the latter will disappear at the end of the reaction, which is thus rendered very distinct (GINTL†).

To determine the amount of real potassium ferrocyanide contained in any given sample of the commercial article, dissolve 5 grm. to 250 c.c.; take 10 c.c. of this solution, and examine as just directed. Suppose, in determining the strength of the permanganate, you have used 20 c.c., and you find now that 19 c.c. is sufficient, the simple rule-of-three sum,

$$20 : \cdot 2 :: 19 : x$$

will inform you how much pure potassium ferrocyanide ·2 grm. of the analyzed salt contains. And even this small calculation may be dispensed with, by diluting the permanganate so that exactly 50 c.c. correspond to ·2 of potassium ferrocyanide, as, in that case,

* If you wish at first for some additional evidence besides the change of color, add to a drop of the mixture on a plate, a drop of solution of sesquichloride of iron: if this fails to produce a blue tint, the conversion is accomplished.

† Zeitschr. f. anal. Chem. 6, 446.

the number of half-c.c. consumed expresses directly the percentage of pure ferrocyanide.

Instead of determining the strength of the permanganate by means of pure potassium ferrocyanide, which is unquestionably the best way, one of the methods given in § 112, 2, may also be employed; bearing in mind, in that case, that 2 mol. potassium ferrocyanide = 885·52, 2 at. iron = 112, and 1 mol. oxalic acid = 126 are equivalent in their action upon solution of potassium permanganate.

The analysis of soluble ferricyanides by this method is effected by reducing them to ferrocyanides, acidifying, and then proceeding in the way described. The reduction is effected as follows: Mix the weighed ferricyanide with a solution of soda or potassa in excess, boil and add concentrated solution of ferrous sulphate gradually, and in small portions, until the color of the precipitate appears black, which is a sign that protosesquioxide of iron has precipitated. Dilute now to 300 c.c., mix, filter, and proceed to determine the ferrocyanide in portions of 50 or 100 c.c. of the fluid. As the space occupied by the precipitate is not taken into account in this process, the results are not absolutely accurate; the difference is so very trifling, however, that it may safely be disregarded. GINTL (*loc. cit.*) suggests to put the neutral or alkaline fluid in a tall vessel and add a few lumps of sodium amalgam as big as peas: in ten minutes the reduction will be effected and without the aid of heat.

Insoluble ferro- or ferricyanides, decomposable by boiling solution of potassa (as are most of these compounds), are analyzed by boiling a weighed sample sufficiently long with an excess of solution of potassa (adding, in the case of ferricyanides, ferrous sulphate), and then proceeding as directed above.

β. After E. BOHLIG.*

In the case of a fluid containing potassium ferrocyanide, and also sulphocyanide (for instance, the red liquor of the prussiate works), the method given in *α* cannot be employed, as the hydrosulphocyanic acid also reduces permanganic acid. The following method—depending on the precipitation of the ferrocyanogen with solution of cupric sulphate—may then be used; it is accurate enough for technical purposes. Dissolve 10 grm. pure cupric sul-

* Polytechn. Notizblatt, 16, 81.

phate to 1 litre, also 4 grm. pure dry potassium ferrocyanide to 1 litre. Add to 50 c.c. of the latter solution (which contain ·2 grm. potassium ferrocyanide) copper solution from a burette to complete precipitation of the ferrocyanogen. In order to hit this point exactly, from time to time dip a strip of filter-paper into the brownish-red fluid which will imbibe the clear filtrate, leaving the precipitate of copper ferrocyanide behind. At first the moist strips of paper, when touched with ferric chloride, become dark blue, the reaction gradually gets weaker and weaker, and finally vanishes altogether. We now know the value of the copper solution with reference to its action on potassium ferrocyanide, and can, therefore, by its means test solutions containing unknown amounts of ferrocyanogen. If alkali sulphides are present, they are first removed by boiling with lead carbonate. After filtering off the lead sulphide, acidify with dilute sulphuric acid, and then proceed.

§ 148.

5. Sulphur.

I. *Determination.*

To determine hydrogen sulphide *in a mixture of gases* confined over mercury* it may be absorbed by a ball made of 2 parts precipitated lead phosphate and 3 parts plaster of Paris. The mixture is made into a paste with water, and pressed into a bullet mould in which the platinum wire is inserted. The mould should previously be oiled. The balls are dried at 100°, saturated with concentrated phosphoric acid, and are then ready for use (Ludwig†).

To determine sulphuretted hydrogen *dissolved in water* the following methods are in use:

a. The method of determining hydrogen sulphide volumetrically by solution of iodine, was employed first by Dupasquier; it is very convenient and accurate. That chemist used alcoholic solution of iodine. But as the action of the iodine upon the alcohol alters the composition of this solution somewhat rapidly, it is better to use a solution of iodine in potassium iodide. The decomposition is as follows:

$$H_2S + 2I = 2HI + S$$

* When this gas remains long in contact with mercury, sulphide of mercury is liable to be formed.

† Annal. d. Chem. u. Pharm. 162, 55.

2 at. I = 253·70 correspond, therefore, to 1 mol. H_2S = 34. However, this exact decomposition can be relied upon with certainty only if the amount of hydrogen sulphide in the fluid does not exceed ·04 per cent. (BUNSEN). Fluids containing a larger proportion of hydrogen sulphide must therefore first be diluted to the required degree with boiled water cooled out of the contact of air.

The iodine solution of § 146 may be used for the estimation of larger quantities of hydrogen sulphide; for weak solutions, *e.g.*, sulphuretted mineral water, it is advisable to dilute the iodine solution 5 times, so that 1 c.c. may contain ·001 grm. iodine.

The process is conducted as follows:

Measure or weigh a certain quantity of the sulphuretted water, dilute, if required, in the manner directed, add some thin starch-paste, and then solution of iodine, with constant shaking or stirring, until the permanent blue color begins to appear. The result of this experiment indicates approximately, but not with positive accuracy, the relation between the examined water and the iodine solution. Suppose you have consumed, to 220 c.c. of the sulphuretted water, 12 c.c. of a solution of iodine containing ·000918 grm. iodine in the c.c.* Introduce now into a flask nearly the quantity of iodine solution required, add the sulphuretted water in quantity either already determined, or to be determined, by weight or measure;† then to the colorless fluid add thin starch-paste, and after this iodine solution until the blue color just begins to show. By this course of proceeding, you avoid the loss of hydrogen sulphide which would otherwise be caused by evaporation and oxidation. In my analysis of the Weilbach water, 256 c.c. of the water required, in my second experiment, 16·26 c.c. of iodine solution, which, calculated to the quantity of sulphuretted water used in the first experiment, viz., 220 c.c., makes 13·9 c.c., or 1·9 c.c. more. But even now the experiment cannot yet be considered quite conclusive, when made with a solution of iodine so dilute; it being still necessary to ascertain how much iodine solution is required to impart the same blue tint to the same quantity of ordinary water mixed with starch-paste, of the same temperature,‡ and as nearly as possible in the same condition § as the analyzed sulphuretted

* The numbers here stated are those which I obtained in the analysis of the Weilbach water.

† Compare Experiment No. 82.

‡ Annal. d. Chem. u. Pharm. 102, 186.

§ In this connection I would recommend, in cases where the sulphuretted

water, and to deduct this from the quantity of iodine solution used in the second experiment. Thus in the case mentioned, I had to deduct ·5 c.c. from the 16·26 c.c. used. If the instructions here given are strictly followed, this method gives very accurate results.

b. Mix the sulphuretted fluid with an excess of solution of sodium arsenite, add hydrochloric acid, allow to deposit, and determine the arsenious sulphide as directed § 127, 4. The results are accurate unless the solution is very dilute, in which case the slight solubility of arsenious sulphide occasions loss.

c. If the hydrogen sulphide is evolved in the gaseous state, and large quantities are to be determined, the best way is to conduct it first through several bulbed U-tubes (fig. 64, p. 435), containing an alkaline solution of sodium arsenite, then through a tube connected with the exit of the last U-tube, which contains pieces of glass moistened with solution of soda; to mix the fluids afterwards, and proceed as in *b.* If, on the other hand, we have to determine small quantities of hydrogen sulphide contained in a large amount of air, etc., it is well to pass the gaseous mixture in separate small bubbles through a very dilute solution of iodine in potassium iodide, of known volume and strength, which is contained in a long glass tube fixed in an inclined position and protected against sunlight. The free iodine remaining is finally estimated by means of a solution of sodium thiosulphate (§ 146); the difference gives us the quantity of iodine which has been converted by hydrogen sulphide into hydriodic acid, and consequently corresponds to the amount of the hydrogen sulphide present. The volume of the gaseous mixture may be known by measuring the water which has escaped from the aspirator used. The arrangement of the absorption tube is the same as is figured in connection with the Determination of Carbonic Acid in Air (§ 221). The thin glass tube conducting the gas into the absorption tube, however, must not be provided with an india-rubber elongation.

From my own experiments* it appears that sulphuretted hydrogen whether in small or large quantities may be also estimated by the increase in weight of absorption tubes. We have only to take care that the mixture of gases is first thoroughly dried by passing over calcium chloride. To take up the hydrogen sulphide

water contains bicarbonate of soda, to add to the ordinary water an equal quantity of this salt, as its presence has a slight influence on the appearance of the final reaction.

* Zeitschr. f. anal. Chem. 10, 75.

we use U-tubes, five sixths filled with copper sulphate on pumice, one sixth at the exit containing calcium chloride. To prepare the pumice with copper sulphate, proceed as follows. Treat 60 grm. pumice in lumps the size of peas in a small porcelain dish with a hot concentrated solution of 30 or 35 grm. copper sulphate, dry the whole with constant stirring, place the dish in an air or oil bath of the temperature of 150° to 160°, and allow to remain therein four hours. A tube containing 14 grm. of this prepared pumice will absorb about ·2 grm. hydrogen sulphide. It is well always to employ two such tubes. If the prepared pumice is dried at a lower temperature it takes up much less of the gas, if dried at a higher temperature the gas is decomposed and sulphurous acid is formed.

Finally, small quantities of hydrogen sulphide mixed with other gases may be estimated by passing through bromine water and converting into sulphuric acid.

II. *Separation and Determination of Sulphur in Sulphides.*

A. Methods based on the Conversion of the Sulphur into Sulphuric Acid.

1. *Methods in the Dry Way.*

a. Oxidation by Alkali Nitrates (applicable to all compounds of sulphur). If the sulphides do not lose any sulphur on heating, mix the pulverized and weighed substance with 6 parts of anhydrous sodium carbonate and 4 of potassium nitrate, with the aid of a rounded glass rod, wipe the particles of the mixture which adhere to the rod carefully off against some sodium carbonate, and add this to the mixture. Heat in a platinum or porcelain crucible (which, however, is somewhat affected by the process), at a gradually increased temperature to fusion;* keep the mass in that state for some time, then allow it to cool, heat the residue with water, filter the fluid, boil the residue with a solution of pure sodium carbonate, filter, wash, remove all nitric acid from the filtrate by repeated evaporation with pure hydrochloric acid, and determine the sulphuric acid as directed in § 132. The metal, metallic oxide, or carbonate, which remains undissolved, is determined, according to circumstances, either by direct weighing or in some other suitable way. In the presence of lead, before filtering, pass carbonic

* If gas not free from sulphur is used for heating, some sulphur is likely to be absorbed—Price, Journ. Chem. Soc., (2) 2, 51. If a platinum crucible is used

acid through the solution of the fused mass, to precipitate the small quantity of that metal which has passed into the alkaline solution.

Should the sulphides, on the contrary, lose sulphur on heating, the finely powdered compound is mixed with 4 parts sodium carbonate, 8 parts nitre, and 24 parts pure and perfectly dry sodium chloride, and the process otherwise conducted as already given.

b. Oxidation by Chlorine Gas (after BERZELIUS and H. ROSE especially suitable for sulphosalts of complicated composition).

The following apparatus (fig. 65), or one of similar construction, is used; corks should be used, not india-rubber stoppers, and wherever there is an india-rubber connection, the glass tubes should be close to each other.

The flask *a* is completely filled with pieces of pyrolusite (native manganese dioxide) of the size of hazelnuts, strong hydrochloric acid is poured in till the spaces between the pieces of pyrolusite are filled up to half the height of the body of the flask. The upper layer of pyrolusite, which should be rinsed with a little water after pouring in the hydrochloric acid, serves to purify the evolved chlorine almost completely from hydrochloric acid. When the stopcock in one of the tubes provided for conducting the chlorine is closed, the chlorine passes down into the cylinder *b* filled with rather dilute soda solution, by which it is completely absorbed. When the stopcock is opened the chlorine is conducted by a tube to the bottom of *c* into a layer of concentrated sulphuric acid, which serves to indicate the rapidity of the current; *c* is moreover completely filled with fragments of pumice-stone moistened with concentrated sulphuric acid, for the purpose of drying the chlorine. The tube with the bulb *d* must be made of glass which is not too easily fusible, and must be adjusted, not horizontally, but a little inclined, so that heavy vapors may not pass back against the slow current of chlorine. The danger that vapors may pass back is further lessened by making the end of the bulb-tube at which chlorine enters no wider than is necessary for the introduction of the substance by means of a long, narrow, thin weighing tube. The part of the tube on the other side of the bulb should have a greater diameter, since it might otherwise be choked up by a sublimate, especially if the substance contains much antimony. It is narrowed at one point to facilitate subsequent fusion and drawing asunder. The downward bent end is fitted into the receiver *e* by means of a cork, or a piece of rubber tubing drawn over it. The receiver contains water or, if antimony is present in the substance, dilute hydrochloric acid to which is also added a little tartaric acid (free from sulphuric acid). The volume of liquid should be only so large as to cause the passing gas to bubble through it in the narrow spaces at each end of the lower bulb, which should be large enough to hold 25

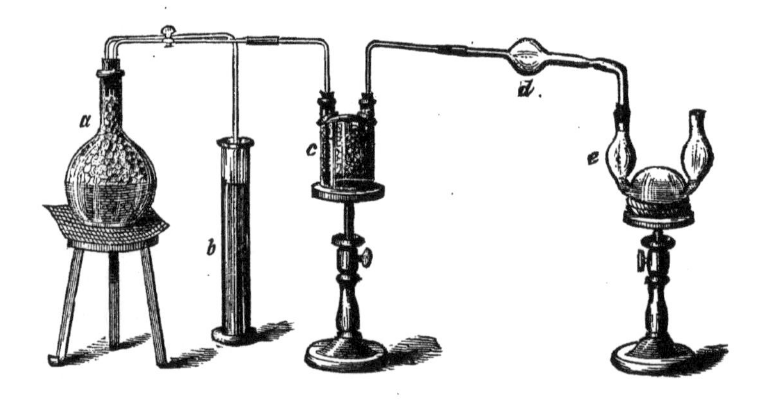

Fig. 65

to 30 c.c. when thus charged. It is well also to attach to the receiver a small U-tube charged with a small volume of the liquid absorbent in such a manner as to increase as little as possible the pressure in the interior of the apparatus. Finally, a long, light glass tube may be attached to the last U-tube for conducting the escaping chlorine into the open air or into a flue.

When the substance has been introduced into the bulb-tube, and the whole apparatus is connected, the stopcock is first closed and evolution of chlorine is produced by application of gentle heat. As soon as gas-bubbles following each other in quick succession appear in the soda solution, the heat is withdrawn. A constant evolution of chlorine will then go on for a long time without further application of heat. When the gas-bubbles are nearly completely absorbed by the soda solution, the stopcock is opened so wide that a slow current of gas enters *c* and after a while reaches the bulb *d*. If the substance is decomposed at the ordinary temperature (*e.g.*, antimony sulphide), care must be taken to diminish the rapidity of the chemical action and consequent elevation of temperature, by partial closing of the stopcock, so that sulphur chloride may not distil over into the receiver at this stage of the process. For if sulphur chloride reaches the liquid in the receiver which is not yet saturated with chlorine, it is decomposed with separation of sulphur which is afterwards not easily converted into sulphuric acid by chlorine. When the action of chlorine ceases to produce elevation of temperature or any apparent change of the substance, and the absorbing liquid has become charged with chlorine, the current is slightly increased and gentle, very gradually increased heat is applied to the bulb, which, however, is not even at the end of the operation brought to redness. During this operation the flow of chlorine must not be so rapid as to carry visible fumes through the absorbing apparatus, and sulphur chloride must be distilled over so slowly that the absorbing liquid remains throughout well charged with chlorine. If the latter precaution is neglected, unoxidized sulphur will remain at the close of the operation, which will render the subsequent part of the process more troublesome and probably less accurate. Besides sulphur chloride, the volatile metallic chlorides distil over. The portion of the tube beyond the bulb may be kept moderately heated so as to prevent it from being stopped up by a sublimate, especially at the narrowed part. When by gradually increased temperature no more volatile products arise from the mass in the bulb and condense in the cooler portion of the tube beyond it, except perhaps ferric chloride (giving a dark brown sublimate), the complete expulsion of which need not be awaited, the heating is extended so that the sublimate in the tube is gradually driven as far as practicable into the receiver, or at least beyond the narrowed part. The stopcock is then closed while the bulb is still warm. When, after a few minutes, the liquid in the receiver has receded somewhat, soften the narrow part of the tube with the flame of a Bunsen burner aided by a blowpipe having a rather large jet, and at the same time draw the tube asunder.

The drawn-off end of the tube containing anhydrous chlorides, which volatilize on exposure to the air, must not be withdrawn from the receiver until the chlorides are dissolved or have by long standing absorbed moisture. Their solution is easily effected provided the tube extends well down into the receiver by inclining the latter so that liquid comes in contact with the end of the tube. The liquid then gradually rises in the tube, absorbing the chlorine gas and dissolving the chlorides in it; meantime, if necessary, the cork may be slightly loosened to admit a little air and prevent the liquid from reaching it by absorption of chlorine. If one fails to effect a solution in the manner above described, the whole may be allowed to stand 24 hours, during which time the chlorides in the tube absorb moisture from the liquid in the receiver, so that the tube can then be withdrawn and the chlorides may be dissolved out with diluted hydrochloric acid and added with rinsings of the tube to the solution in the receiver. Finally, if it is intended to adopt this latter mode of proceeding, the tube may be *cut off* and immediately closed with a cork instead of being fused and drawn off. The solution of the chlorides obtained from the end of the tube, the solution in the receiver and that in the appended U-tube being united, a very gentle heat is applied until the free chlorine is expelled, and the fluid is then allowed to stand until the sulphur, if any is present, has solidified. The sulphur is filtered off on a weighed filter, washed, dried, and weighed. The filtrate is precipitated with barium chloride (§132), by which operation the amount of that portion of the sulphur is determined which has been converted into sulphuric acid. The fluid filtered from the barium sulphate contains, besides the excess of barium chloride added, also the volatile metallic chlorides; which latter are finally determined in it by the proper methods, which will be found in Section V.

The chloride remaining in the bulb-tube is either at once weighed as such (silver chloride, lead chloride), or where this is impracticable—as in the case of copper, for instance, which remains partly as cuprous, partly as cupric chloride—it is dissolved in water, hydrochloric acid, nitrohydrochloric acid, or some other suitable solvent, and the metal or metals in the solution are determined by the methods already described, or which will be found in Section

V. To be enabled to ascertain the weight of the bulb-tube containing silver chloride, it is advisable to reduce the chloride by hydrogen gas, and then dissolve the metal in nitric acid.

In cases where you have only to estimate the sulphur, say in substances containing also sulphuric acid, O. LINDT* recommends conducting the chloride of sulphur and the volatile metallic chlorides into pure solution of soda, when decomposition immediately takes place, producing sodium sulphide, sodium thiosulphate, sodium chloride, and hypochlorite. When the decomposition is over, continue passing the chlorine for two hours through the soda, evaporate then to dryness, ignite the residue cautiously to destroy the sodium chlorate, dissolve in water, and estimate the sulphuric acid according to § 132.

c. Oxidation by Oxide of Mercury (*after* BUNSEN).

This method, which will be found in detail, § 186, is particularly suited to the estimation of sulphur in volatile compounds, or in substances which when heated lose sulphur.

2. *Methods in the Wet Way.*

a. Oxidation of the Sulphur by Acids yielding Oxygen, or by Halogens.†

α. Weigh the finely pulverized sulphide in a small glass tube sealed at one end, and drop the tube into a tolerably capacious strong bottle with glass stopper, which contains red fuming nitric acid (perfectly free from sulphuric acid‡) in more than sufficient quantity to effect the decomposition of the sulphide. Immediately after having dropped in the tube, close the bottle. When the action, which is very impetuous at first, has somewhat abated, shake the bottle a little; as soon as this operation ceases to cause renewed action, and the fumes in the flask have condensed, take out the stopper, rinse this with a little nitric acid into the bottle, and then heat the latter gently.

aa. The whole of the Sulphur has been oxidized, the Fluid is

---

* Zeitschr. f. anal. Chem. 4, 370.

† In presence of lead, barium, strontium, calcium, tin, and antimony, method *b* is preferable to *a*.

‡ To test for sulphuric acid in nitric or hydrochloric acid, it is necessary to evaporate on a water-bath nearly to dryness and take up with water before adding barium chloride. When the acid cannot be got pure, determine the sulphuric acid and allow for it.

perfectly clear :* Evaporate with some sodium chloride, towards the end adding pure hydrochloric acid repeatedly, cooling the dish each time before adding the acid. Dilute with much water, and determine the sulphuric acid as directed § 132. Make sure that the precipitate is pure; if it is not, purify it according to § 132. Separate the bases in the filtrate from the excess of the barium salt by the methods given in Section V.

bb. Undissolved Sulphur floats in the Fluid: Add potassium chlorate in small portions, or strong hydrochloric acid, and digest some time on a water-bath. This process will often succeed in dissolving the whole of the sulphur. Should this not be the case, and the undissolved sulphur appear of a pure yellow color, dilute with water, collect on a weighed filter, wash carefully, dry, and weigh. After weighing, ignite the whole, or a portion of it, to ascertain whether it is perfectly pure. If a fixed residue remains (consisting commonly of quartz, gangue, &c., but possibly also of lead sulphate, barium sulphate, &c.), deduct its weight from that of the impure sulphur. In the filtered fluid determine the sulphuric acid as in *aa*, calculate the sulphur in it, and add the amount to that of the undissolved sulphur. If the residue left upon the ignition of the undissolved sulphur contains an insoluble sulphate, decompose this as directed in § 132, and add the sulphur found in it to the principal amount.

In the presence of bismuth, the addition of potassium chlorate or of hydrochloric acid, is not advisable, as chlorine interferes with the determination of bismuth.

β. Mix the finely pulverized metallic sulphide in a dry flask, by shaking, with powdered potassium chlorate (free from sulphuric acid), and add moderately concentrated hydrochloric acid in small portions. Cover the flask with a watch-glass, or with an inverted small flask. After digestion in the cold for some time, heat gently, finally on the water-bath, until the fluid smells no longer of chlorine. Proceed now as directed in *α*, *aa*, or *bb*, according as the sulphur is completely dissolved or not. In the latter case you must of course immediately dilute and filter. The oxidation of the sulphur may be usually effected more quickly and completely by

* This can of course be the case only in absence of metals forming insoluble salts with sulphuric acid. If such metals are present, proceed as in *bb*, as it is in that case less easy to judge whether complete oxidation of the sulphur has been attained.

warming with nitric acid of 1·36 sp. gr. on a water-bath, and adding potassium chlorate in small portions. Compare STORER,* PEARSON, and BOWDITCH.†

γ. Aqua regia is also frequently used. J. LEFORT‡ recommends a mixture of 1 part strong hydrochloric acid and 3 parts strongest nitric acid. Complete conversion of sulphur into sulphuric acid, however, is rarely effected by aqua regia.

δ. Bromine may also be used. Pyrites or blende is digested at a gentle heat with water, and bromine gradually added. If the sulphides have been prepared in the wet way, good bromine water is sufficient to oxidize them. P. WAAGE§ prefers bromine to all other wet agents, and advises its purification by distillation in an apparatus from which all caoutchouc connections are excluded.

b. Oxidation of the Sulphur by Chlorine in Alkaline Solution, after RIVOT, BEUDANT, *and* DAGUIN.‖ (Suitable also for determining the sulphur in the crude article.)

Heat the very finely pulverized sulphide or crude sulphur for several hours with solution of potassa free from sulphuric acid (which dissolves free sulphur, as well as the sulphides of arsenic and antimony), and then conduct chlorine into the fluid. This speedily oxidizes the sulphur; the sulphuric acid formed combines with the potassa to sulphate, which dissolves in the fluid, whilst the metals converted into oxides remain undissolved. Filter, acidify the alkaline filtrate, and precipitate the sulphuric acid by barium chloride (§ 132). Arsenic and antimony pass into the alkaline solution in the form of acids, but not so lead, which is converted into binoxide, and remains completely undissolved. This method is, therefore, particularly suitable in presence of lead sulphide. In presence of iron sulphide, potassium sulphate is formed at first, and ferric hydroxide, which, if the action of the chlorine is allowed to continue, begins to be converted into potassium ferrate. As soon, therefore, as the fluid commences to acquire a red tint the transmission of chlorine must be discontinued, and the fluid gently heated for a few moments with powdered quartz, to decompose the ferric acid.

It occasionally happens, more particularly in presence of sand, iron pyrites, cupric oxide, &c., that the process is attended with impetuous disengagement of oxygen, which almost completely pre-

* Zeitschr. f. anal. Chem. 9, 71. † *Ib.* 9, 82. ‡ *Ib.* 9, 81.
§ *Ib.* 10, 206. ‖ Compt. Rend. 1835, 865; Journ. f. prakt. Chem. 61, 134.

vents the oxidizing action of the chlorine. However, this accident may be guarded against by reducing the substance to the very finest powder.

B. METHODS BASED ON THE CONVERSION OF THE SULPHUR INTO HYDROGEN SULPHIDE, OR A METALLIC SULPHIDE.

a. The determination of the sulphur in the sulphides of the metals of the alkalies and alkaline earths soluble in water is best effected—provided they are free from excess of sulphur—by I., *b.* In the absence of acids of sulphur you may also convert the sulphur into sulphuric acid by bromine water. The bases are conveniently estimated in a separate portion, which is decomposed by evaporation with hydrochloric or sulphuric acid, or—when none but alkali-metals are present—by ignition with 5 parts of ammonium chloride in a porcelain crucible. If the compounds contain excess of sulphur, they should be oxidized either by chlorine in alkaline solution or treated according to *B*, *c*; if they contain thiosulphate or sulphite, proceed according to § 168.

b. The sulphur contained in alkaline fluids as monosulphide or hydrosulphate of the sulphide may also be determined directly by volumetric analysis, by means of a standard ammoniacal silver or copper solution. In using the former, mix the solution with ammonia, heat and add the standard fluid till, on filtering off a small portion and adding silver solution, a mere opalescence is produced (LESTELLE*). In using the copper solution, mix the fluid to be tested with ammonia, heat to 50° or 60°, and add the standard solution, frequently shaking and boiling till no further precipitation of $CuO, 5CuS$ is produced, and the solution begins to be blue (VERSTRAET†). To make a standard copper solution, 1 c.c. of which shall equal .01, Na_2S, dissolve 9.754 pure copper in 40 grm. nitric acid, boil, add 180 to 200 c.c. ammonia and water to 1 litre. These methods are well adapted for technical purposes, for the estimation of sulphide in soda lies for instance. It need hardly be added that precipitated silver, copper, or lead sulphide (if you have used a solution of oxide of lead in potash) may be estimated gravimetrically.

c. If *all* the sulphur can be expelled from the substance in the form of sulphuretted hydrogen by heating with hydrochloric acid, the sulphide may be heated in a small flask with the concentrated

* Zeitschr. f. anal. Chem. 2, 94. † *Ib.* 4, 216.

acid to complete decomposition and expulsion of the hydrogen sulphide—the latter being determined according to I. In the case of polysulphides, the sulphur separated in the evolution flask is collected on a filter dried at 100°, washed, dried first at 70°, then for a short time at 100°, and weighed.

Third Group.

NITRIC ACID.—CHLORIC ACID.

§ 149.

1. NITRIC ACID.

I. *Determination.*

Free nitric acid in a solution containing no other acid is determined most simply in the volumetric way, by neutralizing with a dilute solution of soda or ammonia of known strength (comp. Special Part, "Acidimetry"). The following method also effects the same purpose: Mix the solution with baryta-water, until the reaction is just alkaline, evaporate slowly in the air, nearly to dryness, dilute the residue with water, filter the solution which has ceased to be alkaline, wash the barium carbonate formed by the action of the carbonic acid of the atmosphere upon the excess of the baryta-water, add the washings to the filtrate, and determine in the fluid the barium as directed in § 101. Calculate for each 1 at. barium 2 mol. nitric acid. Lastly, free nitric acid may also be determined in a simple manner by supersaturating with ammonia, evaporating in a weighed platinum dish, drying the residue at 110° to 120°, and weighing the NH_4NO_3 (SCHAFFGOTSCH).

II. *Separation of nitric acid from the basic radicals, and determination of the acid in nitrates.*

a. Methods based on the decomposition of Nitrates in the Dry Way.

α. In anhydrous metallic nitrates which leave upon ignition a metallic oxide of known and definite composition, the nitric acid may be determined by ignition and calculation from the weight of the residue.

β. In the case of nitrates, whose residue on ignition has no constant composition, or by whose ignition the crucible is much attacked (alkali and alkali-earth nitrates), fuse the substance (which must be anhydrous and also free from organic and other volatile bodies) with a non-volatile flux, and estimate the nitric acid from the loss. Silicic acid is the best flux, as it may be readily procured, and the execution is the most easy and the most certain to succeed. I shall describe the method in its application to potassium or sodium nitrate.

Fuse the latter at a low temperature, pour out on to a warm porcelain dish, powder, and dry again before weighing. Now transfer to a platinum crucible 2 to 3 grm. powdered quartz, ignite well, and weigh after cooling. Add about 0·5 grm. of the salt prepared as above, mix well, and convince yourself by the balance that nothing has been lost during mixing. The covered crucible is then exposed to a low red heat (just visible by day) for half an hour, and weighed after cooling with the cover. The loss of weight represents the quantity of N_2O_5. Sulphates or chlorides are not decomposed at the given temperature; if a higher heat be applied, the latter may volatilize. The action of reducing gases must be avoided. The test-analyses, communicated by REICH,* as well as those performed in my own laboratory,† gave very satisfactory results.

b. Method based on the distillation of Nitric Acid.

All nitrates may be decomposed by distillation with moderately dilute sulphuric acid. The nitric acid passing into the receiver may then be determined, according to I., volumetrically or gravimetrically. 1 to 2 grm. of the nitrate should be treated with a cooled mixture of 1 volume concentrated sulphuric acid and 2 volumes water. For 1 grm. nitre take 5 c.c. sulphuric acid and 10 c.c. water. The distillation may be performed either with a thermometer at 160° to 170° in a paraffin or sand bath (duration of the distillation for 1 to 2 grm. nitre, 3 to 4 hours), or *in vacuo*, with the use of a water-bath. The latter process is the best. In the former, the neck of the tubulated retort (which is drawn out and bent down) is connected with a bulbed U-tube‡ containing a measured quantity of standard soda or potassa solution (§ 192). The distillation *in vacuo* may be conducted, without the use of an air-pump, according to FINKENER,‖ as follows: Transfer the measured

* Berg- und Hüttenmännische Zeitschrift, 1861, No. 21; Zeitschrift f. analyt. Chem. 1, 86.
† Zeitschr. f. anal. Chem. 1, 181.
‡ The bulbed U-tube will be found figured § 185.
‖ Zeitschrift f. analyt. Chem. 1, 309.

quantity of water and concentrated sulphuric acid to the tubulated retort, and the necessary quantity of standard potassa or soda solution, diluted to 30 c.c., to a flask with a narrow neck of about 200 c.c. capacity. Then, by means of an india-rubber tube, connect the flask with the retort air-tight, so that the drawn-out point of the latter may extend to the body of the flask, and—with tubulure open—heat the contents of the retort and of the flask to boiling. When the air has been expelled from the apparatus by long boiling, transfer the salt (weighed in a small tube) to the retort through the tubulure, close the latter immediately, and at the same time take away the lamp. The retort is then heated with a water-bath, the flask being kept cool. The quantity of nitric acid that has passed over is finally ascertained by determining the still free alkali with standard acid. If it is suspected that all the nitric acid has not been driven into the receiver by one distillation, you may—by heating the flask and cooling the retort—distil the water back into the latter, and then the distillation from the retort may be repeated. The distillate thus obtained is always free from sulphuric acid, hence the results are very exact. The base remains as sulphate in the retort. In the presence of chloride add to the contents of the retort a sufficiency of dissolved silver sulphate, or—when much chloride is present—moist silver oxide. The nitric acid is then obtained entirely free from chlorine.

c. Methods based on the decomposition of Nitrates by Alkalies, &c.

α. Nitrates of metals which are completely precipitated by alkali hydroxides or carbonates—provided basic salts are not precipitated at the same time—may be analyzed by simple boiling with an excess of standard potassa or soda or their carbonates. After cooling, dilute to ¼ or ½ litre, mix, allow to settle, draw off a portion of the supernatant clear fluid, determine the free alkali remaining in it, and calculate therefrom the amount which has been converted into nitrate. HAYES obtained with silver and bismuth nitrates good results; but with mercurous nitrate (using sodium carbonate) the results were not so satisfactory.*

β. In nitrates from which the basic metals are precipitated by barium or calcium hydroxides or their carbonates (or by barium sulphide), the nitric acid may be estimated with great accuracy by

* H. Rose, Zeitschrift f. analyt. Chem. 1, 306.

filtering, after precipitation has been effected, warm or cold, passing carbonic acid through the filtrate, if necessary, till all the barium is precipitated, warming, filtering, and determining the barium in the filtrate by sulphuric acid. 1 at. of the same corresponds to 1 mol. nitric anhydride (N_2O_5). [In case of bismuth-salts, boil until the separated oxide is perfectly yellow. PAIGE.]

γ. In many nitrates whose bases are precipitable by sulphuretted hydrogen the nitric acid may be determined according to GIBBS by adding to the salt in solution about its own weight of some neutral organic salt, *e.g.*, Rochelle salt, and throwing down the metal by H_2S. The filtrate and washings are brought to a definite bulk, and the free acid is determined in aliquot portions alkalimetrically.*

d. Methods based upon the decomposition of Nitric Acid by Ferrous Chloride.

Method of PELOUZE† and FRESENIUS. The decomposition is a follows:

$$6FeCl_2 + 2KNO_3 + 8HCl = 3Fe_2Cl_6 + 2KCl + 4H_2O + N_2O_2.$$

α. Select a tubulated retort of about 200 c.c. capacity, with a long neck, and fix it so that the latter is inclined a little upwards. Introduce into the body of the retort about 1·5 grm. fine pianoforte wire, accurately weighed, and add about 30 or 40 c.c. pure fuming hydrochloric acid. Conduct now through the tubulure, by means of a glass tube reaching only about 2 cm. into the retort, hydrogen gas washed by solution of potassa, or pure carbonic acid, and connect the neck of the retort with a U-tube containing some water. Place the body of the retort on a water-bath, and heat gently until the iron is dissolved. Let the contents of the retort cool in the current of hydrogen gas or carbonic acid; increase the latter, and drop in, through the neck of the retort, into the body, a small tube containing a weighed portion of the nitrate under examination, which should not contain more than about 0·200 grm. of N_2O_5. After restoring the connection between the neck and the U-tube, heat the contents of the retort in the water-bath for about a quarter of an hour, then remove the water-bath, heat with the lamp to boiling, until the fluid, to which the nitric oxide had imparted a dark tint, shows the color of ferric chloride, and con-

* Am. Jour. Sci., xliv. 209. † Journ. f. prakt. Chem. 40, 324.

tinue boiling for some minutes longer. Care must be taken to give the fluid an occasional shake, to prevent the deposition of dry salt on the sides of the retort. Before you discontinue boiling, increase the current of hydrogen or carbonic acid gas, that no air may enter through the U-tube when the lamp is removed. Let the contents cool in the current of gas, dilute copiously with water, and determine the iron still present as ferrous chloride volumetrically by potassium dichromate—336 of iron converted by the nitric acid from ferrous to ferric chloride correspond to 108 (N_2O_5). My test-analyses of pure potassium nitrate gave 100·1—100·03—100·03, and 100·05, instead of 100.* [The iron remaining as ferric chloride may also be determined by sodium thiosulphate.]

β. Schulze's Method† modified by Tiemann.‡

The solution containing the nitrate is concentrated if necessary to a volume of about 50 c.c. and introduced into the flask A, which should have a capacity of about 200 c.c. This flask is provided with a rubber stopper, through which pass two bent tubes *a b c* and *e f g*. The first is drawn out to a point (not too small) at *a*, and projects through the stopper about 2 cm.; the second terminates without diminution of size exactly at the lower surface of the stopper. These two tubes are connected by rubber tubes (bound with thread) at *c* and *g* with the glass tubes *c d* and *g h*. A rubber tube is drawn over the lower end of *g h* to protect it from fracture. B is a glass vessel containing 10 per cent. soda solution. A measuring tube graduated to 0·1 c.c., of not too great diameter, filled with previously boiled soda solution, is supported so that its open end is under the surface of the liquid in B.

The solution of the nitrate in the flask is further concentrated by boiling, and finally the lower end of the tube *e f g h* is brought into the soda solution so that a part of the steam escapes through it. After a few minutes the rubber tube at *g* is pressed together with the fingers; if the air has been completely displaced from the flask by boiling, the soda solution will rise suddenly in the tube as in a vacuum, and a slight blow against the finger will be perceptible. In this case, the rubber tube at *g* is closed with a clamp and the steam is allowed to escape through *a b c d* until only 10 c.c.

* Annal. d. Chem. u. Pharm. 106, 217.

† Zeitschr. fur anal. Chem. 1870, 400.

‡ Anleitung zur Untersuchung von Wasser, von W. Kuhel, Zweite Auflage von F. Tiemann, Braunschwerg bei Fr. Vieweg u. Sohn. 1870, s. 55.

of fluid remain in the flask. The lamp is now removed and the rubber tube at *c* is closed with a clamp, and the tube *c d* filled by a jet of water. If an air bubble remains in the rubber tube at *c*, it must be removed by pressure with the fingers. The graduated measuring tube is now brought over the upcurved end of the evolution tube *e f g h* so that the end rises in it 2–3 cm. The flask must next be allowed to stand a few minutes until a partial vacuum is produced in it, which is manifested by a contraction of the rubber tubes at *c* and *g*. A nearly saturated solution of ferrous chloride is poured into a small beaker, the upper part of which is marked

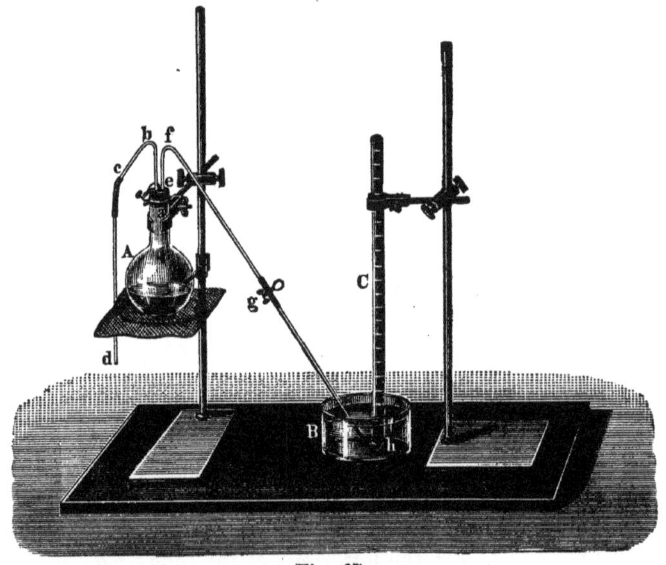

Fig. 67.

so as to show the space occupied by 20 c.c.; two other beakers must also be at hand partly filled with concentrated hydrochloric acid. The tube *c d* is now dipped into the ferrous chloride solution, and the clamp at *c* is loosened until 15–20 c.c. are drawn into the flask. The ferrous chloride remaining in the tube is next removed by drawing in a small quantity of hydrochloric acid in two successive portions. Small bubbles may frequently be observed at *b*, occasioned by evolution of hydrochloric gas caused by diminished pressure in the flask. They disappear almost completely so soon as the pressure rises.

Heat is applied, at first very gently, until the rubber tubes at *c* and *g* are slightly expanded; then the rubber tube at *g* is held compressed by the fingers, the clamp being removed, until the pressure becomes stronger, when the gas is allowed to pass over to the graduated tube. Toward the end of the operation heat is increased and distillation continued until the volume of gas in the measuring tube no longer increases. The hydrochloric gas, abundantly evolved in the last part of the process, is absorbed with violence by the soda solution with a peculiar clattering sound; there is no danger, however, of breaking the evolution tube if care has been taken to enclose the lower end with a rubber tube as above directed.

The measuring tube is brought into a large cylinder containing cold water, best of 15–18° C., and by means of some suitable fixture held wholly submerged in the same. The transfer is effected with the help of a small porcelain dish filled with soda solution.

After 15–20 minutes, the temperature of the water in the cylinder is ascertained with a sensitive thermometer, and the state of the barometer is also observed. Then the tube is taken hold of at the upper end with a strip of paper or cloth, in order to avoid imparting heat to it by direct contact of the hand, and drawn up perpendicularly so far that the level of the fluids within and without it exactly coincide, and the volume of the gas is read off. From the data thus obtained, the volume which the dry gas would occupy at 0° C. and 760 mm. bar. pressure is to be computed. (See p. 836, on Calculation of Analyses.) 1 c.c. N_2O_2 at 0° C. and 760 mm. bar. pressure corresponds to ·002413 grm. N_2O_5.

A condition indispensable for the success of the operation is the complete expulsion of air from the apparatus in the beginning. When an abundant quantity of nitric acid is present in the substance, enough to produce about 80 c.c. nitrogen dioxide is a suitable quantity to use for its determination, and a somewhat larger quantity of ferrous chloride and hydrochloric acid than above indicated may be used. An unnecessary amount of these reagents should, however, be avoided, since it is difficult to boil a small quantity of nitrogen dioxide out of a large volume of liquid.

This method is easy to carry out and gives satisfactory results. It has been selected for description and recommendation here out of a great number of methods, not mentioned in this volume, which have been proposed and more or less used for determination of nitric acid.

e. Methods in which the Nitrogen of the Nitric Acid is separated and measured in the gaseous form.

These methods are more particularly suitable for analyzing nitrates which are decomposed by ignition into oxide or metal and oxides of nitrogen; they will be found in the Section on the Ultimate Analysis of Organic Bodies, § 184. MARIGNAC employed them to analyze mercurous nitrates. BROMEIS analyzed nitrite, &c., of lead by a similar method, recommended by BUNSEN. In cases where it is intended to determine the water of the analyzed nitrate in the direct way, such methods are almost indispensable.*

§ 150.

2. CHLORIC ACID.

I. *Determination.*

Free chloric acid in aqueous solution may be determined by converting it into hydrochloric acid by the agency of nascent hydrogen (II., *b*), and determining the acid formed, as directed in § 141; or by saturating with solution of soda, evaporating the fluid, and treating the residue as directed in II., *a* or *c*.

II. *Separation of Chloric Acid from the Bases and Determination of the Acid in Chlorates.*

a. After BUNSEN.† When warm hydrochloric acid acts upon chlorates, the latter are reduced; as this reduction is not attended with separation of oxygen, the following decompositions may take place:

$$\begin{matrix} Cl_2O_5 \\ 2HCl \end{matrix} \left\{ \begin{matrix} Cl_2O \\ Cl_2O_3 \\ H_2O \end{matrix} \right. \quad \begin{matrix} Cl_2O_5 \\ 4HCl \end{matrix} \left\{ \begin{matrix} 3\,Cl_2O \\ 2\,H_2O \end{matrix} \right. \quad \begin{matrix} Cl_2O_5 \\ 6HCl \end{matrix} \left\{ \begin{matrix} 2Cl_2O \\ 4Cl \\ 3H_2O \end{matrix} \right. \quad \begin{matrix} Cl_2O_5 \\ 8HCl \end{matrix} \left\{ \begin{matrix} Cl_2O \\ 8Cl \\ 4H_2O \end{matrix} \right. \quad \begin{matrix} Cl_2O_5 \\ 10HCl \end{matrix} \left\{ \begin{matrix} 12Cl \\ 5H_2O \end{matrix} \right.$$

Which of these products of decomposition may actually be formed, whether all or only certain of them, cannot be foreseen. But no matter which of them may be formed, they all of them agree in this, that, in contact with solution of potassium iodide, they liberate for every 2 mol. chloric acid ($HClO_3$), or 1 mol. Cl_2O_5 in the chlorate, 12 at. iodine. 1522·2 of iodine liberated correspond accordingly to 150·92 Cl_2O_5. The analytical process is conducted as described § 142, 1.

* See also Gibbs, Am. Journ. Sci., xxxvii. 350.

† Annal. d. Chem. u. Pharm. 86, 282.

b. After SESTINI.* To the concentrated aqueous solution of the weighed chlorate add a piece of zinc and then some pure dilute sulphuric acid and allow to stand for some time (with 0.1 grm. potassium chlorate half an hour is sufficient). By the nascent hydrogen the chloric acid is converted into hydrochloric acid, which, after removal and rinsing of the zinc, is determined according to § 141. To use the volumetric method (§ 141, *b*, *α*), the sulphuric acid is first precipitated with barium nitrate, then the zinc and excess of barium with sodium carbonate, the liquid is filtered and neutralized, then potassium chromate is added, and finally standard silver solution.

c. The basic radicals are determined with advantage in a separate portion, by converting the chlorate either by very cautious ignition, or by warming with hydrochloric acid into chloride.

The estimation of *hypochlorous acid* will be described in the Special Part, article " Chlorimetry."

---

‡ Zeitschrift f. analyt. Chem. 1, 500.

SECTION V.

SEPARATION OF BODIES.

§ 151.

When only one basic or one acid radical is present, the method of its determination has been considered in the previous Section. When more than one basic or more than one acid radical is present, the methods of separating and determining them will be described in the present Section.

The separation of bodies may be effected in three ways: viz., *a*, by *direct analysis; b*, by *indirect analysis; c*, by *estimation by difference.*

By *direct analysis*, we understand the actual separation of radicals or elements. Thus, we separate potassium from sodium by platinic chloride; copper from tin by nitric acid; arsenic from iron by hydrogen sulphide; iodine from chlorine by palladious nitrate; carbon from potassium nitrate by water, &c., &c. In direct analysis we render one body insoluble, while the others remain in solution, or *vice versa*, or we volatilize one body, leaving the others behind, or we effect actual separation in some other manner. This is the mode of analysis most frequently employed. It generally deserves the preference where choice is permitted.

We term an analysis *indirect* if it does not effect the actual separation of the bodies, but causes certain changes which enable us to calculate their quantity. Thus, the quantity of potassium and sodium in a mixture of compounds of the two may be determined by converting them into chlorides, weighing the latter, and determining the chlorine (§ 152, 3).

Finally, if we weigh two bodies together, determine one of them, and subtract its weight from that of the two, we shall find the weight of the other body. In this case the second body is said to be *estimated by difference.* Thus, aluminium may be determined when its oxide is mixed with ferric oxide, by weighing the mixture and determining the iron volumetrically.

Indirect analysis and estimation by difference may be employed in an exceedingly large number of cases; but their use is as a rule only to be recommended where good methods of true separation are wanting. The special cases in which they are preferable to direct analysis cannot be all foreseen; those alone are pointed out which are of more frequent occurrence. As regards the calculations required in indirect analysis, I have given general directions under "the Calculation of Analysis;" wherever it appeared judicious, I have added the necessary directions to the description of the method itself.

I have retained our former subdivision into groups, and, as far as practicable, systematically arranged, first, the general separation of all the bodies belonging to one group from those of the preceding groups; secondly, the separation of the individual bodies of one group from all or from certain bodies of the preceding groups; and finally, the separation of bodies belonging to one and the same group from each other. I think I need scarcely observe that the general methods which serve to separate the whole of the bodies of one group from those of another group are also applicable to the separation of every individual body of the one group from one or several bodies of the other group. It must not be understood that the more special methods are necessarily in all cases preferable to the more general ones. As a rule, it must be left to individual chemists to decide for themselves in each special case which method should be adopted. With respect to the general methods for separating one group from another, I would observe that those adduced appeared to me more adapted to the purpose than others, but still there may be others that are equally suitable, and in special cases even more so. A wide field is here open to the ingenuity of the analyst.

The methods given for the separation of both basic and acid radicals are generally based upon the supposition that they are in the form of free acids or bases, or in the form of salts soluble in water. Wherever this is not the case, special mention is made of the circumstance.

From among the host of proposed methods, I have, as far as practicable, chosen those which have been sanctioned by experience and are distinguished for accurate results. In cases where two methods were on a par with each other as regards these two points, I have either given both or selected the more simple one. Methods which experience has shown to be defective or fallacious have been altogether omitted. I have endeavored to point out, as far as possible, the particular circumstances under which either the one or the other of several methods deserves the preference.

Where the accuracy of an analytical method has been established already, in Section IV., no furthur statements are made on the subject here. Paragraphs of former Sections deserving particular attention are referred to in parentheses.

The extension of chemical science introduces almost every day new analytical methods of every description, which are, rightly or wrongly, preferred to the older methods; the present time may therefore be looked upon in this, as in so many other respects, as a period of transition, in which the new strives more than ever to overcome and supplant the old. I make this remark to show the impossibility of always adding to the description of a method an opinion of its usefulness and accuracy, and also to point out the importance, under such circumstances, of a proper systematic arrangement. I have in this Section generally arranged the various analytical methods upon the bases of their scientific principles, firmly persuaded that this will greatly tend to facilitate the study of the science, and will lead to endeavors to apply known principles to the separation of other bodies besides those to which they are already applied, or to apply new principles where experience has proved the old ones fallacious, and the methods based on them defective.

I conclude these introductory remarks with the important caution to the student *never to look upon a separation as successfully accomplished before he has convinced himself that the weighed precipitates, &c., are pure and more particularly free from those bodies from which it was intended to separate them.*

I. SEPARATION OF THE BASIC RADICALS FROM EACH OTHER.

First Group.

POTASSIUM—SODIUM—AMMONIUM—(LITHIUM).

§ 152.

INDEX. The numbers refer to those in the margin.

1. ***Methods based upon the different degrees of Solubility in Alcohol, of Sodium Platinic Chloride, and Potassium Platinic Chloride.***

a. POTASSIUM FROM SODIUM.

1 It is an indispensable condition in this method that the two alkalies should exist in the form of chlorides. If, therefore, they are present in any other form, they must be first converted into chlorides, which in most cases may be effected by evaporation with hydrochloric acid in excess; in the case of nitrates, the evaporation with hydrochloric acid must be repeated 4—6 times till the weight of the gently ignited mass ceases to diminish. In presence of sulphuric acid, phosphoric acid, and boracic acid, this simple method will not answer. For the methods of separating the alkalies from the two latter acids and converting them into chlorides, see §§ 135 and 136. The presence of sulphuric acid being a circumstance of rather frequent occurrence, the way of meeting this contingency is given below (**2**).

Determine the total quantity of the sodium chloride and potassium chloride* (§§ 97, 98), dissolve in the least quantity

* Never take the weight of the alkali chlorides without convincing yourself of their purity by dissolving them in water, which should give a clear solution, and testing the solution with ammonia and ammonium carbonate, which must throw down no precipitate. It may be thought, perhaps, that a matter so simple need not be mentioned here; still I have found that neglect in this respect is by no means uncommon.

of water, add to the fluid in a porcelain dish an excess of a strong aqueous solution of platinic chloride as neutral as possible. Enough platinum solution should be added to convert the sodium as well as the potassium into platinochloride. It is best to use a solution of known strength and to calculate roughly how much should be added. Evaporate on the water-bath nearly to dryness (the water in the bath should never actually boil, and the sodium platinic chloride should not lose its water of crystallization), treat the residue with alcohol of from ·86 to ·87 sp. gr., cover the dish with a glass plate, and allow to stand a few hours, with occasional stirring. If the supernatant fluid is not deep yellow, this is a proof that the quantity of platinic chloride used is insufficient. When the precipitate has settled, pour off the clear fluid through a filter and examine the precipitate most minutely, if necessary with the aid of a microscope. If it is a heavy yellow powder (sufficiently magnified, small octahedral crystals) it is the pure potassium platinic chloride.* Then transfer it—best with the aid of the filtrate—to the filter, wash it with spirit of ·86 to ·87 sp. gr. and proceed according to § 97, 3 α. (Instead of weighing the double chloride or the platinum obtained from it, you may ignite gently in hydrogen, extract the potassium chloride with water, and weigh this or titrate the chlorine in it by § 141, I., *b*, α). If, on the contrary, white saline particles (sodium chloride) are to be seen mixed with the yellow crystalline powder, platinic chloride has been wanting, the whole of the sodium chloride not having been completely converted into sodium platinic chloride. In this case the precipitate in the dish must be treated with some water, till all the sodium chloride is dissolved, a fresh portion of platinic chloride is added, the whole evaporated nearly to dryness, and the above examination repeated. The quantity of the sodium is usually estimated by subtracting from the united weight of the sodium chloride and potassium chloride the weight of the latter, calculated from that of the potassium platinic chloride.

To make quite sure that the potassium has completely sep-

---

* If small tesseral crystals are visible of a dark orange-yellow color, and relatively large size, and appearing transparent by transmitted light, then the double chloride contains lithium platinic chloride (JENZSCH, Pogg. Ann. 104, 102).

arated, it is advisable to add to the filtrate some water, some more platinic chloride, and if the quantity of sodium is only small, also some sodium chloride; evaporate on the water-bath nearly to dryness, at a temperature not exceeding 75° (BISCHOF), and treat the residue in the manner just described. In order to diminish the solvent action of the alcohol on the potassium platinic chloride, ¼ ether may be now mixed with it. Should this operation again leave a small undissolved residue of potassium platinic chloride, it is filtered off, best on a separate filter, and first washed with alcohol and ether. As, however, this remainder of the double salt is generally impure, dissolve it on the filter with boiling water, evaporate with a few drops of platinic chloride, treat the residue with alcohol, and if any potassium salt remains, determine it either with the principal quantity or by itself.

If you are not satisfied with an indirect estimation of the sodium, one of the following direct methods may be employed. *α*. Evaporate the filtrate till the spirit has gone off, dilute, digest the solution with small pure iron filings till the platinum is all thrown down, filter, add chlorine water till the ferrous is converted into ferric chloride, precipitate with ammonia, filter off the ferric hydroxide, and determine the sodium chloride in the filtrate. *β*. Evaporate the filtrate, finally in a porcelain crucible, to dryness, heat the residue to low redness in a current of hydrogen, extract with water, and determine the sodium chloride in the solution. For small quantities of fluid this method will be found convenient. *γ*. A. MITSCHERLICH recommends to mix the filtrate with sulphuric acid, evaporate to dryness, ignite the residue, extract the sodium sulphate with water, and determine it according to § 98, 1. These methods, of course, yield the sodium salt in a pure condition only when the separation of the potassium has been perfect. They present the advantage that the sodium salt is brought under one's eyes and may be tested after weighing.

2 Should the solution contain sulphuric acid, it may be in presence of hydrochloric acid or of some volatile acid, convert the alkalies first into normal sulphates (§§ 97, 98), and weigh them as such. For the estimation of the potassium, one of the two following methods may be used:

α. First convert the sulphates into chlorides and then proceed as above. For this purpose barium salts were formerly employed, or, better, an alcoholic solution of strontium chloride. The barium sulphate, however, carries down considerable quantities of alkali salt, and the strontium sulphate noticeable quantities; hence the employment of these reagents, more particularly barium, cannot be recommended. H. ROSE advises repeated ignition of the alkali sulphates with ammonium chloride till the weight remains constant; this process is simple and well adapted for small quantities; no loss of alkali need be feared if the heat is not unnecessarily raised. L. SMITH advises the use of lead salts. Dissolve the alkali sulphate, precipitate with pure neutral lead acetate, avoiding a large excess, add some alcohol, filter, precipitate the excess of lead with sulphuric acid, and evaporate to dryness with addition of sulphuric acid. This method, when carefully conducted, yields excellent results.

β. Precipitate the potash directly out of the solution of the sulphates. R. FINKENER* gives the following process: To the rather dilute solution of the salts in a capacious porcelain dish add platinic chloride in quantity more than sufficient to throw down all the potassium, evaporate on a water-bath down to a few c.c., allow to cool, add, at first in small quantities, 20 times the volume of a mixture of 2 parts absolute alcohol and 1 part ether, with stirring; filter after a short time, and wash the precipitate with alcohol and ether till the washings are colorless. If, when the alcohol and ether are first added, a strong aqueous solution of sodium sulphate separates, add some hydrochloric acid till the fluids mix. Dry the precipitate consisting of potassium platinic chloride and sodium sulphate, heat with the filter in a porcelain crucible till the filter is carbonized, then in a current of hydrogen to scarcely visible redness extract the residue with hot water, ignite the platinum in the air, weigh and calculate from the weight the quantity of potassium.

The separation of potassium from sodium by platinic chloride gives results which are fully satisfactory, and at all events far more exact than any method depending on another principle; provided that the platinum solution is pure and the operations have been carefully performed in accordance with the directions. If you have any occasion to doubt the perfect

* H. ROSE, Handbuch der anal. Chem. 6 Aufl. von FINKENER, ii. 923.

purity of the weighed double salt, you may always dissolve it in boiling water, evaporate with addition of a little platinum solution, and reweigh the salt thus purified.

b. Ammonium from Sodium.

3 The process is conducted exactly as in *a*, when the alkalies are present as chlorides. See also § 99, 2. If potassium also is present, the precipitate produced by platinic chloride is a mixture of ammonium platinic chloride and potassium platinic chloride; in which case the weighed precipitate is cautiously ignited for a sufficient length of time, but not too strongly, until the ammonium chloride is expelled, the gentle ignition continued in a stream of hydrogen or with addition of oxalic acid, the residue extracted with water, a few drops of hydrochloric acid added if oxalic acid was employed, and the potassium chloride in the solution determined as directed § 97, 2. The weight found is calculated into potassium platinic chloride, and the result deducted from the weight of the whole precipitate: the difference gives the ammmonium platinic chloride. The weighing of the separated platinum affords a good control. The method is seldom employed, as that given in 2 yields more exact results.

2. *Methods based upon the Volatility of Ammonium Salts and Ammonia.*

Ammonium from Potassium and Sodium.

4 *a. The salts of the alkalies to be separated contain the same volatile acid, and admit of the total expulsion of their water by drying at* 100°, *without losing ammonia* (*e.g.*, the chlorides).

Weigh the total quantity of the salts in a platinum crucible, and heat, with the lid on, gently at first, but ultimately for some time to faint redness; let the mass cool, and weigh. The decrease of weight gives the quantity of the ammonium salt. If the acid present is sulphuric acid, you must, in the first place, take care to heat very gradually, as otherwise you will suffer loss from the decrepitation of ammonium sulphate; and, in the second place, bear in mind that part of the sulphuric acid of the ammonium sulphate remains with the fixed alkali sulphates, and that you must accordingly convert them into normal salts, by ignition in an atmosphere of ammonium carbonate, before proceeding to determine their weight (compare §§ 97 and 98). Ammonium chloride cannot be separated in this manner from fixed alkali sulphates, as it converts them, upon ignition, partly or totally into chlorides.

b. Some one or other of the conditions given in "a" is not fulfilled.

If it is impracticable to alter the circumstances by simple means, so as to make the method *a* applicable, the fixed alkalies and the ammonium must be determined separately in different portions of the substance. The portion in which it is intended to determine the potassium and sodium is gently ignited until ammonium is completely expelled. The fixed alkalies are converted, according to circumstances, into chlorides or sulphates, and treated as directed in **1**, **2**, or **6**. The ammonium is estimated in another portion according to § 99, 3. **5**

3. *Indirect Methods.*

Of course, a great many of these may be devised; but the following is the only one in general use. **6**

POTASSIUM FROM SODIUM.

Convert both alkalies into chlorides (§§ 97 and 98), and weigh as such; estimate chlorine (§ 141); and from the amount of this calculate the quantities of the sodium and potassium (see "Calculation of Analysis" *).

The indirect method of determining sodium and potassium is applicable only in the analysis of mixtures containing tolerably large quantities of both bases; but where this is the case, the process answers very well, affording also, more particularly, the advantage of expedition, if the chlorine in the weighed chlorides is titrated (§ 141, I., *b*).

Supplement to the First Group.

SEPARATION OF LITHIUM FROM THE OTHER ALKALIES.

Lithium may be separated from *potassium* and *sodium* in the indirect way, and by two direct methods: **7**

a. Treat the nitrates or the chlorides, dried at 120°, with a mixture of equal volumes of absolute alcohol and anhydrous ether, digest at least for 24 hours, with occasional shaking (the

* Other methods are given by STOLBA (Zeitschr. f. anal. Chem. 2, 397) and MOHR (*Ib.* 7, 173).

salts must be completely disintegrated), decant rapidly on to a filter covering the funnel, and treat the residue again several times with smaller portions of the mixture of alcohol and ether. Determine, on the one part, the undissolved potassium and sodium salts; on the other, the dissolved lithium salt, by distilling the fluid off, and converting the residue into sulphate. This method is apt to give too much lithium, as the potassium and sodium salts, especially the chlorides, are not absolutely insoluble in a mixture of alcohol and ether. The results may be rendered more accurate by treating the impure lithium salt, obtained by distilling off the ether and alcohol, once more with alcohol and ether, with addition of a drop of nitric or hydrochloric acid, adding the residue left to the principal residue, and then converting the lithium salt into sulphate. If the salts, which it is intended to treat with alcohol and ether, have been ignited, however so gently, caustic lithia is formed—in the case of the chloride by the action of water—and lithium carbonate by attraction of carbonic acid; in that case it is necessary, therefore, to add a few drops of nitric or, as the case may be, hydrochloric acid, in the process of digestion.

If we have to separate the sulphates, they must be converted into nitrates or chlorides before they can be subjected to the above method. This conversion is best effected by means of lead salts, see **2**. Ignition with ammonium chloride does not answer for lithium sulphate, nor can the sulphuric acid be removed by barium, or strontium, as the precipitated sulphates would contain lithium (DIEHL*).

b. Weigh the mixed alkalies, best in form of sulphates, and then determine the lithium as phosphate according to § 100. If the quantity of lithium is relatively very small, convert the weighed sulphates into chlorides (**7**), separate, in the first place, the principal amount of the potassa and soda by means of alcohol (§ 100), and then determine the lithium (MAYER †). **8**

c. When exact results are required, the indirect method is to be preferred. Proceed first according to *a*, evaporate the spirituous solution of the lithium chloride containing the remainder of the other chlorides to dryness, heat moderately, weigh, dissolve in water, estimate the chlorine, and calculate therefrom **9**

* Annal d. Chem. u. Pharm. 121, 98. † *Ib*. 98, 193.

the lithium and sodium or potassium. BUNSEN * also applied the method to the indirect estimation of lithium in presence of potassium and sodium by removing the silver from the filtrate, and separating the potassium with platinum. But I must here point out, that according to JENZSCH † the potassium double salt will contain lithium apparently in the form of the platino-chloride of potassium and lithium.

The sulphuric acid in weighed quantities of the sulphates of lithium, and of potassium and sodium, cannot be determined as barium sulphate (see end of **7**).

The separation of lithium from ammonium may be effected like that of potassium and sodium from ammonium (**4** and **5**).

Second Group.

BARIUM—STRONTIUM—CALCIUM—MAGNESIUM.

I. SEPARATION OF THE BASIC RADICALS OF THE SECOND GROUP FROM THOSE OF THE FIRST.

§ 153.

A. General Method.

1. THE WHOLE OF THE ALKALI-EARTH METALS FROM POTASSIUM AND SODIUM.

10 *Principle on which the method is based: Ammonium carbonate precipitates, from a solution containing ammonium chloride, only barium, strontium, and calcium.*

Mix the solution, in which the metals are assumed to be contained in the form of chlorides, with a sufficient quantity of

* Annal. d. Chem. u. Pharm. 122,348. † Pogg. Annal. 104, 102.

ammonium chloride to prevent the precipitation of the magnesium by ammonia; dilute rather considerably, add some ammonia, then ammonium carbonate in slight excess, let the mixture stand covered for an hour in a moderately warm place, filter, and wash the precipitate with water to which a few drops of ammonia have been added.

The *precipitate* contains the *barium*, *strontium*, and *calcium;* the *filtrate* the *magnesium* and the *alkalies*. So at least we may assume in cases where the highest degree of accuracy is not required. Strictly speaking, however, the solution still contains exceedingly minute traces of calcium and somewhat more considerable traces of barium, as the carbonates of these two metals are not absolutely insoluble in a fluid containing ammonium chloride; the precipitate also may contain possibly a little ammonium magnesium carbonate. Treat the precipitate according to § 154, and the filtrate—in rigorous analyses—as follows: Add 3 or 4 drops (but not much more) of dilute sulphuric acid, then ammonium oxalate, and let the fluid stand again for 12 hours in a warm place. If a precipitate forms, collect this on a small filter, wash, and treat on the filter with some dilute hydrochloric acid, which dissolves the calcium oxalate, and leaves the barium sulphate undissolved. Since a little magnesium oxalate may have separated with the former, add some ammonia to the hydrochloric solution, filter after the precipitate has settled, and mix the filtrate with the principal filtrate.

Evaporate the fluid containing the *magnesium* and the *alkalies* to dryness, and remove the ammonium salts by gentle ignition in a covered crucible, or in a small covered dish of platinum or porcelain.* In the residue, separate the magnesium from the alkalies by one of the methods given **15**—18.

2. THE WHOLE OF THE ALKALI-EARTH METALS FROM AMMONIUM.—The same principle and the same process as in the separation of potassium and sodium from ammonium (**4** and **5**). **11**

* This operation effects also the removal of the small quantity of sulphuric acid added to precipitate the traces of barium, as sulphates of the alkalies are converted into chlorides upon ignition in presence of a large proportion of ammonium chloride.

B. Special Methods.

SINGLE ALKALI-EARTH METALS FROM POTASSIUM AND SODIUM.

1. BARIUM FROM POTASSIUM AND SODIUM.

12 Precipitate the barium with dilute sulphuric acid (§ 101, 1, *a*), evaporate the filtrate to dryness, and ignite the residue, with addition towards the end of ammonium carbonate (§ 97, 1 and § 98, 1). Take care to add a sufficient quantity of sulphuric acid to convert the alkalies also completely into sulphates. In exact analyses, in order to save the alkali salts adhering to the barium sulphate, remove the dry barium sulphate from the filter, heat it with a sufficient quantity of pure strong sulphuric acid to dissolve it completely, allow to cool, dilute largely, collect the barium sulphate (now almost absolutely pure) on the first filter, ignite, and weigh. Evaporate the filtrate in a platinum dish, drive off the sulphuric acid, and estimate the traces of the alkalies.

This method is, on account of its greater accuracy, preferable to the one in *A*, in cases where the barium has to be separated only from one of the two fixed alkalies; but if both alkalies are present, the other method is more convenient, since the alkalies are then obtained as chlorides.

2. STRONTIUM FROM POTASSIUM AND SODIUM.

13 Strontium may be separated from the alkalies like barium, by means of sulphuric acid; but this method is not preferable to the one in **10**, in cases where the choice is permitted (comp. § 102).

3. CALCIUM FROM POTASSIUM AND SODIUM.

14 Precipitate the calcium with ammonium oxalate (§ 103, 2, *b*, *α*), evaporate the filtrate to dryness, and determine the alkalies in the ignited residue. In determining the alkalies, dissolve the residue, freed by ignition from the ammonium salts, in water, filter if necessary, acidify the filtrate, according to circumstances, with hydrochloric acid or sulphuric acid, and then evaporate to dryness; this treatment of the residue is necessary, because ammonium oxalate partially decomposes chlorides of the alkali metals upon ignition with formation of alkali carbonates, except in presence of a large proportion of ammonium chloride. The results are still more accurate than in *A*, except where ammonium oxalate has been used, after the precipitation by ammonium carbonate, to remove the minute traces of lime from the filtrate.

4. MAGNESIUM FROM POTASSIUM AND SODIUM.*

a. Methods based upon the sparing solubility of Magnesium Hydroxide in Water.

15 α. Make the solution as neutral as possible, and free from ammonium salts (it is a matter of indifference whether the magnesium and alkali metals are present as sulphates, chlorides, or nitrates), add baryta-water* as long as a precipitate forms, heat to boiling, filter, and wash the precipitate with boiling water. The precipitate contains the magnesium as hydroxide. Dissolve it in hydrochloric acid, precipitate the barium with sulphuric acid, and then the magnesium as ammonium-magnesium phosphate (§ 104, 2). The alkalies, which are contained in the solution, according to circumstances, as chlorides, nitrates, or caustic alkalies, are separated from the barium as directed in **10** or **12**. LIEBIG, who was the first to employ this method, proposes crystallized barium sulphide as precipitant. The method is not very exact, as magnesium is somewhat more soluble in solutions of alkali salts than in water. On this account the weighed alkali salt must always be tested for magnesium, and the latter determined if required.

16 β. Precipitate the solution with a little pure milk of lime, boil, filter, and wash. Separate the calcium and magnesium in the precipitate according to **24**; the calcium and the alkalies in the filtrate according to **10** or **14**. This method may be employed when magnesium has to be removed from a fluid containing calcium and alkalies, provided the alkalies alone are to be determined. Minute quantities of magnesium also in this case remain with the alkali salt from the cause mentioned in α.

17 γ. Evaporate the solution of the chlorides (which must contain no other acids) to dryness, and if ammonium chloride is present, ignite; warm the residue with a little water (this will dissolve it with the exception of some magnesium oxide, which separates). Add mercuric oxide shaken up with water,

* The methods *a*, α and β, are suitable for the separation of magnesium from lithium.

evaporate to dryness on the water-bath with frequent stirring, dry thoroughly, ignite with increasing temperature till all the resulting mercuric chloride is volatilized. (Avoid inhaling the fumes.) There is no need to continue the ignition until the whole of the mercuric oxide is expelled; on the contrary, part of it may be filtered off together with the magnesium oxide, and subsequently volatilized upon the ignition of the latter. Treat the residue with small quantities of hot water, filter off rapidly, and wash the magnesium oxide with hot water, using small quantities at a time, and not continuing the operation unnecessarily. The solution contains the alkalies in form of chlorides. This method, proposed by BERZELIUS, gives satisfactory results, and, as far as my experience goes, is the best of those given under *a*. Take care to add the mercuric oxide only in proper quantity, and always test the alkali chlorides for magnesium, a trace of which will generally be found.

b. Method based on the Precipitation of the Magnesium as Ammonium Magnesium Carbonate.

18 Mix the solution of sulphates, nitrates, or chlorides (it must be very concentrated) with an excess of a concentrated solution of sesquicarbonate of ammonia in water and ammonia (230 grm. of the salt, 360 c.c. solution of ammonia sp. gr. ·96, and water to 1 litre). After twenty-four hours filter off the precipitate ($MgCO_3 \cdot (NH_4)_2CO_3 + 4H_2O$), wash it with the solution of ammonia and ammonium carbonate used for the precipitation, dry, ignite strongly and for a sufficient length of time, and weigh the magnesium oxide. Evaporate the filtrate to dryness (keeping the heat at first under 100°, expel the ammonium salts, and determine the alkalies as chlorides or sulphates. When sodium alone is present the results are tolerably satisfactory. In the presence of potassium the ignited magnesium oxide must be extracted with water, before weighing, as it contains an appreciable quantity of potassium carbonate; the washings are to be added to the principal filtrate. This last measure is unnecessary in the absence of potassium. The magnesium is always a little too low. Mean error $\frac{9}{1000}$ (F. G. SCHAFFGOTSCH,* H. WEBER †).

---

* Pogg. Annal. 104, 482. † Vierteljahresschrift f. prakt. Pharm. 8, 161.

II. Separation of the Basic Radicals of the Second Group from each other.

§ 154.

Index. (The numbers refer to those in the margin.)

A. *General Method.*

The whole of the Alkali-earth Metals from each other.

19 Proceed as in **10**. The magnesium is precipitated from the filtrate as ammonium magnesium phosphate. The precipitated carbonates of barium, strontium, and calcium are dissolved in hydrochloric acid, and the bases separated as directed in **20**. The traces of magnesium, which may be present in the ammonium carbonate precipitate, are obtained by evaporating the filtrate from the strontium or calcium sulphate to dryness, taking up the residue with water, and precipitating the solution with sodium phosphate and ammonia.

B. *Special Methods.*

1. *Methods based upon the Insolubility of Barium Silicofluoride.*

Barium from Strontium and from Calcium.

20 Mix the neutral or slightly acid solution with hydrofluosilicic acid* in excess, add one third of the volume of alcohol of ·81 sp. gr., let the mixture stand twelve hours, collect the precipitate of *barium silicofluoride* on a weighed filter, wash with

* If not kept in a gutta-percha bottle it should be freshly prepared.

a mixture of equal parts of water and alcohol until the washings cease to show even the least trace of acid reaction (but no longer), and dry at 100°. Precipitate the strontium or calcium from the filtrate by dilute sulphuric acid (§ 102, 1, *a*, and § 103, 1). The results are satisfactory. For the properties of barium silicofluoride, see § 71. If both strontium and calcium are present, the sulphates are weighed, and then separated according to **26**, or they are converted into carbonates (§ 132, II., *b*), and separated according to **31** or **30**.

2. *Methods based upon the Insolubility of Barium Sulphate or Strontium Sulphate, as the case may be, in Water and in Solution of Sodium Thiosulphate.*

a. BARIUM AND STRONTIUM FROM MAGNESIUM.

21 Precipitate the barium and strontium with sulphuric acid (§ 101, 1, *a* and § 102, 1, *a*), and the magnesium from the filtrate with ammonia and sodium ammonium phosphate (§ 104, 2).

b. BARIUM FROM CALCIUM.

22 Mix the solution with hydrochloric acid, then with highly dilute sulphuric acid (1 part acid to 300 water), as long as a precipitate forms; allow to deposit, and determine the barium sulphate as directed § 101, 1, *a*. Concentrate the washings by evaporation and add them to the filtrate, neutralize the acid with ammonia, and precipitate the calcium as oxalate (§ 103, 2, *b*, *α*). The method is principally to be recommended when small quantities of barium have to be separated from much calcium. If we have to separate calcium sulphate from barium sulphate, the salts may (in the absence of free acids) be treated repeatedly with a solution of sodium thiosulphate at a gentle heat. The barium sulphate remains undissolved, the calcium sulphate dissolves. The calcium is precipitated from the filtrate by ammonium oxalate (DIEHL*).

3. *Method based upon the different deportment with Alkali Carbonates of Barium Sulphate on the one hand, and Strontium and Calcium Sulphates on the other.*

BARIUM FROM STRONTIUM AND CALCIUM.

23 Digest the three precipitated sulphates for twelve hours at

* Journ. f. prakt. Chem. 79, 430.

the common temperature (15°—20°), with frequent stirring, with a solution of ammonium carbonate, decant the fluid on to a filter, treat the residue repeatedly in the same way, wash finally with water, and in the still moist precipitate, separate the undecomposed barium sulphate by means of cold dilute hydrochloric acid from the strontium and calcium carbonates formed. To hasten the separation you may boil the sulphates for some time with a solution of potassium (not sodium) carbonate, to which ⅓ the amount of the carbonate, or more, of potassium sulphate has been added. By this process, also, the strontium and calcium sulphates are decomposed, the barium sulphate remaining unacted on. If the basic metals are in solution, the above solution of potassium carbonate and sulphate is added in excess at once, and the whole boiled. The precipitate, consisting of barium sulphate and strontium and calcium carbonates, is to be treated as above with cold hydrochloric acid (H. Rose*).

4. *Methods based on the Insolubility of Calcium Sulphate in Alcohol.*

CALCIUM FROM MAGNESIUM.

a. Remove water and free hydrochloric from a solution of **24** the chlorides by evaporation, dissolve the residue in strong (but not absolute) alcohol, add a slight excess of pure strong sulphuric acid, digest in the cold, allow to stand for some hours, transfer the precipitate consisting of calcium sulphate and some magnesium sulphate to a filter, wash away the acid thoroughly with nearly absolute alcohol, and then continue the washing with alcohol sp. gr. ·96—·95 till a few drops of the washings give no residue on evaporation. Weigh the calcium sulphate according to § 103, 1. Evaporate the alcohol from the filtrate, and determine the magnesium according to § 104, 2. The method is in itself not new, but A. CHIZYNSKI,† adopting the precautions here given, has obtained excellent results, even in the presence of phosphoric acid.

b. SMALL QUANTITIES OF CALCIUM FROM MUCH MAGNESIUM. **25** Convert into neutral sulphates, dissolve the mass in water, and add alcohol, with constant stirring, till a slight permanent tur-

* Pogg. Annal. 95, 286, 299, 427.

† Zeitschr. f. anal. Chem. 4, 348.

bidity is produced, Wait a few hours and then filter, wash the precipitated calcium sulphate with alcohol which has been diluted with an equal volume of water, and determine it after § 103, 1, *a* (in which case the weighed sulphate must be tested for magnesium), or dissolve the precipitate in water containing hydrochloric acid and separate the calcium from the small quantity of magnesium possibly coprecipitated according to **28** (SCHEERER*).

5. *Methods based on the Insolubility of Strontium and Barium Sulphates in solution of Ammonium Sulphate.*

STRONTIUM FROM CALCIUM.

26 If the mixture is soluble, dissolve in the smallest quantity of water, add about 50 times the quantity of the substance of ammonium sulphate dissolved in four times its weight of water, and either boil for some time with renewal of the water that evaporates and addition of a very little ammonia (as the solution of ammonium sulphate becomes acid on boiling), or allow to stand at the ordinary temperature for twelve hours. Filter and wash the precipitate, which consists of strontium sulphate and a little ammonium strontium sulphate, with a concentrated solution of ammonium sulphate, till the washings remain clear on addition of ammonium oxalate. The precipitate is cautiously ignited, moistened with a little dilute sulphuric acid (to convert the small quantity of strontium sulphide into sulphate), reignited and weighed. The highly dilute filtrate is precipitated with ammonium oxalate, and the calcium determined according to § 103, 2, *b*, *α*. If you have the solid sulphates to analyze, they are very finely powdered and boiled with concentrated solution of ammonium sulphate with renewal of the evaporated water and addition of a little ammonia. Results very close, *e.g.*, 1·048 $Sr(NO_3)_2$ instead of 1·053, and ·497 $CaCO_3$, instead of ·504 (H. ROSE‡).

27 BARIUM may be separated FROM CALCIUM in the same way.

6. *Methods based upon the Insolubility of Calcium Oxalate in Ammonium Chloride and in Acetic Acid.*

CALCIUM FROM MAGNESIUM.

28 *a.* Mix the properly diluted solution with sufficient ammo-

* Annal. d. Chem. u. Pharm. 110, 237.

‡ Pogg. Annal. 110, 296.

nium chloride to prevent the formation of a precipitate by ammonia, which is added in slight excess; add ammonium oxalate as long as a precipitate forms, then a further portion of the same reagent, about sufficient to convert the magnesium also into oxalate (which remains in solution). This excess is absolutely indispensable to insure complete precipitation of the calcium, as calcium oxalate is slightly soluble in magnesium chloride not mixed with ammonium oxalate (Expt. No. 92). Let the mixture stand twelve hours, decant the supernatant clear fluid, as far as practicable, from the precipitated calcium oxalate, mixed with a little magnesium oxalate, on to a filter, wash the precipitate once in the same way by decantation, then dissolve in hydrochloric acid, add water, then ammonia in slight excess, and a little ammonium oxalate. Let the fluid stand until the precipitate has completely subsided, then pour on to the previous filter, transfer the precipitate finally to the latter, and proceed exactly as directed § 103, 2, *b*, *α*. The first filtrate contains by far the larger portion of the magnesium, the second the remainder. Evaporate the second filtrate, acidified with hydrochloric acid, to a small volume, then mix the two fluids, and precipitate the magnesium with sodium ammonium phosphate $(HNaNH_4)PO_4$,* as directed § 104, 2. If the quantity of ammonium salts present is considerable, the estimation of the magnesium is rendered more accurate by evaporating the fluids in a large platinum or porcelain dish to dryness, and igniting the residuary saline mass, in small portions at a time, in a smaller platinum dish, until the ammonium salts are expelled. The residue is then treated with hydrochloric acid and water, warmed, allowed to cool, and rendered just alkaline with ammonia. If enough ammonium chloride is present, no magnesium hydroxide will fall down, but occasionally small flocks of silica or alumina are to be seen. Filter them off and finally precipitate with ammonia and $(HNaNH_4)PO_4$. If the precipitate produced by ammonia is at all considerable, dissolve it in hydrochloric acid, evaporate the solution on a water-bath to dryness, treat the residue with hydrochloric acid and water, render alkaline with ammonia, filter, and add the filtrate to the principal solution.

* This is preferable to sodium phosphate as a precipitant, see MOHR, Zeitschr. f. anal. Chem. 12, 36.

Numerous experiments have convinced me that this method, which is so frequently employed, gives accurate results only if the foregoing instructions are strictly complied with. It is only in cases where the quantity of magnesium present is relatively small that a single precipitation with ammonium oxalate may be found sufficient (comp. Expt. No. 93*).

29 *b.* In the case of calcium and magnesium phosphates, dissolve in the least possible quantity of hydrochloric acid, add ammonia until a copious precipitate forms; redissolve this by addition of acetic acid, and precipitate the calcium with an excess of ammonium oxalate. To determine the magnesium, precipitate the filtrate with ammonia and $(HNaNH_4)PO_4$. As free acetic acid by no means prevents the precipitation of small quantities of magnesium oxalate, the precipitate contains some magnesium, and as calcium oxalate is not quite insoluble in acetic acid, the filtrate contains some calcium; these two sources of error compensate each other in some measure. In accurate analysis, however, these trifling admixtures of magnesium and calcium are afterwards separated from the weighed precipitates of calcium carbonate or oxide and magnesium pyrophosphate respectively.

7. *Method based upon the Insolubility of Strontium Nitrate in Alcohol and Ether.*

STRONTIUM FROM CALCIUM (*after* STROMEYER).

30 Digest the perfectly dry nitrates in a closed flask with absolute alcohol, to which an equal volume of ether should be added (H. ROSE). Filter off the undissolved strontium nitrate in a covered funnel, wash with the mixture of alcohol and ether, dissolve in water, and determine as strontium sulphate (§ 102, 1). Precipitate the calcium from the filtrate by sulphuric acid. The results are satisfactory.

8. *Indirect Method.*

STRONTIUM FROM CALCIUM.

31 Determine both bases first as carbonates or oxides, precipi-

* Further experiments will be found in Zeitschr. f. anal. Chem. 7, 310. Compare also WITTSTEIN, Zeitschr. f. anal. Chem. 2, 318, and COSSA, *Ib.* 8, 141. According to HAGER, *Ib.* 9, 254, the precipitate of calcium oxalate will be free from magnesium if filtered off immediately; however, I fear that a little calcium might in this case be left in solution.

tating them either with ammonium carbonate or oxalate (§§ 102, 103); then estimate the amount of carbonic acid in them, and calculate the amount of strontium and calcium as directed in "Calculation of Analyses." The determination of the carbonic acid may be effected by fusion with vitrified borax (§ 139, II., *c*), but the application of a moderate white heat, such as is given by a good gas blowpipe without the use of a crucible jacket, is alone sufficient to drive out all the carbonic acid from both the carbonates (F. G. SCHAFFGOTSCH*). I can strongly recommend this method. It is well to precipitate the carbonates hot, to press the precipitate cautiously down in the platinum crucible and turn over the agglomerated cake every now and then till, after repeated ignitions, the weight has become constant. The results are good if neither of the bases is present in too minute quantity.

32 The indirect separation may, of course, be effected by means of other salts, and can be used also for the determination of CALCIUM IN PRESENCE OF BARIUM or of BARIUM IN PRESENCE OF STRONTIUM. In the expulsion of carbonic acid from barium carbonate vitrified borax must be used (§ 139, II., *c*).

Third Group.

ALUMINIUM—CHROMIUM.

I. SEPARATION OF ALUMINIUM AND CHROMIUM FROM THE ALKALIES.

§ 155.

1. FROM AMMONIUM.

33 *a.* Aluminium and chromium salts may be separated from ammonium salts by ignition. However, in the case of aluminium, this method is applicable only in the absence of chlorine (volatilization of aluminium chloride). The safest way, therefore, is to mix the compound with sodium carbonate before igniting.

34 *b.* Determine the ammonium by one of the methods given in § 99, 3, using solution of potassa or soda to effect the expulsion of ammonia. The aluminium and chromium are then determined in the residue in the same way as in **35**.

* Pogg. Annal. 113, 615.

2. FROM POTASSIUM AND SODIUM.

35 *a* Precipitate and determine the chromium and aluminium with ammonia as directed in § 105, *a*, and § 106, 1, *a*. The filtrate contains the alkalies, which are then freed from the ammonium salt formed, by evaporation to dryness and ignition. In the presence of large quantities of alkali salts it is well to dissolve the moderately ignited precipitate in hydrochloric acid, and reprecipitate with ammonia.

36 *b*. *Aluminium* may be separated also from potassium and sodium by heating the nitrate (see **38**).

II. SEPARATION OF ALUMINIUM AND CHROMIUM FROM THE ALKALI-EARTH METALS.

§ 156.

INDEX. (The numbers refer to those in the margin.)

SEPARATION OF ALUMINIUM FROM THE ALKALI-EARTH METALS.

A. *General Methods.*

THE WHOLE OF THE ALKALI-EARTH METALS FROM ALUMINIUM.

1. *Method based upon the Precipitation of Aluminium Hydroxide by Ammonia, and upon its solution in Soda.*

37 Put the solution in a platinum dish or, with less advantage, a porcelain dish. Let it be dilute and warm. Add a tolerable quantity of ammonium chloride, if such be not already present, then very gradually, almost drop by drop (WRINKLE*), ammonia as free as possible from carbonic acid, in moderate excess, and boil till no more free ammonia is observable. Under these circumstances, a little magnesium hydroxide, and also a small quantity of calcium, barium, or strontium carbonates are at first precipitated along with the aluminium hydroxide; on the boil-

* Zeitschr. f. anal. Chem. 10, 96.

ing with ammonium chloride, the coprecipitated alkali-earth metal compounds redissolve, so that the aluminium hydroxide finally retains only an unweighable or scarcely weighable trace of them. Allow to deposit, and proceed with the aluminium determination according to § 105, *a*. In very exact analysis it is well, after moderately washing the aluminium precipitate, to redissolve it in hydrochloric acid, and reprecipitate with ammonia as above. In separations of aluminium from calcium or magnesium this double precipitation is especially necessary in the presence of sulphates. After the aluminium oxide has been weighed, fuse it for a long time with potassium disulphate, dissolve the fused mass in water, and determine any silicic acid* that may remain. The solution, when mixed with potassa in excess, will often not appear perfectly clear, but will contain a few flocks of magnesium hydroxide (perhaps also traces of barium, strontium, or calcium carbonates). If there is any amount of the latter, filter it off, dissolve in nitric acid, precipitate with ammonia, boil till the fluid ceases to smell of ammonia, filter, evaporate the small quantity of fluid in a platinum capsule, ignite, weigh the residual magnesium oxide (which may contain traces of other alkali-earth metals), deduct it from the aluminium oxide, dissolve it in hydrochloric acid, and add to the original filtrate. In order to the further separation of the alkali-earth metals, acidify the fluid containing them with hydrochloric acid, evaporate (preferably in a platinum dish) to a small bulk, and while still warm add ammonia just in excess. A small precipitate of aluminium hydroxide is sometimes formed at this stage; filter off, wash, and weigh with the principal precipitate. In the filtrate determine the alkali-earth metals according to § 154.

[The difficulty of washing aluminium hydroxide usually increases with lapse of time between precipitation and filtration. This difficulty may be to some extent obviated by the following slight modification of the above-described manipulation. Add ammonia to the solution, which may occupy a volume of 400 c.c. for ·2 gr. Al_2O_3, until free acid is partially neutralized, but not until a permanent precipitate is formed; add

* A small quantity will always be found if you have boiled in a glass or porcelain vessel.

also ammonium chloride if but little free acid was present. Heat nearly to boiling, and add ammonia slowly until a permanent precipitate begins to form, then drop by drop until a slip of red litmus-paper dipped into the fluid changes to blue and the odor of ammonia becomes perceptible on boiling. *Carefully avoid* the use of more ammonia than is sufficient to produce these indications of a *slight* excess. Boil rapidly 7 to 10 minutes, allow the precipitate to settle 5 to 10 minutes, filter and wash the precipitate moderately upon the filter. Remove the filter with the moist precipitate from the funnel, and unfold it upon the side of a beaker having a height exceeding the diameter of the filter, so that the latter may not extend to the bottom of the beaker. Rinse the precipitate from the filter down to the bottom of the beaker with a strong jet of water and dissolve (completely or nearly) by adding concentrated hydrochloric acid. Moisten also the filter with acid somewhat diluted, and rinse the small amount of aluminium chloride solution thus formed out of the paper with a jet of water. Push up the filter now, if necessary with a rod, so that it may be above the solution, and allow it to remain adhering to the side of the beaker. The solution need not, for this second precipitation, occupy a volume above 200—250 c.c. Precipitate the aluminium precisely as before, moistening also the filter with ammonia solution. Immediately after boiling pour the solution with the precipitate upon a filter. Push the old filter down to the bottom of the beaker, wash it by adding and decanting small successive portions of hot water, stirring and pressing the paper with a rod and pouring the water upon the precipitate, until a few drops of the decanted water give no turbidity with silver nitrate. Next complete the washing of the precipitate on the filter with hot water. After the washing is complete, beat up the old filter in the beaker with a glass rod and rinse it out upon the top of the washed precipitate—the old filter must on no account be thrown away, since it may retain a little aluminium hydroxide which treatment with hydrochloric acid failed to dissolve. Add to the united filtrates ammonia to decided alkaline reaction; heat until the solution becomes neutral. If more aluminium hydroxide separates, collect it on a small filter.]

2. *Method based upon the unequal Decomposability of the Nitrates at a Moderate Heat* (DEVILLE*).

38 To make this simple and convenient method applicable, the basic metals must be present as pure nitrates. Evaporate to dryness in a platinum dish, and heat gradually, with the cover on, in the sand- or air-bath—or, better still, on a thick iron disk, with two cavities, one for the platinum dish, the other, filled with brass turnings, for inserting a thermometer—to from 200° to 250°, until a glass rod moistened with ammonia ceases to indicate further evolution of nitric-acid fumes. You may also, without risk, continue to heat until nitrous-acid vapors form. The residue consists of aluminium oxide, barium, strontium and calcium nitrates, and normal and basic magnesium nitrates.

Moisten the mass with a concentrated solution of ammonium nitrate, and heat gently, but do not evaporate to dryness. Repeat this operation until no further evolution of ammonia is perceptible. (The basic magnesium nitrate, insoluble in water, dissolves in nitrate of ammonia, with evolution of ammonia, as normal magnesium nitrate.) Add water, and digest at a gentle heat.

[If the ammonium nitrate has evolved only imperceptible traces of ammonia, pour hot water into the dish, stir, and add a drop of dilute ammonia; this must cause no turbidity in the fluid; should the fluid become turbid, this proves that the heating of the nitrates has not been continued long enough; in which case you must again evaporate the contents of the dish, and heat once more.]

The aluminium oxide remains undissolved in the form of a dense granular substance. Decant after digestion, and wash with boiling water; ignite strongly in the same vessel in which the separation has been effected, and weigh. Test the weighed aluminium oxide according to **37**. Separate the alkali-earth metals as directed § 154.

In the same way aluminium may be separated also from potassium and sodium (**36**.)

3. *Method in which the processes of* 1 *and* 2 *are combined.*

39 Precipitate the aluminium as in **37**, wash in the same way

* Journ. f. prakt. Chem. 1853, 60, 9.

as there directed, then treat, while still moist, with nitric acid, and proceed according to **38**, to remove the trifling amount of magnesium, etc., coprecipitated; add the solution obtained to the principal solution of the alkaline earths, and treat the fluid as directed in **37**. This method may be employed also in the case of chlorides; it will be sometimes found useful.

4. *Method based upon the Precipitation of Aluminium by Sodium Acetate or Formate upon boiling.*

40 The same process as for the separation of ferric iron from the alkali-earth metals. The method is employed more particularly when both aluminium and ferric iron have to be separated from alkali-earth metals at the same time. The precipitation of the aluminium is usually not quite complete, so that it will be necessary to separate the aluminium which remains in solution from the filtrate (**37**).

5. *Method based on the Precipitation of Aluminium by Ammonium Succinate.*

41 Proceed as for the precipitation of ferric iron by the same reagent (§ 159); especially to be employed when aluminium and ferric iron are both to be separated from alkali-earth metals at the same time. The filtrate must be tested according to **40.**

B. *Special Methods.*

SOME OF THE ALKALI-EARTH METALS FROM ALUMINIUM.

1. *Methods based upon the Precipitation of some of the Salts of the Alkali-earth Metals.*

a. BARIUM AND STRONTIUM FROM ALUMINIUM.

43 Precipitate the barium and strontium with *sulphuric acid* (§§ 101 and 102), and the aluminium from the filtrate as directed § 105, *a.* This method is especially suited for the separation of barium from aluminium. In accurate analyses the barium sulphate must be purified according to **12.**

b. CALCIUM FROM ALUMINIUM.

44 Add ammonia to the solution until a permanent precipitate forms, then acetic acid until this precipitate is redissolved, then ammonium acetate, and finally *ammonium oxalate* in slight excess (§ 103, 2, *b*, *β*); allow the precipitated calcium oxalate to deposit in the cold, then filter, and precipitate the aluminium from the filtrate as directed § 105, *a*.

2. Method based upon the Precipitation of Aluminium by Barium Carbonate.

ALUMINIUM FROM MAGNESIUM AND SMALL QUANTITIES OF CALCIUM.

46 Mix the slightly acid dilute fluid in a flask, with a moderate excess of barium carbonate shaken up with water; cork the flask and let the mixture stand in the cold until the aluminium hydroxide has subsided, wash by decantation three times, filter, and then determine the aluminium in the precipitate as directed **43**; in the filtrate, first precipitate the barium by sulphuric acid (**22**), and then separate the calcium and magnesium according to § 154.

SEPARATION OF CHROMIUM FROM THE ALKALI-EARTH METALS.

1. The best way to separate THE WHOLE OF THE ALKALI-EARTH METALS from chromium at the same time is to convert the latter into chromic acid. This may be done in the dry or the wet way.

47 *a. Dry way.* Mix the powdered substance with about 8 times its weight of a mixture of 2 parts of sodium carbonate and 1 part of nitre, and fuse in a platinum crucible. On treating the fused mass with hot water, the chromium dissolves as alkali chromate (to be determined according to § 130), while the alkali-earth metals remain in the residue as carbonates or oxides (magnesium oxide). If the residue is not perfectly white, extract the remainder of the chromic acid from it by boiling with solution of sodium carbonate.

48 *b. Wet way.* Suitable for separating chromium from barium, strontium, and calcium.

Nearly neutralize the acid fluid with sodium carbonate, add excess of sodium acetate, warm and pass chlorine, adding sodium carbonate occasionally to keep the fluid nearly neutral. As soon as all the chromium is oxidized, precipitate with sodium carbonate by the aid of heat, and proceed for the rest according to **47** (GIBBS*). Bromine instead of chlorine may be used;

* Zeitschr. f. anal. Chem. 3, 328.

however, the oxidation is but tardily effected by the mere addition of bromine water.

2. CHROMIUM FROM BARIUM, STRONTIUM, AND CALCIUM. To **49** separate barium and strontium, precipitate the moderately acid, hot, dilute solution with sulphuric acid—in the presence of strontium, allow to cool and add alcohol—and when the precipitate has settled, filter. Chromium cannot be separated by ammonia from the alkali-earth metals, since, even though carbonic acid be completely excluded, they are partially precipitated along with the chromic hydroxide. From solutions containing a salt of chromium, calcium cannot be precipitated completely by ammonium oxalate; but it may be by sulphuric acid and alcohol (§ 103, 1).

3. CHROMIUM may also be separated from MAGNESIUM and **50** small quantities of CALCIUM by means of barium carbonate. See **46**.

III. SEPARATION OF CHROMIUM FROM ALUMINIUM.*

§ 157.

a. Fuse the oxides with 2 parts of potassium nitrate and 4 **51** parts of sodium carbonate in a platinum crucible, treat the fused mass with boiling water, rinse the contents of the crucible into a porcelain dish or beaker, add a somewhat large quantity of potassium chlorate, supersaturate slightly with hydrochloric acid, evaporate to the consistence of syrup, and add, during the latter process, some more potassium chlorate in portions, to remove the free hydrochloric acid. Dilute now with water, and separate the aluminium and chromium as directed § 130, II., *c*, *α*. If you omit the evaporation with hydrochloric acid and potassium chlorate, part of the chromic acid will be reduced by the nitrous acid in the fluid, and chromic hydroxide will accordingly, upon addition of ammonia, be precipitated with the aluminium hydroxide (DEXTER†).

b. Dissolve the oxides in hydrochloric acid, add soda or **52** potassa solution in sufficient excess and saturate the clear green solution with chlorine gas. The chromium will be converted

---

* The separation of aluminium from titanic acid will be given under the Analysis of Silicates.

† Pogg. Anal. 89, 142.

into chromic acid, and the aluminium partially separated. When the fluid has become of a pure yellow color, heat to remove the excess of chlorine, add ammonium carbonate, and digest to destroy the hypochlorous acid and precipitate the still dissolved aluminium, and proceed according to § 130, II., *c*, *α* (WÖHLER*).

c. Nearly neutralize the acid solution with sodium carbonate, add sodium acetate in excess, pass chlorine or add bromine and warm. The chromium will readily be converted into chromic acid, especially if sodium carbonate is added every now and then to keep the fluid nearly neutral. As soon as this is effected proceed according to § 130, II., *c*, *α* (GIBBS†). **53**

Fourth Group.

ZINC—MANGANESE—NICKEL—COBALT—FERROUS IRON—FERRIC IRON—(URANIUM).

I. SEPARATION OF THE METALS OF THE FOURTH GROUP FROM THE ALKALIES.

§ 158.

A. *General Methods.*

1. ALL METALS OF THE FOURTH GROUP FROM AMMONIUM.

Proceed as for the separation of chromium and aluminium from ammonium, **33**. It must be borne in mind that the oxides of the fourth group comport themselves, upon ignition with ammonium chloride, as follows: Ferric oxide is partly converted into ferric chloride which volatilizes; the oxides of manganese are converted into manganous chloride and manganous oxide with volatilization of some of the former;‡ the oxides of nickel and cobalt are reduced to the metallic state, no chloride being lost by volatilization;§ oxide of zinc is converted into chloride which volatilizes. It is, therefore, generally the safest way to add sodium carbonate. The ammonium is determined in a separate portion. **54**

* Anal. d. Chem. u. Pharm. 106, 121.
† Zeitschr. f. anal. Chem. 3. 327.
‡ Zeitschr. f. anal. Chem. 11, 424.
§ *Ib.* 12, 73.

2. All Metals of the Fourth Group from Potassium and Sodium.

55 Mix the solution in a flask with ammonium chloride if necessary, add ammonia till neutral or slightly alkaline, then yellow ammonium sulphide saturated with hydrogen sulphide, fill the flask nearly to the top with water, cork it, allow the precipitated sulphides to subside, and then filter them off from the fluid containing the alkalies. In performing this process the precautionary rules given under the heads of the several metals in question (§§ 108—113) must be borne in mind.* (If, notwithstanding, the filtrate is brownish, acidify it with acetic acid, pass hydrogen sulphide, boil and filter off the small quantity of the nickel sulphide which then separates.) Acidify the filtrate with hydrochloric acid, evaporate, filter off the sulphur if necessary, continue the evaporation to dryness, ignite the residue to remove ammonium salts, and determine the alkalies by the methods given § 152.

B. *Special Methods.*

56 1. Zinc from Potassium and Sodium, by precipitating the zinc from the solution of the acetates with hydrogen sulphide (see **73**).

2. Ferric Iron from Potassium and Sodium, by precipitating with ammonia; or by heating the nitrates (see **37** and **38**).

57 3. Manganese from the Alkalies. Mix the neutral or slightly acid solution with ammonium chloride and precipitate the manganese with a slight excess of ammonium carbonate. Allow the precipitate to settle in a warm place, filter through a thick filter, wash with hot water and weigh as protosesquioxide (H. Tamm†) In the filtrate separate the alkalies from ammonium salts by gentle ignition. The separation of manganese as hydrated peroxide cannot be recommended, as the precipitate retains alkali.‡

---

* Manganese may be separated from the alkalies according to § 109, 1, *c.* 2, *b.* Nickel and cobalt may be separated from the alkalies according to **58**, substituting ammonium acetate for sodium acetate.

† Zeitschr. f anal. chem. 11, 425. ‡ *Ib.* 11, 298.

II. Separation of the Metals of the Fourth Group from those of the Second.

§ 159.

A. *General Method.*

All Metals of the Fourth Group from the Alkali-earth Metals.

Add ammonium chloride, and, if acid, also ammonia, and **58** precipitate with ammonium sulphide, as in **55**. Take care to use slightly yellow ammonium sulphide, perfectly saturated with hydrogen sulphide, and free from ammonium carbonate and sulphate, and to employ it in sufficient excess. Insert the cork, and let the flask stand for some time to allow the precipitate to subside, then wash quickly, and as far as practicable out of the contact of air, with water to which some ammonium sulphide has been added. Acidify the filtrate with hydrochloric acid, heat, filter from the sulphur, and separate the alkali-earth metals, as directed § 154. If the filtrate is brownish from a little dissolved nickel sulphide, make it *slightly* acid with acetic acid instead of with hydrochloric acid, add some alkali acetate, pass hydrogen sulphide, boil, and filter.

If the quantity of the alkali-earth metals is rather considerable, it is advisable to treat the slightly washed precipitate once more with hydrochloric acid (in presence of nickel or cobalt, it is not necessary to effect complete solution), heat the solution gently for some time, and then reprecipitate in the same way.

[If we have merely to effect removal of nickel and cobalt, we may also, after making neutral or slightly alkaline with ammonia, acidify *slightly* with acetic acid, add alkali acetate, heat, and while boiling pass H_2S gas through the solution. Presence of ammonium salts facilitates separation of the nickel and cobalt. Compare § 79, *e*.]

B. *Special Methods.*

1. BARIUM AND STRONTIUM FROM THE WHOLE OF THE METALS OF THE FOURTH GROUP. **59**

Precipitate the barium and strontium from the slightly acid solution with sulphuric acid (§§ 101, 102). The barium sulphate should first be washed with water acidified with hydrochloric acid, but even then you cannot be sure of getting it free from iron. The sulphates after weighing must therefore always be tested for iron, etc.

2. ZINC FROM THE ALKALI-EARTH METALS. **60**

Convert the basic metals into acetates, and precipitate the zinc from the solution according to § 108, 1, *b*.

3. FERRIC IRON FROM THE ALKALI-EARTH METALS. **61**

a. Mix the somewhat acid solution with enough ammonium chloride, boil, add slight excess of ammonia, boil till the excess of the latter is nearly expelled, and filter. The solution is free from iron, the precipitate is free from calcium, barium, and strontium, but contains a very slight trace of magnesium (H. ROSE *). In delicate analyses, after moderately washing the ferric hydroxide, redissolve it in hydrochloric acid, and repeat the precipitation.

b. Precipitate the iron as basic ferric acetate or formate, compare **71**. The method is good, and can frequently be employed.

c. Decompose the nitrates by heat (**38**). A good method.†

4. MANGANESE FROM CALCIUM AND MAGNESIUM. **62**

[The solution must not contain ammonium salts. The manganese, calcium, and magnesium may be present as chlorides, nitrates, or acetates (or sulphates if but little calcium is in the solution and care be taken to avoid deposition of calcium sulphate). Neutralize any free acid which may be present by adding sodium carbonate till a slight precipitate is formed. Redissolve this precipitate by the addition of just sufficient HCl. Add next sodium acetate to the solution, then aqueous solution

---

* Pogg. Annal. 100, 300.

† Compare LATSCHINOW, Zeitschr. f. anal Chem. 7, 213.

of bromine. The solution should at this point be rather dilute.' Expose to a temperature of 50° to 70° a few hours, till free bromine is all or nearly all expelled from the solution, and filter. Test the filtrate by adding more sodium acetate and more bromine water, and warming. The manganese is precipitated as hydrated dioxide which is liable to contain soda. If the quantity is very small, it may, unless great accuracy is required, be converted by ignition, after careful washing with hot water, directly into Mn_3O_4, and weighed. If, however, the quantity is considerable, it should be dissolved in HCl and converted into some other suitable form for weighing.

According to FINKENER,* manganese dioxide precipitated as above described (except using chlorine instead of bromine) from a solution containing the alkali-earth metals, will not be entirely free from the latter, especially from barium if that is present. He recommends to dissolve the manganese precipitate and reprecipitate boiling hot with ammonium sulphide, by which means pure manganese sulphide is obtained. GIBBS† observes that when manganese is separated from zinc, calcium, and magnesium by the above process (precipitation as binoxide), a repetition of the process is necessary to secure good results; but in case manganese is to be separated only from calcium and magnesium, the second treatment may be omitted.‡]

5. COBALT, NICKEL, AND ZINC, FROM BARIUM, STRONTIUM, AND CALCIUM.

63 Mix with sodium carbonate in excess, add potassium cyanide, heat very gently, until the precipitated carbonates of cobalt, nickel, and zinc are redissolved; then filter the alkali-earth carbonates from the solution of the cyanides in potassium cyanide. The former are dissolved in dilute hydrochloric acid, and separated according to § 154; the latter are separated according to § 160 (HAIDLEN and FRESENIUS§).

* Handbuch d. anal. Chem. v. H. ROSE, 6 Aufl. v. FINKENER, 2, 925.

† Zeitschr. f. anal. Chem. 3, 321.

‡ E. A. COLBY (priv. contrib.) finds, by experiments made in the Sheffield Laboratory on the separation of Ca from Mn, that by proceeding as above directed only a slight unweighable trace of Ca goes down with the Mn; while if the amount of free acetic acid is moderately increased, the manganese is precipitated *entirely free* from calcium. Too much acetic acid, however, prevents or delays precipitation of Mn.

§ Annal. d. Chem. u. Pharm. 43, 140.

III. Separation of the Metals of the Fourth Group from those of the Third, and from each other.

§ 160.

INDEX. (The numbers refer to those in the margin.)

A. *General Methods.*

1. *Method based upon the Precipitation of some Basic Radicals by Barium Carbonate.*

FERRIC IRON, ALUMINIUM, AND CHROMIUM, FROM ALL OTHER BASIC RADICALS OF THE FOURTH GROUP.

64 Mix the sufficiently dilute solution of the chlorides or nitrates, but not sulphates, which must contain a little free acid,* in a flask, with a moderate excess of barium carbonate diffused in water; cork, and allow to stand some time in the cold, with occasional shaking. The ferric iron, aluminium, and chromium are completely separated,† whilst the other basic radicals remain in solution, with the exception perhaps of traces of cobalt and nickel, which will generally fall down with the precipitate. This may be prevented, at least as regards nickel, by addition of ammonium chloride to the fluid to be precipitated (SCHWARZENBERG‡). Decant, stir up with cold water, allow to deposit, decant again, filter, and wash with cold water. The precipitate contains, besides the precipitated metals, barium carbonate; and the filtrate, besides the non-precipitated metals, a barium salt.

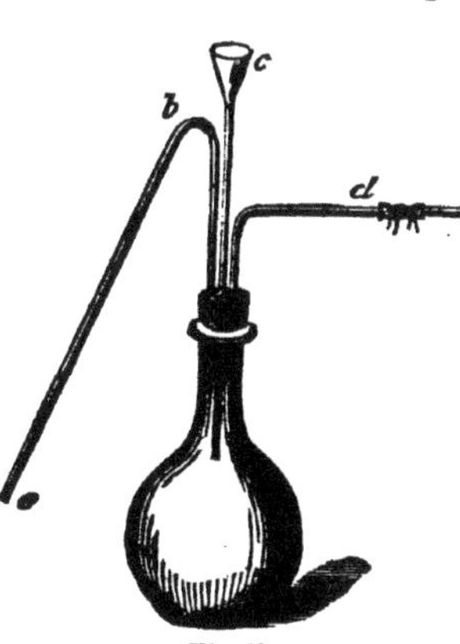

Fig. 68.

If ferrous iron is present, and it is wished to separate it by this method from ferric iron, etc., the air must be excluded during the whole of the operation. In that case, the solution of the substance, the precipitation, and the washing by decantation, are effected in a flask (*A*, fig. 68), through which carbonic acid is transmitted (*d*). The washing water, boiled free from air, and cooled out of contact of air (preferably in a current of carbonic acid), is poured in through a funnel tube (*c*), and the fluid drawn off by means of

---

* If there is much free acid, the greater part of it must first be saturated with sodium carbonate.

† The separation of the chromium requires the most time.

‡ Annal. d. Chem. u. Pharm. 97, 216.

a movable syphon (*b*); all the tubes are fitted air-tight into the cork; they are smeared with tallow.

2. *Method based upon the Precipitation of the Metals of the Fourth Group by Sodium Sulphide or Ammonium Sulphide, from Alkaline Solution effected with the aid of Tartaric Acid.*

ALUMINIUM AND CHROMIUM FROM THE METALS OF THE FOURTH GROUP.

Mix the solution with pure normal potassium tartrate,* then with pure solution of soda or potassa until the fluid has cleared again;† add sodium sulphide as long as a precipitate forms, allow it to deposit until the supernatant fluid no longer exhibits a greenish or brownish tint; decant, stir the precipitate up with water containing sodium sulphide, decant again, transfer the precipitate, which contains all the metals of the fourth group, to a filter, wash with water containing sodium sulphide, and separate the metals as directed in B. Add to the filtrate potassium nitrate, and evaporate to dryness; fuse the residue in a platinum dish, and separate the aluminium from the chromic acid formed as directed § 157. If you have merely to separate aluminium from the metals of the fourth group, it is better, after addition of potassium tartrate, to supersaturate with ammonia, add ammonium chloride, and precipitate in a flask with ammonium sulphide. When the precipitate has settled it is filtered off and washed with water containing ammonium sulphide. The filtrate is evaporated in a platinum dish with sodium carbonate and potassium nitrate to dryness, fused, and the aluminium determined in the residue. **65**

B. *Special Methods.*

1. *Methods based upon the Solubility of Aluminium Hydroxide in Caustic Alkalies.*

a. ALUMINIUM FROM FERROUS AND FERRIC IRON, AND SMALL QUANTITIES OF MANGANESE (but not from nickel and cobalt).

Mix the hydrochloric solution with sodium carbonate or **66**

---

* Tartaric acid often contains aluminium, therefore this is best made from the acid tartrate.

† Chromium and zinc cannot be obtained together in alkaline solution (CHANCEL, Compt. rend. 43, 927; Journ. f. prakt. Chem. 70, 378).

pure potash till the greater part of the free acid is neutralized, and pour the solution gradually into excess of pure potash heated nearly to boiling in a platinum or silver dish, stirring all the while. Porcelain does not answer so well, and glass should on no account be used. The iron, if present as ferric chloride, separates as ferric hydroxide, while the aluminium remains in solution as alkali aluminate. Hydrated protosesquioxide of iron is more easy to wash than ferric hydroxide, hence when much iron is present it is better to reduce a part by cautiously adding sodium sulphite and heating, so that when the solution is added to the boiling potash a black granular precipitate may be formed. The iron precipitate is sure to contain alkali, and must be dissolved in hydrochloric acid, the solution boiled with nitric acid if necessary, and reprecipitated with ammonia.

To the alkaline filtrate add a few drops of hydrochloric acid. If the potash was present in sufficient excess the precipitate will redissolve readily on stirring. Continue adding hydrochloric acid till in excess, boil with a little potassium chlorate (to destroy traces of organic matter), concentrate by evaporation, and throw down the aluminium according to § 105, *a*. The above is the best method of procedure, but it is always to be feared that small quantities of aluminium will be retained by the iron precipitate.

b. Aluminium from Ferrous and Ferric Iron, Cobalt, and Nickel.

67 Fuse the oxides with potassium hydroxide in a silver crucible, boil the mass with water, and filter the alkaline fluid, which contains the aluminium, from the oxides, which are free from aluminium, but contain potassa (H. Rose).

2. *Methods based on the different behavior of Ammonia or Ammonium Carbonate in the presence of Chloride with solutions of certain basic radicals.*

a. Aluminium and Ferric Iron from Cobalt and Nickel.

68 Ferric iron may be completely separated from these metals by mixing the hot solution with ammonium chloride, and then with excess of ammonia, digesting for several hours, washing the precipitate, redissolving in hydrochloric acid, reprecipitating with ammonia, and repeating the operation a third time. Nickel and cobalt are to be precipitated from the filtrate after concentration to a small volume, as directed in § 110, 1, *b*, *β*.

In separating iron and aluminium from nickel and cobalt, it is well to substitute ammonium carbonate for ammonia, so as to insure the complete precipitation of the aluminium.

b. MANGANESE FROM NICKEL AND ZINC.

69 The solution should be slightly acid and contain ammonium chloride. Precipitate the manganese as white carbonate with ammonium carbonate, allow to settle in a warm place, filter through a thick paper, if necessary double, wash with hot water, dry the precipitate and convert it into protosesquioxide by ignition with access of air. This excellent method was proposed by TAMM,* and has given me good results.† It is not adapted to the separation of cobalt from manganese, as the former is partly precipitated with the latter.

3. *Method based upon the different deportment of neutralized Solutions at boiling heat.*

a. FERRIC IRON FROM MANGANESE, NICKEL AND COBALT, AND OTHER STRONG BASIC METALS, AFTER HERSCHEL,‡ SCHWARZENBERG,§ AND MY OWN EXPERIMENTS.

70 Mix the dilute solution largely with ammonium chloride (at least 40 of NH_4Cl to 1 of MnO,NiO, &c.), add ammonium carbonate in small quantities, at last drop by drop and in very dilute solution, as long as the precipitated iron redissolves, which takes place promptly at first, but more slowly towards the end. As soon as the fluid has lost its transparency, without showing, however, the least trace of a distinct precipitate in it, and fails to recover its clearness after standing some time in the cold, but, on the contrary, becomes rather more turbid than otherwise, the reaction may be considered completed. When this point has been attained, heat slowly to boiling, and keep in ebullition for a short time after the carbonic acid has been entirely expelled. The iron separates as a basic ferric salt, which rapidly settles, if the solution was not too concentrated. Pour off the hot fluid through a filter and wash by decantation combined with filtration with boiling water containing a little

* Chem. News, 26, 37.

† Zeitschr. f. anal. Chem. 11, 425.

‡ Annal. de Chim. et de Phys. 49, 306.

§ Annal. d. Chem. u. Pharm. 97, 216.

ammonium chloride. It is well to redissolve the precipitate in hydrochloric acid, and throw down the iron with ammonia. The first filtrate should be mixed with excess of ammonia. If a small portion of ferric hydroxide is thrown down here, filter it off, dissolve in hydrochloric acid, precipitate with ammonia and thus free the small quantity of iron entirely from the strong basic metals; if, on the other hand, a larger quantity of iron is thrown down, this is a sign that the operation has been conducted improperly, and the hydrochloric solution of the precipitate must be reprecipitated as above. The fluid should not contain more than 2 or 3 grm. of iron in the litre, and should be tolerably free from sulphuric acid, as when this is present it is impossible to hit the exact point of saturation.

4. *Method based on the behavior of the Acetates at a boiling heat.*

FERRIC IRON AND ALUMINIUM FROM MANGANESE, ZINC, COBALT, NICKEL, AND FERROUS IRON.

The metals should be present in the form of chlorides. The solution should be in a flask. If much free acid is present first nearly neutralize with sodium or ammonium carbonate; the solution should remain clear, but if there is much ferric chloride it should be of a deep red color. Add a concentrated solution of neutral sodium or ammonium acetate, not in large excess, and boil for a short time—long-continued boiling would make the precipitate slimy. When the lamp is removed the precipitate should settle rapidly, leaving the supernatant fluid clear. Wash the precipitate immediately by decantation and filtration with boiling water containing a little sodium or ammonium acetate. In very particular analyses it would be well after washing the precipitate a little to redissolve it in hydrochloric acid and reprecipitate. 71

In separating ferric from ferrous iron REICHARDT* recommends a slight addition of ammonium chloride or of sodium chloride to prevent oxidation of the ferrous salt.

The precipitate of basic ferric acetate or basic aluminium acetate is best dissolved in hydrochloric acid, in order to precipitate the basic metals from this solution again by ammonia.

* Zeitschr. f. anal. Chem. 5, 64.

This method is more suitable to the separation of ferric iron or ferric iron and aluminium from the strong basic metals than to the separation of aluminium alone. It is a good method, and is very generally used.

[The results obtained by this method depend greatly on the proper adjustment of free acetic acid to the volume of the solution which is boiled. The solution at this point may contain about four per cent. (by volume) of acetic acid sp. gr. 1·044 (JEWETT*). If too little acetic is present, zinc, manganese, nickel, and cobalt are precipitated in notable quantity along with the iron. If too much is present the precipitation of iron is incomplete. The operator may control the amount of acid within narrow limits by proceeding as follows. Add the alkali carbonate to the *cold* and preferably concentrated solution until a slight precipitate forms which no longer redissolves in four or five minutes with occasional shaking, but imparts a turbidity to the deep red solution; HCl is then added without further delay, slowly, drop by drop, until the fluid, though still dark, becomes clear. Next the amount of acetic acid required to form four per cent. of the final volume is added, then sodium acetate (about ten times as much of the crystallized salt as there is iron present, or more if but little iron is present). Dilute now to the final volume, which should amount to at least 100 c.c. per ·1 grm. iron and heat to boiling. After boiling two or three minutes only, allow the iron precipitate to settle. Pour the clear liquid through a filter, then bring the precipitate upon the filter at once and wash as above directed. The iron precipitate contains no zinc and but an inappreciable trace of manganese. Small quantities of cobalt and still more nickel will, however, be precipitated with the iron. When these two metals are present in considerable quantity a repetition of the process is indispensible when accuracy is required. Coprecipitation of nickel is lessened but not entirely prevented by presence of ammonium chloride.†

In carrying out the process according to this plan great care must be taken in the preliminary neutralization with alkali carbonate to leave as little free mineral acid as possible without formation of a permanent precipitate, otherwise this free acid

* Am. Chem. Jour. I. 251. † *Loc. cit.*

will liberate enough acetic acid from the sodium acetate to prevent (with that intentionally added) the precipitation of iron in a form easy to wash.

In separating large quantities of iron from small quantities of manganese the addition of 2 or 3 per cent. of acetic acid will secure a separation satisfactory enough for most purposes (e.g. in iron and iron ores), and the danger that the acetic acid present may accidentally exceed the proper limit will of course be lessened.]

5. *Method based on the different behavior of the Succinates.*

FERRIC IRON (AND ALUMINIUM) FROM ZINC, MANGANESE, NICKEL, AND COBAT.

72 The solution should contain no considerable quantity of sulphuric acid. If acid, as is usually the case, add ammonia till the color is reddish brown, then sodium or ammonium acetate (H. ROSE) till the color is deep red, finally precipitate with neutral alkali succinate at a gentle heat, and when cool filter the ferric succinate from the solution which contains the rest of the metals. Wash the precipitate first with cold water, then with warm ammonia, which removes the greater part of the acid, leaving it darker in color. Dry and ignite, moisten with a little nitric acid, and ignite again. With proper care the separation is complete, and especially to be recommended when a relatively large quantity of iron is present. The method may also be used in the presence of aluminium. The latter falls down completely with the iron (E. MITSCHERLICH, PAGELS*).

6. *Methods based upon the different deportment of the several Sulphides with Acids, or of Acid Solutions with Hydrogen Sulphide.*

a. ZINC FROM ALUMINIUM AND MANGANESE.

73 The solution of the acetates, which must be free from inorganic acids, and must contain a sufficient excess of acetic acid, is precipitated with hydrogen sulphide, which throws down the zinc only (§ 108, *b*). The metals are usually most readily obtained in the form of acetates, by converting them into

* Jahresber. v. KOPP u. WILL. 1858, 617.

sulphates, and adding a sufficient quantity of barium acetate. Hydrogen sulphide is then conducted, without application of heat, into the unfiltered fluid, to which, if necessary, some more acetic acid has been added. The precipitate, which consists of a mixture of zinc sulphide and barium sulphate, is washed with water containing hydrogen sulphide. It is then heated with dilute hydrochloric acid, the solution filtered, and the zinc in the filtrate determined as directed § 108, *a*. The other metals are determined in the fluid filtered from the zinc sulphide, after removal of the barium by precipitation. BRUNNER† has proposed a modification of this process, especially for the separation of zinc from nickel.

b. ZINC FROM NICKEL, COBALT, AND MANGANESE.

74 To the hydrochloric solution add sodium carbonate till a permanent precipitate just forms, and then a drop or two of hydrochloric acid to redissolve the precipitate. Now pass hydrogen sulphide till the precipitate of zinc sulphide ceases to increase. Add a few drops of a very dilute solution of sodium acetate, and continue passing the gas for some time. When all the zinc is precipitated, allow to stand for twelve hours, filter, wash with hydrogen sulphide water, and determine the nickel and cobalt in the filtrate (SMITH and BRUNNER*) A good method; compare KLAYE and DEUS.† The method is also adapted for separating zinc from manganese.

[*Precautions.*—Bear in mind that Zn can be precipitated from solutions containing free HCl, but only in case the amount of the latter is very small.‡ When ZnS is precipitated the amount of HCl set free may be sufficient to prevent complete precipitation of the Zn. Addition of sodium acetate converts this HCl into NaCl, and allows the formation of ZnS to continue. Care must be taken not to add enough sodium acetate to convert *all* the HCl into NaCl, for in that case NiS and CoS will be precipitated.]

c. ZINC FROM NICKEL COBALT, AND MANGANESE.

75 [Zinc can be precipitated by hydrogen sulphide from a cold solution containing a sufficient amount of free acetic acid to

* Dingler's polyt. Journ. 150, 369; Chem. Centralbl. 1859, 26.

† Zeitschr. f. anal. Chem. 10, 200.

‡ STORER and ELIOT, Mem. Am. Acad. Arts and Sciences, viii. 95.

prevent precipitation of nickel and cobalt. To effect separation by this means* add sodium or potassium carbonate to the solution till it is slightly alkaline. If a large quantity of any free volatile acid is present it may be previously removed by evaporation. Dissolve the precipitate produced by the alkali carbonate (without filtering) in acetic acid, and add a large quantity more of acetic acid. Precipitate the zinc by passing H_2S through the cold moderately diluted solution. Wash the sulphide of zinc with water to which hydrogen sulphide and a little ammonium acetate has been added. The zinc sulphide should not be dark-colored. This will only be the case when not enough acetic is present to prevent precipitation of nickel or cobalt. Cobalt and nickel may be best separated from the filtrate by evaporating till the greater part of the acetic acid is removed, then adding some ammonium chloride and ammonia to slight alkaline reaction, evaporating further till the reaction becomes acid, heating finally to boiling, and passing hydrogen sulphide through the solution, as directed in § 110, 1, *b*, *β*.—A good method.]

7. *Different deportment of the several Oxides with Hydrogen Gas at a red heat.*

FERRIC IRON FROM ALUMINIUM AND CHROMIUM.

[Precipitate with ammonia, heat, filter, ignite, and weigh. **76** Triturate, and weigh off a portion in a porcelain crucible. Ignite to redness in a stream of hydrogen gas as long as water forms (about one hour). Then ignite over the blast-lamp in a current of mixed hydrogen and hydrochloric acid gases.

This leaves aluminium and chromium oxides in a state of purity; the iron volatilizes as ferrous chloride, and is determined by the loss. (Method of RIVOT and DEVILLE modified.) This method is further modified by COOKE,† who by means of a platinum boat in a platinum tube ignites the mixed oxides over a Bunsen lamp half an hour in a current of hydrogen, then alternately in HCl gas and hydrogen till the light color shows that iron has been removed.]

* ROSE and FINKENER, Anal. Chem. ii. 129 and 143.
† Zeitschr. f. anal. Chem. 6, 226.

8. *Methods based upon the different capacity of the several Oxides to be converted by Oxidizing Agents into higher Oxides, or by Chlorine into higher Chlorides.*

a. CHROMIUM FROM ALL THE METALS OF THE FOURTH GROUP, AND FROM ALUMINIUM.

77 Fuse the oxides with potassium nitrate and sodium carbonate (comp. **51**), boil the mass with water, add a small quantity of alcohol, and heat gently for several hours. Filter and determine in the filtrate the chromium as directed § 130, and in the residue the metals of the fourth group. The following is the theory of this process: the oxides of zinc, cobalt, nickel, iron, and partly that of manganese, separate upon the fusion, whilst, on the other hand, potassium manganate (perhaps also some ferrate) and chromate are formed. Upon boiling with water, part of the manganic acid of the potassium manganate is converted into permanganic acid at the expense of the oxygen of another part, which is reduced to the state of binoxide; the latter separates, whilst the potassium salts are dissolved. The addition of alcohol, with the application of a gentle heat, effects the decomposition of the potassium manganate and permanganate, manganese binoxide being separated. Upon filtering the mixture, we have therefore now the whole of the chromium in the filtrate as alkali chromate, and all the oxides of the fourth group on the filter. Aluminium, if present, will be found partly in the residue, partly as alkali aluminate in the filtrate; proceed with the latter according to **51.**

If you have to deal with the native compound of sesquioxide of chromium with protoxide of iron (chromic iron) the above method does not answer. This substance may be decomposed by fusion with cryolite and potassium disulphate.*

78 *b.* The radicals to be separated may be in the form of a solution of their salts; nearly neutralize the solution, add sodium acetate, heat and convert the chromium into chromic acid by passing chlorine, compare **53.** If ferric iron and aluminium are present, they will separate during boiling by the action of the sodium acetate, while the chromic acid and any zinc will remain in solution. If manganese, nickel, and cobalt are present, the method loses its simplicity; the manganese is precipitated as hydrated peroxide with a portion of the cobalt,

* GIBBS and CLARK Am. Jour. Sci. II ser. 48 198.

almost the whole of the nickel and some zinc, while the chromic acid remains in solution with the principal amount of the zinc and the rest of the cobalt and nickel (W. GIBBS).

9. *Method based upon the different deportment of the Nitrites.*

COBALT FROM NICKEL, ALSO FROM MANGANESE AND ZINC.

79 The separation of cobalt as tripotassium cobaltic nitrite was recommended first by FISCHER,* afterwards by A. STROMEYER,† GENTH and GIBBS,‡ H. ROSE,§ FR. GAUHE,‖ and myself (compare last edition of this work). The results are quite satisfactory both in presence of much cobalt and little nickel, and in the presence of little cobalt and much nickel; but the process is peculiarly good for the latter case. However, it is absolutely necessary that barium, strontium, and calcium should be absent, as in their presence nickel is thrown down as triple nitrite of nickel, potassium, and alkali-earth metal (KÜNZEL, O. L. ERDMANN¶). The best way of proceeding is as follows: The solution (from which any iron must first be separated) is evaporated to a small bulk, and then, if much free acid is present, neutralized with potassa. Then add a concentrated solution of potassium nitrite (previously neutralized with acetic acid and filtered from any flocks of silica and alumina that may have separated) in sufficient quantity, and finally acetic acid, till any flocculent precipitate that may have formed from excess of potassa has redissolved and the fluid is decidedly acid. Allow it to stand at least for twenty-four hours in a warm place, take out a portion of the supernatant fluid with a pipette, mix it with more potassium nitrite, and observe whether a further precipitation takes place in this after long standing. If no precipitate is formed the whole of the cobalt has fallen down, otherwise the small portion must be returned to the principal solution, some more potassium nitrite added, and after long standing the same test applied. Thus, and thus alone, can the analyst be sure of the complete precipitation of the cobalt. Finally filter and treat the precipitate according to § 111, 1, *d*. Boil the filtrate

* Pogg. Annal. 72, 477.

† Annal. d. Chem. u. Pharm. 96, 218.

‡ *Ib.* 104, 309.

§ Pogg. Annal. 110, 412.

‖ Zeitschr. f. anal. Chem. 5, 74.

¶ Zeitschr. f. anal. Chem. 3, 161; Journ. f. prakt. Chem. 97, 387.

with excess of hydrochloric acid, precipitate with potash, redissolve the precipitate in hydrochloric acid, throw down the nickel according to § 110, 1, *b*, α or β, as sulphide, and then convert into metal. In this manner alone can the nickel be obtained pure, as the original filtrate contains so much alkali salt and also generally alumina and silica.

[When nickel and cobalt are obtained in the form of sulphides in the process of separation from other metals, the mixed sulphides may be converted into metals without previous separation, by the same process that is described for nickel sulphide § 110, 1, *b*, and 2. Cobalt may then be separated from a nitric acid solution of the two metals and nickel estimated by difference.]

10. *Methods based upon the different deportment with Potassium Cyanide.*

a. Aluminium from Zinc, Cobalt, and Nickel.

80 Mix the solution with sodium carbonate, add potassium cyanide in sufficient quantity, and digest in the cold, until the precipitated zinc, cobalt, and nickel carbonates are redissolved. Filter off the undissolved aluminium precipitate, wash, and remove the alkali which it contains, by resolution in hydrochloric acid and reprecipitation by ammonia (Fresenius and Haidlen *).

b. Cobalt from Nickel.

81 Liebig's method,† which depends upon the conversion of the cobalt into potassium cobalticyanide, and of the nickel into double nickel and potassium cyanide, has been carefully studied in my laboratory by Fr. Gauhe.‡ It has been thus found that boiling the solution containing potassium cyanide and hydrocyanic acid (Liebig's first method) does not completely convert the double cobalt and potassium cyanide first formed into potassium cobalticyanide, but that passing chlorine (Liebig's second method) effects a ready and thorough conversion. The method then gives a very excellent separation, and is more particularly to be recommended where the quantity of nickel is small in proportion to the cobalt. We proceed as follows,

* Annal. d. Chem. u. Pharm. 43, 129. † *Ib.*, 65, 244, and 87, 128.
‡ Zeitschr. f. anal. Chem. 5, 75.

taking a hydrochloric solution of the metals: Remove the greater part of the free acid by evaporation or neutralize it by potash, add pure potassium cyanide till the precipitate first formed has redissolved; then add more cyanide, dilute, boil for some time or not, as you like, pass chlorine through the cold fluid, adding potash or soda occasionally, so that the fluid may remain strongly alkaline to the end. Bromine may be used instead of chlorine, and indeed is far more convenient. In the course of an hour the whole of the nickel will have precipitated as black hydrate of the sesquioxide. Having taken out a portion and satisfied yourself of this by addition of a further quantity of chlorine or bromine, filter, and wash with boiling water. The precipitate always retains alkali, and must be redissolved in hydrochloric acid, and estimated according to § 110, 1, *a*, or 2.

As regards the cobalt, it is most convenient to estimate it by difference. But if you wish to make a direct estimation, it will be advisable, in consequence of the large quantity of salts present in solution, first to evaporate to dryness with excess of hydrochloric acid, to take up the residue with a little water, and to heat in a large platinum dish, with the addition of excess of pure concentrated sulphuric acid till the greater part of the sulphuric acid has escaped. The red mass, consisting principally of alkali disulphate, is then treated with water, and the cobalt estimated according to § 111, 1, *c*.

c. Cobalt from Zinc.

82 Add to the solution of the two metals, which must contain some free hydrochloric acid, common potassium cyanide (prepared by Liebig's method), in sufficient quantity to redissolve the precipitate of cobalt cyanide and zinc cyanide which forms at first; then add a little more potassium cyanide, and boil some time, adding occasionally one or two drops of hydrochloric acid, but not in sufficient quantity to make the solution acid. After cooling add some chlorine or bromine, and digest for some time to complete the conversion of the cobalt into potassium cobalticyanide. Mix the solution with hydrochloric acid in an obliquely placed flask, and boil until the zinc cobalticyanide which precipitates at first is redissolved, and the hydrocyanic acid completely expelled. Add solution of soda or potassa in excess, and boil until the fluid is clear; the solution may now be assumed to contain all the cobalt as potassium cobalticyanide, and all the zinc as a compound of oxide of zinc and alkali. Precipitate the zinc by hydrogen sulphide (§ 108). Filter, and determine the cobalt in the filtrate as in **81**. The process is simple and the separation complete (FRESENIUS and HAIDLEN).

11. *Methods based upon the Volumetric Determination of one of the Metals, and the finding of the other from the difference.*

a. FERRIC IRON FROM ALUMINIUM.

83 Precipitate both metals with ammonia (§ 105, *a*, and § 113, 1). Dissolve the weighed residue, or an aliquot part of it, by digestion with concentrated hydrochloric acid, or by fusion with bisulphate of potassa and treatment with water containing sulphuric acid, and determine the iron volumetrically as directed § 113, 3, *a* or *b*. The alumina is found from the difference. This is an excellent method, and to be recommended more particularly in cases where the relative amount of iron is small. If you have enough substance, it is of course much more convenient to divide the solution, by weighing or measuring, into 2 portions, and determine in the one the sesquioxide of iron + alumina, in the other the iron.

b. FERRIC IRON FROM FERROUS IRON (ZINC AND NICKEL).

84 *α*. Determine in a portion of the substance the total amount of the iron as sesquioxide, or by the volumetric way. Dissolve another portion by warming with sulphuric acid in a flask through which carbonic acid is conducted, to exclude the air; dilute the solution, and determine the protoxide of iron volumetrically (§ 112, 2, *a*). The difference gives the quantity of the sesquioxide. Or, dissolve the compound in like manner in hydrochloric acid, and determine the ferric chloride with sodium thiosulphate according to § 113, 3, *b*. In this case the difference gives the ferrous iron. If it is desired to determine the ferrous chloride in the hydrochloric solution directly, it will be well to use PENNY's method (§ 112, 2, *b*). If the compound in which the ferrous and ferric basic radicals are to be estimated is decomposed by acids with difficulty, heat it with a mixture of 4 parts sulphuric acid and 1 part water (or with hydrochloric acid) in a sealed tube for 2 hours at 210° (MITSCHERLICH†). Or, if this is not enough, fuse it with borax (1 part mineral, 5 to 6 vitrified borax) in a small retort, connected with a flask containing nitrogen (produced by combustion of phosphorus in air); an atmosphere of carbonic acid is less suitable. Triturate the fused mass with the glass, and dissolve in boiling hydrochloric acid in an atmosphere of carbonic acid (HERMANN v. KOBELL). Or, as will generally be the best way, you may dissolve the substance in a mixture of hydrofluoric and hydrochloric or hydrofluoric and sulphuric acids with exclusion of

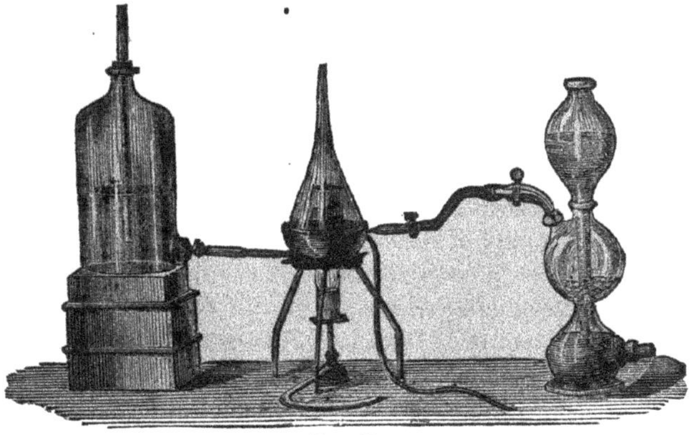

FIG. 69.

air. COOKE* dissolves silicates in a mixture of sulphuric and hydrofluoric acids in an atmosphere of steam and carbonic acid, and determines the ferrous iron by means of potassium permanganate.

Fig. 69 exhibits his apparatus. To the sides of a copper water-bath are attached three tubes. The tube on the left connects with a Mariotte's flask to maintain the water at a constant level. The upper tube on the right connects with a carbonic acid gas generator, while the third tube carries off any overflow of water to the sink.

On the cover of the water-bath close to the rim is a circular groove, which receives the edge of an inverted glass funnel. When the apparatus is in use this groove is kept full of water by the spray from the boiling liquid, and thus forms a perfect

* Am. Jour. Science, 2d ser., 44, 347.

† Jour. f. prakt. Chem., 81, 108 and 83, 455.

water-joint; but in order to secure this result the bath must be kept nearly full of water, and holes for the ready escape of the steam and spray should be provided in the rings, which cover the bath and adapt it for vessels of various sizes. By this arrangement the funnel may be kept filled with an atmosphere of steam or of carbonic acid for an indefinite period. Moreover, we can either pour in fresh quantities of solvent, or we can stir up the material, in the vessel within, introducing a tube-funnel or stirrer through the spout of the covering funnel.

The finely pulverized substance (½ to 1 grm.) is placed in a large platinum crucible. Upon it pour a mixture of dilute sulphuric acid (sp. gr. 1·5) with as little hydrofluoric acid as experience may show is required to dissolve or decompose the substance, stirring up the material with a platinum spatula. The crucible is next transferred to the water-bath, the covering funnel put in place, water poured into the groove, the interior filled with carbonic acid, and the lamp lighted. As soon as the water boils, the supply of carbonic acid is stopped; and if the water-level has been properly adjusted, the apparatus will take care of itself, the groove will be kept full of water, and the interior of the funnel full of steam. If the materials cake on the bottom of the crucible, as is not unfrequently the case when a large amount of insoluble sulphate is formed, the lamp may be removed, the apparatus again filled with carbonic acid, and the contents of the crucible stirred up by aid of a stout platinum wire about two inches long, fused to the end of a glass tube. Anything adhering to the rod can easily be washed back into the crucible by directing the jet from the wash-bottle down the throat of the covering funnel. The lamp may then be replaced, the current of carbonic acid interrupted, and the process of digestion continued. When the decomposition is complete, the current of carbonic acid gas is re-established, the lamp extinguished, and the air-tube of the Mariotte's flask raised until its lower end is above the level of the overflow. A slow current of water is thus caused to flow through the bath, which soon cools down the whole apparatus. The crucible may now be removed, its contents washed into a beaker-glass, and the solution diluted with pure water until the volume is about 500 c.c., when the amount of ferrous iron present can be determined with a solution of potassium permanganate in the usual way.

Many iron compounds in fine powder are completely decomposed by boiling a few minutes only with the mixed acids above mentioned. If a preliminary experiment shows this to be the case, a simple and satisfactory way of effecting a solution is to boil the substance with the solvent acids in a platinum crucible of 40 to 50 c.c. capacity, provided with a well-fitting concave cover. By watching the escaping vapor, one can regulate the boiling so as to prevent access of air without appreciable mechanical loss. If on removing the cover the decomposition is complete, the operation may be considered successful. Put the crucible and its contents at once into cold water in a beaker and titrate with permanganate (or thiosulphate if HCl has been used).

Iron may also be determined volumetrically in presence of zinc, nickel, etc. It is, indeed, often the better way, instead of effecting the actual separation of the oxides, to determine in one portion of the solution the iron + zinc or + nickel, in another portion the iron alone, and to find the quantity of the other metal by the difference. However, this can be done only in cases where the quantity of iron is relatively small.

12. *Cobalt and Nickel from Manganese.*

To the acid solution add sodium carbonate in excess, then acetic acid in liberal excess, then to the clear fluid, containing say a grm. of nickel or cobalt, 30 to 40 c.c. of sodium acetate solution (1 in 10), and pass hydrogen sulphide to saturation, keeping at 70°. Filter off the precipitated nickel or cobalt sulphide, wash and dry it. Concentrate the filtrate by evaporation, add ammonium sulphide, and then acetic acid, thus obtaining a second precipitate of nickel or cobalt sulphide. Test the filtrate again in the same manner. In the united precipitates determine the nickel or cobalt according to § 110, 1, *b*, or § 111, 1, *c*; in the filtrate, the manganese according to § 109, 2. **85**

IV. Separation of Iron, Aluminium, Manganese, Calcium, Magnesium, Potassium and Sodium.

§ 161.

As these metals are found together in the analysis of most silicates, and also in many other cases, I devote a distinct paragraph to the description of the methods which are employed to effect their separation.

1. *Method based upon the employment of Barium Carbonate* (particularly applicable in cases where the mixture contains only a small proportion of calcium).

The solution should contain no free chlorine, and the iron should be all in the form of ferric salt. Precipitate the iron and aluminium by barium carbonate * (**46** and **64**), dissolve the precipitate in hydrochloric acid, throw down the barium with sulphuric acid, filter, and estimate the iron and aluminium according to one of the methods given § 160, by preference **83**, at least when the quantity of aluminium is not too small. **86**

To the filtrate from the barium carbonate precipitate add hydrochloric acid, heat, throw down the barium with sulphuric acid, added just in excess. Filter off the precipitate, wash till free from soluble sulphate, concentrate if necessary, precipitate, and determine the manganese as sulphide (§ 109, 2). To the filtrate add hydrochloric acid, heat, filter off the sulphur, precipitate the lime with oxalate of ammonia, and finally separate the magnesia from the alkalies by one of the methods given § 153.

2. *Method based upon the application of Alkali Acetates or Formates.*

Remove by evaporation any very considerable excess of acid which may be present, dilute, add sodium carbonate,† until the fluid is nearly neutral, then sodium acetate (or sodium formate) and precipitate iron and aluminium, observing all directions given in **71**. Wash the precipitate well, dissolve in hydrochloric acid, precipitate the solution with ammonia (**37**), dry, ignite, and weigh. Dissolve in concentrated hydrochloric acid, or digest it with 16 times its weight of a mixture of 8 parts sulphuric acid and 3 parts water, or fuse it for a long time with **87**

* Before adding the barium carbonate, it is *absolutely indispensable* to ascertain whether a solution of it in hydrochloric acid is completely precipitated by sulphuric acid, so that the filtrate leaves no residue upon evaporation in a platinum dish.

† In cases where it is intended to estimate the alkalies in the filtrate, ammonium salts must be used instead of the sodium salts. If, however, it is intended to precipitate manganese subsequently with bromine, ammonium salts must not be introduced into the solution.

bisulphate of potassa, dissolve in water, and determine the iron volumetrically as in § 113, 3, *a* or *b*. The difference gives the quantity of the aluminium. If any silicic acid remains behind on dissolving the precipitate, it is to be collected on a filter, ignited, weighed, and deducted from the alumina. The filtrate contains the manganese, the alkali-earth metals, and the alkalies. Precipitate the manganese with ammonium sulphide (§ 109, 2), boil with hydrochloric acid and filter off the sulphur, precipitate the calcium, after addition of ammonia, with ammonium oxalate, and lastly, after removing the ammonium salts by ignition, precipitate the magnesium from the hydrochloric acid solution of the residue with ammonium sodium phosphate. However, if it is intended to estimate the alkalies, the magnesium must be separated by one of the processes in § 153, 4. This method is convenient, and gives good results, especially in the presence of much iron and little aluminium. Since aluminium is not precipitated by alkali acetates or formates with the same certainty as iron, it is necessary to test the weighed manganese sulphide for aluminium.

[This method is to be recommended when manganese is present with iron, or with iron and a moderate proportion of aluminium. If, however, the amount of aluminium is large in proportion to the iron, it is difficult to precipitate it completely with sodium acetate. Instead of precipitating manganese with ammonium sulphide it may be separated from calcium and magnesium by precipitation with bromine. Add aqueous solution of bromine to the filtrate from the iron precipitate without previous concentration of the filtrate, unless its volume exceeds 600 or 700 c.c., and proceed according to § 159, **62**.]

3. *Method based upon the application of Ammonium Sulphide.*

88 Mix the fluid in a flask with ammonium chloride, then with ammonia, until a precipitate just begins to form, then with yellow ammonium sulphide, fill the flask nearly up to the top with water, cork it, allow to settle in a warm place, filter, and wash the precipitate—consisting of iron and manganese sulphides and aluminium hydroxide—without interruption with water containing ammonium sulphide. Separate the calcium, magnesium, and alkalies in the filtrate as in **87**. Dissolve the precipitate in hydrochloric acid, and separate the aluminium from the iron and manganese according to **65** or **66**, and then the iron from the manganese, say by **70** or **71**.

The following method is particularly suitable in cases where no manganese is present, or only inappreciable traces:

4. *Method based upon the application of Ammonia.*

89 The solution must contain all the iron in the state of a ferric salt. Add a relatively large quantity of ammonium chloride, and—observing the precautions indicated in **37**—precipitate with ammonia. The precipitate contains the whole of the iron and aluminium; at most an inappreciable amount of the latter remains in solution if the free ammonia has been almost but not entirely driven off by heat, if the solution was diluted sufficiently, and if enough ammonium chloride was present. It may also contain small quantities of calcium and magnesium and a little manganese. It is well, therefore, usually to redissolve the washed precipitate in hydrochloric acid, and reprecipitate with ammonia. In this way the precipitate will be got free from alkali-earths and manganese. Wash the precipitate completely, dry, ignite, and treat according to **87**. If silicic acid remains undissolved, it is to be determined and deducted. The solution filtered from the aluminium and ferric hydroxide is concentrated by evaporation, the manganese is precipitated and determined according to § 109, 2, as sulphide, the alkali-earth metals and alkalies in the filtrate are determined according to **87**. The weighed sulphide of manganese is digested with dilute hydrochloric acid, any residue that may remain fused with bisulphate of potassa, dissolved in water, and tested for alumina.

Supplement to the Fourth Group.

To §§ 158, 159, 160.

SEPARATION OF URANIUM FROM THE OTHER METALS OF GROUPS I.—IV.

90 It has already been stated, in § 114, that uranium in uranyl compounds cannot be completely separated from the *alkalies* by means of ammonia, as the precipitated ammonium uranate is likely to contain also fixed alkalies. The precipitate should therefore be dissolved in hydrochloric acid, the solution evaporated in the platinum crucible, the residue gently ignited in a current of hydrogen gas, the chlorides of the alkali metals extracted with water, and the uranous oxide (UO_2) ignited in hydrogen, in order to its being weighed as such, or in the air, whereby it is converted into uranous uranate, $U(UO_4)_2$. Instead of dissolving the precipitate in hydrochloric acid and treating the solution as directed, you may heat the precipitate cautiously* with ammonium chloride, and treat the residue with water (H. Rose). Uranium may be completely separated from the alkalies also by ammonium sulphide as H. Rose found. Remelé† has examined this subject with great care and recommends the following method of precipitation:—The solution being *neutral or slightly acid*, add an excess of yellow ammonium sulphide and keep nearly boiling for an hour to convert the first formed precipitate of uranium oxysulphide entirely into a mixture of uranous oxide and sulphur. The fluid, at first dark from presence of dissolved uranium, will now appear yellow and transparent. Filter off the precipitate containing all the uranium and wash it with cold or warm water, first by decantation, finally on the filter. It is well to mix a little ammonium sulphide or chloride with the water, as when pure water is used the last filtrate is apt to be turbid. The dried precipitate is roasted and then converted into uranous uranate by ignition in the air, or into uranous oxide by ignition in hydrogen (§ 114).

91 From *barium* uranyl may be separated by sulphuric acid, from strontium and calcium by sulphuric acid and alcohol. Ammonia fails to effect complete separation of uranyl from the alkali-earth metals, the precipitate always containing not inconsiderable quantities of the latter. In such precipitates, however, the uranium and the alkali-earth metals may likewise be separated by gentle ignition with ammonium chloride and treatment of the residue with water.

92 Uranyl may be separated from strontium and calcium also by precipitation with ammonium sulphide by the method given above in the separation from the alkalies. As carbonates of the alkali-earth metals may be coprecipitated, treat the washed pre-

* Strong ignition would occasion the volatilization of uranium chloride.
† Zeitschr. f. anal. Chem. 4, 379.

cipitate of uranous oxide and sulphur *in the cold* with dilute hydrochloric acid which will not dissolve uranous oxide. Ammonium sulphide will not answer for the separation of uranium from *barium* (REMELÉ*).

Magnesium may be separated from uranyl not only by **93** ammonium sulphide in presence of ammonium chloride, but also by ammonia. Add enough ammonium chloride to the solution, heat to boiling, supersaturate with ammonia, continue boiling till the odor of ammonia is but slight, filter the hot fluid, and wash the precipitate, which is free from magnesium, with hot water containing ammonia (H. ROSE). It is always well to test the uranous oxide obtained by ignition in hydrogen for magnesium by treating with dilute hydrochloric acid.

Aluminium is best separated from uranyl by mixing the somewhat acid fluid with ammonium carbonate in excess. The uranyl passes completely into solution, while the aluminium remains absolutely undissolved. Filter, evaporate, add hydrochloric acid to resolution of the precipitate produced, heat till all the carbonic acid is expelled, and precipitate with ammonia (§ 114).

Uranyl is best separated from *chromium* (W. GIBBS†) by adding to the solution soda in slight excess, heating to boiling and adding bromine water, when the chromium is rapidly converted into chromic acid. Filter the solution containing sodium chromate from the precipitate which has a deep orange-red color and consists of a compound of soda and uranic oxide mixed with some uranyl chromate. Wash the precipitate with hot water containing a little soda, dissolve it in hot nitric acid, boil the solution a few minutes to drive off any nitrous acid, and precipitate the chromic acid according to § 130, I., *a*, *β* with mercurous nitrate (according to GIBBS at a boiling heat). The filtrate now contains the whole of the uranium, of course in presence of mercury.

The separation of uranyl from the metals of the *fourth* **94** *group* may be based simply on the fact that ammonium carbonate prevents the precipitation of uranyl, but not that of the other metals by ammonium sulphide. Mix the solution with a mixture of ammonium carbonate and ammonium sulphide, allow

* Zeitschr. f. anal. Chem. 4,383. † *Ib.* 12, 310.

to-subside in a closed flask, and wash the precipitate with water containing ammonium carbonate and ammonium sulphide.

Remove the greater part of the excess of ammonium carbonate from the filtrate by a very gentle heat, acidify with hydrochloric acid, warm, filter off the separated sulphur, and throw down the uranium either by ammonium sulphide (see above, *Separation of Uranium from the Alkalies*) or by heating with nitric acid and then adding ammonia (H. ROSE,* REMELÉ†). The method is not so suitable in presence of nickel, as a little of this metal is very liable to pass into the filtrate on precipitation with ammonium carbonate and ammonium sulphide.

Ferric iron may be also separated from uranyl by means of an excess of ammonium carbonate. The small quantity of iron which passes with the uranium into solution will fall down on allowing the solution to stand for several hours, or it may be precipitated with ammonium sulphide, before the uranium is thrown down (PISANI‡).

From *nickel*, *cobalt*, *manganese*, *zinc*, and *magnesium* the uranyl may also be separated by barium carbonate. The fluid, which should contain a little free acid, is mixed with the precipitant in excess, and allowed to stand in the cold for 24 hours with frequent shaking (**64**).

95 From *cobalt*, *nickel*, and *zinc* uranyl may also be separated (GIBBS and PERKINS§) by taking the neutral or slightly acid solutions of the chlorides, adding sodium acetate in excess and a few drops of acetic acid, and passing a rapid current of hydrogen sulphide for half an hour through the boiling fluid. The uranium remains dissolved while the other metals are precipitated. I should advise testing the filtrate with a mixture of ammonium carbonate and ammonium sulphide to see if any nickel, cobalt, or zinc remain in solution.

---

* Zeitschr. f. anal. Chem. 1, 412.

† *Ib.* 4, 385.

‡ Compt. rend. 52, 106.

§ Zeitschr. f. anal. Chem. 3, 334.

Fifth Group.

SILVER—MERCURY (IN MERCUROUS AND MERCURIC COMPOUNDS)—LEAD —BISMUTH—COPPER—CADMIUM.

I. SEPARATION OF THE METALS OF THE FIFTH GROUP FROM THOSE OF THE FIRST FOUR GROUPS.

§ 162.

INDEX. (The numbers refer to those in the margin.)

A. *General Method.*

ALL THE METALS OF THE FIFTH GROUP FROM THOSE OF THE FIRST FOUR GROUPS.

Principle: Hydrogen Sulphide precipitates from Acid Solutions the Metals of the Fifth Group, but not those of the first Four Groups.

The following points require especial attention in the execution of the process:

96 *α.* To effect the separation of the metals of the fifth group from those of the first three groups, by means of hydrogen sulphide, it is necessary simply that the reaction of the solution should be acid, the nature of the acid to which the reaction is due being of no consequence. But, to effect the separation of the metals of the fifth group from those of the fourth, the presence of a free mineral acid is indispensable; otherwise zinc and, under certain circumstances, also cobalt and nickel may be coprecipitated.

β. But even the addition of hydrochloric acid to the fluid will not always entirely prevent the coprecipitation of the zinc. RIVOT and BOUQUET* declare a complete separation of copper

* Annal. d. Chem. u. Pharm. 80, 364.

from zinc by means of hydrogen sulphide altogether impracticable. CALVERT * states that he has arrived at the same conclusion. On the other hand, SPIRGATIS† concurs with H. ROSE in maintaining that the complete separation of copper from zinc may be effected by means of hydrogen sulphide in presence of a sufficient quantity of free acid.

In this conflict of opinions, I thought it necessary to subject this method once more to a searching investigation. I therefore instructed one of the students in my laboratory, Mr. GRUNDMANN, to make a series of experiments in the matter, with a view to settling the question. ‡

The following process is founded on the results which we obtained:

Add to the COPPER and ZINC solution a large amount of hydrochloric acid (*e.g.*, to ·4 grm. oxide of copper in 250 c.c. of solution, 30 c.c. hydrochloric acid of 1·1 sp. gr.), conduct into the fluid at about 70° hydrogen sulphide largely in excess, filter before the excess of hydrogen sulphide has had time to escape or become decomposed, wash with hydrogen sulphide water, dry, roast, redissolve in nitrohydrochloric acid, evaporate nearly to dryness, add water and hydrochloric acid as above, and precipitate again with hydrogen sulphide. This second precipitate is free from zinc; it is treated as directed in § 119, 3.

If CADMIUM is present, it is well to have less acid present, *e.g.*, to ·4 grm. oxide of cadmium in 250 c.c. of solution add 10 c.c. hydrochloric acid of 1·1 sp. gr. If the quantity of zinc is considerable, dissolve the first precipitate of cadmium sulphide in hot hydrochloric acid, evaporate nearly to dryness, add 10 c.c. hydrochloric acid and about 250 c.c. water, and precipitate again. In this way the results are quite satisfactory.

γ. The other metals of the fifth group comport themselves in this respect similarly to cadmium, *i.e.*, they are not completely precipitated by hydrogen sulphide in presence of too much free acid in a concentrated solution. Lead requires the least amount of free acid to be retained in solution; then follow in order of succession, cadmium, mercury, bismuth, copper, silver (M. MARTIN§). A portion of the filtrate should, if neces-

* Journ. f. prakt. Chem. 71, 155.
† *Ib.* 57, 184.
‡ *Ib.* 73, 241.
§ *Ib.* 67, 371.

sary, be tested by addition of a large quantity of hydrogen sulphide to see if the precipitation of the fifth group was complete.

δ. If hydrochloric acid produces no precipitate in the solution, it is preferred as acidifying agent, otherwise sulphuric or nitric acid must be used. In the latter case the fluid must be rather largely diluted. ELIOT and STORER * arrived at the same conclusion as ourselves, and showed that the cause of CALVERT'S unfavorable results was the too large dilution of his solutions. For to prevent the precipitation of zinc you have not merely to preserve a certain proportion between the zinc and the free acid, but also a certain degree of dilution. Although I agree with the above-named chemists in the opinion that it is possible to produce a condition of the fluid, under which one precipitation will effect complete separation, still it appears to me better, for practical purposes, to precipitate twice, as this is sure to lead to the desired result.

ε. A somewhat copious experience in the separation of COPPER from NICKEL (and COBALT) which so frequently occurs has led me to the opinion that a double precipitation is unnecessary. If the solution which is to be treated with hydrogen sulphide contains enough free hydrochloric acid and not too much water, the copper falls down absolutely free from nickel, while, on the other hand, if the quantity of free acid is not too large, the filtrate will be quite free from copper.

ζ. CADMIUM and ZINC may, according to FOLLENIUS, also be completely separated by a single precipitation, if the metals are present in a sulphuric acid solution containing 25 or 30 per cent. of dilute acid of 1·19 sp. gr. Precipitate with hydrogen sulphide at 70°. Collect the precipitate on a weighed asbestos filter, dry in a current of heated air, ignite gently in a stream of pure hydrogen sulphide (to convert small quantities of cadmium sulphate into sulphide), remove the small quantity of separated sulphur by gentle ignition in a current of air, and weigh.

---

* On the Impurities of Commercial Zinc, &c.—Memoirs of the American Academy of Arts and Sciences. New Series. Vol. 8.

B. *Special Methods.*

SINGLE METALS OF THE FIFTH GROUP FROM SINGLE OR MIXED METALS OF THE FIRST FOUR GROUPS.

1. SILVER is most simply and completely separated from the METALS OF THE FIRST FOUR GROUPS by means of hydrochloric acid. The hydrochloric acid must not be used too largely in excess, and the fluid must be sufficiently dilute; otherwise a portion of the silver will remain in solution. Care must be taken also not to omit the addition of nitric acid, which promotes the separation of the silver chloride. The latter should be treated according to § 115, 1, *a*. **97**

2. The separation of MERCURY from the METALS OF THE FIRST FOUR GROUPS may be effected also by ignition, which will cause the volatilization of the mercury or the mercurial compound, leaving the non-volatile bodies behind. The method is applicable in many cases to alloys, in others to oxides, chlorides, or sulphides. If the mercury is estimated only from the loss, the operation is conducted in a crucible; otherwise in a bulb-tube, or a wide glass tube with porcelain boat. In the latter case it is well to use a current of hydrogen (compare § 118, 1, *a*). **98**

The precipitation of mercury as mercurous chloride with phosphorous acid, according to § 118, 2, is also well adapted for its separation from metals of the first four groups. If the mercury is already present as a mercurous salt, it may be separated and determined in a simple manner, by precipitation with hydrochloric acid (§ 117, 1).

3. FROM THOSE BASIC RADICALS WHICH FORM SOLUBLE SALTS WITH SULPHURIC ACID, LEAD may be readily separated by that acid. The results are very satisfactory, if the rules given in § 116, 3 are strictly adhered to. **99**

If you have lead in presence of barium, both in form of sulphates, digest the precipitate with a solution of ordinary ammonium sesquicarbonate, without application of heat. This decomposes the lead salt, leaving the barium salt unaltered. Wash, first with solution of ammonium carbonate, then with water, and separate finally the lead carbonate from the barium sulphate, by acetic acid or dilute nitric acid (H. ROSE*). The

* Journ. f. prakt. Chem. 66, 166.

same object may also be attained by suspending the washed insoluble salts in water and digesting with a clear concentrated solution of sodium thiosulphate at 15—20° (not higher). The barium sulphate remains undissolved, the lead sulphate dissolves. Determine the lead in the filtrate (after § 116, 2) as lead sulphide (J. Löwe*).

4. Copper from all Metals of the first Four Groups.

a. Free the solution as far as possible from hydrochloric **100** and nitric acids by evaporation with sulphuric acid. Dilute if necessary, boil, and add sodium thiosulphate † as long as a black precipitate continues to form. As soon as this has deposited, and the supernatant fluid contains only suspended sulphur, the whole of the copper is precipitated. The precipitate is cuprous sulphide (Cu_2S), and may be readily washed without suffering oxidation. Convert it into anhydrous cuprous sulphide by ignition in hydrogen (§ 119, 3). The other metals are in the filtrate and washings. Evaporate with some nitric acid, filter, and determine the metals in the filtrate. ‡ Results good. The method requires practice, as the end of the precipitation of the copper is not so easy to hit as when hydrogen sulphide is employed.

If the solution contained hydrochloric or nitric acid, and this was not first removed before the addition of the thiosulphate, the precipitant would be required in much larger quantity; in the presence of hydrochloric acid because the cuprous chloride produced is only decomposed by a large excess of thiosulphate, in the presence of nitric acid because the thiosulphate does not begin to act on the copper salt till all the nitric acid is decomposed.

---

* Journ. f. prakt. Chem.

† The commercial salt is often not sufficiently pure; in which case some sodium carbonate must be added to its solution, and the mixture filtered.

‡ As far back as 1842, C. Himly made the first proposal to employ sodium thiosulphate for the precipitation of many metals as sulphides (Annal. d. Chem. u. Pharm. 43, 150). The question, after long neglect, was afterwards taken up again by Vohl (Annal. d. Chem. u. Pharm. 96, 237), and Slater (Chem. Gaz. 1855, 369). Flajolot, however, made the first quantitative experiment (Annal. des. Mines, 1853, 641; Journ. f. prakt. Chem. 61, 105). The results obtained by him are perfectly satisfactory.

b. Precipitate the copper as cuprous sulphocyanate according to § 119, 3, *b*; the other metals remain in solution (RIVOT). If alkalies were present and it were desired to determine them in the filtrate, ammonium sulphocyanate must be used instead of the potassium salt usually employed. This method is particularly well adapted for the separation of copper from zinc. The zinc can be precipitated at once from the filtrate by sodium carbonate. The method is also suitable for separating copper from iron (H. ROSE*); in this case it is unnecessary that ferric salts be completely reduced by the sulphurous acid added; the separation may be effected, even if the solution becomes blood-red on the addition of the precipitant. **101**

c. The solution should be free from hydrochloric acid, and should contain a certain quantity of free nitric acid (20 c.c. nitric acid of 1·2 sp. gr. to 200 c.c.), and some sulphuric acid. Throw down the copper by a galvanic current, so that the metal may be firmly deposited on a platinum vessel (preferably a platinum cone), which forms the negative pole. Take care that the current is strong enough, and, without interrupting it, remove the cone from the fluid occasionally to see when the copper is all precipitated. With proper execution the separation of copper from all metals of Groups 1–4 is thorough. All metals of Groups 1–4 remain dissolved, except manganese, which separates as binoxide at the positive pole. The method requires practice and strict attention to the conditions which have been determined by a long course of experiments. It is particularly suited for mining assays and manufactures. The electrolytic method of separating copper was, I believe, first recommended by GIBBS,† and afterwards improved by LUCKOW.‡ LECOQ DE BOISBAUDRAN,§ ULLGREN,‖ and MERRICK¶ have also written on this subject. Finally the method was very accurately and minutely described by the Mansfelder Ober-Berg- und Hüttendirection at Eisleben,** who, after giving a prize to LUCKOW's method, afterwards adopted it, and still further improved it. I must refer the reader for details to the last mentioned memoir and LUCKOW's paper. **102**

* Pogg. Annal. 110, 424. † Zeitschr. f. anal. Chem. 3, 334.
‡ Dingler's polyt. Journ. 177, 296, and (in detail) Zeitschr. f. anal. Chem. 8, 25.
§ Zeitschr. f. anal. Chem. 7, 253, and 9, 102. ‖ *Ib.* 7, 255.
¶ American Chemist, 2, 136. ** Zeitschr. f. anal. Chem. 11, 1.

COPPER FROM ZINC.

103 BOBIERRE* employed the following method with satisfactory results in the analysis of many alloys of zinc and copper: The alloy is put into a porcelain boat lying in a porcelain tube, and heated to redness for three-quarters of an hour at the most, a rapid stream of hydrogen gas being conducted over it during the process. The zinc volatilizes, the copper remains behind. If the alloy contains a little lead (under 2 to 3 per cent.) this goes off entirely with the zinc, and is partly deposited in the porcelain tube in front of the boat; if more lead is present part only is volatilized, the rest remaining with the copper (M. BURSTYN†).

6. BISMUTH FROM THE METALS OF THE FIRST FOUR GROUPS, WITH THE EXCEPTION OF FERRIC IRON.

104 Precipitate the bismuth according to § 120, 4, as basic chloride, and determine it as metal; all the other basic metals remain completely in solution. Results very satisfactory (H. ROSE‡).

7. CADMIUM FROM ZINC.

105 Prepare a hydrochloric or nitric acid solution of the two oxides, as neutral as possible, add a sufficient quantity of tartaric acid, then solution of potassa or soda, until the reaction of the clear fluid is distinctly alkaline. Dilute now with a sufficient quantity of water, and boil for 1½–2 hours. All the cadmium precipitates as hydroxide, free from alkali (to be determined as directed § 121), whilst the whole of the zinc remains in solution; the latter metal is determined as directed in § 108, 1, *b* (AUBEL and RAMDOHR§). The test-analyses communicated are satisfactory. As the separation only succeeds when the substances are present in correct proportions, I will add the quantities employed by AUBEL and RAMDOHR with especially good effect. About 1 grm. oxide of zinc and 1 grm. oxide of cadmium were dissolved in hydrochloric acid, 30 grm. solution of tartaric acid (containing ·23 grm. acid in 1 grm.), 50 grm. soda solution of 1·16 sp. gr., and 120 grm. water added, and the whole boiled 2 hours. (The boiling must on no account be done in glass; a platinum or silver dish should be used.)

* Compt. rend. 36, 224; Journ. f. prakt. Chem. 58, 380.
† Zeitschr. f. anal. Chem. 11, 175.
‡ Pogg. Annal. 110, 429.
§ Annal. d. Chem. u. Pharm. 103, 33.

II. Separation of the Metals of the Fifth Group from each other.*

§ 163.

Index. (The numbers refer to those in the margin.)

Silver from copper, 106, 111, 113, 124, 125.
" cadmium, 106, 111, 113.
" bismuth, 106, 110, 113, 122.
" mercuricum,* 106, 111, 113, 119, 121.
" lead, 106, 109, 110, 113, 124.

*Mercuricum** from silver, 106, 111, 113, 119, 121.
" mercurosum,* 107.
" lead, 108, 109, 110, 113, 119, 121.
" bismuth, 108, 110, 113, 114, 119.
" copper, 108, 112, 113, 119, 121.
" cadmium, 108, 113, 119.

*Mercurosum** from mercuricum, 107.
" copper, 107.
" cadmium, 107.
" lead, 107, 109.

Compare also mercuricum from other metals.

Lead from silver, 106, 110, 113, 124, 125.
" mercuricum, 108, 109, 110, 113, 119, 121.
" mercurosum, 107, 109.
" copper, 109, 110, 113, 115.
" bismuth, 109, 115, 122, 123.
" cadmium, 109, 110, 113.

Bismuth from silver, 106, 110, 113, 122.
" lead, 109, 115, 122, 123.
" copper, 110, 113, 114, 116, 122.
" cadmium, 110, 113, 114, 115, 118.
" mercuricum, 108, 110, 113, 114, 119.

Copper from silver, 106, 111, 113, 124.
" lead, 109, 110, 113, 115.
" bismuth, 110, 113, 114, 116, 122.
" mercuricum, 108, 112, 113, 119, 121.
" mercurosum, 107.
" cadmium, 112, 113, 117, 120.

Copper in cupric from copper in cuprous compounds, 125.

Cadmium from silver, 106, 111, 113.
" lead, 109, 110, 113.
" bismuth, 110, 113, 114, 115, 118.
" copper, 112, 113, 115, 117, 120.
" mercuricum, 108, 113, 119.
" mercurosum, 107.

* For the sake of brevity the terms "mercuricum" and "mercurosum" are used to designate respectively mercury in mercuric and mercurous compounds.

1. *Methods based upon the Insolubility of certain of the Chlorides in Water or Alcohol.*

a. SILVER FROM COPPER, CADMIUM, BISMUTH, MERCURICUM, AND LEAD.

α. To separate *silver* from *copper*, *cadmium*, and *bismuth*, **106** add to the nitric acid solution containing excess of nitric acid, hydrochloric acid as long as a precipitate forms, and separate the precipitated silver chloride from the solution which contains the other metals, as directed § 115, 1, *a*. In the presence of bismuth, after pouring off the supernatant fluid, heat again with nitric acid, and wash with dilute nitric acid before washing with water.

β. If you wish to separate *mercuricum* from *silver* by hydrochloric acid, special precautions must be taken, as a solution of mercuric nitrate possesses the property of dissolving silver chloride (WACKENRODER, v. LIEBIG,* H. DEBRAY†). Although the silver chloride in solution for the most part separates on the addition of enough hydrochloric acid to convert the mercuric nitrate into chloride, or on addition of sodium acetate, still we cannot depend upon the complete precipitation of the silver. On this account, mix the nitric acid solution—which must not contain any mercurous salt, and is to be in a sufficiently dilute condition and acidified with nitric acid—with hydrochloric acid, as long as a precipitate forms. Allow to deposit, filter off the clear fluid, heat the precipitate —to free it from any possibly coprecipitated basic mercuric salts—with a little nitric acid, add water, then a few drops of hydrochloric acid, and filter off the silver chloride. In the filtrate determine the mercury as sulphide (§ 118, 3), and finally test this for silver, by ignition in a stream of hydrogen —any silver that may happen to be present will remain behind in the metallic state.

γ. In the separation of *silver* from *lead*, the precipitation is advantageously preceded by addition of sodium acetate. The fluid must be hot and the hydrochloric acid rather dilute; no more must be added of the latter than is just necessary. In this manner, the separation may be readily effected, since

* Annal. d. Chem. u. Pharm. 81, 128.
† Compt. rend. 70, 847; Zeitschr. f. anal. Chem. 13, 349.

lead chloride dissolves in sodium acetate (ANTHON). The silver chloride is washed with hot water. The lead is thrown down from the filtrate with hydrogen sulphide. If you desire to prevent the occasionally injurious influence of sodium acetate, great care must be given to the washing of the silver chloride. It is also well to reduce the weighed chloride by gentle ignition in a current of hydrogen, and to test the silver obtained for lead.

δ. The volumetric method (§ 115, 5) is usally resorted to in mints to determine the *silver in alloys*. In presence of a mercuric salt, sodium acetate is mixed with the fluid, immediately before the addition of the solution of chloride of sodium. In the East India mint the silver is separated and weighed as chloride.*

b. MERCUROSUM FROM MERCURICUM, COPPER, CADMIUM, AND LEAD.

107 Mix the highly dilute cold solution with hydrochloric acid as long as a precipitate (mercurous chloride) forms; allow this to deposit, filter on a weighed filter, dry at 100°, and weigh. The filtrate contains the other metals. If you have to analyze a solid body, insoluble in water, either treat directly, in the cold, with dilute hydrochloric acid, or dissolve in highly dilute nitric acid, and mix the solution with a large quantity of water before proceeding to precipitate. Care must always be taken that the mode of solution is such as not to convert mercurous into mercuric compounds. If lead is present the washing of the mercurous chloride must be executed with special care with water of 60—70°, till the filtrate ceases to be colored with hydrogen sulphide. As an additional security, it is well to test at last whether the weighed mercurous chloride leaves no lead sulphide behind on cautious ignition with sulphur in a stream of hydrogen.

c. MERCUROSUM AND MERCURICUM FROM COPPER, CADMIUM, AND (but less well) FROM BISMUTH AND LEAD.

108 If mercury is present as a mercuric compound, or partly in a mercuric and partly in a mercurous compound, it is precipitated according to § 118, 2, by means of hydrochloric acid

* Chem. Centralbl. 1872, 202.

and phosphorous acid as mercurous chloride. The precipitate, particularly when bismuth is present, is first washed with water containing hydrochloric acid, then with pure water, till the washings are no longer colored with hydrogen sulphide (H. ROSE*). In the presence of lead, the remarks in **107** must be attended to.

2. *Methods based upon the Insolubility of Lead Sulphate.*

LEAD FROM ALL OTHER METALS OF THE FIFTH GROUP.

109 Mix the nitric acid solution with pure sulphuric acid in not too slight excess, evaporate until the sulphuric acid begins to volatilize, allow the fluid to cool, add water (in which, if there is a sufficient quantity of free sulphuric acid present, the mercuric and bismuth sulphates dissolve completely), and then filter the solution, which contains the other metals, *without delay* from the undissolved lead sulphate. If it is feared that the residue no longer contains enough free sulphuric acid, add some dilute acid to it before adding the water. Wash the precipitate with water containing sulphuric acid, displace the latter with alcohol, dry, and weigh (§ 116, 3). Precipitate the other metals from the filtrate by hydrogen sulphide. If *silver* is present in any notable quantity, this method cannot be recommended, as the silver sulphate is not soluble enough. In this case you may follow ELIOT and STORER,† viz., mix the solution with ammonium nitrate, warm, precipitate the greater portion of the silver with ammonium chloride, evaporate the filtrate, remove the ammonium salts by ignition, and in the residue separate the small remainder of the silver from the lead with sulphuric acid as just directed. For the separation of *lead from bismuth*, on the above principle, H. ROSE‡ gives the following process as the best. If both oxides are in dilute nitric acid solution, as is usually the case, evaporate to small bulk, and add enough hydrochloric acid to dissolve all the bismuth; the lead separates partially as chloride. Should a portion of the clear fluid poured off become turbid on the addition of a drop of water, you must add some more hydro-

* Pogg. Annal. 110, 534.

† Proceedings of the American Academy of Arts and Sciences, Sept. 11, 1860, p. 52; Zeitschr. f. anal. Chem. 1, 389.

‡ Pogg. Annal. 110, 432.

chloric acid, till no permanent turbidity is produced unless several drops of water are added. The turbid fluids should all be returned, and the glasses rinsed with alcohol. Add now dilute sulphuric acid, allow to stand some time with stirring, add spirit of wine of ·8 sp. gr., stir well, allow to settle for a long time, filter, wash the lead sulphate first with alcohol mixed with a small quantity of hydrochloric acid, then with pure alcohol. Determine it after § 116, 3. Mix the filtrate at once with a large quantity of water, and proceed with the precipitated basic bismuth chloride according to § 120, 4.

3. *Methods based upon different deportment with Cyanide of Potassium* (FRESENIUS and HAIDLEN*).

a. LEAD AND BISMUTH FROM ALL OTHER METALS OF THE FIFTH GROUP.

110 Mix the *dilute* solution with sodium carbonate in *slight* excess, add solution of potassium cyanide (free from sulphide), heat gently for some time, filter and wash. On the filter you have lead and bismuth carbonates (containing alkali); the filtrate contains the other metals as cyanides in combination with potassium cyanide. The method of effecting their further separation will be learnt from what follows. In very accurate analyses bear in mind that the filtrate generally contains traces of bismuth, which may be precipitated by ammonium sulphide.

b. SILVER FROM MERCURICUM, COPPER, AND CADMIUM.

111 Add to the solution, which, if it contains much free acid, must previously be nearly neutralized with soda, potassium cyanide until the precipitate which forms at first is redissolved. The solution contains the cyanides of the metals in combination with potassium cyanide as soluble double salts. Add dilute nitric acid in excess, which effects the decomposition of the double cyanides; the insoluble silver cyanide precipitates permanently, whilst the mercuric cyanide remains in solution, and the cyanides of copper and cadmium redissolve in the excess of nitric acid. Treat the silver cyanide as directed § 115, 3. If the filtrate contains only mercury and cadmium, precipitate at once with hydrogen sulphide, which completely

* Annal. d. Chem. u. Pharm. 43, 129.

throws down the sulphides of the two metals; but if it contains copper, you must first heat with sulphuric acid, until the odor of hydrocyanic acid is no longer perceptible, and then precipitate with hydrogen sulphide (§ 119, 3).

c. Copper from Mercuricum and Cadmium.

112 Mix the solution, as in *b*, with potassium cyanide until the precipitate which is first thrown down redissolves; add some more potassium cyanide, then hydrogen sulphide water or ammonium sulphide, as long as a precipitate forms. The cadmium and mercury sulphides are completely thrown down, whilst the copper remains in solution, as sulphide dissolved in potassium cyanide. Allow the precipitate to subside, decant repeatedly, treat the precipitate, for security, once more with solution of potassium cyanide, heat gently, filter, and wash the sulphides of the metals. To determine the copper in the filtrate, evaporate the latter, with addition of nitric and sulpuric acids, until there is no longer any odor of hydrocyanic acid, and then precipitate with hydrogen sulphide (§ 119, 3).

d. All the Metals of the Fifth Group from each other.

113 Mix the dilute solution with sodium carbonate, then with potassium cyanide in excess, digest some time at a gentle heat, and filter. On the filter you have lead carbonate and bismuth carbonate (containing alkali); separate the two metals by a suitable method. Add to the filtrate dilute nitric acid in excess, warm gently till the cuprous sulphocyanate first precipitated with the silver cyanide has redissolved, and filter off the undissolved silver salt, which is to be determined as directed § 115, 3. Neutralize the filtrate with sodium carbonate, add potassium cyanide, and pass hydrogen sulphide in excess. Add now some more potassium cyanide, to redissolve the copper sulphide which may have fallen down, and filter the fluid, which contains the whole of the copper, from the precipitated sulphides of mercury and cadmium. Determine the copper as directed in *c*, and separate the mercury and cadmium as in **108**.

4. *Methods based on the Formation and Separation of insoluble Basic Salts.*

a. BISMUTH FROM COPPER, CADMIUM, AND MERCURICUM (also from the basic radicals of the first four groups, with the exception of ferric iron).

Precipitate the bismuth as basic chloride according to § 120, 4, and throw down the copper, &c., in the filtrate by hydrogen sulphide. Results thoroughly satisfactory (H. ROSE*). **114**

b. BISMUTH FROM LEAD AND CADMIUM.

Separate the bismuth according to § 120, 1, *c*, as basic nitrate, and precipitate the lead and cadmium in the filtrate by hydrogen sulphide. Results very satisfactory (J. LÖWE†). **115**

c. BISMUTH AND COPPER FROM LEAD AND CADMIUM.

Separate the bismuth after § 120, 1, *c*, as basic nitrate, then heat the dish on the water-bath till the normal copper nitrate is completely converted into bluish-green basic salt and no blue solution is produced on addition of water. Allow to cool, treat with an aqueous solution of ammonium nitrate (1 in 500), filter, wash with the same solution, and separate in the solution lead from cadmium; in the residue copper from bismuth. Results very satisfactory (J. LÖWE, *loc. cit.*).

5. *Method based upon the Precipitation of some of the Metals by Ammonia or Ammonium Carbonate.*

COPPER FROM BISMUTH.

Mix the (nitric acid) solution with ammonium carbonate in excess, and warm gently. The bismuth separates as carbonate, whilst the copper carbonate is redissolved by the excess of ammonium carbonate. As the precipitate, however, generally retains a little copper, it is necessary to redissolve it, after washing, in nitric acid, and precipitate again with ammonium carbonate; the same operation must be repeated a third time if required. Some solution of ammonium carbonate may be added to the water used for washing. Apply heat to the filtrate that the ammonium carbonate may volatilize, acidify cautiously with hydrochloric acid, and determine the copper as cuprous sulphide (§ 119, 3). The oxide of bismuth thus **116**

* Pogg. Annal. 110, 430.

† Journ. f. prakt. Chem. 74, 345.

obtained is quite copper-free, but a little bismuth passes into the copper solution, hence the separation does not give such exact results as that in **114** (H. Rose*).

6. *Method based on the Precipitation of the Copper as Cuprous Sulphocyanate.*

Copper from Cadmium (and the metals of Groups I.—IV., comp. **101**).

117 Precipitate the copper according to § 119, 3, *b*, as cuprous sulphocyanate (Rivot), and the cadmium from the filtrate as sulphide. Results good (H. Rose). Palladium may also be separated from copper in this way (Wöhler†).

7. *Method based upon the different deportment of the Chromates.*

Bismuth from Cadmium.

118 Precipitate the bismuth as directed § 120, 2. The filtrate contains the whole of the cadmium. Concentrate by evaporation, and then precipitate the cadmium by the cautious addition of sodium carbonate, as directed § 121, 1, *a* (J. Löwe,‡ W. Pearson§). The results given are satisfactory.

8. *Method based upon the different deportment of the Sulphides with Acids.*

a. Mercuricum from Silver, Bismuth, Copper, Cadmium, and (but less well) from Lead.

119 Boil the thoroughly washed precipitated sulphides with perfectly pure moderately dilute nitric acid. The mercuric sulphide is left undissolved, the other sulphides are dissolved. No chlorine may be present, and it is necessary that the mercuric sulphide should be pure, that is, free from finely divided mercury, which, as is well known, is precipitated when mercurous salts are treated with hydrogen sulphide. G. v. Rath‖ employed this method, which is so universally used in qualitative analysis; with perfect success for the separation of mercury from bismuth.

* Pogg. Annal. 110, 430.
† Annal. d. Chem. u. Pharm. 140, 144; Zeitschr. f. anal. Chem. 5, 403.
‡ Journ. f. prakt. Chem. 67, 439.
‖ Pogg. Annal. 96, 322.
§ Phil. Mag. 11, 204.

b. COPPER FROM CADMIUM.

120 Boil the well-washed precipitate of the sulphides with dilute sulphuric acid (1 part concentrated acid and 5 parts water), and, after some time, filter the undissolved copper sulphide, to be determined according to § 119, 3, from the solution containing the whole of the cadmium (A. W. HOFMANN*).

9. *Methods based upon the Volatility of some of the Metals, Oxides, Chlorides, or Sulphides at a high Temperature.*

a. MERCURY FROM SILVER, LEAD, COPPER (in general from the metals forming non-volatile chlorides).

121 Precipitate with hydrogen sulphide, collect the precipitated sulphides on a weighed filter, dry at 100°, weigh, and mix uniformly. Introduce an aliquot part into the bulb *d*,

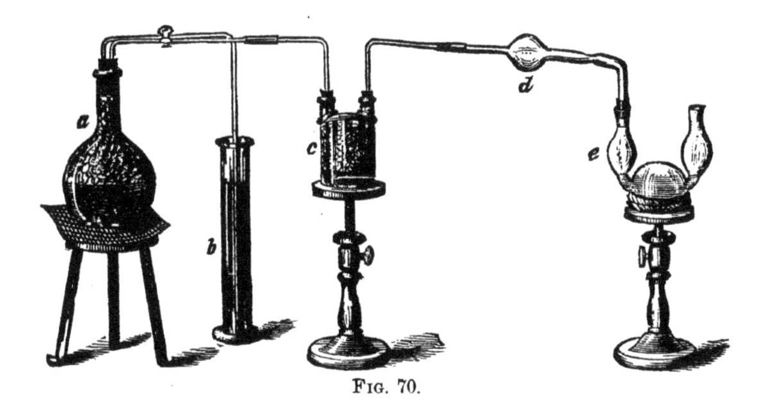

FIG. 70.

(fig. 70), pass a slow stream of chlorine gas, and apply a gentle heat to the bulb, increasing this gradually to faint redness. To ensure complete absorption it is well to have another small U-tube connected with *e*. The excess of chlorine escaping from the latter during the operation may be conducted into a flue or into a carboy containing moist slaked lime. First sulphur chloride distils over, which decomposes with the water in *e* (p. 463); then the mercuric chloride formed volatilizes, condensing partly in *e*, partly in the hind part of *d*. Cut off

* Annal. d. Chem. u. Pharm. 115, 286.

that part of the tube, rinse the sublimate with water into *e*, and mix the contents of the latter with the water in the second U-tube (not shown by the figure). Mix the solution with excess of ammonia, warm gently till no more nitrogen is evolved, acidify with hydrochloric acid, and then determine in the fluid filtered from the sulphur, which may still remain undissolved, the mercury as directed § 118, 3. If the residue consists of silver chloride alone, or lead chloride alone, you may weigh it at once; but if it contains several metals, you must reduce the chlorides by ignition in a stream of hydrogen, and dissolve the reduced metals in nitric acid, for their ulterior separation. Bear in mind that, in presence of lead, the sulphides and the chlorides must be heated *gently*, in the chlorine and hydrogen respectively, otherwise some lead chloride might volatilize.

In alloys or mixtures of oxides the mercury may usually be determined with simplicity from the loss on ignition in the air or in hydrogen.

b. BISMUTH FROM SILVER, LEAD, AND COPPER.

122 The separation is effected exactly in the same way as that of mercury from the same metals (**121**). The method is more especially convenient for the separation of the metals in alloys. Care must be taken not to heat too strongly, as otherwise lead chloride might volatilize; nor to discontinue the application of heat too soon, as otherwise bismuth would remain in the residue. AUG. VOGEL * gives 360° to 370° as the best temperature. Put water containing hydrochloric acid in U-tubes, which serve as receivers (fig. 70), and determine the bismuth therein according to § 120.

10. *Precipitation of one Metal in the Metallic State by another or the lower Oxide of another.*

a. LEAD FROM BISMUTH.

123 Precipitate the solution with ammonium carbonate (§ 116, 1, *a* and § 120, 1, *a*), wash the precipitated carbonates, and dissolve in acetic acid, in a flask; place a weighed rod of pure lead in the solution, and nearly fill up with water, so that the

* Zeitschr. f. anal. Chem. 13, 61.

rod may be entirely covered by the fluid; close the flask, and let it stand for about 12 hours, with occasional shaking. Wash the precipitated bismuth off from the lead rod, collect on a filter, wash, and dissolve in nitric acid; evaporate the solution, and determine the bismuth as directed § 120. Determine the lead in the filtrate as directed § 116. Dry the leaden rod, and weigh; subtract the loss of weight which the rod has suffered in the process from the amount of the lead obtained from the filtrate (ULLGREN*). PATERA† recommends precipitating from dilute nitric solution, washing the precipitated bismuth first with water, then with alcohol, transferring to a small filter, drying and weighing. If it is feared that the finely divided bismuth has undergone oxidation, it is well to fuse it with potassium cyanide (§ 120, 4).

11. *Separation of Silver by Cupellation.*

124 CUPELLATION was formerly the universal method of determining SILVER in alloys with COPPER, LEAD, etc. The alloy is fused with a sufficient quantity of pure lead to give to 1 part of silver 16 to 20 parts of lead, and the fused mass is heated, in a muffle, in a small cupel made of compressed bone-ash. Lead and copper are oxidized, and the oxides absorbed by the cupel, the silver being left behind in a state of purity. One part by weight of the cupel absorbs the oxide of about 2 parts of lead; the quantity of the sample to be used in the experiment may be estimated accordingly. This method is only rarely employed in laboratories;‡ I have given it a place here, however, because it is one of the safest processes to determine very small quantities of silver in alloys.§

12. *Methods depending on the Volumetric Estimation of one Metal.*

a. COPPER OF CUPROUS COMPOUNDS IN PRESENCE OF CUPRIC COMPOUNDS.‖

125 Dissolve the substance, if necessary in a current of carbonic

* BERZELIUS' Jahresber. 21, 148. † Zeitschr. f. anal. Chem. 5, 226.

‡ For details of this process consult "Bodemann and Kerl's Assaying," translated by GOODYEAR; or "Notes on Assaying," by P. DE RICKETTS.

§ Compare MALAGUTI and DUROCHER, Comp. rend. 29, 689; DINGLER, 115, 276. Also W. HAMPE, Zeitschr. f. anal. Chem. 11, 221.

‖ The method of COMMAILLE (Comp. rend. 56, 309) can no longer be relied

acid, in hydrochloric acid, and add ferric chloride. A volumetric determination of the amount of iron reduced to ferrous salt affords a basis for calculating the amount of copper present originally as a cuprous compound. Or if a known quantity of ferric chloride is used, a determination of the iron remaining in the state of a ferric salt suffices equally well.

b. SILVER IN PRESENCE OF LEAD AND COPPER.

Small quantities of silver may be estimated by PISANI'S method, § 115, II.

Sixth Group.

GOLD—PLATINUM—TIN—ANTIMONY—(ANTIMONIC ACID)—ARSENIOUS ACID—ARSENIC ACID.

I. SEPARATION OF THE METALS OF THE SIXTH GROUP FROM THOSE OF THE FIRST FIVE GROUPS.

§ 164.

INDEX. (The numbers refer to those in the margin.)

upon, since STAS has shown that the finely divided silver thrown down by ammoniacal solution of cuprous chloride dissolves largely in ammonia with access of air.

Tin from zinc, 126, 128, 133, 135.
" manganese, 126, 128, 135.
" nickel and cobalt, 126, 128, 133, 135, 140.
" iron, 126, 128.
" silver, 127, 128, 133, 140.
" mercury, 127, 128, 133.
" lead, 127, 128, 133, 140.
" copper, 127, 128, 133, 135, 180.
" bismuth, 127, 128.
" cadmium, 127, 128, 133, 135.

Antimony from the metals of Groups I. and II., 126.
" Group III., 126.
" zinc, 126, 128, 134.
" manganese, 126, 128.
nickel and cobalt, 126, 128, 134, 139, 140.
iron, 126, 128, 138.
silver, 127, 128, 134, 140.
mercury, 127, 128, 134, 136, 148.
lead, 127, 128, 134, 140, 150.
copper, 127, 128, 134, 138, 140, 151.
bismuth, 127, 128.
" cadmium, 127, 128, 134.

Arsenic from the metals of Group I., 126, 145, 146.
" " II., 126, 137, 145, 146.
" " III., 126, 144, 145.
zinc, 126, 128, 137, 143, 145, 146.
manganese, 126, 128, 137, 143, 144, 145, 146.
nickel and cobalt, 126, 128, 137, 139, 140, 143, 144, 145, 146.
iron, 126, 128, 137, 138, 143, 144, 145.
silver, 127, 128, 137, 140, 145.
" mercury, 127, 128, 145, 148.
" lead, 127, 128, 137, 140, 145, 149.
copper, 127, 128, 137, 138, 140, 143, 144, 145, 151.
" bismuth, 127, 128, 137, 145.
" cadmium, 127, 128, 137, 144, 145.

A. *General Methods.*

1. *Method based upon the Precipitation of Metals of the Sixth Group from Acid Solutions by Sulphuretted Hydrogen.*

ALL METALS OF THE SIXTH GROUP FROM THOSE OF THE FIRST FOUR GROUPS.

Conduct into the acid* solution hydrogen sulphide in excess, and filter off the precipitated sulphides (corresponding to the oxides of the sixth group). **126**

* Hydrochloric acid answers best as acidifying agent.

The points mentioned **96**, α, β, and γ, must also be attended to here. As regards γ, antimony and tin are to be inserted between cadmium and mercury, in the order of metals there given. With respect to the particular conditions required to secure the proper precipitation of certain metals of the sixth group, I refer to Section IV. I have to remark in addition:

α. That hydrogen sulphide fails to separate arsenic acid from zinc, as, even in presence of a large excess of acid, the whole or at least a portion of the zinc precipitates with the arsenic (Wöhler). To secure the separation of the two bodies in a solution, the arsenic acid must first be converted into arsenious acid, by heating with sulphurous acid, before the hydrogen sulphide is conducted into the fluid.

β. That in presence of antimony, tartaric acid should be added, as otherwise the sulphide of antimony will contain chloride; and that sulphide of antimony, when thrown down from a boiling solution by hydrogen sulphide, becomes black after a time, and so dense that it is deposited like sand, whereby the filtration and washing are much facilitated (S. P. Schäfeler *).

2. *Method based upon the Solubility of the Sulphides of the Metals of the Sixth Group in Sulphides of the Alkali Metals.*

a. The Metals of Group VI. (with the exception of Gold and Platinum) from those of Group V.

127 Precipitate the acid solution with hydrogen sulphide, paying due attention to the directions given in Section IV. under the heads of the several metals, and also to the remarks in **126**. The precipitate consists of the sulphides of the metals of Groups V. and VI. Wash, and treat at once with yellow ammonium sulphide in excess. (It is usually best to spread out the filter in a porcelain dish, add the ammonium sulphide, cover with a large watch-glass, and place on a heated water-bath. Unnecessary exposure to air should be avoided.) Add some water, filter off the clear fluid, treat the residue again

* Berichte der deutschen chem. Gesellsch. 1871, 279. I have myself confirmed these observations.

with ammonium sulphide, digest a short time, repeat the same operation, if necessary, a third and fourth time, filter, and wash the residuary sulphides of Group V. with water containing ammonium sulphide. If stannous sulphide is present, some flowers of sulphur must be added to the ammonium sulphide, unless the latter be very yellow. In presence of copper, the sulphide of which is a little soluble in ammonium sulphide, sodium sulphide should be used instead. However, this substitution can be made only in the absence of mercury, since the sulphide of that metal is soluble in sodium sulphide.

Add to the alkaline filtrate, gradually, hydrochloric acid in small portions, until the acid predominates; allow to subside, and then filter off the sulphides of the metals of the sixth group, which are mixed with sulphur.

If a solution contains much arsenic acid in presence of small quantities of copper, bismuth, &c., it is convenient to precipitate these metals (together with a very small amount of arsenious sulphide) by a brief treatment with hydrogen sulphide. Filter, extract the precipitate with ammonium sulphide (or potassium sulphide), acidify the solution obtained, mix it with the former filtrate containing the principal quantity of the arsenic, and proceed to treat further with hydrogen sulphide (§ 127, 4, *b*).

b. The Metals of Group VI. (with the exception of Gold and Platinum) from those of Groups IV. and V.

α. Neutralize the solution with ammonia, add ammonium chloride, if necessary, and then yellow ammonium sulphide in excess; digest in a closed flask, for some time at a moderate heat, and then proceed as in **127**. Repeated digestion with fresh quantities of ammonium sulphide is indispensable. On the filter, you have the sulphides of the metals of Groups IV and V. Wash with water containing ammonium sulphide. In presence of nickel, this method offers peculiar difficulties; traces of mercuric sulphide, too, are liable to pass into the filtrate. In presence of copper (and absence of mercury), soda and sodium sulphide are substituted for ammonia and ammonium sulphide.* **128**

* The accuracy of this method has been called in question by Bloxam (Quart. Journ. Chem. Soc. 5, 119). That chemist found that ammonium sulphide fails

β. In the analysis of solid compounds (oxides or salts), it is in most cases preferable to fuse the substance with 3 parts of dry sodium carbonate and 3 of sulphur, in a covered porcelain crucible. When the contents are completely fused, and the excess of sulphur is volatilized, the mass is allowed to cool, and then treated with water, which dissolves the sulphosalts of the metals of the sixth group, leaving the sulphides of Groups IV. and V. undissolved. By this means, even ignited stannic oxide may be readily tested for iron, &c., and the amount of the admixture determined (H. ROSE). The solution of the sulphosalts is treated as in **127**. In the presence of copper, traces of the sulphide may be dissolved with the sulphides of Group VI. Occasionally a little ferrous sulphide dissolves, coloring the solution green. In that case add some ammonium chloride, and digest till the solution has turned yellow. Instead of the mixture of sodium carbonate and sulphur you may also use already prepared *hepar sulphuris*, or, as FRÖHDE* says, you may fuse the substance with 4 or 5 parts of sodium thiosulphate.

B. *Special Methods.*

1. *Methods based upon the Insolubility of some Metals of the Sixth Group in Acids.*

a. GOLD FROM METALS OF GROUPS IV. AND V. IN ALLOYS.

129 *α*. Boil the alloy with pure nitric acid (not too concentrated), or, according to circumstances, with hydrochloric acid. The other metals dissolve, the gold is left. The alloy must be reduced to filings, or rolled out into a thin sheet. If the alloy were treated with concentrated nitric acid, and at a temperature below boiling, a little gold might dissolve in consequence of the co-operation of nitrous acid. In the presence of silver and lead, this method is only applicable when they

---

to separate small quantities of stannic sulphide from large quantities of mercuric sulphide or cadmium sulphide (1 : 100); and that more especially the separation of copper from tin and antimony (also from arsenic) by this method is a failure, as nearly the whole of the tin remains with the copper. The latter statement I cannot confirm, for Mr. LUCIUS, in my laboratory, has succeeded in separating copper from tin by means of yellowish sodium sulphide completely; but it is indispensable to digest three or four times with sufficiently large quantities of the solvent, as stated in the text.

* Zeitschr. f. anal. Chem. 5, 405.

amount to more than 80 per cent., since otherwise they are not completely dissolved. Alloys of silver and gold containing less than 80 per cent. of silver are therefore fused with 3 parts of lead, before they are treated with nitric acid. The residuary gold is weighed; but its purity must be ascertained, by dissolving in cold dilute nitrohydrochloric acid, not in concentrated hot acid, as silver chloride also is soluble in the latter. In the presence of silver, a small quantity of its chloride is usually obtained here. If it can be weighed, it should be reduced and deducted.

At the Mint Conference held at Vienna in 1857, the following process was agreed upon for the mints in the several states of Germany. Add to 1 part of gold, supposed to be present, 2½ parts of pure silver; wrap both the alloy and the silver in a paper together, and introduce into a cupel in which the requisite amount of lead is just fusing.* After the lead has been absorbed,† the button is flattened by hammering or rolling, then ignited and rolled. The rolls are treated first with nitric acid of 1·2 sp. gr., afterwards with nitric acid of 1·3 sp. gr., rinsed, ignited, and weighed.‡ Even after boiling again with nitric acid of 1·3 sp. gr., they retain ·75 to ·001 of silver which will remain as chloride if the rolls are treated with cold dilute aqua regia (H. Rössler, *loc. cit.*).

β. Heat the alloy (previously filed or rolled) in a capacious platinum dish with a mixture of 2 parts pure concentrated sulphuric acid and 1 part water, until the evolution of gas has ceased and the sulphuric acid begins to volatilize; or fuse the alloy with potassium disulphate (H. Rose). Separate the gold from the sulphates of the other metals, by treating the mass with water which should finally be boiling. It is advisable to repeat the operation with the separated gold, and ultimately

* If the weighed sample, say ·25 grm., contains 98–92‰ gold, 3 grm. of lead are required; if 92–87·5, 4 grm.; if 87·5–75, 5 grm.; if 75–60, 6 grm.; if 60–35, 7 grm.; if less than 35, 8 grm.

† A small quantity of gold—from one to three thousandths—is always lost in cupellation. The loss increases with the amount of lead, and is also dependent on the proportion of silver to gold. The more silver present the less is the loss of gold. In large buttons the loss is less than in small ones (H. Rössler, Ding. polyt. Journ. 206, 185; Zeitschr. f. anal. Chem. 13, 87).

‡ Kunst-und Gewerbeblatt f. Baiern, 1857, 151; Chem. Centralbl. 1857, 307 Polyt. Centralbl. 1857, 1151, 1471, 1639.

test the purity of the latter. In presence of lead this method is not good.

γ. The methods given in α and β may be united, *i.e.*, the cupelled and thinly-rolled metal may be first warmed with nitric acid of 1·2 sp. gr., then thoroughly washed, the gold boiled 5 minutes with concentrated sulphuric acid, washed again, and ignited (MASCAZZINI, BUGATTI).

b. PLATINUM FROM METALS OF GROUPS IV. AND V. IN ALLOYS.

130 The separation is effected by heating the alloy in filings or foil with pure concentrated sulphuric acid, with addition of a little water, or by fusing with potassium disulphate (**129**, β); but not with nitric acid, as platinum in alloys will, under certain circumstances, dissolve in that acid.

2. *Method based upon the Separation of Gold in the metallic state.*

GOLD FROM ALL METALS OF GROUPS I.—V., with the exception of LEAD, MERCURY, AND SILVER.

131 Precipitate the hydrochloric acid solution with oxalic acid as directed § 123 *b*, γ, or with ferrous sulphate, § 123, *b*, α, and filter off the gold when it has completely separated. Take care to add a sufficient quantity of hydrochloric acid after the reduction to insure solution of any oxalates. In the presence of copper the addition of hydrochloric acid does not suffice, since the coprecipitated cupric oxalate will dissolve with difficulty in this acid. E. PURGOTTI* recommends in this case, after precipitation, adding potash cautiously to the boiling hot fluid till it is neutral, and then if necessary some normal potassium oxalate. Double oxalate of copper and potash will be formed which dissolves with a blue color. The gold after washing will now be pure.

3. *Method based upon the Precipitation of Platinum as Potassium Platinic, or Ammonium Platinic Chloride.*

PLATINUM FROM THE METALS OF GROUPS IV. AND V., with the exception of MERCURY IN MERCUROUS COMPOUNDS, LEAD, AND SILVER.

132 Precipitate the platinum with potassium chloride or

* Zeitschr. f. anal. Chem. 9, 128.

ammonium chloride as directed § 124, and wash the precipitate thoroughly with alcohol. The platinum prepared from the precipitated ammonium or potassium salt is to be tested after being weighed, to see whether it yields any metal (especially iron) to fusing potassium disulphate.

4. *Methods based upon the Separation of Oxides insoluble in Nitric Acid.*

a. TIN FROM METALS OF GROUPS IV. AND V. (not from Bismuth, Iron, or Manganese*) IN ALLOYS.

133 Treat the finely divided alloy, or the metallic powder obtained by reducing the oxides in a stream of hydrogen with nitric acid, as directed § 126, 1, *a*. The filtrate contains the other metals as nitrates. As stannic oxide is liable to retain traces of copper and lead and iron, you must, in an accurate analysis, test an aliquot part of it for these bodies, and determine their amount as directed **128**, *β*.

BRUNNER recommends the following course of proceeding, by which the presence of copper in the tin may be effectually guarded against. Dissolve the alloy in a mixture of 1 part of nitric acid, 4 parts of hydrochloric acid, and 5 parts of water; dilute the solution largely with water, and heat gently. Add crystals of sodium carbonate until a distinct precipitate has formed, and boil. (In presence of copper, the precipitate must, in this operation, change from its original bluish-green to a brown or black tint.) When the fluid has been in ebullition some 10 or 15 minutes, allow it to cool, and then add nitric acid, drop by drop, until the reaction is distinctly acid; digest the precipitate for several hours, when it should have acquired a pure white color. The stannic oxide thus obtained is free from copper; but it may contain some iron, which can be removed as directed in **128**, *β*.

Before the stannic oxide can be considered pure, it must be tested also for silicic acid, as it frequently contains traces of this substance. To this end, an aliquot part is fused in plati-

* If the alloy of tin contains bismuth or manganese, there remains with the stannic oxide, bismuth trioxide or manganese sesquioxide, which cannot be extracted by nitric acid; if it contains iron, on the contrary, some stannic oxide always dissolves with the iron, and cannot be separated even by repeated evaporation (H. ROSE, Pogg. Annal. 112, 169, 170, 172).

num with 3—4 parts of sodium and potassium carbonate, the fused mass boiled with water, and the solution filtered; hydrochloric acid is then added to the filtrate, and, should silicic acid separate, the fluid is filtered off from this substance. The tin is then precipitated by hydrogen sulphide, and the silicic acid still remaining in the filtrate is determined in the usual way (§ 140). If hydrochloric acid has produced a precipitate of silicic acid, the last filtration is effected on the same filter (KHITTEL*).

b. ANTIMONY FROM THE METALS OF GROUPS IV. AND V. IN ALLOYS (not from Bismuth, Iron and Manganese).

134 Proceed as in **133**, filter off the precipitate, and convert it by ignition into antimony tetroxide according to § 125, 2. Results only approximate, as a little antimony dissolves. Alloys of antimony and lead, containing the former metal in excess, should be previously fused with a weighed quantity of pure lead (VARRENTRAPP†).

5. *Methods based on the Precipitation of Tin in Stannic Salts by Normal Salts* (e.g., *Sodium Sulphate*) *or by Sulphuric Acid.*

TIN FROM THE METALS OF GROUPS I., II., III,; ALSO FROM MANGANESE, ZINC, NICKEL AND COBALT, COPPER, CADMIUM (GOLD).

135 Precipitate the hydrochloric acid solution, which must contain the tin entirely as stannic chloride, according to § 126, 1, *b*, by ammonium nitrate or sodium sulphate (LÖWENTHAL), or by sulphuric acid, which, H. ROSE says, answers equally well. Alloys are always treated as follows: First, oxidize by digestion with nitric acid; when no more action takes place, evaporate the greater portion of the nitric acid in a porcelain dish, moisten the mass with strong hydrochloric acid, and after half an hour add water, in which the metastannic chloride and the other chlorides dissolve. Alloys of tin and gold are dissolved in aqua regia, the excess of acid evaporated, and the solution diluted with much water, before precipitating with sulphuric acid.

It must be remembered that in this process any phosphoric

* Chem. Centralbl. 1857, 929. † Dingler's polyt. Journ. 158, 316.

acid that may be present is precipitated entirely or partially with the tin. After the precipitate has been well washed by decantation, LÖWENTHAL recommends to boil with a mixture of 1 part nitric acid (sp. gr. 1·2) and 9 parts water, then to transfer to the filter, and wash thoroughly. Results very satisfactory. If the fluid contains a ferric salt, a portion of the iron always falls down with the tin. Hence the stannic oxide must be tested for iron according to **128**, *β*, which, if present, must be determined and deducted.

6. *Method based on the Insolubility of Mercuric Sulphide in Hydrochloric Acid.*

MERCURY FROM ANTIMONY.

136 Digest the precipitated sulphides with moderately strong hydrochloric acid in a distilling apparatus. The sulphide of antimony dissolves, while the mercuric sulphide remains behind. Expel all the hydrogen sulphide, then add tartaric acid, dilute, filter, mix the filtrate with the distillate which contains a little antimony, and precipitate with hydrogen sulphide. The mercuric sulphide may be weighed as such (F. FIELD*).

7. *Methods based upon the Conversion of Arsenic and Antimony into Alkali Arsenate and Antimonate.*

a. ARSENIC FROM THE METALS OF GROUPS II., IV., AND V.

137 If you have to do with arsenites or arsenates, fuse with 3 parts of sodium and potassium carbonates and 1 part of potassium nitrate; if an alloy has to be analyzed it is fused with 3 parts of sodium carbonate and 3 parts of potassium nitrate. In either case the residue is boiled with water, and the solution, which contains the arsenates of the alkalies, filtered from the undissolved oxides or carbonates. The arsenic acid is determined in the filtrate as directed § 127, 2. If the quantity of arsenic is only small, a platinum crucible may be used, otherwise a porcelain crucible must be used, as platinum would be seriously injured. In the latter case, bear in mind that the fused mass is contaminated with silicic acid and alumina. If the alloy contains much arsenic a small quantity may be readily lost by volatilization, even though the operation be cautiously

* Quart. Journ. Chem. Soc. 12, 32.

conducted. In such a case, therefore, it is better first to oxidize with nitric acid, then to evaporate, and to fuse the residue as above directed with sodium carbonate and potassium nitrate.

b. ARSENIC AND ANTIMONY FROM COPPER AND IRON, especially in ores containing sulphur.

138 Diffuse the very finely pulverized mineral through pure solution of potassa, and conduct chlorine into the fluid (comp. p. 467). The iron and copper separate as oxides, the solution contains sulphate, arsenate, and antimonate of potassium (RIVOT, BEUDANT, and DAGUIN*).

c. ARSENIC AND ANTIMONY FROM COBALT AND NICKEL.

139 Dilute the nitric acid solution, add a large excess of potassa, heat gently, and conduct chlorine into the fluid until the precipitate is black. The solution contains the whole of the arsenic and antimony, the precipitate the nickel and cobalt as sesquioxides (RIVOT, BEUDANT, and DAGUIN, *loc. cit.*)

8. *Methods based upon the Volatility of certain Chlorides or Metals.*

a. TIN, ANTIMONY, ARSENIC FROM COPPER, SILVER, LEAD, COBALT, NICKEL.

140 Treat the sulphides with a stream of perfectly dry chlorine, proceeding exactly as directed in **121**. In presence of antimony, fill the receiver *e* (fig. 70) with a solution of tartaric acid in water, mixed with hydrochloric acid. The metals may be also separated by this method in alloys. The alloy must be very finely divided. Arsenical alloys are only very slowly decomposed in this way. In separating arsenic and copper the temperature must not exceed 200°, and chlorine water should be put into the receiver (PARNELL†). If tin and copper are separated in this manner, according to the experience of H. ROSE,‡ a small trace of tin remains with the copper chloride.

b. STANNIC OXIDE, ANTIMONIOUS OXIDE (AND ALSO ANTIMONIC ACID), ARSENIOUS AND ARSENIC ACIDS, FROM ALKALIES AND ALKALINE EARTHS.

141 Mix the solid compound with 5 parts of pure ammonium chloride in powder, in a porcelain crucible, cover this with a

* Compt. rend. 1853, 835; Journ. f. prakt. Chem. 61, 133.
† Chem. News, 21, 133. ‡ Pogg. Annal. 112, 169.

concave platinum lid, on which some ammonium chloride is sprinkled, and ignite gently until all ammonium chloride is driven off; mix the contents of the crucible with a fresh portion of that salt, and repeat the operation until the weight remains constant. In this process, the chlorides of tin, antimony, and arsenic escape, leaving the chlorides of the alkalies and alkali-earth metals. The decomposition proceeds most rapidly with alkali salts. With regard to salts of alkali-earth metals it is to be observed that those which contain antimonic acid or stannic oxide are generally decomposed completely by a double ignition with ammonium chloride (magnesium alone cannot be separated perfectly from antimonic acid by this method). The arsenates of the alkali-earth metals are the most troublesome to decompose; barium, stronium, and calcium salts usually require to be subjected 5 times to the operation, before they are free from arsenic, and magnesium arsenate it is impossible thoroughly to decompose in this way (H. ROSE*). According to SALKOWSKI† barium arsenate may be converted into chloride quite free from arsenic by one ignition with ammonium chloride; however calcium arsenate was found to leave a residue containing arsenic acid even after six ignitions with ammonium chloride.

c. MERCURY FROM GOLD (SILVER, AND GENERALLY FROM THE NON-VOLATILE METALS).

142 Heat the weighed alloy in a porcelain crucible, ignite till the weight is constant, and determine the mercury from the loss. If it desired to estimate it directly, the apparatus, p. 307, fig. 54, may be used. In cases where the separation of mercury from metals that oxidize on ignition in the air is to be effected by this method, the operation must be conducted in an atmosphere of hydrogen (p. 251, fig. 50).

9. *Methods based on the Volatility of Arsenious Sulphide.*

ARSENIC ACID FROM THE OXIDES OF MANGANESE, IRON, ZINC, COPPER, NICKEL, COBALT (NOT SO WELL FROM OXIDE OF LEAD, AND NOT FROM OXIDES OF SILVER, ALUMINUM, OR MAGNESIUM).

143 Mix the arsenic acid compound (no matter whether it has

* Pogg. Annal. 73, 582; 74, 578; 112, 173. † Journ. f. prakt. Chem. 104, 138.

been air-dried or gently ignited) with sulphur, and ignite under a good draught in an atmosphere of hydrogen (p. 251, fig. 50; the perforated lid must in this case be of porcelain; platinum would not answer). The whole of the arsenic volatilizes, the sulphides of manganese, iron, zinc, lead, and copper remain behind; they may be weighed directly. After weighing, add a fresh quantity of sulphur to the residue, ignite as before, and weigh again; repeat this operation until the weight remains constant. Usually, if the compound was intimately mixed with the sulphur, the conversion of the arsenate into sulphide is complete after the first ignition. Results very good.

In separating *nickel* the analyst will remember that the residue cannot be weighed directly, since it does not possess a constant composition; hence the ignition in hydrogen may be saved; nickel arsenate loses all its arsenic on being simply mixed with sulphur and heated. The heat should be moderate and continued till no more red sulphide of arsenic is visible on the inside of the porcelain crucible. It is advisable to repeat the operation. The separation of arsenic from *cobalt* cannot be completely effected in this manner even by repeated treatment with sulphur, but it can be effected by oxidizing the residue with nitric acid, evaporating to dryness, mixing with sulphur, and reigniting. Smaltine and cobaltine must be treated in the same manner (H. Rose*). I should not forget to mention that Ebelmen,† a long while ago, noticed the separation of arsenic acid from sesquioxide of iron by ignition in a stream of hydrogen sulphide.

10. *Method based upon the Separation of Arsenic as Ammonium Magnesium Arsenate.*

Arsenic Acid from Copper, Cadmium, Ferric Iron, Manganese, Nickel, Cobalt, Aluminium.

144 Mix the hydrochloric acid solution, which must contain the whole of the arsenic in the form of arsenic acid, with enough tartaric acid to prevent precipitation by ammonia, precipitate the arsenic acid according to § 127, 2, as ammonium magnesium arsenate, allow to settle, filter, wash once with a

* Zeitschr. f. anal. Chem. 1, 413.

† Annal. de Chim. et de Phys. (3) 25, 98.

mixture of 3 parts water and 1 part ammonia, redissolve in hydrochloric acid, add a very minute quantity of tartaric acid, supersaturate again with ammonia, add some more magnesium chloride and ammonium chloride, allow to deposit, and determine the now pure precipitate according to § 127, 2. In the filtrate the bases of Groups IV. and V. may be precipitated by ammonium sulphide; if aluminium is present, evaporate the filtrate from the sulphides with addition of sodium carbonate and a little nitre to dryness, fuse, and estimate the aluminium in the residue. The method is more adapted to the separation of rather large than of very small quantities of arsenic from the above-named metals, since in the case of small quantities the minute portions of ammonium magnesium arsenate that remain in solution may exercise a considerable influence on the accuracy of the result.

11. *Method based upon the Separation of Arsenic as Ammonium Arsenio-molybdate.*

Arsenic Acid from all Metals of Groups I.—V.

145 Separate the arsenic acid as directed in § 127, 2, *b*; long continued heating at 100° is indispensable. The determination of the basic metals is most conveniently effected in a special portion.

12. *Method based upon the Insolubility of Ferric Arsenate.*

Arsenic Acid from the Metals of Groups I. and II., and from Zinc, Manganese, Nickel, and Cobalt.

146 Mix the hydrochloric solution with a sufficient quantity of pure ferric chloride, neutralize the greater part of the free acid with sodium carbonate, and precipitate the iron and arsenic acid together with barium carbonate in the cold or with sodium acetate at a boiling heat. The precipitate should be so basic as to have a brownish-red color. The method is especially suitable for the separation of arsenic acid when its estimation is not required. However, the precipitate may be dissolved in hydrochloric acid and the arsenic determined by precipitation with hydrogen sulphide.

13. *Methods based upon the Insolubility of some Chlorides.*

a. SILVER FROM GOLD.

Treat the alloy with cold dilute nitrohydrochloric acid, **147** dilute, and filter the solution of auric chloride from the undissolved silver chloride. This method is applicable only if the alloy contains less than 15 per cent. of silver; for if it contains a larger proportion, the silver chloride which forms protects the undecomposed part from the action of the acid. In the same way silver may be separated also from *platinum.*

b. MERCURY FROM THE OXYGEN COMPOUNDS OF ARSENIC AND ANTIMONY.

Precipitate the mercury from the hydrochloric solution by **148** means of phosphorous acid as mercurous chloride (§ 118, 2). The tartaric acid, which in the presence of antimony must be added, does not interfere with the reaction (H. ROSE*).

14. *Methods based upon the Insolubility of certain Sulphates in Water or Alcohol.*

a. ARSENIC ACID FROM BARIUM, STRONTIUM, CALCIUM, AND LEAD.

Proceed as for the separation of phosphoric acid from the **149** same metals (§ 135, *b*). The compounds of these basic radicals with arsenious acid are first converted into arsenates, before the sulphuric acid is added; this conversion is effected by heating the hydrochloric acid solution with potassium chlorate or by means of bromine.

b. ANTIMONY FROM LEAD.

Treat the alloy with a mixture of nitric and tartaric acids. **150** The solution of both metals takes place rapidly and with ease. Precipitate the greater part of the lead as sulphate (§ 116, 3), filter, precipitate with hydrogen sulphide, and treat the sulphides according to **128**, with ammonium sulphide, in order to separate the antimony from the lead left unprecipitated by the sulphuric acid (A. STRENG†).

15. *Method based upon the Separation of Copper as Cuprous Sulphocyanate.*

COPPER FROM ARSENIC AND ANTIMONY.

From the properly prepared solution precipitate the cop- **151**

* Pogg. Annal. 110, 536. † Ding. polyt. Journ. 151, 389.

per by § 119, 3, *b*, as cuprous sulphocyanate, allow to settle, filter, wash with water containing ammonium nitrate (to prevent the washings being muddy), and determine antimony and arsenic in the filtrate, precipitating first with hydrogen sulphide. Results good.

16. *Method based upon the different deportment with Cyanide of Potassium.*

GOLD FROM LEAD AND BISMUTH.

152 These metals may be separated in solution by potassium cyanide in the same way in which the separation of mercury from lead and bismuth is effected (see **110**). The solution of the double cyanide of gold and potassium is decomposed by boiling with aqua regia, and, after expulsion of the hydrocyanic acid, the gold determined by one of the methods given in § 123.

II. SEPARATION OF THE METALS OF THE SIXTH GROUP FROM EACH OTHER.

§ 165.

INDEX. (The numbers refer to those in the margin.)

1. *Method based upon the Precipitation of Platinum as Potassium Platinic Chloride.*

PLATINUM FROM GOLD.

153 Precipitate from the solution of the chlorides the platinum as directed § 124, *b*, and determine the gold in the filtrate as directed § 123, *b*.

2. *Methods based upon the Volatility of the Chlorides of the inferior Metals.*

a. PLATINUM AND GOLD FROM TIN, ANTIMONY, AND ARSENIC.

154 Heat the finely divided alloy or the sulphides in a stream of chlorine gas. Gold and platinum are left, the chlorides of the other metals volatilize (compare **121**).

b. ANTIMONY FROM TIN.

155 The tin should be present wholly as a stannous salt. Precipitate with hydrogen sulphide, filter (preferably through an asbestos filtering tube), dry the precipitate, and pass through it a current of dry hydrochloric gas at the ordinary temperature. The sulphides are converted into the corresponding chlorides; the chloride of antimony alone escapes, and may be received in water. Dissolve the residual stannous chloride in water containing hydrochloric acid, and estimate the tin according to § 126 (C. TOOKEY*). The method can only be used in rare cases, as it is difficult to obtain a precipitate quite free from stannic sulphide.

c. ARSENIOUS ACID FROM ARSENIC ACID.

156 The amount of substance taken should not contain more than ·2 grm. arsenious acid. Heat with 45 grm. sodium chloride, 135 grm. sulphuric acid (free from arsenic) of 1·61 sp. gr., and 30 grm. water in a tubulated retort containing a spiral of platinum, and provided with a thermometer. The temperature should rise to about 125°. In order to condense the arsenious chloride in the products of distillation, a LIEBIG's condenser is connected with the retort; a tubulated receiver is connected with the condenser; a U-tube is connected with the receiver, and finally a calcium chloride tube containing fragments of glass moistened with weak soda solution is fixed

* Journ. Chem. Soc. 15, 462.

upright in the exit end of the U-tube. In the receiver and U-tube water is placed. At the end of the operation rinse the calcium chloride tube, and mix with the contents of the receiver. Determine the arsenic in the distillate according to § 127, 4, *a*, in the residue according to § 127, 4, *b*. The sulphide obtained from the former corresponds to the arsenious acid, from the latter to the arsenic acid. Results satisfactory (RIECKHER*). If the substance given is a dilute fluid, render slightly alkaline with sodium carbonate, and concentrate to about 20 c.c., finally in a tubulated retort.

3. *Methods based upon the Volatility of Arsenic and Arsenious Sulphide.*

a. ARSENIC FROM TIN (H. ROSE).

157 Convert into sulphides or oxides, dry at 100°, and heat a weighed portion with addition of a little sulphur in a bulb-tube, gently at first, but gradually more strongly, conducting a stream of dry hydrogen sulphide gas through the tube during the operation. Sulphur and arsenious sulphide volatilize; sulphide of tin is left. The arsenious sulphide is received in U-tubes containing dilute ammonia, which are connected with the bulb-tube in the manner described in **121**. When upon continued application of heat no sign of further sublimation is observed in the colder part of the bulb-tube, drive off the sublimate which has collected in the bulb, allow the tube to cool, and then cut it off above the coating. Divide the separated portion of the tube into pieces, and heat these with a little solution of soda until the sublimate is dissolved; unite the solution with the ammoniacal fluid in the receivers, add hydrochloric acid, then, without filtering, potassium chlorate, and heat gently until the arsenious sulphide is completely dissolved. Filter from the sulphur, and determine the arsenic acid as directed § 127, 2. The quantity of tin cannot be calculated at once from the blackish-brown sulphide of tin in the bulb, since this contains more sulphur than SnS. It is therefore weighed, and the tin determined in a weighed portion of it, by converting it into stannic oxide, which is effected by moistening with nitric acid, and roasting (§ 126, 1, *c*).

* Pharm. Centralhalle, 11, 92.

Tin and arsenic in alloys are more conveniently converted into oxides by cautious treatment with nitric acid. If, however, it is wished to convert them into sulphides, this may readily be effected by heating 1 part of the finely divided alloy with 5 parts of sodium carbonate and 5 parts of sulphur, in a covered porcelain crucible until the mass is in a state of calm fusion. It is then dissolved in water, the solution filtered from the ferrous sulphide, &c., which may possibly have formed, and then precipitated with hydrochloric acid.

If the tin only in the alloy is to be estimated directly, while the arsenic is to be found from the difference, convert as above directed into sulphides or oxides, mix with sulphur and ignite in a porcelain crucible with perforated cover in a stream of hydrogen sulphide. The residual arsenic-free stannous sulphide is to be converted into stannic oxide and weighed as such.

4. *Methods based upon the Insolubility of Sodium Metantimonate.*

a. Antimony from Tin and Arsenic (H. Rose).

158 If the substance is metallic, oxidize the finely divided weighed sample, in a porcelain crucible, with nitric acid of 1·4 sp. gr., adding the acid gradually. Dry the mass on the water-bath, transfer to a silver crucible, rinsing the last particles adhering to the porcelain into the silver crucible with solution of soda, dry again, add eight times the bulk of the mass of solid sodium hydroxide, and fuse for some time. Allow the mass to cool, and then treat with hot water until the undissolved residue presents the appearance of a fine powder; dilute with some water, and add one-third the volume of alcohol of ·83 sp. gr. Allow the mixture to stand for 24 hours, with frequent stirring; then filter, transfer the last adhering particles from the crucible to the filter by rinsing with dilute alcohol (1 vol. alcohol to 3 vol. water), and wash the undissolved residue on the filter, first with alcohol diluted with twice its volume of water, then with a mixture of equal volumes of alcohol and water, and finally with a mixture of 3 vol. alcohol and 1 vol. water. Add to each of the alcoholic fluids used for washing a few drops of solution of sodium carbonate. Continue the washing until the color of a portion of the fluid running off remains unaltered upon being acidified with hydrochloric acid and mixed with hydrogen sulphide water.

Rinse the sodium metantimonate from the filter, wash the latter with a mixture of hydrochloric and tartaric acids, dissolve the metantimonate in this mixture, precipitate with hydrogen sulphide, and determine the antimony as directed § 125, 1. In presence of much tin it is well to fuse the metantimonate again with caustic soda, &c.

To the filtrate, which contains the tin and arsenic, add hydrochloric acid, which produces a precipitate of stannic arsenate; conduct now into the unfiltered fluid hydrogen sulphide for some time, allow the mixture to stand at rest until the odor of that gas has almost completely gone off, and separate the weighed sulphides of the metals which contain free sulphur, as in 157.

If the substance contains only *antimony* and *arsenic*, the alcoholic filtrate is heated, with repeated addition of water, until it scarcely retains the odor of alcohol; hydrochloric acid is then added, and the arsenic acid determined as magnesium pyroarsenate (§ 127, 2), or as arsenious sulphide (§ 127, 4, *b*).

159 *b*. Small quantities of the sulphides of arsenic and antimony mixed with sulphur are often obtained in mineral analysis. The two metals may in this case be conveniently separated as follows:—Exhaust the precipitate with bisulphide of carbon, oxidize with chlorine-free red fuming nitric acid, evaporate the solution nearly to dryness; mix the residue with a copious excess of sodium carbonate, add some sodium nitrate, and treat the fused mass as given in *a*, **158**. If, on the other hand, you have a mixture of sulphides of tin and antimony to analyze, oxidize it with nitric acid of 1·5 sp. gr., and treat the residue obtained on evaporation as given in *a*, **158**.

5. *Methods based upon the Precipitation of Arsenic as Ammonium Magnesium Arsenate.*

a. Arsenic from Antimony.

160 Oxidize the metals or sulphides with nitrohydrochloric acid, with hydrochloric acid and potassium chlorate, with bromine dissolved in hydrochloric acid, or with chlorine in alkaline solution; add tartaric acid, a large quantity of ammonium chloride, and then ammonia in excess. (Should the addition of the latter reagent produce a precipitate, this is a proof that an insufficient quantity of ammonium chloride or of tartaric acid has been used, which error must be corrected before proceeding with the analysis.) Then precipitate the arsenic acid as directed § 127, 2, and determine the antimony in the filtrate as directed in § 125, 1. As basic magnesium tartrate might precipitate with the ammonium magnesium arsenate, the precipitate should always, after slight washing, be redissolved in hydrochloric acid, and reprecipitated with ammonia with addition of a little magnesia mixture—an excellent method.

b. ARSENIOUS ACID FROM ARSENIC ACID.

161 Mix the sufficiently dilute solution with a large quantity of ammonium chloride, precipitate the arsenic acid as directed § 127, 2, and determine the arsenious acid in the filtrate by precipitation with hydrogen sulphide (§127, 4). LUDWIG* has observed that if the solution is too concentrated, magnesium arsenite falls down with the ammonium magnesium arsenate, hence it is necessary to dissolve the weighed magnesium precipitate in hydrochloric acid and test the solution with hydrogen sulphide. The presence of arsenious acid will be betrayed by the immediate formation of a precipitate.

c. TIN AND ANTIMONY FROM ARSENIC ACID.

162 LENSSEN† separated tin from arsenic acid with good results by digesting the oxides obtained by oxidation with nitric acid with ammonia and yellow ammonium sulphide, and precipitating the arsenic acid from the clear solution according to 127, 2, as ammonium magnesium arsenate. On acidifying the filtrate the tin separates as stannic sulphide. The method can only give good results when the whole of the arsenic was present as arsenic acid before the addition of ammonium sulphide, for the arsenic in a solution of arsenious acid in yellow ammonium sulphide is not thrown down by magnesia mixture. The method also answers for separating antimony from arsenic.

---

* Archiv für Pharm. 97, 24.

† Annal. d. Chem. u. Pharm. 114, 116.

6. *Methods based on the different behavior of the freshly Precipitated Sulphides towards Solution of Potassium Hydrogen Sulphite or Oxalic Acid.*

a. ARSENIC FROM ANTIMONY AND TIN (BUNSEN*).

If freshly precipitated arsenious sulphide is digested with **163** sulphurous acid and potassium sulphite, the precipitate is dissolved; on boiling, the fluid becomes turbid from separated sulphur, which turbidity for the most part disappears again on long boiling. The fluid contains, after expulsion of the sulphurous acid, potassium arsenite and thiosulphate. The sulphides of antimony and tin do not exhibit this reaction. Both therefore may be separated from arsenious sulphide by diluting the solution of the three sulphides in potassium sulphide to about 500 c.c. and precipitating with a large excess (about a litre) of saturated aqueous sulphurous acid, digesting the whole for some time in a water-bath, and then boiling till one-third of the water and the whole of the sulphurous acid are expelled and the sulphur has disappeared; this will take about an hour and a half. The residuary sulphide of antimony or tin is arsenic-free, the filtrate contains the whole of the arsenic and may be immediately precipitated with hydrogen sulphide. BUNSEN determines the arsenic by oxidizing the dried sulphide together with the filter with *fuming* nitric acid, diluting the solution a little, warming *gently* with a little potassium chlorate (in order to oxidize more fully the substances formed from the paper), and finally precipitating as ammonium magnesium arsenate.

With regard to the separation of stannic sulphide from the solution of potassium arsenite, it is to be observed that the stannic sulphide must be washed with concentrated solution of sodium chloride, as, if water were used, the fluid would run through turbid. As soon as the precipitate is thoroughly washed with the sodium chloride, the latter is displaced by solution of ammonium acetate, containing a slight excess of acetic acid. These last washings must not be added to the first, as the ammonium acetate hinders the complete precipita tion of the arsenious acid by hydrogen sulphide.

* Annal. d. Chem. u. Pharm. 106, 3.

The test-analyses adduced by BUNSEN show very satisfactory results.

b. TIN FROM ARSENIC AND ANTIMONY (F. W. CLARKE*).

164 Moist freshly precipitated bisulphide of tin completely dissolves on boiling for a moderate length of time with excess of oxalic acid, and therefore tin in the form of bichloride is not thrown down by hydrogen sulphide from a hot solution containing excess of oxalic acid. The sulphides of arsenic are barely affected by boiling with oxalic acid, and hydrogen sulphide immediately reprecipitates the traces dissolved. Sulphide of antimony dissolves more copiously on boiling with oxalic acid, but hydrogen sulphide reprecipitates the antimony from the solution.

[These reactions form the basis of CLARKE'S method, which, with some important modifications, has been successfully applied to the separation of tin from antimony in alloys by F. P. DEWEY,† who proceeds as follows:

Dissolve with a mixture of 1 part strong nitric acid, 4 parts strong hydrochloric acid, and 5 parts water. Since even small quantities of free mineral acids prevent complete precipitation of antimony, they are removed by evaporating to dryness on a water-bath, with previous addition of enough potassium chloride to form double salts with the tin and antimony chlorides present. The presence of the potassium chloride entirely prevents loss of tin and antimony by volatilization as chlorides during the evaporation. Add to the salts thus obtained a large quantity of pure oxalic acid (at least 20 parts crystallized acid to 1 part tin), and dilute with water to about 125 c.c. per ·1 grm. antimony present. The salts dissolve readily. Boil and pass H_2S through the boiling solution half an hour. Filter immediately while hot, and wash the greater part of the soluble matter out of the precipitate with hot water. The precipitated antimonious sulphide will contain a little stannic sulphide. Dissolve in ammonium sulphide, avoiding an unnecessary quantity of the solvent, and pour the solution into a strong hot solution of oxalic acid. A liberal excess of oxalic acid should be present after decomposition of the sulphur salts. Heat the oxalic solution with the suspended precipitate of antimonious

* Chem. News, 21, 124. † Am. Chem. Journ. i. 244.

sulphide to boiling, and pass H_2S gas ten minutes. Collect the Sb_2S_3 now free from tin on a weighed filter, wash with hot water, and proceed to determine the antimony as directed in § 125, 1, *b*. To recover tin from the filtrate, evaporate nearly to dryness, add strong sulphuric acid, and heat till all the oxalic acid present is decomposed and removed. Dilute largely, and precipitate the tin with hydrogen sulphide according to § 126, 1, *c*.]

7. Methods based upon the Separation of the Metals themselves, or, as the case may be, on the different deportment of the same with Acids.

a. TIN FROM ANTIMONY (TOOKEY,* improvements by CLASEN (*loc. cit.*) and ATTFIELD†).

165 The hydrochloric solution should be oxidized if necessary with a few drops of nitric acid or a little potassium chlorate. Heat nearly to boiling and add iron as long as it dissolves. Either hoop-iron or fine bright wire will answer the purpose; it should dissolve in dilute hydrochloric acid, leaving little or no residue. The antimony will be thrown down, the tin reduced to stannous chloride. As soon as all antimony appears to be precipitated and the iron to be dissolved, add more hydrochloric acid, allow to deposit, decant and test whether iron produces any further precipitate. In this way you will ensure the absence of any metallic iron and the complete precipitation of the antimony. Wash the antimony with hot water, which should be at first acidified, then with alcohol, finally with ether, drying at 100°. Throw down the tin with hydrogen sulphide (§ 126, 1, *c*). With care the results are good; compare CLASEN (*loc. cit.*).

b. MUCH TIN FROM LITTLE ANTIMONY AND ARSENIC.

166 If an alloy of the three metals is treated in a very finely divided condition in a stream of carbonic acid with strong hydrochloric acid, the whole of the tin dissolves to stannous chloride. A part of the arsenic and antimony escapes as arsenetted and antimonetted hydrogen, whilst the rest remains behind in the state of metal, or, as the case may be, of a solid combination with hydrogen. Conduct the gas through several

* Journ. Chem. Soc. 15, 402.

† Zeitschr. f. anal. Chem. 9, 107.

U-tubes, containing a little chlorine-free red fuming nitric acid, whereby the arsenic and antimony will be oxidized. When the solution is effected, dilute the contents of the flask with air-free water to a certain volume, mix, allow to settle, and determine the tin in an aliquot part, either gravimetrically or volumetrically. Filter the rest of the fluid, wash the precipitate thoroughly, dry the filter with its contents in a porcelain crucible, add the contents of the U-tubes, evaporate to dryness, and in the residue separate the antimony and arsenic as directed **158**. It is well to treat an aliquot part of the hydrochloric solution with iron (**165**) to find, and, if necessary, estimate traces of antimony which may have passed into the hydrochloric acid solution.

c. Tin from Gold.

167 Gold may be separated from excess of tin by boiling the finely divided alloy with only slightly diluted sulphuric acid, to which hydrochloric acid has been cautiously added. The tin dissolves as stannous chloride. Heat is applied till the sulphuric acid begins to volatilize copiously. Stannic oxide is formed which dissolves in the concentrated sulphuric acid, while the gold remains behind. On addition of much water, the stannic oxide falls, mixed with finely divided gold, in th form of a purple-red precipitate. On warming with concentrated sulphuric acid, the stannic oxide finally redissolves, while the gold is left pure (H. Rose*).

d. Platinum from Gold.

168 The aqua regia solution is freed as far as possible from nitric acid by evaporation with hydrochloric acid, and treated with a solution of ferrous chloride, the gold being determined as directed § 123, *b*. The platinum may be precipitated from the filtrate by hydrogen sulphide according to § 124, *c*.

8. *Method based upon the Precipitation of Tin as Stannic Arsenate.*

Tin from Arsenic.

169 E. Häffely† has proposed the following method of determining both the tin and the arsenic in commercial sodium stannate, which often contains a large admixture of sodium

* Pogg. Annal. 112, 172. † Phil. Mag. 10, 220.

arsenate. Mix a weighed sample with a known quantity of sodium arsenate in excess, add nitric acid also in excess, boil, filter off the precipitate, which has the composition $2SnO_2,As_2O_5 + 10H_2O$, and wash; expel the water by ignition, and weigh the residue, which consists of $2SnO_2,As_2O_5$. In the filtrate determine the excess of arsenic acid as directed § 127, 2. The amount of the stannic oxide is found from the weight of the precipitate, that of the arsenic acid is obtained by adding the quantity in the precipitate to the quantity in the filtrate, and deducting the quantity added.

9. *Volumetric Methods.*

a. Arsenious from Arsenic Acid.

170 Convert the whole of the arsenic in a portion of the substance into arsenic acid and determine the total amount of this as directed § 127, 2; determine in another portion the arsenious acid as directed in § 127, 5, *a*, and calculate the arsenic acid from the difference.

b. Antimony of Antimonious Compounds from Antimonic Acid.

171 Determine in a sample of the substance the total amount of the antimony as directed § 125, 1, in another portion estimate the antimony present as an antimonious compound as directed § 125, 3, and calculate the antimonic acid from the difference.

c. Tin of Stannous, from Tin of Stannic Compounds.

172 In one portion of the substance convert the whole of the stannous into stannic salts by digestion with chlorine water or some other means, and determine the total quantity of tin as directed § 126, 1, *b*; in another portion, which, if necessary, is to be dissolved in hydrochloric acid in a stream of carbonic acid, determine the stannous tin according to § 126, 2.

II. SEPARATION OF THE ACIDS FROM EACH OTHER.

It must not be forgotten that the following methods of separation proceed generally upon the assumption that the acids exist either in the free state, or as alkali salts; compare the introductory remarks, p. 479. Where several acids are to be determined in one and the same substance, we very often use a separate portion for each. The methods here given do not embrace every imaginable case, but only the most important cases, and those of most frequent occurrence.

First Group.

ARSENIOUS ACID—ARSENIC ACID—CHROMIC ACID—SULPHURIC ACID—PHOSPHORIC ACID—BORACIC ACID—OXALIC ACID—HYDROFLUORIC ACID—SILICIC ACID—CARBONIC ACID.

§ 166.

1. ARSENIOUS ACID AND ARSENIC ACID FROM ALL OTHER ACIDS.

Precipitate the arsenic from the solution by hydrogen sulphide (§ 127, 4, *a* or *b*), filter, and determine the other acids in the filtrate. It must be remembered, that the arsenious sulphide will be obtained mixed with sulphur if chromic acid, ferric salts, or any other substances which decompose hydrogen sulphide are present. The estimation of sulphuric acid in the filtrate cannot be accurate unless air is excluded, and oxidizers such as chromic acid are absent; sulphuric acid is, therefore, best estimated in a separate portion (**174**). From those acids which form soluble magnesium salts, arsenic acid may be separated also by precipitation as ammonium magnesium arsenate (§ 127, 2). **173**

2. SULPHURIC ACID FROM ALL THE OTHER ACIDS.*

a. From Arsenious, Arsenic, Phosphoric,† Boracic, Oxalic, and Carbonic Acids.

Acidify the dilute solution strongly with hydrochloric acid, mix with barium chloride, and filter the barium sulphate from the solution, which contains all the other acids. Determine the barium sulphate as directed § 132. If acids are present with which barium forms salts insoluble in water but soluble in acids, the barium sulphate is apt to carry down with it such salts, and this is all the more liable to happen, the longer the **174**

* With respect to the separation of sulphuric acid from selenic acid, comp. WOHLWILL (Annal. d. Chem. u. Pharm. 114, 183).

† If metaphosphoric acid is present, it must first be converted into orthophosphoric by fusion with alkali carbonate.

precipitate is allowed to settle. This remark applies especially to barium oxalate, and tartrate, and the barium salts of other organic acids (H. Rose). In such cases I would recommend, after washing, to stop up the neck of the funnel, and digest the precipitate with a solution of hydrogen sodium carbonate, then to wash with water, with dilute hydrochloric acid, and again with water. In every case, however, the purity of the weighed barium sulphate must be tested as directed § 132, 1.

In the fluids filtered from the barium sulphate the other acids are determined according to the directions of the Fourth Section, after the removal of the excess of barium chloride. Or the other acids may be estimated in separate portions of the substance, which is indeed usually the best way, and for carbonic acid is of course the only way.

b. From Hydrofluoric Acid.

175 α. When sulphuric acid and hydrofluoric acid are present in the free state in aqueous solution, it is best to estimate the acidity in one portion by means of standard soda (§ 192), and the sulphuric acid in another (§ 132, I., 1), finding the hydrofluoric acid by difference. The barium sulphate should be purified by fusion with sodium carbonate (§ 132, I., 1).

176 β. To estimate both acids in minerals or other dry substances, it is safest, provided the fluoride can be decomposed by sulphuric acid, to determine the fluorine in one portion according to § 138, 3, *a*, and to fuse another portion for a long time with four times its amount of sodium carbonate, which will decompose the sulphate thoroughly, the fluoride generally but partially. The fused mass is soaked in water, the solution filtered, acidified with hydrochloric acid and precipitated with barium chloride. The barium sulphate thus obtained generally contains barium fluoride, and must be purified according to § 132, I., 1, by fusion with sodium carbonate, &c.

177 γ. An actual separation of both acids may be effected, when both are in the form of alkali salts, by adding sodium carbonate if necessary, and then precipitating the fluorine according to § 138, I., adding the calcium chloride cautiously in very slight excess. The sulphuric acid is for the most part found in the filtrate from the calcium carbonate and fluoride, a very small part is generally also found in the calcium acetate filtered from the calcium fluoride. Both filtrates are acidified and precipitated with barium chloride (§ 132, I., 1. H. Rose).

178 δ. Insoluble compounds may also be decomposed by fusion with six parts of sodium and potassium carbonates, and two parts of silica. The fused mass, after cooling, is treated with water, the solution is mixed with ammonium carbonate, and heated, more ammonium carbonate is added to replace what evaporates, the silicic acid thrown down is filtered off and washed with water containing ammonium carbonate, a solution of zinc oxide in ammonia is added to precipitate the remaining silica, the fluid is evaporated till all ammonia is driven off, filtered and the process concluded as in γ. The precipitate produced by the zinc should be tested for sulphuric acid.

c. From Chromic Acid.

179 Boil the dry compound with strong hydrochloric acid (p. 357, β) and estimate the chromic acid from the evolved chlorine. Neutralize some of the acid with ammonia, dilute and precipitate the sulphuric acid by long boiling with excess of barium chloride. The barium sulphate thus obtained retains chromic oxide (H. Rose) and must always be fused with sodium carbonate, &c. (p. 367).

d. From Hydrofluosilicic Acid.

180 First throw down the hydrofluosilicic acid according to § 133, as potassium silicofluoride, then the sulphuric acid in the filtrate with barium chloride.

e. From Silicic Acid.

Compare **192**.

3. Phosphoric Acid from the other Acids.

181 *a.* From the *acids of arsenic*, see **173**: from *sulphuric acid*, see **174**; from *silicic acid*, see **192**.

b. From Chromic Acid.

Precipitate the phosphoric acid by adding ammonium nitrate and ammonia, and then magnesium nitrate, and determine the chromic acid in the filtrate as directed § 130, I., *a*, *β* or I., *b*.

c. From Boracic Acid.

Precipitate the phosphoric acid with a solution of double chloride of magnesium and ammonium (§ 134, *b*, *α*), wash the precipitate partially, redissolve it in hydrochloric acid, reprecipitate with ammonia, adding a little magnesium and ammonium chloride, and estimate the phosphoric acid as magnesium pyrophosphate. In the filtrate estimate the boracic acid as magnesium borate (§ 136, I., 1, *d*). **182**

d. From Oxalic Acid.

α. If the two acids are to be determined in one portion, the aqueous or hydrochloric solution is mixed with sodium auric chloride in excess, heat applied, and the oxalic acid calculated from the reduced gold (§ 137, *c*). The gold added in excess is separated from the filtrate by hydrogen sulphide, and the phosphoric acid then precipitated by double chloride of magnesium and ammonium. **183**

β. If there is enough of the substance, the oxalic acid is determined in one portion according to § 137, *b*, or *d*, and the phosphoric acid in another portion. If the substance is soluble in water, and the quantity of oxalic acid inconsiderable, the phosphoric acid may be precipitated at once with magnesium chloride, ammonium chloride, and ammonia: if not, the substance is ignited with potassium carbonate and sodium carbonate, and the oxalic acid being thus destroyed, the phosphoric acid is determined in the nitric acid solution of the residue according to § 134, I., *b*, *β*. **184**

e. From Hydrofluoric Acid.

α. Phosphates and fluorides are frequently found together in minerals. In the analysis of phosphorites, for instance, we have to estimate small quantities of fluorine, often too in the presence of aluminium and iron, which increase the difficulty. According to my own experience,* it is always safest in such cases to estimate in one portion the fluorine as silicon fluoride (§ 138, II., 3, *a*), and in another portion the phosphoric acid. Regarding the first estimation, it must be mentioned that car- **185**

* Zeitschr. f. anal. Chem. 5, 190, and 6, 403.

bonic acid if present must first be removed. To this end heat the finely powdered weighed substance with water, add acetic acid in slight excess, and also, if the fluoride present is soluble in water, some calcium acetate; evaporate to dryness on a water bath, treat with water, filter, wash the insoluble matter, dry, separate as far as possible from the filter, add the filter ash, weigh, test a small portion for carbonic acid by heating with hydrochloric acid, and weigh the rest for the fluorine estimation. For the estimation of the phosphoric acid, dissolve the finely powdered substance in hydrochloric acid, evaporate to dryness on a water-bath, moisten with a little hydrochloric acid, add nitric acid, warm, dilute, filter, evaporate filtrate and washings to dryness, dissolve in nitric acid, and proceed according to § 134, I., *b*, *β*.

186 *β*. Where you have an alkali phosphate and an alkali fluoride together in aqueous solution the phosphoric acid may be separated according to § 135, II., *d*, *β*, as silver phosphate, or according to § 135, II., *k*, as mercurous phosphate. The fluoride will be all in the filtrate. If the former method is adopted the silver is removed from the filtrate by sodium chloride, and the fluorine estimated as calcium salt (§ 138, I.). If the latter method is adopted, as the solution is always acid, the use of glass and porcelain must be avoided. The mercury is removed from the filtrate by neutralizing with sodium carbonate and—without filtering—passing hydrogen sulphide. The fluorine is estimated in the filtrate as calcium salt, according to § 138, I. (H. Rose).

187 *γ*. Substances which are insoluble in water, and cannot be decomposed by acids, are fused with sodium carbonate and silica (**178**), the fused mass is treated with water, and the solution with ammonium carbonate. In this way all the fluorine and all, or nearly all, the phosphoric acid will be brought into solution. The solution is treated as in **186**, and any remainder of phosphoric acid in the undissolved residue is estimated according to **185**.

4. Hydrofluoric Acid from other Acids.

a. Fluorides from Borates.

188 Mix the solution containing alkali borate and fluoride with some sodium carbonate, and add calcium acetate in excess. A precipitate is formed, which contains the whole of the fluorine as calcium fluoride, and besides this, calcium carbonate and some calcium borate; the greater portion of the latter having been redissolved by the excess of the calcium salt added. Determine the calcium fluoride in the precipitate as directed § 138, I. The small quantity of boracic acid in the precipitate is, in this process, partly volatilized, partly dissolved after evaporating the mass with acetic acid and extracting with water. It is therefore necessary to determine the boracic acid in a separate portion of the substance, according to § 136, I., 2 (A. STROMEYER).*

b. Fluorides from Silicic Acid and Silicates.

A great many native silicates contain fluorides: care must, therefore, always be taken, in the analysis of minerals, not to overlook the latter. If the silicates containing fluoride are decomposable by acids—which is only rarely the case—and the silicic acid is separated in the usual way by evaporation, the whole of the fluorine may volatilize.

189 α. BERZELIUS's method. Fuse the elutriated substance with 4 parts of sodium carbonate for some time at a strong red heat, digest the mass in water, boil, filter, and wash, first with boiling water, then with ammonium carbonate. The filtrate contains all the fluorine as sodium fluoride, and, besides this, sodium carbonate, silicate, and aluminate. Mix the filtrate with ammonium carbonate and heat the mixture, replacing the ammonium carbonate, which evaporates. Filter off the precipitate of hydrate of silicic acid and aluminium hydroxide, and wash with ammonium carbonate. To separate the last portions of silica from the filtrate add a solution of zinc oxide in ammonia, evaporate till no more ammonia escapes, and filter off the precipitate of zinc silicate and oxide. Determine the silica in this precipitate by dissolving in nitric acid, evaporating to dryness, taking up with nitric acid, and filtering off the undissolved silica. In the alkaline filtrate estimate the fluorine as calcium salt (§ 138, I.). The residue, insoluble in water, and the precipitate produced by ammonium carbonate are finally treated with hydrochloric acid according to § 140, II., *a*, in order to separate the silica.

---

* Annal d. Chem. u, Pharm. 100, 91

β. In substances readily decomposed by sulphuric acid you **190** may also separate and weigh the silica according to **189** in one portion, and determine the fluorine in another portion according to § 138, II., 3, *a*.

c. Fluorides, Silicates and Phosphates together.

Compounds of this kind are not rare in nature, and may **191** be decomposed according to **189**. We cannot always rely on complete decomposition of the phosphate, as, for instance, calcium phosphate is but partially decomposed on fusion with sodium carbonate. The solution, obtained after separation of the silica by ammonium carbonate and the zinc solution, is made up to a definite volume, and a portion is tested for phosphoric acid with molybdic solution. If none is present the fluorine is estimated in the measured remainder of the fluid as fluoride of calcium (§ 138, I.). If on the other hand phosphoric acid is still present, treat the measured remainder of the fluid according to **186**. In the original residue and the ammonium carbonate precipitate estimate the principal amounts of the silicic and phosphoric acids and the basic metals. In the zinc precipitate estimate the remainder of the silicic acid, and in the filtrate from the latter estimate the portion of the phosphoric acid which was thrown down by zinc oxide.

As the phosphoric acid is so divided by this method, it is well to make a direct estimation of it in another portion of the substance, especially when only a small quantity is present. For this purpose decompose the silicate with hydrofluoric and hydrochloric acids, add enough but not too large an excess of sulphuric acid, and evaporate till all the fluorine has escaped as silicon fluoride and hydrofluoric acid. Do not increase the heat to the escape of sulphuric acid, or phosphoric acid may be lost. Take up the residue with nitric acid, dilute, filter, and estimate the phosphoric acid in the filtrate by the molybdic method.

If the substance can be easily decomposed with sulphuric acid, the fluorine may of course also be expelled as silicon fluoride and estimated according to § 138, II., 3, *a*.

5. Silicic Acid from all other Acids.

a. In compounds which are decomposed by hydrochloric acid.

Decompose the substance by digestion with hydrochloric **192** or nitric acid, evaporate the whole *on the water bath* to dryness (§ 140, II., *a*), treat with water, hydrochloric acid or nitric acid according to circumstances, filter off the silica, and estimate the other acids in the filtrate. The following points require attention.

α. In the presence of borates or fluorides this method cannot be used, employ **193**.

β. In the presence of phosphoric acid the silica always retains a small portion, which cannot be extracted by washing with acidified water (H. Rose, W. Skey*). After washing the silica with water, treat it repeatedly with ammonia, which will leave only a very minute quantity of the phosphoric acid. Evaporate the ammoniacal fluid, finally adding a little hydrochloric acid, dissolve in water with addition of a little nitric acid, filter off the small amount of silica which was taken up by the ammonia, and estimate the remainder of the phosphoric acid in the filtrate.

b. In compounds which are not decomposed by hydrochloric acid.

193 Fuse with carbonate of potash and soda (p. 422), and treat the residue either at once cautiously with dilute hydrochloric or nitric acid, in order to proceed with the solution according to **192** (not applicable in presence of boracic acid or fluorine); or taking the fluid obtained by boiling the residue with water, precipitate the dissolved silica by warming with ammonium carbonate, and throw down the last portion of silica from the filtrate by zinc oxide dissolved in ammonia (**189**).

The silicic acid is then found partly in the residue left undissolved by water, partly in the precipitate produced by ammonium carbonate, and partly in the precipitate produced by the zinc solution. Separate it according to § 140, II., *a*. Boracic acid and fluorine will be found entirely in the last alkaline filtrate (**189**). Regarding phosphoric acid see **191**. Sulphuric acid passes for the most part into the last alkaline filtrate, yet it is well also to examine the acid filtrates from the silica.

6. Carbonic Acid from all other Acids.

194 When carbonates are heated with stronger acids, the car-

* Zeitschr. f. anal. Chem. 8, 70.

bonic acid is expelled; the presence of carbonates, therefore, does not interfere with the estimation of most other acids. And as, on the other hand, the carbonic acid is determined by the loss of weight or by combination of the expelled gas, the presence of salts of non-volatile acids does not interfere with the determination of the carbonic acid. Accordingly, with compounds containing carbonates, sulphates, phosphates, &c., either the carbonic acid is determined in one portion, and the other acids in another, or both estimations are performed on one portion. In the latter case the process described p. **412**, *e*, may be used with advantage, the other acids being determined in the solution remaining in the decomposing flask. In presence of fluorides, one of the weak non-volatile acids, such as tartaric acid or citric acid, must be employed to expel the carbonic acid; since, were sulphuric or hydrochloric acid used, part of the liberated hydrofluoric acid would escape with the carbonic acid. If, as will occasionally happen in an analysis, a mixed precipitate of calcium fluoride and calcium carbonate is thrown down from a solution, the two salts may be separated by evaporating with acetic acid to dryness, and extracting the residue with water; the calcium acetate formed from the carbonate is dissolved the calcium fluoride is left behind.

Second Group.

CHLORINE—BROMINE—IODINE—CYANOGEN—SULPHUR.

I. SEPARATION OF THE ACIDS OF THE SECOND GROUP FROM THOSE OF THE FIRST.

§ 167.

a. All the Acids of the Second Group from those of the First.

Mix the dilute solution with nitric acid, add silver nitrate in excess, and filter off the insoluble chloride, bromide, iodide, &c., of silver. The filtrate contains the whole of the acids of the first group, the silver salts of these acids being soluble in water or nitric acid. Carbonic acid must, under all circumstances, be determined in a separate portion (§ 139, *e*). **195** *b. Some of the Acids of the Second Group from Acids of the First Group.*

As it is often inconvenient for the further separation of the acids of the second group to have them all in the form of insoluble silver compounds, the analysis is sometimes effected by separating first the acid of the first group, then that of the second. If the quantity of substance is large enough, the most convenient way generally is to determine the several acids, *e.g.*, sulphuric acid, phosphoric acid, chlorine, sulphur, &c., in separate portions. **196**

Of the infinite number of combinations that may present themselves we will here consider only the most important.

1. SULPHURIC ACID may be readily separated from chlorine, bromine, iodine, and cyanogen, by precipitation with a barium salt. If the acids of the second group are to be determined in the same portion, barium nitrate or acetate is used instead of barium chloride. In presence of hydrogen sulphide, sulphuric acid cannot be determined in this way, as part of the hydrogen sulphide would be converted into sulphuric acid by the oxygen of the air. The error thus introduced into the process may be very considerable (FRESENIUS*) The hydrogen sulphide must, therefore, first be removed by cupric chloride, and the sulphuric acid determined in the filtrate; or, the hydrogen sulphide must be completely oxidized into sulphuric acid by chlorine or bromine, and a corresponding deduction afterwards made in calculating the quantity of the sulphuric acid. In other cases it is well to expel the hydrogen sulphide according to p. 468. § 148, *c*, by heating with hydrochloric acid, and to estimate the sulphuric acid in the residual fluid. **197**

2. PHOSPHORIC ACID may be precipitated by ammonium magnesium nitrate, after addition of ammonium nitrate; OXALIC ACID by calcium nitrate; chlorine, bromine, iodine, &c.. are determined in the filtrate. **198**

3. CHLORINE IN SILICATES.

a. If the silicates dissolve in dilute nitric acid, precipitate the highly dilute solution with silver nitrate, without applying heat, remove the excess of silver from the filtrate by dilute **199**

---

* Journ. f. prakt. Chem. 70, 9.

hydrochloric acid, still without applying heat, and then separate the silicic acid in the usual way.

b. If the silicate becomes gelatinous upon decomposition with nitric acid, dilute, allow to deposit, filter, wash the separated silicic acid, and treat the filtrate as in *a.*

In the processes *a* and *b* the silver chloride may contain silica. Reduce the weighed silver salt by hydrogen and treat with nitric acid, the silica will remain behind.

c. If nitric acid fails to decompose the silicates, mix the substance with sodium and potassium carbonates, moisten the mass with water, dry in the crucible, fuse, boil with water, remove the dissolved silicic acid by ammonium carbonate and zinc oxide dissolved in ammonia (**189**), and then precipitate, after addition of nitric acid, with silver nitrate.

4. Chlorides in presence of Fluorides.

200 If the substance is soluble in water, the separation may be effected as directed **195**; but it is more convenient to precipitate the fluorine with calcium nitrate, and the chlorine in the filtrate with silver nitrate. Insoluble compounds are fused with sodium carbonate and silicic acid, and treated as in **201.**

5. Chlorides in presence of Fluorides in Silicates.

201 Proceed as directed **189.** Saturate the alkaline filtrate nearly with nitric acid, precipitate with calcium nitrate, separate the calcium fluoride and carbonate as directed in **194**, and precipitate the chlorine in the filtrate by silver nitrate.

6. Sulphides in Silicates.

202 If the substance is decomposable by acids, reduce it to the very finest powder, and treat with fuming nitric acid free from sulphuric acid (§ 148, II., 2, *a*), or with rather dilute nitric acid in sealed tubes at 120—150° (Carius). When the sulphur is completely oxidized, rinse the contents of the flask or tube into a dish, evaporate on the water bath, treat with hydrochloric or nitric acid, dilute, filter off the silica, and determine in the filtrate the sulphuric acid formed. If, on the contrary, the substance is not decomposable by acids, fuse with 4 parts of sodium carbonate and 1 part of potassium nitrate, boil the fused mass with water, filter, remove the dissolved silicic acid from the filtrate by acidifying with hydrochloric or nitric acid and evaporating, and proceed as above directed.

7. SULPHIDES IN PRESENCE OF CARBONATES.

If you have to estimate sulphur in sulphides, which can easily be decomposed by acids (*e.g.*, calcium sulphide), in presence of carbonates, decompose the substance by heating with hydrochloric acid, dry the evolved mixture of hydrogen sulphide and carbonic acid, take up the hydrogen sulphide by tubes filled with pumice prepared with cupric sulphate (p. 410), and the carbonic acid by soda-lime tubes (p. 631). **203**

Supplement.

ANALYSIS OF COMPOUNDS, CONTAINING ALKALI SULPHIDES, CARBONATES, SULPHATES, AND THIOSULPHATES.

§ 168.

The following method was first employed by G. WERTHER* in the examination of gunpowder residues. N. FEDOROW† has shown that the original process included an error, which has been put right in the method described below. **204**

Put the substance into a flask, add water, in which a sufficient quantity of cadmium carbonate‡ is suspended; cork, and shake the vessel well. The alkali sulphide decomposes completely with the cadmium carbonate. Filter the yellowish precipitate off, and treat it with dilute acetic acid (not with hydrochloric acid); the cadmium carbonate dissolves, the cadmium sulphide is left undissolved. Oxidize the latter with potassium chlorate and nitric acid (p. 466), or with bromine (p. 467), and precipitate with barium chloride the sulphuric acid formed from the sulphide.

Heat the fluid filtered from the yellow precipitate, and mix with solution of neutral silver nitrate. The precipitate consists of silver carbonate and silver sulphide ($K_2S_2O_3 + 2AgNO_3 + H_2O = K_2SO_4 + Ag_2S + 2HNO_3$). Filter it off, and wash with carbonic acid water, then remove the silver

* Journ. f. prakt. Chem. 55, 22.

† Zeitschr. f. anal. Chem. 9, 127.

‡ To obtain the cadmium carbonate free from alkali, ammonium carbonate must be used as precipitant.

carbonate by ammonia and precipitate the silver from the ammoniacal solution by acidifying with nitric acid and adding sodium chloride. 2 mol. silver chloride so obtained correspond to 1 mol. carbonate.* Dissolve the silver sulphide in dilute boiling nitric acid, determine the silver in the solution as silver chloride, and calculate from the result the quantity of the thiosulphuric acid; 1 mol. AgCl corresponds to 1 at. sulphur in thiosulphuric acid, or 2 AgCl correspond to $K_2S_2O_3$.

From the fluid filtered from the silver sulphide and carbonate remove first the excess of silver by means of hydrochloric acid, and then precipitate the sulphuric acid by a barium salt. From the sulphuric acid found you have, of course, to deduct the quantity of that acid resulting from the decomposition of the thiosulphuric acid, and accordingly for 1 part of silver chloride formed from the sulphide, ·279 parts of sulphuric anhydride (SO_3). The difference gives the amount of sulphuric acid originally present in the analyzed compound. By way of control, you may determine, in the fluid filtered from the barium sulphate, the alkali as sulphate (§ 97 or § 98).

II. Separation of the Acids of the Second Group from each other.

§ 169.

1. Chlorine from Bromine.

All the methods of direct analysis hitherto proposed to effect the separation of chlorine from bromine are defective. The bromine is therefore always determined in a more indirect way.

a. Precipitate with silver nitrate, wash the precipitate, **205** wash it from the filter into a porcelain dish, extract the filter with hot ammonia, evaporate the ammonia in a weighed porcelain crucible, add the principal quantity of the precipitate, dry, fuse, and weigh. Transfer an aliquot part of the mixed

* A quantity equivalent to the sulphur found existing as sulphide has to be deducted from this ($K_2S+CdCO_3=CdS+K_2CO_3$). On the other hand, a quantity equivalent to the sulphide of silver precipitated by the thiosulphate must be added, for each mol. of sulphide of silver from the thiosulphate gives 2 mol. HNO_3, which decomposes 1 mol. carbonate of silver. This correction was overlooked by Werther.

silver chloride and bromide to a light weighed bulb-tube of hard glass,* fuse in the bulb, let the mass cool, and weigh. This operation gives both the total weight of the tube with its contents, and the weight of the portion of mixed silver chloride and bromide in the bulb. The greatest accuracy in the several weighings is indispensable. Now transmit through the tube a slow stream of dry pure chlorine gas, heat the contents of the bulb to fusion, and shake the fused mass occasionally about in the bulb. After the lapse of about 20 minutes, take off the tube, allow it to cool, hold it in an oblique position, that the chlorine gas may be replaced by atmospheric air, and then weigh. Heat once more for about 10 minutes in a stream of chlorine gas, and weigh again. If the two last weighings agree, the experiment is terminated; if not, the operation must be repeated once more. The loss of weight suffered, multiplied by 4·22297 (which may be taken as 4·223), gives the quantity of the silver bromide decomposed by the chlorine. For the proof of this rule, see "Calculation of Analyses."

This method gives very accurate results, if the proportion of bromine present is not too small; but most uncertain results in cases where mere traces of bromine have to be determined in presence of large quantities of chlorides, as, for instance, in salt-springs. To render the method available in such cases, the great point is to produce a silver compound containing all the bromine, and only a small part of the chlorine. This end may be attained in several ways. In these processes the quantity of chlorine is found by completely precipitating a separate portion with silver solution, and deducting the silver bromide found from the weight of the precipitate.

α. Mix the solution with sodium carbonate in excess (if a precipitate is formed, do not filter), evaporate to dryness, powder the residue, extract with hot absolute alcohol; the solution contains the whole of the alkali bromide, and only a small portion of the alkali chloride; add a drop of soda solution, and evaporate, dissolve the residue in water, acidify with nitric acid, and precipitate with silver solution.

* The best way of effecting the removal of the fused mass from the crucible is to fuse again, and then pour out.

β. FEHLING'S method.* Mix the solution *cold* with a quantity of solution of silver nitrate not nearly sufficient to effect complete precipitation, shaking the mixture vigorously, and leave the precipitate for some time in the fluid, with repeated shaking. If the amount of the precipitate produced corresponds at all to the quantity of bromine present, the whole of the latter substance is obtained in the precipitate. **206**

FEHLING gives the following rule:

If the fluid contains 1 bromine to 1000 chlorine use $\frac{1}{5}$ or $\frac{1}{6}$ the quantity of silver nitrate that would be required to effect complete precipitation; if the fluid contains 10,000 times as much chlorine as bromine, use $\frac{1}{10}$; if 50,000, use $\frac{1}{30}$; if 100,000, use $\frac{1}{60}$.

Wash the mixed precipitate of silver chloride and bromide *thoroughly*, dry, ignite, weigh, and treat with chlorine as above.

γ. MARCHAND† has slightly modified FEHLING'S method. He reduces with zinc the mixed precipitate of silver chloride and bromide obtained by FEHLING'S fractional precipitation, decomposes the solution of zinc chloride and bromide with sodium carbonate, evaporates to dryness, and extracts the residue with absolute alcohol, which dissolves all the sodium bromide with only a little of the sodium chloride; he then evaporates the solution to dryness, takes up the residue with water, precipitates again with silver nitrate, and subjects a part of the weighed precipitate to the treatment with chlorine. **207**

δ. If a fluid containing chlorides in presence of some bromide is heated in a retort with hydrochloric acid and manganese dioxide, the whole of the bromine passes over before any of the chlorine. Under this circumstance MOHR‡ bases the following method for effecting the concentration of bromine:—Distil as stated, and conduct the vapors, through a doubly bent tube, into a wide WOULF'S bottle, which contains some strong ammonia. Dense fumes form in the bottle, filling it gradually. Conduct the excess of vapors from the first into a second bottle, with narrow neck, containing ammoniated water. Both bottles must be sufficiently large to allow no vapors to escape. When the whole of the bromine is evolved,

* Journ. f. prakt. Chem. 45, 269.

† *Ib.* 47, 363.

‡ Annal. d. Chem. u. Pharm. 93, 80.

which may be distinctly seen by the color of the space above the liquid in the retort and tubes, raise the cork of the flask to prevent the receding of ammonium bromide fumes. Let the apparatus cool, and unite the contents of the two bottles; the fluid contains the whole of the bromine, with a relatively small portion of the chlorine.

b. Determine in a portion of the solution the chlorine + bromine (by precipitating with silver), either gravimetrically or volumetrically; in another portion the bromine, either by the colorimetric method (§ 143, I., *b*, *β*) or volumetrically (§ 143, I., *b*, *α*). Calculate the chlorine from the difference. The method is very suitable for an expeditious analysis of mother-liquors. **208**

2. Chlorine from Iodine.

a. Mix the solution with palladious nitrate, and determine the precipitated palladious iodide as directed § 145, I., *a*, *β*. Conduct hydrogen sulphide into the filtrate to remove excess of the palladium, destroy the excess of hydrogen sulphide by solution of ferric sulphate, and precipitate the chlorine finally with solution of silver. It is generally found more simple and convenient to precipitate from one portion the iodine, by means of palladious chloride, as directed § 145, I., *a*, *β*, from another portion the chlorine and iodine jointly with silver nitrate, and to calculate the chlorine from the difference. If you have no solution of palladious nitrate ready, and the chlorine and iodine must be determined in one portion of the solution under examination, add a measured quantity of a solution of palladious chloride, determine the amount of chlorine in this in another exactly equal portion of the same solution, and deduct it. The results are accurate. In the case of fluids containing a large proportion of alkali chlorides to a small quantity of iodide—and such cases often occur—the iodide is concentrated by adding sodium carbonate to the fluid, evaporating to dryness, extracting the residue with hot alcohol, evaporating the alcoholic solution with addition of a drop of solution of soda, and taking the residue up with water. **209**

b. Proceed exactly as for the indirect determination of bromine in presence of chlorine (**205**). The greatest care must be taken that as little as possible of the mixed silver chlo- **210** ride and iodide adheres to the filter, for silver iodide dissolves only very slightly in ammonia. Any particles of silver iodide remaining attached to the filter may be saved by incinerating the filter and evaporating the ash with a drop of nitric acid and a drop of hydriodic acid. The loss of weight suffered by the silver precipitate on fusion in chlorine multiplied by 2·569 gives the amount of silver iodide present. This method gives still better results than in the separation of bromine from chlorine, inasmuch as the difference between the atomic weights of iodine and chlorine is far greater than the difference between those of bromine and chlorine. Regarding the concentration of the iodide, if necessary, see **209**.

211 *c.* Liberate the iodine by nitrous acid, take it up with carbon disulphide, wash the latter, and then estimate the iodine in it by sodium thiosulphate (p. 440, *β*).

In this process the chlorine is determined, either in the fluid separated from the violet carbon disulphide, or with greater accuracy by precipitating the chlorine + iodine in a second portion with silver, and deducting the weight of silver iodide corresponding to the iodine already found from the weight of the precipitate. A good and approved method.

If the quantity of iodine is small, the following method may also be used with advantage for estimating it:

The carbon disulphide should be thoroughly washed, covered with a layer of water, and in a stoppered bottle. Add drop by drop, with shaking, dilute chlorine water (of unknown strength), till the coloration has *just* vanished, and all the iodine is consequently converted into ICl_5. Separate the solution from the disulphide, add potassium iodide solution in sufficient excess, and determine the free iodine after § 146. Six parts of the iodine found correspond to 1 part originally present. If the analyst would avoid the trouble of pouring off the fluid from the disulphide, and of washing the latter, he may transfer the mixture, after the addition of chlorine to decoloration, to a somewhat narrow measuring cylinder, note the volume occupied by the iodine pentachloride solution, take out a portion with a pipette, and proceed as above directed.

212 *d.* For technical purposes the following method is also suitable. It was recommended by WALLACE and LAMONT* for

* Chem. Gaz. 1859, 137.

the estimation of iodine in kelp. The kelp-lie is nearly neutralized with nitric acid, evaporated to dryness, and the residue fused in a platinum vessel to oxidation of all the sulphides. Treat with water, filter, add silver nitrate till the precipitate appears perfectly white, wash, digest with strong ammonia, and weigh the residual silver iodide. Finally, add to the weight of the latter the amount which passes into solution in the ammonia; it is $\frac{1}{2493}$ of the aqueous ammonia (sp. gr. ·89) used.

3. Chlorine, Bromine, and Iodine from each other.

a. The three acid radicals are determined jointly in a portion of the fluid by precipitating with solution of silver nitrate (§ 141, I., *a* or *b*, *α*). To determine the iodine, another portion is precipitated with palladious chloride in the least possible excess (§ 145, I., *a*, *β*). The fluid filtered from the precipitate is freed from palladium by hydrogen sulphide and the excess of the latter removed by means of ferric sulphate; the chlorine and bromine are then precipitated jointly either completely or partially with silver nitrate, and the bromine determined as directed **205**. **213**

If the compound contains a large proportion of chlorine to a small proportion of bromine, the iodine may be precipitated also by palladious nitrate, as there is no danger, in that case, of palladious bromide being coprecipitated. The filtrate is treated as above.

These methods give accurate results; but they are applicable only if the quantity of iodide present is somewhat considerable.

b. Mix the neutral dilute and cold solution containing alkali iodide with alkali chloride or alkaki bromide, or both, with a saturated neutral solution of thallium nitrate, stirring well till, on repeated trial, you obtain a transient white precipitate—the first and permanent precipitate being yellow. It is best to have the thallium solution in a burette, so that you can easily add it by drops. If the white precipitate of thallium chloride or bromide does not at once disappear on stirring, add more water, but not an unnecessary quantity, or some of the thallium iodide will remain in solution. **214**

Allow to stand eight or twelve hours in a cold place, pour off the clear fluid through a weighed filter dried at 100°, wash the filter a little so that no more water than necessary may pass through the precipitate, turn the precipitate on to the filter, wash with as little water as you can, dry at 100°, and weigh. Precipitate the chlorine and bromine in the filtrate by silver solution. If they are both present, the mixed silver precipitate is to be treated according to **205**. Results quite satisfactory (HÜBNER and SPEZIA,* and HÜBNER and FRERICHS†).

215 *c*. Remove the iodine from the solution by carbon disulphide or chloroform, as in **211**. In the fluid separated from the iodized carbon disulphide determine the chlorine and bromine as directed **205**, and in the iodized carbon disulphide, the iodine as directed § 145, I., *b*, *β*. This method is particularly recommended for the separation of small quantities of iodine, and in this respect is supplementary to **213**.

216 *d*. Determine in a portion of the compound the chlorine, bromine, and iodine jointly by adding a known quantity of standard silver solution in slight excess, filtering and determining the small excess of silver in the filtrate by iodide of starch (p. 295). The precipitate is weighed, compare **210**. We now know the tatal of the chloride, bromide, and iodide of silver and also the silver therein contained.

Determine the iodine separately as in **215**, calculate the quantity of silver iodide and of silver corresponding to the amount found, deduct the calculated amount of silver iodide from the mixed iodide, chloride, and bromide of silver, that of the silver from the known quantity of the metal contained in the mixed compound; the remainders are respectively the joint amount of chloride and bromide of silver, and the quantity of the metal contained therein; these are the data for calculating the chlorine and bromine.

4. ANALYSIS OF IODINE CONTAINING CHLORINE.

217 *a*. Dissolve a weighed quantitity of the dried iodine in cold sulphurous acid, precipitate with silver nitrate, digest the precipitate with nitric acid, to remove the silver sulphite which may have coprecipitated, and weigh. The calculation of the iodine and chlorine is made by the following equations, in which A represents the quantity of iodine analyzed, x the

* Zeitschr. f. anal. Chem. 11, 397. † *Ib.* 11, 400.

iodine contained in it, y the chlorine contained in it, and B the amount of silver chloride and iodide obtained:

$$x + y = A$$

and

$$\frac{Ag + I}{I} x + \frac{Ag + Cl}{Cl} y = B$$

Now as

$$\frac{Ag + I}{I} = 1{\cdot}8508$$

and

$$\frac{Ag + Cl}{Cl} = 4{\cdot}0437$$

we have

$$y = \frac{B - 1{\cdot}851\,A}{2{\cdot}1929}$$

b. If you have free iodine and free chlorine in solution, determine in one portion, after heating with sulphurous acid, the iodine as palladium iodide (§ 145, I., *a*, *β*), and treat another portion as directed § 146. Deduct from the apparent amount of iodine found by the latter process, the actual quantity calculated from the palladium iodide: the difference expresses the amount of iodine equivalent to the chlorine contained in the substance. **218**

5. Analysis of Bromine containing Chlorine.

a. Proceed exactly as in **217**, weighing the bromine in a small glass bulb. Taking A to be equal to the analyzed bromine, B to the silver bromide and chloride obtained, x to the bromine contained in A, y to the chlorine contained in A, the calculation is made by the following equations: **219**

$$x + y = A$$

and

$$y = \frac{B - 2{\cdot}34997\,A}{1{\cdot}69374}$$

b. Mix the weighed anhydrous bromine with solution of iodide of potassium in excess, and determine the separated iodine as directed § 146. **220**

From these data, the respective quantities of bromine and chlorine are calculated by the following equations. Let A represent the weighed bromine, i the iodine found, y the chlorine contained in A, x the bromine contained in A, then

$$x + y = A$$
$$y = \frac{i - 1\cdot5866\,A}{1\cdot9907}$$

Bunsen, the originator of methods 4 and 5, has experimentally proved their accuracy.*

6. Cyanogen from Chlorine, Bromine, or Iodine.

221 *a.* Precipitate with solution of silver, collect the precipitate upon a weighed filter, and dry in the water-bath until the weight remains constant; then determine the cyanogen by the method of organic analysis; the quantity of the chlorine, bromine, or iodine is found by difference.

222 *b.* Precipitate with solution of silver as in *a*, dry the precipitate at 100° and weigh. Heat the precipitate, or an aliquot part of it, in a porcelain crucible, with cautious agitation of the contents, to complete fusion; add dilute sulphuric acid to the fused mass, then reduce by zinc, filter the solution from the metallic silver and silver paracyanide, and determine the chlorine, iodine, or bromine in the filtrate, in the usual way by silver. The silver cyanide is the difference. Neubauer and Kerner† obtained very satisfactory results by this method.

223 *c.* Precipitate with solution of silver as in *a*, weigh the precipitate and heat it, or an aliquot part, with nitric acid of 1·2 sp. gr. in a sealed tube at 100° for several hours, or at 150° for one hour. The silver cyanide is completely decomposed, while the chloride, bromide, or iodide are unaffected. Filter the contents of the tube, wash the precipitate and weigh it, the loss indicates the amount of silver cyanide (K. Kraut‡).

224 *d.* Determine the radicals jointly in a portion by precipitating with solution of silver, and the cyanogen in another portion, in the volumetric way (§ 147, I., *b*).

---

* Annal. d. Chem. u. Pharm. 86, 274, 276.

† *Ib.* 101, 344.

‡ Zeitschr. f. anal. Chem. 2, 243.

7. Ferro- or Ferricyanogen from Hydrochloric Acid.

225 To analyze say potassium ferro- or ferricyanide, mixed with an alkali chloride, determine in one portion the ferro- or ferricyanogen as directed § 147, II., *g*; acidify another portion with nitric acid, precipitate with solution of silver, wash the precipitate, fuse with 4 parts of sodium carbonate and 1 part of potassium nitrate, extract the fused mass with water, and determine the chlorine in the solution as directed in § 141.

8. Sulphur (in Sulphides) from Chlorine.

226 The old method of separating the two radicals by means of a metallic salt is liable to give false results, as part of the chlorine may fall down as chloride with the sulphide. We, therefore, precipitate both as silver compounds, dry the precipitate at 100°, weigh it, and determine the sulphur in a weighed portion; or—and this is usually preferred—determine in a portion of the solution the sulphur as directed § 148, I., *a*, or *b*, in another portion the sulphur + chlorine in form of silver salts. If you employ a solution of silver nitrate mixed with excess of ammonia, for the determination of the sulphur, you may, after filtering off the silver sulphide, estimate the chlorine directly as silver chloride, by adding nitric acid, and, if necessary, more neutral silver solution. In this case you must take care that the silver sulphide is pure; should it contain calcium carbonate, which is not unlikely if calcium is present, you remove this with dilute acetic acid. The weighed silver sulphide should be reduced by hydrogen, and then weighed again by way of control. To remove hydrogen sulphide from an acid solution, in order that chlorine may be determined in the latter by means of silver nitrate, H. Rose recommends to add solution of ferric sulphate, which will effect the separation of sulphur alone; the separated sulphur is allowed to deposit, and then filtered off.

Third Group.

NITRIC ACID—CHLORIC ACID.

I. Separation of the Acids of the Third Group from those of the first two Groups.

§ 170.

a. If you have a mixture of nitric acid or chloric acid with another free acid in a fluid containing no bases, determine in one portion the joint amount of the free acid, by the acidimetric method (see Special Part), in another portion the acid mixed with the chloric or nitric acid, and calculate the amount of either of the latter from the difference. **227**

b. If you have to analyze a mixture of a nitrate or chlorate with some other salt, determine in one portion the nitric or chloric acid volumetrically (§ 149, II., *d*, *α*, or *β*, or II., *e*, and § 150), or the nitric acid by § 149, II., *a*, *β*; and in another portion the other acid. I think I need hardly remark that no substances must be present which would interfere with the application of these methods. **228**

c. From the chlorides of many metals whose carbonates or normal phosphates are insoluble, chlorates and nitrates may be separated also by digesting the solution with recently precipitated thoroughly washed silver carbonate or normal silver phosphate in excess, and boiling the mixture. In this process, the chlorides react with the carbonate or phosphate—silver chloride and carbonate or phosphate of the metal with which the chlorine was originally combined being formed, which both separate, together with the excess of the silver carbonate or phosphate, whilst the chlorates and nitrates remain in solution (H. Rose, Chenevix, Lassaigne*). **229**

d. The estimation of an alkaline chlorate, in presence of a chloride, may be effected also by precipitating one portion at once, and another portion after gentle ignition, with solution of silver, and calculating the chloric acid from the difference between the two precipitates. **230**

e. Where you have nitrate of soda or potash in presence of **231**

* Journ. de Pharm. 16, 289; Pharm. Centralbl. 1850, 121.

nitrite or carbonate, as for instance in the commercial alkali nitrites, estimate in one portion the carbonate by standard acid (see Special Part),* in another portion the nitrous acid by permanganate or chromate of potash (p. 365). The nitrate is found by difference.

II. Separation of the Acids of the Third Group from each other.

232 We have as yet no method to effect the direct separation of nitric acid from chloric acid; the only practicable way, therefore, is to determine the two acids jointly in a portion of the compound, by the method described for nitric acid, § 149, II., *d*, *α*, bearing in mind that 12 atoms iron are converted from a ferrous to a ferric salt by 2 mol. chloric acid ($HClO_3$) or 1 mol. chloric anhydride (Cl_2O_5). In another portion estimate the chloric acid, by adding sodium carbonate in excess, evaporating to dryness, fusing the residue until the chlorate is completely converted into chloride, and then determining the chlorine in the latter, taking care that the silver chloride contains no difficultly soluble nitrite. 2 mol. silver chloride produced from this corresponds to $2HClO_3$ or Cl_2O_5, provided there was no chloride originally present.

* The alkali nitrites have no alkaline reaction.

SECTION VI.

ORGANIC ANALYSIS.

§171.

ORGANIC compounds contain comparatively only few of the elements. A small number of them consist simply of 2 elements, viz.,

C and H;

the greater number contain 3 elements, viz., as a rule,

C, H, and O;

most of the rest 4 elements, viz., generally,

C, H, O, and N;

a small number 5 elements, viz.,

C, H, O, N, and S;

and a few, 6 elements, viz.,

C, H, O, N, S, and P.

This applies to all the natural organic compounds which have as yet come under our notice. But we may artificially prepare organic compounds containing other elements besides those enumerated; thus we know many organic substances, which contain chlorine, iodine, or bromine; others which contain arsenic, antimony, tin, zinc, platinum, iron, cobalt, etc.; and it is quite impossible to say which of the other elements may not be similarly capable of becoming more remote constituents of organic compounds (constituents of organic radicals).

With these compounds we must not confound those in which organic acids are combined with inorganic bases, or organic bases with inorganic acids, such as tartrate of lead, for instance, silicic ether, borate of morphia, etc.; since in such bodies any of the elements may of course occur.

Organic compounds may be analyzed either with a view simply to resolve them into their proximate constituents; thus, for instance, a gum-resin into resin, gum, and ethereal oil; or the analysis may have for its object the determination of the ultimate constituents (the elements) of the substance. The simple resolution of organic compounds into their proximate constituents is effected by methods perfectly similar to those used in the analysis of inorganic compounds; that is, the operator endeavors to separate (by solvents, application of heat, etc.) the individual constituents from one another, either directly, or after having converted them into appropriate forms. We disregard here altogether this kind of organic analysis—of which the methods must be nearly as numerous and varied as the cases to which they are applied—and proceed at once to treat of the second kind, which may be called *the ultimate analysis of organic bodies.*

The ultimate analysis of organic bodies (*here termed simply, organic analysis*) has for its object, as stated above, the determination of the elements contained in organic substances. It teaches us how to isolate these elements or to convert them into compounds of known composition, to separate the new compounds formed from one another, and to calculate from their several weights, or volumes, the quantities of the elements. Organic analysis, therefore, is based upon the same principle upon which rest most of the methods of separating and determining inorganic compounds.

The conversion of most organic substances into distinctly characterized and readily separable products, the weights of which can be accurately determined, offers no great difficulties, and organic analysis is therefore usually one of the more easy tasks of analytical chemistry; and as, from the limited number of the elements which constitute organic bodies, there is necessarily a great sameness in the products of their decomposition, the analytical process is always very similar, and a few methods suffice for all cases. It is principally ascribable to this latter circumstance that organic analysis has so speedily attained its present high degree of perfection: the constant examination and improvement of a few methods by a great number of chemists could not fail to produce this result.

An organic analysis may have for its object either simply to ascertain the relative quantities of the constituent elements of a substance—thus, for instance, woods may be analyzed to ascertain their heating power, fats to ascertain their illuminating power—or to determine not only the relative quantities of the constituent elementary atoms, but also the number of atoms of carbon, hydrogen, oxygen, &c., which constitute 1 molecule of the analyzed compound. In scientific investigations we have invariably the latter object in view, although we are not yet able to achieve it in all cases. These two objects cannot well be attained by one operation; each requires a distinct process.

The methods by which we ascertain the proportions of the constituent elements of organic compounds may be called collectively *the ultimate analysis of organic bodies*, in a more restricted sense; whilst the methods which reveal to us the absolute number of elementary atoms constituting the molecule of the analyzed compound may be styled *the determination of the molecular weight of organic bodies.*

The success of an organic analysis depends both upon the method and its execution. The latter requires patience, circumspection, and skill; whoever is moderately endowed with these gifts will soon become a proficient in this branch. The selection of the method depends upon the knowledge of the constituents of the substance, and the method selected may require certain modifications, according to the properties and state of aggregation of the same. Before we can proceed, therefore, to describe the various methods applicable in the different cases that may occur, we have first to occupy ourselves here with the means of testing organic bodies qualitatively.

I. Qualitative Examination of Organic Bodies.

§ 172.

It is not necessary, for the correct selection of the proper method, to know all the elements of an organic compound, since, for instance, the presence or absence of oxygen makes not the slighest difference to the method. But with regard to other elements, such as nitrogen, sulphur, phosphorus, chlorine, iodine, bromine, &c., and also the various metals, it is absolutely indispensable that the operator should know positively whether either of them is present. This may be ascertained in the following manner:

1. *Testing for Nitrogen.*

Substances containing a tolerably large amount of nitrogen exhale upon combustion, or when intensely heated, the well-known smell of singed hair or feathers. No further test is required if this smell is distinctly perceptible; otherwise one of the following experiments is resorted to:

a. The substance is mixed with hydrate of potassa in powder or with soda-lime (§ 66, 4 or 5), and the mixture heated in a test-tube. If the substance contains nitrogen, ammonia will be evolved, which may be readily detected by its odor and reaction, and by the formation of white fumes with volatile acids. Should these reactions fail to afford positive certainty, every doubt may be removed by the following experiment: Heat a somewhat larger portion of the substance, in a short tube, with an excess of soda-lime, and conduct the products of the combustion into dilute hydrochloric acid; evaporate the acid on the water-bath, dissolve the residue in a little water, and mix the solution with platinic chloride and alcohol. Should no precipitate form, even after the lapse of some time, the substance may be considered free from nitrogen.

b. LASSAIGNE has proposed another method, which is based upon the property of potassium to form potassium cyanide when ignited with a nitrogenous organic substance. The following is the best mode of performing the experiment:

Heat the substance under examination, in a test-tube, with a small lump of potassium, and after the complete combustion of the potassium, treat the residue with a little water (cautiously); filter the solution, add 2 drops of solution of ferrous sulphate containing some ferric sulphate, digest the mixture a short time, and add hydrochloric acid in excess. The formation of a blue or bluish-green precipitate or coloration proves the presence of nitrogen.

Both methods are delicate: *a* is the more commonly employed, and suffices in almost all cases; *b* does not answer so well in the case of alkaloids containing oxygen (*e.g.* morphia, brucia).

c. In organic substances containing oxides of nitrogen, the presence of nitrogen cannot be detected with certainty by either *a* or *b*, but it may be readily discovered by heating the substance in a tube, when red acid fumes, imparting a blue tint to iodide of starch paper, will be evolved, accompanied often by deflagration.

2. *Testing for Sulphur.*

a. Solid substances are fused with about 12 parts of pure hydrate of potassa and 6 parts of potassium nitrate. Or they are intimately mixed with some pure potassium nitrate and sodium carbonate; potassium nitrate is then heated to fusion in a porcelain crucible, and the mixture gradually added to the fusing mass. The mass is allowed to cool, then dissolved in water, and the solution tested with barium chloride, after acidifying with hydrochloric acid.

b. Fluids are treated with fuming nitric acid, or with a mixture of nitric acid and potassium chlorate, at first in the cold, finally with application of heat; the solution is tested as in *a*.

c. As the methods *a* and *b* serve simply to indicate the presence of sulphur in a general way, but afford no information regarding the state or form in which that element may be present, I add here another method, which serves to detect only the sulphur in the non-oxidized state in organic compounds.

Boil the substance with strong solution of potassa and evaporate nearly to dryness. Dissolve the residue in a little water, and bring the solution into a small flask provided with a loosely-fitting stopper, through which passes a funnel tube reaching nearly to the bottom of the flask. Suspend from the lower surface of the stopper within the flask a strip of paper dipped first in lead acetate, then in ammonium carbonate solution. Add slowly dilute sulphuric acid, and observe whether the lead paper becomes brown; or test the first alkaline solution by means of a polished surface of silver, or by nitroprusside of sodium, or by just acidifying the dilute solution with hydrochloric acid, and adding a few drops of a mixture of ferric chloride and potassium ferricyanide (see "Qual. Anal." § 159).

3. *Testing for Phosphorus.*

The methods described in 2, *a* and *b*, may likewise serve for phosphorus. The solutions obtained are tested for phosphoric acid with magnesium sulphate or chloride; or with ferric chloride, with addition of sodium acetate; or with solution of molybdic acid (comp. "Qual. Anal."). In method *b*, the greater part of the excess of nitric acid must first be removed by evaporation.

4. *Testing for Inorganic Substances.*

A portion of the substance is heated on platinum foil, to see whether or not a residue remains. When acting upon difficultly combustible substances, the process may be accelerated by heating the spot which the substance occupies on the platinum foil to the most intense redness, by directing the flame of the blow-pipe upon it from below. The residue is then examined by the usual methods. That volatile metals in volatile organic compounds —*e.g.*, arsenic in kakodyl—cannot be detected by this method need hardly be mentioned.

These preliminary experiments should never be omitted, since neglect in this respect may give rise to very great errors. Thus, for instance, taurin, a substance in which a large proportion of sulphur was afterwards found to exist, had originally the formula $C_4N_2H_{14}O_{10}$ assigned to it. The preliminary examination of organic substances for chlorine, bromine, and iodine is generally unnecessary, as these elements do not occur in native organic compounds, and as their presence in compounds artificially produced by the action of the halogens requires generally no further proof. Should it, however, be desirable to ascertain positively whether a substance does or does not contain chlorine, iodine, or bromine, this may be done by the methods given § 188.

II. Determination of the Elements in Organic Bodies.*

§ 173.

A. Analysis of Compounds which consist simply of Carbon and Hydrogen, or of Carbon, Hydrogen, and Oxygen.

The principle of the method which serves to effect the quantitative analysis of such compounds is exceedingly simple. The substance is burned to carbonic acid and water; these products are separated from each other and weighed, and the carbon of the substance is calculated from the weight of the carbonic acid, the hydrogen from that of the water. If the sum of the carbon and hydrogen is equal to the original weight of the substance, the substance contains no oxygen; if it is less than the weight of the substance, the difference expresses the amount of oxygen present.

The combustion is effected either by igniting the organic substance with oxygenized bodies which readily part with their oxygen (cupric oxide, lead chromate, &c.); or at the expense both of free and combined oxygen.

a. Solid Bodies.

* [For Prof. Warren's admirable methods we must refer to his original papers in Am. Journ. Sci., 2d ser., vol. 38, p. 387, vol. 41, p. 40, and vol. 42, p. 156.]

COMBUSTION WITH CUPRIC OXIDE.

Applicable (with modification described in § 176) to non-volatile organic compounds not containing chlorine, bromine, iodine, alkali metals, alkali-earth metals, nitrogen, or sulphur.

§ 174.

I. APPARATUS AND PREPARATIONS REQUIRED FOR THE ANALYSIS.

1. THE SUBSTANCE.—This must be most finely pulverized and perfectly pure and dry;—for the method of drying, I refer to § 26.

2. A TUBE IN WHICH TO WEIGH THE SUBSTANCE, made of thin glass about 20 cm. long, and of 7 mm. internal diameter; one end of the tube is closed by fusion; the other, during the operation of weighing, is stopped with a smooth cork.

3. THE COMBUSTION TUBE.—A tube of difficultly fusible glass (potassa glass), about 2 mm. thick in the glass, 80 to 90 cm. in length, and from 12 to 14 mm. inner diameter, is softened in the middle before a glass-blower's lamp, drawn out as represented in fig. 71, and finally apart at *b*. The fine points of the two pieces

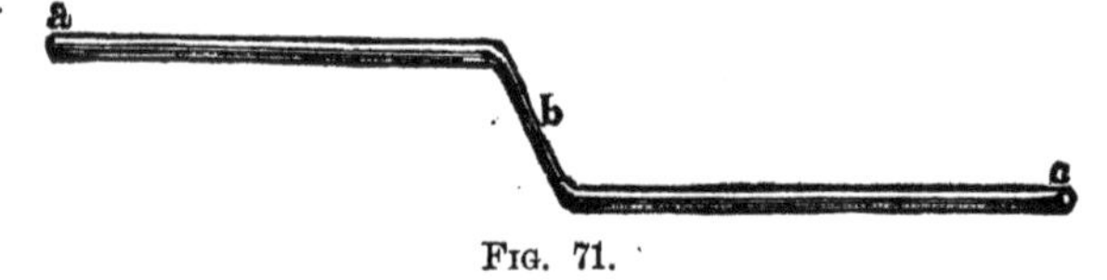

FIG. 71.

are then sealed and thickened a little in the flame, and the sharp edges of the open ends, *a* and *c*, are slightly rounded by fusion, care being taken to leave the aperture perfectly round. The posterior part of the tube should be shaped as shown in fig. 72, and not as in fig. 73.

FIG. 72. FIG. 73.

Two perfect combustion tubes are thus produced. The one intended for immediate use is cleaned with linen or paper attached to a piece of wire, and then thoroughly dried. This is effected either by laying the tube, with a piece of paper twisted over its mouth, for some time on a sand-bath, with occasional removal of the air from it by suction, with the aid of a glass tube, or (rapidly) by moving the tube to and fro over the flame of a gas or spirit lamp, heating its entire length, and continually removing the hot air by suction through the small glass tube (Fig. 74).

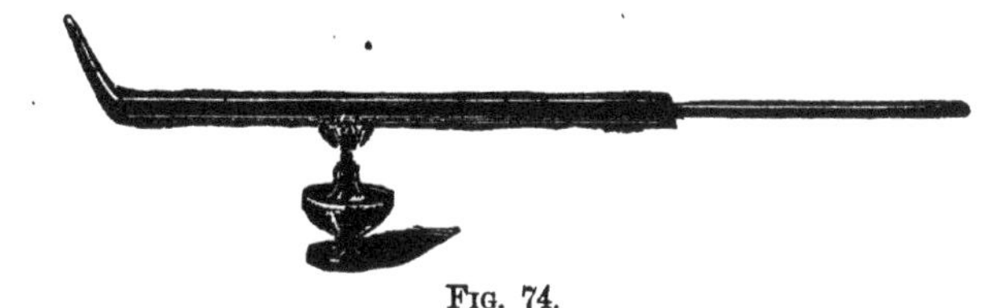

Fig. 74.

The combustion tube, when quite dry, is closed air-tight with a cork, and kept in a warm place until required for use.

In default of glass tubes possessed of the proper degree of infusibility, thin brass or copper foil, or brass gauze, is rolled round the tube, and iron wire coiled round it.

4. The Potash Bulbs (fig. 75).—This apparatus, devised by Liebig, is filled to the extent indicated in the engraving, with a clear solution of caustic potassa of 1·27 sp. gr. (§ 66, 7). The introduction of the solution of potassa into the apparatus is effected by plunging the end *a* into a beaker or dish into which a little of the solution has been poured out, and applying suction to *b*, by means of a caoutchouc tube. The two ends are then wiped perfectly dry with twisted slips of paper, and the outside of the apparatus with a clean cloth.

Fig. 75.

5. The Calcium Chloride Tube (fig. 76) is filled in the following manner:—In the first place, the neck between the two bulbs of the tube is loosely stopped with a small cotton plug; this is effected by introducing a loose cotton plug into the wide tube, and applying a sudden and energetic suction at the other end. The large bulb is then filled with lumps of calcium chloride (§ 66, 8, *a*), and the tube with smaller fragments, intermixed with coarse powder of the same substance; a loose cotton plug is then inserted, and the tube finally closed with a perforated cork, into which a small glass tube is fitted; the protruding part of the cork

is cut off, and the cut surface covered over with sealing-wax; the edge of the little tube is slightly rounded by fusion.

In using this tube a considerable quantity of the water condenses in the empty bulb *a*, and at the close of the experiment

FIG. 76.

may be poured out. The operator is thus enabled to test it as to reaction, &c., and also to use the same tube far oftener without fresh filling than he could otherwise.

6. A SMALL TUBE OF VULCANIZED INDIA-RUBBER.—This must be so narrow that it can only be pushed with difficulty over the tube of the calcium chloride tube on the one hand, and over the end of the potash bulbs on the other hand; in which case there is no need of binding with silk cord. If the rubber tube should be a little too wide, it must be tied round with silk cord, or with ignited piano wire. It is self-evident that the narrow end of the calcium chloride tube should be of the same width as the tube *a* of the potash bulbs. The india-rubber tube is purified from any adherent sulphur, and dried in the water-bath previous to use.

7. CORKS.—These should be soft and smooth, and as free as possible from visible pores. A cork should be selected which, after careful squeezing, fits perfectly tight, and screws with some difficulty to one-third of its length, at the most, into the mouth of the combustion tube; a perfectly smooth and round hole, into which the end *b* of the chloride of calcium tube must fit perfectly air-tight, is then carefully bored through the axis of the cork. The cork is then kept for an hour of two in the water-bath. It is advisable always to have two corks of this description ready. Instead of ordinary corks, caoutchouc stoppers may be used with great advantage.

8. OXIDE OF COPPER.—A Hessian crucible, of about 100 c.c. capacity, is nearly filled with oxide of copper prepared as directed in § 66, 1; the crucible is covered with a well-fitting overlapping lid, and heated to dull redness with charcoal, or in a suitable gas-furnace; it is then allowed to cool, so that by the time the oxide of copper is required for use, the hand can only just bear contact with it.

9. A WIDE GLASS TUBE sealed at one end, or a FLASK (fig. 77), in which the freshly ignited oxide of copper is allowed to cool, and from which it is transferred to the combustion tube, secure from the possible absorption of moisture from the air.

The freshly ignited and still quite hot oxide of copper is transferred direct from the crucible to this filling tube, or flask, which is then closed air-tight with a cork. It saves time to fill in at once a sufficient quantity of oxide to last for several analyses. If the cork fits tight, the contents will remain several days fit for use, even though a portion has been taken out, and the tube repeatedly opened.

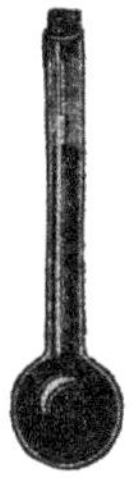

Fig. 77.

10. A MIXING WIRE of copper (fig. 78) with ring at one end

Fig. 78.

for a handle and a single corkscrew turn at the other, which should taper smoothly to a point.

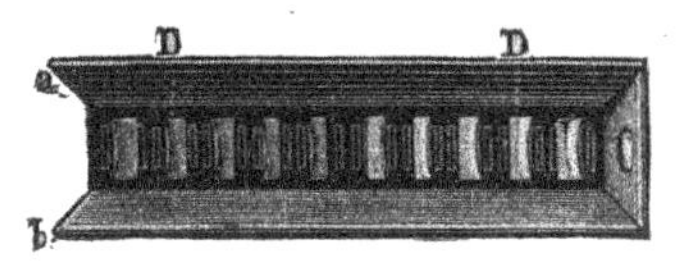

Fig. 79.

11. A COMBUSTION-FURNACE.—Some time ago the only one used was LIEBIG'S, in which charcoal is the fuel. Recently gas combustion furnaces have been introduced into most laboratories, because they are more cleanly and convenient.

a. LIEBIG'S combustion-furnace is of sheet-iron. It has the form of a long box, open at the top and behind. It serves to heat the combustion tube with red-hot charcoal. Fig. 79 represents the furnace as seen from the top.

It is from 50 to 60 cm. long, and from 7 to 8 deep; the bottom, which, by cutting small slits in the sheet-iron, is converted into a grating, has a width of about 7 cm. The side walls are inclined slightly outward, so that at the top they stand about 12

Fig. 80.

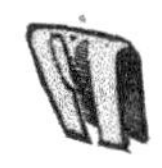

Fig. 81.

cm. apart. A series of upright pieces of strong sheet-iron, having the form shown in *D*, fig. 80, and riveted on the bottom of the furnace at intervals of about 5 cm., serves to support the combustion tube. They must be of exactly corresponding height with the round aperture in the front piece of the furnace (fig. 80, *A*).

This aperture must be sufficiently large to admit the combustion tube easily. Of the two screens used, one has the form shown in fig. 81, the other is a single plate precisely like the end piece of the furnace (fig. 79). The openings cut into the screens must be sufficiently large to receive the combustion tube without difficulty. The furnace is placed upon two bricks resting upon a flat surface, and is slightly raised at the farther end, by inserting a piece of wood between the supports (see fig. 84). The apertures of the grating at the anterior end of the furnace must not be blocked up by the supporting bricks. In cases where the combustion tubes are of good quality, the furnace may be raised by introducing a little iron rod between the furnace and the supporting brick. Placing the tube in a gutter of Russia sheet-iron tends greatly to preserve it, but contact of the glass and iron must be prevented by an intervening layer of asbestos.

b. Gas combustion furnaces of the most various descriptions have been proposed.

§ 175.

II. Performance of the Analytical Process.

a. Weigh first the potash apparatus, then the calcium chloride tube. Introduce about 0·35—0·6 grm. of the substance under examination (more or less, according as it is rich or poor in oxygen) into the weighing tube,* which must be no longer warm, and weigh the latter accurately with its contents. The weight of the empty tube being approximately known, it is easy to take the right quantity of substance required for the analysis. Close the tube then with a smooth cork.

b. The filling of the combustion tube is effected as follows: The perfectly dry tube is rinsed with some oxide of copper; a layer of oxide of copper, about 13 cm. long, is introduced into the posterior end of the combustion tube, by inserting the latter into the filling tube or flask containing the oxide of copper

* Care must be taken that no particles of the substance adhere to the sides of the tube, at least not at the top.

(fig. 82), holding both tubes in an oblique direction, and giving a few gentle taps.

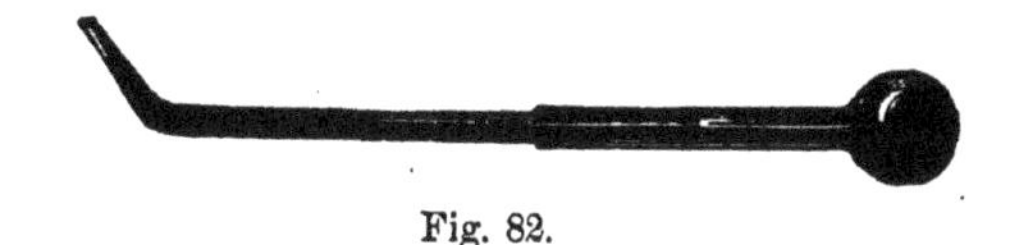

Fig. 82.

From the tube containing the substance remove the cork cautiously, to prevent the slightest loss of substance; insert the open end of the tube as deep as possible into the combustion tube, and pour from it the requisite quantity of substance by giving it a few turns, pressing the rim all the while gently against the upper side of the combustion tube, to prevent its coming into contact with the powder already poured out; the two tubes are, in this manipulation, held slightly inclined (see fig. 83).

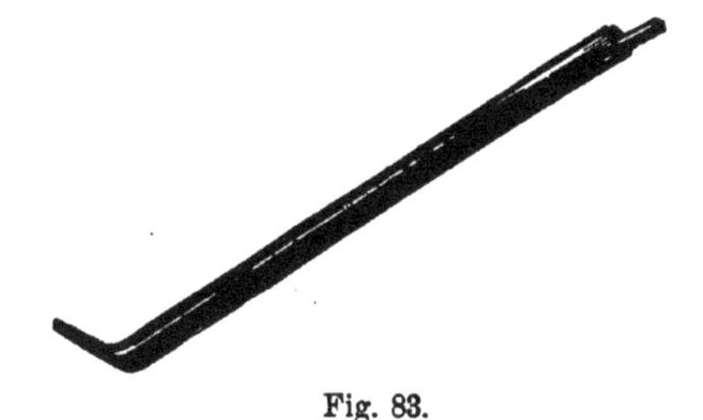

Fig. 83.

When a sufficient quantity of the substance has been thus transferred from the weighing to the combustion tube, the latter is restored to the horizontal position, which gives to the former a gentle inclination with the closed end downwards. If the little tube is now slowly withdrawn, with a few turns, the powder near the border of the opening falls back into it, leaving the opening free for the cork. The tube is then immediately corked and weighed, the combustion tube also being meanwhile kept closed with a cork. The difference between the two weighings shows the quantity of substance transferred from the weighing to the combustion tube. The latter is then again opened, and a quantity of oxide of copper, equal to the first, transferred to it from the filling tube, or flask, taking care to rinse down with this the particles of the substance still adhering to the sides of the tube.

There is now in the hind part of the tube a layer of oxide of copper, about 25 cm. long, with the substance in the middle.

The next operation is the mixing: this is performed with the aid of the wire (fig. 78), which is pushed down to within 3 to 4 cm. of the end, and rapidly moved about in all directions until the mixture is complete and uniform, the tube being held nearly horizontal.

Oxide of copper is then poured in to within 5 to 6 cm. of the open end, and the tube is corked.

c. A few gentle taps on the table will generally suffice to shake together the contents of the tube, so as to completely clear the tail from oxide of copper, and leave a free passage for the evolved gas from end to end. Should this fail, as will occasionally happen, owing to malformation of the tail, the object in view may be attained by striking the mouth of the tube several times against the side of a table.

d. Connect the end *b* (fig. 84) of the weighed calcium chloride tube with the combustion tube by means of a dried perforated cork, lay the furnace upon its supports, with a slight inclination forward, and place the combustion tube in it; connect the end

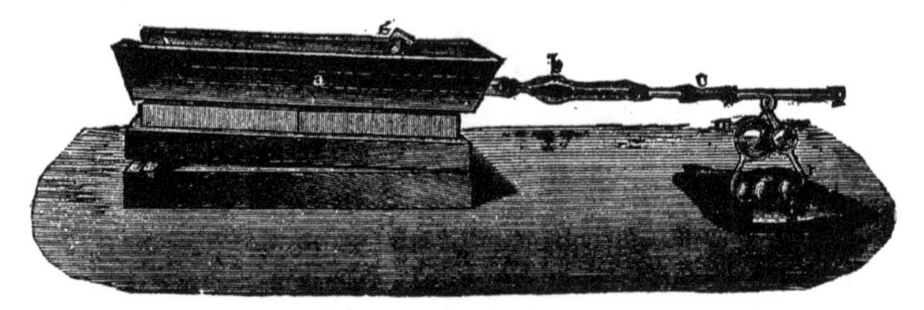

Fig. 84.

of the calcium chloride tube, by means of a vulcanized india-rubber tube, with the end *m* of the potash apparatus, and, if necessary, secure the connection with silk cord, taking care to press the joint of the two thumbs close together whilst tightening the cords, since otherwise, should one of the cords happen to give way, the whole apparatus might be broken. Rest the potash apparatus upon a folded piece of cloth. Fig. 84 shows the whole arrangement.

e. To ascertain whether the joinings of the apparatus fit air-tight, put a piece of wood about the thickness of a finger (*s*), or a cork or other body of the kind, under the bulb *r* of the potash apparatus, so as to raise that bulb slightly (see fig. 84). Heat the bulb *m*, by holding a piece of red-hot charcoal near it, until a certain amount of air is driven out of the apparatus; then remove the piece of wood (*s*), and allow the bulb *m* to cool. The solution of potassa will now rise into the bulb *m*, filling it more or less; if the liquid in *m* preserves, for the space of a few minutes, the same level which it has assumed after the perfect cooling of the bulb, the joinings may be considered perfect; should the fluid, on the other hand, gradually regain its original level in both limbs of the apparatus, this is a positive proof that the joinings are not air-tight. (The few minutes which elapse between the two observations may be advantageously employed in reweighing the little tube in which the substance intended for analysis was originally weighed.)

f. Let the mouth of the combustion tube project a full inch beyond the furnace; suspend the single screen over the anterior end of the furnace, as a protection to the cork; put the double screen over the combustion tube about two inches farther on (see fig. 84), replace the little piece of wood (*s*) under *r*, and put small pieces of red-hot charcoal first under that portion of the tube which is separated by the screen; surround this portion gradually altogether with ignited charcoal, and let it get red-hot; then shift the screen an inch farther back, surround the newly exposed portion of the tube also with ignited charcoal, and let it get red-hot; and proceed in this manner slowly and gradually extending the application of heat to the tail of the tube, taking care to wait always until the last exposed portion is red-hot before shifting the screen, and also to maintain the whole of the exposed portion of the tube before the screen in a state of ignition, and the projecting part of it so hot that the fingers can hardly bear the shortest contact with it. The whole process requires generally from $\frac{3}{4}$ to 1 hour. It is quite superfluous, and even injudicious, to fan the charcoal constantly; this should be done however when the process is drawing to an end, as we shall immediately have occasion to notice.

The liquid in the potash bulbs is gradually displaced from the bulb *m* upon the application of heat to the anterior portion of the combustion tube, owing simply to the expansion of the heated air. The evolution of gas proceeds with greater briskness when the heat begins to reach the actual mixture; the first bubbles are only partly absorbed, as the carbonic acid contains still an admixture of air; but those which follow are so completely absorbed by the potassa, that a solitary air-bubble only escapes from time to time through the liquid. The process should be conducted in a manner to make the gas-bubbles follow each other at intervals of from $\frac{1}{2}$ to 1 second. Fig. 85 shows the proper position of the potash bulbs during the operation.

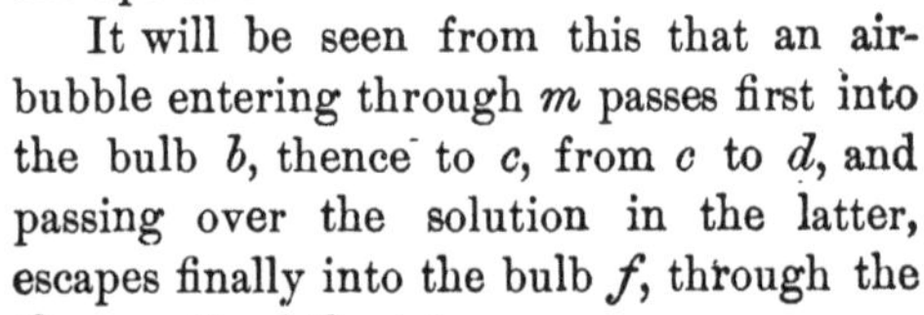

Fig. 85.

It will be seen from this that an air-bubble entering through *m* passes first into the bulb *b*, thence to *c*, from *c* to *d*, and passing over the solution in the latter, escapes finally into the bulb *f*, through the fluid which just covers the mouth of the tube *e*.

g. When the tube is in its whole length surrounded with red-hot charcoal, and the evolution of gas has relaxed, fan the burning charcoal gently with a piece of pasteboard. When the evolution of gas has entirely ceased, adjust the position of the potash bulbs to a level, remove the charcoal from the farther end of the tube, and place the screen before the tail. The ensuing cooling of the tube on the one hand, and the absorption of the carbonic acid in the potash bulbs on the other, cause the solution of potassa in the latter to recede, slowly at first, but with increased rapidity from the moment the liquid reaches the bulb *m*. (If you have taken care to adjust the position of the potash bulbs correctly, you need not fear that the contents of the latter will recede to the calcium chloride tube.) When the bulb *m* is about half filled with solution of potassa, break off the point of the combustion tube with a pair of pliers or scissors, whereupon the fluid in the potash bulbs will immediately resume its level. Restore the potash bulbs now again to their original oblique position, join a caoutchouc tube to the potash bulbs, and slowly apply suction until the last bubbles no longer diminish in size in passing through the latter. It is better to employ a small aspirator instead of sucking with the mouth. You then know the volume of air that has passed through the apparatus.

This terminates the analytical process. Disconnect the potash bulbs and remove the calcium chloride tube, together with the cork, which must not be charred, from the combustion tube;

remove the cork also from the calcium chloride tube, and place the latter upright, with the bulb upwards. After the lapse of half an hour, weigh the potash bulbs and the calcium chloride tube, and then calculate the results obtained. They are generally very satisfactory. As regards the carbon, they are rather somewhat too low (about 0·1 per cent.) than too high. The carbon determination, indeed, is not free from sources of error; but none of these interfere materially with the accuracy of the results, and the deficiency arising from the one is partially balanced by the excess arising from the other. In the first place, the air which passes through the solution of potassa during the combustion, and finally during the process of aspiration, carries away with it a minute amount of moisture. The loss arising from this cause is increased if the evolution of gas proceeds very briskly, since this tends to heat the solution of potassa; and also if nitrogen or oxygen passes through the potash bulbs (compare § 176 and § 178). This may be remedied, however, by fixing to the exit end of the latter a tube, either straight or U-formed (see fig. 86 or fig. 60); the tube may be filled with small fragments of potassa, or one-half may be filled with soda lime (§ 66, 4) and the other half with calcium chloride, the end containing soda lime being connected to the potash apparatus, which is always weighed along with the appended tube. In the second place, traces of carbonic acid from the atmosphere are carried into the potash apparatus during the final aspiration; this may be avoided by connecting the tail of the combustion tube during the aspiration with a tube filled with potassa crushed to small lumps, by means of a flexible tube. In the third place, it may happen in the analysis of substances containing a considerable proportion of water or hydrogen, that the carbonic acid is not completely dried in passing through the calcium chloride; this may be avoided by using instead of the calcium chloride tube, or in conjunction with it, a U-tube filled with fragments of pumice stone and H_2SO_4; but usually a calcium chloride tube, if filled for about 12 c.c. of its length with not too coarsely granulated calcium chloride, will suffice, provided the combustion is not pushed too rapidly. Finally, if the mixture was not sufficiently intimate, traces of carbon will remain unconsumed. It is therefore better to complete the combustion in oxygen gas. See below.

§ 176.

[*Completion of the Combustion by Oxygen Gas.* To insure the oxidation of the last traces of carbon and to leave the oxide of copper ready for use again, it is advisable to finish the combustion in a stream of oxygen. For this purpose the tail of the combustion tube must be made rather stout and long. When the potash-lye recedes, slip tightly over the suitably cooled tail a caoutchouc tube connected with a source of pure and dry oxygen gas, nip off the tip within this tube by help of a pliers, and cautiously let on the oxygen until the reduced copper is oxidized and the gas traverses the potash bulbs. Then replace the stream of oxygen by one of pure and dry air, to remove all oxygen from the bulbs. To prevent loss by evaporation from the potash-lye, append to the potash bulb the additional absorbing apparatus above mentioned (in § 175).

The oxygen and purified air are supplied as in the process described in § 178.]

COMBUSTION WITH LEAD CHROMATE, OR WITH LEAD CHROMATE AND POTASSIUM DICHROMATE.

§ 177.

This is not only a good method for the analysis of compounds mentioned in § 174, but is especially resorted to in the analysis of salts of organic acids with alkalies or alkali-earth metals (as the chromic acid completely displaces carbonic acid from their carbonates), and of bodies containing sulphur, chlorine, bromine, or iodine, and also for the combustion of substances containing carbon in a difficultly oxidizable form—*e.g.*, graphite.

Of the apparatus, &c., enumerated in § 174, all are required except oxide of copper, which is here replaced by lead chromate (§ 66, 2). A narrow combustion tube may be selected, as lead chromate contains a much larger amount of available oxygen in an equal volume than oxide of copper. A quantity of the chromate, more than sufficient to fill the combustion tube, is heated in a platinum or porcelain dish over a gas or BERZELIUS lamp, until it begins to turn brown; before filling it into the tube, it is allowed to cool down to 100°; and even below. The process is conducted as the one described in § 174.

If the substance analyzed contains a large proportion of sulphur, use a rather long combustion tube (60–70 cm.) and place in front of the mixture 10–20 cm. pure lead chromate, which should be kept only at a dull red heat during the combustion (CARIUS).

One of the principal advantages which lead chromate has over oxide of copper as an oxidizing agent being its property of fusing at a high heat, the temperature must, in the last stage of the process of combustion, be raised (by fanning the charcoal, &c.) sufficiently high to fuse the contents of the tube completely, as far as the substance extends. To heat the *anterior* end of the tube to the same degree of intensity would be injudicious, since the lead chromate in that part would thereby lose all porosity, and thus also the power of effecting the combustion of the products of decomposition which may have escaped oxidation in the other parts of the tube.

As the lead chromate, even in powder, is, on account of its density, by no means all that could be desired in this latter respect, it is preferable to fill the anterior part of the tube, instead of with lead chromate, with coarsely pulverized strongly ignited oxide of copper, or with copper turnings which have been superficially oxidized by ignition in a muffle or in a crucible with access of air.

In the case of very difficultly combustible substances—*e.g.*, graphite—it is desirable that the mass should not only readily cake, but also, in the last stage of the process, give out a little more oxygen than is given out by lead chromate. It is therefore advisable in such cases to add to the latter one-eighth of its weight of fused and powdered potassium dichromate. With the aid of this addition, complete oxidation of even very difficultly combustible bodies may be effected (LIEBIG).

COMBUSTION WITH OXIDE OF COPPER IN A STREAM OF OXYGEN GAS.

§ 178.

Many chemists effect combustion with oxide of copper in a stream of oxygen supplied by a gasometer. The methods based upon this principle are employed not only for the analysis of difficultly combustible bodies, but also to effect the determination of the carbon and hydrogen in organic substances in general.

These methods require a gasometer filled with oxygen, and another with air, together with certain arrangements to dry the oxygen and air completely, and to free them from carbonic acid. They are resorted to in cases where a number of ultimate analyses have to be made in succession; and also more particularly in the analysis of substances which cannot be reduced to powder, and do not admit therefore of intimate mixture with oxide of copper, &c.

The heating may be effected with the charcoal combustion furnace, but a gas furnace is most convenient.

Fig. 86 represents the manner in which the several requisite pieces of apparatus are arranged and connected. The combustion tube rests in a gutter of sheet iron, but the glass is kept from contact with the metal by a layer of asbestos. It is well to secure the tube to the gutter by binding it with copper wire. At its anterior end the combustion tube is connected in the usual manner with the absorbing apparatus, consisting of a calcium chloride tube, potash bulb, and additional absorbing tube. To the latter, which is prepared as described in § 175, p. 619, must also be attached an unweighed calcium chloride tube to prevent moist air from entering during the process. B is a bell jar standing in open vessel. By opening the stop-cock in the tube which enters the top of the bell jar, the pressure within the combustion tube caused by the liquid in the potash bulb may be removed; provided the water levels within and without the bell are properly adjusted. (This arrangement for relieving the pressure within the combustion tube is, however, usually quite unnecessary.) It is hardly necessary to state that any or even all of the several parts of the apparatus here represented may be replaced by other forms. An Erlenmeyer combustion furnace 80 cm. long, with 25 burners, is quite satisfactory. Many forms of apparatus have been devised for drying and purifying the air and oxygen which are used in the process. Fig. 87 shows one which is durable and efficient. The bulb tube entering the bottle *d* is connected with the gasometer by means of a rubber tube. The bottle *d* is half filled with concentrated sulphuric acid, through which the gas or air passes in bubbles and enters the bottom of the cylinder *c*. The lower half of this cylinder is filled with fragments of fused potash; the upper half with calcium chloride, which is separated from the potash by a layer of asbestos. Glass tubes provided with glass stopcocks enter top of each cylinder through rubber stoppers, and are connected by means of strong rubber tubes to the two limbs of the forked tube *b*, so that a regulated current of either air or oxygen can be made to enter the combustion tube through *a* at will.

In carrying out the process, the substance to be analyzed may be mixed directly with the oxide of copper, or it may be placed in a porcelain or platinum tray and kept out of contact with the oxide. The latter mode of proceeding affords an opportunity for obtaining inorganic constituents—*e.g.*, ashes in coal—at the same time, and for this and other reasons is oftenest preferable, and will be first described under *a*.

a. Combustion in a platinum or porcelain tray. The combustion tube should be about 85 cm. in length and open at both ends. Fix a plug of asbestos, or better fine copper gauze, at a point 40 cm. from the rear end of the tube, and fill with coarsely granulated oxide of copper up to within 16 cm. of the front end, and secure the oxide in place by another plug of copper gauze, thus leaving about 15 cm. at the front end free. Into this space, a roll of copper gauze, 8 to 10 cm. long, is pushed up to the oxide. This copper roll should be in the metallic state if the substance to be analyzed contains nitrogen, chlorine, or bromine; otherwise it should previously be superficially oxidized by passing air or oxygen over it in a tube at a red heat. Provide another similar copper roll (metallic for all cases) to be placed behind the tray containing the substance, next connect an unweighed calcium chloride tube to the front end of the combustion tube, and fit a cork or rubber stopper, through which passes a small strong glass tube, closely into the rear end.

Now, in order to thoroughly expel all hygroscopic moisture from the copper rolls and the oxide, place the combustion tube in the furnace, and connect the tube inserted in the rear end with the tube *a*, fig. 87, of the drying and purifying apparatus, which must stand at the end of the furnace connected with two gasometers, one holding air, the other oxygen prepared as directed in § 66, 3. The contents of the tube are then brought gradually to a dull red heat, while a slow current of dry air is passed through it. Next, after allowing the whole to cool, remove the stopper from the rear end, draw out the copper roll with a piece of wire, introduce the weighed substance in a tray, pushing it nearly up to the oxide, replace immediately the copper roll, leaving it midway between the tray and the end of the tube, replace also without delay the

Fig. 86.

stopper, and connect as before with the drying apparatus. The unweighed calcium chloride tube is now removed from the front end of the combustion tube, and the weighed absorbing apparatus is attached. All being thus made ready, with stopcocks both for air and oxygen closed, heat the front roll of copper and oxide of copper, so far as possible without affecting the substance in the tray, and also the rear copper roll, all to dull redness. Then turn on a very slow current of oxygen and commence cautiously to heat the tray. The oxygen is not designed to aid the combustion at this

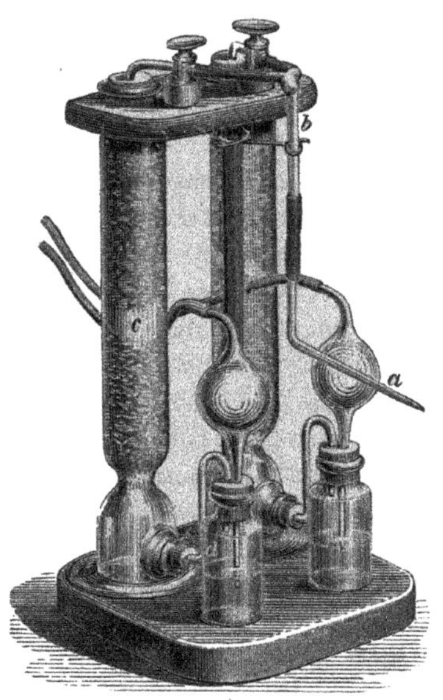

Fig. 87.

stage: it is taken up by the metallic copper; but serves to prevent products of distillation from receding and being deposited in the end of the tube. The heat which the substance itself will bear, or the rate at which the heat may be increased, depends upon its volatility, and is best regulated by observing the rate at which gas bubbles enter the first bulb of the potash apparatus. When the evolution of gas is much diminished and only fixed carbon remains in the tray, the rear copper roll is allowed to cool, and the combustion is completed by increasing the current of oxygen. When this is effected, and it is observed that the greater part of the reduced copper before the tray has been reoxidized, shut off the oxygen and supply instead a current of air until the oxygen is expelled from the apparatus. The operation is now at an end. The stopcock admitting air is shut, and the furnace allowed to cool.

Several combustions can usually be made in the same tube by this method. If the rear roll of copper becomes too much oxidized by repeated use, it may be reduced to the metallic state by heating in a current of hydrogen. The oxide of copper and the front copper roll will remain in order, unless it is desired to have the front roll in the metallic state for substances containing nitrogen, &c., in which case it may also be deoxidized by hydrogen. If one combustion is to be made immediately after another in the same tube, the preliminary heating to expel hygroscopic moisture is, of course, unnecessary.

b. Combustion of the substance mixed with oxide of copper. The combustion tube is prepared and ignited to expel hygroscopic moisture from its contents as above described in *a*, in all particulars except the following: It may be somewhat shorter—the rear copper roll is omitted, a few cm. of the rear portion of the copper oxide should be in a fine state of division instead of coarsely granulated. After the preliminary ignition and sufficient cooling, the tube is taken from the furnace, and the substance is introduced from the long narrow tube in which it has been weighed, and quickly mixed with the oxide by means of a copper wire with twisted end (see fig. 78, § 174); next, without delay, oxide of copper which has been ignited and cooled in a tube or flask (see fig. 77, § 174) is poured into the rear end of the tube—enough to occupy about 12 cm. in length of the space behind the mixture—while about an equal space at the very end should remain empty. A few gentle taps on the table by the tube in a horizontal position will serve to shake the contents down a little, so as to leave a narrow clear passage above. The tube is now laid in the furnace, the unweighed calcium chloride tube is removed from the front end, the proper absorbing apparatus is attached, connection of the after-end with the drying and purifying apparatus is made, and finally the whole remaining part of the process is conducted as directed above in *a*, except that a very slow stream of oxygen may be used at an earlier stage of the process, or at least before signs of moisture or receding gases appear in the empty space at the rear end of the combustion tube.

The combustion tube may be used a second time, or so long as it remains uninjured, removing the fine oxide of copper from the after-part with a wire each time, and allowing the granulated to remain.

Volatile Substances, or Bodies undergoing Alteration at 100° (losing Water, for instance).

§ 179.

The process is conducted either according to § 174, or as directed § 178. Ignited chromate of lead, cooled in a closed tube, may also be employed as oxidizing agent.

b. FLUID BODIES.

α. Volatile liquids (*e.g.*, ethereal oils, alcohol, etc.).

§ 180.

1. The analysis of organic volatile fluids requires the objects enumerated in § 174. The combustion tube should be somewhat longer than there mentioned; it should have a length of 50 or 60 cm., according as the substance is less or more volatile. The process requires besides several small glass bulbs for the reception of the liquid to be analyzed. These bulbs are made in the following manner:

A glass tube, about 30 cm. long and about 8 mm. wide, is drawn out as shown in fig. 88, fused off at *d*, and *A* expanded into a bulb, as shown in fig. 89. The bulbed part is then cut off at *β*. Another bulb is then made in the same way, and a third and fourth, &c., as long as sufficient length of tube is left to secure the bulb from being reached by the moisture of the mouth.

Fig. 88.

Fig. 89.

Two of these bulbs are accurately weighed; they are then filled with the liquid to be analyzed, closed by fusion, and weighed again. The filling is effected by slightly heating the bulb over a lamp and immersing the point into the liquid to be analyzed, part of which will now, upon cooling, enter the bulb. If the fluid is highly volatile, the portion entering the still warm bulb is converted into vapor, which expels the fluid again; but the moment the vapor is recondensed, the bulb fills the more completely. If the liquid is of a less volatile nature, a small portion only will enter at first; in such cases the bulb is heated again, to convert what has entered into vapor, and the point is then again immersed into the fluid, which will now readily enter and fill the bulb. The excess of fluid is ejected from the neck of the little tube by a sudden jerk; the point of the capillary neck is then sealed in the blowpipe flame. The combustion tube is now prepared for the process by introducing into it from the filling-tube or flask (§ 174) a layer of oxide of copper occupying about 6 cm. in length. The middle of the neck of one of the bulbs is slightly scratched with a file, the pointed end is quickly broken off, and the bulb and end are dropped into the combustion tube (see fig. 90). Another layer of oxide of copper, about 6—9 cm. long, is then filled in, and the other bulb introduced in the same manner as the first. The tube is finally nearly filled with oxide of copper. A few gentle taps upon the table suffice to clear a free passage for the gases evolved. (It is advisable to place in the anterior half of the combustion tube small lumps of oxide of copper [comp. § 66, 1], or superficially oxidized copper turnings, which will permit the free passage of the gases, even with a narrow channel, or no channel at all; since with a wide channel there is the risk of vapors passing unconsumed through the tube.)

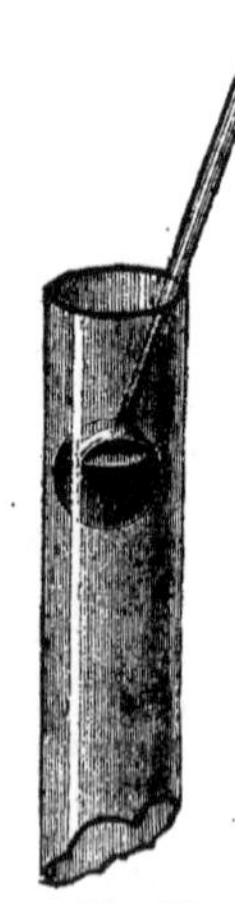

Fig. 90.

The combustion of highly volatile substances demands great care, and requires certain modifications of the common method. The operation commences by heating to redness the anterior half of the tube, which is separated from the rest by a screen, or in the case of highly volatile substances, by two screens; ignited charcoal is then placed behind the tube to heat the tail and prevent the condensation of vapor in that part. A piece of red-hot charcoal is now applied to that part of the tube which is occupied by the first bulb; this causes the efflux and evaporation of the contents of the latter; the vapor passing over the oxide of copper suffers combustion, and thus the evolution of gas commences, which is then maintained by heating very gradually the first, and after this the second bulb; it is better to conduct the operation too slowly than too quickly. Sudden heating of the bulbs would at once cause such an impetuous rush of gas as to eject the fluid from the potash bulbs. The tube is finally in its entire length surrounded with ignited charcoal, and the rest of the operation conducted in the usual way. If the air drawn through the apparatus tastes of the analyzed substance, this is a sure sign that complete combustion has not been effected.

2. In the combustion of liquids of high boiling point and abounding in carbon, *e.g.*, ethereal oils, unconsumed carbon is apt to deposit on the completely reduced copper near the substance; it is therefore advisable to distribute the quantity intended for analysis (about 0·4 grm.) in 3 bulbs, separated from each other in the tube by layers of oxide of copper.

3. In the combustion of less volatile liquids, it is advisable to empty the bulbs of their contents before the combustion begins: this is effected by connecting the filled tube with an exhausting syringe, and rarefying the air in the tube by a single pull of the handle; this will suffice to expand the air-bubble in each bulb sufficiently to eject the oily liquid from it, which is then absorbed by the oxide of copper.

4. If there is reason to apprehend that the oxide of copper may not suffice to effect the complete combustion of the carbon, the process is terminated in a stream of oxygen gas (compare § 176).

5. If it is intended to effect the combustion in the apparatus described in § 178 (in a current of oxygen gas), the bulb must be drawn out to a fine long point, and filled almost completely with the fluid. The point is then sealed in the blowpipe flame, and the bulbs are transferred in that state to the combustion tube. When the anterior and the farther end of the tube are red-hot, a piece of ignited charcoal is put to the part occupied by the first bulb, when the expansion of the liquid will cause it to burst. When the contents of the first bulb are consumed, the second, and after this the third, are treated in the same way. This method will not answer, however, for very volatile liquids, as, *e.g.*, ether, on account of the explosion which would inevitably take place.

β. Non-volatile Liquids (*e.g.*, fatty oils).

§ 181.

The combustion of non-volatile liquids is effected either, 1, with chromate of lead, or oxide of copper and oxygen; 2, in the apparatus described § 178.

1. The operation is conducted in general as directed § 175 or § 176. The substance is weighed in a small tube, placed for that purpose in a tin foot (see fig. 91), and the mixing effected as follows: Introduce into the combustion tube first a layer, about 6 cm. long, of chromate of lead, or of oxide of copper; then drop in the small cylinder with the substance, and let the oil completely run out into the tube; make it spread about in various directions, taking care, however, to leave the upper side (intended for the channel) and the forepart, to the extent of $\frac{1}{4}$ or $\frac{1}{8}$ of the length of the tube, entirely clean. Fill the tube now nearly with chromate of lead or oxide of copper—which has previously been cooled in the filling tube or flask—taking care that the little cylinder which contained the oil be completely filled with the oxidizing agent. Place the tube in hot sand, which, imparting a high degree of fluidity to the oil, leads to the perfect absorption of the latter by the oxidizing agent, and proceed with the combustion in the usual way. It is advisable to select a tolerably long tube. Chromate of lead is usually to be preferred. If it is used, a very intense heat, sufficiently strong to fuse the contents of the tube, is cautiously applied in the last stage of the process.

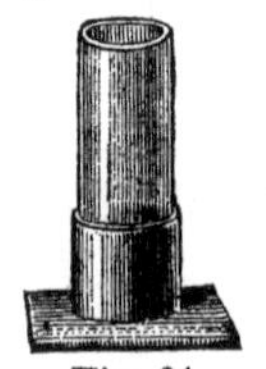
Fig. 91.

Solid fats or waxy substances which, not being reducible to powder, cannot be mixed with the oxidizing agent in the usual way, are treated in a similar manner to fatty oils. They are fused in a small weighed glass boat, made of a tube divided lengthwise; when cold, the little boat with its contents is weighed, and then dropped into the combustion tube, which has been previously filled to the extent of about 6 cm. with chromate of lead, or with oxide of copper. The substance is then fused by the application of heat, and made to spread about in the tube in the same manner as is done with fatty oils; the rest of the operation also being conducted exactly as in the latter case. If chromate of lead is employed, it will be found advantageous to add some potassium dichromate

(§ 177). If oxide of copper be used, finish in a stream of oxygen (§ 176).

2. If it is intended to effect the combustion of fatty substances or other bodies of the kind in a tray, in a current of oxygen gas, by means of the apparatus described in § 178, the combustion must be conducted with great care. As soon as the oxide of copper in the anterior and the copper roll in the posterior parts of the tube are red-hot, carefully regulated heat is applied to the part occupied by the tray. The volatile products generated by the dry distillation of the substance burn at the expense of the oxide of copper. When it is perceived that the surface layer of the oxide of copper is reduced, the application of heat to the substance is suspended for a time, and resumed only after the reduced copper is reoxidized in the stream of oxygen gas. Care is finally taken to insure the complete combustion of the carbon remaining in the boat.

Supplement to A., §§ 174—181.

§ 182.

Modified Apparatus for the Absorption of Carbonic Acid.

G. J. Mulder* has replaced the potash bulbs altogether by a totally different absorption apparatus. The calcium chloride tube is immediately connected with the system of U-tubes, fig. 92; *a* contains small pieces of glass, 6 to 10 drops concentrated sulphuric acid, and at the top asbestos plugs. *b* is filled to $\frac{7}{8}$ with granulated soda lime (say 20 grm. prepared as directed in § 66, 4), the remaining $\frac{1}{8}$ (in the 2d limb) contains calcium chloride (say 3 grm.). Lastly, *c* is filled with lumps of potassa. *a* and *b* are weighed together, *c* serves as a guard to *b*, and is not weighed. The sulphuric acid tube serves to show the rate of the evolution of gas; it contains enough sulphuric acid,

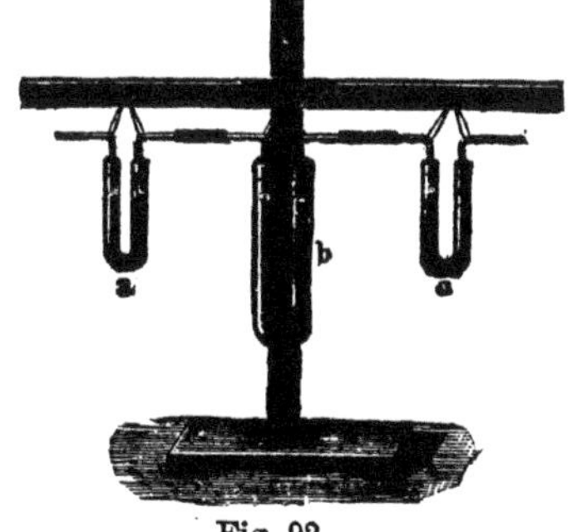

Fig. 92.

* Zeitschrift f. analyt. Chem. 1, 2.

when the lower part is just stopped up. If the process goes on properly, the weight of the tube does not increase more than 1 mgrm.; generally the increment is unweighable. If the tube is closed after use with caoutchouc caps, it may be used over and over again. The sulphuric acid possesses the advantage over other fluids that it indicates whether the combustion was complete or not; for in the first case it remains colorless, in the second it becomes brown from the escaping hydrocarbons, and then the results cannot be expected to be perfectly accurate. The absorption of the carbonic acid by the soda-lime tube is as rapid as it is complete; even when a stream of carbonic acid is passing, with ten times the rapidity usual in organic analysis, no trace of the acid makes its escape. The absorption of the carbonic acid is attended with warming of the soda-lime; if any water evaporates from the soda-lime, it is retained by the calcium chloride in the second limb. The corks of the absorption tubes are, like the others, coated with sealing-wax. A filled soda-lime tube weighs about 40 grm. The first time it is used alone; the second time the same tube is used, but as a precautionary measure a second similarly filled and separately weighed tube is placed in front of it. The second tube rarely increases in weight, and unless it does, the first tube can be used a third time, but of course in connection with the second. If the second tube has gained in the third operation, the first tube is rejected at the fourth operation, and the second is now used alone, &c. If after the combustion a stream of oxygen is transmitted through the combustion tube, the tubes are of course at the end full of oxygen. If, then, care be taken that the tubes are full of oxygen before weighing, the trouble of the final transmission of air may be saved. For weighing, MULDER closes the ends of the glass tubes with caps made out of india-rubber tube. According to DIBBITS,* however, this is not to be recommended.

MULDER's absorption apparatus is peculiarly suitable, when the carbonic acid is mixed with another gas. It insures complete absorption, precludes the evaporation of any water, and offers perfect security in case of the sudden occurrence of a too rapid evolution of gas.

---

* Zeit. f. anal. Chem. 15, 157.

B. Analysis of Compounds consisting of Carbon, Hydrogen, Oxygen, and Nitrogen.

The principle of the analysis of such compounds is in general this: in *one portion* the carbon and the hydrogen are determined as carbonic acid and water respectively; in *another portion*, the nitrogen is determined either in the gaseous form, or as ammonium platinic chloride, or by determining volumetrically the ammonia formed from the nitrogen; the oxygen is calculated from the loss.

As the presence of nitrogen exercises a certain influence upon the estimation of carbon and hydrogen, we have here to consider not only the method of determining the nitrogen, but also the modifications which the presence of the nitrogen renders necessary in the usual method of determining the carbon and hydrogen.

a. Determination of the Carbon and Hydrogen in Nitrogenous Substances.

§ 183.

1. When nitrogenous substances are ignited with oxide of copper or with lead chromate, a portion of the nitrogen present escapes in the gaseous form, together with the carbonic acid and aqueous vapor; whilst another portion, minute indeed, still, in bodies abounding in oxygen, not quite insignificant, is converted into nitric oxide gas, which is subsequently transformed wholly or partially into nitrous acid by the air in the apparatus. The application of the methods described in §§ 174, etc., in the analysis of nitrogenous substances would accordingly give too much carbon; since the potash bulbs would retain, besides the carbonic acid, also the nitrous acid formed and a portion of the nitric oxide (which in the presence of potassa decomposes slowly into nitrous acid and nitrous oxide). This defect may be remedied by selecting a combustion tube about 12—15 cm. longer than those commonly employed, filling this in the usual way, but finishing with a loose layer, about 9—12 cm. long, of clean, fine copper turnings (§ 66, 6), or a compact roll of copper wire-gauze. The roll of copper gauze in front of the oxide should not be previously oxidized (as is recommended for substances free from nitrogen chlorine and

bromine), but should be in the metallic state*. The process is commenced by heating these copper turnings to redness, in which state they are maintained during the whole course of the operation. These are the only modifications required to adapt the methods above described for the analysis of nitrogenous substances. The use of the metallic copper depends upon its property of decomposing, when in a state of intense ignition, all the oxides of nitrogen into oxygen, with which it combines, and into pure nitrogen gas. As the metal exercises this action only when in a state of intense ignition, care must be taken to maintain the anterior part of the tube in that state throughout the process. As metallic copper recently reduced retains hydrogen gas, and, when kept for some time, aqueous vapor condensed on the surface, the copper turnings intended for the process must be introduced into the tube hot as they come from the drying closet (which is heated to 100°). v. LIEBIG recommends to compress the hot turnings in a tube into a cylindrical form, to facilitate their rapid introduction into the combustion tube.

2. If it is intended to burn nitrogenous bodies in the apparatus described in § 178, care must be taken to keep at least the anterior half of the roll from oxidizing, both during the ignition in the current of air and during the actual process of combustion. When the operation is terminated, and the oxidation of the metallic copper is visibly progressing, the oxygen is turned off, and the cock of the air gasometer opened a little instead, to let the tube cool in a slow stream of atmospheric air.

3. Since the metallic copper is usually oxidized during each combustion and must be reduced again, STEIN† uses silver instead of copper. Silver has the additional advantage that it retains also chlorine. According to the investigations of CALBERLA, silver at a red heat reduces oxides of nitrogen completely, while it does not exercise the least influence on carbonic acid.

b. DETERMINATION OF THE NITROGEN IN ORGANIC COMPOUNDS.

As already indicated, two essentially different methods are in

* The copper turnings or gauze cannot be replaced by the metallic powder obtained by the reduction of the oxide with hydrogen, as this obstinately retains hydrogen, and consequently decomposes appreciable quantities of carbonic acid with formation of carbonic oxide. Schrötter, Lautemann, Journ. f. prakt. Chem. 77, 316.

† Zeitschrift f. anal. Chem. 8, 83.

use for effecting the determination of the nitrogen in organic compounds; viz., the nitrogen is either separated in the pure form and its volume measured, or it is converted into ammonia, and this is determined either as ammonium platinic chloride, or volumetrically by neutralization.

α. *Determination of the Nitrogen from the Volume.*

§ 184.

aa. Dumas' Method, modified by Schiel.

This method may be employed in the analysis of all organic compounds containing nitrogen. It requires a graduated glass cylinder of about 200 c.c. capacity, with a ground-glass plate to cover it.

The combustion tube should be 60 or 70 cm. long, and drawn out at the posterior end to a stout open tail, which should have a small bulb or swell for the better fastening of a rubber tube to it. Introduce into it near the tail a plug of newly ignited asbestos,

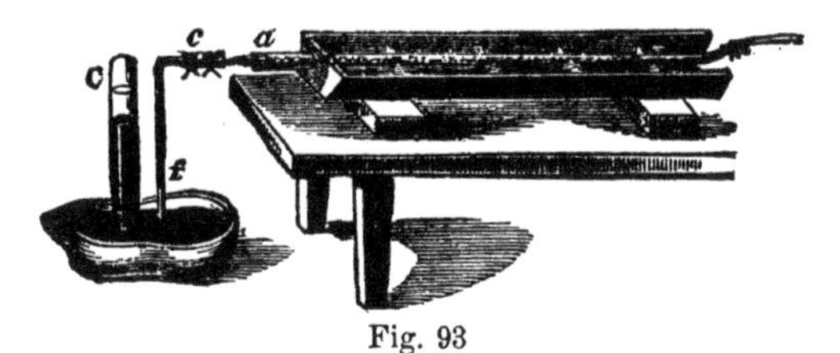

Fig. 93

then a layer of oxide of copper, 4 cm. long; after this the *intimate mixture* of an accurately weighed portion of the substance (0·3—0·6 grm., or, in the case of compounds poor in nitrogen, a somewhat larger quantity) with oxide of copper, then the oxide which has served to rinse the mortar, followed by a layer of pure oxide, and, lastly, a layer of copper turnings, about 15 cm. long. Make a channel along the top of the tube by gentle tapping. Connect the tube with the bent delivery tube *c f* (fig. 93), and place in the furnace. Connect the tail by means of a *stout tube* of india-rubber with an apparatus for giving a continuous stream of washed carbonic acid gas. Transmit this slowly through the tube for half an hour, then immerse the end of the bent delivery tube under mercury, and invert over it a test tube filled with solution of potassa. If the gas bubbles entering the cylinder are completely absorbed by the solution of potassa, this is a proof that the air is thoroughly expelled from the tube. But should this not be the case, the evolution of carbonic acid must be continued until the desired point is attained. When the gas is completely absorbed, close the communication between the CO_2 generator and the combustion tube by a screw clamp or stopcock, invert the graduated cylinder, filled $\frac{2}{3}$ with mercury, $\frac{1}{3}$ with concentrated solution of potassa, over the end of the delivery tube, with the aid of a ground-glass plate,* and proceed with the combustion in the usual way, heating first the anterior end of the tube to redness, and advancing gradually towards the farther end. In the last stage of the process, communication is reëstablished with the CO_2 generator, and thus the whole of the nitrogen gas which still remains in the tube is forced into the cylinder. Wait now until the volume of the gas in the cylinder no longer decreases, even upon shaking the latter (consequently, until the whole of the carbonic acid has been absorbed), then place the cylinder in a large and deep glass vessel filled with water, the transport from the mercurial trough to this vessel being effected by keeping the aperture closed with a small dish filled with mercury. The mercury and the solution of potassa sink to the bottom, and are replaced by water. Immerse the cylinder, then raise it again until the water is inside and outside on an exact level; read off the volume of the gas and mark the temperature of the water and the state of the barometer; calculate the weight of the nitrogen gas from its volume, after reduction to the normal temperature and pressure, and with due regard to the tension of the aqueous vapor (comp. "Calculation of Analyses"). The results are generally somewhat too high, viz., by about 0·2—0·5 per cent.; this is owing to the circumstance that even long-continued transmission of carbonic acid through the tube fails to expel every trace of atmospheric air adhering to the oxide of copper.

* The following is the best way of filling the cylinder and inverting it over the opening of the bent delivery tube:—The mercury is introduced at first, and the air-bubbles which adhere to the walls of the vessel are removed in the usual way. The solution of potassa is then poured in, leaving the top of the cylinder free to the extent of about 2 lines; this is cautiously filled up to the brim with pure water, and the ground-glass plate slided over it. The cylinder is now inverted, and the opening placed under the mercury in the trough; the glass plate is then withdrawn from under the cylinder. In this manner the operation may be performed easily, and without soiling the fingers.

It is highly advisable, before making any nitrogen determinations with this method, to subject a non-nitrogenous substance, *e.g.*, sugar, to the same process. The analyst thereby acquaints himself with the extent of the error to which he will be exposed. In such an experiment the quantity of unabsorbed gas should not exceed 1 or 1½ c.c.

To insure complete combustion of difficultly combustible bodies, STRECKER recommends the addition of arsenious oxide in powder to the oxide of copper with which the substance is to be mixed; the arsenious oxide is volatilized by the action of the heat, the fumes burning the whole of the carbon like a current of oxygen. The arsenious oxide sublimes in the anterior part of the tube, arsenic remains in the copper.

bb. By exhaustion of the combustion tube with an air pump, and measurement of nitrogen in Schiff's Azotometer.

A process capable of giving much more accurate results than the preceding (*aa*) has been developed by FRANKLAND and ARMSTRONG,* GIBBS † and JOHNSON. It is described ‡ as follows:

REAGENTS.

Cupric oxide.—"Copper scale," which may contain cuprous oxide, coal dust, oil, &c., is mixed in an iron pot with 10 per cent. of potassium chlorate and enough water to make a thin paste. The mass is heated and stirred till dry, the heat is then raised to the point of ignition, and until the mass does not glow nor sparkle when stirred.

The potassium chloride is washed out by decantation and the cupric oxide is dried and moderately ignited.

Metallic copper.—Granular copper oxide, or fine copper gauze, is suitable for its preparation. The granular copper is most convenient; copper gauze must be made into rolls adapted to the combustion tube. The copper is reduced and cooled as usual in a stream of hydrogen.

Potassium chlorate.—Commercial potassium chlorate is fused in porcelain and pulverized.

Sodium bicarbonate must contain no organic matter.

* Jour. Chem. Soc. [ii], vol. vi. p. 77.
† Am.Journ.Sci. and Arts, vol. xlviii.
‡ By JOHNSON and JENKINS, American Chemical Journal, ii. 27.

Solution of Caustic Potash.—Dissolve commercial "stick potash" in less than its weight of water, making a solution so concentrated that, on cooling, it deposits crystals of potassium hydrate.

The same clear solution may be used for a number of combustions or until the absorption of carbonic acid gas is not quite prompt.

APPARATUS.

The Combustion tube should be of the best hard Bohemian glass, about 2 feet 4 inches long. The rear end is bent and sealed as in fig. 96.

It is best to protect the horizontal part with thin copper foil. The tube is connected with the pump by a close fitting rubber cork, smeared with glycerine.

Azotometer.—This is a modification of the apparatus invented and described by Schiff, *Fres. Zeitschrift*, Bd. 7, p. 430.

It is represented in fig. 94.

The gas is measured in an accurately calibrated cylinder (burette) A of 120 c. c. capacity, graduated to fifths of cubic centimetres, and closed at the upper end by a glass stopcock. The lower end is connected, by means of a perforated rubber stopper about $1\frac{3}{4}$ inches long and $1\frac{1}{2}$ inches diameter, with another tube having two arms, one, D, to receive the delivery tube from the pump, the other connected by a rubber tube with a bulb of 200 c. c. capacity, F, through which potash solution is supplied. The graduated tube is enclosed in a water-jacket with an external diameter of about $1\frac{3}{4}$ inches. Its lower end is closed by the caoutchouc stopper that connects the two parts of the azotometer described above. The upper end of the jacket is closed by a thin rubber disc, slit radially and having four perforations: one in the centre, through which the neck of the graduated tube passes, and three others near the circumference.

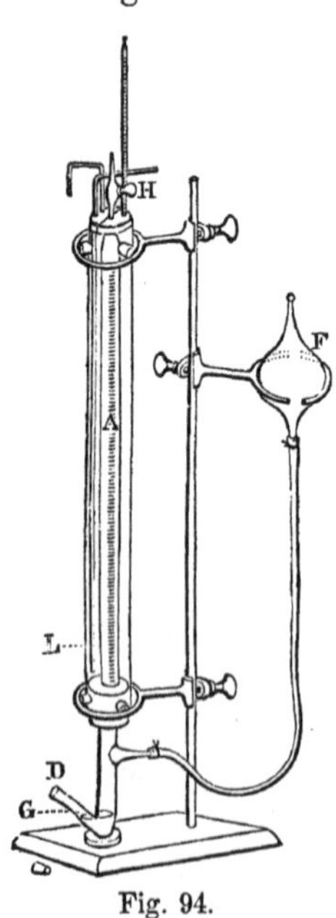

Fig. 94.

Through one of the latter, a glass tube, L, bent as in the figure, reaches to the bottom of the jacket, another short tube just passes through the disc, and the third hole is for supporting a thermometer. The azotometer is held upright and firm on a stand by rings fitting around the jacket and by cork wedges.

The bulb for potash solution rests in a slotted, sliding ring.

The air pump used is the Sprengel mercury pump, modified merely so as to be easily constructed and durable. Its essential parts are sketched in fig. 95. Some of them are exaggerated in order to show their construction more plainly. Through a rubber stopper wired into the nozzle of the mercury reservoir, A, passes a glass tube, B, 4 inches long; this connects by a caoutchouc tube with the straight tube D, 3 feet long. The rubber tube E, 6 inches long, connects D with a straight glass tube, F, of about the same length as D.

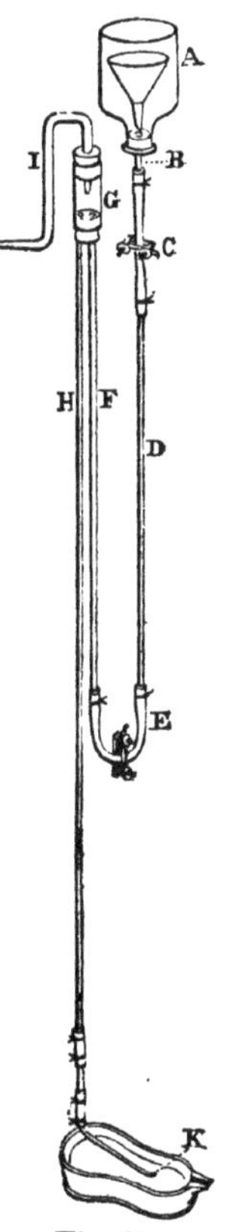

Fig. 95.

G is a piece of combustion tube $1\frac{1}{2}$ inches long, closed below by a doubly perforated soft rubber stopper admitting the tubes F and H, and above by a singly perforated rubber stopper into which a tube, I, is fitted. The tube H has a length of 45 inches. At the bottom it is connected by rubber with a straight tube of 3 inches, and this again with a tube, K, of 7 inches. The tubes H K should have an internal diameter of $1\frac{1}{2}$ millimetres, F may be 2 millimeters, and D still larger.

We have used for H and F slender Bohemian glass tubes of 4 millimetres exterior diameter. Their elasticity compensates for their slenderness. If heavy barometer tubes be used, the stoppers and G must be of correspondingly larger dimensions.

The joints at G must be made with the greatest care.

It is best to insert the lower stopper for half its length into G, having the dimensions of the parts so related that it requires considerable effort to force the slightly greased tubes F and H to their places just through the stopper. The tube I must be of *stout glass*—a decimetre in diameter. It is drawn out at either end to a long taper, and bent as in the figure, in order to bring its free extremity to the level of the combustion furnace. The hole in the upper rubber stopper has a diameter of 5 mm., just sufficient to admit the narrowed end of the tube, which, after greasing or moistening with glycerine, is "screwed down" into the stopper.

These three joints are the only ones belonging to the pump which have to resist diminished pressure, and require extreme care in making.

If not entirely secure they are to be trapped with glycerine. For this purpose it is needful to pass F and H through a stopper of half an inch greater diameter than G and correspondingly perforated before entering the latter. Then, previous to inserting I, a tube 4 inches long is slipped over G upon this wider stopper. When I has been inserted and the tubes have been secured to their support, the space between G and the outer tube is filled with the most concentrated glycerine, which is prevented from absorbing moisture by corking above.

The two rubber tubes are both provided with stout screw clamps, to admit of exactly regulating the flow of mercury. The tubes D, F, H, and I are secured to a vertical plank framed below into a heavy horizontal wooden foot on which rests the mercury trough, and having above a horizontal shelf through an aperture of which passes the neck of A.

The tubes D, F, H, and I are secured to the plank at several points by wooden or cork clamps, clasping the tubes and fastened by screws or wires.

These fastenings are made elastic by the intervention of a thick rubber tube between the glass and wood. The connections C and E should be made of stout vulcanized rubber, those at the base of H K of fine black rubber.

The latter should be soaked in melted tallow previous to use, all excess being carefully removed from the interior. The joints should be wound with waxed silk.

A glass funnel is placed within A to prevent spattering of the mercury when it is filled.

OPERATION.

From 3 to 4 grams of potassium chlorate, according to the amount of carbon to be burned, are put into the tail of the combustion tube, fig. 96, followed by an asbestos plug just at the bend. The substance to be analyzed (0·6—0·8 grams) is well mixed in a mortar with enough cupric oxide that has been *freshly ignited* and allowed to cool to make a layer 11 or 12 inches long in the tube. The mixture is introduced through a funnel and rinsed with enough cupric oxide to make a layer of 3 inches, a second asbestos plug, and upon it a layer of reduced copper of 4 or 5 inches long are put in, then a third asbestos plug, then 2 inches of cupric oxide, a fourth asbestos plug, then ·8 to 1· grams of sodium bicarbonate. The remaining space in the tube is loosely filled with asbestos, to absorb the water which is formed during combustion,

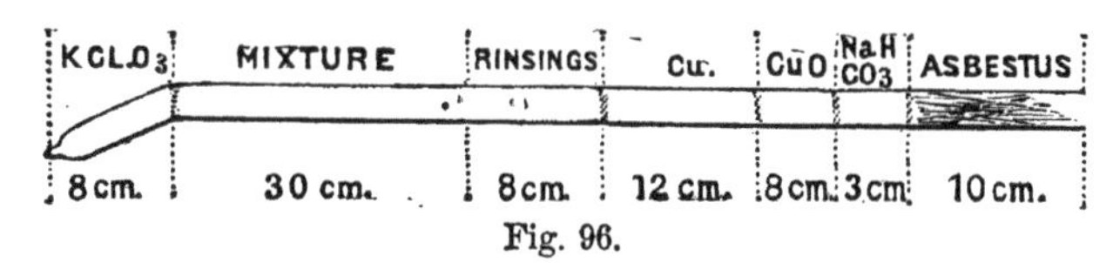

Fig. 96.

and prevent it from flowing back upon the heated glass. The anterior part of the tube containing the cupric oxide and reduced copper is wound with copper foil, leaving, however, a little of the copper (Cu. in fig. 96) visible at its rear. The combustion tube is placed in the furnace at the bend of the tube I, and connected with the latter by a close-fitting rubber stopper smeared with glycerine.

Care must be taken to make the joint perfectly tight. The combustion tube has its conical rubber stopper partly inserted, and is then forced and rotated upon the tapering and stout end of the tube I, the latter being supported by one hand applied at the lower bend.

PREPARATION OF THE AZOTOMETER.

Fill the bottom of the azotometer to about the level indicated by the dotted line G, with mercury. Close the arm D securely with a rubber stopper. Grease the stop-cock H and insert the plug, leaving the cock open.

Pour potash solution into F till A is nearly full, and there is still some solution in the bulb F. Raise the bulb cautiously with one hand, holding the stop-cock H in the other hand. When the solution in A has risen *very nearly* to the glass cock, close the latter, avoiding contact of the alkali with the ground glass bearings. Replace the bulb in the ring and lower it as far as may be. If the level of the solution in the azotometer does not fall in 15 or 20 minutes, it is tight. Place the delivery tube of the pump K in a mercury trough.

Supply the vessel A with at least 500 c. c. of mercury. Cautiously open the clamps C and E. If the mercury does not start at once pinch the rubber at E repeatedly. The mercury should flow nearly as fast as it can be discharged at K, without filling the cylinder G. Five to ten minutes working of the pump will generally suffice to make a complete exhaustion of the combustion tube. If most of the mercury runs out before exhaustion is complete, close the clamp C, return the mercury to A, and repeat the operation. When there is a complete exhaustion, the mercury falls with a rattling or clicking sound. After it has been distinctly heard for half a minute, close the clamp C. If the mercury column in H remains stationary for some minutes, the connections are proved to be tight.

ADJUSTING THE AZOTOMETER.

Remove the mercury trough, placing K in a capsule.

Heat the part of the tube containing sodium bicarbonate. Water vapor and carbon dioxide are evolved, which fill the vacuum in H and expel the mercury. While this is being done place the azotometer near by, remove the bulb F from the ring and support it in a box near the level of D, so that the stopper may be removed from D without greatly changing the level of the mercury G, and so that the azotometer can be moved freely without disturbing it. When the cork in D has been removed fill D half full or more with water.

As soon as the mercury has fully escaped from K insert the latter in D. Let a few bubbles escape through the water and then pass the tube K down so that the escaping gas enters the azotometer. It will much facilitate the delivery of gas if the extremity of the tube K just touches the inside of the azotometer tube, and is kept, as near as possible, to the surface of the mercury.

The carbon dioxide is absorbed in passing through the caustic potash solution. In spite of all precautions very minute bubbles of permanent gas will occasionally ascend, but, as will be seen on observing the amount of potash solution thus displaced, the error thereby occasioned is extremely small.

THE COMBUSTION.

First heat the anterior cupric oxide to full redness, and afterwards the copper. The fine gauze or pulverulent copper very completely reduces any oxides of nitrogen which might be produced in the combustion, and also retains any excess of oxygen which is evolved at the close of the process.

The anterior cupric oxide burns the traces of hydrogen which may be held by the reduced copper, even when the tube is exhausted, and also destroys the carbon monoxide which is usually formed when steam and carbon dioxide pass together over reduced copper, if iron or carbon be present. Go on with the combustion as usual, bringing the heat up to a fair redness. The flow of gas may be made quite rapid, say one bubble a second, or a little faster.

When the horizontal part of the tube has all been heated, and the evolution of gas has nearly ceased, heat the potassium chlorate so that it boils vigorously from evolution of oxygen. The reoxidization of the reduced copper oxide and of any unburned carbon proceeds rapidly.

When the oxygen, whose flow admits of easy regulation, begins to attack the anterior layer of reduced copper, stop its evolution and lower the flames all along the tube, keeping the reduced copper still faint red.

After a few minutes start the pump, slowly at first, having some vessel under the tube D of the azotometer to receive the mercury. A few minutes pumping suffices to clear the tube. Remove the azotometer, close the tube D with its rubber stopper, then raise the bulb into its ring to such a height that the potash solution in it shall be at about the same level as that in the graduated tube. Connect L at its upper end with a water supply, insert a thermometer in the top of the water jacket and let the water run, until the temperature and the volume of gas are constant.

Read off the volume of gas and temperature, after having accurately adjusted the level of the solution in the bulb to that in the azotometer.

Read the barometer and make the calculations in the usual way. When 50 per cent. potash solution is used, no correction need be made for tension of aqueous vapor, as SCHIFF has shown.

The calculation is somewhat shortened by the use of the table in Jour. of Chem. Soc., Vol. XVIII. (1865) p. 212.

Very fair results are got by employing, with suitable precaution, a stream of carbon dioxide to displace the air of the combustion tube, but the process is very tedious, the sources of error are more numerous, and the results are apt to be higher and not so accordant as when the mercury pump is used to evacuate the tube.

The pump above described has been in use for eighteen months without any repairs, and by its help two or even three analyses may be performed in a day.

β. Determination of Nitrogen by conversion into Ammonia.

VARRENTRAPP and WILL'S Method.

§ 185.

This method may be applied to all nitrogenous compounds, except those containing the nitrogen in the form of nitric acid, hyponitric acid, &c.* It is based upon the same principle as the method of examining organic bodies for nitrogen (§ 172, 1, *a*), viz., upon the circumstance that, when nitrogenous bodies are ignited with an alkali hydroxide, the latter is decomposed, yielding water, the oxygen of which combines with carbon to CO_2, which remains in combination with the alkali as carbonate, whilst the hydrogen at the moment of its liberation combines with the whole of the nitrogen present to form ammonia.

In the case of substances abounding in nitrogen, such as uric acid, mellon, &c., the whole of the nitrogen is not at once converted into ammonia in this process; a portion of it combining with part of the carbon of the organic matter to cyanogen, which then combines, either in that form with the alkali metal, or in the form of cyanic acid with the alkali. Direct experiments have proved, however, that even in such cases the whole of the nitrogen is ultimately obtained as ammonia, if the alkali hydroxide is present in excess, and the heat applied sufficiently intense.

As in all organic nitrogenous compounds the carbon preponderates over the nitrogen, the oxidation of the former, at the expense of the water, will invariably liberate a quantity of hydrogen more than sufficient to convert the whole of the nitrogen present into ammonia; for instance,

$$CN + 2H_2O = CO_2 + NH_3 + H.$$

[* Vegetable matters, as dried plants, containing not more than 3 per cent. of NO_5 may be analyzed by this method. In a case where 6 per cent. of N_2O_5 was present, a loss of 0·2 per cent. of N took place in the experiments of E. Schulze.—Fres. Zeitschrift vi. 387.]

The excess of the liberated hydrogen escapes either in the free state, or in combination with the not yet oxidized carbon, according to the relative proportions of the two elements and the temperature, as marsh gas, olefiant gas, or vapor of readily condensible hydrocarbons, which gases serve in a certain measure to dilute the ammonia. As a certain dilution of that product is necessary for the success of the operation, I will here at once state that substances rich in nitrogen should be mixed with more or less of some non-nitrogenous body—sugar, for instance—so that there may be no deficiency of diluent gas.

The ammonia is determined volumetrically, see § 196.

aa. Requisites.

1. The objects enumerated § 174, and a Porcelain Mortar for mixing the weighed substance.

2. A Combustion Tube of the kind described § 174, 3; length about 40 cm., width about 12 mm. The combustion is effected in an ordinary combustion furnace.

3. Soda-Lime (§ 66, 5).—It is advisable to gently heat in a platinum or porcelain dish, a quantity of the soda-lime sufficient to fill the combustion tube, so as to have it perfectly dry for the process of combustion. In the analysis of non-volatile substances, the best way is to use the soda-lime while still warm.

4. Asbestos.—A small portion of this substance is ignited in a platinum crucible previous to use.

5. A Verrentrapp and Will's Bulb Apparatus.—This may be obtained from the shops. Fig. 97 shows its form. It is filled to the extent indicated in the drawing with standard sulphuric or

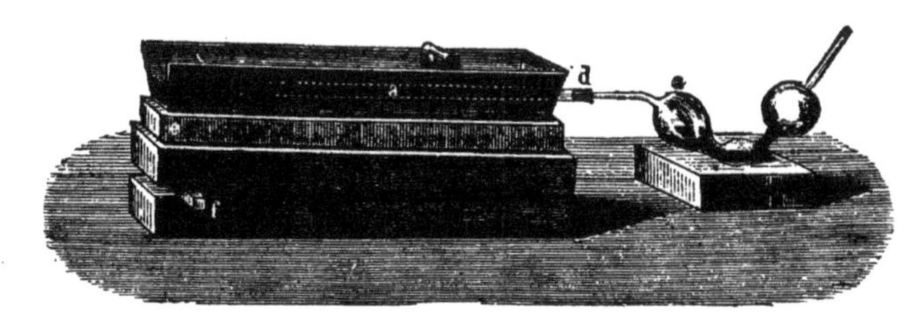

Fig. 97.

hydrochloric acid § 192, of which 20 c.c. should be employed. The acid is introduced either by dipping the point into the acid, and applying suction to *d*, or by means of a burette.

In order to guard against the receding of the acid into the combustion tube, ARENDT and KNOP have suggested the form indicated fig. 98.

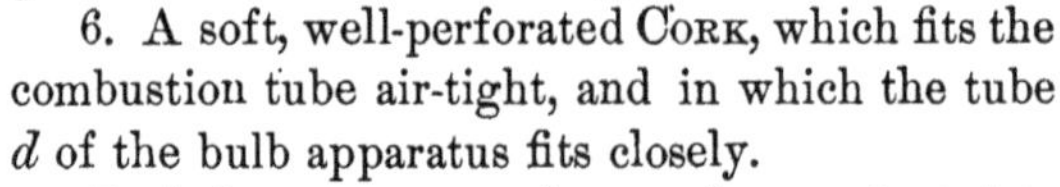

Fig. 98.

6. A soft, well-perforated CORK, which fits the combustion tube air-tight, and in which the tube *d* of the bulb apparatus fits closely.

7. A SUCTION-TUBE of caoutchouc adapted to the point of the bulb apparatus.

bb. The Process.

The combustion tube is half filled with soda-lime, which is then gradually transferred to the perfectly dry, and, if the nature of the substance permits, rather warm mortar, where it is most intimately mixed with the weighed substance, forcible pressure being carefully avoided; a layer of soda-lime, occupying about 3 cm., is now introduced into the posterior part of the combustion tube, and the mixture filled in after; the latter, which will occupy about 20 cm., is followed by a layer of about 5 cm. of soda-lime, which has been used to rinse the mortar, and this again by a layer of 12 cm. of pure soda-lime, leaving thus about 4 cm. of the tube clear. The tube is then closed with a loose plug of asbestos, and a free passage for the evolved gases formed by a few gentle taps; it is then connected with the bulb apparatus by means of the perforated cork, and finally placed in the combustion furnace (see fig. 97).

To ascertain whether the apparatus closes air-tight, some air is expelled by holding a piece of red hot charcoal to the bulb *a*, and the apparatus observed, to see whether the liquid will, upon cooling, permanently assume a higher position in *a* than in the other limb. The tube is then gradually surrounded with ignited charcoal, commencing at the anterior part, and progressing slowly towards the tail, the operation being conducted exactly as in an ordinary combustion (§ 175). Care must be taken to keep the anterior part of the tube tolerably hot throughout the process, since this will almost entirely prevent the passage of liquid hydrocarbons, the presence of which in the standard acid would be inconvenient. The asbestos should be kept sufficiently hot to guard against its retaining water, and with this, ammonia. The combustion should be conducted so as to maintain a steady and uninterrupted evolution of gas; there is no fear of any ammonia escaping unabsorbed, even if the evolution is rather brisk; but the operator must constantly be on his guard against the receding of the acid, which takes place the moment the evolution of gas ceases, and this, in some instances, with such impetuosity as to force the acid into the combustion tube, which of course spoils the whole analysis. This difficulty may be readily met, however, by mixing with the substance an equal quantity of sugar, which will give rise to the evolution of more permanent gases diluting the ammonia.

When the tube is ignited in its whole length, and the evolution of gas has *totally* ceased,* the point of the combustion tube is broken off, and air to the extent of several times the volume of the gas in the tube is sucked through the apparatus, to force all the rest of the ammonia into the acid.

Liquid nitrogenous compounds are weighed in small sealed glass bulbs, and the process is conducted as directed § 180, with this difference, that soda-lime is substituted for oxide of copper. It is advisable to employ tubes of greater length for the combustion of liquids than are required for solid bodies. The best method of conducting the operation, is to heat first about one-third of the tube at the anterior end, and then to force the liquid from the bulbs into the tube by heating the hinder end of the latter; the expelled liquid will thus become diffused in the central part of the tube, without being decomposed. By a progressive application of heat, proceeding slowly from the anterior to the posterior end, a steady and uniform evolution of gas may be easily maintained.

When the combustion is terminated, the bulb apparatus is emptied, through the opening at the point, into a beaker, and rinsed with water until the rinsings cease to manifest acid reaction.

The excess of acid is determined by means of standard potash or ammonia solution and cochineal tincture, or, if the acid is so colored that the point of neutralization cannot readily be decided by cochineal, employ slips of turmeric paper (see § 196).

It is advantageous to use a rather dilute acid, 1 c. c. = 0·005 grm. of nitrogen. The receiver (fig. 99) may be advantageously substituted for the bulb-tube. The tube *a*—previously provided with the caoutchouc stopper *b*—is first connected by the aid of a good cork with the combustion tube, and then the U-tube *c*—having been charged with the proper quantity of acid from a MOHR's burette —is added. At the termination of the combustion, when air has

* This is indicated by the white color which the mixture reassumes when all the carbon deposited on the surface is oxidized.

been drawn through the apparatus, the tube *a* is rinsed into the apparatus *c*, some tincture of cochineal added, and standard alkali run into the tube from a second burette, until the acid is almost neutralized. Now pour the contents of the apparatus into a beaker, rinse with water, and complete the neutralization. With this receiver neither receding nor spirting is possible. By not pouring out the fluid till the point of saturation is nearly attained, you require less water for rinsing the tube. This method is rapid and accurate.

Fig. 99.

[From the results of a critical investigation of this method by JOHNSON and JENKINS* the following facts may be here added:

1. The efficiency of the "soda-lime" mixture described § 66, 5, is fully confirmed. It is easier to prepare than the mixture of caustic lime and soda (§ 66,4) formerly used for this purpose, and does not, like the latter, attract moisture readily from the air, and is not liable to swell and choke the tube during combustion.

2. Neither the highest heat possible to obtain in an ERLENMEYER gas combustion furnace, nor a long layer of strongly heated soda-lime, nor these two conditions united, occasion any appreciable dissociation of the ammonia formed in combustion.

3. A suitable length of the anterior layer of soda-lime must be secured in order to get a good result. With 0·5 gram of substances, such as are encountered in agricultural chemistry, containing less than 8 per cent. of nitrogen, a glass tube of 12 to 14 inches is long enough. As the content of nitrogen increases to 10 per cent or over, the tubes should be made several inches longer. In the combustion of dried blood or egg-albumin a tube 20—25 inches long is preferred, and the mixture of soda-lime and substance should occupy rather less than half the tube, a layer of pure soda-lime of 12 or more inches long being essential for perfectly destroying the volatile organic matters.

4. The long anterior layer of pure soda-lime must be brought to a *full red heat* before heating the mixture, and must be so kept throughout the combustion.

5. No fumes or tarry matters, indicative of incomplete combustion, should appear in bulb-tube or receiver.

* Report of Connecticut Agr. Exp. Station, 1878, p. 111.

6. When the combustion proper is begun under the conditions above described, it can be carried on quite rapidly until completed. The contents of the tubes then show no sign of unburned carbon.

7. Equally good results are obtained whether the mixture is made intimately in a mortar, or more roughly by stirring with a spatula in a metallic capsule or scoop, or by mixing in the tube with a wire.]

Iron gas tubes may be substituted for glass tubes. They are closed at the rear with a cork, carrying a bit of glass tube drawn out to a sealed tail. The mixture is confined to its place by loose asbestos plugs. The corks are kept from charring by wrapping the end of the tube with two or three thicknesses of filter-paper, which is kept wet by a wash-flask, or by dipping the depending end into a vessel of water. The tubes should be 45 cm. long, and 5 cm. at each end should project from the fire and be protected with wet paper.

C. Analysis of Organic Compounds containing Sulphur.*

§ 186.

The usual method of determining the carbon in organic bodies —viz., by combustion with oxide of copper or lead chromate— would give results too high in the analysis of compounds containing sulphur, since—more especially if oxide of copper is used—a portion of the sulphur would be converted in the process into sulphurous acid, which would be absorbed with the carbonic acid in the potash bulbs. Carius recommends to burn substances containing sulphur in a tube 60—80 cm. long, with lead chromate, care being taken that the anterior 10—20 cm., which contains pure lead chromate, are never heated above low redness. The lead chromate may be used again three or four times without refusion; and, finally, if treated by Vohl's method (p. 124), it is just as fit for use as if it had not been employed for the combustion of a substance containing sulphur.

The presence of sulphur demands no modification in the process described §§ 184 and 185 for the determination of nitrogen. In substances containing oxygen in presence of sulphur, the oxygen is estimated from the loss.

[* Warren's method of determining carbon, hydrogen, and sulphur in one operation is described in Am. Journ. Sci., vol. 41, 2d ser., p. 40.]

As regards the *estimation of the sulphur* in organic compounds, that element is invariably weighed in the form of barium sulphate, into which it may be converted either in the dry or in the wet way.

a. Methods in the Dry Way.

1. *Method suitable, more particularly, to determine the sulphur in non-volatile Substances poor in Sulphur, e.g., in the so-called Protein Compounds* (v. LIEBIG).

Put some lumps of potassa, free from sulphuric acid (§ 66, 7, *c*) into a capacious silver dish, add $\frac{1}{8}$ of pure potassium nitrate, and fuse the mixture, with addition of a few drops of water. When the mass is cold, add to it a weighed quantity of the finely pulverized substance, fuse over the lamp, stir with a silver spatula, and increase the heat, continuing the operation until the color of the mass shows that the carbon separated at first has been completely consumed. Should this occupy too much time, you may accelerate it by the addition of potassium nitrate in small portions. Let the mass cool, then dissolve in water, supersaturate the solution with hydrochloric acid in a capacious beaker covered with a glass dish, and precipitate with barium chloride. Wash the precipitate well with boiling water, first by decantation, then on the filter. Dry and ignite. Treat the ignited barium sulphate as directed p. 367; if this latter operation is omitted, the result is almost always too high.

A suitable alcohol lamp is preferable to a gas flame, since the latter may communicate sulphur to the fused mass. As it is by no means easy to obtain the required reagents perfectly free from sulphur, it is well to try a parallel experiment, using the same quantities of each that is used for the analysis, and if an appreciable amount of barium sulphate is obtained, make the necessary correction in the analysis.

2. *Method adapted more particularly for the Analysis of non-volatile or difficultly volatile Substances containing more than* 5 *per cent. of Sulphur* (KOLBE *).

Introduce into the posterior part of a straight combustion tube,† 40—45 cm. long, a layer, 7—8 cm. long, of an intimate mixture of 8 parts of pure anhydrous sodium carbonate, and 1 part of pure

* Supplemente zum Handwörterbuch, 205.
† Sealed and rounded at the end like a test tube.

potassium chlorate; after this introduce the weighed substance, then another layer, 7 or 8 cm. long, of the same mixture; mix the organic compound intimately with the sodium carbonate and potassium chlorate, by means of the mixing wire (fig. 78, p. 613); fill up the still vacant part of the tube with anhydrous sodium carbonate or potassium carbonate mixed with a little potassium chlorate. Clear a *wide* passage from end to end by a few gentle taps, place the tube in a combustion furnace, heat the anterior part to redness, and then, progressing slowly toward the posterior part, proceed to surround with red-hot charcoal the part occupied by the mixture. In the analysis of substances abounding in carbon, it is advisable to introduce into the posterior part of the tube a few lumps of pure potassium chlorate, to insure complete combustion of the carbon, and perfect conversion into sulphates of the compounds of potassa with the lower oxides of sulphur that may have formed. The sulphuric acid in the contents of the tube is determined as in 1.

3. *Method adapted for the Analysis both of non-volatile and volatile Substances, but more especially the latter* (DEBUS *).

Dissolve 149 parts of potassium dichromate purified by recrystallization, and 106 parts anhydrous sodium carbonate in water, evaporate the solution to dryness, reduce the lemon-colored saline mass to powder, heat to intense redness in a Hessian crucible, and transfer still hot to a filling-tube (fig. 77, p. 613).† When the powder is cold, introduce a layer of it, 7—10 cm. long, into a common combustion tube; then introduce the substance, and after this another layer, 7—10 cm. long, of the powder. Mix intimately by means of the mixing wire, then fill the still unoccupied part of the tube with the saline mixture, and apply heat as in an ordinary ultimate analysis. When the entire mass is heated to redness, conduct a slow stream of dry oxygen gas over it for $\frac{1}{2}$—1 hour. When cold, wipe the ash off the tube, cut the latter into several pieces over a sheet of paper, and treat them in a beaker with a sufficient quantity of water to dissolve the saline mass. Add hydrochloric acid in tolerable excess, then some alcohol, and apply a

* Annal. d. Chem. u. Pharm. 76, 90.

† The saline mass must always first be tested for sulphur. For this purpose a small portion of it is reduced with hydrochloric acid and alcohol, barium chloride added, and the mixture allowed to stand 12 hours at rest. No trace of a precipitate should be discernible.

gentle heat until the solution shows a beautiful green color; filter off the chromic oxide produced by the combustion (this contains sulphuric acid); wash first with water containing hydrochloric acid, then with alcohol, dry, and transfer to a platinum crucible; add the filter-ash, mix with 1 part of potassium chlorate and 2 parts of potassium (or sodium) carbonate, and ignite until the chromic oxide is completely converted into alkalin chromate. Dissolve the fused mass in dilute hydrochloric acid, and reduce by heating with alcohol; add the solution to the fluid filtered from the chromic oxide, heat the mixture to boiling, and precipitate the sulphuric acid with barium chloride. DEBUS's test-analyses were very satisfactory; thus he obtained 99·76 and 99·50 of sulphur for 100, again 30·2 of sulphur in xanthogenamide for 30·4, &c.

4. *Method equally adapted for the Analysis of Solid and Liquid Volatile Compounds.* (W. J. RUSSELL;* suggested by BUNSEN.)

Introduce into a combustion tube, 40 cm. long, sealed at the posterior end, first 2—3 grm. pure mercuric oxide, then a mixture of equal parts of mercuric oxide and pure anhydrous sodium carbonate, mixed with the substance, and fill up the tube with sodium carbonate mixed with a little mercuric oxide. Connect the open end of the tube with a gas delivery tube dipping under water, to effect the condensation of the mercurial fumes. Place a screen in front of the part of the tube occupied by the substance, then heat the anterior part to bright redness, and maintain this temperature during the entire process. At the same time, heat another portion of the tube, nearer to the end, but not to the same degree of intensity, so that there may be alternate parts in the tube in which the mercuric oxide is left undecomposed. When the part before the screen is at bright redness, remove the screen, heat the mixture containing the substance, regulating the application of heat so as to insure complete decomposition in the course of 10—15 minutes, and heat at the same time the still unheated parts of the tube, and lastly also the pure oxide of mercury at the extreme end. The gas must be tested from time to time, to ascertain whether it contains free oxygen. Dissolve the contents of the tube in water, add some mercuric chloride to decompose the sodium sulphide which may have formed, acidify with hydrochloric acid, oxidize the

* Quart. Journ. Chem. Soc. 7, 212.

mercuric sulphide which may have formed with potassium chlorate, and finally precipitate the sulphuric acid with barium chloride. W. J. RUSSELL obtained by this method very satisfactory results in the analysis of pure sulphur, potassium sulphocyanate, and carbon disulphide.

*b. Method in the Wet Way.**

According to RIVOT, BEUDANT, and DAGUIN,† the sulphur in organic compounds may be readily determined by heating with pure solution of potassa, adding 2 volumes of water and conducting chlorine into the fluid. When the oxidation is effected, the solution is acidified and freed from the excess of chlorine by application of heat, then filtered, and the filtrate precipitated by barium chloride. Mr. C. J. MERZ, in my laboratory, has employed both this method and v. LIEBIG'S (*a*, 1) in the analysis of fine horn shavings. This process appears convenient and exact.‡

Substances leaving an ash on incineration, and which may therefore be presumed to contain sulphates, are boiled with hydrochloric acid; the solution obtained is filtered, and the filtrate tested with barium chloride. If a precipitate of barium sulphate forms, the sulphur contained in it is deducted from the quantity found by one of the methods described above; the difference gives the quantity of the sulphur which the analyzed substance contains in organic combination.

[*c. Methods depending on Combustion in a Current of Oxygen.*

When organic compounds containing sulphur are burned in a current of oxygen gas in a combustion tube, usually not only SO_3 or sulphuric acid, but SO_2 also is formed. Additional means of completing the oxidization of the sulphur and absorbing the sulphuric acid are required. Several methods have been proposed and used for attaining the desired end. C. M. WARREN § conducts the products of combustion over heated lead dioxide, thus obtaining

[* For the excellent processes of Carius, see Annal. d. Chem. u. Pharm. 116, 11.]

† Comp. rend. 1853, 835; Journ. f. prakt. Chem. 61, 135.

‡ Two experiments were made with each method, on horn dried at 100°. The percentages obtained were as follows: By v. Liebig's method, 3·37 and 3·345; by the present method, 3·31 and 3·33.

§ Zeitschr. f. anal. Chem. 5, 169.

lead sulphate mixed with lead dioxide. BRÜGLEMANN * conducts the products of combustion over quicklime (or soda-lime), obtaining calcium sulphate mixed with quicklime. MIXTER† uses bromine and water to complete the oxidization and absorb the sulphuric acid. SAUER‡ uses also bromine (a hydrochloric solution) for the same purpose.

These methods are applicable to all classes of organic compounds containing sulphur. The two last mentioned possess the

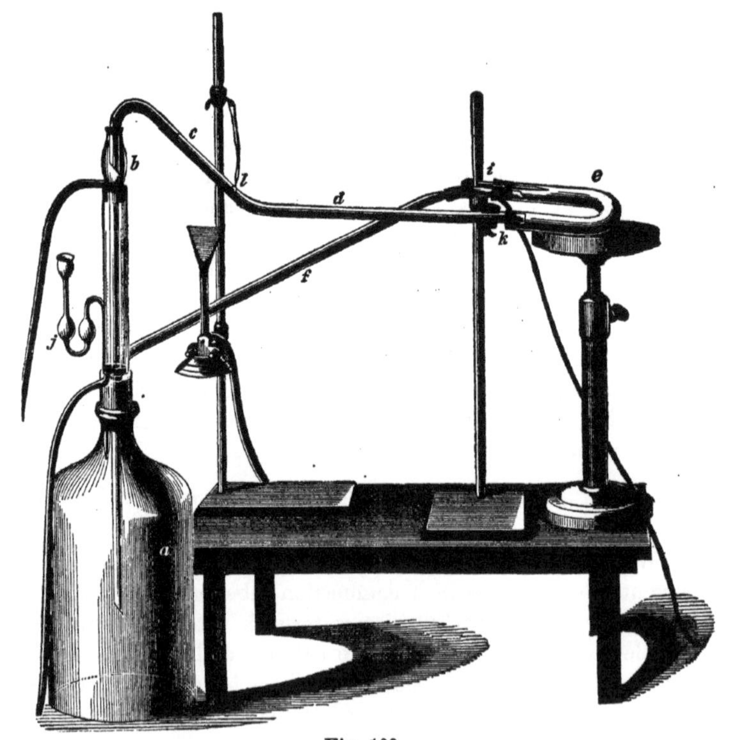

Fig. 100.

advantage that the sulphur is obtained as free sulphuric acid in a solution containing no fixed matter, and consequently in a condition to be easily and accurately determined.

---

* Zeitschr. f. anal. Chem. 15, 1, and 15, 175.

† American Jour. Sci. and Arts, iv. 90.

‡ Zeitschr. f. anal. Chem. 12, 32, and 12, 178.

1. MIXTER's method is described * as follows:

The apparatus (fig. 100) is designed to effect the combustion in a confined volume of gas; a device resorted to on account of the difficulty of completely condensing by liquid absorbents in U-tubes the dense white fumes of sulphuric acid produced by combustion. The bottle (*a*) has a capacity of from 4 to 10 litres, according to the amount of oxygen required. The neck should be large enough for a stopper 35 to 40 mm. in diameter. The condenser *b* is made of rather thin tubing 14 mm. in diameter; at the upper end it is expanded to a bulb in order to admit some motion to the tube *c d*. Below the bulb it is surrounded by a water-jacket 22 cm. high: from the point where it enters the stopper of the bottle it is narrowed somewhat for convenience of fitting. The combustion tube *c d* is made of hard glass of 12—15 mm. internal diameter; the portion *c* is 18 cm. from curve to curve, and is protected by a sheet-iron trough lined with asbestos; the part *d* is from 35 to 45 cm. in length. The wire attached at *l* is to sustain *c* in case *d* breaks; *c* is joined to *b* by a collar of black rubber. The U-tube *e* is connected with *d* by a rubber collar drawn over the latter at *k*; this U-tube is slightly inclined, that no liquid may run against the rubber connectors. The tube *f* connects *a* with *e*; it is narrowed at both ends to 10 mm. diameter. Near the upper end it is jointed by a piece of black-rubber tubing in order that the apparatus may be easily disconnected at *k*. The ends of *f* extend 2 cm. or more beyond the stoppers. Through the rubber stopper *i* a small glass tube passes beyond the end of *f*, where it is narrowed to an opening of 1 mm. The double bulb tube *j* is to accommodate variations of pressure, and to admit air as the original volume of gas diminishes during the combustion. The tubes *b*, *c*, *d*, and *f* should at no point have an internal diameter less than 8 mm.—10 mm. is preferable—and the narrowed ends should be cut obliquely that drops of water may not obstruct the circulation. The rubber stoppers and connections should be freed from adhering sulphur by heating in a solution of soda. The joints of the apparatus are sufficiently tight when water will stand in one limb of the safety tube.

The bottle *a* is filled over water with oxygen, and, if necessary, rinsed with distilled water; a few drops of bromine are poured in, the tubes adjusted, and a slow stream of water made to flow through

---

* American Jour. Sci. and Arts, iv. 90.

the water-jacket. The assay, if not volatile, is introduced into the tube *d* in a platinum tray,* which should not fill more than half the bore of *d*, leaving space enough for the free circulation of the oxygen. The part *c* is gradually heated and kept hot during the combustion. This hot inclined tube acts as a chimney; the heated gases rise in it, pass into the cold tube *b* and fall, thus causing a constant stream of gas to pass over the assay. It is important to ignite the assay without distilling off any considerable portion. To do this a small splinter of wood may be placed in contact with that part of the substance nearest *l*, or that end of the tray may hold a thin layer of the assay, which is heated as rapidly as safety allows by a lamp held in the hand. To insure a full supply of gas in the tube *d* at the commencement of the combustion, oxygen is passed from a gasometer through the tube *i* till the white fume which appears in the condenser *b* passes into *a*. The products of combustion being denser fall to the bottom of the bottle, and for a while displace the oxygen, thus increasing the circulation. After the substance is ignited, the fire passes to the other end of the tray. The part of the tube about the tray is heated by a lamp as is required to keep up the combustion. At the end of the operation the heat is increased. If drops of liquid collect in *c*, and are liable to run down to the hotter parts of the tube, they should be driven off by heat. If carbonic acid be the principal product of the combustion, there is little change in the volume of gases in the apparatus; but if water and sulphuric acid are formed in much quantity, the volume is diminished and air enters through the safety tube.

Most solid substances heated alone in the open tray yield volatile products too rapidly for entire combustion, but if mixed with sand in suitable proportion they burn slowly and completely. Liquids should be enclosed in narrow tubes sealed at one end and drawn out at the other to a capillary bore for two or three inches of length. Upon the point of the tube a bit of platinum sponge is fixed to assist the oxidation. The liquid should not fill more than two thirds of the wider part of the tube.

Before introducing very volatile substances, the 10 cm. of the

* A platinum tray which answers well may be made 10 to 20 cm. long, 10 mm. wide, and 7 to 10 mm. deep by bending thin foil over a glass tube. The ends may be roughly bent together or left open.

combustion tube *l d* should be heated to dull redness. Oxygen is passed in at *i*, the tubes are disjointed at *k*, and the tube holding the assay is then pushed in, till the platinum just reaches the heated zone. The apparatus being connected at *k*, slow volatilization of the liquid is effected by cautiously applying a flame under the empty portion of the tube containing the substance, so as to maintain the platinum sponge in a steady glow. As soon as a cloud of combustion-products appears in the vessel *a*, oxygen is shut off from *i*. When all the liquid has distilled from the interior tube, the tube *c d* is cooled slowly and the apparatus is left for two hours or until the fume has entirely subsided. If no odor of bromine be perceptible when the apparatus is disconnected at *k* to remove the tray or tube, a few drops of it should be poured through a funnel-tube put in the place of *j*, and the whole allowed to stand some time to insure complete oxidation of the sulphur-compounds and deposition of the sulphuric acid.

The tubes *d* and *e* are then rinsed into a beaker, this water is poured into *b*, which is then thoroughly washed by the aid of the wash-bottle; the large rubber stopper is lifted from the bottle and the lower part of *b* rinsed; without removing the tube *f* from the stopper, it is rinsed into a beaker, and finally the bottle is carefully washed. The solution obtained, which need not exceed 500 c. c., is evaporated to a small volume, filtered if necessary, and the sulphuric acid is determined by precipitation with barium chloride, observing all precautions mentioned in § 132, 1. In case the substance leaves an ash or residue in the tray, this must be dissolved in aqua regia, the nitric acid removed by evaporation with strong chlorhydric acid, and any sulphuric acid it may contain separated in the usual manner. In the use of this apparatus there is no danger from explosions if care be taken to have the combustion tube hot enough to ignite combustible vapor. Before attempting to burn a substance in the apparatus, it is best to try it in a large inclined tube open at both ends, or with oxygen supplied at the lower end. Such a preliminary trial will usually indicate the precautions necessary in burning the substance in the apparatus.

For the determination of sulphur in substances rich in sulphur, ·5 to ·75 grm., requiring about 4 litres of oxygen, may be used. When but little sulphur is present, a combustion of 2· grms. may be effected with 9 litres of oxygen. External heat is best applied to the part of the tube containing the substance by a Bunsen burner held in the hand. The length of time required for the actual combustion seldom exceeds 20 minutes.

This method gives very accurate results.

2. SAUER'S *Method, modified by* MIXTER.*

α. If the substance gives off but little volatile matter on heating, e.g., coke, anthracite coal, &c.

A combustion tube 30 to 40 cm. in length is drawn out quite narrow at one end, and the drawn out narrow part is bent downward at a right angle and fitted by means of a perforated stopper into the U-tube A, fig. 101, containing aqueous solution of bro-

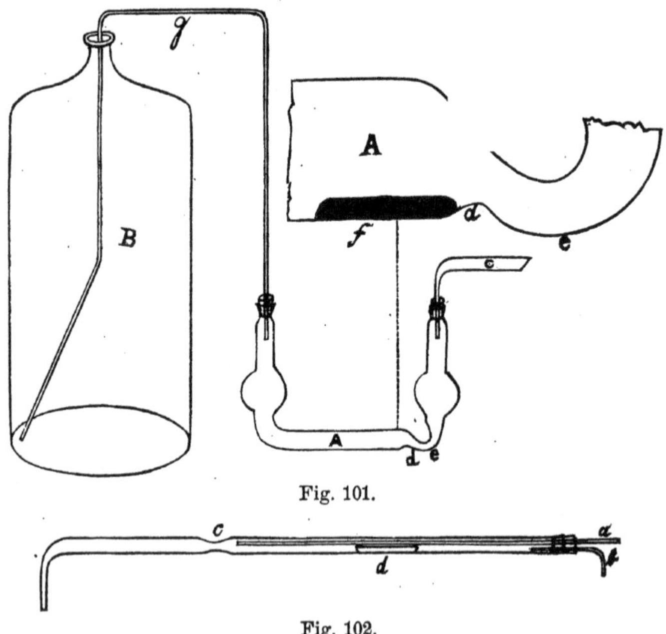

Fig. 101.

Fig. 102.

mine and also a large drop of undissolved bromine. The globule of bromine is made to rest at the point *f* by giving the apparatus a suitable inclination. The combustion tube is laid in a combustion furnace, and the substance contained in a tray is pushed into

*Am. Journ. Chem., 2, 396.

the open end about 15 cm. This end is then closed with a stopper, through which passes a glass tube. Pure oxygen gas is then conducted into the combustion tube, and the part containing the tray is heated to redness. If during the process the bromine in solution becomes nearly exhausted by the action of sulphurous acid, a portion of the undissolved globule is shaken over into the narrow part *e* of the U-tube, where it is rapidly dissolved by the agitation caused by the passing gas-bubbles. In order to complete the condensation of fumes of sulphuric acid which may pass through the U-tube, they are conducted by means of the tube *g* to the bottom of the bottle B, which has a capacity of about ·8 litres. The bottom of the bottle should be barely covered with water. During the process of combustion a cloud of fumes may be observed in the lower part of the bottle, while the air in the upper part remains perfectly clear. After combustion is completed, the tube *g* is removed, and the bottle with its mouth closed is allowed to stand until the visible fumes are absorbed. The combustion tube is rinsed to remove sulphuric acid which may have condensed in the part near the U-tube. The rinsings are added to the united solutions obtained in A and B. The solution containing the sulphuric acid is now heated to remove free bromine, and concentrated if the volume appears too great. The sulphuric acid in it is determined as in the similar solution obtained by the process described above in 1.

If the operator cannot procure a U-tube of the form represented by A, the more common form shown by fig. 99 may be used. In that case it is best to use a saturated solution of bromine in hydrochloric acid, of which the U-tube should contain 12 to 15 c. c. when filled to extent indicated in fig. 99. On account of the small volume of liquid which can be used in such tube, an aqueous solution would hardly suffice. The free hydrochloric acid should be nearly all removed by evaporation from the final solution of sulphuric acid before proceeding to precipitate the latter with barium chloride.

If inorganic matter remains in the tray after completing the combustion, it is to be treated as directed in *c*, 1.

β. The substance gives off volatile matter at a high temperature.

A combustion tube about 85 cm. long, narrowed at the point indicated by *c* in fig. 102, is employed. Having introduced the substance in a tray (or if volatile at the ordinary temperature in bulb tube with capillary orifice), the narrow part of the combustion tube and also a portion beyond extending to within 10 or 15 cm. of the end entering the U-tube is heated to dull redness in a combustion furnace. Oxygen gas is now conducted by means of the hard glass tube *a* to the point *c* beyond the tray. At the same time a very slow current of carbon dioxide is made to enter through the tube *b* in order to prevent vapors from receding. Now, by a cautious application of heat the volatile matter in the tray is first distilled off and burned by the constantly supplied current of oxygen. Next the combustion of any fixed residue remaining in the tray is effected by transferring the supply of oxygen from *a* to *b*, and that of carbon dioxide from *b* to *a*. The only use of carbon dioxide at this stage is to prevent products of combustion from entering the tube *a*. The combustion tube during the process is connected with the same absorbing apparatus as used in 2, *α*. The remaining part of the process is also indicated as in 2, *α*.

Mixter* obtained quite satisfactory results with this process. When very volatile substances, *e.g.*, carbon disulphide, are to be burned, it is necessary to apply heat very cautiously to the part of the tube containing the substance, so that the flame produced by the meeting of the combustible vapor with oxygen shall be a few millimetres *back* of the end of the tube delivering the oxygen.]

D. Determination of Phosphorus in Organic Compounds.

§ 187.

The phosphorus in organic compounds is determined by methods similar to those employed for determination of sulphur in organic compounds, *i.e.*, the organic substance is oxidized either in the wet or dry way, and a solution is obtained in which the phosphoric acid formed by oxidization is determined.

For oxidation the methods given in § 186, 1, 2, 4, are suitable.

From the solution obtained phosphoric acid is precipitated, either directly with ammonium chloride, magnesium chloride, and ammonia mixture, or with molybdic acid solution, after removing hydrochloric acid by repeated evaporation with nitric acid.

The phosphorus cannot be determined by incineration of the

*Am. Journ. Chem., 2, 396.

substance and examination of the ash. Vitellin, which when treated with nitric acid gives 3 per cent. of phosphoric acid, yields barely 0·3 per cent. of ash (V. BAUMHAUER).

If a substance contains phosphorus both in an unoxidized state and in the form of phosphates, treat a separate portion with hydrochloric acid, filter if necessary and determine the phosphoric acid in the solution. The quantity thus found is deducted from the total phosphoric acid found in the portion submitted to oxidation in order to find the amount which existed in the compound in an unoxidized state.

E. ANALYSIS OF ORGANIC SUBSTANCES CONTAINING CHLORINE, BROMINE, OR IODINE.

§ 188.

Substances containing Bromine and Iodine are analyzed generally in the same manner as those containing Chlorine.

Those portions of the following § which are enclosed between square brackets refer exclusively to combinations of Iodine or Bromine, as the case may be.

The combustion of organic substances containing chlorine with oxide of copper gives rise to the formation of cuprous chloride, which, were the process conducted in the usual manner, would condense in the calcium chloride tube, and would thus vitiate the determination of the hydrogen. This and every other error may be prevented by the employment of lead chromate (§ 177). The chlorine is, in that case, converted into lead chloride, and retained in that form in the combustion tube.

If the combustion is effected with oxide of copper in a current of oxygen, the cuprous chloride is decomposed by the oxygen, oxide of copper and free chlorine being formed; the latter is retained partly in the calcium chloride tube, partly in the potash bulbs. To remedy this defect, STAEDELER* proposes to fill the anterior part of the tube with clean copper turnings; these must be kept red-hot during the combustion, and the current of oxygen must be arrested the moment they begin to oxidize. K. KRAUT† observes with reference to this process that it is well to place a roll of silver foil,

* Annal. d. Chem. u. Pharm. 69, 335. † Zeitschr. f. anal. Chem. 2, 242.

about 5 inches long, in front of the layer of metallic copper. In the absence of the silver the transmission of oxygen has to be conducted with caution, in order that no chlorine may be expelled from the cuprous chloride first formed, but by adopting KRAUT's recommendation we may continue passing the gas without fear till it escapes free from the potash tube. [In the case of substances containing iodine, it is needless to employ metallic copper as well as silver foil.] The silver may be used over and over again, but at last requires ignition in a stream of hydrogen. According to A. VÖLCKER,* the evolution of chlorine may be prevented by mixing the oxide of copper with ⅛ lead oxide.

[In the analysis of bodies containing bromine the above methods do not always answer. v. GORUP-BESANEZ† satisfied himself of this by analyzing dibromotyrosin. Whether this body was burnt with lead chromate, with a mixture of lead chromate and potassium chromate, with oxide of copper and oxygen and an anterior layer of lead chromate, with an anterior layer of copper turnings, whether mixed or in the platinum boat, in whichever way the analysis was performed the carbonic acid always came out several per-cents. too low, because metallic bromide was formed, which fused and enclosed carbon, thereby preventing its oxidization. The following process, on the contrary, yielded good results: Into a combustion tube drawn out to a long point, introduce first a three-inch layer of oxide of copper, then a plug of asbestos, then a mixture of the substance (finely powdered) with about an equal weight of well-dried lead oxide in a porcelain boat; again a plug of asbestos, then granulated oxide of copper, then lead chromate or copper turnings. First heat the anterior and then the posterior layers to ignition, and warm the part where the boat is very cautiously and gradually; everything combustible distils over, arrives at the oxide of copper in the form of vapor, and is there burnt. In the boat nothing remains but a mixture of lead bromide and oxide. Complete the combustion with oxygen, taking care not to heat the point where the boat is too strongly, nor continue the transmission of oxygen longer than necessary. Observe also that no copper bromide sublimes into the calcium chloride tube.]

As regards the determination of the *chlorine itself*, this is usually effected either (*a*) by igniting the substance with alkalies or

* Chem. Gaz. 1849, 245, 29. † Zeitschr. f. anal. Chem. 1, 439.

alkali-earths, by which process all the chlorine is obtained as chloride, or (*b*) by oxidizing the substance with nitric acid, &c., in a sealed tube.

a. As chlorine-free lime is easily obtainable (by burning marble), this body is usually preferred to effect the decomposition. It must always be tested for chlorine previous to use.

Introduce into a combustion tube, about 40 cm. long, the posterior end of which is sealed and rounded like a test tube, a layer of lime, 6 cm. long, then the substance, after this another layer of lime, 6 cm. long, and mix with the wire; fill the tube almost to the mouth with lime, clear a free passage for the evolved gases by a few gentle taps, and apply heat in the usual way. Volatile fluids are introduced into the tube in small glass bulbs. When the decomposition is terminated, dissolve in dilute nitric acid, and precipitate with solution of silver nitrate (§ 141). KOLBE recommends the following process to obtain the contents of the combustion tube:—When the decomposition is completed, remove the charcoal, insert a cork into the open end of the tube, remove every particle of ash, and immerse the tube, still hot, with the sealed end downwards, into a beaker filled two-thirds with distilled water; the tube breaks into many pieces, and the contents are then more readily acted upon. As in this method the ignition of compounds abounding in nitrogen may be attended with formation of calcium cyanide or sodium cyanide,* the separation of the silver chloride and the cyanide, if required, is to be effected by the process given in § 169, 6, *b* (NEUBAUER and KERNER†). [In determining iodine by this method, a little iodine set free by action of nitric acid must be converted in hyriodic acid by addition of a little sulphurous acid before precipitating with silver nitrate.] In the analysis of acid organic compounds containing chlorine (*e.g.*, chlorospiroylic acid), the chlorine may often be determined in a simpler manner, viz., by dissolving the substance under examination in an excess of dilute solution of potassa, evaporating to dryness, and igniting the residue, by which means the whole of the chlorine present is converted into a soluble chloride (LÖWIG).

* The formation of cyanides may be prevented by using, instead of lime, a mixture of lime and soda, obtained by slaking 3 parts quicklime in a solution of 1 part sodium hydroxide (free from chlorine) and heating the mixture to dryness in a silver dish. *Rose Handb. der Anal. Chem., Ed. 6 by Finkener,* ii. 735.

† Annal. d. Chem. u. Pharm. 101, 324, 344.

b. In more readily decomposable compounds, *e.g.*, in the substitution products of acids, the halogen may also be determined by decomposing the substance by contact during several hours with water and sodium amalgam, acidifying the fluid with nitric acid, and precipitating with silver solution (KEKULÉ*).

F. ANALYSIS OF ORGANIC COMPOUNDS CONTAINING INORGANIC BODIES.

§ 189.

In the analysis of organic compounds containing inorganic bodies, it is, of course, necessary first to ascertain the quantity of the latter before proceeding to the determination of the carbon, &c., as otherwise the amount of the organic body whose constituents have furnished the carbonic acid, water, &c., not being known, it would be impossible to estimate the oxygen from the loss.

If the substances in question are salts or similar compounds, their basic radicals are determined by the methods given in the Fourth Section; but in cases where the inorganic bodies are of a nature to be regarded more or less as impurities (*e.g.*, the ash in coal), they may usually be determined with sufficient accuracy by the combustion of a weighed portion of the substance, in an obliquely placed platinum crucible, or in a platinum dish. In the analysis of substances containing fusible salts, even long-continued ignition will often fail to effect complete combustion, as the carbon is protected by the fused salt from the action of the oxygen. In such cases, the best way to effect the purpose is to carbonize the substance, treat the mass with water, and incinerate the undissolved residue; the aqueous solution is, of course, likewise evaporated to dryness, and the weight of the residue added to that of the ash.

If organic compounds whose ash contains potassium, sodium, barium, strontium, or calcium are burnt with oxide of copper, part of the carbonic acid evolved remains as carbonate of these metals. As, in many cases, the amount of carbonic acid thus retained is not constant, and the results are, moreover, more accurate if the whole amount of the carbon is expelled and weighed as carbonic acid, the combustion is effected with lead chromate, with addition of $\frac{1}{8}$ of potassium dichromate, according to the directions given in § 177.

* Jahresb. v. Kopp. u. Will. 1861, 832.

Accurate experiments have shown that in this case not a trace of carbonic acid remains with the bases.

If the substance is weighed in a porcelain or platinum boat, and the combustion is effected according to § 178, the ash, carbon, and hydrogen may be determined in one portion. The amount of carbonic acid contained in the ash is added to that found by the process of combustion; if the carbonic acid in the ash cannot be calculated, as in the case of alkali carbonates, it may be determined by means of fused borax (§ 139, II., *c*).

In burning substances containing mercury, the arrival of any of the metal at the calcium chloride tube may be prevented by having a layer of copper-turnings in the anterior part of the combustion tube, and by not allowing the foremost portion to get too hot.

PART II.

SPECIAL PART.

1. ANALYSIS OF FRESH WATER (SPRING-WATER, RIVER-WATER, &C.).*

§ 190.

THE analysis of the several kinds of fresh water is *usually* restricted to the quantitative estimation of the following substances:

a. Basic metals: Sodium, calcium, magnesium.

b. Acids: Sulphuric acid, nitric acid, silicic acid, carbonic acid, chlorine.

c. Mechanically suspended Matters: Clay, &c.

We confine ourselves, therefore, here to the estimation of these bodies.

I. *The Water is clear.*

1. *Determination of the Chlorine.*—This may be effected either, *a*, in the gravimetric, or, *b*, in the volumetric way.

a. Gravimetrically.

Take 500—1000 grm. or c. c.† Acidify with nitric acid, and precipitate with silver nitrate. Filter when the precipitate has *completely* subsided (§ 141, I., *a*). If the quantity of the chlorine is so inconsiderable that the solution of silver nitrate produces only a slight turbidity, evaporate a larger portion of the water to ½, ¼, ⅛, &c., of its bulk, filter, wash the precipitate, and treat the filtrate as directed.

b. Volumetrically.

Evaporate 1000 grm. or c. c. to a small bulk, and determine the chlorine in the residual fluid, without previous filtration, by solution of silver nitrate, with addition of potassium chromate (§ 141, I., *b*, *α*).

* Compare Qualitative Analysis, p. 320 et seq. See a paper recently read before the Chemical Society by Dr. Miller—the Society's Journal (2), iii. 117 et seq.; also, Frankland, idem (2), iv. 239, and vi. 77; and Wanklyn, Chapman, and Smith, idem, vi. 152.

† As the specific gravity of fresh water differs but little from that of pure water, the several quantities of water may safely be measured instead of weighed. The calculation is facilitated by taking a round number of c. c.

2. *Determination of the Sulphuric Acid.*—Take 1000 grm. or c. c. Acidify with hydrochloric acid and mix with barium chloride. Filter after the precipitate has *completely* subsided (§ 132, I., 1). If the quantity of the sulphuric acid is very inconsiderable, evaporate the acidified water to ½, ¼, ⅙, &c., of the bulk, before adding the barium chloride.

3. *Determination of Nitric Acid.*—If, on testing the residue on evaporation of a water for nitric acid, such a strong reaction is obtained that the presence of a determinable quantity of the acid may be inferred, evaporate according to the apparent quantity of nitric acid indicated by qualitative testing 500 to 1000 or 2000 c. c. of the water in a porcelain dish, wash the residue into a flask (it is immaterial whether any solid matter which may have separated goes partially or not at all into the flask), evaporate in the flask still further, if necessary, and in the small quantity of residual fluid determine the nitric acid according to § 149, *d*, *β*.

4. *Determination of the Silicic Acid, Calcium, and Magnesium.*

Evaporate 1000 grm. or c. c. to dryness—after addition of some hydrochloric acid—preferably in a platinum dish, treat the residue with hydrochloric acid and water, filter off the separated silicic acid, and treat the latter as directed § 140, II., *a*. Determine calcium and magnesium in the filtrate as directed § 154, 6, *a* (**28**).

5. *Determination of the total Residue and of the Sodium.*

a. Evaporate 1000 grm. or c. c. of the water, with proper care, to dryness in a weighed platinum dish, first over a lamp, finally on the water-bath. Expose the residue, in the air-bath, to a temperature of about 180°, until no further diminution of weight takes place. This gives the *total amount of the salts.*

b. Treat the residue with water, and add, cautiously, pure dilute sulphuric acid in moderate excess; cover the vessel during this operation with a dish, to avoid loss from spirting; then place on the water-bath, without removing the cover. After ten minutes, rinse the cover by means of a washing bottle, evaporate the contents of the dish to dryness, expel the free sulphuric acid, ignite the residue, in the last stage with addition of some ammonium carbonate (§ 97, 1), and weigh. The residue consists of sodium sulphate, calcium sulphate, magnesium sulphate, and some separated silica. It must not redden moist litmus paper. The quantity of the sodium sulphate in the residue is now found by subtracting from the weight of the latter the known weight of the silica and the weight of the calcium and magnesium sulphates calculated from the quantities of these earths found in 4.

6. *Direct Determination of the Sodium.*

The sodium may also be determined in the direct way, with comparative expedition, by the following method:—

Evaporate 1250 grm. or c. c. of the water, in a dish, to about $\frac{1}{6}$, and then add 2—3 c. c. of thin pure milk of lime, so as to impart a strongly alkaline reaction to the fluid; heat for some time longer, then wash the contents of the dish into a quarter-litre flask. (It is not necessary to rinse every particle of the precipitate into the flask; but the whole of the fluid must be transferred to it, and the particles of the precipitate adhering to the dish well washed, and the washings also added to the flask.) Allow the contents to cool, dilute to the mark, shake, allow to deposit, filter through a dry filter, measure off 200 c. c. of the filtrate, corresponding to 1000 grm. of the water, transfer to a quarter-litre flask, mix with ammonium carbonate and some ammonium oxalate, add water up to the mark, shake, allow to deposit, filter through a dry filter, measure off 200 c. c., corresponding to 800 grm. of the water, add some ammonium chloride,* evaporate, ignite, and weigh the residual sodium chloride as directed § 98, 2.†

Or by the following method:—

Evaporate the filtrate from the barium sulphate obtained in 2 to dryness in a platinum dish (or if nitrates are present in porcelain) to remove free hydrochloric acid and separate silica. Digest the residue with a few c. c. water, and precipitate magnesium without previous filtration by addition of solution of barium hydroxide, avoiding a large excess. Enough has been added if a pellicle of barium carbonate forms upon the surface of the liquid on exposure a short time to the air. Filter and wash the usually slight precipitate. Heat the filtrate, and add ammonium carbonate to precipitate

* To convert the still remaining sodium sulphate, on ignition, into sodium chloride.

† This process, which entirely dispenses with washing, presents one source of error—viz., the space occupied by the precipitates is not taken into account. The error resulting from this is, however, so trifling, that it may safely be disregarded, as the excess of weight amounts to $\frac{1}{500}$ at the most.

the barium introduced and the calcium originally present, filter from the precipitated carbonates, and evaporate the filtrate to dryness, and remove by heating the ammonium chloride completely. Dissolve the sodium chloride in the residue with 4 or 5 c. c. water, warm, and add a few drops of ammonium carbonate and ammonia to separate possibly remaining traces of barium and calcium, filter again into a weighed platinum dish, evaporate to dryness, heat nearly to fusion, and weigh the sodium chloride. The sodium chloride obtained by either process will contain the potassium (as chloride) if any is present in the water. If enough alkali chloride is obtained it may, after weighing, be examined for potassium according to § 152, 1, *a*.

7. Calculate the numbers found in 1—6 to 1000 parts of water, and determine from the data obtained the amount of *carbonic acid* in combination, as follows :—

Add together the quantities of SO_3 corresponding to the basic oxides found, and subtract from the sum, first, the amount of sulphuric acid SO_3 precipitated from the water by barium chloride (2), secondly, the amount equivalent to the nitric acid found, and thirdly, the amount equivalent to the chlorine found; the remainder is equivalent to the carbonic acid combined with the bases in the form of *normal* carbonates. 80 parts of SO_3 remaining after subtracting the quantities just stated, correspond accordingly to 44 parts of CO_2.

If, by way of control, you wish to determine the combined carbonic acid in the direct way, evaporate 1000 grm. or c. c. of the water, in a flask, to a small bulk; add tincture of cochineal, then standard nitric acid, and proceed as directed p. 698.

8. *Control.*

If the quantities of the Na_2O, CaO, MgO, SO_3, N_2O_5, SiO_2, CO_2 and Cl are added together, and an amount of oxygen equivalent to the chlorine (since this latter is combined with metal and not with oxide) is subtracted from the sum, the remainder must nearly correspond to the total amount of the salts found in 5, *a*. *Perfect* correspondence cannot be expected, since, 1, upon the evaporation of the water magnesium chloride is partially decomposed, and converted into a basic salt; 2, the silicic acid expels some carbonic acid; and 3, it being difficult to free magnesium carbonate from water without incurring loss of carbonic acid, the residue remaining upon the evaporation of the water contains the magnesium carbonate as a basic salt, whereas, in our calculation, we have assumed the quantity of carbonic acid corresponding to the normal salt.

9. *Determination of the free Carbonic Acid.*

In the case of well-water this may be conveniently executed by the process described § 139, *β* (p. 405). We here obtain the carbonic acid which is contained in the water over and above the quantity corresponding to the normal carbonates, or in other words, the carbonic acid which is free and which is combined with the carbonates to bicarbonates.

10. *Determination of the Organic Matter.*

Many fresh waters contain so much organic matter as to be quite yellow, others contain traces, and many again may be said to be free from such substances. The exact estimation of organic matter is by no means an easy task, and the method usually adopted—viz., ignition of the residue of the water dried at 180°, treatment with ammonium carbonate, gentle ignition again, and calculation of the organic matter from the loss of weight—yields merely an approximate result, since we can never be sure as to the condition of the magnesium carbonate in the residue dried at 180° and in the same after ignition, and since the silicic acid expels some carbonic acid, which is not taken up again on treatment with ammonium carbonate, &c.

[This approximation, however, will generally suffice, if the purpose of the analysis is to enable one to judge of the quality of the water with reference to its use in steam-boilers and for most manufacturing processes. But if it is desired to learn by analysis whether the water is fit or unfit for drinking, the case is quite different, for its quality as a potable water doubtless depends greatly on the amount, and still more on the *kind*, of organic matter present. Detailed descriptions of methods used in the examination of the organic matter contained in water are to be found in "Water Analysis," by J. A. WANKLYN and E. T. CHAPMAN, third edition, London, Truebner & Co., 1874; Anleitung zur Untersuchung von Wasser, von Kubel und Tiemann, 2 Aufl.; also in several articles on the subject by WANKLYN, CHAPMAN, and SMITH, in Journal of the Chem. Soc.]

II. *The water is not clear.*

Fill a large flask of known capacity with the water, close with a glass stopper, and allow the flask to stand in the cold until the suspended matter is deposited; draw off the clear water with a siphon as far as practicable, filter the bottoms, dry or ignite the contents of the filter, and weigh. Treat the clear water as directed in I.

Respecting the calculation of the analysis, I remark simply that the results are *usually** arranged upon the following principles:—

The *chlorine* is combined with sodium; if there is an excess, this is combined with calcium. If, on the other hand, there remains an excess of sodium, this is combined with sulphuric acid. The *sulphuric acid*, or the remainder of the sulphuric acid, as the case may be, is combined with calcium. The *nitric acid* is, as a rule, to be combined with calcium. The *silicic acid* is put down in the free state, the remainder of the calcium and the magnesium as carbonates, either normal or acid, according to circumstances.

It must always be borne in mind that the results of the qualitative analysis may render another arrangement of the acids and bases necessary. For instance, if the evaporated water reacts strongly alkaline, sodium carbonate is present, generally in company with sodium sulphate and sodium chloride, occasionally also with sodium nitrate. The calcium and magnesium are then to be entirely combined with carbonic acid.

In the report, the quantities are represented in parts per 1000 (or 1000,000), and also in grains per gallon.

For technical purposes, it is sometimes sufficient to estimate the *hardness* of the water (the relative amount of calcium and magnesium in it) by means of a standard solution of soap. A detailed description of this method, which was first employed by CLARK, may be found in BOLLEY & PAUL'S Handbook of Technical Analysis. See also SUTTON'S Volumetric Analysis.

---

* A certain latitude is here allowed to the analyst's discretion

2. ACIDIMETRY.

A. Estimation by Specific Gravity.

§ 191.

Tables, based upon the results of exact experiments, have been drawn up, expressing in numbers the relation between the specific gravity of the aqueous solution of an acid, and the amount of real acid contained in it. Therefore, to know the amount of real acid contained in an aqueous solution of an acid, it suffices, in many cases, simply to determine its specific gravity. Of course the acids must, in that case, be free, or at least nearly free from admixtures of other substances dissolved in them. Now, as most common acids are volatile (sulphuric acid, hydrochloric acid, nitric acid, acetic acid), any non-volatile admixture may be readily detected by evaporating a sample of the acid in a small platinum or porcelain dish.

The determination of the specific gravity is effected either by comparing the weight of equal volumes of water and acid, or by means of a good hydrometer. The estimations must, of course, be made at the temperature to which the Tables refer.

The following Tables on pages 676—679 give the relations between the specific gravity and the strength for sulphuric acid, hydrochloric acid, nitric acid, and acetic acid.

In all cases in which the determination of the specific gravity fails to attain the end in view, or which demand particular accuracy, the volumetric method described under B, is employed.

B. Estimation by Saturation with an Alkaline Fluid of known strength.*

§ 192.

1. This method requires:—

A dilute acid of known strength. Sulphuric or hydrochloric acid may be used. Nitric and oxalic acids are less frequently employed.

* According to Nicholson and Price (Chem. Gaz., 1856, p. 30) the common method of acidimetry is not suited for determining free acetic acid, on account of the alkaline reaction of neutral sodium acetate; however, Otto (Annal. d. Chem. u. Pharm. 102, 69) has clearly demonstrated that the error arising from this is so inconsiderable that it may safely be disregarded.

TABLE I.

Showing the percentages of Acid (H_2SO_4) *and Anhydride* (SO_3) *corresponding to various specific gravities of aqueous Sulphuric Acid by* BINEAU; *calculated for* **15°**, *by* OTTO.

Specific gravity.	Percentage of H_2SO_4.	Percentage of SO_3.	Specific gravity.	Percentage of H_2SO_4.	Percentage of SO_3.
1 8426	100	81·63	1·398	50	40·81
1·842	99	80·81	1·3886	49	40·00
1·8406	98	80·00	1·379	48	39·18
1·840	97	79·18	1·370	47	38·36
1·8384	96	78·36	1·361	46	37·55
1·8376	95	77·55	1·351	45	36·73
1·8356	94	76·73	1·342	44	35·82
1·834	93	75·91	1·333	43	35·10
1·831	92	75·10	1·324	42	34·28
1·827	91	74·28	1·315	41	33·47
1·822	90	73·47	1·306	40	32·65
1·816	89	72·65	1·2976	39	31·83
1·809	88	71·83	1·289	38	31·02
1·802	87	71·02	1·281	37	30·20
1·794	86	70·10	1·272	36	29·38
1·786	85	69·38	1·264	35	28·57
1·777	84	68·57	1·256	34	27·75
1·767	83	67·75	1·2476	33	26·94
1·756	82	66·94	1·239	32	26·12
1·745	81	66·12	1·231	31	25·30
1·734	80	65·30	1·223	30	24·49
1·722	79	64·48	1·215	29	23·67
1·710	78	63·67	1·2066	28	22·85
1·698	77	62·85	1·198	27	22·03
1·686	76	62·04	1·190	26	21·22
1·675	75	61·22	1·182	25	20·40
1·663	74	60·40	1·174	24	19·58
1·651	73	59·59	1·167	23	18·77
1·639	72	58·77	1·159	22	17·95
1·627	71	57·95	1·1516	21	17·14
1·615	70	57·14	1·144	20	16·32
1·604	69	56·32	1·136	19	15·51
1·592	68	55·59	1·129	18	14·69
1·580	67	54·69	1·121	17	13·87
1·568	66	53·87	1·1136	16	13·06
1·557	65	53·05	1·106	15	12·24
1·545	64	52·24	1·098	14	11·42
1·534	63	51·42	1·091	13	10·61
1·523	62	50·61	1·083	12	9·79
1·512	61	49·79	1·0756	11	8·98
1·501	60	48·98	1·068	10	8·16
1·490	59	48·16	1·061	9	7·34
1·480	58	47·34	1·0536	8	6·53
1·469	57	46·53	1·0464	7	5·71
1·4586	56	45·71	1·039	6	4·89
1·448	55	44·89	1·032	5	4·08
1·438	54	44·07	1·0256	4	3·26
1·428	53	43·26	1·019	3	2·445
1·418	52	42·45	1·013	2	1·63
1·408	51	41·63	1·0064	1	0·816

TABLE II.

Showing the percentages of Anhydrous Acid (HCl) *corresponding to various specific gravities of aqueous solutions of Hydrochloric Acid, by* URE. *Temperature* 15°.

Specific gravity.	Percentage of hydrochloric acid gas (HCl).	Specific gravity.	Percentage of hydrochloric acid gas (HCl).
1·2000	40·777	1·1000	20·388
1·1982	40·369	1·0980	19·980
1·1964	39·961	1·0960	19·572
1·1946	39·554	1·0939	19·165
1·1928	39·146	1·0919	18·757
1·1910	38·738	1·0899	18·349
1·1893	38·330	1·0879	17·941
1·1875	37·923	1·0859	17·534
1·1857	37·516	1·0838	17·126
1·1846	37·108	1·0818	16·718
1·1822	36·700	1·0798	16·310
1·1802	36·292	1·0778	15·902
1·1782	35·884	1·0758	15·494
1·1762	35·476	1·0738	15·087
1·1741	35·068	1·0718	14·679
1·1721	34·660	1·0697	14·271
1·1701	34·252	1·0677	13·863
1·1681	33·845	1·0657	13·456
1·1661	33·437	1·0637	13·049
1·1641	33·029	1·0617	12·641
1·1620	32·621	1·0597	12·233
1·1599	32·213	1·0577	11·825
1·1578	31·805	1·0557	11·418
1·1557	31·398	1·0537	11·010
1·1537	30·990	1·0517	10·602
1·1515	30·582	1·0497	10·194
1·1494	30·174	1·0477	9·786
1·1473	29·767	1·0457	9·379
1·1452	29·359	1·0437	8·971
1·1431	28·951	1·0417	8·563
1·1410	28·544	1·0397	8·155
1·1389	28·136	1·0377	7·747
1·1369	27·728	1·0357	7·340
1·1349	27·321	1·0337	6·932
1·1328	26·913	1·0318	6·524
1·1308	26·505	1·0298	6·116
1·1287	26·098	1·0279	5·709
1·1267	25·690	1·0259	5·301
1·1247	25·282	1·0239	4·893
1·1226	24·874	1·0220	4·486
1·1206	24·466	1·0200	4·078
1·1185	24·058	1·0180	3·670
1·1164	23·650	1·0160	3·262
1·1143	23·242	1·0140	2·854
1·1123	22·834	1·0120	2·447
1·1102	22·426	1·0100	2·039
1·1082	22·019	1·0080	1·631
1·1061	21·611	1·0060	1·124
1·1041	21·203	1·0040	0·816
1·1020	20·796	1·0020	0·408

TABLE III.

Showing the percentages of Nitric Anhydride (N_2O_5) *corresponding to various specific gvavities of aqueous Nitric Acid, by* URE. *Temperature* 15°.

Specific gravity.	Percentage of N_2O_5.	Specific gravity.	Percentage of N_2O_5.	Specific gravity.	Percentage of N_2O_5.	Specific gravity.	Percentage of N_2O_5.
1·500	79·7	1·419	59·8	1·295	39·8	1·140	19·9
1·498	78·9	1·415	59·0	1·289	39·0	1·134	19·1
1·496	78·1	1·411	58·2	1·283	38·3	1·129	18·3
1·494	77·3	1·406	57·4	1·276	37·5	1·123	17·5
1·491	76·5	1·402	56·6	1·270	36·7	1·117	16·7
1·488	75·7	1·398	55·8	1·264	35·9	1·111	15·9
1·485	74·9	1·394	55·0	1·258	35·1	1·105	15·1
1·482	74·1	1·388	54·2	1·252	34·3	1·099	14·3
1·479	73 3	1·383	53·4	1·246	33·5	1·093	13·5
1·476	72·5	1·378	52·6	1·240	32·7	1·088	12·7
1·473	71·7	1·373	51·8	1·234	31·9	1·082	11·9
1·470	70·9	1·368	51·1	1·228	31·1	1·076	11·2
1·467	70·1	1·363	50 2	1·221	30·3	1·071	10·4
1·464	69·3	1·358	49·4	1 215	29·5	1·065	9·6
1·460	68·5	1·353	48·6	1·208	28·7	1·059	8·8
1·457	67·7	1·348	47·9	1·202	27·9	1·054	8·0
1·453	66·9	1·343	47·0	1·196	27·1	1·048	7·2
1·450	66·1	1·338	46·2	1·189	26·3	1·043	6·4
1·446	65·3	1·332	45·4	1·183	25·5	1·037	5·6
1·442	64·5	1·327	44·6	1·177	24·7	1·032	4·8
1·439	63·8	1·322	43·8	1·171	23·9	1·027	4·0
1·435	63·0	1·316	43·0	1·165	23·1	1·021	3·2
1·431	62·2	1·311	42·2	1·159	22·3	1·016	2·4
1·427	61·4	1·306	41·4	1·153	21·5	1·011	1·6
1·423	60·6	1·300	40·4	1·146	20·7	1·005	0·8

2. An alkaline fluid of known strength. Potassa or ammonia may be employed.

a. PREPARATION OF THE SOLUTIONS.

The solutions should be of suitable strength. As the first step in the preparation of a dilute sulphuric acid, of convenient strength for ordinary use, dilute 20 cubic centimetres of oil of vitriol with water to the volume of 2 litres.

The standard alkali is made from commercial caustic potash; this is dissolved in water and diluted until a given volume, *e.g.*, 5. c. c., neutralizes 4 to 5 c. c. of the standard acid, as is determined by a few rough trials.

The alkali solution thus obtained is heated to boiling in a flask, and a little freshly-slaked lime is added to decompose any potassium carbonate. The boiling is continued a few minutes and,

TABLE IV.

Showing the percentages of Acetic Acid ($HC_2H_3O_2$) *corresponding to various specific gravities of aqueous solutions of Acetic Acid, by* MOHR.

Specific gravity.	Percentage of acetic acid ($HC_2H_3O_2$).	Specific gravity.	Percentage of acetic acid ($HC_2H_3O_2$).	Specific gravity.	Percentage of acetic acid $HC_2H_3O_2$).	Specific gravity.	Percentage of acetic acid ($HC_2H_3O_2$).	Specific gravity.	Percentage of acetic acid ($HC_2H_3O_2$).
1·0635	100	1·0735	80	1·067	60	1·051	40	1·027	20
1·0555	99	1·0735	79	1·066	59	1·050	39	1·026	19
1·0670	98	1·0732	78	1·066	58	1·049	38	1·025	18
1·0680	97	1·0732	77	1·065	57	1·048	37	1·024	17
1·0690	96	1·0730	76	1·064	56	1·047	36	1·023	16
1·0700	95	1·0720	75	1·064	55	1·046	35	1·022	15
1·0706	94	1·0720	74	1·063	54	1·045	34	1·020	14
1·0708	93	1·0720	73	1·063	53	1·044	33	1·018	13
1·0716	92	1·0710	72	1·062	52	1·042	32	1·017	12
1·0721	91	1·0710	71	1·061	51	1·041	31	1·016	11
1·0730	90	1·0700	70	1·060	50	1·040	30	1·015	10
1·0730	89	1·0700	69	1·059	49	1·039	29	1·013	9
1·0730	88	1·0700	68	1·058	48	1·038	28	1·012	8
1·0730	87	1·0690	67	1·056	47	1·036	27	1·010	7
1·0730	86	1·0690	66	1·055	46	1·035	26	1·008	6
1·0730	85	1·0680	65	1·055	45	1·034	25	1·007	5
1·0730	84	1·0680	64	1·054	44	1·033	24	1·005	4
1·0730	83	1·0680	63	1·053	43	1·032	23	1·004	3
1·0730	82	1·0670	62	1·052	42	1·031	22	1·002	2
1·0732	81	1·0670	61	1·051	41	1·029	21	1·001	1

finally, the lye is poured upon a filter, and the filtrate is collected in the bottle from which it is to be used. Care should be taken to bring upon the filter some of the excess of lime that is suspended in the liquid, so that the latter may acquire no carbonic acid from the air. This clear liquid thus obtained is a potash-lye containing lime in solution. If exposed to the air, the carbonic acid that is absorbed separates as calcium carbonate, leaving the liquid perfectly caustic.

It now remains to determine with the greatest accuracy, 1st, the volume of alkali which neutralizes a cubic centimetre of the acid, and, 2d, the amount of SO_3 contained in a cubic centimetre of the latter.

As a means of recognizing the point of neutralization, tincture of cochineal possesses great advantages over solution of litmus. The knowledge of this fact is due to LUCKOW, who has detailed its application in *Journ. für prakt. Chem.*, lxxxiv. p. 424. Tincture of cochineal is prepared by digesting and frequently agitating three grammes of pulverized cochineal in a mixture of 50 cubic centimetres of strong alcohol with 200 c. c. of distilled water, at ordinary temperatures, for a day or two. The solution is decanted, or filtered through Swedish paper.

The tincture thus prepared has a deep ruby-red color. On gradually diluting with pure water (free from ammonia), the color becomes orange and finally yellowish-orange. Alkalies and alkali-earths as well as their carbonates change the color to a carmine or violet-carmine. Solutions of strong acid and acid salts make it orange or yellowish-orange.

To determine the volumetric relation of the alkali and acid, a given volume of the latter, *e.g.*, 20 c. c., is measured off into a wide-mouthed flask, 10 drops of cochineal-tincture, and about 150 c.c. of water are added; the alkali is now allowed to flow in from a burette, until the yellowish liquid in the flask, suddenly, and by a single drop, acquires a violet-carmine tinge.

In nicer determinations, it is important to bring the liquid each time to a given volume, by adding water after the neutralization is nearly finished. For this purpose, two or more flasks of equal capacity are selected, and on the outside of each a strip of paper is gummed to indicate the level of the proper amount of liquid, *e.g.*, 200 c. c. The same amount of coloring matter being thus always diffused in the same volume of the same water, the errors of varying dilution and varying amount of ammonia (which is rarely absent from distilled water) are avoided. The contents of one flask, in which the neutralization has been satisfactorily effected, may be kept as a standard of color for the succeeding trials, as the tint remains constant for hours, being unaffected by the absorption of carbonic acid. The greatest convenience and accuracy of measurement are obtained by using burettes provided with ERDMANN'S swimmer (see p. 40).

When three or four accordant results have been obtained, the average is taken as expressing the relative strength of the acid and alkali.

To ascertain the absolute standard, weigh off in a small platinum crucible about 0·8 grm. of pure sodium carbonate, ignite to dull redness, cool and weigh accurately: bring the crucible with its contents into one of the wide-mouthed flasks and let flow from the burette a slight excess, *e.g.* 50 c. c., of standard acid. The solution of sodium carbonate is facilitated by warming, and, finally, the contents of the flask are gently boiled for several minutes to expel carbonic acid. The solution is now allowed to become *perfectly cold*, then add ten drops of cochineal and lastly the standard alkali to neutralization, diluting to the proper volume.

To illustrate the accuracy of the process and the calculations employed, the following actual data may be useful. The acid solution was made by diluting 50 c. c. of oil of vitriol to the volume of ten litres and had half the strength above recommended. The alkali was from a stock on hand and more dilute than necessary.

Relation of acid to alkali.

Exp. I., 20 c. c. H_2SO_4 = 32·8 c. c. KOH, or 1 : 1·64
Exp. II., 20 c. c. H_2SO_4 = 32·8 c. c. KOH, or 1 : 1·64
Exp. III., 40 c. c. H_2SO_4 = 65·7 c. c. KOH, or 1 : 1·6425.

We have accordingly:

1 c. c. H_2SO_4 = 1·64 c. c. KOH and 1 c. c. KOH = 0·60976 c. c. H_2SO_4.

Absolute strength of acid and alkali.

Exp. I. 0·4177 grm. of sodium carbonate were treated with 44·2 c. c. of H_2SO_4 solution. To neutralize the excess of the acid were required 3·8 c. c. KOH, which correspond to 2·32 c. c. H_2SO_4 (3·8 × 0·60976). Deducting this from the total amount of acid (44.2 — 2·32) we have 41·88 c. c. of acid, neutralized by the sodium carbonate taken.

41·88 c. c. solution of H_2SO_4 = 0·4177 grm. Na_2CO_3.

Exp. II. 0.4126 grm. Na_2CO_3 treated with 44 c. c. H_2SO_4 required 4·28 c. c. KOH. 4·28 × 0·60976 = 2·61 c. c. H_2SO_4. 44 — 2·61 = 41·39 c. c. H_2SO_4.

41·39 c. c. solution of H_2SO_4 = 0·4126 grm. Na_2CO_3.

Now, from the data obtained by each of these experiments, the absolute strength of the sulphuric acid solution may be calculated; for when sodium carbonate and sulphuric acid exactly neutralize each other, one molecule or 106·08 pts. Na_2CO_3 reacts with 1 molecule or 98 pts. H_2SO_4.

$$\overbrace{H_2SO_4}^{\text{mol. weight 98}} + \overbrace{Na_2CO_3}^{\text{mol. weight 106·08}} = Na_2SO_4 + CO_2 + H_2O.$$

Consequently that volume of a sulphuric acid solution which is found to exactly neutralize 1·0608 grm. Na_2CO_3 *must* contain ·98 grm. H_2SO_4.

The volume of the sulphuric acid solution in the present case which would neutralize 1·0608 grm. Na_2CO_3 is found by calculation from the data furnished by the experiments to be:—

$$\begin{array}{lcc} & \overbrace{\text{grms. } Na_2CO_3} & \overbrace{\text{c. c. } H_2SO_4 \text{ solution}} \\ \text{I.,} & \cdot 4177 : 1\cdot 0608 :: & 41\cdot 88 : 106\cdot 35 \\ \text{II.,} & \cdot 4126 : 1\cdot 0608 :: & 41\cdot 39 : 106\cdot 41 \end{array}$$

According to Exp. I., 106·35 c. c.; according to Exp. II., 106·41 c. c.—mean, 106·38 c. c.

This volume therefore contains ·98 grm. H_2SO_4. By dividing ·98 grm. by 106·38 the weight of H_2SO_4 in 1 c. c. would of course be found.

But the already found volume of the sulphuric acid which contains a weight of H_2SO_4 corresponding to its molecular weight is a more convenient basis for calculating the weight of any alkali neutralized by 1 c. c. Suppose it is desired, for instance, to find the weight of NH_4OH, NH_3, or N which corresponds to 1 c. c. of the acid solution. One molecule of sulphuric acid neutralizes 2 mol. NH_4OH:—

$$\overbrace{H_2SO_4}^{\text{mol. weight } 98} + \overbrace{2NH_4OH}^{\text{mol. weight } 35 \times 2 = 70} = (NH_4)_2SO_4 + (H_2O)_2.$$

Hence 106·38 c. c. = ·98 grm. H_2SO_4 neutralize ·70 grm. NH_4OH, and further, observing that 35 parts (1 mol.) of NH_4OH contain 14 pts. N or 17 NH_3,—

$$\begin{array}{lr} N & 14 \\ H_3 & 3 \\ H_2 & 2 \\ O & 16 \\ \hline & 35 \end{array} \quad \left.\begin{array}{l} \\ \end{array}\right\} 17 \text{ (N + } H_3\text{)}$$

and 70 pts. (2 mol.) twice these quantities—

106·38 c. c. acid solution correspond to ·34 grm. NH_3,
" " " " " ·28 " N.

Finally find the weight of the substance which corresponds to

1 c. c. of the standard solution. For nitrogen, *e.g.*, this is ·28 grm. ÷ 106·38 = 0·002632 grm.

We may then write on the label of the acid bottle the following data for calculation :—

$$1 \text{ c. c. KOH } = 0\cdot60976 \text{ c. c. } H_2SO_4,$$
$$1 \text{ c. c. } H_2SO_4 = 1\cdot64 \text{ c. c. KOH},$$
$$1 \text{ c. c. } H_2SO_4 = 0\cdot002632 \text{ grm. N}.$$

In a like manner, we may calculate the weight of any base, or any constituent part of a base corresponding to 1 c. c. of the standard acid, being careful to observe whether *one* or *two* molecules of the base are neutralized by one mol. H_2SO_4.

To ascertain the absolute strength of the alkali solution. No further experimental work is required for this purpose. For since ·98 grm. H_2SO_4 neutralize **1·1226** grm. KOH, as clearly seen from the formulæ with appended molecular weights expressing the reaction,

$$\overbrace{H_2SO_4}^{\text{mol. weight 98}} + \overbrace{2KOH}^{\text{mol. weight } 56\cdot13 \times 2 = 112\cdot26} = K_2SO_4 + (H_2O)_2,$$

it follows that a volume of potash solution which exactly neutralizes 106·38 c. c. sulphuric acid solution (*i.e.*, the volume already found to contain ·98 grm. H_2SO_4) *must* contain 1·1226 grm. KOH. This volume of potash solution may be calculated from the already determined volumetric relation of the acid and alkali solutions, viz.:—

$$1 \text{ c. c. } H_2SO_4 \text{ sol.} = 1\cdot64 \text{ c. c. KOH sol.}$$
$$106\cdot38 \text{ c. c.} \times 1\cdot64 = 174\cdot46 \text{ c. c.}$$

Accordingly 174·46 c. c. potash solution contain **1·1226** grm. KOH, or 112·26 *centigrammes* a number equal to *twice* the number which expresses the molecular weight of KOH. The weight of any acid neutralized by 1 c. c. of this alkali solution may now be readily calculated, bearing in mind that 2 mol. KOH neutralize 2 mol. of any monobasic and 1 mol. of any dibasic acid. For hydrochloric acid, *e.g.*:

$$\overbrace{(KOH)_2}^{\text{mol. weight } 56\cdot13 \times 2 = 112\cdot26} \text{ neutralize } \overbrace{(HCl)_2}^{\text{mol. weight } 36\cdot46 \times 2 = 72.92}$$

176·46 c. c. sol. = 1·1226 grm. KOH neutralize ·7292 grm. HCl.

$$\frac{\cdot 7292}{176 \cdot 46} = \cdot 00418.$$

1 c.c. alkali solution = ·00418 grm. HCl.

b. The actual analysis. It is only necessary to weigh or measure off the acid to be examined, dilute to about 150 c. c. and ascertain how much of the standard alkali is required for its neutralization, proceeding just as detailed for ascertaining the volumetric relation of the acid and alkali solutions. It is best to use for determination a quantity which will require 15 to 30 c. c., but not over a burette full of the standard alkali. It is often convenient in case of strong acids to weigh off about five times the amount required for a single trial, dilute to exactly 500 c. c. and make two or more determinations, using for each 100 c. c.

α. If the color of a fluid conceals the change of the dissolved cochineal, or if salts of iron be present, we use red litmus or turmeric paper to hit the point of neutralization, *i.e.*, we add alkali till a strip of test paper dipped in just indicates a weak alkaline reaction. In this case more alkali will be employed than when cochineal can be used in solution, and in exact determinations it may be worth while to rectify the error by a correction. This may be done by taking a like quantity of water and adding alkali solution, till the fluid just gives a reaction on the test paper in question, as strong as was obtained at the close of the first experiment. The quantity of alkali used is, of course, to be deducted from the quantity employed in the first experiment.

β. Determination of Acids by means of normal sodium carbonate solution. See *c*, *δ*, page 687.

γ. Application of the Acidimetric principle to the determination of combined acids.

The acidimetric principle may often be employed also for the determination of acids in combination with bases, if solution of soda or sodium carbonate precipitates the latter completely, and in a state of purity. For instance, acetic acid in iron mordant, or in verdigris, may be estimated in this way, by the following process: Precipitate with a measured quantity of standard solution of soda or sodium carbonate in excess, boil, filter, wash, concentrate the filtrate which contains the excess of alkali used. Determine the number of c. c. of this excess by means of standard acid solution. Subtract the number of c. c. thus found from the c. c of soda solution consumed in the experiment; the difference expresses the quantity of soda solution neutralized by the acid contained in the substance, in combination as well as in the free state. Of course, correct results can be expected only if no basic salt has been thrown down by the soda solution.

δ. Determination of combined acids by Gibbs' method. See § 149, ii., *c*, *γ*, p. 472.

c. Deviations from the preceding method.

α. Different acids and alkalies for standard solutions.

Hydrochloric may be used instead of sulphuric acid, and ammonia instead of potash solution. A hydrochloric acid solution containing 13 grm. HCl per 1000 c. c. will have about the same neutralizing power as the sulphuric acid solution recommended in the preceding (made by diluting 20 c. c. oil of vitriol to 2000 c. c.). Find the specific gravity of a sample of pure aqueous hydrochloric acid and calculate by means of Table II., page 677, the volume required to contain 13 grm. HCl, and dilute to 1000 c. c. Prepare the ammonia solution by diluting pure ammonia solution until it is found by trial with a few c. c. that it neutralizes nearly an equal volume of the HCl solution. Ascertain next the exact volumetric relation of the acid to alkali solution as before directed.

The ammonia should be nearly or quite free from ammonium carbonate. Ammonia from a freshly opened bottle which gives a very slight or no immediate precipitate with calcium chloride is suitable.

The absolute strength of both solutions may be found by the same method that is applied to sulphuric acid and potash solutions.

To insure accuracy it is well also to determine chlorine in the acid solution by precipitation with silver nitrate. These solutions affect the burettes less than potash and sulphuric acid solutions, and are more readily prepared.

β. Different indicators. Many kinds have been proposed. Litmus solution is preferable to cochineal when organic acids or aluminium salts are present. Carbonic acid, however, renders the indication which it affords quite indistinct, while it interferes with the action of cochineal far less.

γ. *To save time in computations.* When many successive determinations are to be made, it is convenient to have an alkali solution that will neutralize exactly an equal volume of the standard acid solution. This may be obtained as follows: After ascertaining the volumetric relation of the two solutions, calculate how many c. c. of water must be added to the stronger to make equal the weaker; *e.g.*, if 1 c. c. acid solution = 1·132 c. c. alkali solution, ·132 c. c. water must be added to each c. c. of the acid solution, or 132 c. c to 1000 c. c. Measure accurately 1000 c. c. of the acid solution in a litre flask, and pour it into a dry bottle designed for keeping the standard solution, then measure by means of a burette 132 c. c. of water, and allow it to run into the flask. Shake the water about in the flask and pour into the bottle, pour a part of the contents of the bottle into the flask and return it again to the bottle, to insure a uniform mixing of all the water with the acid.

Calculations may be still further shortened by using for each determination such quantities of the substance to be analyzed that each c. c. of the standard solution shall correspond to 1 per cent. of the constituent determined. It is only necessary to observe the following rule: Take of the substance to be analyzed a weight equal to the weight of the constituent to be determined which corresponds to 100 c. c. of the standard solution.

Suppose, for instance, it is desired to determine the percentage of H_2SO_4 in several samples of sulphuric acid which may be concentrated or more or less dilute by means of a standard alkali solution, of which

$$1 \text{ c. c.} = 0·0142 \text{ grm. } H_2SO_4$$

or

$$100 \text{ c. c.} = 1·1420 \text{ grm. } H_2SO_4$$

If now 1·1420 grm. of a sample be accurately weighed off, and it is found that 100 c. c. of the alkali solution are required to neutralize it, the sample must be pure H_2SO_4, or contain 100 per cent. of H_2SO_4. If 1·1420 grm. of another sample requires for neutralization 40 c. c. of the standard alkali, it is clear that it contains 40 per cent. of H_2SO_4, and, finally, that the percentage will equal the number of c. c. used, whatever the number may be. This mode of proceeding is of course applicable in the determination of alkalies by means of standard acid solutions, or in determination of elements which form a part of the neutralized body; *e.g.*, nitrogen in ammonia.

When the substance analyzed contains above 50 per cent. of the constituent to be determined, it is preferable, in order to avoid using too large a quantity of standard solution, to weigh off just half the quantity required by the preceding rule; then each c. c. of the standard solution will indicate *two* per cent., or the percentage is found by doubling the number of c. c. used. On the other hand, it is sometimes advantageous to use twice or perhaps five times the quantity demanded by the first rule, in which case the percentage is found by dividing the number of c. c. by 2 or 5.

Often when the substance to be analyzed is in solution, or soluble, it is advisable to weigh five times the amount required, dilute in a half-litre flask to 500 c. c. and take out $\frac{1}{5}$ by means of a 100 c. c. pipette for the determination.

δ. *Use of normal solutions.*

Solutions may be made of such strength that 1000 c. c. contain an amount of acid or base *equivalent* to 1 gramme of hydrogen; one molecule of an acid or base being equivalent to the number of H atoms corresponding to the *quantivalence* of its radical. Such solutions are called *normal* solutions.

The following examples show the relation of molecular weights and quantivalence to the actual weights of acids and bases (or salts) in normal solutions:—

	Mol. weight.	Radical.	Quantivalence.	Weight in 1000 c. c.
HCl	36·46	Cl	1	36·46 grms.
H_2SO_4	98·	SO_4	2	49· "
NH_4OH	35·	NH_4	1	35· "
NaOH	40·04	Na	1	40·04 "
(Na_2CO_3	106·08	Na_2	2	53·04 ")

From this relation it follows that a given volume of any normal acid solution exactly neutralizes an equal volume of any normal alkali solution. Moreover, the weight of any acid or alkali which 1 c. c. of a normal solution neutralizes can be very easily calculated.

The method generally used for preparing normal solutions is to first make solutions of acid and alkali somewhat stronger than required, ascertain their volumetric relation and actual strength, as in *a*, page 680 ; and next dilute to the volume required to make the solutions normal by addition of the calculated and accurately measured volume of water as described above under γ. The trouble involved in these operations overbalances the advantages of normal solutions except for some special purposes. By using *sodium carbonate*, however, for the alkali, normal solutions may be prepared in a comparatively simple manner, since sodium carbonate, unlike the caustic alkalies, can easily be procured pure, and can be accurately weighed. 53·08 grm. of pure Na_2CO_3 previously ignited to dull redness are dissolved in water and the solution is diluted to exactly 1000 c. c. To make a normal acid solution mix 60 grm. concentrated sulphuric acid with 1050 c. c. water, let cool, and ascertain how many c. c. of this acid neutralize 50 c. c. of the normal sodium carbonate solution.

Suppose 48·6 c. c. acid sol. = 50 c. c. alkali sol.;

then 1 " " $= \frac{50}{48 \cdot 6} = 1 \cdot 0288$ " "

or 1000 " " = 1028·8 " "

Accordingly 28·8 c.c. of water must be added to 1000 c. c. of the acid solution to make it normal. The acid solution and water are measured and mixed as described above under γ. Test finally the acid against the alkali to be certain that equal volumes neutralize each other. In neutralization it is not necessary to expel carbonic acid by boiling. Tincture of cochineal must be used as an indicator. Litmus is quite unsuitable, since in the presence of carbonic acid the change of color which it undergoes during neutralization is gradual and indistinct. Even cochineal is not quite indifferent to carbonic acid, the slight excess of acid or alkali required to produce a distinct change of color being perceptibly increased by its presence.

The error thus caused by the disturbing influence of carbonic acid, which is slight when normal solutions are used, is increased when more dilute standard solutions of acid and alkali are used, for the more dilute the solution is, the greater the volume required to supply the slight unavoidable excess of acid or alkali. A standard solution of caustic alkali is therefore to be recommended for nicer investigation, and for all purposes when dilute standard solutions are required.

The common mineral acids in a free state, in not too dilute solutions may be determined with sufficient accuracy for many technical purposes without the use of a standard acid solution, by simply adding tincture of cochineal solution to a suitable weighed amount previously diluted to about 150 c. c., and neutralizing cold with the normal sodium carbonate solution.

Modification of the common Acidimetric Method (Kiefer*).

§ 193.

Instead of estimating free acid by a solution of soda of known strength, and determining the neutralization point by means of cochineal tincture, an ammoniacal solution of copper may be used for the purpose, in which case the neutralization point is known by the turbidity observed as soon as the free acid present is completely neutralized. The copper solution is prepared by adding to an aqueous solution of cupric sulphate, solution of ammonia until the precipitate of basic salt which forms at first is just redissolved. After determining the strength of the solution by standard sulphuric or hydrochloric acid (not oxalic), it may be employed for the estimation of all the stronger acids (with the exception of oxalic acid), provided the fluids are clear. The basic cupric salt, in the precipitatiou of which the final reaction consists, is not insoluble in the ammonium salt formed, and its solubility depends on the degree of concentration, and on the presence of other salts, especially of ammonium salts (Carey Lea†). Hence the method cannot boast of scientific accuracy, but as the variations occasioned by the causes mentioned are inconsiderable,‡ the process retains its applicability to technical purposes, for which, indeed, it was originally proposed. This method is of especial value in cases in which free acid is to be determined in presence of a normal metallic salt with acid reaction; *e.g.*, free sulphuric acid in mother-liquors of cupric sulphate or zinc sulphate, &c. It is advisable to determine the strength of the ammoniacal copper solution anew before every fresh series of experiments.

---

* Annal. d. Chem. u. Pharm. 93, 386.

† Chem. News, 4, 195.

‡ Compare my experiments on the subject in the Zeitschrift f. analyt. Chem. 1, 108.

TABLE I.

Percentages of Anhydrous Potassa (K_2O) *corresponding to different specific gravities of solution of potassa.*

Dalton.		Tünnermann (at 15°).			
Specific gravity.	Percentage of anhydrous potassa.	Specific gravity.	Percentage of anhydrous potassa.	Specific gravity.	Percentage of anhydrous potassa.
1·60	46·7	1·3300	28·290	1·1437	14·145
1·52	42·9	1·3131	27·158	1·1308	13·013
1·47	39·6	1·2966	26·027	1·1182	11·882
1·44	36·8	1·2803	24·895	1·1059	10·750
1·42	34·4	1·2648	23·764	1·0938	9·619
1·39	32·4	1·2493	22·632	1·0819	8·487
1·36	29·4	1·2342	21·500	1·0703	7·355
1·33	26·3	1·2268	20·935	1·0589	6·224
1·28	23·4	1·2122	19·803	1·0478	5·002
1·23	19·5	1·1979	18·671	1·0369	3·961
1·19	16·2	1·1839	17·540	1·0260	2·829
1·15	13·0	1·1702	16·408	1·0153	1·697
1·11	9·5	1·1568	15·277	1·0050	0·5658
1·06	4·7				

TABLE II.

Percentages of Anhydrous Soda (Na_2O) *corresponding to different specific gravities of solution of soda.*

Dalton.		Tünnermann (at 15°).					
Specific gravity.	Percentage of anhydrous soda.	Specific gravity.	Percentage of anhydrous soda.	Specific gravity.	Percentage of anhydrous soda.	Specific gravity.	Percentage of anhydrous soda.
1·56	41·2	1·4285	30·220	1·2982	20·550	1·1528	10·275
1·50	36·8	1·4193	29·616	1·2912	19·945	1·1428	9·670
1·47	34·0	1·4101	29·011	1·2843	19·341	1·1330	9·066
1·44	31·0	1·4011	28·407	1·2775	18·730	1·1233	8·462
1·40	29·0	1·3923	27·802	1·2708	18·132	1·1137	7·857
1·36	26·0	1·3836	27·200	1·2642	17·528	1·1042	7·253
1·32	23·0	1·3751	26·594	1·2578	16·923	1·0948	6·648
1·29	19·0	1·3668	25·989	1·2515	16·319	1·0855	6·044
1·23	16·0	1·3586	25·385	1·2453	15·714	1·0764	5·440
1·18	13·0	1·3505	24·780	1·2392	15·110	1·0675	4·835
1·12	9·0	1·3426	24·176	1·2280	14·506	1·0587	4·231
1·06	4·7	1·3349	23·572	1·2178	13·901	1·0500	3·626
		1·3273	22·967	1·2058	13·297	1·0414	3·022
		1·3198	22·363	1·1948	12·692	1·0330	2·418
		1·3143	21·894	1·1841	12·088	1·0246	1·813
		1·3125	21·758	1·1734	11·484	1·0163	1·209
		1·3053	21·154	1·1630	10·879	1 0081	0·604

TABLE III.

Percentages of Ammonia (NH_3) *corresponding to different specific gravities of solution of ammonia at* 16° (J. OTTO).

Specific gravity.	Percentage of ammonia.	Specific gravity.	Percentage of ammonia.	Specific gravity.	Percentage of ammonia.
0·9517	12·000	0·9607	9·625	0·9697	7·250
0·9521	11·875	0·9612	9·500	0·9702	7·125
0·9526	11·750	0·9616	9·375	0·9707	7·000
0·9531	11·625	0·9621	9·250	0·9711	6·875
0·9536	11·500	0·9626	9·125	0·9716	6·750
0·9540	11·375	0·9631	9·000	0·9721	6·625
0·9545	11·250	0·9636	8·875	0·9726	6·500
0·9550	11·125	0·9641	8·750	0·9730	6·375
0·9555	11·000	0·9645	8·625	0·9735	6·250
0·9556	10·950	0·9650	8·500	0·9740	6·125
0·9559	10·875	0·9654	8·375	0·9745	6·000
0·9564	10·750	0·9659	8·250	0·9749	5·875
0·9569	10·625	0·9664	8·125	0·9754	5·750
0·9574	10·500	0·9669	8·000	0·9759	5·625
0·9578	10·375	0·9673	7·875	0·9764	5·500
0·9583	10·250	0·9678	7·750	0·9768	5·375
0·9588	10·125	0·9683	7·625	0·9773	5·250
0·9593	10·000	0·9688	7·500	0·9778	5·125
0·9597	9·875	0·9692	7·375	0·9783	5·000
0·9602	9·750				

3. ALKALIMETRY.

A. Estimation of Potassa, Soda, or Ammonia, from the Specific Gravity of their Solutions.

§ 194.

In pure or nearly pure solutions of hydrated soda or potassa, or of ammonia, the percentage of alkali may be estimated from the specific gravity of the solution.

B. Estimation of the total Amount of Carbonated and Caustic Alkali in crude Soda and in Potashes.

The "soda ash" of commerce is a crude sodium carbonate—the "potashes" and "pearlash" a crude potassium carbonate. The commercial value of these articles depends on the percentage of alkali carbonate (or caustic alkali) that they contain, which is very variable.

I. Volumetric Methods.

Method of Descroizilles *and* Gay-Lussac, *slightly modified.*

§ 195.

The principle of this method is the converse of that on which the acidimetric method described § 192 is based, *i.e.*, if we know the quantity of an acid of known strength required to saturate an unknown quantity of caustic potassa or soda, or of potassium carbonate or sodium carbonate, we may readily calculate from this the amount of alkali present.

For technical analyses we may employ normal sulphuric acid. For the method of preparing and using it, see p. 687—8.

For the analysis we may conveniently weigh off such a quantity of the substance that the number of c. c. of acid required to neutralize it shall directly express its percentage of the alkali or carbonate sought.

Since 100 c. c. of the normal solution contain $\frac{1}{20}$ of 98 grm. H_2SO_4, the proper quantities of the sodium and potassium compounds to employ are $\frac{1}{20}$ of the weight of the compound required to neutralize 98 grm. H_2SO_4 viz.:

Potassa, K_2O	4·713	grm.
Potassium hydroxide, KOH	5·613	"
Potassium carbonate, K_2CO_3	6·913	"
Hydrogen potassium carbonate, $KHCO_3$	10·013	"
Soda, Na_2O	3·104	"
Sodium hydroxide, NaOH	4·004	"
Sodium carbonate (dry), Na_2CO_3	5·304	"
Sodium carbonate crystallized, $Na_2CO_3 \cdot 10\,H_2O$	14·304	"
Hydrogen sodium carbonate, $NaHCO_3$	8·404	"

With regard to the examination of pearlash by this method, the following points deserve attention :—

The various sorts of potash of commerce contain, besides potassium carbonate (and caustic potassa):

a. Normal salts (*e.g.*, potassium sulphate, potassium chloride).

b. Salts with alkaline reaction (*e.g.*, potassium silicate, potassium phosphate).

c. Admixtures insoluble in water, more especially calcium carbonate, phosphate, and silicate.

The salts named in *a* exercise no influence upon the results, but not so those named in *b* and *c*. Those in *c* may be removed by filtration; but the admixture of the salts named in *b* constitutes an irremediable though slight source of error; that is to say, if it is desired to confine the determination to the caustic and carbonated alkali. But as regards the estimation of the value of pearlash for many purposes, the term error cannot be applied; as, for instance, in the preparation of caustic potassa, by boiling the solution with lime, the alkali combined with silicic acid and with phosphoric acid is converted, like the carbonate, into the caustic state.

If you are not satisfied with finding the percentage of available alkali, but desire also to know whether the remainder consists simply of foreign salts, or whether water is also present, the determination of the latter substance must precede the alkalimetric examination. The same remark applies also to soda.

With regard to the examination of soda by this method, the following points deserve attention:—

The soda of commerce, prepared by LEBLANC's method, contains, besides sodium carbonate, always, or at least generally, sodium hydroxide, sodium sulphate, sodium chloride, sodium silicate and aluminate, and not seldom also sodium sulphide, sodium thiosulphate and sulphite.*

The three last-named substances impede the process, and interfere more or less with the accuracy of the results. Their presence is ascertained in the following way:—

a. Mix with sulphuric acid; a smell of hydrogen sulphide reveals the presence of *sodium sulphide*, with which sodium thiosulphate is also invariably associated.

b. Color dilute sulphuric acid with a drop of solution of potassium permanganate or chromate, and add some of the soda under examination, but not sufficient to neutralize the acid. If the solution retains its color, this proves the absence of both sodium sulphite and thiosulphate; but if the fluid loses its color, or turns green, as the case may be, one of these salts is present.

c. Whether the reaction described in *b* proceeds from sodium

* Traces of sodium cyanide are also occasionally found.

sulphite or thiosulphate is ascertained by supersaturating a clear solution of the sample under examination with hydrochloric acid. If the solution, after the lapse of some time, becomes turbid, owing to the separation of sulphur (emitting at the same time the odor of sulphurous acid), this may be regarded as a proof of the presence of sodium thiosulphate; however, the solution may, besides the thiosulphate, also contain sodium sulphite. With respect to the detection of sodium sulphite in the presence of thiosulphate, comp. "Qual. Anal.," p. 204.

The defects arising from the presence of the three compounds in question may be remedied in a measure, by igniting the weighed sample of the soda with potassium chlorate, before proceeding to saturate it. This operation converts the sodium sulphide, thiosulphate, and sulphite into sodium sulphate. But if sodium thiosulphate is present, the process serves to introduce another source of error, as that salt, upon its conversion into sulphate, decomposes a molecule of sodium carbonate, and expels the carbonic acid of the latter [$Na_2S_2O_3 + 4O$ (from the potassium chlorate) $+ Na_2CO_3 = 2(Na_2SO_4) + CO_2$].

The presence of sodium silicate and of sodium aluminate may be generally recognized by the separation of a precipitate as soon as the solution is saturated with acid. If you intend the result to express the quantity of carbonated and caustic alkali only, the presence of these two bodies becomes a slight source of error; but if you wish to estimate the value of the soda for many purposes, no error will be caused.

Method of FR. MOHR, *modified.*

§ 196.

Instead of estimating the alkalies in the direct way by means of an acid of known strength, we may estimate them also, as proposed first by FR. MOHR,* by supersaturating with standard acid, expelling the carbonic acid by boiling, and finally by determining by standard alkali solution the excess of standard acid added.

This process gives very good results, and is therefore particularly suited for scientific investigations. It requires the standard fluids mentioned in § 192, viz., a standard acid and standard solu-

* Annal. d. Chem. u. Pharm. 86, 129.

tion of potassium or sodium hydroxide. Each of these fluids is filled into a MOHR's burette.

The process is as follows:—

Dissolve the alkali in water, and add a measured quantity of tincture of cochineal; run in now as much of the standard acid as will suffice to impart an orange tint to the fluid; then boil, and remove the last traces of carbonic acid, by boiling, shaking, blowing into the flask, and finally sucking out the air.

Now add standard solution of potash, drop by drop, until the color just appears violet. There is no difficulty in determining the exact point at which the reaction is completed.

If the standard solutions of potash and acid are of corresponding strength, the number of c. c. used of the potash solution is simply deducted from the number of c. c. used of the acid. The remainder expresses the volume of the acid solution neutralized by the alkali in the examined sample. If the two standard fluids are not of corresponding strength, the excess of acid added, and subsequently neutralized by the potash solution, is calculated from the known relation the one bears to the other.

For the method of calculating the weight of any alkali corresponding to 1 c. c. of a standard solution is given in § 192, p. 682.

With regard to certain variations from the ordinary course which are occasionally convenient, comp. p. 685.

§ 197.

There now still remain two questions to be considered, which are of importance for the estimation of the commercial value of potash and soda. The first concerns the separate determination of the caustic alkali, which the sample under examination may contain besides the carbonate; the second, the determination of sodium carbonate in presence of potassium carbonate.

The product may be tested qualitatively for caustic alkali by adding to a solution barium chloride so long as a precipitate forms; if the solution still remains alkaline, presence of caustic alkali is indicated.

C. Determination of the Caustic Alkali which Commercial Alkali may contain beside the Carbonate.

Many kinds of potashes and crude soda, more especially the latter, contain, besides alkali carbonate, also caustic alkali; and the chemist is often called upon to determine the amount of the latter; as it is, for instance, by no means a matter of indifference to the soap-boiler how much of the soda is supplied to him already in the caustic state. This may be effected as follows:—

a. Determine in a portion of the substance carbonic acid directly: convert the caustic alkali in another weighed portion into carbonate by mixing with about an equal quantity of sand and about $\frac{1}{3}$ its weight of ammonium carbonate, adding as much water as the mass will absorb, evaporating and igniting to expel water, and determine carbonic acid in the residue. Deduct the percentage of carbonic acid found in the original substance from that found after treatment with ammonium carbonate. An amount of caustic alkali must be present equivalent to this difference (for 44 pts. CO_2, 62·08 pts. Na_2O, or 94·26 pts. K_2O).

b. Determine (according to § 196) the total amount of alkali existing both as carbonate and caustic, and reckon it all as carbonate. In another portion determine the caustic alkali as follows: Dissolve a suitable weighed portion in a measuring flask, add barium chloride so long as a precipitate forms, fill to the mark with water, shake and allow to settle clear without exposure to the air, take out a known portion of the supernatant clear fluid with a pipette, and determine volumetrically the caustic alkali or equivalent barium hydroxide which it contains. The amount actually existing as carbonate can now be found by calculating the carbonate equivalent to the caustic thus found, and deducting it from the total amount of alkali found and reckoned as carbonate in the other portion.

D. Estimation of Sodium Carbonate in presence of Potassium Carbonate.

Soda, being much cheaper than potash, is occasionally used to adulterate the latter. The common alkalimetric methods not only fail to detect this adulteration, but they give the admixed sodium carbonate as potassium carbonate. Many processes* have been proposed for estimating in a simple way the soda contained in potash, but not one of them can be said to satisfy the requirements of the case.

The following tolerably expeditious process, however, gives

* Comp. Handwörterbuch der Chemie, 2 Aufl., I. 443.

accurate results: Dissolve 6·25 grm. of the gently-ignited pearlash in water, filter the solution into a quarter-litre flask, add acetic acid in slight excess, apply a gentle heat until the carbonic acid is expelled, then add to the fluid, while still hot, lead acetate, drop by drop, until the formation of a precipitate of lead sulphate *just* ceases; allow the mixture to cool, add water up to the mark, shake, allow to deposit, filter through a dry filter, and transfer 200 c. c. of the filtrate, corresponding to 5 grm. of pearlash, to a ¼-litre flask. Add hydrogen sulphide water up to the mark, and shake. If the lead acetate has been carefully added, the fluid will now smell of hydrogen sulphide, and no longer contain lead; in the contrary case, hydrogen sulphide gas must be conducted into it. After the lead sulphide has subsided, filter through a dry filter. Evaporate 50 c. c. of the filtrate (corresponding to 1 grm. of pearlash) with addition of 10 c. c. hydrochloric acid, of 1·10 sp. gr., in a weighed platinum dish, to dryness, then cover the dish, heat, and weigh; the weight found expresses the total quantity of potassium and sodium chlorides given by 1 grm. of the pearlash. Estimate the potassium and sodium now severally in the indirect way, by determining the chlorine volumetrically (§ 141, I., *b*). For the calculation of the results, see "Calculation of Analyses" in the Appendix.

4. ESTIMATION OF ALKALI-EARTH METALS BY THE ALKALIMETRIC METHOD.

§ 198.

The alkali-earth metals, when in the form of oxides, hydroxides, or carbonates, may also be determined volumetrically by means of standard acid and alkali solutions. Standard sulphuric acid may be used for magnesium; standard nitric acid for barium, strontium, and calcium. The only advantage which these acids possess over hydrochloric is that there is less liability of loss on heating solutions containing them in the free state, which is necessary when carbonic is present. Hydrochloric acid can, however, be used with safety if precaution be taken to avoid the presence of an unnecessary quantity when the solution is heated.

If an oxide or hydroxide free from carbonic acid is to be examined, add some water to a weighed quantity, and allow the standard acid to flow in from a burette until solution is effected and the solution colored with litmus or cochineal gives an acid reaction. Then determine the excess of acid used by means of the standard alkali solution.

In case of a carbonate, dissolve in a flask, adding first water, then standard hydrochloric acid from a burette in small successive portions, until solution is complete; next add the indicator (litmus or cochineal) and allow the standard alkali solution to run in from a burette until the free acid is nearly neutralized. Now remove the carbonic acid by boiling a few minutes, and complete the neutralization with the standard alkali.

To calculate the amount of the alkali-earth metal, deduct from the volume of standard acid used the amount neutralized by the total quantity of standard alkali used (which is calculated from the known volumetric relation of the two solutions); the remainder has formed a normal salt with the alkali-earth metal. The quantity found to be required, its absolute strength, molecular weight, and the atomic weight of the alkali-earth metal, are data for calculating the amount of the latter, in the manner explained on pp. 681–4. If hydrochloric acid is used for a standard acid, bear in mind that 2 mol. correspond to 1 at. of an alkali-earth metal.

5. CHLORIMETRY.

§ 199.

The "chloride of lime," or bleaching powder" of commerce, contains calcium hypochlorite, calcium chloride, and calcium hydroxide. The two latter ingredients are for the most part combined with one another as calcium oxychloride. In freshly prepared and perfectly normal chloride of lime, the quantities of calcium hypochlorite and calcium chloride present stand to each other in the proportion of their mol. weights. When such chloride of lime is brought into contact with dilute sulphuric acid, the whole of the chlorine it contains is liberated in the elementary form, in accordance with the following equation:—

$$CaCl_2O_2 + CaCl_2 + 2H_2SO_4 = 2CaSO_4 + 2\,H_2O + 4Cl.$$

On keeping chloride of lime, however, the proportion between calcium hypochlorite and calcium chloride gradually changes—the former decreases, the latter increases. Hence, from this cause alone, to say nothing of original difference, the commercial article is not of uniform quality, and on treatment with acid gives sometimes more and sometimes less chlorine.

As the value of this article depends entirely upon the amount of chlorine set free on treatment with acid, chemists have devised various simple methods of determining the available amount of chlorine in any given sample. These methods have collectively received the name of Chlorimetry. We describe a few of the best.

Preparation of the Solution of Chloride of Lime.

The solution is prepared alike for all methods, and best in the following manner:—

Weigh off 10 grm., triturate finely with a little water, add gradually more water, pour the liquid into a litre flask, triturate the residue again with water, and rinse the contents of the mortar carefully into the flask; fill the latter to the mark, shake the milky fluid, and examine it at once in that state, *i.e.*, without allowing it to deposit; and every time, before measuring off a fresh portion, shake again. The results obtained with this turbid solution are much more constant and correct than when, as is usually recommended, the fluid is allowed to deposit, and the experiment is made with the supernatant clear portion alone. The truth of this may readily be proved by making two separate experiments, one with the decanted clear fluid, and the other with the residuary turbid mixture. Thus, for instance, in an experiment made in my own laboratory, the decanted clear fluid gave 22·6 of chlorine, the residuary mixture 25·0, the uniformly mixed turbid solution 24·5.

1 c. c. of the solution of chloride of lime so prepared corresponds to 0·01 grm. chloride of lime.

A. Penot's *Method.**

§ 200.

This method is based upon the conversion of arsenious acid into arsenic acid, or more strictly, an arsenite into an arsenate, since the conversion is effected in an alkaline solution. Potassium iodide-starch paper is employed to ascertain the exact point when the reaction is completed.

* Bulletin de la Société Industrielle de Mulhouse, 1852, No. 118.—Dingler's Polytech. Journal, 127, 134.

a. Preparation of the Potassium Iodide-Starch Paper.

The following method is preferable to the original one given by PENOT:—

Stir 3 grm. of potato starch in 250 c. c. of cold water, boil with stirring, add a solution of 1 grm. potassium iodide and 1 grm. crystallized sodium carbonate, and dilute to 500 c. c. Moisten strips of fine white unsized paper with this fluid, and dry. Keep in a closed bottle.

b. Preparation of the solution of Arsenious Acid.

Dissolve 4·436 grm. of pure arsenious oxide (As_2O_3) and 13 grm. pure crystallized sodium carbonate in 600—700 c. c. water, with the aid of heat, let the solution cool, and then dilute to 1 liter. Each c. c. of this solution contains an amount of sodium arsenite equivalent to 0·004436 grm. arsenious oxide (As_2O_3), which corresponds to 1 c. c. chlorine gas of 0° and 760 mm. atmospheric pressure.*

As sodium arsenite in alkaline solution is liable, when exposed to access of air, to be gradually converted into sodium arsenate, PENOT's solution should be kept in small bottles with glass stoppers, filled to the top, and a fresh bottle used for every new series of experiments. According to Fr. MOHR† the solution keeps unchanged, if the arsenious oxide and the sodium carbonate are both *absolutely* free from oxidizable matters (arsenious sulphide, sodium sulphide, and sodium sulphite).

---

* Penot gives the quantity of arsenious oxide as 4·44; but I have corrected this number to 4·436, in accordance with the now received atomic weights of the substances and specific gravity of chlorine gas—after the following proportion:—

141·84 (4 at. Cl) : 198 (1. mol. As_2O_3) :: 3·17763 (weight of 1 litre of chlorine gas) : x; $x = 4·436$, *i.e.*, the quantity of arsenious oxide which 1 litre of chlorine gas converts into arsenic acid.

This solution is arranged to suit the foreign method of designating the strength of chloride of lime—viz., in chlorimetrical degrees (each degree represents 1 litre chlorine gas at 0° and 760 mm. pressure in a kilogramme of the substance). This method was proposed by Gay-Lussac. The degrees may readily be converted into per cents, and *vice versa*, thus: A sample of chloride of lime of 90° contains $90 \times 3·17763 = 285·986$ grm. chlorine in 1000 grm. or 28·59 in 100; and a sample containing 34·2 per cent. chlorine is of 107·6°, for 100 grm. of the substance contain 34·2 grm. chlorine; ∴ 1000 grm. of the substance contain 342 grm. chlorine, but 342 grm. chlorine $= \frac{342}{3·17763}$ litres $= 107·6$ litres; ∴ 1000 grm. of the substance contain 107·6 litres chlorine.

† His Lehrbuch der Titrirmethode, 2 Aufl., S. 290.

c. The Process.

Measure off, with a pipette, 50 c. c. of the solution of chloride of lime prepared according to the directions of § 199, transfer to a beaker, and from a 50 c. c. burette add slowly, and at last drop by drop, the solution of arsenious acid, with constant stirring, until a drop of the mixture produces no longer a blue-colored spot on the iodized paper ; it is very easy to hit the point exactly, as the gradually increasing faintness of the blue spots made on the paper by the fluid dropped on it indicates the approaching termination of the reaction, and warns the operator to confine the further addition of the solution of arsenious acid to a single drop at a time. The number of ½ c. c. used indicates directly the number of chlorimetrical degrees (see note), as the following calculation shows: Suppose you have used 40 c. c. of solution of arsenious acid, then the quantity of chloride of lime used in the experiment contains 40 c. c. of chlorine gas. Now, the 50 c. c. of solution employed correspond to 0·5 grm. of chloride of lime; therfore 0·5 grm. of chloride of lime contain 40 c. c. chlorine gas, therefore 1000 grm. contain 80000 c. c. = 80 litres. This method gives very constant and accurate results, and appears to be particularly well suited for use in manufacturing establishments where there is no objection, on the score of danger, to the employment of arsenious acid. (Expt. No. 99.)

B. Otto's *Method.*

§ 201.

The principle of this method is as follows:—

Two molecules of ferrous sulphate when brought in contact with chlorine in presence of water and free sulphuric acid, give 1 mol. ferric sulphate and 2 mol. HCl, the process consuming 2 at. chlorine.

$$2FeSO_4 + H_2SO_4 + 2Cl = Fe(SO_4)_3 + 2HCl.$$

1 mol. crystallized ferrous sulphate:—

$$(FeSO_4 \cdot 7H_2O) = 278$$

correspond to 35·46 of chlorine, or, in other terms, 0·7839 grm. crystallized ferrous sulphate correspond to 0·1 grm. chlorine.

The ferrous sulphate required for these experiments is best prepared as follows:—

Take iron nails, free from rust, and dissolve in dilute sulphuric acid, applying heat in the last stage of the operation: filter the solution, still hot, into about twice its volume of common alcohol. The precipitate consists of

$$FeSO_4 + 7H_2O.$$

Collect upon a filter, wash with common alcohol, spread upon a sheet of blotting paper, and dry in the air. When the mass smells no longer of alcohol, transfer to a bottle and keep this well corked. Instead of ferrous sulphate, ammonium ferrous sulphate (p. 118) may be used. 0·1 grm. of chlorine reacts with 1·1055 grm. of this double sulphate.

The Process.

Dissolve 3·1356 grm. (4 × ·7839 grm.) of the precipitated ferrous sulphate, or 4·422 grm. (4 × 1·1055 grm.) of ammonium ferrous sulphate, with addition of a few drops of dilute sulphuric acid, in water, to 200 c. c.; take out, with a pipette, 50 c. c., corresponding to 0·7839 grm. ferrous sulphate, or 1·1055 grm. ammonium ferrous sulphate, dilute with 150—200 c. c. water, add a sufficiency of pure hydrochloric acid, and run in from a 50 c. c. burette the freshly shaken solution of chloride of lime, prepared according to § 199, until the ferrous sulphate is completely converted into ferric sulphate. To know the exact point when the reaction is completed, place a number of drops of a solution of potassium ferricyanide on a plate, and when the operation is drawing to an end apply some of the mixture with a stirring-rod to one of the drops on the plate, and observe whether it produces a blue precipitate; repeat the experiment after every fresh addition of two drops of the solution of chloride of lime. When the mixture no longer produces a blue precipitate in the solution of potassium ferricyanide on the plate, read off the number of volumes used of the solution of chloride of lime.

The amount of solution of chloride of lime used contained 0·1 grm. of chlorine. Suppose 40 c. c. have been used: as every c. c. corresponds to 0·10 grm. of chloride of lime, the percentage by weight of available chlorine in the chloride of lime is found by the following proportion :—

$$0{\cdot}40 : 0{\cdot}10 :: 100 : x; \; x = 25;$$

or, by dividing 1000 by the number of c. c. used of the solution of chloride of lime.

This method also gives very satisfactory results, provided always that the ferrous salt is perfectly dry and free from ferric salt.

Modification of the preceding Method.

Instead of the solution of ferrous sulphate, a solution of ferrous chloride, prepared by dissolving pianoforte wire in hydrochloric acid (according to p. 268), may be used with the best results. If 0·6316 of pure metallic iron, *i. e.*, 0·6335 of fine pianoforte wire (which may be assumed to contain 99·7 per cent. of iron), are dissolved to 200 c. c., the solution so prepared contains exactly the same amount of iron as the solution of ferrous sulphate above mentioned—that is to say, 50 c. c. of it correspond to 0·1 grm. chlorine. But as it is inconvenient to weigh off a definite quantity of iron wire, the following course may be pursued in preference: Weigh off accurately about 0·15 grm., dissolve, dilute the solution to about 200 c. c., convert the ferrous into ferric chloride with the solution of chloride of lime, prepared according to the directions of § 199, and calculate the chlorine by the proportion

$$56 : 35{\cdot}46 :: \text{the quantity of iron used} : x;$$

the x found corresponds to the chlorine contained in the amount used of the solution of chloride of lime. This calculation may be dispensed with by the application of the following formula, in which the carbon in the pianoforte wire is taken into account:—

Multiply the weight of the pianoforte wire by 6313, and divide the product by the number of c. c used of the solution of chloride of lime; the result expresses the percentage of chlorine by weight.

This method gives very good results. I have described it here principally because it dispenses altogether with the use of standard fluids. It is therefore particularly well adapted for occasional examinations of samples of chloride of lime, and also by way of control. (See Expt. No. 99.)

C. Bunsen's *Method.*

Pour 10 c. c. of the solution of chloride of lime, prepared according to the directions of § 199 (containing 0·1 chloride of lime), into a beaker, and add about 6 c. c. of the solution of potassium iodide, prepared according to p. 445 (containing 0·6 KI); dilute the mixture with about 100 c. c. of water, acidify with hydrochloric acid, and determine the liberated iodine as directed § 146. As 1 at. iodine corresponds to 1 at. chlorine, the calculation is easy. This method gives excellent results. (Compare Expt. No 99.)

6. EXAMINATION OF BLACK OXIDE OF MANGANESE.

§ 202.

The native black oxide of manganese (as also the regenerated artificial product) is a mixture of manganese dioxide with lower oxides of that metal, and with ferric oxide, clay, &c., ; it also invariably contains moisture, and frequently chemically combined water. The commercial value of the article depends entirely upon the amount of dioxide (or, more correctly expressed, of available oxygen) which it contains. By "available oxygen" we understand the excess of oxygen contained in a manganese, over the 1 at. combined with the metal to monoxide; upon treating the ore with hydrochloric acid, an amount of chlorine is obtained equivalent to this excess of oxygen. This available oxygen is always expressed in the form of manganese dioxide. 1 at. corresponds to 1 mol. manganese dioxide, since $MnO_2 = MnO + O$.

I. Drying the Sample.

All analyses of manganese proceed, of course, upon the supposition that the sample operated upon is a fair average specimen of the ore. A portion of a tolerably finely powdered average sample is generally sent for analysis to the chemist; in the case of new lodes, however, a number of samples, taken from different parts of the mine, are also occasionally sent. If, in the latter case, the average composition of the ore is to be ascertained, and not simply that of several samples, the following course must be resorted to: Crush the several samples of the ore, in an iron mortar, to coarse powder, and pass the whole of this through a rather coarse sieve. Mix uniformly, then remove a sufficiently large portion of the coarse powder with a spoon, reduce it to powder in a steel mortar, passing the whole of this through a fine sieve. Mix the powder obtained by this second process of pulverization most intimately; take about 8—10 grm. of it, and triturate this, in small portions at a time, in an agate mortar, to an impalpable powder. Average samples are generally already sufficiently fine to require only the last operation.

As regards the temperature at which the powder is to be dried, if you desire to expel the whole of the moisture without disturbing any of the water of hydration, the temperature adopted must be 120° (this is the result of my own experiments, see Expt. No. 100). But, as there appears to be at present an almost universal understanding, in the manganese trade, to limit the drying temperature to 100°, the fine powder is exposed, in a shallow copper or brass pan, for 6 hours, to the temperature of boiling water, in a water bath (p. 53, fig. 23.)

When the samples have been dried, they are introduced, still hot, into glass tubes 12—14 cm. long and 8—10 mm. wide, sealed at one end; these tubes are then corked and allowed to cool.

In laboratories where whole series of analyses of different ores are of frequent occurrence, it is advisable to number the drying-pans and glass tubes, and to transfer the samples always from the pan to the tube of the corresponding number.

II. Determination of the Manganese Dioxide.

§ 203.

Of the many methods that have been proposed for the valuation of manganese ores, I select three as the most expeditious and accurate. The first is more particularly adapted for technical purposes.

A. Fresenius *and* Will's *Method.*

a. If oxalic acid (or an oxalate) is brought into contact with manganese dioxide in presence of water and excess of sulphuric acid, manganous sulphate is formed, and carbon dioxide evolved, while the oxygen, which we may assume to exist in the manganese dioxide in combination with the monoxide, combines with the elements of the oxalic acid, and thus converts the latter into carbon dioxide.

$$MnO_2 + H_2SO_4 + H_2C_2O_4 = MnSO_4 + 2H_2O + 2CO_2.$$

Each atom of available oxygen, or, what amounts to the same, each mol. binoxide of manganese = 87, gives 2 mol. carbon dioxide = 88.

b. If this process is performed in a weighed apparatus from which nothing except the evolved carbonic acid can escape, and which, at the same time, permits the complete expulsion of that acid, the diminution of weight will at once show the amount of carbonic acid which has escaped, and consequently, by a very simple calculation, the quantity of dioxide contained in the analyzed manganese ore. As 88 parts, by weight, of carbon dioxide correspond to 87 of manganese dioxide, the carbon dioxide found need simply be multiplied by 87, and the product divided by 88, or the carbon dioxide may be multiplied by

$$\frac{87}{88}=0{\cdot}9887,$$

to find the corresponding amount of manganese dioxide.

c. But even this calculation may be avoided by simply using in the operation the exact weight of ore which, if the latter consisted of pure dioxide, would give 100 parts of carbon dioxide.

The number of parts evolved of carbon dioxide expresses, in that case, directly the number of parts of dioxide contained in 100 parts of the analyzed ore. It results from *b* that 98·87 is the number required. Suppose the experiment is made with 0·9887 grm. of the ore, the number of centigrammes of carbon dioxide evolved in the process expresses directly the percentage of dioxide contained in the analyzed manganese ore. Now, as the amount of carbon dioxide evolved from 0·9887 grm. of manganese would be rather small for accurate weighing, it is advisable to take a multiple of this weight, and to divide afterwards the number of centigrammes of carbon dioxide evolved from this multiple weight by the same number by which the unit has been multiplied. The multiple which answers the purpose best for superior ores is the triple, = 2·966; for inferior ores, I recommend the quadruple, = 3·955, or the quintuple, = 4·9435.

The analytical process is performed in the apparatus illustrated in fig. 58, p. 409.

The flask *A* should hold, up to the neck, about 120 c. c.; *B* about 100 c. c. The latter is half filled with sulphuric acid; the tube *a* is closed at *b* with a little wax ball, or a very small piece of caoutchouc tubing, with a short piece of glass rod inserted in the other end.

Place 2·966, or 3·955, or 4·9435 grm.—according to the quality of the ore—in a watch-glass, and tare the latter most accurately on a delicate balance; then remove the weights from the watch-glass, and replace them by manganese from the tube, very cautiously, with the aid of a gentle tap with the finger, until the equilibrium is exactly restored. Transfer the weighed sample, with the aid of a card, to the flask *A*, add 5—6 grm. normal sodium oxalate, or about 7·5 grm. normal potassium oxalate, in powder, and as much water as will fill the flask to about one third. Insert the cork into *A*, and tare the apparatus on a strong but delicate balance, by means of shot, and lastly, tinfoil, not placed directly on the scale, but in an appropriate vessel. The tare is kept under a glass bell. Try whether the apparatus closes air-tight. Then make some sulphuric acid flow from *B* into *A*, by applying suction to *d*, by means of a caoutchouc tube. The evolution of carbon dioxide commences immediately in a steady and uniform manner. When it begins to slacken, cause a fresh portion of sulphuric acid to pass into *A*, and repeat this until the manganese ore is completely decomposed, which, if the sample has been very finely pulverized, requires at the most about five minutes. The complete decomposition of the analyzed ore is indicated, on the one hand, by the cessation of the disengagement of carbon dioxide, and its non-renewal upon the influx of a fresh portion of sulphuric acid into *A*; and, on the other hand, by the total disappearance of every trace of black powder from the bottom of *A*.*

Now cause some more sulphuric acid to pass from *B* into *A*, to heat the fluid in the latter, and expel the last traces of carbon dioxide therein dissolved; remove the wax stopper, or india-rubber tube, from *b*, and apply gentle suction to *d* until the air drawn out tastes no longer of carbon dioxide. Let the apparatus cool completely in the air, and place it on the balance, with the tare on the other scale, and restore equilibrium. The number of centigramme weights added, divided by 3, 4, or 5, according to the multiple of 0·9887 grm. used, expresses the percentage of dioxide contained in the analyzed ore.

In experiments made with definite quantities of the ore, weighing in an open watch-glass cannot well be avoided, and the dried manganese is thus exposed to the chance of a reabsorption of water

* If the manganese ore has been pulverized in an iron mortar, a few black spots (particles of iron from the mortar) will often remain perceptible.

from the air, which of course tends to interfere, to however so trifling an extent, with the accuracy of the results. In very precise experiments, therefore, the best way is to analyze an indeterminate quantity of the ore, and to calculate the percentage as shown above. For this purpose, one of the little corked tubes, filled with the dry pulverized ore, is accurately weighed, and about 3 to 5 grm. (according to the quality of the ore) are transferred to the flask *A*. By now reweighing the tube, the exact quantity of ore in the flask is ascertained. To facilitate this operation, it is advisable to scratch on the tube, with a file, marks indicating approximately the various quantities which may be required for the analysis, according to the quality of the ore.

With proper skill and patience on the part of the operator, a good balance and correct weights, this method gives most accurate and corresponding results, differing in two analyses of the same ore barely to the extent of 0·2 per cent.

If the results of two assays differed by more than 0·2 per cent., a third experiment should be made. In laboratories where analyses of manganese ores are matters of frequent occurrence, it will be found convenient to use an aspirator for sucking out the carbon dioxide. In the case of very moist air, the error which proceeds from the fact that the water in the air drawn through the apparatus is retained, and which is usually quite inconsiderable, may now be increased to an important extent. Under such circumstances, connect the end of the tube *b* with a calcium chloride tube during the suction.

Very accurate determinations may also be made by weighing the evolved carbon dioxide. For this purpose the apparatus described on page 414, fig. 61, is well adapted. From ·5 to 1. grm. ore should be used for a determination. Introduce the ore and oxalic acid or oxalate into the decomposing flask, fill the flask about one third with water, connect the several parts of the apparatus as for the determination of carbonic acid, decompose the ore by admitting gradually strong sulphuric acid, remove the evolved CO_2 completely from the unweighed portion of the apparatus into the potash bulbs as described for the determination of CO_2.

Some ores of manganese contain *carbonates of the alkali-earth metals*, which of course necessitates a modification of the foregoing process. To ascertain whether carbonates of the alkali-earth metals are present, boil a sample of the pulverized ore with water, and add nitric acid. If any effervescence takes place, the process is modified as follows (Röhr*):

After the weighed portion of ore has been introduced into the flask *A*, treat it with water, so that the flask may be about ¼ full, add a few drops of dilute sulphuric acid (1 part, by weight, sulphuric acid, to 5 parts water) and warm with agitation, preferably in a water bath. After some time dip a rod in and test whether the fluid possesses a strongly acid reaction. If it does not, add more sulphuric acid. As soon as the whole of the carbonates are decomposed by continued heating of the acidified fluid, completely neutralize the excess of acid with soda solution free from carbonic acid, allow to cool, add the usual quantity of sodium oxalate, and proceed as above.

If you have no soda solution free from carbonic acid at hand, you may place the sodium oxalate or oxalic acid (about 3 grm.) in a small tube, and suspend this in the flask *A* by means of a thread fastened by the cork. When the apparatus is tared, and you have satisfied yourself that it is air-tight, release the thread and proceed as above.

B. Bunsen's *Method.*

Reduce the ore to the very finest powder, weigh off about 0·4 grm., introduce this into the small flask *a*, illustrated in fig. 64, p. 435, and pour pure fuming hydrochloric acid over it; conduct the process exactly as in the analysis of chromates. Boil until the ore is completely dissolved and all the chlorine expelled, which is effected in a few minutes. 2 atoms of iodine separated correspond to 2 at. chlorine evolved, and accordingly to 1 mol. of manganese dioxide. For the estimation of the separated iodine, the method § 146 may be employed. Results most accurate.

C. *Estimation of the Manganese Dioxide by means of Iron.*

Dissolve, in a small long-necked flask, placed in a slanting position, about 1 grm. pianoforte wire, accurately weighed, in moderately concentrated pure hydrochloric acid; weigh off about 0·6 grm. of the sample of manganese ore in a little tube, drop this into the flask, with its contents, and heat cautiously until the ore is dissolved. 1 mol. of manganese dioxide converts 2 at. of dissolved iron from the state of ferrous to ferric chloride. When complete

* Zeitschr. f. anal. Chem. 1, 48.

solution has taken place, dilute the contents of the flask with water, allow to cool, rinse into a beaker, and determine the iron still remaining in the state of ferrous chloride with potassium dichromate (p. 274). Deduct this from the weight of the wire employed in the process; the difference expresses the quantity of iron which has been converted by the oxygen of the manganese from ferrous to ferric chloride.* This difference multiplied by $\frac{43.5}{56}$ or 0·7768 gives the amount of manganese dioxide in the analyzed ore. This method also, if carefully executed, gives very accurate results.

The main reason why this method is less suitable for industrial use than the first lies in the fact that the analyst must work with much smaller quantities of substance. Hence to obtain results equally accurate with those yielded by A, far greater nicety in weighing and manipulating is required. Instead of metallic iron, weighed quantities of pure ferrous sulphate or ferrous ammonium sulphate may be used.

III. Estimation of Moisture in Manganese.

§ 204.

In the purchase and sale of manganese, a certain proportion of moisture is usually assumed to be present, and often a percentage is fixed within which the moisture must be confined. In estimating the moisture the same temperature should be employed, at which the drying for the purpose of determining the dioxide is effected (§ 202, I.).

As the amount of moisture in an ore may be altered by the operations of crushing and pulverizing, the experiment should be made with a sample of the mineral which has not yet been subjected to these processes. The drying must be continued until no further diminution of weight is observed; at 100°, this takes about 6 hours; at 120°, generally only 1½ hours. If the moisture in a manganese ore is not to be estimated on the spot, but in the laboratory, a fair average sample of the ore should be forwarded to the chemist in a strong, perfectly dry, and well-corked bottle.

---

* In very precise experiments, the weight of the iron must be multiplied by 0·997, since pianoforte wire may always be assumed to contain about 0·003 impurities.

IV. Estimation of the Amount of Hydrochloric Acid required for the complete Decomposition of a Manganese.

§ 205.

Different manganese ores, containing the same amount of available oxygen, or, as it is usually expressed, of binoxide of manganese, may require very different quantities of hydrochloric acid to effect their decomposition and solution, so as to give an amount of chlorine corresponding to the available oxygen in them; thus, an ore consisting of 60 per cent. of binoxide of manganese and 40 per cent. of sand and clay requires 4 mol. hydrochloric acid to 1 at. of available oxygen; whereas an equally rich ore containing lower oxides of manganese, ferric oxide, or calcium carbonate requires a much larger proportion of hydrochloric acid.

The quantity of hydrochloric acid in question may be determined by the following process:—

Determine volumetrically the strength of a moderately strong hydrochloric acid (of, say, 1·10 sp. gr.). Warm 10 c. c. of the same acid with a weighed quantity (about 1 grm.) of the manganese, in a small, long-necked flask, with a glass tube, about 3 feet long, fitted into the neck. Fix the flask in a position that the tube is directed obliquely upwards, and then gently heat the contents. As soon as the manganese is decomposed, apply a somewhat stronger heat for a short time, to expel the chlorine which still remains in solution; but carefully avoid continuing the application of heat longer than is absolutely necessary, as it is of importance to guard against the slightest loss of hydrochloric acid. Let the flask cool, dilute the contents with water, and determine the free hydrochloric acid remaining. Deduct the quantity found from that originally added; the difference expresses the amount of hydrochloric acid required to effect the decomposition of the manganese ore.

7. ANALYSIS OF COMMON SALT.

§ 206.

I select this example to show how to analyze, with accuracy and tolerable expedition, salts which, with a predominant principal ingredient, contain small quantities of other substances.

a. Reduce the salt by trituration to a uniform powder, and put this into a stoppered bottle.

b. Weigh off 10 grm. of the powder, and dissolve in a beaker by digestion with water; filter the solution into a ½-litre flask, and thoroughly wash the small residue which generally remains. Finally, fill the flask with water up to the mark, and shake the fluid.

If small white grains of calcium sulphate are left on dissolving the salt, reduce them to powder in a mortar, add water, let the mixture digest for some time, decant the clear supernatant fluid on to a filter, triturate the undissolved deposit again, add water, &c., and repeat the operation until complete solution is effected.

c. Ignite and weigh the dried insoluble residue of *b*, and subject it to a qualitative examination, more especially with a view to ascertain whether it is perfectly free from calcium sulphate.

d. Of the solution *b*, measure off successively the following quantities:

For *e.*	50 c. c.	corresponding	to	1	grm. of	common salt.
" *f*	150 c. c.	"	"	3	"	"
" *g.*	150 c. c.	"	"	3	"	"
" *h.*	50 c. c.		"	1		

e. Determine in the 50 c. c. measured off, the *chlorine* as directed § 141, I., *a* or *b*.

f. Determine in the 150 c. c. measured off, *sulphuric acid* as directed § 132, I., 1.

g. Determine in the 150 c. c. measured off, the *calcium* and *magnesium*, as directed p. 496, 28.

h. Mix the 50 c. c. measured off in a platinum dish, with about ½ c. c. of pure concentrated sulphuric acid, and proceed as directed § 98, 1. The neutral residue contains the sulphates of sodium, calcium, and magnesium. Deduct from this the quantity of the two latter substances as resulting from *g*; the remainder is *sodium* sulphate.

i. Determine in another weighed portion of the salt, the *water* as directed § 35, *a*, *α*, at the end.

k. Bromine and other bodies, of which only very minute traces are found in common salt, are determined by the methods described in Part I.

8. ANALYSIS OF GUNPOWDER.*

§ 207.

Gunpowder, as is well known, consists of nitre, sulphur, and charcoal, and, in the ordinary condition, invariably contains a small quantity of moisture. The analysis is frequently confined to the determination of the three constituents and the moisture, but often the examination is extended to the nature of the charcoal, and the carbon, hydrogen, oxygen, and ash therein are estimated.

a. Determination of the Moisture.

Weigh 2—3 grm. of the substance (not reduced to powder) between two well-fitting watch-glasses, and dry in the desiccator, or at a gentle heat, not exceeding 60°, till the weight remains constant.

b. Determination of the Nitre.

Place an accurately weighed quantity (about 5 grm.) on a filter, moistened with water; saturate with water, and, after some time, repeatedly pour small quantities of hot water upon it until the potassium nitrate is completely extracted. Receive the first filtrate in a small weighed platinum dish, the washings in a beaker or small flask. Evaporate the contents of the platinum dish cautiously, adding the washings from time to time, heat the residue cautiously to incipient fusion, and weigh it.†

c. Determination of the Sulphur.

Oxidize 2—3 grm. of the powder with pure concentrated nitric acid and potassium chlorate, the latter being added in small portions, while the fluid is maintained in gentle ebullition. If the operation is continued long enough, it usually happens that both the charcoal and sulphur are fully oxidized, and a clear solution is

---

* As regards the determination of the sp. gr. of gunpowder, I refer to HEEREN's paper on the subject, in Mittheilungen des Gewerbevereins für Hannover, 1856, 168—178; Polyt. Centralbl. 1856, 1118.

† The potassium nitrate may also be estimated in an expeditious manner, and with sufficient accuracy for technical purposes, by means of a hydrometer, which is constructed to indicate the percentage of this ingredient when floated in water containing a certain proportion of gunpowder in solution. A method based upon the same principle, proposed by Uchatius, is given in the Wiener akad. Ber. X. 748; also Annal. d. Chem. u. Pharm. 88, 395.

finally obtained. Evaporate with excess of pure hydrochloric acid on a water-bath to dryness, filter, if undissolved charcoal should render it necessary, and determine the sulphuric acid after § 132, I., 1.

d. Determination of the Charcoal.

Digest a weighed portion of the powder repeatedly with ammonium sulphide, till all sulphur is dissolved, collect the charcoal on a filter dried at 100°, wash it first with water containing ammonium sulphide, then with pure water, dry at 100°, and weigh.

The charcoal so obtained must, under all circumstances, be tested for sulphur by the method given under *c*, and if occasion require, the sulphur must be determined in an aliquot part. The charcoal may also be examined as regards its behavior to potash solution (in which "red charcoal" * is partially soluble) and an aliquot part may be subjected to elementary analysis according to § 177 or § 178. For this latter purpose take a portion of the charcoal dried at 100°, and dry at 190° (WELTZIEN). If the charcoal, on this second drying, suffers a diminution of weight, calculate the latter into per cents. of the gunpowder, deduct it from the charcoal, and add it to the moisture.

9. ANALYSIS OF SILICATES AND SILICEOUS ROCKS.

§ 208.

The separation of silica in silicates which are decomposable by acids has been described in § 140, II., *a.*; and in silicates which are not thus decomposed in § 140, II., *b*. For determination of the alkalies, see page 426, *γ*. Methods for separating the other basic metals more commonly occurring in silicates are given in § 161. Some silicates contain water, fluorine, and iron both in the ferrous and ferric state, while siliceous rocks may contain in addition small quantities of carbonic acid, titanic acid, sulphur, phosphoric acid, &c., due to admixture of various minerals.

Below are some remarks respecting processes which may be required in such cases, more especially in the analysis of rocks.

1. *Decomposition.* If the greater part of the rock mass is undecomposable by acids, fuse directly with sodium carbonate, and

* Incompletely carbonized wood.

separate silica according to § 140, II., *b*. If the greater part is decomposable by hydrochloric acid, treat with hydrochloric acid and evaporate to dryness as described in § 140, II., *a*, for separation of silica. Treat the residue with strong hydrochloric acid 5 to 10 minutes, add water and filter. The insoluble portion, consisting of silica and the undecomposed part of the rock, is ignited with the filter in a platinum crucible and fused with sodium carbonate. Silica is then separated in the usual manner, and the second solution of basic metals thus obtained is added to the first. Alkalies are determined in another portion by the method of J. L. Smith (page 426.)

2. *Water*. Silicates dried at 100° occasionally contain *water*. This is determined by taking a weighed portion dried at 100° and igniting in a platinum crucible, or—in presence of carbon, carbonates, or ferrous iron—in a tube, through which a stream of dry air is drawn, the moisture expelled from the substance being retained by a weighed calcium chloride tube.

If the escaping aqueous vapors manifest acid reaction, owing to disengagement of *hydrofluoric acid* or *silicon fluoride*, mix the substance with 6 parts of finely triturated recently ignited lead oxide in a small retort, weigh, ignite, and weigh again. This method, however, cannot be used if carbonic acid as well as fluorrine is present. In that case the method employed by L. Sipöcz* may be used. Ignite 4 parts sodium carbonate in a platinum crucible till water is completely expelled, allow to cool to 50° or 60°, mix intimately with a platinum wire with 1 part of the pulverized dried silicate, introduce the mixture into a capacious platinum boat, rinsing out the last adhering portions with sodium carbonate. The boat, provided with a cover, is now placed in the middle of a porcelain tube (glazed inside) and heated in an air bath an hour to 120° or 130° C. During this time every trace of moisture should be removed from the mixture by passing dried air by means of a gasometer through the tube. The end of the tube through which the current of air makes its exit is provided with a calcium chloride tube, which at the end of the drying process is replaced by a weighed U tube, containing glass beads moistened with pure strong sulphuric acid. The substance is now brought to a red heat in a combustion furnace, and a regulated current of air (dried by sul-

* Zeitschr. f. anal. Chemie. 17, 207.

phuric acid) is passed over it about half an hour to carry the expelled water vapor into the absorbing apparatus. (It is obvious that this method can be used in any case instead of ignition with lead oxide).

3. *Carbonic acid and water.* If it can be proved by a preliminary experiment, that carbonic acid, as well as water, can be completely removed by intense ignition, and no other constituents (ferrous iron, manganese, fluorine, sulphides, alkali fluorides, and chlorides, &c.) are present which will cause change of weight on ignition; the joint amount of water and carbonic acid may be determined by loss of weight on ignition, and carbonic acid in another portion according to § 139, II., *e*. The amount of water present equals the difference between the two quantities thus found. If, as is usually the case, the nature of the substance does not allow the joint amount of water and carbonic acid to be determined by loss on ignition, water may be determined by the method recommended by SIPÖCZ (described above in 2). A much simpler and sufficiently accurate method, however, is to determine both water and carbonic acid at once by ignition of the substance with lead chromate mixed with $\frac{1}{10}$ its weight of potassium dichromate in a combustion tube, collecting and weighing the evolved water and carbonic acid. The process is conducted precisely as in the combustion of organic compounds (see § 177) except that it is not necessary (unless sulphides are present) to place lead chromate in front of the mixture of chromates with the pulverized rock. Heat should be applied toward the end of the process sufficient to fuse the mixture. It is desirable to use 2 grm. or even more of the substance for the determination, and to take great care to avoid presence of hygroscopic moisture in the chromates. Carbonaceous matter (rarely present in rocks) would of course interfere with the determination of CO_2 by this process.*

4. *Ferrous iron* is most readily determined by decomposing with a mixture of sulphuric and hydrofluoric acids and titration with potassium permanganate according to § 160, **84**, page 526. With a little skill and proper attention, the simple method of effecting the

* Very satisfactory results were obtained both by myself and Mr. W. J. Comstock in the determination of CO_2 in Iceland spar by this method. Mr. Comstock also obtained equally accurate results where a considerable quantity of pulverized fluor spar was added to the weighed Iceland spar, showing that presence of fluorides does not interfere with the process.—O. D. A.

decomposition described at the end of **84** can safely be employed, unless the substance is unusually difficult to decompose.

5. *Titanic acid* is frequently present in siliceous rocks in quantity sufficient to be determined. In the ordinary process of analysis, a part of it remains with the separated and weighed silica, while the remainder is precipitated and weighed along with the ferric and aluminium oxides. Both portions may be brought together, and determined as follows: Dissolve the silica in the platinum crucible in which it has been weighed with hydrofluoric acid, adding also a few drops of pure dilute sulphuric acid, and evaporate on a water-bath; add to the residue 2 or 3 c. c. of hydrofluoric acid, and evaporate again to ensure complete removal of silica. Ignite the residue strongly and weigh, in order to be able to deduct from the weight of the silica the amount of impurities thus found in it. Add next to the residue a little sodium carbonate and fuse. After cooling, add strong sulphuric acid drop by drop till with the aid of heat the mass is dissolved. It is best to use so much sulphuric acid that the mass will just remain liquid on cooling. It will then easily dissolve in a small quantity of water. Dissolve it in water and reserve the solution.

Dissolve the weighed ferric and aluminium oxides by prolonged digestion in concentrated hydrochloric acid; 12 to 24 hours will usually suffice for the solution if the oxides have not been too strongly ignited. Some flocks of silica, however, and possibly titanic acid may remain In order to ensure solution of the latter, add 20 c. c. dilute sulphuric acid and evaporate till fumes of sulphuric acid appear; cool, add a little water, digest till sulphates, &c., are dissolved, filter off and weigh the traces of silica, add the filtrate to the solution of titanic acid previously obtained. Add next to the solution, sodium carbonate until a *slight* precipitate is formed which does not redissolve on stirring. Next add 4 c. c. pure dilute sulphuric acid, which is designed to dissolve the slight precipitate and prevent precipitation of iron along with titanium in the subsequent part of the process (too much free acid would prevent complete precipitation of the titanic acid). Solution of sulphurous acid is then added to reduce the iron to ferrous sulphate, the solution being exposed to a very gentle heat till it becomes nearly colorless, when it should be diluted to a volume of 700 to 800 c. c. and boiled two hours with occasional addition of a few c. c. dilute, previously heated, solution of sulphurous acid. Titanic acid if present will be precipitated. After filtering, iron may be determined in the filtrate by concentrating, reducing with H_2S, boiling out excess of H_2S and titrating with standard potassium permanganate according to § 113, 3, *a*.

It may here be observed that by proceeding as above directed, in separating titanic acid from the silica the traces of alumina and possibly other basic oxides which may be retained by the silica are lost. This defect may be remedied, at the expense of some delay, by reserving the solution containing all the basic metals as first filtered from the separated silica until the sulphuric acid solutions of the titanic acid and other impurities possibly present in the silica can be obtained and added to it. The iron and alumina precipitate will then contain all the titanic acid, and the traces of basic metals recovered from the silica will be united to the main portion. If this course is adopted, the use of an unnecessary amount of sodium carbonate and sulphuric acid in obtaining a solution of the residue from the silica should be avoided. The precipitate of aluminium and ferric hydroxides should, in order to eliminate basic sulphates, be dissolved and reprecipitated—a proceeding which is always advisable, even in the absence of sulphates, when alkali-earth metals are present.

Sulphur. If sulphides are present determine sulphur as in iron ores (see page 745). It must be borne in mind, however, that if barium, strontium, or lead is present a portion of the sulphuric acid formed may be retained in the undissolved residue. By prolonged boiling of this residue with sodium carbonate and filtering, the sulphuric acid may be brought into solution as sodium sulphate. This solution may also contain silica and lead, from which the sulphuric must be separated.

If sulphates are present in the original material along with sulphides, the sulphuric acid may be determined by boiling a separate portion a long time with sodium carbonate, filtering, acidifying the filtrate with HCl, and precipitating with $BaCl_2$. The sulphur in the sulphuric acid thus found, deducted from the total amount existing in both sulphides and sulphates, leaves that belonging to the sulphides.

Phosphoric acid may be determined as in iron ores (see p. 741). In careful investigations the residue insoluble in acids should be examined also by fusing it with sodium carbonate, separating silica by drying down with nitric acid, redissolving with nitric acid and adding molybdic acid solution. The reagents used in this process must be free from the least trace of phosphoric acid.]

10. SEPARATION OF SILICATES DECOMPOSABLE FROM THOSE UNDECOMPOSABLE BY ACIDS.

§ 209.

After the silicate has been very finely pulverized and dried at 100° it is usually treated for some time, at a gentle heat, with moderately concentrated hydrochloric acid, evaporated to dryness on the water-bath, the residue moistened with hydrochloric acid, water added, and the solution filtered; it is often preferable, however, to digest the powder with dilute hydrochloric acid (of about 15 per cent.) for some days at a gentle heat, and then at once filter the solution. Which of the two ways it is advisable to adopt, and indeed whether the method here described (which was first employed by Chr. Gmelin in the analysis of phonolites) may be resorted to, depends upon the nature of the mixed minerals. The more readily decomposable the one of the constituent parts of the mixture is, and the less readily decomposable the other, the more constant the proportion between the undissolved and the dissolved part is found to remain in different experiments; in other words, the less the undissolved part is affected by further treatment with hydrochloric acid, the more safely may this method of decomposition be resorted to.

The process gives:

a. A *hydrochloric acid solution*, containing, besides a little silicic acid, the basic metals of the decomposed silicate in the form of metallic chlorides, which are separated and determined by the proper methods.

b. An *insoluble residue*, which contains, besides the undecomposed silicate, the silica separated from the decomposed silicate.

After the latter has been well washed with water, to which a few drops of hydrochloric acid have been added, transfer it, still moist, in small portions at a time, to a boiling solution of sodium carbonate (free from silicic acid) contained in a platinum dish; boil for some time, and filter off each time, still very hot, through a weighed filter. Finally, rinse the last particles of the residue which still adhere to the filter completely into the dish, and proceed as before. Should this operation not fully succeed, dry and incinerate the filter, transfer the ash to the platinum dish, and boil repeatedly with the solution of sodium carbonate till a few drops of the fluid finally passing through the filter remain clear on warming with excess of ammonium chloride. Wash the residue, first with hot water, then—to insure the removal of every trace of sodium carbonate which may still adhere to it—with water slightly acidified with hydrochloric acid, and finally again with pure water. Collect the washings in a separate vessel (H. Rose).

Acidify the alkaline filtrate with hydrochloric acid, and determine in it the silicic acid which belongs to the silicate decomposed by hydrochloric acid, as directed § 140, II., *a*. To ascertain how much *water* the part decomposed by acid contains, the following data are required: The percentage which the decomposed part is of the whole, the percentage of water in the undecomposed part, the percentage of water in the original mixture of silicates. Dry the undissolved silicate at 100° and weigh. Then calculate by difference the quantity of the dissolved silicate. Treat the undissolved silicate as directed § 140, II., *b*. For determination of water, ferrous iron, titanic acid, and other minor constituents, see § 208.

11. ANALYSIS OF LIMESTONES, DOLOMITES, MARLS, &c.

As the minerals containing calcium and magnesium carbonates play a very important part in manufactures and agriculture, the chemist is often called upon to analyze them. The analytical process differs according to the different object in view. For technical purposes, it is sufficient to determine the principal constituents; the geologist takes an interest also in the matter present in smaller proportions; whilst the agricultural chemist seeks a knowledge not only of the constituents, but also of the state of solubility, in different menstrua, in which they are severally present.

I will give, in the first place, a process for effecting a complete and accurate analysis; and, in the second place, the volumetric methods by which the calcium and magnesium carbonates may be determined. An accurate qualitative examination should always precede the quantitative analysis.

A. Method of Effecting the Complete Analysis.

§ 210.

a. Reduce a large piece of the mineral to powder, mix this uniformly, and dry at 100°.

b. Treat about 2 grm., in a covered beaker, with dilute hydrochloric acid in excess, evaporate to dryness in a platinum or porcelain dish, moisten the residue with hydrochloric acid, heat with water, filter on a dried and weighed filter, wash the insoluble residue, dry at 100°, and weigh. It generally consists of separated *silica*, *clay*, and *sand:* but it often contains also *humus-like matter.* Opportunity will be afforded in *g* for examining this residue.

c. Mix the hydrochloric acid solution with chlorine water [or aqueous solution of bromine], then with ammonia in slight excess, and let the mixture stand at rest for some time, in a covered vessel, at a gentle heat. Filter off the precipitate, which contains—besides the hydrate of sesquioxide of manganese, ferric and aluminium hydroxides—the phosphoric acid which the analyzed compound may contain, and, moreover, invariably traces of calcium and magnesium; wash slightly, and redissolve in hydrochloric acid; heat the solution, add chlorine [or bromine] water, and then precipitate again with ammonia; filter off the precipitate, wash, dry, ignite, and weigh.

For the estimation of the several components of the precipitate, viz., iron, manganese, aluminium, and phosphoric acid, opportunity will be afforded in *g.*

d. Unite the fluids filtered from the first and second precipitates produced by ammonia, and determine the calcium and magnesium as directed in § 154, 6 (**28**).

e. If the limestone dried at 100° still gives *water* upon ignition, this is estimated best as directed § 36.

f. Determine carbonic acid by one of the methods described in § 139: most accurately by absorption and weighing of liberated carbonic acid, II., *e*, page 412, or with simpler apparatus by loss of weight on decomposition with acid, II., *d*, *bb*, page 410; or, in absence of water and notable quantities of ferrous iron and manganese, by fusion with vitrified borax, II., *c*, page 408.

g. To effect the estimation of the constituents present in smaller proportion, as well as the analysis of the residue insoluble in hydrochloric acid, and of the precipitate produced by ammonia, dissolve 20—50 grm. of the mineral in hydrochloric acid. As the evaporation to dryness of large quantities of fluid is always a tedious operation, gently heat the solution for some time, to expel the carbonic acid; then filter through a weighed filter into a litre flask, wash the residue, dry, and weigh it. (The weight will not quite agree with that of the residue in *b*, as the latter contains also that part of the silicic acid which here still remains in solution.)

α. Analysis of the insoluble Residue.

aa. Treat a portion with boiling solution of pure sodium carbonate (§ 209, *b*), and separate the silicic acid from the solution (§ 140, II., *a*); this process gives the quantity of that portion of the *silicic acid* contained in the residue, which is *soluble in alkalies.*

bb. Treat another portion, by the usual process for silicates (§ 140, II., *b*), and deduct from the *silicic acid* found the amount obtained in *aa.*

cc. If the residue contains *organic matter* (humus), determine, in a portion, the carbon by the method of ultimate analysis (§ 177). PETZHOLDT,* who determined by this method the coloring organic matter of several dolomites, assumes that 58 parts of carbon correspond to 100 parts of organic substance (humic acid).

dd. If the residue contains *pyrites*,† fuse another portion of it with sodium carbonate and potassium nitrate; macerate in water, add hydrochloric acid, evaporate to dryness, moisten with hydrochloric acid, gently heat with water, filter, determine the sulphuric acid in the filtrate, and calculate from the result the amount of pyrites present.‡

β. Analysis of the Hydrochloric Acid Solution.

Make the solution up to 1 litre.

aa. For the determination of the *silicic acid* that has passed into solution, and of the *barium, strontium, aluminium, manga-*

* Journ. f. prakt. Chem. 63, 194.

† Compare PETZHOLDT, *loc. cit.;* EBELMEN (Compt. rend. 33, 681); DEVILLE (Compt. rend. 37, 1001; Journ. f. prakt. Chem. 62, 81); ROTH (Journ. f. prakt. Chem. 58, 84).

‡ If the residue contains barium or strontium sulphate, these compounds are formed again upon evaporating the soaked mass with hydrochloric acid; they remain accordingly on the filter, whilst the sulphuric acid formed by the sulphur of the pyrites passes into the filtrate.

nese, *iron*, and *phosphoric acid*, evaporate 500 c. c., and dry the residue at 100°—110°. Treat the dry mass, in order to separate silicic acid, &c. (precipitate I.), with hydrochloric acid and water, boil the solution with nitric acid, add ammonia, boil till the excess of ammonia has escaped, filter, wash slightly, dissolve on the filter with hydrochloric acid, reprecipitate in the same manner with ammonia, and filter off precipitate II., which contains ferric hydroxide, &c. Digest the united filtrates in a nearly filled and closed flask with sulphide of ammonium in a slightly warm place for 24 hours, then filter off precipitate III. This consists principally of manganese sulphide; it is to be washed with water containing ammonium sulphide. Precipitate the filtrate with ammonium carbonate and ammonia, allow to stand 24 hours, and then filter off precipitate IV., which consists for the most part of calcium carbonate, and is to be washed with water containing ammonia. Evaporate the filtrate in a porcelain dish to dryness, project the residue, little by little, into a red hot platinum dish, drive off the ammonium salts, moisten the residue with hydrochloric acid, dissolve it in water, and boil, with addition of pure milk of lime, to strongly alkaline reaction. Filter off precipitate V., which is composed of magnesium hydroxide and the excess of calcium hydroxide, wash it, precipitate the filtrate with ammonium carbonate and ammonia, and, after long standing, filter off precipitate VI., which is to be washed with water containing ammonia.

Precipitate I. consists principally of silicic acid. It may also contain barium and strontium sulphates. Treat it in a platinum dish with hydrofluoric acid and a little sulphuric acid, evaporate to dryness, and, if necessary, repeat this operation. Should a residue remain, fuse it with a small quantity of sodium carbonate, treat with water, filter, wash, dissolve in hydrochloric acid, and precipitate the solution with sulphuric acid. When the precipitate has settled filter it from solution *a*, and wash. Stop up the tube of the funnel, and fill the latter with solution of ammonium carbonate, allow to stand 12 hours, open the funnel tube, wash the residue first with water, then with hydrochloric acid (solution *b*), finally again with water, and then weigh the pure residual *barium* sulphate. Mix the united solutions *a* and *b* with ammonium carbonate and ammonia, allow to stand some time; if a precipitate forms (which may contain strontium carbonate) filter it off, dry, and add to precipitate IV.

Precipitate II. consists principally of ferric hydroxide; it contains also the aluminium, and, provided there is enough iron, the whole of the phosphoric acid. Dissolve in hydrochloric acid, add pure tartaric acid, and then ammonia. Having fully convinced yourself that no precipitate is formed, precipitate the iron with ammonium sulphide in a small flask, which must be nearly filled and closed, allow to stand till the fluid appears of a pure yellow color, filter, wash with water containing ammonium sulphide, and determine the *iron* after § 113, 2. To the filtrate add a little pure sodium carbonate and pure potassium nitrate, evaporate to dryness, and ignite till the residue is white. Add water and hydrochloric acid till the whole is dissolved,* and precipitate the clear fluid with ammonia. If a precipitate forms (aluminium hydroxide or phosphate, or a mixture of both), filter it off, and weigh. Mix the filtrate with a little magnesium sulphate. If another precipitate forms, this time consisting of ammonium magnesium *phosphate* (which is to be determined after § 134, I., *b*, *α*), the aluminium precipitate may be calculated as *aluminium phosphate* (Al_2O_3, P_2O_5). If, on the contrary, no precipitate is formed, the phosphoric acid must be determined in the alumina precipitate as directed § 134, I., *b*, *β*.

Precipitate III. consists principally of manganese sulphide. It may also contain traces of nickel, cobalt, and zinc sulphides, calcium carbonate, &c. Treat with moderately dilute acetic acid, heat the filtrate, to remove any carbonic acid, add ammonia, precipitate with ammonium sulphide, allow to stand 24 hours, and determine the *manganese* as sulphide (§ 109, 2). If any residue was left insoluble in acetic acid, test it for the above-mentioned metals. The fluid filtered from the pure manganese sulphide is to be mixed with ammonium carbonate. If a precipitate forms it is to be treated with precipitate IV.

Precipitates IV., V., VI. The united mass of these precipitates, together with the small portions of alkali-earth carbonates obtained during the treatment of precipitates I. and III. contain the whole of the strontium and the whole of the barium which originally passed into the hydrochloric acid solution. Ignite the dried precipitate (if necessary in portions) in a platinum crucible, most intensely over the gas blowpipe. By this means any barium

* I may remind the operator that the residue, which contains nitric acid, cannot be heated with hydrochloric acid in a platinum dish.

and strontium carbonates, and a part, at all events, of the calcium carbonate, are converted into oxides (ENGELBACH*). Boil the residue 5 or 6 times with small portions of water, pouring off the solution through a filter; neutralize the solution with hydrochloric acid, evaporate to dryness, and test a minute portion with the spectroscope—this minute portion is afterwards added to the rest. If calcium and *strontium* alone are present, separate according to **26**. If barium is present, separate the three alkali-earth metals after **23**.

bb. Although it is possible in *aa* to test for metals precipitable by hydrogen sulphide from acid solution, *e.g.*, copper, and if required to determine them, still it is more convenient to employ a fresh quarter of the hydrochloric acid solution for this purpose. The precipitate obtained by passing the gas into the warm dilute solution is washed, dried, and treated with carbon disulphide. If a residue remains it is to be examined.

cc. The remaining quarter of the dilute hydrochloric acid solution is used for the estimation of the *alkalies*.† Mix with chlorine water, then with ammonia and ammonium carbonate; after allowing the mixture to stand for some time, filter off the precipitate, evaporate the filtrate to dryness, ignite the residue in a platinum dish to remove the ammonium salts, and finally separate the magnesium from the alkalies as directed p. 491, **15**. The reagents must be most carefully tested for fixed alkalies, and the use of glass and porcelain vessels avoided as far as practicable.

Should the limestone contain a sulphate soluble in hydrochloric acid, precipitate the sulphuric acid by a small excess of barium chloride, allow to settle, and filter off the barium sulphate (which is to be determined in the usual manner) before proceeding as above to the estimation of the alkalies.

h. As calcite and aragonite may contain *fluorides* (JENZSCH‡),

---

* Zeitschr. f. anal. Chem. 1, 474.

† The simplest way of ascertaining whether and what alkalies are present in a limestone, is the process given by ENGELBACH (Annal. d. Chem. u. Pharm. 123, 260)—viz., ignite a portion of the triturated mineral strongly in a platinum crucible over the blast, boil with a little water, filter, neutralize with hydrochloric acid, precipitate with ammonia and ammonium carbonate, filter, evaporate the filtrate to dryness and examine with the spectroscope. The ammonium carbonate precipitate may be evaporated with hydrochloric acid to dryness, and examined in like manner for barium and strontium.

‡ Pogg. Annal. 96, 145.

the possible presence of fluorine must not be disregarded in accurate analyses of limestones. Treat, therefore, a larger sample of the mineral with acetic acid until the calcium and magnesium carbonates are decomposed; evaporate to dryness until the excess of acetic acid is completely expelled, and extract the residue with water (§ 138, I.). We have the fluorine in the residue. If it can be distinctly detected in a portion of the same,* the determination may be attempted after § 138, II., 3, *a*.

i. If the limestone under examination contains *chlorides*, treat a large sample with water and nitric acid, at a very gentle heat; filter, and precipitate the chlorine from the filtrate by solution of silver nitrate.

k. It is often interesting for agriculturists to know the degree of solubility of a sample of limestone or marl in the weaker solvents. This may be ascertained by treating the sample first with water, then with acetic acid, finally with hydrochloric acid, and examining each solution and the residue. The analysis of marls made by C. STRUCKMANN† were done in this manner.

l. To effect the separation of the caustic or carbonated lime, in hydraulic limes, from the silicates, DEVILLE‡ proposed to boil with solution of ammonium nitrate, which he stated would dissolve the caustic lime and carbonate of lime, without exercising a decomposing action on the silicates. GUNNING§ found, however, that by this process the double silicates of aluminium and calcium are more or less decomposed, with separation of silicic acid. As yet no method is known by which the object here stated can be accomplished with absolute accuracy; the best way, perhaps, is treating the sample with dilute acetic acid; C. KNAUSZ‖ recommends hydrochloric acid.

B. VOLUMETRIC DETERMINATION OF CALCIUM CARBONATE AND MAGNESIUM CARBONATE (for technical purposes).

§ 211.

a. If a mineral contains only calcium carbonate, the amount of the latter may be estimated from the quantity of acid required to

---

* See Qual. Anal. § 146, 6. † Annal. d. Chem. u. Pharm. 74, 170.

‡ Compt. rend. 37, 1001; Journ. f. prakt. Chem. 62, 81.

§ Journ. f. prakt. Chem. 62, 318.

‖ Gewerbeblatt aus Würtemberg, 1855, Nr. 4; Chem. Centralbl., 1855, 244.

effect its decomposition, the method described in § 198 being employed for the purpose. Or the carbonic acid in the mineral may be determined, and 1 mol. calcium carbonate = 100 calculated for each mol. carbonic acid = 44.

b. But if the mineral contains, besides calcium carbonate, also magnesium carbonate, the results obtained by either process give the quantity of calcium carbonate + magnesium carbonate, the latter being expressed by its equivalent quantity of calcium carbonate (*i.e.*, 100 of calcium carbonate for 84 of magnesium carbonate). If, therefore, you desire to know the actual amount of each, you must, in addition to the above determination, determine one of the alkali-earth metals separately. For this purpose one of the two following methods may be employed :—

1. Mix the dilute solution of 2—5 grm. of the mineral with ammonia and ammonium oxalate in excess, allow to stand for 12 hours and then filter. Ignite the precipitate of calcium oxalate, together with the filter, and treat the calcium carbonate produced as directed § 198. This process gives the amount of calcium contained in the analyzed mineral; the difference between this and the former result gives the calcium carbonate which is equivalent to the amount of magnesium carbonate present. To obtain perfectly accurate results by this method, repeated precipitation is indispensable (see § 154, 6, *a*).

2. Dissolve 2—5 grm. of the mineral in the least possible excess of hydrochloric acid, and add a solution of lime in sugar water as long as a precipitate forms. By this operation the magnesia only is precipitated. Filter, wash, and treat the precipitate as directed § 198; the result represents the quantity of the magnesium. Deduct the quantity of calcium carbonate equivalent thereto from the result of the total determination; the remainder is the amount of calcium carbonate present.

The method 2 is only suitable when the proportion of magnesium is small.

[12. ASSAY OF COPPER ORES.

§ 212.

For the assayer who has occasion to determine the amount of copper in ores very frequently, the method of LUCKOW* depending on the electrolytic deposition of copper, and the method of STEINBECK† (volumetric determination by means of potassium cyanide), are to be recommended. For the analytical chemist, who is only occasionally called upon to assay a copper ore, the processes below described may be useful, as they require no preparation of special reagents or apparatus.

1. *When Arsenic, Antimony, Bismuth, (Cadmium, Tin) are not present.*

Take 1 grm. of the pulverized ore if rich, 3 to 5 if poor. Put it in a dry beaker (best of a diameter of 8—10 centimetres at the bottom), cover with a watch-glass. Mix in another glass vessel, which should be dry to avoid dilution, 1 pt. sulphuric with 3 to 4 pts. nitric acid, both concentrated. Pour the mixed acids upon the ore. 40—50 c. c. will suffice if not more than 2 grm. ore are taken. The action is less violent if the requisite amount of acid is added at once than it is if added gradually. Heat the covered beaker on a sand-bath, to near 100° C. an hour or two, or until the action of the acids on the ore has apparently nearly ceased, raise then the watch-glass covering the beaker so far as to allow vapors to pass off freely by interposing between it and the beaker a triangle made of thick glass rod, and evaporate until nitric acid is completely removed. Copper exists in the residue as sulphate, mixed usually with other metallic sulphates, together with such constituents of the ore as are not decomposed by the acids. A small quantity of liquid (sulphuric acid) will still remain with the solid residue, and should not be removed by further elevation of temperature. After cooling, add about 100 c. c. of water to the residue and keep it warm on the sand-bath an hour to ensure solution of the anhydrous copper sulphate. Complete solution of the copper in the residue may be effected in a few minutes by adding with the water a little hydrochloric acid; but since hydrochloric acid dissolves notable quantities of lead sulphate, it should not be used for this purpose unless the ore is known to be free from lead. Filter from the undissolved

* Zeitschr. f. anal. Chem. 8, 23, and 11, 1.

† *Ib.* 8, 9.

part of the residue and wash it with hot water. If the residue contains lead sulphate, avoid the use of an unnecessary volume of water in washing, or use water acidified with sulphuric acid to prevent lead sulphate from being dissolved. Dilute the filtrate to about 500 c. c. If hydrochloric acid has not previously been used, add now 1 or 2 c. c. and heat to boiling and pass hydrogen sulphide through the solution until the copper is all precipitated, the temperature meanwhile being maintained nearly or quite to the boiling point. If a tolerably rapid current of H_2S be passed into the solution through a tube contracted at the orifice to a diameter of 1 to 2 millimetres, so as to produce numerous and small bubbles, the precipitation is usually complete in course of 15 to 25 minutes. Wash the precipitate, dry, ignite in an atmosphere of hydrogen, and weigh the resulting Cu_2S according to directions in § 119, 3, *a*, page 316.

2. *When Antimony, Arsenic, (Bismuth, Cadmium, Tin) are present.*

Arsenical or antimonial minerals are occasionally present in copper ores, bismuth compounds very rarely, while appreciable amounts of cadmium or tin can usually safely be assumed to be absent. If the ore to be assayed has not been pulverized and can be examined in large fragments, the assayer, with sufficient knowledge of mineralogy and experience in the use of the blowpipe, can usually decide whether arsenic, antimony, or even bismuth are present. If this cannot be done, qualitative testing in the wet way should be resorted to. If any of the metals which interfere with the process described in 1 are present, decompose the ore with aqua regia, adding also enough sulphuric acid to convert into sulphates, on evaporation, any nitrates and chlorides which may be formed. An unnecessary excess of sulphuric should be avoided, as it is difficult to remove by evaporation and, if allowed to remain in large amount, may render the subsequent precipitation of copper less perfect. After removing the nitric and hydrochloric acids by evaporation, dissolve the copper sulphate in the residue by digestion with water, filter, add solution of sulphurous acid (100—200 c. c.), set in a warm place; if the solution ceases to smell of SO_2 after half an hour, add more solution of sulphurous acid. Finally, after allowing the solution to stand in a warm place 4 or 5 hours at least, precipitate the copper with ammonium or potassium sulphocyanate and determine it as Cu_2S as directed in § 119, 3, *b*.]

[13. ASSAY OF LEAD ORES

§ 213.

The commercial value of lead ore is often estimated from assays made in the dry way. These assays are not very exact even in rich ore, and still less so in poor ores. If the means for making a dry assay are not at hand, or a more accurate determination is required, the following method applicable to all ores (rich or poor, galenas or carbonates), not containing arsenic or antimony may be used.

Decompose the finely pulverized ore—2 grm. if rich, 5 grm. if poor—with a mixture of concentrated nitric and sulphuric acid precisely in the manner described for copper ores (§ 212, 1). After free nitric acid has been removed by evaporation, the lead exists in the residue as lead sulphate, mixed, according to the character of the ore, with different kinds and quantities of other metallic sulphates, and also such constituents of the ore as are not decomposed by the acids. Sulphuric acid, which may be used freely in the decomposing mixture, should remain also with this residue and not be driven off by evaporation. When, toward the close of the process of evaporation by gradually increasing temperature, dense white fumes of sulphuric acid begin to appear, or when, at a temperature little exceeding 100° C., no odor of nitric acid is perceptible, heat is removed and the residue allowed to cool. About 100 c. c. of water are then added. After digesting 2 hours to dissolve such metallic sulphates as are soluble in water, the residue, containing lead sulphate and other insoluble matter (possibly, for example, quartz, native silicates, barium sulphate, calcium sulphate), is collected on a moderate sized filter and washed with water to which a little sulphuric acid has been added. The filter and its contents, without previous drying, are then placed in the bottom of a large beaker. Pour first ammonia (20—30 c. c.) upon the filter and residue, and next acetic acid to decided acid reaction, keep warm some 20 minutes with occasional stirring. The lead sulphate readily dissolves. Filter, wash the still remaining undissolved residue and filter, chiefly by decantation, adding to the wash water at first a little ammonium acetate. The filtrate may contain, besides lead, also calcium sulphate. To prevent the possible deposition of the latter during the subsequent precipitation of the lead, dilute the filtrate by adding $\frac{1}{10}$ to $\frac{1}{5}$ its volume of water, and add 1 or 2 c. c. dilute hydrochloric acid. Precipitate now the lead, either in the cold or slightly heated solution, with a current of hydrogen sulphide. Treat the precipitate according to § 116, 2, page 299. In igniting in hydrogen, take care that only the bottom of the porcelain crucible is faintly red. Too high heat will cause loss by volatilization.

If this ore contains arsenic and antimony, these elements, or at least the latter, will cause a small quantity of lead to remain undissolved in the residue remaining after the treatment with ammonium acetate. In such a case, lead may be separated from the residue by means of methods of separation described in § 164, or a wholly different course may be devised and applied to the original substance.]

[14. DETERMINATION OF NICKEL AND COBALT IN ORES, SPEISS* AND MATTE.†

§ 214.

The separation of nickel and cobalt from other metals which accompany them in ores and determination of their joint amount precedes the separation of the two metals. The process requires great care throughout. A method is here given for determining these metals, and also copper and lead, in a "matte" containing, besides nickel and cobalt, copper, lead, iron, and accidental adhering particles of sand and earthy silicates. To the description of this method will be appended modifications necessary or admissible in the examination of other products.

1. *Decomposition and Separation of Lead.*

Decompose 2 gr. of the very finely pulverized material with a mixture of concentrated nitric and sulphuric acid, proceeding precisely as recommended for decomposition of copper ores. (See § 212, 1.) It is, however, safer, after the nitric acid first employed

* A product consisting chiefly of metallic arsenides obtained by smelting ores is called "speiss."

† A product consisting chiefly of metallic sulphides obtained by smelting ores is called "matte."

has been expelled, to add to the residual sulphuric acid a fresh portion of concentrated nitric acid; keep hot for some time, and finally evaporate off all the nitric acid a second time, allowing at least 8 or 10 hours for the whole process of decomposition. Add water (100 to 150 c.c.) to the cooled residue and digest 3 or 4 hours, and filter the solution of metallic sulphates obtained from the insoluble residue, which contains the lead as sulphate, together with particles of sand, &c. Dissolve the lead sulphate in this residue with ammonium acetate and determine the lead as in the process of assaying lead ores, § 213. Incinerate the filter containing the portion of the residue which ammonium acetate fails to dissolve in a porcelain crucible, and add to it in the crucible aqua regia; evaporate to a few drops. If the few drops of remaining liquid show no greenish color, it may be assumed that the residue so treated contains no nickel or cobalt. If the color indicates presence of these metals, rinse the contents of the crucible into the filtrate from the lead sulphate. Before throwing away the filtrate from the final lead precipitate (lead sulphide), prove that it contains no nickel or cobalt by adding ammonia to neutral reaction, and a few drops of ammonium sulphide.

2. *Separation of Copper.*

Dilute the filtrate from the lead sulphate and other insoluble matter to about 500 c.c., and precipitate copper with hydrogen sulphide and determine it, proceeding as directed in "Assay of Copper Ores," § 212.

3. *Separation of Iron.*

Concentrate the filtrate from the copper sulphide, add nitric acid enough to convert the iron into a ferric salt, and boil a few minutes. Allow the solution, which should now occupy a volume of about 300 c.c., to become nearly cold, and add a *large excess* of ammonia at once, and let it stand in a warm place (50° to 70° C.) half an hour. Filter then into a porcelain casserole and wash the greater part of the saline matter out of the ferric hydroxide with hot water. Complete washing is needless. Although this filtrate contains the greater part of the nickel and cobalt, a very considerable quantity is retained by the ferric hydroxide. Even a second precipitation with ammonia cannot be relied upon for effecting a satisfactory separation. Dissolve, therefore, the ferric hydroxide in hydrochloric acid, which may be used freely for this purpose, since it will be next converted into ammonium chloride, which acts favorably, rather than otherwise, in the subsequent precipitation of iron (§ 160, 71). Wash the filter from which the iron precipitate has been dissolved, neutralize the greater part of the free hydrochloric acid in the solution with ammonia, proceed then carefully to prepare the solution for precipitation of iron with *sodium acetate* by neutralizing properly with sodium or ammonium carbonate and addition of acetic acid, as described in § 160, 71, p. 517. The iron precipitated thus a second time is still not absolutely free from a trace of nickel or cobalt, which may, however, be neglected unless extraordinary accuracy is required, or unless the original substance was comparatively rich in these metals (containing over 20 per cent). In that case, dissolve the iron precipitate in hydrochloric acid and precipitate the iron again as basic ferric acetate and filter. Add the filtrate or filtrates, as the case may be, to the first ammoniacal filtrate, which should meanwhile have been concentrated to expel free ammonia and reduce its volume. The mixed filtrates will now contain some free acetic acid. Concentrate in a porcelain dish to 300 or 400 c.c., and add ammonia to alkaline reaction. This will usually throw down a little ferric or aluminium hydroxide, which is to be filtered off. If the precipitate is very slight it may be thrown away. If considerable, dissolve in HCl, reprecipitate with ammonia, filter, and add the filtrate to the main filtrate.

4. *Precipitation of Nickel and Cobalt.*

Concentrate the now alkaline solution to a volume of about 250 c.c. During the concentration the solution will sometimes become slightly acid, as may be shown by testing with litmus paper. If acid, it is in just the right condition for precipitation of nickel and cobalt as sulphides. If alkaline, add cautiously HCl till the fluid gives with litmus paper a distinct but *slight* acid reaction. Precipitate next nickel and cobalt in the solution previously heated to gentle ebullition in a beaker covered with a watch-glass, with hydrogen sulphide, according to § 110, *b*, *β*, p. 261. Dry the sulphides on the filter and remove them from the filter to a beaker. Incinerate the filter, add the ash and adhering nickel and cobalt sulphides to the portion in the beaker, or if the whole quantity of sulphides is small, calcine the whole with the filter. Treat with aqua regia until only a residue of yellow sulphur remains undissolved, evaporate, and expose the residue to a heat of 100° C., to make traces of silica insoluble; then moisten with a few drops of hydrochloric acid, add 15—20 c.c. of water to dissolve nickel and cobalt salts, add without previous filtration some fresh solution of H_2S as long as it produces a dark-colored precipitate or turbidity (indicating traces of still unremoved copper or lead), filter into a porcelain dish (having a capacity of 250 c.c.), and concentrate the solution to about 100 c.c. Boil gently, and during the boiling add gradually *pure* sodium carbonate solution to alkaline reaction, avoiding much excess; continue the boiling a few minutes to remove free carbonic acid. Next dissolve in a platinum dish a few decigrammes of pure sodium hydroxide (prepared from metallic sodium*), add the solution, and heat again to boiling to precipitate the nickel and cobalt that may still remain in solution. Wash the precipitate with boiling hot water, first by decantation, and finally on the filter (best with the BUNSEN filtering apparatus) until a drop of the washings evaporated on polished platinum foil gives no more residue than distilled water, and then wash still a little longer. After drying the precipitate, remove from the filter to a piece of glazed paper, and cover it immediately with a bell-glass† and incinerate the filter with the small portion adhering to it until carbon is completely removed, and expose for some time longer to the air at a red heat, to remove traces of carbon which metallic nickel reduced by the paper may have absorbed. Transfer the main portion of the oxides to the crucible, cover and heat to redness, finally reduce the oxides to metals by ignition in hydrogen, according to § 110, 2.

If the preceding operations have been conducted with due care the metals thus obtained are free from appreciable quantities of impurities. Test them, however, first for soda by adding in the crucible a few drops of water, allowing it to remain in contact with the metal 10 minutes or longer, and applying reddened litmus paper. If the experiment shows presence of soda it can be removed (probably but incompletely) by prolonged digestion with water,

---

* If pure sodium hydroxide is not at hand, a small quantity can easily be prepared by dissolving a fragment of sodium in water in a platinum dish.

† Strongly dried nickelous hydroxide exposed to moist air will often decrepitate projecting particles a considerable distance.

removing the water with a pipette, drying and igniting the metal again in hydrogen. Test also for silica by dissolving in nitric acid. If an appreciable amount of silica remains undissolved,* collect on a small filter, weigh, and make the necessary deduction from the first weight of the metals.

5. *Separation of Cobalt from Nickel.*

Evaporate the nitric acid solution of the metals nearly to dryness, or until free nitric is almost completely removed; add 4 or 5 c.c. of water to dissolve the nickel and cobalt salts, add drop by drop solution of potassium carbonate until a permanent precipitate just begins to be formed. (If free nitric has been removed by evaporation completely, the addition of potassium carbonate is not required.) Next add 6 to 8 grm. potassium nitrite dissolved in 10 to 15 c.c. of hot water. This usually produces a flocculent precipitate containing both nickel and cobalt, on account of the potassium carbonate which the nitrite commonly used as a reagent contains. Add then a little acetic acid. Any flocculent precipitate which may have been previously formed now disappears, and a precipitate of tripotassium cobaltic nitrite, either simultaneously or after a short time, is deposited. Much acetic acid hinders the complete separation of cobalt. If a flocculent precipitate has been formed, a few drops usually suffice to dissolve it; 2 or 3 c.c. more may then be added. The whole volume of the solution should then amount to only 15 to 20 c.c. Cover the beaker containing it with a watch-glass and allow it to stand in a warm place at least 24 hours. Filter; wash with a solution of potassium acetate.† Test the filtrate for cobalt as follows: Dilute, heat, precipitate with sodium hydrox-

* According to my own experience it is not difficult to obtain the nickel and cobalt free from iron and aluminium. If the operator wishes to assure himself on this point, an excess of ammonia may be added to the nitric acid solution before filtering off the silica. If a precipitate is formed it must be collected on a filter, washed slightly, dissolved on the filter (which retains the accompanying silica), reprecipitated with ammonia, collected again on the same filter, washed, ignited, and weighed, when its weight jointly with the silica may be deducted from the first weight of the cobalt and nickel. If cobalt is then to be separated from nickel, both metals may be precipitated from the filtrates as sulphides, dissolved in aqua regia, when after removing free acids, the process of separating cobalt may be applied.—O. D. A.

† A suitable solution may be prepared by neutralizing acetic acid with crystallized potassium bicarbonate, leaving the solution slightly acid.

ide, wash the greater part of the saline matter out of the precipitate, dissolve it in nitric acid, evaporate to dryness on a water-bath, add two or three drops of nitric acid to the residue, dissolve in a small volume of water, filter, concentrate, and repeat the process of separation with potassium nitrite as before. Put the filter containing the cobalt precipitate, still moist, into a beaker, add about 100 c.c. of water, heat, add hydrochloric acid until solution is effected, separate the filter paper by filtration, evaporate the solution on a water-bath, allowing the residue to remain on the water-bath 2 or 3 hours to render traces of silica insoluble, moisten with hydrochloric acid, add water, filter, and convert the cobalt in the solution into the metallic form by the same process as before employed for obtaining metallic nickel and cobalt from a similar solution; test also the weighed cobalt for impurities in the same manner. The amount of *nickel* in the mixture of the two metals may now be calculated by difference.

It has been assumed that lead and copper are determined in the process here described. No essential modification, however, can be made in the method of separating them from nickel and cobalt when their determination is not required.

6. *Modifications admissible or required for ores, or other products of different compositions.*

If only very *little copper* is present (less than 1 per cent.), and no other metals of the fifth and sixth groups, the treatment of the first solution with H_2S may be omitted. The copper then remains in solution until after the iron is separated and is precipitated along with nickel and cobalt as sulphide. After the mixed sulphides are dissolved as before described, the copper may be separated from the slightly acid solution by hydrogen sulphide.

When no *lead* is present, the treatment of the first insoluble residue with ammonium acetate is omitted.

If other metals of the fifth and sixth groups are present no modification is required. In case, however, arsenic is present (as in speiss and some ores), it is advisable to convert the resulting arsenic acid into arsenious by means of sulphurous acid before treatment with H_2S.

No modifications are required on account of the presence of any metals of second, third, or fourth groups, except zinc or a notable amount of manganese. In respect to manganese, it can safely be assumed when not more than a trace is present (as in matte or speiss and *most* ores) that the precipitation of nickel and cobalt as directed in a solution containing a small quantity of free acetic acid will effect its separation. But if a considerable amount is present a small portion is liable to be precipitated along with the nickel and cobalt sulphides. This can be removed when the mixed sulphides are brought into solution, and excess of acid is removed by adding ammonium chloride freely, also a little ammonium acetate, and repeating the precipitation of nickel and cobalt in same manner as before, as sulphides, taking care to have the solution always contain a small quantity of free acetic acid. Or the manganese may be separated from the solution of the three metals according to § 160, **85**, page 529.

If zinc is present it all goes down with the nickel and cobalt when they are precipitated as sulphides. After the mixed sulphides have been dissolved and free acid has been removed by evaporation, zinc may then be separated according § 160, **75**, p. 520.

15. ASSAY OF ZINC ORES.

§ 215.

Method of SCHAFFNER,* *modified by* C. KÜNZEL,† *as employed in the Belgian zinc-works; described by* C. GROLL.‡

a. Solution of the ore and preparation of the ammoniacal solution.

Powder and dry the ore.

Take 0·5 grm. in the case of rich ores, 1 grm. in the case of poor ores, transfer to a small flask, dissolve in hydrochloric acid with addition of some nitric acid by the aid of heat, expel the excess of acid by evaporation, add some water, and then excess of ammonia. Filter into a beaker, and wash the residue with lukewarm water and ammonia, till ammonium sulphide ceases to produce a white turbidity in the washings. The oxide of zinc remaining in the ferric hydroxide is disregarded. Its quantity,

* Journ. f. prakt. Chem. 73, 410.

† *Ib.* 88, 486.

‡ Zeitschr. f. anal. Chem. 1, 21.

according to GROLL, does not exceed 0·3—0·5 per cent. This statement probably has reference only to ores containing relatively little iron; where much iron is present the quantity of zinc left behind in the precipitate may be not inconsiderable. The error thus arising may be greatly diminished by dissolving the slightly washed iron precipitate in hydrochloric acid and adding excess of ammonia. But the surer mode of proceeding is to add to the original solution—after evaporating off the greater part of the free acid as above, and allowing to cool—dilute sodium carbonate nearly to neutralization, then to precipitate the iron, after p. 517, with sodium acetate, boiling, to filter, and wash. The washings, after being concentrated by evaporation, are added to the filtrate and the whole is then mixed with ammonia, till the first-formed precipitate is redissolved.

If the ore contains manganese—provided approximate results will suffice—digest the solution of the ore in acids, after the addition of excess of ammonia and water, at a gentle heat for a long time, and then filter off, with the iron precipitate, the hydrated protosesquioxide of manganese which has separated from the action of the air. The safer course—though undoubtedly less simple—is, after separating the iron with sodium acetate, to precipitate the manganese by passing chlorine, as directed p. 510, or by adding bromine and heating.

If lead is present, it is separated by evaporating the aqua regia solution with sulphuric acid, taking up the residue with water and filtering; then proceed as directed.

b. Preparation and standardizing of the sodium sulphide solution.

The solution of sodium sulphide is prepared either by dissolving crystallized sodium sulphide in water (about 100 grm. to 1000—1200 water), or by supersaturating a solution of soda, free from carbonic acid, with hydrogen sulphide, and subsequently heating the solution in a flask to expel the excess of hydrogen sulphide. Whichever way it is prepared, the solution is afterwards diluted, so that 1 c.c. may precipitate about 0·01 grm. zinc. Prepare a solution of zinc, by dissolving 10 grm. chemically pure zinc in hydrochloric acid, or 44·122 grm. dry crystallized zinc sulphate in water, or 68·133 grm. dry crystallized potassium zinc sulphate in water, and making the solution in either case up to 1 litre with water.

Each c.c. of this solution corresponds to 0·01 grm. zinc. Now measure off 30—50 c.c. of this zinc solution into a beaker, add ammonia till the precipitate is redissolved, and then 400—500 c.c. distilled water. Run in sodium sulphide as long as a distinct precipitate continues to be formed, then stir briskly, remove a drop of the fluid on the end of a rod to a porcelain plate, spread it out so that it may cover a somewhat large surface, and place in the middle a drop of pure dilute solution of nickel chloride. If the edge of the drop of nickel solution remains blue or green, proceed with the addition of sodium sulphide, testing from time to time, till at last a blackish gray coloration appears surrounding the nickel solution. The reaction is now completed, the whole of the zinc is precipitated, and a slight excess of sodium sulphide has been added. The precise depth of color of the nickel must be observed and remembered, as it will have to serve as the stopping signal in future experiments. To make sure that the zinc is really quite precipitated, you may add a few tenths of a c.c. more of the reagent, and test again, of course the color of the nickel-drop must be darker. Note the number of c.c. used, and repeat the experiment, running in at once the necessary quantity of the reagent, *less* 1 c.c., and then adding 0·2 c.c. at a time, till the end-reaction is reached. The last experiment is considered the more correct one. The sodium sulphide solution must be restandardized before each new series of analyses.

c. Determination of the zinc in the solution of the ore.

Proceed in the same way with the ammoniacal solution prepared in *a* as with the known zinc solution in *b*. Here also repeat the experiment, the second time running in at once the required number of c.c., less 1, of sodium sulphide, and then adding 0·2 c.c. at a time, till the end-reaction makes its appearance. The second result is considered the true one. There are three different ways in which this repetition of the experiment may be made. You may either weigh out at the first two portions of the zinc ore, or you may weigh out double the quantity required for one experiment, make the ammoniacal solution up to 1 litre and employ ½ litre for each experiment, or lastly, having reached the end-reaction in the first experiment, you may add 1 c.c. of the known zinc solution, which will destroy the excess of sodium sulphide, and then run in sodium sulphide in portions of 0·2 c.c., till the end-reaction is again attained. Of course, in this last process to obtain the second result, you deduct from the whole quantity of sodium sulphide used the amount of the same, corresponding to 1 c.c. of the zinc solution.

If the ore contains copper, remove it from the acid solution by hydrogen sulphide, evaporate the filtrate with nitric acid, dilute, treat with ammonia, and determine the zinc as above.

16. PARTIAL ANALYSIS OF IRON ORES.

§ 216.

For the purpose of ascertaining the quantity and more especially the *quality* of the iron, which can be produced from an ore, determinations of iron, phosphorus, sulphur, and manganese, without regard to other constituents, are often required. An examination of the ore for titanic acid, especially if it is a magnetic iron ore, is also often demanded. Frequently a determination of only two or three of these substances, as iron and phosphorus, or phosphorus and sulphur, is required. In any case it is most convenient to use a separate portion of the ore for the determination of each substance.

Iron.

Iron may be determined volumetrically many ways. For the present purpose, either of the two following methods may be used.

1. Decompose 2 grm. with concentrated hydrochloric acid by heating gently on a sand-bath in a covered beaker, having a capacity of not less than half a litre. If the ore contains carbonaceous matter, the weighed portion should first be ignited with exposure to the air until the carbon is burned out. Two or three hours suffice for the decomposition of most limonites and magnetites, but some varieties of hematite resist the action of acid for a long time. The appearance of the still undissolved residue, which may consist of quartz, clay, hornblende, or other silicates, will indicate when the

oxide (or carbonate) of iron in the ore is dissolved.* Dilute somewhat and filter, add to the filtrate 20 to 30 c.c. of strong sulphuric acid, or an equivalent amount of dilute, and evaporate till hydrochloric acid is all removed, avoiding a heat much above 100° C. After cooling, add about 100 c.c. of water and digest till the ferric sulphate has gone into solution. Then pour the solution into a half-litre flask, without filtering from a slight residue of silica which may have separated, dilute up to the mark, mix thoroughly, and make two determinations of iron, each in 100 c.c. of this solution, by reduction to ferrous sulphate and titration with potassium permanganate according to § 113, 3, *a*, p. 278.

2. Proceed as above until the decomposition with hydrochloric acid has been effected. Now, if no *ferrous iron* is present, concentrate, if the volume of the liquid is greater, to 20—30 c.c. and dilute without filtering in a half-litre flask to 500 c.c., and determine iron in portions of 100 c.c. each, according to 113, 3, *b* (titration with standard sodium thiosulphate and iodine solutions). If ferrous iron is present, add, a little at a time, potassium chlorate (less than ½ gr. for 2 grm. ore usually suffices) until a minute portion of the solution taken out with pointed glass rod and tested after dilution on a watch-glass with fresh solution of potassium ferricyanide gives no blue color. The unavoidable excess of potassium chlorate must now be decomposed by heating with hydrochloric acid, and the liberated chlorine removed by evaporating off about half the solution. Next dilute to 500 c.c. and determine the iron by the process just mentioned.

Phosphorus.

Take 5 grm. of the ore, unless it is known to contain a rather large quantity of phosphoric acid, in which case 2 or 3 grm. should be used. Decompose with concentrated hydrochloric acid in a beaker having a diameter at the bottom of about 9 centimetres. The solution, without filtering from the insoluble residue, may be allowed to evaporate in a sand-bath nearly to dryness, but before the liquid has been all removed, transfer the beaker to an air-

* Some iron ores, especially magnetites, contain a small quantity of iron existing as a constituent of a silicate (e.g., hornblende or garnet), undecomposable by hydrochloric acid. In the *assay* of iron ores, the slight inaccuracy which on this account results is usually disregarded.

bath* provided with a thermometer, and continue the evaporation at a temperature of 130° to 140° C. until the mass is quite dry and appears no longer sticky, but brittle when touched with a glass rod. Then, after cooling, add concentrated nitric acid (40 to 50 c.c.) and heat on a sand-bath until the iron is again dissolved and only a residue similar in color and appearance to the original insoluble residue remains. This solution of the iron in nitric acid is easily effected, provided the heat used in the preceding drying operation has not exceeded the prescribed limit. The excess of nitric acid, however, which it is usually necessary to use for this purpose, if allowed to remain in the free state, retards the subsequent precipitation of phosphoric acid by molybdic acid solution, and may even cause an appreciable error by retaining a portion permanently in solution. Evaporate off, therefore, a part of the nitric acid, reducing the volume to about 25 c.c. If the evaporation proceeds too far, basic ferric salts containing phosphoric acid will remain undissolved on subsequent addition of water. After proper concentration, add 100 c.c. of cold water, stir, and allow the insoluble residue, which must not show evidence of containing basic ferric salts, to settle. Filter into a tall narrow beaker, or better still into a cone-shaped flask. To the filtrate and washings, which need hardly exceed 200 c.c., add 100 c.c. of molybdic acid solution (prepared as directed in "Qual. Anal.," p. 72). Add next gradually ammonia so long as the reddish-brown precipitate, which it forms, dissolves *very readily* on stirring. A few cubic centimetres can usually be added at this point, without danger of forming a permanent iron precipitate, on account of the free nitric acid which the added molybdic acid solution contained. Stir well the solution, which should now occupy a volume of 300 to 350 c.c., and let it stand at a temperature of 40° to 50° C. at least 24 hours. The greater part of the solution standing over the precipitated ammonium phosphomolybdate can be removed perfectly clear by means of a siphon.

---

* A suitable air-bath may be easily constructed as follows: Procure an iron pot having its diameter greatest at the rim (about 12 inches), fit a sheet tin cover to it, cut circular holes (2 or 4) in the cover 10 centimetres in diameter to receive the beaker (which must be selected of a proper size), and also a small hole in the centre for inserting a thermometer. The pot is heated by setting it into a sheet-iron cylinder, made to fit it, down to the rim, and placing a Bunsen burner under it within the cylinder.

Collect the precipitate on a small filter (2½ inches in diameter) and wash it with the same molybdic acid solution that is used for precipitation, diluted with an equal volume of water. Allow the filtrate and washings to stand in a warm place several hours to ascertain whether any more phosphoric acid can be precipitated. The moist precipitate is to be dissolved on the filter with ammonia. It is advisable to have the ammonia used for this purpose in a small graduated glass cylinder so that the quantity used may be observed. Pour 2 or 3 c.c. into the flask in which the precipitation has been effected in order to dissolve what may adhere to it, then pour from the flask upon the filter, and at the same time stir up the precipitate with a jet of hot water. Repeat this operation till complete solution takes place. By cautious use of ammonia solution (sp. gr. ·95) its volume should be restricted to about 10 c.c. for small quantities of the precipitated phosphomolybdate, while for comparatively large quantities, such as are obtained from 4 gr. of ore containing upwards of ·5 per cent. of phosphoric acid, more may be used. Usually the solution, after passing through the filter remains clear or at most exhibits but a slight opalescence. Occasionally it is turbid to such extent that it is advisable to pass it through the same filter again. Wash the solution out of the filter paper with the smallest sufficient volume of hot water. Add now, according to the quantity of the dissolved precipitate, 6 to 12 c.c. of hydrochloric acid (sp. gr. 1·1). If this occasions (by supersaturation of the ammonia) a permanent precipitate of ammonium phosphomolybdate, redissolve it with a slight excess of ammonia, adding enough to give a perceptible odor of ammonia to the solution when cold. This addition of a measured volume of hydrochloric acid is designed to form a moderate quantity of ammonium chloride—not enough to have a sensible solvent effect on the ammonium magnesium phosphate which is next to be precipitated, but sufficient to prevent the coprecipitation of other magnesium compounds. Next precipitate the phosphoric acid with "magnesium mixture" (see p. 113). An excess of this solution is required to effect complete precipitation of phosphoric acid; 8 to 10 c.c. may be used in any case, while more may be required if the ore is rich in phosphoric acid. Finally, to render the separation of ammonium magnesium phosphate complete, add to the solution about one-tenth its volume of ammonia solution, and stir well. The preceding operations should be conducted in such a manner as not to unnecessarily increase the volume of the solution. The final volume after addition of all reagents may amount to 40 to 60 c.c. in ordinary cases, or to 100 c.c. for ores containing an unusually large quantity of phosphorus. Filter the solution after standing 6 to 12 hours in the cold, wash the precipitate with dilute ammonia, dry, detach from the filter (unless the quantity is very small), incinerate the filter in an open platinum crucible, add the precipitate, ignite, weigh, and calculate the amount of phosphorus (or if required P_2O_5) in the ore.

In view of the importance of *accurate* determinations of phosphorus in iron ores, pig-iron, &c., for technical purposes, some further explanation may here be properly given of the causes of the possible errors which the above directions are intended to obviate. If, in the beginning of the process, HCl is not removed by evaporation to dryness, it may prevent complete precipitation of phosphoric acid by the molybdic acid solution.* The presence of a large quantity of free nitric acid also prevents precipitation of the last traces of phosphoric acid by the molybdic solution. If, in attempting to obviate this cause of error by evaporating the nitric acid, the evaporation is carried *too far*, basic ferric nitrate will be formed, which will retain phosphoric acid. If too great heat is used in precipitating the ammonium phosphomolybdate, free molybdic acid will be deposited along with it. A slight deposition of molybdic acid, provided the precipitate remains pulverulent, may have no sensible injurious effect; but a larger amount, especially if deposited in the form of a crust, will retain iron which cannot be washed out on the filter. If then ammonia is applied to dissolve the precipitate on the filter, a ferric compound containing phosphoric acid will remain on the filter undissolved.

In order to insure the complete precipitation of phosphoric acid, it is necessary to use not only enough molybdic solution to

---

* It is true that after evaporation and drying at a temperature between 130° and 140° C. some chlorine still remains as ferric chloride, which might be further decreased or entirely removed by evaporating again the nitric acid solution to dryness. I have repeatedly taken this course and compared the results with those obtained without evaporating a second time; but do not thereby obtain a larger amount of phosphorus, and conclude that this extra precaution is unnecessary.—O. D. A.

convert it into ammonium phosphomolybdate, but a liberal excess proportional to the volume of the solution. It may occasionally happen, in case of an ore or a sample of iron unexpectedly rich in phosphorus, that 100 c.c. will not suffice. But since more molybdic solution is added in the process of washing the precipitate, the formation of an additional precipitate in the filtrate, which should be kept warm 6 hours, will indicate any deficiency in the quantity of molybdic solution first used.

If, in the final precipitation as ammonium magnesium phosphate, a large amount of free ammonia, and no ammonium chloride, is present when the magnesia mixture is added, it is possible that magnesium oxide or basic magnesium phosphate may be coprecipitated; while, on the other hand, a very large amount of ammonium chloride may retard the precipitation of phosphoric acid. If the precipitation is attempted in too large a volume of solution there is more danger that it may not be complete, and also more difficulty in removing the precipitate from the sides of the vessel, to which it may adhere in the form of minute transparent crystals.

Finally, notwithstanding the use of all due precaution, the weighed magnesium pyrophosphate may contain a trace of silica —so slight that it may in most cases be neglected. But if a very accurate determination of minute quantities of phosphorus in the purer kinds of ore, iron, &c., is required, it is advisable to dissolve the weighed precipitate in the crucible by warming with nitric or hydrochloric acid, collect any remaining residue on a very small filter, wash, return to the crucible, ignite, weigh, and deduct the weight of the crucible + silica from its weight + first ignited pyrophosphate.

Sulphur. If any considerable quantity of metallic sulphides is visible on close inspection of the ore, take 5 grm. *finely* pulverized. If no sulphides can be seen take 7 to 10 grm. Add to the ore in a large beaker about 20 c. c. of aqua regia for each gramme taken. Allow it to stand at common temperature of the room 6 hours, then 12 hours longer at 40° to 50° C. Finally evaporate to dryness, treat the residue with strong hydrochloric acid, dilute to 200 to 300 c.c., filter, concentrate to about 100 c.c., transfer to a small beaker, add while hot a few c.c. of barium chloride solution. If the ore contains much sulphur, the greater part of the sulphuric acid produced from it will be at once precipitated; a quantity far too great to neglect, however, will remain in solution on account of the presence of free acids and ferric salts. If the ore contains a comparatively small though still determinable amount of sulphur, it may happen that no precipitate will appear at this stage. In either case, therefore, remove the free acid by evaporation, after the addition of barium chloride so far as it can be removed without the formation of basic ferric salts insoluble in water. The last part of the evaporation is carried on best by heating the small beaker in an iron plate. The solution may usually be brought thus to a volume of 10 or 15 c.c. The formation of a dark pellicle on the surface of the liquid at this stage can usually be observed, and is a sure indication that further evaporation would render the iron insoluble in water. After cooling, add cold water (about 100 c.c.) and 1 c.c. dilute HCl to dissolve the soluble saline matter. If the ore contained sulphur, a residue of barium sulphate will now appear.* (If the preceding evaporation has been carried too far a bulky mass of ferric salts will remain undissolved, in which case add hydrochloric acid freely till it dissolves, and repeat the evaporation.) The barium sulphate thus obtained usually contains iron and other impurities. Collect it on a filter, wash till the greater part of the saline matter is removed, ignite in an open platinum crucible till carbon is burned away, add a little sodium carbonate, fuse, warm the fused mass with water in the crucible until it becomes disintegrated; pour the contents upon a small filter, wash the sodium sulphate out of the insoluble part, add HCl to the filtrate till it gives, after boiling, an acid reaction with test paper (avoiding much excess of acid), and precipitate while boiling with barium chloride. The barium sulphate thus obtained, after washing first by decantation 2 or 3 times, and afterwards on a filter with boiling water, may be assumed to be sufficiently pure. Weigh it and calculate percentage of sulphur in the ore.

* The nitric and hydrochloric acids used for the examination must always be tested for sulphuric acid as follows: Evaporate 200 c.c. (or the same volume used in analysis) of the mixed acids, with addition of a few centigrammes of pure Na_2CO_3 till only some half dozen drops remain, dilute, add barium chloride while hot. If a weighable amount of $BaSO_4$ is formed, weigh it. If the weight of $BaSO_4$ from the 200 c.c. does not exceed ·002 or ·003 gr., the acids may be used with the required correction of result.

Manganese.

1. *Method suitable for ores not unusually rich in manganese.*

The most reliable methods of determining manganese in iron ores involve the precipitation of iron as basic ferric acetate. In order to avoid the tedious operation of washing a large quantity of iron precipitated in this form, the whole volume of the solution in which the precipitate is formed may be measured, and after the precipitate has settled a measured portion of the nearly clear supernatant liquid may be taken for the determination of manganese. A wide graduated cylinder of thin glass holding 1200 to 1400 c.c. is required for measuring the solution.* Take 4 or 5 grm. ore, decompose with strong hydrochloric acid, evaporate to dryness (with addition of nitric acid if ferrous iron is present), redissolve with strong hydrochloric acid, evaporate off the greater part of the excess of acid used for redissolving, dilute and filter into a flask capable of holding at least 1500 c.c., previously marked at a height corresponding to 1000 c.c.

Precipitate now the iron by the successive addition of sodium carbonate, a little hycrochloric acid, acetic acid, sodium acetate, and boiling; according to directions given in § 160, **71**. The final volume to which the solution is brought before boiling must in this case be limited to about 1000 c.c. After precipitation, pour the contents of the flask immediately without cooling into the graduated vessel, rinse the flask with a small volume of water which must be carefully mixed from top to bottom with the main solution by stirring with a long glass rod. When the precipitate has settled to such an extent that at least half of the solution can be drawn off nearly free from suspended matter, note the volume which it occupies, and siphon off the nearly clear solution. Note the volume remaining. Suppose, having used 5 gr. ore, the whole volume was 1140 c.c., and the remaining volume 420 c.c. The

* If a suitable measuring vessel is not at hand, one which will suffice may be prepared in the following manner: Procure a tall narrow beaker (9—10 in. in height, 3—3½ in. in diameter). Run 50 c.c. of water into it from a burette; mark the side of the beaker at the surface of the liquid with a writing diamond (or mark with a pencil a vertical strip of paper fastened to the beaker with shellac). Continue adding portions of 50 c.c. and marking till the vessel is filled. Afterwards graduate the portion between 1000 c.c. and 1200 c.c., also between 300 c.c. and 500 c.c., into spaces corresponding each to 10 c.c.

volume drawn off is then 1140 — 420 = 720 c.c., corresponding to $\frac{720}{1140} \times 5$ gr. ore; or rather the 720 c.c. corresponds *approximately* to that amount of ore. For no account is taken of the volume occupied by the solid ferric acetate, nor can very accurate measurements be made in wide graduated glass vessels. But these sources of error have no appreciable influence on the final result unless the ore is comparatively rich in manganese (containing over 2 or 3 per cent.). For such ores it is, in fact, preferable to use 1 grm. for the determination, and wash the basic ferric acetate in the ordinary manner.

The solution which is drawn off by means of a siphon may contain, besides a little suspended iron precipitate, a trace of iron still in solution, calcium and magnesium, and a large amount of saline matter. Concentrate without filtration by evaporation in a beaker, or, more expeditiously. by boiling in a flask to about 300 c.c.; add sodium carbonate to alkaline reaction, boil and add a little sodium hydroxide. Manganese is thus precipitated along with iron calcium, &c.; collect the precipitate on a filter, wash slightly and dissolve on the filter with the smallest possible quantity of hydrochloric acid. If the precipitate dissolves with difficulty on account of the presence of higher oxides of manganese, add a few drops of solution of sulphurous acid. Boil the filtrate to expel chlorine, or if sulphurous acid has been used, boil with addition of a few drops of nitric acid. Add sodium carbonate solution till a slight deepening of color, due to presence of ferric chloride, indicates that the solution is nearly neutral. (If sufficient iron is not already present to give this indication, two or three drops of ferric chloride solution may be added.) Add next sodium acetate and boil to precipitate the slight quantity of iron present, and filter the hot solution. If the preceding operations have been properly conducted, the filtrate and washings need rarely exceed 200 c.c. Precipitate the manganese in it by adding aqueous solution of bromine and keeping it warm a few hours. When the excess of bromine has escaped, filter and wash with hot water. Test the filtrate for manganese by adding a little more bromine solution and also more sodium acetate. Ignite the precipitate and weigh as Mn_3O_4. The manganese protosesquioxide thus obtained may contain a trace of soda; but when the quantity does not amount to more than 2 or 3 per cent. of the ore, it is not probable that greater accuracy would be attained by dissolving it and converting the manganese into another form for weighing. But it should, after weighing, be examined to ascertain whether it contains enough cobalt to cause an appreciable error in the estimation of manganese, since traces of cobalt are frequently present in iron ores, more especially in brown hematites. Dissolve it in hydrochloric acid, evaporate to a few drops. If the bright green color, which even a very small amount of cobalt would occasion, does not appear, it may be assumed that cobalt is not present in sufficient quantity to require any change in the percentage of manganese calculated from the weighed protosesquioxide. But if the color indicates presence of cobalt, continue the evaporation with heat not exceeding 100° C. until free acid is *completely* removed. Dissolve the residue in about 20 c.c. of water and acidify with not more than one or two drops of acetic acid, add sodium acetate, heat and pass H_2S through the solution. Cobalt will then be precipitated as sulpide. If the quantity is sufficient, the cobalt may be determined by converting it into cobalt sulphate (see p. 265); or manganese may be determined in the filtrate from the cobalt sulphide, by precipitating (after boiling out H_2S) with sodium carbonate, igniting and weighing again as Mn_3O_4.

2. *Method suitable for Ores containing larger quantities of Manganese.*

Weigh out from ·75 to 1· gr. and proceed as in the above-described process so far as the precipitation of iron as basic ferric acetate. Filter, wash the precipitate (best collected on *two* filters) with hot water containing 1 or 2 per cent. of sodium acetate. Boil the filtrate and washings, and filter again if any additional flocks of basic ferric acetate separate. Concentrate the filtrate to 600—800 c.c., transfer about one-half of the solution into another beaker, nearly or quite neutralize the acetic acid which it contains with sodium carbonate, add to it the remainder of the solution which still contains free acid, and should dissolve any slight precipitate caused by sodium carbonate. Precipitate next manganese with bromine and treat the precipitate as above directed in 1, not omitting examination for cobalt; or, if great accuracy is desired, the manganese may be converted into pyrophosphate for weighing.

TITANIC ACID.

Qualitative examination. Fuse about 1 grm. of the *finely* pulverized ore with potassium disulphate in the manner described below under "Quantitative determination." Treat the fused mass with boiling dilute hydrochloric acid, which readily dissolves the iron and titanic acid. Boil the solution, without filtering from the insoluble residue which usually remains, in a porcelain casserole with granulated tin. If the violet color indicating titanium does not sooner appear, concentrate by rapid boiling, with addition of more tin in case the first portion has dissolved, until saline matter begins to be deposited and but some half-dozen c. c. of liquid remain. If no decided violet color now appears it may be concluded that either no titanium or but *very little* is present. The only certain way to detect minute quantities is to proceed as in quantitative determinations, which indeed requires but little more time if the method of decomposing the ore with hydrofluoric acid is employed.

Quantitative determination. Fuse 1 grm. of the very finely pulverized ore with potassium disulphate. The potassium disulphate has but little effect on the ore until the temperature approaches dull redness. If it contains too much sulphuric acid it will froth and occasion mechanical loss before the proper temperature is reached. If prepared strictly according to directions in § 64, 7, p. 115, this trouble will be obviated. After the bottom of the crucible is faint red, apply no more heat than is just sufficient to maintain the mass in a state of fusion. The temperature must be gradually increased, since the sulphate becomes more and more infusible as fumes of sulphuric acid escape with formation of normal sulphate. When the mass is no longer fluid at a full red heat, allow it to cool; add concentrated sulphuric acid (2 or 3 c. c.), heat very gradually until, aided by stirring with a platinum wire, the fused mass becomes disintegrated and mixed with the acid. The temperature may then be gradually increased as before.

The progress of the decomposition may be ascertained by dipping a thick cold platinum wire or spatula to the bottom of the crucible, allowing it to cool and repeating the dipping a few times. By inspection of the sample thus taken up with a lens one can see whether undecomposed particles of magnetite are present. One addition of fresh sulphuric acid often suffices, but the addition of acid and reheating may be repeated as often as required. It is advisable at the end of the operation, after the decomposition appears complete, to incorporate a liberal amount of sulphuric acid uniformly with the mass, and allow the greater part to remain in order to facilitate subsequent solution of the mass in water. After cooling, digest with 300 c. c. of cold water until all soluble matter (ferric sulphate, titanic acid) is taken up. This often requires a long time, usually 24 to 48 hours. If the ore contains quartz or silicates an insoluble residue is sure to remain, possibly retaining a small quantity of titanic acid. Collect it on a filter, incinerate the filter, and fuse the residue with a small quantity of potassium disulphate, and at the end of the operation add, after the mass has sufficiently cooled, concentrated sulphuric acid enough to retain the potassium salt and the titanic acid permanently in solution. Heat till the whole is liquid with exception of the undecomposable parts of the ore, cool, and put the crucible with its still liquid contents into a small beaker containing just sufficient water to cover it. If titanic is present it will now readily go into solution. The filtered solution can be treated for titanic separately like the main solution, or it may be added to the main solution. To separate titanic acid from the first solution, or from the two mixed solutions, add first sodium carbonate so long as it can be added without producing a permanent precipitate, then 3 c. c. of pure dilute sulphuric acid and 100 to 150 c. c. strong solution of sulphurous acid; expose to heat of 40° to 50° C. an hour. If the solution continues to smell of sulphurous acid enough of that reagent has been added, otherwise more should be added. Dilute to 700 to 800 c. c. in a large beaker and boil steadily 2 hours, covered with a watch-glass. A moderate quantity of free acid must be present to prevent iron from being precipitated. The iron must also be in the state of *ferrous* sulphate. Too much free acid prevents precipitation of titanic acid. When a considerable amount of titanic acid is present the formation of a precipitate on heating the solution, a little before the actual boiling begins, is an indication that the free acid present does not exceed the proper amount. To compensate for the water lost by evaporation during the boiling, add from time to time *hot* water so gradually as not to check the boiling. A little solution of sulphurous acid should be mixed with the water thus added to keep the iron in the state of ferrous sulphate.

Allow the precipitated titanic acid to settle till the solution above it is perfectly clear (12 to 24 hours). Filter (not with a Bunsen pump) through a filter carefully fitted to the funnel and stir the precipitate as little as possible during the washing, as it is somewhat inclined to pass through the pores of the filter paper; if necessary, ammonium sulphate may be added to the water used for washing to prevent this tendency. Ignite the precipitate strongly, let the crucible partially cool, throw in a small lump of clean ammonium carbonate, and heat rapidly again to bright redness in order to remove traces of sulphuric acid. The weighed titanic acid, notwithstanding all precautions, is likely to contain a little ferric oxide. Fuse it with sodium carbonate, add gradually to the cold fused mass in the crucible 5 or 6 c. c. strong sulphuric acid, heat till evolution of CO_2 has ceased and the mass has dissolved. Add then more strong sulphuric acid (6—10 c. c.) and dilute with about 100 c. c. of water. Determine the iron in this solution by titration with potassium permanganate, with previous reduction by H_2S according to § 113, 3. (Zinc cannot in this case be employed for the reduction since it reduces also titanic acid.) Calculate the iron found as ferric oxide and subtract it from the impure titanic acid weighed.

If the analyst has at hand hydrofluoric acid and a platinum dish capable of holding 100 to 200 c. c., the following method of decomposing the ore may be substituted, with great saving of time, for the fusion with disulphate. Heat the ore nearly to boiling in the platinum dish with a mixture of hydrofluoric and strong hydrochloric acids. Magnetic iron ores, in which it is oftenest required to determine titanic acid, are thus usually decomposed in a few minutes. Add 20 to 25 c. c. concentrated sulphuric acid diluted with half its volume of water, and concentrate by means of a carefully adjusted flame till fumes of sulphuric acid begin to escape. It is of utmost importance to remove every trace of hydrofluoric acid. The appearance of fumes of sulphuric acid can be considered as proof that this has been effected only when means are employed to protect the sides of the dish above the liquid from heat sufficient to volatilize sulphuric acid, since the mixed acids are attracted upward along the surface of the platinum. After cooling add nearly 100 c. c. of water. Either at once or in a few hours the whole dissolves, with exception perhaps of a slight residue, which may, if it appears too considerable to be neglected, be subjected again to the same treatment.* The solution of the ore obtained in this way is neutralized with sodium carbonate and further treated in the same manner as a solution obtained by decomposing with potassium disulphate.]

[17. COMPLETE ANALYSIS OF IRON ORES.

§ 217.

(*Process adapted to all iron ores except such as contain a large amount of titanic acid.*)

1. *Silica, iron, aluminium, manganese, calcium, magnesium.* Take about 1 grm. ore. Add concentrated hydrochloric acid and heat in a water bath or sand bath nearly to boiling one or two hours. Evaporate finally to dryness and expose the residue to a heat slightly exceeding 100° C. in order to render insoluble any silica which may have been dissolved. A porcelain dish may be used in this operation, but a beaker of 200 to 300 c.c. capacity is more convenient, especially if a suitable air bath is at hand for raising the temperature at the end to 120 to 130° C. Add two or three c.c. concentrated hydrochloric acid to the residue and warm till the iron is redissolved, and filter at once† after suitable dilution, carefully removing every particle of the residue to the filter; wash and reserve the filtrate; ignite the filter and its contents till carbon is burned away; add then to the residue 5 or 6 times its weight of pure sodium carbonate and fuse; disintegrate the fused mass by heating with water, acidify with hydrochloric acid, and separate silica by evaporation and drying in the usual manner. The filtrate from the silica is now added to the other reserved solution of basic metals.

Another method of decomposing the ore and separating silica is to fuse directly with sodium carbonate, without previous treat-

* If the ore contains much calcium, a residue of calcium sulphate insoluble in the limited amount of water above recommended must be expected.

† By prolonged digestion of this residue with hydrochloric acid, traces of silica might be taken up from certain silicates which being very slowly acted on by acid may have escaped complete decomposition by the first treatment with acid.

ment with hydrochloric acid, and separate silica from the fused mass in the ordinary manner. More time, however, is required to disintegrate the mass and separate the silica, more saline matter is introduced into the solution, and the platinum crucible used for the fusion is likely to become permeated with iron to such an extent that for a long time it is unsuitable for most other uses. These disadvantages overbalance the apparent greater simplicity of this mode of proceeding.

From the solution of basic metals precipitate first, the iron as basic ferric acetate according to direction given in § 160, **71**, p. 517. The iron may be precipitated without concentration of the two mixed filtrates if care has been taken to avoid too large a volume and the presence of an unnecessary amount of free acid, otherwise the solution should be concentrated until the greater part of the free acid is removed. It is usually best to collect the precipitated ferric acetate on *two* filters. Wash at first with boiling hot water containing 1 or 2 per cent. of sodium acetate until a few drops of the washings give but a slight turbidity when tested with silver nitrate. Reserve the filtrate and washings which contain manganese, calcium and magnesium, and continue without interruption to wash the precipitate with hot water to which a little ammonium acetate has been added until a drop of the washings leaves no residue on evaporation on platinum foil. The last washings containing ammonium acetate are thrown away. Dry the precipitate, which contains besides iron the aluminium and phosphoric acid of the ore, detach from the filters, incinerate the latter with prolonged exposure to the air at a full red heat in order to convert into ferric oxide the lower oxides of iron formed by reducing action of the filter paper. Add the precipitate and moisten it in the crucible with concentrated nitric acid, dry with gentle heat, repeat the moistening with nitric acid and drying, ignite and weigh. The weighed precipitate should exhibit no magnetic attraction when a magnet is applied *externally* to the bottom of the crucible.*

Dissolve the weighed Fe_2O_3, Al_2O_3 and P_2O_5 in strong hydro-

* If the treatment with nitric acid is omitted, lower oxides of iron are usually formed by ignition of basic ferric acetate which are converted into ferric oxide with great difficulty by prolonged ignition. Even treatment with nitric acid has but little oxidizing effect after the precipitate has once been ignited.

chloric acid, add 10 to 15 c.c. of pure dilute sulphuric acid, remove all the hydrochloric acid by evaporation, and dilute moderately with water. Occasionally, but not often, a residue of silica may be observed at this point, so considerable in quantity as to render it advisable to collect it in a small filter and deduct its weight from the precipitate which contained it and add it to that of the main portion of silica. It is necessary now, in order to estimate satisfactorily the comparatively small amount of aluminium usually present, to determine very accurately the iron. Titration with potassium permanganate in the sulphuric acid solution is the best of all volumetric methods; previous reduction with hydrogen sulphide according to directions on p. 729 is to be recommended as a method of reduction involving least sources of error. The Al_2O_3 is calculated by deducting Fe_2O_3 found, and also P_2O_5* (determined in another portion of the ore) from the joint weight of the three substances.

Concentrate the filtrate from the basic ferric acetate to about 600 c.c. and precipitate manganese with bromine water after partial neutralization of the free acetic acid as directed on p. 749 (second method of determining manganese). Neutralize the filtrate from the manganese dioxide with ammonia and precipitate calcium with ammonium oxalate. In the filtrate from calcium oxalate precipitate (without concentration unless the volume exceeds 600 c.c.) the magnesium with sodium phosphate adding a liberal quantity of ammonia and allowing 24 hours for complete separation of the ammonium magnesium phosphate.

2. *Alkalies.*

Small quantities of potash or soda are sometimes found in magnetic iron ores owing to the presence of felspars. More rarely an appreciable amount of potash may be found in brown

* If the amount of P_2O_5 is very small, not exceeding say 0·1 per cent., the precipitate produced by boiling with sodium acetate, instead of being treated as here recommended, may be washed sufficiently to free it from appreciable quantities of manganese and alkali-earth metals, dissolved in hydrochloric acid and reprecipitated with ammonia. The ferric and aluminium hydroxides thus precipitated are easily washed free from saline matter, ignited and weighed. The phosphoric acid, however, is liable to be but partially precipitated by ammonia along with the iron, so that an error (not exceeding the amount of P_2O_5) will result in calculating the Al_2O_3 by difference.

hematite on account of intermixed micaceous minerals. For qualitative or quantitative examination use the method of J. L. SMITH. See p. 426.

3. *Ferrous and Ferric Oxides.*

Decompose ·5 grm. by boiling in a large covered platinum crucible with a mixture of sulphuric and hydrofluoric acids, dilute and determine ferrous iron by titration with potassium permanganate. (See p. 529). The amount of ferric oxide can of course then be calculated, the total iron having being previously determined.

4. *Carbonic Acid.*

Determine carbonic acid in 1 to 5 grm. according to the amount present by decomposing with hydrochloric acid and weighing the evolved CO_2 by the process described on p. 413. Or determine carbonic acid and water at the same time by igniting in a combustion tube with lead chromate and potassium chromate. (See Analysis of Silicates and Siliceous Rocks, § 208, p. 716). The latter method, however, cannot be used when the ore contains carbonaceous matter.

5. *Water.*

When carbonic acid and ferrous oxide are absent, water is determined by loss on ignition. Water should be determined in the same pulverized sample, kept carefully corked in a tube, which has been used for the determination of the chief constituents. If the sample has not been previously dried at 100° C., hygroscopic and combined water may be determined separately (if desired) by drying a weighed portion (about 1 grm.) at 100° C. to constant weight in a platinum crucible and afterwards igniting.

In the presence of carbonic acid or ferrous oxide determine water by igniting 1 or 2 grm. in a platinum boat in a combustion tube and collecting the water in a calcium chloride tube, a slow current of dried air being meanwhile drawn through the apparatus by an aspirator. Or the water and at the same time carbonic acid may be determined as suggested above (under 4. Carbonic Acid).

6. *Presence of carbonaceous or bituminous matter* interferes with determination of water by either of the above methods, and also requires a modification of the processes used for determining some of the other constituents. *Sulphur* should be determined in

the ore without previous ignition, as directed in § 216; *carbonic acid* by decomposition with hydrochloric acid, and weighing the evolved CO_2, as described p. 413. In determining *phosphorus* according to § 216, ignite the portion weighed out in an open crucible till carbon is burned out before dissolving it. Treat also in the same manner the weighed portion in which silica and the other chief constituents are to be determined.

For the estimation of *ferrous* and *ferric oxides* decompose ·5 gr. with a mixture of hydrochloric and hydrofluoric acids, proceeding as in 3. Since presence of organic matter may interfere with the volumetric determination of either ferrous or ferric iron, separate ferric iron by barium carbonate with exclusion of air, according to § 160, p. 513. Determine the amount of ferric iron thus precipitated by dissolving in sulphuric acid, reduction to ferrous sulphate, and titration with potassium permanganate.

From the results of these several operations the composition of the ore in its original state can now be calculated, with the exception of water and carbonaceous matter.

7. *Titanic acid.*

It is rarely or never necessary to make complete analyses of iron ores containing over 5 or 6 per cent. of titanic acid, since such ores are usually rejected as unsuitable for smelting. When ores containing this amount or less are subjected to the above-described process of analysis, a portion of the titanic acid follows the silica and is weighed along with it. The remainder is precipitated with the basic ferric acetate and is weighed with ferric oxide. A method of separating the titanic acid from these two products is described in § 208 (Analysis of Silicates and Siliceous Rocks), p. 717. In washing silica which contains titanic acid, the latter sometimes passes through the pores of the paper, making the filtrate turbid. This, however, will occasion no error if the filter retains the silica.

8. *Phosphoric acid* must be determined in a separate portion of the ore as in § 216.

9. *Sulphur* must also be determined in a separate portion as in § 216.

[18. ANALYSIS OF PIG-IRON, STEEL, AND WROUGHT IRON.

§ 218.

I. PIG-IRON.

Preparation of the sample.

The chemist usually receives for analysis a short section broken from a pig. If the iron is hard and brittle (white pig or spiegel), procure, by breaking on an anvil with a heavy hammer, some fragments free from the outer surface, to which sand or other impurities may adhere. Pulverize these fragments to a coarse powder in a mortar of the hardest steel. If the sample is too tough to be crushed, it must be reduced to a suitable condition by drilling.* To obtain the necessary quantity (40 to 50 grm.) bore one or more holes in the clean broken end of the sample, at a distance half way between the centre and outside. Use no oil or water in the process, and cleanse the drill from oil before beginning. The borings may be taken up from time to time during the operation with a magnet and transferred to a bottle provided with a glass stopper.

1. *Determination of the total amount of carbon.*

Method of BERZELIUS (somewhat modified).

The determination of carbon by the method here recommended requires the use of a special reagent, viz., a strong solution of cupric ammonium chloride containing *no free acid.* This solution may be prepared as follows: Dissolve common blue vitriol (crystallized cupric sulphate) in 10 to 15 times its weight of water, filter the solution, and heat to boiling in a copper kettle. Add solution of common sal soda gradually to alkaline reaction, keeping up meanwhile the boiling. Basic cupric carbonate (not entirely free from basic sulphate) is precipitated in a dense form easy to wash by decantation. Wash it by decantation until the sodium sulphate is nearly all removed. Transfer to a glass or porcelain vessel. Reserve about one tenth, and dissolve the remainder in concentrated hydrochloric acid; add the reserved portion which is

* It is best if possible to employ the assistance of a machinist who can use a drill press run by steam.

designed to neutralize the acid completely. Let the solution stand cold several hours with occasional agitation. A portion of the basic carbonate should remain permanently undissolved. Filter, and add, for five parts blue vitriol used, two parts ammonium chloride previously dissolved in a small volume of hot water and filtered. Care should be taken to conduct the above operations so that the final solution obtained may not be too dilute. If its volume does not exceed twice the volume of the concentrated hydrochloric acid used in dissolving the carbonate, it is satisfactory in respect to strength.

The Process. Pour at once, not gradually, at least 200 c.c. of the cupric ammonium chloride solution upon the iron borings (3 or 4 grm. may be taken) in a large beaker. The beaker should be set in a vessel of cold water, and the contents should be frequently stirred during the first 15 or 20 minutes to prevent too great elevation of temperature by the chemical action; otherwise a slight evolution of hydrogen might take place, carrying off with it some hydrocarbon compound. (Evolution of hydrogen and loss of carbon is sure to result if the cupric ammonium chloride contains free acid.) Afterwards the beaker is allowed to remain at the common temperature of the room. The iron dissolves as ferrous chloride with deposition of metallic copper, which, in presence of excess of cupric chloride, is converted into cuprous chloride. The latter is soluble in the cupric ammonium chloride. After the metallic iron has all dissolved, leaving a residue which crumbles under pressure (6 to 12 hours may be required according to fineness of the borings), add a few c.c. of hydrochloric acid to dissolve ferric compounds which may be deposited by the action of the air on the ferrous chloride in solution. If, after standing several hours longer, an accumulation of metallic copper or cuprous chloride is observed remaining persistently undissolved, more of the double chloride may be added. The complete solution of iron and copper is generally effected in 48 hours, and often much sooner. When nothing remains undissolved except a black carbonaceous residue, filter through an asbestos filter prepared by packing well-disintegrated asbestos, neither too closely nor too loosely,* in a tube 8 or 9 inches

* Let the first 2 cm. of the filtering tube at the very bottom have a diameter of ½ cm. and the next 2 cm., above a diameter of about 1 cm. Leave the lower 2 cm. empty and fill with asbestos to a point a little above where the tube has its full diameter. After filling, pour water into the tube; if it runs through in a continuous stream, the packing is too loose, but it should drop rapidly.

long and ⅜ of an inch in diameter, narrowed at one end. Wash the carbon residue until the copper solution is completely removed. It is not safe to apply an exhausting apparatus to hasten the filtration with a filter prepared in this manner.* Dilute the filtrate with distilled (or perfectly clear) water, and observe whether particles of the carbon residue have passed through. Dry the residue in the tube at 100°, and determine the amount of carbon in it by combustion with lead chromate mixed with potassium dichromate according to § 177. Remove, for this purpose, the carbon residue together with the asbestos from the tube with the aid of a steel wire slightly curved at the end, introducing it through the narrow end of the tube, loosening and pushing the whole mass out into a small porcelain mortar already containing some of the chromates. Rinse out the tube with the remainder of that portion of the chromates which is to be mixed with the substance, and mix with a pestle in the mortar till the asbestos is broken up to such an extent that the mixture can be introduced into the combustion tube through a funnel.

2. *Determination of the graphite.*

Treat 4 grm. with moderately concentrated hydrochloric acid, at a gentle heat, until no more gas is evolved; filter the solution through an asbestos filter prepared as in 1; wash the undissolved residue, first with boiling water, then with solution of potassa, after this with alcohol, and lastly with ether; then dry, and burn after § 177. Deduct the graphite obtained here from the total amount of carbon found in 1; the difference gives the combined carbon.

3. *Determination of Sulphur.*

The general plan adopted in all good methods of determining sulphur in iron is to dissolve the metal as completely as is practicable in hydrochloric acid, whereby the greater part of the sulphur, being converted into hydrogen sulphide, passes off along with a large volume of hydrogen which is conducted through some liquid capable of absorbing the H_2S. For this purpose bromine dissolved in hydrochloric acid, potassium permanganate solution,

* J. CREAGH SMITH has devised, and described in the Americal Chemical Journal, vol. i., p. 368, an asbestos filter for filtering carbon residues, which is simple in construction, and can be used with the BUNSEN pump.

alkaline lead solution, ammoniacal cadmium solution, ammoniacal silver solution have all been employed, some suitable method in each case being devised for bringing the absorbed sulphur into a weighable form. Only the method in which an ammoniacal silver solution is used will here be described in detail.

Apparatus. A flask of 300 to 350 c.c. capacity is provided with a doubly perforated rubber stopper. Through one hole passes a funnel tube for the introduction of acid. The end of this tube, which should reach nearly to the bottom of the flask, is drawn out narrower and bent upward with a short curve to prevent gas bubbles from entering it and escaping. For absorbing the hydrogen sulphide from the evolved gas a pair of connected U-tubes are used like those in fig. 64, p. 435. The absorbing tubes are connected with the flask by a strong (not too narrow) tube about 8 inches long, bent downward, and contracted if necessary, at each end, so as to fit into the perforation in the stopper.

Treat the rubber stoppers used in making connection with warm soda solution, and carefully rub the loosened sulphur from the surface, not neglecting the perforations, till a clean black surface is obtained.

The process. Dissolve a gramme or more of silver nitrate in 15 to 20 c.c. of ammonia solution. Pour at least 10 c.c. of this solution into the first U-tube and the remainder into the second. A little water may be added if the size and form of the U-tubes require it, in order to secure proper contact with the gas bubbles which are to pass through them. Introduce 10 grm. of the iron and 40—50 c.c. water into the flask. Adjust the funnel tube so that the lower end may be under the surface of the water. Connect the several parts of the apparatus, and add concentrated hydrochloric acid in small portions at a time so as to produce as nearly as practicable a constant evolution of gas. The addition of acid may be regulated according to the appearance of the second U-tube. The first tube should absorb the hydrogen sulphide. If a blackening of the solution in the second tube begins to appear, add the hydrochloric acid more gradually. When (usually after 4 or 5 hours) the addition of more hydrochloric acid fails to increase the very slow evolution of gas, heat the flask gently, but not to boiling, 20 or 30 minutes, with addition of more hydrochloric acid, taking care not to distil over enough acid to neutralize the ammonia in the first U-tube. Collect the precipitate formed in the first tube on a small filter, wash slightly, and dry at 100° C. Dry also the U-tube, to which a portion of the precipitate invariably adheres. The second tube will not contain an appreciable quantity of silver sulphide unless too rapid a current of gas has been unintentionally produced. Place the dry filter and its contents in a small dry beaker. Dissolve or loosen the sulphide of silver from the U-tube by shaking with successive portions of aqua regia and pouring into a small beaker, using in all about 20 c.c. Then put into the beaker the dried silver sulphide with the filter.

The insoluble residue in the flask, consisting chiefly of graphite and silica, often contains sulphur, and should never be neglected in the analysis of pig-irons. Collect it on a filter and wash out the free acid, dry on the filter thoroughly at 100°, detach from the filter carefully, rub the mass to a powder in a beaker with a glass rod, and add aqua regia.

Allow the aqua regia to act on the two products at the common temperature 6 hours, and afterwards 12 to 24 hours, at 40° to 54° C. Then concentrate to one third the first volume, dilute, and filter each through separate filters and unite the filtrates. After concentrating to about 50 c.c. add barium chloride, and continue the concentration not quite to dryness, but till only liquid enough remains to moisten the residue. Add a small volume of water and 5 or 6 drops of hydrochloric acid. Treat the residue of impure barium sulphate thus obtained as in the determination of sulphur in iron ores (p. 746).

The aqua regia used for oxidizing the silver sulphide and the insoluble residue must be tested for sulphuric acid as directed in "Analysis of Iron Ores," p. 746. The process of dissolving the iron in the flask should be carried on without interruption.

This method gives results agreeing with remarkable closeness when repeated determinations are made in the same sample.

The substitution of a hydrochloric acid solution of bromine for the solution of silver nitrate in ammonia requires no essential change in the details of the process. The apparatus, however, must be modified so as to avoid much contact of rubber stoppers with the bromine vapor. The bromine solution at the close of the operation contains the sulphur which has been evolved at H_2S already in the form of sulphuric acid, which can be determined simply by precipitation with barium chloride after evaporating off the hydrochloric acid to the proper extent. But since the insoluble residue, when accurate results are desired,* must be treated for sulphur, as before described, there is little saving of time or trouble by this shorter method of determining the sulphur which passes into the U-tube.

4. *Determination of Phosphorus.*

If the iron is known to contain over 0·5 per cent. of phosphorus 2 grm. will suffice. If less is present 4 grm. may be taken for the determination.

Dissolve with a mixture of equal parts of concentrated nitric and hydrochloric acids, using about 30 c.c. per gramme of iron taken, and pouring the whole quantity upon the iron at once.†

Proceed further in all details precisely as directed for determination of phosphorus in iron ores.

5. *Determination of Silicon.*

The residue from the solution used for determining phosphorus may be used for determining silicon. Ignite it without separation from the filter until the graphite is partially burned away. Fuse with sodium carbonate mixed with a little potassium nitrate, sufficient to effect complete combustion of the carbon still present. Treat the fused mass first with boiling water, in which it readily dissolves, except some silica in light flocculent form, and traces of metallic oxides. Acidify with hydrochloric acid, or nitric acid in case the solution is to be in contact with platinum, and separate silica as usual. When the quantity of silica is not over 1 per cent., these operations may be most conveniently performed in a large platinum crucible without transferring the substance to any other vessel.

6. *Determination of Manganese.*

Dissolve 3 grm. in aqua regia, evaporate to dryness to separate

---

* I have frequently determined separately the sulphur remaining in the insoluble residue obtained by treating pig-iron as described in this process, and seldom find it to be free from a weighable quantity of sulphur; in some cases amounting even to one third of the total amount found.—O. D. A.

† If the mixture of acids is gradually added to the iron, especially if a larger proportion of hydrochloric is used, a possible escape of phosphoretted hydrogen may be apprehended.

silica, redissolve with hydrochloric acid, filter, and determine manganese in the solution as in iron ores. Method 1, p. 747.

In spiegel-iron the manganese may be more accurately determined by dissolving ·5 grm., evaporating to dryness, redissolving with hydrochloric acid, and proceeding with the solution as in Method 2 for iron ores (p. 749).

7. *Determination of Copper.*

If a determination of the minute quantity of copper sometimes present in pig-iron is required, it may be done in the same portion used for sulphur. Dilute the filtrate from the first insoluble residue and pass hydrogen sulphide through it nearly an hour. More or less sulphur separates. Allow it several hours to settle. If the deposit is darker in color than pure sulphur, presence of copper is indicated. In that case collect it on a filter and wash with a dilute solution of hydrogen sulphide. Copper is also often found in the insoluble residue. When this residue is treated with aqua regia to extract the sulphur possibly retained by it, the copper is dissolved and goes finally into the filtrate from the impure barium sulphate first obtained. Pass H_2S through this filtrate and filter off any precipitate which may result.

Incinerate the two filters containing the copper precipitates in a porcelain crucible. Treat the residue in the crucible with aqua regia, add finally a few drops of sulphuric acid, remove the other acids by evaporation, take up the cupric sulphate in a small volume of water, filter and precipitate the copper again with H_2S, and weigh it as cuprous sulphide.

8. *Presence of Other Elements in Pig-Iron.*

Besides the above-mentioned elements, sodium, potassium, lithium, calcium, magnesium, aluminium, chromium, titanium, zinc, cobalt, nickel, tin, arsenic, antimony, vanadium, and, according to some authorities, nitrogen, may occur in minute quantities in pig-iron. Their determination, however, is rarely undertaken; partly because it is not known whether they have any influence, good or bad, on the quality of the iron when present in such minute proportions, and partly because it is very difficult to determine them accurately on account of lack of sufficiently pure reagents, the action of solutions on the vessels used in the process, &c.

II. Steel and Wrought Iron.

Determine carbon, silicon, sulphur, phosphorus, and manganese as in pig-iron, with the following modifications only of the processes used for *carbon*, *silicon*, and *sulphur*.

Silicon is best determined in a separate portion, since the quantity used for phosphorus does not afford enough silica to weigh accurately; 10 grm. will suffice. Place the weighed quantity in a platinum (or porcelain) dish, add first 30 to 40 c. c. water; next, gradually, concentrated hydrochloric acid until with aid of heat the metal is dissolved, leaving a residue of more or less carbonaceous matter. Evaporate to dryness, expose to a temperature of 120° to 150° C. in an air-bath, redissolve the iron by adding first concentrated hydrochloric acid, and next water. Filter through a small filter, incinerate the filter and burn the carbon out of the residue, fuse with sodium carbonate, disintegrate the fused mass with water, acidify with hydrochloric acid, and separate silica by evaporating in the crucible. The now pure silica is collected in a very small filter, washed and weighed in the usual manner.

In the analysis of Bessemer steel, or any steel or iron which has been melted, it may be assumed that the silica thus obtained is formed by oxidation, in the process of analysis, of *silicon* existing in the metal. In the analysis of ordinary wrought iron the silica obtained may come partly from *silicon* and partly from mechanically mixed particles of slag in which it existed as silica.

Carbon. Use for determination 6 to 10 grm. of steel or 10 grm. of wrought iron.

Sulphur. Treat the comparatively small quantity of insoluble residue collected on a small filter, washed and dried, directly with aqua regia without removing from the filter.]

19. ANALYSIS OF COAL AND PEAT.

§ 219.

For technical purposes, estimations of moisture, ash, coke, and volatile matters usually suffice. Determination of sulphur is less frequently required, and ultimate analysis is only resorted to in special cases.

a. Moisture. The finely pulverized coal (3—5 grm.) is heated to 110°—115° for an hour or more or until it ceases to lose weight (see § 29). Many bituminous coals gain weight after a time from oxidation of sulphides or hydrocarbons (WHITNEY). According to HINRICHS,* drying the coal for one hour effects the maximum loss.

b. Coke and *volatile matters.* The dried coal of *a* is sharply heated in a closed platinum, or, in presence of sulphides, in a porcelain crucible as long as combustible matters issue from it. It is then cooled quickly. The loss is set down as volatile matters. The residue, less the ash, is coke.

c. Ash. The residue of *b* is incinerated in a crucible placed aslant.

d. Carbon and *hydrogen* are determined by combustion with chromate of lead and bichromate of potash, § 177.

e. Sulphur is best determined according to § 186, *c*, 2, *α*, p. 658. The method thus described gives the amount of ash as well as sulphur.

Or the following simple method recommended by ESCHKA† may be employed. About 1 grm. of the finely pulverized substance is intimately mixed by stirring with a platinum wire with 1 grm. burned magnesia (MgO) and ·5 grm. dry sodium carbonate in a platinum crucible. The uncovered crucible is then heated in an inclined position with an alcohol lamp so that only the lower half becomes red hot. In order to facilitate combustion, which requires, according to the nature of the substance, ¾ to 1 hour, the mixture is frequently stirred with a platinum wire. After the carbon is consumed and the color of the mass has changed to brownish or yellowish, ½ to 1 grm. of pulverized anhydrous ammonium nitrate is added and intimately mixed with the contents of the crucible. The mixture is then ignited again, in the covered crucible, from 5 to 10 minutes. Any sulphites which may have been formed at first are hereby converted into sulphates. The mixture, which retains its pulverulent form, is next transferred to a beaker and warmed with 150 c. c. of water. The solution is filtered and acidified with hydrochloric acid. Sulphuric acid is then precipitated with barium chloride.

* Chemical News, 19, 282.

† Zeitschr. f. anal. Chem. 13, 344.

All the sulphur, whether existing in the form of calcium sulphate or pyrites, in the coal is obtained by this method.

The sulphur of calcium sulphate in coal may be separately determined by boiling 24 hours the finely powdered coal with an equal weight of sodium carbonate dissolved in water, filtering, acidifying with hydrochloric acid, and precipitating with barium chloride.

The calcium sulphate is decomposed by the sodium carbonate, while sulphides of iron are not attacked.

[20. ANALYSIS OF COMMERCIAL FERTILIZERS.

§ 220.

1. *Preparation of the sample.* Mix the sample uniformly and, if need be, take a portion of 20—50 grms. which shall accurately represent the whole, for further pulverization. Bone, dried blood, guano, &c., should be ground or pounded fine enough to pass through sieve meshes of $\frac{1}{25}$ in. diameter.

Superphosphates should be merely rubbed in a mortar to crush lumps and secure uniformity. Grinding of superphosphates may occasion a further action of the acid on the undissolved phosphate and increase the per cent. of soluble phosphoric acid.

If the substance is very moist and coarse dry 20 to 50 grms. at 100°, with addition of a weighed amount of oxalic acid if ammonia is likely to escape, till it can be easily handled, grind fine and weigh. Make nitrogen determinations in this portion and reckon the results back to the original material.

Analysis of Superphosphate.

2. *Soluble Phosphoric Acid.* Bring 20 grm. into a litre flask with about 800 c. c. of water and shake frequently (every 10 minutes) for 2 hours: then make up to volume of 1 litre; mix thoroughly, pour on dry filter and measure off 100 c. c. = 2 grm. substance.

3. The determination of phosphoric acid in the solution thus obtained may be made most accurately by the molybdic method. (See p. 375).

4. A simpler, more rapid and for most purposes sufficiently

accurate process is the following "citric method," first published in its present form by PETERMANN, but worked out independently in the Connecticut Agricultural Experiment Station, as follows:

To the solution add 55 c. c. solution of ammonium citrate,* (equivalent to 10 grm. of crystallized citric acid), 40 c. c. of magnesia mixture†—in all cases use about four times as much as would be required to combine with the phosphoric acid—and then add to the solution 75 c. c. of water and 90 c. c. of ammonia solution of sp. gr. 0·96. The precipitate should be distinctly crystalline; a flocculent precipitate indicates that insufficient ammonium citrate has been added.

Stir vigorously and repeatedly and after 12 hours filter, wash with dilute ammonia, ignite and weigh. The use of GOOCH's asbestus filter greatly facilitates the work.

5. *Reverted Phosphoric Acid.* Place 2 grm. of substance in a mortar. Take 100 c. c. of neutral or slight alkaline ammonium citrate solution, sp. gr. 1.09, (the commercial citrate is strongly acid), pour 50 c. c. on the substance, add dilute ammonia to slight alkaline reaction, pulverize the substance, let the coarser parts settle, pour off the turbid liquid into a flask, grind the residue to the finest powder and wash it with the remaining citrate solution into the flask, keep the contents of the latter at 30°—40° for half an hour, with very frequent shaking, then dilute to 200 c. c., pour upon a dry filter, take 100 c. c. of filtrate=1 grm. substance, add 40 c. c. magnesia mixture, 120 c. c. water and 100 c. c. ammonia, stir, filter, ignite, etc., as under 4.

Deducting the soluble phosphoric acid from that here found gives the amount of "*reverted phosphoric acid.*"

6. *Insoluble Phosphoric Acid.* 5 grm. of the superphosphate are wet with 5 c. c. solution of magnesium nitrate, sp. gr. 1.355,‡

---

* Neutralize 185 grm. citric acid with ammonia or ammonium carbonate, in very slight excess, and bring to a volume of 1000 c. c.

† 110 grm. crystallized $MgCl_2 6H_2O$, 140 grm. NH_4Cl, 700 c. c. solution of ammonia sp. gr. 0·96, and water to make 2 litres. Instead of $MgCl_2 6H_2O$, 22 grm. of calcined magnesia may be dissolved in the equivalent quantity of HCl, the solution boiled with a little calcined magnesia in excess and filtered.

‡ Dissolve 160 grm. calcined MgO in the equivalent quantity of HNO_3, boil with a little excess of MgO, filter and bring to volume of 1 litre. 5 c. c. of this solution is enough to prevent formation of pyrophosphate in 5 grm. of any commercial superphosphate. If not enough to destroy organic matters, moisten the residue of ignition with HNO_3 and heat again.

evaporated to dryness and gently ignited. The residue is digested with hydrochloric acid, diluted to 500 c. c. and filtered on a dry filter. To 100 c. c. of filtrate (= 1 grm. substance) are added 40 c. c. ammonium citrate solution, 25 c. c. magnesium mixture, 100 c. c. water and 90 c. c. ammonia. The precipitate is treated as under 4. Subtracting the result of 5 from that of 6, gives the "insoluble phosphoric acid."

7. To apply the molybdic method to the analysis of superphosphates, determine total phosphoric acid in 2 grm., first ignited with addition of magnesium nitrate, then treated with nitric acid to *complete* solution of the phosphates and diluted to 500 c. c. 100 c. c. of this solution are used. Determine "insoluble phosphoric acid" in a suitable aliquot of the *nitric solution* of the insoluble residue of 5. Reverted phosphoric acid is found indirectly by subtracting from the total the sum of the soluble and insoluble.

8. *Potash.* Boil 10 grm. with water for 10 minutes, dilute the solution to 1000 c.c. and filter through a dry filter.

The error of measurement due to the presence of undissolved matters is inconsiderable and may be neglected. Heat 100 c.c. of the filtrate to boiling, precipitate sulphuric acid by barium chloride and magnesium iron, &c., together with phosphoric acid by barium hydroxide and filter. In the filtrate, heated nearly to boiling, precipitate the barium by ammonium carbonate and filter. Evaporate the filtrate to dryness, expel ammonium salts by ignition, dissolve the residue in a little water and determine the potash by excess of platinic chloride in the usual way.

When the substance contains much soluble organic matters it is better to destroy these at the outset by heat, which should be very gentle at first and may finally reach faint redness.

Nitrogen may exist in superphosphates either in organic combination, as ammonium salts, or as nitrates.

9. The nitrogen of ammonium salts is determined in all cases by distilling with calcined magnesia—proceeding as directed p. 220, except that *magnesia* must be used instead of potash or lime.

10. The nitrogen in organic combination when alone or together with ammonium salts is determined by combustion with soda lime (p. 644), in the latter case subtracting from the result the amount of nitrogen already found to exist in ammonium salts.

11. Nitrogen in the form of nitrates is determined by SCHULZE's method as described on p. 473.

12. When nitrates and nitrogenous organic matters occur together, it is necessary to determine the total nitrogen by the absolute method as described on p. 637*bb*.

The determination of soluble, reverted, and insoluble phosphoric acid, of potash and of nitrogen, is usually sufficient to fix the commercial value of a superphosphate. It is sometimes required, however, to determine water, sulphuric acid and chlorine.

13. *Water.* Dry one gramme for three hours at 100°. It is often impracticable, and for commercial purposes is unnecessary, to make an accurate water determination. Gypsum, which most superphosphates contain in considerable quantity, does not part with all its water readily or completely at 100°, while a higher heat to some extent decomposes the organic matters.

14. *Sulphuric acid.* Boil one grm. with water acidulated with hydrochloric acid, filter, and determine sulphuric acid in the filtrate in the usual way. It is advisable in all cases to purify the precipitate as described on p. 366.

15. *Chlorine* is estimated by VOLHARD's method, or by precipitation with silver nitrate in the clear hot water extract of 1 grm.

GUANO.

16. The determinations of *phosphoric acid*, soluble, "reverted," and insoluble, are made precisely as in the case of a superphosphate. The soluble phosphoric acid consists of phosphates of the alkalies, and the washings, except in the case of "rectified" guanos which have been treated with oil of vitriol, are alkaline.

17. Determine *nitrogen* as in superphosphates. Many guanos contain ammonium carbonate and therefore require care in manipulation to prevent its escape. If nitrates are present, add to the 0·5 gr. taken for combustion with soda lime an equal weight of pure sugar or oxalic acid. The quantity of nitrate is so small that with this precaution accurate results are obtained without resorting to the absolute method.

18. *Potash* is determined as in superphosphates.

19. If a determination of *water* is required, weigh the guano in a boat and introduce it into a tube which is heated to 100° in an air or water bath. One end of this tube is connected with a drying apparatus containing oil of vitriol or calcium chloride. The other is provided with a U-tube and standard acid for receiving

ammonia, and an aspirator to maintain a current of dry air. The volatilized ammonia is measured with a standard alkali and taken into account in reckoning the loss of weight.

BONE.

20. *Water.* Dry 1 grm. at 100°, and determine water by loss.

21. *Fat.* Transfer the dry bone to an extraction apparatus and extract with absolute ether as long as anything is removed.

Evaporate the ether extract, dry at 100° for two hours and weigh.

22. *Carbonic acid.* Determine carbonic acid in 1 grm. by the method described on p. 412*e*.

23. *Ash.* Incinerate 1 grm. till the ash is white or light gray. Moisten with ammonium carbonate solution, dry, ignite gently and weigh.

24. *Phosphoric acid.* Dissolve the ash, prepared as above, in hydrochloric acid, filter, dilute to 250–300 c.c., add 12–15 grm. of citric acid as ammonium citrate and precipitate with magnesia mixture in the manner previously described, 4, or dissolve in nitric acid and proceed by the molybdic process.

25. *Nitrogen.* Determine nitrogen in 1 grm. by combustion with soda lime.

For most purposes the determination of phosphoric acid and nitrogen in sufficient.

POTASH SALTS.

26. Boil 5 grm. with water for 10 minutes, dilute to 1000 c.c. and determine potash in 100 c.c. as described under 8 or 29.

27. In another portion of 100 c.c., determine sulphuric acid by barium chloride, and in a third portion chlorine may be determined by precipitation with silver nitrate, or more conveniently by VOLHARD's method.

28. Determine water by heating 2–5 grm. in a platinum capsule to dull redness.

29. *Potash.* STOHMANN directs to boil 10 grm. of substance with about 300 c.c. of water, and to add dropwise $BaCl_2$ solution until no further precipitate appears, to let cool and dilute to 1000 c.c., and after subsidence or filtration to take 100 c.c. of the clear solution, add a large excess of $PtCl_4$ (equivalent to about 2 grm.

Pt), evaporate and proceed as usual with the precipitate. As the alkali-earth platinchlorides are all soluble in alcohol, the results are good.

21. ANALYSIS OF ATMOSPHERIC AIR.

§ 221.

In the analysis of atmospheric air we usually confine our attention to the following constituents: oxygen, nitrogen, carbonic acid, and aqueous vapor. It is only in exceptional cases that the exceedingly minute quantities of ammonia and other gases—many of which may be assumed to be always present in infinitesimal traces—are also determined.

It does not come within the scope of the present work to describe all the methods which have been employed in the capital investigations made in the last few years by BRUNNER, BUNSEN, DUMAS and BOUSSINGAULT, REGNAULT and REISET, and others. To these methods we are indebted for a more accurate knowledge of the composition of our atmosphere, and excellent descriptions of them will be found in the works below.*

I confine myself to those methods which are found most convenient in the analysis of the air for medical or technical purposes.

A. DETERMINATION OF THE WATER AND CARBONIC ACID.

§ 222.

It was formerly the custom to effect these determinations by BRUNNER'S method, which consisted in slowly drawing, by means of an aspirator, a measured volume of air through accurately weighed apparatuses filled with substances having the property of retaining the aqueous vapor and the carbonic acid, and estimating these two constituents by the increased weights of the apparatuses.

Fig. 103 represents the arrangement recommended by REGNAULT.

* Ausführliches Handbuch der analytischen Chemie, von H. Rose, II. 853; Graham-Otto's Ausführliches Lehrbuch der Chemie, Bd. II. Abth. 1, S. 102 *et seq.*; Handwörterbuch der Chemie, von Liebig, Poggendorff und Wöhler, 2 Aufl. Bd. II. S. 431 *et seq.*; and Bunsen's Gasometry.

The vessel *V* is made of galvanized iron, or of sheet zinc; it holds from 50 to 100 litres, and stands upon a strong tripod in a trough large enough to hold the whole of the water that *V* contains. At *a* a brass tube, *c*, with stopcock, is firmly fixed in with cement. Into the aperture *b*, which serves also to fill the apparatus, a thermometer reaching down to the middle of *V* is fixed air-tight by means of a perforated cork soaked in wax.

The efflux tube, *r*, which is provided with a cock, is bent slightly upward, to guard against the least chance of air entering the vessel

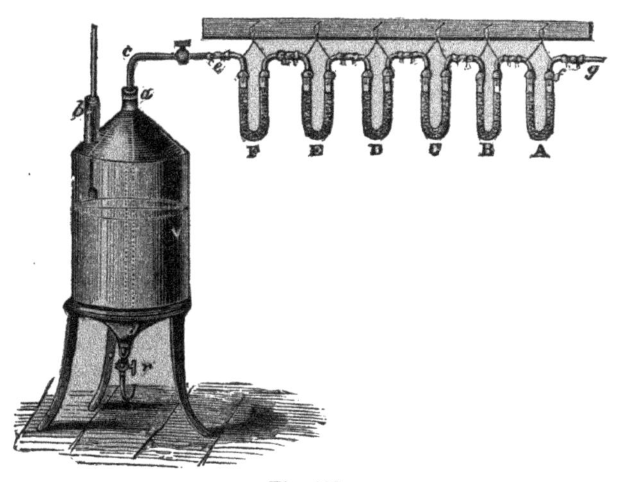

Fig. 103.

from below. The capacity of the vessel is ascertained by filling it completely with water, and then accurately measuring the contents in graduated vessels. The end of the tube *c* is connected air-tight with *F*, by means of a caoutchouc tube; the tubes *A*—*F* are similarly connected with one another. *A*, *B*, *E*, and *F* are filled with small pieces of glass moistened with pure concentrated sulphuric acid, *C* and *D* with moist slaked lime.* Finally, *A* is also con-

* With regard to *C* and *D*, I have returned to lime, preferring it to pumice saturated with solution of potash, because, as Hlasiwetz (Chem. Centralbl. 1856, 575) has shown, the solution of potash absorbs not only carbonic acid, but also oxygen. Indeed, H. Rose had previously made a similar observation. With respect to the other tubes, I prefer the concentrated sulphuric acid to calcium chloride as the absorbent for water (see Pettenkofer, Sitzungsber. der bayer

nected with a long tube leading to the place from which the air intended for analysis is to be taken. The corks of the tubes are coated over with sealing-wax. The tubes *A* and *B* are intended to withdraw the moisture from the air; they are weighed together. *C*, *D*, and *E* are also weighed jointly. *C* and *D* absorb the carbonic acid; *E* the aqueous vapor which may have been withdrawn from the hydrate of lime by the dry air. *F* need not be weighed; it simply serves to protect *E* against the entrance of aqueous vapor from *V*.

The aspirator is completely filled with water; *c* is then connected with *F*, and thus with the entire system of tubes; the cock *r* is opened a little, just sufficiently to cause a slow efflux of water. As the height of the column of water in *V* is continually diminishing, the cock must from time to time be opened a little wider, to maintain as nearly as possible a uniform flow of water. When *V* is completely emptied, the height of the thermometer and that of the barometer are noted, and the tubes *A* and *B*, and *C*, *D*, and *E* weighed again.

As the increase of weight of *A* and *B* gives the amount of water, that of *C*, *D*, and *E* the amount of carbonic acid, in the air which has passed through them; and as the volume of the latter (freed from water and carbonic acid) is accurately known from the ascertained capacity of *V*,* the calculation is in itself very simple; but it involves, at least in very accurate analyses, the following corrections:—

α. Reduction of the air in *V*, which is saturated with aqueous vapor, to dry air; since the air which penetrates through *c* is dry.

β. Reduction of the volume of dry air so found to 0°, and 760 mm.

When these calculations have been made (see "Calculations of Analyses," in Appendix), the weight of the air which has penetrated into *V* is readily found from the datum in Table V. at the

Akad. 1862, II. Heft 1, S. 59). Hlasiwetz's statement, that concentrated sulphuric acid also takes up carbonic acid, I have found to be unwarranted. Calcium chloride does not dry the air completely, and, besides, Hlasiwetz says that when it is used a trace of chlorine is carried away corresponding to the amount of ozone in the air (op. cit. p. 517).

* Or from the quantity of water which has flown from *V*, as the experiment may be altered in this way, that a portion only of the water is allowed to run out, and received in a measuring vessel.

end of the volume; and as the carbonic acid and water have also been weighed, the respective quantities of these constituents of the air may now be expressed in per cents. by weight, or, calculating the weights into volumes, in per cents. by measure.

Considering the great weight and size of the absorption apparatus, in comparison to the increase of weight by the process, at least 25,000 c. c. of air must be passed through; the air inside the balance-case must be kept as dry as possible by means of a sufficient quantity of calcium chloride, and the apparatus left for some time in the balance-case before proceeding to weigh. Neglect of these measures would lead to considerable errors, more particularly as regards the carbonic acid, the quantity of which in atmospheric air is, on an average, about 10 times less than that of the aqueous vapor (comp. HLASIWETZ, *loc. cit.*).

For the *exact determination of the carbonic acid* one of the following methods is far better suited:—

a. Process suggested by FR. MOHR, *applied and carefully tested by* H. v. GILM.* VON GILM employed in his experiments an aspirator holding at least 30 litres, which was arranged like that shown in fig. 103, but had a third aperture, bearing a small manometer. The air was drawn through a tube, 1 metre long and about 15 mm. wide; this tube was drawn out thin at the upper end, and at the lower end bent at an angle of 140°—150°. It was more than half filled with coarse fragments of glass and perfectly clear baryta water, and fixed in such a position that the long part of it was inclined at an angle of 8°—10° to the horizontal. A narrow glass tube, fitted into the undrawn-out end of the tube by means of a cork, served to admit the air. Two small flasks, filled with baryta water, were placed between the absorption tube and the aspirator; these were intended as a control, to show that the whole of the carbonic acid had been retained. When about 60 litres of air had slowly passed through the absorption tube, the barium carbonate formed was filtered off out of contact of air, and the tube as well as the contents of the filter washed, first with distilled water saturated with barium carbonate, then with pure boiled water. The barium carbonate in the filter and in the tube was then dissolved in dilute hydrochloric acid, the solution evaporated to dry-

* Chem. Centralbl. 1857, 760.

ness, the residue gently ignited, and the chlorine of the barium chloride determined as directed § 141, *b*, *α*. 2 atoms of chlorine represent 1 mol. carbonic acid. It is obvious that one may also determine the barium in the hydrochloric acid solution by precipitating with sulphuric acid. For filtering the barium carbonate, v. GILM employed a double funnel (fig. 104); the inner cork has, besides the perforation through which the neck of the funnel passes, a lateral slit, which establishes a communication between the air in the outer funnel and the air in the bottle.

Fig. 104.

As, with the absorption apparatus arranged as described, the air has to force its way through a column of fluid, the manometer is required to determine the actual volume of the air; the height indicated by this instrument being deducted from the barometric pressure observed during the process.

FR. MOHR * now recommends as the absorbent fluid a solution of barium hydroxide in potash. This is prepared by dissolving crystals of barium hydroxide in weak solution of potash with the aid of heat, and filtering off the barium carbonate, which invariably forms in small quantity. The clear filtrate is accordingly saturated with barium carbonate. MOHR now leaves out the fragments of glass.

This method afforded v. GILM very harmonious results. Nevertheless, it involves one source of error. If clear baryta water is passed through paper with the most careful possible exclusion of air, and the filter is washed till the washings are free from baryta, and dilute hydrochloric acid is then poured upon the filter, and the filtrate thus obtained is evaporated, a small quantity of barium chloride will be left, showing that a little baryta was kept back by the paper. AL. MÜLLER † has already called attention to the capacity of filter paper for retaining baryta.

b. M. PETTENKOFER'S *process*.‡

α. *Principle and Requisites*. A known volume of air is made

---

* Lehrbuch der Titrirmethode, 2d ed. 446.

† Journ. f. prakt. Chem. 83, 384.

‡ Abhandl. der naturw. u. techn. Commission der k. bayer. Akad. der Wiss. II. 1; Annal. d. Chem. u. Pharm. II. Supplem. Bd. p. 1.

to act upon a definite quantity of standard baryta water (standardized by oxalic acid solution), in such manner that the carbonic acid is completely bound by the baryta. The baryta water is then poured out into a cylinder, and allowed to deposit with exclusion of air, a part of the clear fluid is then removed, and the baryta remaining in solution is determined. The difference between the oxalic acid required for a certain quantity of baryta water before and after the action of the air represents the barium carbonate formed, and consequently the carbonic acid present.

Two kinds of baryta water are used: one contains 21 grm. and the other 7 grm. crystallized barium hydroxide * in the litre; these serve for the determination of larger and smaller quantities of carbonic acid respectively. 1 c. c. of the stronger corresponds to about 3 mgrm. carbonic acid, of the weaker 1 c. c. corresponds to about 1 mgrm.†

The oxalic acid solution which serves for standardizing the baryta water contains 2·8636 grm. cryst. oxalic acid in 1 litre. 1 c. c. corresponds to 1 mgrm. carbonic acid. The baryta water is standardized as follows:—Transfer 30 c. c. of it to a flask, and then run in the oxalic acid from a MOHR's burette with float; shake the fluid from time to time, closing the mouth of the flask with the thumb. The vanishing point of the alkaline reaction is ascer-

---

* The barium hydroxide must be entirely free from caustic potash, and soda, the smallest quantities of which render the volumetric estimation in the presence of barium carbonate impossible, since the normal alkali oxalates decompose the alkali-earth carbonates. When a trace even of barium carbonate is suspended in the fluid—and this is always the case when a baryta water which has been used for the absorption of carbonic acid is not filtered—the reaction continues alkaline if the smallest trace of potash or soda is present, because the alkali oxalate formed immediately enters into decomposition with the barium carbonate. A fresh addition of oxalic acid converts the alkali carbonate again into oxalate, and the fluid is for a moment neutral, till, on shaking with air, the carbonic acid escapes, and any barium carbonate still present converts the alkali oxalate again into carbonate. To test a baryta water for caustic alkali, determine the alkalinity of a perfectly clear portion, and then of a portion that has been mixed with a little pure precipitated barium carbonate. If you use more oxalic acid in the second than in the first experiment, caustic alkali is present, and some barium chloride must be added to the baryta water before it can be used.

† [The baryta water is kept in a bottle under a thin stratum of kerosene (MOHR). It is drawn off through a siphon supported in the stopper, the outer leg of which is recurved upwards and closed with a bit of rubber tube and clip. By having this leg of the siphon sufficiently long the burette may be filled by inserting its delivery end in the rubber tube and opening both clips.]

tained with delicate turmeric paper.* As soon as a drop of the fluid placed on the paper does not give a brown ring, the end is attained. If you were obliged, in the first experiment, to take out too many drops for testing with turmeric paper, consider the result as only approximate, and make a second experiment, adding at once the whole quantity of oxalic acid to within 1 or ½ c. c. and then beginning to test with paper. A third experiment would be found to agree with the second to $\frac{1}{10}$ c. c. The reaction is so sensitive that all foreign alkaline matter, particles of ash, tobacco smoke, &c., must be carefully guarded against.

β. The actual Analysis. This may be effected in two different ways.

aa. Take a perfectly dry bottle, of about 6 litres capacity, with well-fitting ground glass stopper, and accurately determine the capacity; fill the bottle, by means of a pair of bellows, with the air to be analyzed; add 45 c. c. of the dilute standard baryta water, and cause the baryta water to spread over the inner surface of the bottle by turning the latter about, but without much shaking. In the course of about ½ an hour the whole of the carbonic acid is absorbed. Pour the turbid baryta water into a cylinder, close securely, and allow to deposit; then take out, by means of a pipette, 30 c. c. of the clear supernatant fluid, run in standard oxalic acid, multiply the volume used by 1·5 (as only 30 c. c. of the original 45 are employed in this experiment), and deduct the product from the c. c. of oxalic acid used for 45 c. c. of the fresh baryta water; the difference represents the quantity of baryta converted into carbonate, and consequently the amount of the carbonic acid. If the air is unusually rich in carbonic acid, the concentrated baryta water is employed.

bb. Pass the air through a tube or through two tubes containing measured quantities of standard baryta water and finish the experiment as in *aa.* For passing a definite quantity of air we should generally employ an aspirator (p. 773); PETTENKOFER in his experiments with the respiration apparatus forced the air by means of small mercurial pumps first through the tubes, and then through an apparatus for measuring the gas. The form and arrangement

* Prepared with lime-free Swedish filter paper and tincture of turmeric. The spirit used in making the latter must be free from acid. Dry the paper in a dark room, and keep it protected from the light. It is lemon yellow.

of the tubes is illustrated by fig. 105. Two such tubes were used; the first was 1 metre, the second ·3 metres long; they were filled with baryta water—the former with the stronger solution, the latter with the weaker. The air is introduced through the short limbs of the tubes, and the glass tubes themselves are so inclined

Fig. 105.

that the bubbles of air move on with the necessary rapidity without uniting. The motion of the gas bubbles keeps up a constant mixing of the baryta water.

B. Determination of the Oxygen and Nitrogen.

§ 223.

The method I shall give is that proposed by v. Liebig.* It is based upon the observation made by Chevreul and Döbereiner, that pyrogallic acid, in alkaline solutions, has a powerful tendency to absorb oxygen.

1. A strong measuring tube, holding 30 c. c. and divided into $\frac{1}{5}$ or $\frac{1}{10}$ c. c., is filled to $\frac{2}{3}$ with the air intended for analysis. The remaining part of the tube is filled with mercury, and the tube is inverted over that fluid in a tall cylinder, widened at the top.

2. The volume of air confined is measured (§ 12). If it is intended to determine the carbonic acid—which can be done with sufficient accuracy only if the quantity of the acid amounts to several per cents.—the air is dried by the introduction of a ball of calcium chloride before measuring. If it is not intended to determine

* Annal. d. Chem. u. Pharm. 77, 107.

the carbonic acid this operation is omitted. A quantity of solution of potassa of 1·4 sp. gr. (1 part of dry potassium hydroxide to 2 parts of water), amounting to from $\frac{1}{40}$ to $\frac{1}{50}$ of the volume of the air, is then introduced into the measuring tube by means of a pipette with the point bent upwards (fig. 106), and spread over the entire inner surface of the tube by shaking the latter; when no further diminution of volume takes place, the decrease is read off. If the air has been dried previously with calcium chloride, the diminution of the volume expresses exactly the amount of carbonic acid contained in the air; but if it has not been dried with calcium chloride, the diminution in the volume cannot afford correct information as to the amount of the carbonic acid, since the strong solution of potassa absorbs aqueous vapor.

Fig. 106.

3. When the carbonic acid has been removed, a solution of pyrogallic acid, containing 1 grm. of the acid* in 5 or 6 c. c. of water, is introduced into the same measuring tube by means of another pipette, similar to the one used in 2 (fig. 106); the quantity of pyrogallic acid employed should be half the volume of the solution of potassa used in 2. The mixed fluid (the pyrogallic acid and solution of potassa) is spread over the inner surface of the tube by shaking the latter, and, when no further diminution of volume is observed, the residuary nitrogen is measured.

4. The solution of pyrogallic acid mixing with the solution of potassa of course dilutes it, causing thus an error from the diminution of its tension; but this error is so trifling that it has no appreciable influence upon the results; it may, besides, be readily corrected, by introducing into the tube, after the absorption of the oxygen, a small piece of hydrate of potassa corresponding to the amount of water in the solution of the pyrogallic acid.

5. There is another source of error in this method; viz., on account of a portion of the fluid always adhering to the inner surface of the tube, the volume of the gas cannot be read off with absolute accuracy. In comparative analyses, the influence of this defect upon the results may be almost entirely neutralized, by taking nearly equal volumes of air in the several analyses.†

---

* Liebig has described a very advantageous method of preparing pyrogallic acid. See Annal. d. Chem. u. Pharm. 101, 47.

† Bunsen employs for the absorption of oxygen a papier-mâché ball saturated

6. Notwithstanding these sources of error, the results obtained by this method are very accurate and constant. In eleven analyses which v. LIEBIG reports, the greatest difference in the amount of oxygen found was between 20·75 and 21·03. The numbers given express the actual and uncorrected results.

[22. DETECTION AND ESTIMATION OF ARSENIC IN ORGANIC MATTER.

§ 224.

GAUTIER'S *Method simplified by* JOHNSON AND CHITTENDEN.

The following method for the detection and estimation of arsenic in organic matter is a modification of the process recently described by GAUTIER.* GAUTIER'S method consists in treating the substance with certain quantities of nitric acid, and afterwards of sulphuric acid, at a high temperature, whereby the organic matters are partly destroyed and converted into slightly soluble humus-like bodies, from which all the arsenic may be extracted by boiling water. Gautier treats the brown solution thus obtained with "sodium bi-sulphate, throws down the arsenic in the state of sulphide, with hydrogen sulphide, transforms this sulphide into arsenic acid by known means," treats the solution thus obtained in the Marsh Apparatus, and finally weighs the arsenic in the metallic state as below described.

JOHNSON and CHITTENDEN dispense with the use of all reagents but sulphuric acid, nitric acid, and zinc alloyed with a little platinum, which are not difficult to obtain in a state of absolute freedom from arsenic, and they, together with DONALDSON, have demonstrated that the method, thus essentially simplified, gives exact results. The following account of the process is from a paper by CHITTENDEN and DONALDSON.†

---

with a concentrated alkaline solution of potassium pyrogallate, which he introduces into the gaseous mixture attached to a platinum wire. By adopting this proceeding the source of error mentioned in 5 is avoided. See also Russell, Jour. Chem. Soc. 1868, pp. 130, 131.

* Bulletin de la Société Chimique, 24, 250.

† American Chemical Journal, vol. ii. p. 235.

I. Reagents and Apparatus.

The reagents required are pure granulated zinc alloyed with a small quantity of platinum, pure concentrated nitric and sulphuric acids, and three dilute sulphuric acids of increasing strength, which, for the sake of convenience, may be prepared in considerable quantities.

Acid No. 1. 180 c. c. pure conc. H_2SO_4+1000 c. c. H_2O.
Acid No. 2. 260 c. c. pure conc. H_2SO_4+1000 c. c. H_2O.
Acid No. 3. 425 c. c. pure conc. H_2SO_4+1000 c. c. H_2O.
The form of Marsh apparatus used is shown by fig. 107.
The flask, a Bunsen's wash-bottle of 200 c. c. capacity, is pro-

Fig. 107.

vided with a small separating funnel of 65 c. c. capacity, with glass stopcock. This is a very material aid to the obtaining of a slow and even evolution of gas, and is nearly indispensable in accurate quantitative work. The gas generated is dried by passing through a calcium chloride tube,* and then passes through a tube of hard glass, heated to a red heat by a furnace of three Bunsen lamps with spread burners, so that a continuous flame of six inches is obtained, and with a proper length of

* Otto and also Dragendorff recommend to pass the gas first over fragments of caustic potassa. We find, however, in accordance with Doremus, that arsenic is arrested by caustic alkali.—S. W. J. and R. H. C.

cooled tube not a trace of arsenic passes by. The glass tube where heated is wound with a strip of wire gauze, both ends being supported upon the edges of the lamp frame, so that the tube does not sink down when heated. The small furnace is provided with two appropriate side pieces of sheet metal, so that a steady flame is always obtained. When the quantity of arsenic is very small the tube is naturally so placed that the mirror is deposited in the narrow portion, but when the arsenic is present to the extent of ·005 grm. the tube should be 6 millimetres in inner diameter, and so arranged that fully two inches of this large tube are between the flame and the narrow portion. When the quantity of arsenic is less the tube can naturally be smaller.

II. Process.

a. Method for the complete extraction of arsenic from organic matter.

100 grms of the material to be examined, cut into small pieces, are placed in a porcelain casserole of 600 c. c. capacity and provided with a stirring rod of stout glass. 23 c. c. of pure concentrated nitric acid are added, and the dish placed on a small air-bath * provided with a thermometer and a single Bunsen burner. The mixture is then heated at 150°—160° C., with occasional stirring. At first the tissue takes on a yellowish color, then swells up somewhat, becoming finally quite thick; soon changes again, becoming liquid, and then generally requires heating from 1½ to 2 hours, the temperature sometimes being raised to 180° C.

At this point the mass, being now quite thick again, usually takes on a deeper yellow color or orange shade. When this change of color is noticed the casserole is taken from the bath and 3 c. c. of pure concentrated sulphuric acid added and the mixture stirred vigorously. The addition of concentrated sulphuric acid to the viscid residue rich in nitric acid and nitro-compounds naturally

* For air-bath an ordinary flat-bottomed tin basin, 7 inches in diameter, 3 inches deep, is used with a cover provided with an opening 5 inches in diameter. This bath is set in an iron ring fastened to a stout lamp-stand, while the end of the thermometer passes through a small hole near the edge of the cover a short distance into the bath, so that the temperature can be regulated.

gives rise, especially at this temperature, to a considerable commotion: the mass becomes brown, swells up, nitrous fumes are copiously evolved, immediately followed by dense white fumes of suffocating odor, while the residue in the dish is changed either into a dry carbonaceous mass or a black, sticky, tar-like mass. Although the oxidation is so powerful, no deflagration takes place, and the carbonization is effected in this manner without the volatilization of any arsenic. The casserole is again placed on the bath and heated for a few minutes at 180° C., then, while still on the bath, 8 c. c. of pure concentrated nitric acid are added drop by drop with continual stirring, the object being to destroy more completely the organic matter, and at the same time the nitric acid falling drop by drop on the carbonaceous residue tends to prevent the formation of sulphurous acid and the consequent formation of insoluble arsenious sulphide.

After the addition of the nitric acid the dish is heated at 200° C. for fifteen minutes, and when cold a hard carbonaceous residue is the result, entirely free from nitric acid. In working with different kinds of tissue, slight deviations from the above description will frequently be observed. When much bony matter is present the last residue takes on a somewhat different character, owing to the presence of calcium sulphate, and occasionally when the 3 c. c. of sulphuric acid are added the oxidation does not at once take place, but requires a little longer heating on the air-bath. When such is the case the mixture needs constant watching in order to remove the dish from the bath at the first approach of the oxidation.

The arsenic now exists as arsenic acid, readily soluble in water. The carbonaceous residue is thoroughly extracted with boiling water, and in order to avoid all loss is not previously pulverized, but the casserole in which the oxidation took place is filled with water and heated on the water-bath for several hours. The hard mass soon softens, and by repeated treatment in this manner readily gives up all its arsenic to the aqueous solution; it is, however, better to have the carbonaceous residue in contact with different portions of warm water for about 24 hours to insure the complete extraction of the arsenic.

The reddish-brown fluid containing some organic matter and arsenic acid is now evaporated on the water-bath to dryness, care being taken that the entire residue is finally obtained in one casserole. This residue* of organic matter and arsenic is warmed with 45 c. c. of sulphuric acid No. 1, and the clear solution so obtained, or, as more frequently happens, the fluid with organic matter in suspension, is then ready for introduction into the MARSH apparatus.

b. Method for the conversion of arsenic acid into arsenetted hydrogen and then into metallic arsenic.

25–35 grms. of granulated zinc previously alloyed with a small quantity of platinum are placed in the generator, and everything being in position, the MARSH apparatus is filled with hydrogen by the use of a small quantity of acid No. 1. After a sufficient time has elapsed the gas is lighted at the jet and the glass tube heated to a bright redness. The 45 c. c. of acid No. 1 containing the arsenic is then poured into the separating funnel, from which it is allowed to flow into the generator at such a rate that the entire fluid is introduced in one hour or one hour and a half; 40 c. c. of acid No. 2 are then poured into the casserole, to which considerable organic matter usually adheres, and then transferred to the separating funnel and allowed to flow slowly into the generator, and lastly 45 c. c. of acid No. 3. In this manner we are sure to have all of the arsenic acid dissolved and thus carried into the generator, while at the same time the stronger acids Nos. 2 and 3 serve as a rinse fluid and thereby prevent mechanical loss, while, at the same time, the increasing strength of acid added counteracts the diluting effect of the reaction so that the strength of acid remains about the same during the entire process of 2½ to 3 hours and thereby insures a regular flow of gas. The amount of time required will vary with the amount of arsenic: 2—3 mgrms. of arsenic will require about two to three hours for the entire decomposition, while 4—5

* When the residue left by the evaporation of the water is quite large, it is sometimes better to reoxidize it. This is quickly accomplished by adding a few cubic centimetres of concentrated nitric acid to the contents of the casserole and heating on the air-bath at 150°—180° C. until a reddish solution is obtained. Then 3—5 c. c. of concentrated sulphuric acid are added and the mixture heated at the above temperature until the nitric acid is completely driven off. The thin black fluid is then carefully mixed with the requisite quantity of No. 1 acid, and introduced into the Marsh apparatus. Frequently quite a heavy, flocculent precipitate separates from the sulphuric acid solution. This does not interfere, but is poured, together with the fluid, directly into the receiving bulb, which is purposely provided with a delivery tube of large calibre.

mgrmes. will need perhaps three to four hours. Where the amount of arsenic is small, only 25 grms. of zinc are needed, and but 45 c. c. of acid No. 1, 30 c. c. of acid No. 2, and 30 c. c. of acid No. 3; but when 4—5 mgrms. of arsenic are present it is better to take the first mentioned quantities of zinc and acids.

The arsenic being thus collected as a large or small mirror of metal, the tube is cut at a safe distance from the mirror, so that a tube of perhaps 2—6 grms. weight is obtained. This is carefully weighed and then the arsenic removed by simple heating; or, if the arsenic is to be saved as in a toxical case, dissolved out with strong nitric acid. The tube is then cleaned, dried, and again weighed, the difference giving the weight of metallic arsenic, from which by a simple calculation the amount of arsenious oxide can be obtained. The delicacy of the method is shown by the fact that ·00001 grm. As_2O_3 when introduced into 100 grms. of beef yielded by this method a distinct mirror of metallic arsenic. In a similar manner ·000001 grm. As_2O_3 yielded a faint mirror of arsenic, this amount appearing to be the limit.

In conducting these experiments with organic matter, after the zinc is placed in the generator, 15 drops of olive oil are allowed to flow down the side, and this as the fluid is introduced floats on top and thereby prevents any troublesome frothing. The only other thing to be guarded against is the too rapid introduction of the acids, whereby loss as well as frothing of the mixture may ensue, and secondly the heating of the flask by the chemical reaction. If necessary this latter can be prevented by placing the generator in a glass or other dish so that a stream of cold water can continually play about it, which will keep the flask sufficiently cool to prevent the formation of any hydrogen sulphide which might sometimes show itself in slight quantity.

The following results show the accuracy of the method:—

Quantity of Arsenic introduced.	Wt. of Metallic Arsenic found.	Theoretical Wt. Metallic Arsenic.
100 grms. beefsteak with .004 grm. As_2O_3	·00300	·00303
" " " ·004 "	·00300	·00303
" " " ·004 "	·00290	·00303
" ·003 "	·00219	·00227
" ·005 "	·00369	·00378
" ·005 "	·00372	·00378]

PART III.

EXERCISES FOR PRACTICE.

EXERCISES FOR PRACTICE.

THE principal point kept in view in the selection of these exercises has been that most of them, and more particularly the first, should permit an exact control of the results. This is of the utmost importance for students, since a well-grounded self-reliance is among the most indispensable requisites for a successful pursuit of quantitative investigations, and this is only to be attained by ascertaining for one's self how near the results found approach the truth.

Now a rigorously accurate control is practicable only in the analysis of pure salts of known composition, or of mixtures composed of definite proportions of pure bodies. When the student has acquired, in the analysis of such substances, the necessary self-reliance, he may proceed to the analysis of minerals or products of industry in which such rigorous control is unattainable.

The second point kept in view in the selection of these exercises has been to make them comprise both the more important analytical methods and the most important bodies, so as to afford the student the opportunity of acquiring a thorough knowledge of every branch of quantitative analysis.

Organic analysis offers less variety than the analysis of inorganic substances; the exercises relating to the former branch are therefore less numerous than those relating to the latter.

I would advise the student to analyze the same substance repeatedly, until the results are quite satisfactory. [It is a good habit always to carry on together duplicate analyses. It requires but little more time to make two analyses than to make one, and the operator's experience is thus very economically doubled.]

It is by no means necessary for the student to go through the whole of these examples; the time which he may require to attain proficiency in analysis depends, of course, upon his own abilities. One may be a good analyst without having tried *every* method or determined *every* body. A few substances *well* analyzed yield more profit than can be obtained from going over many processes in a superficial manner.

Finally, the student is warned against prematurely attempting to discover new methods; he should wait until he has attained a good degree of proficiency in general chemistry, and more particularly in practical analysis.

EXERCISES.

A. SIMPLE DETERMINATIONS IN THE GRAVIMETRIC WAY, INTENDED TO PERFECT THE STUDENT IN THE PRACTICE OF THE MORE COMMON ANALYTICAL OPERATIONS.

[WE give here, in the first place, quite full details of all the steps in the estimation of chlorine in sodium chloride, including the preparation of this salt in a state of purity. This, it is hoped, will relieve much of the perplexity which the beginner must at first experience in making out a scheme of operations from the various separate paragraphs where the processes are described. The student should not fail, however, to study carefully the chapter on operations while carrying on the analysis, nor to examine every reference.

1. SODIUM CHLORIDE.

Preparation. Sodium chloride is far less soluble in hydrochloric acid than in water. On account of this property the crude product—common salt—may be purified from the magnesium chloride and calcium sulphate which it contains as follows:—To 100 c. c. of a saturated solution add very gradually an equal volume of pure concentrated hydrochloric acid. Drain the mass of fine crystals which separate on a funnel, the throat of which is loosely closed with filter paper. Wash with a small volume of pure dilute hydrochloric acid, and at last, in order to test the purity of the product, allow 5 or 6 c. c. of distilled water to pass through. Collect the water that runs through in a test tube separately, and add to it barium chloride. If no turbidity results, the sodium chloride is free from sulphates and may be assumed to be pure enough for analysis. Remove it from the funnel and dry it in a porcelain dish. If not free from sulphates, the product may be subjected to a repetition of the process. This, however, will rarely be necessary.*

* When large quantities of pure sodium chloride are required, it is more economical to prepare it from a solution of common salt by saturating the solution with HCl gas.

A portion* of the salt thus obtained is heated in a covered crucible until it ceases to decrepitate, but not to fusion, and preserved in a weighing tube (like a small test tube, but not flared at the mouth) that is closed with a soft, well-fitting, and smooth cork.

ESTIMATION OF CHLORINE.

1. *Weighing out the substance.* The tube containing the prepared salt is wiped, if need be, from dust. The cork is taken out, and by means of a bit of thin paper, or a clean linen handkerchief, any particles of salt adhering to the cork, and to the inside of the tube as far as the cork reaches, are removed. The cork is replaced, and the whole is weighed (see §§ 9 and 10), the weight being immediately recorded in the note-book. A clean beaker or assay-flask, of about 200 c. c. capacity, being ready, the weighing-tube is held over it and the cork carefully removed. A portion of substance is allowed to fall in the vessel, and, the cork being replaced, the tube is again counterpoised. If two to three decigrammes have been emptied, the operator is ready to proceed. If less, more should be transferred from the tube to the vessel. If more, or much more, it is better to begin anew, by weighing off another portion into another beaker or flask. In this manner weigh off two portions in separate vessels, so as to carry together duplicate analyses. Now affix a piece of gummed paper to each vessel, and label them to correspond with their designation in the note-book.

2. *Solution and precipitation.* Dissolve the weighed portions, each in about 100 c. c. of cold distilled water, add a few drops of pure nitric acid, and, lastly, clear solution of silver nitrate† until further addition no longer produces a precipitate.

Agitate the mixture well, but with care to avoid loss. This can be done by shaking, if a flask be in use, or by stirring with a glass rod, if a beaker be employed.

Set the vessel aside in a dark place, covered with paper or a watch-glass to exclude dust, and let stand for about 12 hours, or until the precipitate has subsided and the liquid above it is *perfectly* clear, then add a drop of silver nitrate to make sure that the precipitation is complete (if not complete, add more solution of silver, and let stand again for some hours).

---

* Pure sodium chloride is needed in other analyses, and the chief part of what is thus prepared should be carefully bottled and reserved for future use.

† Solution of a silver coin in nitric acid answers for this purpose as well as pure nitrate, provided it be clear and contain but little free acid.

3. *Filtration.* A filter is placed in a funnel at least ¼ inch deeper than itself, and moistened with water, at the same time being carefully pressed down so that its edges touch the glass at all points. The funnel being supported on a stand, a clean beaker or flask is put beneath it, and the operator proceeds to pour the liquid —on whose surface some particles of silver chloride usually float—into the filter, leaving the bulk of the precipitate undisturbed. To do this without loss the following precautions may be regarded: *a.* Touch the edge or lip of the vessel with a very slight coat of tallow (a small bit of which is kept at hand under the edge of the work-table, and is applied with the finger). *b.* Pour slowly over the greased place, along a glass rod held nearly vertical, so directing the stream that it shall strike against the side, not into the vertex of the filter. *c.* When the filter is filled to within ¼ inch of the top discontinue the pouring, bringing the rod into the vessel containing the precipitate, after it has drained so that nothing will fall from it.

The pouring-rod may be simply straight, and an inch longer than the diagonal of the vessel, or when it is desirable not to disturb a precipitate, it may be 3—4 inches long and bent siphon fashion so as to hang on the edge of a beaker or flask. In either case its end should be rounded by fusion, and those portions along which the liquid flows must not be handled.

The vessel containing the precipitate, as well as that which receives the filtrate, and likewise the funnel, should be kept *covered* as much as possible in all cases when nicety is required, to prevent access of dust, insects, &c.

The most convenient covers are large watch-glasses, but square plates of glass, or even cards, will generally answer.

The filtration of silver chloride should be conducted without exposing it to strong light, whereby it is blackened, with loss of chlorine, p. 168.

4. When all, or nearly all, the liquid has passed the filter, it remains *to wash* and *to transfer* the precipitate.

These operations may be carried on as follows: pour about 100 c.c. of cold distilled water upon the precipitate, which mostly remains in the vessel where it was formed, and agitate vigorously, in order to break up and divide the lumpy silver chloride, and bring every part of it perfectly in contact with the water.

When in a beaker, the agitation must be made with great caution, by means of a glass stirring-rod; when in a narrow-mouthed flanged flask, this may be tightly closed by a perfectly smooth cork (softened for the purpose by squeezing) and then shaken violently.

The water and precipitate are now poured together upon the filter, with the precautions before detailed. The last portions of the precipitate are removed from the beaker or flask by repeated rinsings, in which a wash-bottle like fig. 36, p. 81, may be conveniently employed.

Any portions of precipitate that adhere to the sides of the vessel too strongly to be removed by a stream from the wash-bottle must be *rubbed off*. For this purpose the feather is employed.

It is made from a goose-quill, by cutting off the extreme tip for an inch or so, and smoothly trimming away the beard, except a portion of one half-inch in length on the inside of the curve. The tubular part may be removed or not, to suit the depth of the dish which is to be washed.

The dish being wiped clean, externally, a little water is put in it, and, it being held up to the light, its whole interior surface is gently rubbed with the feather, then rinsed, rubbed again and rinsed, so long as careful inspection discovers any portions of adhering precipitate; finally, the feather is rinsed in a stream of water, the rinsings in each case being poured upon the filter.

The washing is now continued by help of the wash-bottle. A jet of cold water is directed, first, upon the interior of the funnel, just *above* the filter, then upon the edge of the filter itself. If thrown immediately against the paper, this is liable to be perforated. The stream of water is carried around the edge of the filter until the latter is nearly full, and the liquid is then allowed to drain off. This process is repeated until a portion of the wash-water, collected to the depth of an inch in a test tube containing a drop of hydrochloric acid, gives no turbidity of silver chloride. When this is accomplished, the precipitate is washed down into the vertex of the filter. The funnel is then closely covered with paper (p. 85), labelled, allowed to drain thoroughly, and set away in a warm place for drying.

5. *Drying the filter*. In public laboratories a heated closet is usually provided for drying filters. Its temperature should not exceed 100° C. In default of such special arrangement, the drying may be effected over the register of a hot-air furnace, or over a common stove or kitchen range.

The funnel may also be supported on a retort-stand over a sheet of iron, which is heated beneath by a lamp, or may be placed at once in the water-bath. See § 50.

6. When the precipitate is perfectly dry we proceed to ignite it for *weighing.*

A small porcelain crucible (platinum must not be used) is cleaned, gently ignited, and when cool (after 15—20 minutes) weighed.

The work-table being clean, two small sheets of fine and smooth writing or glazed paper are opened and laid down side by side. The filter is removed from the funnel and carefully inverted upon one of the papers. The precipitate is loosened from the filter by squeezing and rubbing gently between the fingers, and when it has mostly separated the filter is lifted, reversed, and any portions of silver chloride still adhering are loosened by rubbing its sides together. What is thus detached is poured or shaken out on the paper.

The filter is now spread out as a half-circle upon the other sheet of paper, and, beginning with the straight edge, is folded up into a narrow flattened roll, the two ends of which are then brought together. In this way those central portions of the filter to which particles of precipitate adhere are thoroughly enveloped by the exterior parts, so that in the subsequent burning nothing can easily escape.

The crucible being placed on the glazed paper, the filter is taken by the two free ends in a clean pincers or tongs, put to the flame of a lamp to set it on fire, and then held over the crucible until it is completely charred. It is then dropped into the crucible and moistened with two or three drops of nitric acid. The crucible is covered and placed over a low flame until its contents are dry; it is then heated somewhat stronger, whereby the carbon is nearly or entirely consumed.

The crucible being allowed to cool, one more drop of nitric acid, and afterwards a drop of hydrochloric acid, is added to the residue, and it is heated cautiously, without the cover, until fumes cease to escape. This treatment with nitric acid serves to destroy carbon and convert any reduced silver to nitrate, which the hydrochloric acid in turn transforms into chloride. When the crucible is cool, it is placed again on the paper, and the precipitate is poured into it from the other sheet, the last particles being detached by cautious tapping with the fingers underneath, or by the use of a clean camel's-hair pencil.

The crucible is now put over a low flame and heated cautiously until the silver chloride begins to fuse on the edges. It is then covered and let cool. When cold it is weighed. Read § 115, 1, and the references there made.

7. *Record and calculation of results.* The amount of silver chloride is learned by subtracting from the total the joint weight of the crucible and filter-ash. The quantity of chlorine is obtained by multiplying the amount of silver chloride by the decimal 0·2473. In order to compare results they are reduced to *per cent.* statements by the following proportion :

Substance : chlorine in substance : : 100 : chlorine in 100; i.e. *per cent.*

The record may be made as follows: It is well to work out the calculations in full in the weight-book, as in case of mistake the data are at hand for revision.

		No. 1.		No. 2.
NaCl and tube		6·615		6·180
" " —substance		6·180		5·765
Substance		·435		·415
Crucible, AgCl and Ash		15·3630		14·3270
Cr.	14·298	14·2995	13·309	13·3105
Ash	·0015		·0015	
AgCl		1·0635		1·0165
		0·2473		0·2473
		31905		30495
		74445		71155
		42540		40660
		21270		20330
Cl. =		·26300355		·25138045

No. 1.	No. 2.
·435) 26·300355 (60·46	·415) 25·138045 (60·57
2610	2490
2003	2380
1740	2075
2635	3054
2610	2905

	Found. No. 1.	Found. No. 2.	Calculated.
Chlorine	60·46	60·57	60·62

We have here employed the simplest arithmetical calculation. It is well to duplicate the calculation with help of the tables given in the Appendix.

The first determination given above is not only fair for this method, but answers all ordinary purposes. The second is very good, though with care still closer accordance with theory can be easily attained.]

2. IRON.

Procure 10—15 grms. of fine bright pianoforte wire, cut it into lengths of about 0·3 grm. and keep it free from rust in a dry bottle.

Weigh, on a watch-glass, for each estimation, about 0·3 grm. of wire, and dissolve in hydrochloric acid, with addition of nitric acid. The acids are diluted with a little water.

The solution is effected by heating in a moderate-sized beaker covered with a watch-glass. When complete solution has ensued, and the color of the fluid shows that all the iron is dissolved as ferric chloride (if this is not the case some more nitric acid must be added), rinse the watch-glass, dilute the fluid to about 150 c.c., heat to incipient ebullition, add ammonia in moderate excess, filter through a filter exhausted with hydrochloric acid, &c. (Comp. § 113, 1, *a.*) If BUNSEN's filtering apparatus is employed, proceed as described on p. 97.

As the ferric oxide generally contains a small quantity of silica partially arising from the silicon in the wire, partially taken up from the glass vessels), after it is weighed, digest with fuming hydrochloric acid for some hours; when the ferric oxide is all dissolved, dilute, collect the silica on a small filter, ignite and weigh. The weight is the silica + the ashes of both filters.

The records are made as follows:—

Watch-glass + iron	10·3192
" empty	9·9750
Iron	·3442
Crucible + ferric oxide + silica + filter ash.	17·0703
" empty	16·5761
	·4942
Ash of large filter	·0008
Ferric oxide + silica	·4934
Crucible + silica + ashes of both filters	16·5809
" empty	16·5761
	·0048
Ashes of the filters	·0014
Silica	·0034

·4934 — ·0034 = ·4900 ferric oxide = ·343 iron which gives 99·65 per cent.

3. Lead Acetate.

Determination of Lead.—Triturate the dry and non-effloresced crystals* in a porcelain mortar, and press the powder between sheets of blotting paper until fresh sheets are no longer moistened by it.

a. Weigh about 1 grm., dissolve in water, with addition of a few drops of acetic acid, and proceed exactly as directed § 116, 1, *a*.

b. Weigh about 1 grm., and proceed exactly as directed § 116, 4.

PbO	223·00	58·84
$(C_2H_3O)_2O$	102·00	26·91
$3H_2O$	54·00	14·25
	379·00	100·00

4. Potash Alum.

Determination of Aluminium.—Press pure triturated potash alum between sheets of blotting paper; weigh off about 2 grm., dissolve in water, and determine aluminium as directed p. 241, *a*.

K_2O	94·26	9·93
Al_2O_3	103·00	10·85
$4SO_3$	320·00	33·71
$24H_2O$	432·00	45·51
	949·26	100·00

5. Potassium Dichromate.

Determination of Chromic Acid.—Fuse pure potassium dichromate at a gentle heat, weigh off ·4—·6 grm., dissolve in water, reduce with hydrochloric acid and alcohol, and proceed as directed § 130, I., *a*, *α*.

K_2O	94·26	31·93
$2CrO_3$	200·96	68·07
	295·22	100·00

* Obtained by dissolving the pulverized commercial salt in hot water nearly to saturation, filtering, adding a drop or two of acetic acid to the solution, and slowly evaporating to crystallization.

6. Arsenious Oxide.

Dissolve about 0·2 grm. pure arsenious oxide in small lumps in a middle-sized flask, with a glass stopper, in some solution of soda, by digesting on the water-bath; dilute with a little water, add hydrochloric acid in excess, and then nearly fill the flask with clear hydrogen sulphide water. Insert the stopper and shake. If the hydrogen sulphide is present in excess, the precipitation is terminated; if not, conduct an excess of hydrogen sulphide gas into the fluid; proceed in all other respects exactly as directed § 127, 4.

As_2	150	75·76
O_3	48	24·24
	198	100·00

B. COMPLETE ANALYSIS OF SALTS IN THE GRAVIMETRIC WAY; CALCULATION OF THE FORMULÆ FROM THE RESULTS OBTAINED (see "Calculation of Analyses," in the Appendix).

7. Calcium Carbonate.*

Heat pure calcium carbonate in powder (no matter whether Iceland spar or the artificially prepared substance, see "Qual. Anal.," Am. Ed., p. 87) gently in a platinum crucible.

a. Determination of Calcium.—Dissolve in a covered beaker about 1 grm. in dilute hydrochloric acid, heat gently until the carbonic acid is completely expelled, and determine calcium as directed § 103, 2, *b*, *α*.

b. Determination of Carbonic Acid.—Determine in about 0·8 grm. the carbonic acid after § 139, II., *e*.

CaO	56	56·00
CO_2	44	44·00
	100	100·00

8. Cupric Sulphate.†

Triturate the pure crystals ‡ in a porcelain mortar, and dry as directed p. 47, *a*.

* $Ca < {O \atop O} > CO$. † $Cu < {O \atop O} > SO_2 + 5H_2O$.

‡ [Boil a solution of commercial blue vitriol with a little pure binoxide of lead to oxidize the iron, then with a little barium carbonate to precipitate it, filter and crystallize.—H. Wurtz, Am. Jour. (2), XXVI. 367.]

a. Determination of Water of Crystallization.—1. Weigh off in a crucible 1—2 grm. of the salt, and, having first heated the air-bath (fig. 22, p. 52) so that the thermometer stands steadily at 120°—140°, introduce the crucible, uncovered, and maintain the heat for two hours. Then cool the crucible in a desiccator and weigh. Heat again as before, for an hour, and weigh. If need be, repeat the heating until no more loss occurs. The loss expresses the amount of water expelled at the temperature of 140°, or four molecules. 2. Raise the temperature of the air-bath to between 250°—260° and proceed as before. The loss is the one molecule of strongly combined water of crystallization, or, as some term it, *water of halhydration.*

b. Determination of Sulphuric Acid.—In another portion of the copper sulphate (about 1·5 grm.) determine the sulphuric acid according to § 132, I., 1.

d. Determination of Copper.—In about 1·5 grm. determine the copper as cuprous sulphide, as directed § 119, 3, *a.*

CuO	79·40	31·83
SO_3	80·00	32·08
H_2O	18·00	7·22
4_2HO	72·00	28.87
	249·40	100·00

9. CRYSTALLIZED HYDROGEN SODIUM PHOSPHATE.*

a. Determination of the Water of Crystallization.—Heat about 1 grm. of the pure uneffloresced salt in a platinum crucible, slowly and moderately, first in the water-bath, then in the air-bath, and finally some distance above the lamp (not to visible redness); the loss of weight gives the amount of water of crystallization.

b. Determination of the Hydrogen in the Anhydrous Salt.—Ignite the residue of *a.* The loss is water.

c. Determination of Phosphoric Acid.

α. Treat 1·5—2 grm. of the salt as directed § 134, *b*, *α*.

β. Treat about 0·2 grm. of the salt as directed § 134, *b*, *β*.

I recommend the student to perform the determination by each of these methods, as they are both in common use in the analytical laboratory.

---

* HO, NaO, NaO > PO + $24H_2O$.

d. Determination of Sodium.—Treat about 1·5 grm. of the salt according to § 135, *a*, *α*. After the excess of lead has been separated with hydrogen sulphide, the fluid is to be evaporated to dryness and weighed in a platinum dish; comp. § 69, *b*, and § 98, 2.

P_2O_5	142·00	19·83
$2Na_2O$	124·16	17·34
H_2O	18·00	2·51
$24H_2O$	432·00	60·32
	716·16	100·00

10. Silver Chloride.

Ignite pure fused silver chloride in a stream of pure dry hydrogen till complete decomposition is effected, and weigh the silver obtained. The ignition may be performed in a light bulb tube, or in a porcelain boat in a glass tube, or in a porcelain crucible with perforated cover (§ 115, 4).

The chlorine may be in this case estimated by difference; if you want to determine it directly, proceed as directed § 141, II., *b*.

Ag	107·93	75·27
Cl	35·46	24·73
	143·39	100·00

11. Mercuric Sulphide.

Reduce to a fine powder, and dry at 100°.

a. Determination of Sulphur.—Treat about 0·5 grm., as directed § 148, *β*, p. 466, using nitric acid and potassium chlorate. Precipitate with barium chloride, and after decanting the clear liquid into a filter, boil the barium sulphate twice with dilute solution of ammonium acetate and finally wash with hot water.

b. Determination of Mercury.—Dissolve about 0·5 grm. as before, dilute, and allow to stand in a moderately warm place until the smell of chlorine has nearly gone off; filter if necessary, add ammonia in excess, heat gently for some time, add hydrochloric acid until the white precipitate of mercuric chloride and amide of mercury is redissolved, and treat the solution, which now no longer smells of chlorine, as directed § 118, 3.

Hg	200·00	86·21
S	32·00	13·79
	232·00	100·00

12. CRYSTALLIZED CALCIUM SULPHATE.*

Select clean and pure cystals of selenite, triturate to a coarse powder, avoiding as much as possible exposure to the air, and cork up in a weighing tube.

a. Determination of Water.—After § 35, *a*, *α*.

b. Determination of Sulphuric Acid and Calcium (§ 132, II., *b*, *α*).

CaO	56	32·56
SO_3	80	46·51
$2H_2O$	36	20·93
	172	100·00

C. SEPARATION OF TWO BASIC OR TWO ACID RADICALS FROM EACH OTHER, AND DETERMINATIONS IN THE VOLUMETRIC WAY.

13. SEPARATION OF IRON FROM MANGANESE.

Dissolve in hydrochloric acid about 0·2 grm. fine pianoforte wire, and about the same quantity of ignited protosesquioxide of manganese (prepared as directed § 109, 1, *a*); heat with a little nitric acid, and separate the two metals by means of sodium acetate (p. 517). Determine the manganese as directed § 109, 3.

14. VOLUMETRIC DETERMINATION OF IRON BY SOLUTION OF POTASSIUM PERMANGANATE.

a. Graduation of the Solution of Potassium Permanganate.

α. By metallic iron (fine piano wire) dissolved in dilute sulphuric acid (p. 268).

β. By ammonium oxalate (p. 270).

b. Determination of Iron in Ammonium Ferrous Sulphate.

In solution acidified with sulphuric acid (p. 272, *β*).

The formula requires 18·37 per cent. of FeO.

---

* $Ca < {}^{O}_{O} > SO_2 + 2H_2O$.

c. Determination of the Iron in a Limonite.

Powder finely, dry at 100°, weigh off 2 grm., heat with strong hydrochloric acid till the ferric oxide is completely dissolved, dilute the acid solution with twice its volume of water, filter, evaporate with sulphuric acid, dilute the ferric sulphate to 500 c.c., and in two or three portions of 100 c.c. each reduce ferric to ferrous sulphate, and determine iron as directed in § 113, *a*, p. 278.

15. Volumetric Determination of Iron with Sodium Thiosulphate.

a. Graduation of the Solution of Sodium Thiosulphate.

α. By solution of ferric chloride (p. 280).

β. By ammonia-iron-alum (p. 119). 2 grms. to be weighed off, dissolved in water with addition of hydrochloric acid.

b. Determination of Iron in Limonite.

Decompose 2 to 3 grms. with concentrated hydrochloric acid. Transfer to a 500 c.c. flask, mix well the solution diluted to 500 c.c., and determine repeatedly iron in portions of 100 c.c. each, after § 113, 3, *b*. (If the ore contains ferrous iron, oxidize the hydrochloric acid solution with potassium chlorate, avoiding needless excess, and concentrate it one half to remove excess of chlorine. See § 112, 1, p. 266.)

16. Determination of Nitric Acid in Potassium Nitrate.

Heat pure nitre, not to fusion, and transfer it to a tube provided with a cork.

Treat 0·5 grm. as directed p. 473, *β*.

K_2O	94·26	46·59
N_2O_5	108·08	53·41
	202·34	100·00

17. Separation of Magnesium from Sodium.

Dissolve about 0·4 grm. pure recently ignited magnesia and about 0·5 grm. pure well-dried sodium chloride in dilute hydrochloric acid (avoiding a large excess), and separate by one of the methods described in § 153, 4.

18. Separation of Potassium from Sodium.

Triturate crystallized sodium potassium tartrate (Rochelle salt), press between blotting paper, weigh off about 1·5 grm., heat in a platinum crucible, gently at first, then for some time to gentle ignition. The carbonaceous residue is first extracted with water, finally with dilute hydrochloric acid, the acid fluid is evaporated in a weighed platinum dish, and the chlorides are weighed together (§ 97, 2). Then separate them by platinic chloride (p. 481, 1), and calculate from the results the quantities of soda and potassa severally contained in the Rochelle salt.

K_2O	94·26	16·70
Na_2O	62·08	11·00
$C_8H_8O_{10}$	264·00	46·78
$8H_2O$	144·00	25·52
	564·34	100·00

19. Volumetric Determination of Chlorine in Chlorides.

a. Preparation and examination of the solution of silver nitrate (§ 141, I., *b*, *α*).

b. Indirect determination of the sodium and potassium in Rochelle salt, by volumetric estimation of the chlorine in the alkali chlorides prepared as in No. 18. For calculation, see "Calculation of Analyses," in the Appendix.

20. Acidimetry.

a. Preparation of standard acid and alkali solutions. Sulphuric acid and potash, or hydrochloric acid and ammonia may be used (§ 192).

b. Determination of acid in hydrochloric acid, by the specific gravity (p. 677).

c. Determination of acid in the same hydrochloric acid, by an alkaline fluid of known strength (p. 684).

d. Determination of acid in colored vinegar, by saturation with a standard alkaline solution. (Application of test-papers, p. 684).

21. Alkalimetry.

a. Preparation of the normal acid after Descroizilles and Gay-Lussac (§ 195).

b. Valuation of a soda-ash after expulsion of the water by gentle ignition.

α. After DESCROIZILLES and GAY-LUSSAC (§ 195).

β. After MOHR (§ 196).

22. DETERMINATION OF AMMONIUM.

Treat about 0·8 grm. chloride of ammonium as directed § 99, 3, *α*.

NH_4	18·04	33·72	NH_3	17·04	31·85
Cl	35·46	66·28	HCl	36·46	68·15
	53·50	100·00		53·50	100·00

D. ANALYSIS OF ALLOYS, MINERALS, INDUSTRIAL PRODUCTS, &c., IN THE GRAVIMETRIC AND VOLUMETRIC WAY.

23. ANALYSIS OF BRASS.

Brass consists of from 25 to 35 per cent. of zinc and from 75 to 65 per cent. of copper. It also contains usually small quantities of tin and lead, and occasionally traces of iron.

Dissolve about 2 grm. in nitric acid, evaporate on the water-bath to dryness, moisten the residue with nitric acid, add some water, warm, dilute still further, and filter off any residual meta-stannic acid (§ 126, 1, *a*). Add to the filtrate, or, if the quantity of tin is very inconsiderable, directly to the solution, about 20 c.c. dilute sulphuric acid; evaporate to dryness on the water-bath, add 50 c.c. water, and apply heat. If a residue remains (lead sulphate), filter it off, and treat it as directed § 116, 3. In the filtrate, separate the copper from the zinc by sodium thiosulphate (p. 540). If the quantity of iron present can be determined, determine it in the weighed oxide of zinc (§ 160).

24. ANALYSIS OF SOLDER (TIN AND LEAD).

Introduce about 1·5 grm. of the alloy, cut into small pieces, into a flask, treat it with nitric acid, and proceed as directed p. 339, to effect the separation and estimation of the tin.

Mix the filtrate in a porcelain dish with pure dilute sulphuric acid, evaporate the nitric acid on the water-bath, and proceed with the lead sulphate obtained as directed § 116, 3. Test the fluid filtered from the lead sulphate with hydrogen sulphide and ammonium sulphide for the other metals which the alloy might contain besides tin and lead. The stannic oxide may contain small quantities of iron or copper; it is tested for these by fusion with sodium carbonate and sulphur (p. 558).

25. Analysis of a Dolomite.

See § 210.

26. Analysis of Felspar.

a. Decomposition by sodium carbonate (§ 140, II., *b*); removal of the silicic acid; precipitation of the aluminium with the small quantity of iron as hydroxides by ammonia (in platinum or Berlin porcelain, not in glass vessels) after § 156, 1, (**37**); separation of barium, if present, from the filtrate with dilute sulphuric acid, and then of calcium with ammonium oxalate, § 154 (**28**). Finally, solution of the weighed alumina in concentrated hydrochloric acid, separation and weighing of traces of silica if present; evaporation with sulphuric acid and volumetric determination of iron, generally present in small quantities after § 113, p. 279.

b. Decomposition by Smith's method, p. 426. Separate the alkalies after § 152, **1.**

c. Determine loss by ignition.

27. Assay of a Calamine or Smithsonite.

After § 215.

Volumetric determination of the zinc.

28. Assay of Galena.

Determination of the lead, as directed § 213.

29. Valuation of Chloride of Lime (§ 199).

a. After Penot (p. 699).

b. After Otto (p. 701).

30. Valuation of Manganese (§ 202).

a. After Fresenius and Will (p. 705). The evolved CO_2 to be weighed (p. 708).

b. After Bunsen (p. 709).

c. By means of iron (p. 709).

31. Complete Analysis of Iron Ore (§ 217).

E. DETERMINATION OF THE SOLUBILITY OF SALTS.

32. Determination of the Degree of Solubility of Common Salt.

a. At boiling heat.—Dissolve perfectly pure pulverized sodium chloride in distilled water, in a flask, heat to boiling, and keep in ebullition until part of the dissolved salt separates. Filter the fluid now with the greatest expedition, through a funnel surrounded with boiling water and covered with a glass plate, into an accurately tared capacious measuring flask. As soon as about 100 c.c. of fluid have passed into the flask, insert the cork, allow to cool, and weigh. Fill the flask now up to the mark with water, and determine the salt in an aliquot portion of the fluid, by evaporating in a platinum dish (best with addition of some ammonium chloride, which will, in some measure, prevent decrepitation); or by determining the chlorine (§ 141).

b. At 14°.—Allow the boiling saturated solution to cool down to this temperature with frequent shaking, and then proceed as in *a.*

100 parts of water dissolve at 109·7°....40·35 of sodium chloride.
100 " " 14° ....35·87 " "

33. Determination of the Degree of Solubility of Calcium Sulphate.

a. At 100°.

b. At 12°.

Digest pure pulverized calcium sulphate for some time with water, in the last stage of the process at 40°—50° (at which temperature sulphate of lime is most soluble); shake the mixture frequently during the process. Decant the clear solution, together with a little of the precipitate, into two flasks, and boil the fluid in one of them for some time; allow that in the other to cool down to 12°, with frequent shaking, and let it stand for some time at that temperature. Then filter both solutions, weigh the filtrates, and determine the amount of calcium sulphate respectively contained in them, by evaporating and igniting the residues.

100 parts of water dissolve at 100° ....0·217 of anhydrous calcium sulphate.
100 " " 12° ....0·233 " " "

34. Analysis of Atmospheric Air.

See § 221.

F. ORGANIC ANALYSIS AND ANALYSES IN WHICH ORGANIC ANALYSIS IS APPLIED.

35. Analysis of Tartaric Acid.

Select clean and white crystals. Powder and dry at 100°.

a. Burn with lead chromate after § 177. For details of manipulation see §§ 174 and 175.

b. Burn with oxygen gas in a tray, § 178.

C_4	48	32
H_6	6	4
O_6	96	64
	150	100

36. Determination of the Nitrogen in Crystallized Potassium Ferrocyanide.

Triturate the perfectly pure crystals, press in blotting paper, and determine the nitrogen as directed § 185. (Combustion with soda lime.) The formula requires 19·93 per cent. of nitrogen.

37. Analysis of Uric Acid (or any other perfectly pure organic compound of carbon, hydrogen, oxygen, and nitrogen).

Dry pure uric acid at 100°.

a. Determination of the carbon and hydrogen (§ 183).

b. Determination of the nitrogen.

α. After § 185.

β. After § 184, *bb.*

C_5	60·00	35·68
N_4	56·16	33·40*
H_4	4·00	2·38
O_3	48·00	28·54
	168·16	100·00

38. Analysis of a Superphosphate (§ 220).

39. Analysis of Coal (§ 219.)

40. Analysis of a Cast Iron.

After § 218.

* Taking 14 for the atomic weight of N gives 33·33 per cent. of nitrogen.

APPENDIX.

ANALYTICAL EXPERIMENTS.*

1. ACTION OF WATER UPON GLASS AND PORCELAIN VESSELS, IN THE PROCESS OF EVAPORATION (to § 41).

A large bottle was filled with water cautiously distilled from a copper boiler with a tin condensing tube. All the experiments in 1 were made with this water.

a. 300 c.c., cautiously evaporated in a platinum dish, left a residue weighing, after ignition, 0·0005 grm.=0·0017 per 1000.

b. 600 c.c. were evaporated, boiling, nearly to dryness, in a wide flask of Bohemian glass; the residue was transferred to a platinum dish, and the flask rinsed with 100 c.c. distilled water, which was added to the residue in the dish; the fluid in the latter was then evaporated to dryness, and the residue ignited.

The residue weighed..	0·0104	grm.
Deducting from this the quantity of fixed matter originally contained in the distilled water, viz............................	0·0012	"
There remains substance taken up from the glass...........	0·0092	"

=0·0153 per 1000.

In three other experiments, made in the same manner, 300 c.c. left, in two 0·0049 grm., in the third 0·0037 grm.; which, calculated for 600 c.c., gives an

average of..	0·0090	grm.
And after a deduction of....................................	0·0012	"
	0·0078	"

=0·013 per 1000.

We may therefore assume that 1 litre of water dissolves, when boiled down to a small bulk in glass vessels, about 14 milligrammes of the constituents of the **glass.**

c. 600 c.c. were evaporated nearly to dryness in a dish of Berlin porcelain, and in all other respects treated as in *b.*

The residue weighed..	0·0015	grm.
Deducting from this the quantity of fixed matter contained in the distilled water, viz....................................	0·0012	"
There remains substance taken up from the porcelain........	0·0003	"

=0·0005 per 1000.

* The experiments are numbered as in the original edition, but some are omitted.

2. Action of Hydrochloric Acid upon Glass and Porcelain Vessels, in the Process of Evaporation (to § 41).

The distilled water used in 1 was mixed with $\frac{1}{10}$ of pure hydrochloric acid.

a. 300 grm., evaporated in a platinum dish, left 0·002 grm. residue.

b. 300 grm., evaporated first in Bohemian glass nearly to dryness, then in a platinum dish, left 0·0019 residue; the dilute hydrochloric acid, therefore, had not attacked the glass.

c. 300 grm., evaporated in Berlin porcelain, &c., left 0·0036 grm., accordingly after deducting 0·002, 0·0016=0·0053 per 1000.

d. In a second experiment made in the same manner as in *c.*, the residue amounted to 0·0034, accordingly after deducting 0·002, 0·0014=0·0047 per 1000.

Hydrochloric acid, therefore, attacks glass much less than water, whilst porcelain is about equally affected by water and dilute hydrochloric acid. This shows that the action of water upon glass consists in the formation of soluble basic silicates.

3. Action of Solution of Ammonium Chloride upon Glass and Porcelain Vessels, in the Process of Evaporation (to § 41).

In the distilled water of 1, $\frac{1}{10}$ of ammonium chloride was dissolved, and the solution filtered.

a. 300 c.c., evaporated in a platinum dish, left 0·006 grm. fixed residue.

b. 300 c.c., evaporated first nearly to dryness in Bohemian glass. then to dry ness in a platinum dish, left 0·0179 grm.; deducting from this 0·006 grm., there remains substance taken up from the glass, 0·0119=0·0397 per 1000.

c. 300 c.c., treated in the same manner in Berlin porcelain, left 0·0178 deducting from this 0·006, there remains 0·0118=0·0393 per 1000.

Solution of ammonium chloride; therefore, strongly attacks both glass and porcelain in the process of evaporation.

4. Action of Solution of Sodium Carbonate upon Glass and Porcelain Vessels (to § 41).

In the distilled water of 1, $\frac{1}{10}$ of pure crystallized sodium carbonate was dissolved.

a. 300 c.c., supersaturated with hydrochloric acid and evaporated to dryness in a platinum dish, &c., gave 0·0026 grm. silica=0·0087 per 1000.

b. 300 c.c. were gently boiled for three hours in a glass vessel, the evaporating water being replaced from time to time; the tolerably concentrated liquid was then treated as in *a*; it left a residue weighing 0·1376 grm.; deducting from this the 0·0026 grm. left in *a*, there remains 0 135 grm.=0·450 per 1000.

c. 300 c.c., treated in the same manner as in *b*, in a porcelain vessel, left 0·0099; deducting from this 0·0026 grm., there remains 0·0073=0·0243 per 1000.

Which shows that boiling solution of sodium carbonate attacks glass very strongly, and porcelain also in a very marked manner.

5. Water Distilled from Glass Vessels (to § 56, 1).

42·41 grm. of water distilled with extreme caution from a tall flask with a Liebig's condenser, left upon evaporation in a platinum dish, a residue weighing, after ignition, 0·0018 grm., consequently $\frac{1}{23561}$.

6. POTASSIUM SULPHATE AND ALCOHOL (to § 68, *a*).

a. Ignited pure potassium sulphate was digested cold with absolute alcohol, for several days, with frequent shaking; the fluid was filtered off, the filtrate diluted with water, and then mixed with barium chloride. It remained perfectly clear upon the addition of this reagent, but after the lapse of a considerable time it began to exhibit a slight opalescence. Upon evaporation to dryness, there remained a very trifling residue, which gave, however, distinct indications of the presence of sulphuric acid.

b. The same salt treated in the same manner, with addition of some pure concentrated sulphuric acid, gave a filtrate which, upon evaporation in a platinum dish, left a clearly perceptible fixed residue of potassium sulphate.

7. DEPORTMENT OF POTASSIUM CHLORIDE IN THE AIR AND AT A HIGH TEMPERATURE (to § 68, *b*).

0·9727 grm. of ignited (not fused) pure potassium chloride, heated for 10 minutes to dull redness in an open platinum dish, lost 0·0007 grm.; the salt was then kept for 10 minutes longer at the same temperature, when no further diminution of weight was observed. Heated to bright redness and semi-fusion, the salt suffered a further loss of weight to the extent of 0·0009 grm. Ignited intensely and to perfect fusion, it lost 0·0034 grm. more.

Eighteen hours' exposure to the air produced not the slightest increase of weight.

8. SOLUBILITY OF POTASSIUM PLATINIC CHLORIDE IN ALCOHOL (to § 68, *c*).

a. In absence of free Hydrochloric Acid.

α. An excess of perfectly pure, recently precipitated potassium platinic chloride was digested for 6 days at 15—20°, with alcohol of 97·5 per cent., in a stoppered bottle, with frequent shaking. 72·5 grm. of the perfectly colorless filtrate left upon evaporation in a platinum dish a residue which, dried at 100°, weighed 0·006 grm.; 1 part of the salt requires therefore 12083 parts of alcohol of 97·5 per cent. for solution.

β. The same experiment was made with alcohol of 76 per cent. The filtrate might be said to be colorless; upon evaporation, slight blackening ensued, on which account the residue was determined as platinum. 75·5 grm. yielded 0·008 grm. platinum, corresponding to 0·02 grm. of the salt. One part of the salt dissolves accordingly in 3775 parts of alcohol of 76 per cent.

γ. The same experiment was made with alcohol of 55 per cent. The filtrate was distinctly yellowish. 63·2 grm. left 0·0241 grm. platinum, corresponding to 0·06 grm. of the salt. One part of the salt dissolves accordingly in 1053 parts of alcohol of 55 per cent.

b. In presence of free Hydrochloric Acid.

Recently precipitated potassium platinic chloride was digested cold with alcohol of 76 per cent., to which some hydrochloric acid had been added. The solution was yellowish; 67 grm. left 0·0146 grm. platinum, which corresponds to 0·0365 grm. of the salt. One part of the salt dissolves accordingly in 1835 parts of alcohol mixed with hydrochloric acid.

9. SODIUM SULPHATE AND ALCOHOL (to § 69, *a*).

Experiments made with pure anhydrous sodium sulphate, in the manner described in 6, showed that this salt comports itself both with pure alcohol, and with alcohol containing sulphuric acid, exactly like potassium sulphate.

10. Deportment of ignited Sodium Sulphate in the Air (to § 69, *a*).

2 5169 grm. anhydrous sodium sulphate were exposed, in a watch-glass, to the open air on a hot summer day. The first few minutes passed without any increase of weight, but after the lapse of 5 hours there was an increase of 0·0061 grm.

12. Deportment of Sodium Chloride in the Air (to § 69, *b*).

4·3281 grm. of chemically pure, moderately ignited (not fused) sodium chloride, which had been cooled under a bell-glass over sulphuric acid, acquired during 45 minutes' exposure to the (somewhat moist) air an increase of weight of 0·0009 grm.

13. Deportment of Sodium Chloride upon Ignition by itself and with Ammonium Chloride (to § 69, *b*).

4·3281 grm. chemically pure, ignited sodium chloride were dissolved in water, in a moderate-sized platinum dish, and pure ammonium chloride was added to the solution, which was then evaporated and the residue gently heated until the evolution of ammonium chloride fumes had apparently ceased. The residue weighed 4·3334 grm. It was then very gently ignited for about 2 minutes, and after this re-weighed, when the weight was found to be 4·3314 grm. A few minutes' ignition at a red heat reduced the weight to 4·3275 grm., and 2 minutes' further ignition at a bright red heat (upon which occasion white fumes were seen to escape), to 4·3249 grm.

14. Deportment of Sodium Carbonate in the Air and on Ignition (to § 69, *c*).

2·1061 grm. of moderately ignited chemically pure sodium carbonate were exposed to the air in an open platinum dish in July in bad weather; after 10 minutes the weight was 2·1078, after 1 hour, 2·1113, after 5 hours, 2·1257.

1·4212 grm. of moderately ignited chemically pure sodium carbonate were ignited for 5 minutes in a covered platinum crucible; no fusion took place, and the weight was unaltered. Heated more strongly for 5 minutes, it partially fused, and then weighed 1·4202. After being kept fusing for 5 minutes, it weighed 1·4135.

15. Deportment of Ammonium Chloride upon Evaporation and Drying (to § 70, *a*).

0·5625 grm. pure and perfectly dry ammonium chloride was dissolved in water in a platinum dish, evaporated to dryness in the water-bath and completely dried; the weight was now found to be 0·5622 grm. (ratio 100:99·94). It was again heated for 15 minutes in the water-bath, and afterwards re-weighed, when the weight was found to be 0·5612 grm. (ratio 100:99·77). Exposed once more for 15 minutes to the same temperature, the residue weighed 0·5608 grm. (ratio 100:99·69).

16. Solubility of Ammonium Platinic Chloride in Alcohol (to § 70, *b*).

a. In absence of free Hydrochloric Acid.

α. An excess of perfectly pure, recently precipitated ammonium platinic chloride was digested for 6 days. at 15—20°, with alcohol of 97·5 per cent., in a stoppered bottle, with frequent agitation.

74·3 grm. of the perfectly colorless filtrate left, upon evaporation and ignition in a platinum dish, 0·0012 grm. platinum, corresponding to 0·0028 of the salt. One part of the salt requires accordingly 26535 parts of alcohol of 97·5 per cent.

β. The same experiment was made with alcohol of 76 per cent. The filtrate was distinctly yellowish.

81·75 grm. left 0·0257 platinum, which corresponds to 0·0584 grm. of the salt. One part of the salt dissolves accordingly in 1406 parts of alcohol of 76 per cent.

γ. The same experiment was made with alcohol of 55 per cent. The filtrate was distinctly yellow. Slight blackening ensued upon evaporation, and 56·5 grm. left 0·0364 platinum, which corresponds to 0·08272 grm. of the salt. Consequently, 1 part of the salt dissolves in 665 parts of alcohol of 55 per cent.

b. In presence of Hydrochloric Acid.

The experiment described in β was repeated, with this modification, that some hydrochloric acid was added to the alcohol. 76·5 grm. left 0·0501 grm. of platinum, which corresponds to 0·1189 grm. of the salt. 672 parts of the acidified alcohol had therefore dissolved 1 part of the salt.

17. SOLUBILITY OF BARIUM CARBONATE IN WATER (to § 71, *b*).

a. In Cold Water.—Perfectly pure, recently precipitated $BaCO_3$ was digested for 5 days with water of 16—20°, with frequent shaking. The mixture was filtered, and a portion of the filtrate tested with sulphuric acid, another portion with ammonia; the former reagent immediately produced turbidity in the fluid, the latter only after the lapse of a considerable time. 84·82 grm. of the solution left, upon evaporation, 0·0060 $BaCO_3$. One part of that salt dissolves consequently in 14137 parts of cold water.

b. In Hot Water.—The same barium carbonate being boiled for 10 minutes with pure distilled water, gave a filtrate manifesting the same reactions as that prepared with cold water, and remaining perfectly clear upon cooling. 84·82 grm. of the hot solution left, upon evaporation, 0·0055 grm. of barium carbonate. One part of that salt dissolves therefore in 15421 parts of boiling water.

18. SOLUBILITY OF BARIUM CARBONATE IN WATER CONTAINING AMMONIA AND AMMONIUM CARBONATE (to § 71, *b*).

A solution of chemically pure barium chloride was mixed with ammonia and ammonium carbonate in excess, gently heated and allowed to stand at rest for 12 hours; the fluid was then filtered off; the filtrate remained perfectly clear upon addition of sulphuric acid; but after the lapse of a very considerable time, a hardly perceptible precipitate separated. 84·82 grm. of the filtrate left, upon evaporation in a small platinum dish, and subsequent gentle ignition, 0·0006 grm. One part of the salt had consequently dissolved in 141000 parts of the fluid.

19. SOLUBILITY OF BARIUM SILICO-FLUORIDE IN WATER (to § 71, *c*).

a. Recently precipitated, thoroughly washed barium silico-fluoride was digested for 4 days in cold water, with frequent shaking; the fluid was then filtered off, and a portion of the filtrate tested with dilute sulphuric acid, another portion with solution of calcium sulphate; both reagents produced turbidity—the former immediately, the latter after one or two seconds—precipitates separated from both portions after the lapse of some time. 84.82 grm. of the filtrate left a residue which, after being thoroughly dried, weighed 0·0223 grm. One part of the salt had consequently required 3802 parts of cold water for its solution.

b. A portion of another sample of recently precipitated barium silico-fluoride was heated with water to boiling, and the solution allowed to cool (upon which a portion of the dissolved salt separated). The cold fluid was left for a considerable time longer in contact with the undissolved salt, and was then filtered off. The filtrate showed the same deportment with solution of sulphate of lime as that of *a.* 84·82 grm. of it left 0·025 grm. One part of the salt had accordingly dissolved in 3392 parts of water.

20. Solubility of Barium Silico-Fluoride in Water acidified with Hydrochloric Acid (to § 71, *c*).

a. Recently precipitated pure barium silico-fluoride was digested with frequent agitation for 3 weeks with cold water acidified with hydrochloric acid. The filtrate gave with sulphuric acid a rather copious precipitate. 84·82 grm. left 0·1155 grm. of thoroughly dried residue, which, calculated as barium silico-fluoride, gives 733 parts of fluid to 1 part of that salt.

b. Recently precipitated pure barium silico-fluoride was mixed with water very slightly acidified with hydrochloric acid, and the mixture heated to boiling. Cooled to 12°, 84·82 grm. of the filtrate left a residue of 0·1322 grm., which gives 640 parts of fluid to 1 part of the salt.

N.B. The solution of barium silico-fluoride in hydrochloric acid is not effected without decomposition; at least, the residue contained, even after ignition, a rather large proportion of barium chloride.

21. Solubility of Strontium Sulphate in Water (to § 72, *a*).

a. In Water of 14°.

84·82 grm. of a solution prepared by 4 days' digestion of recently precipitated strontium sulphate with water at the common temperature, left 0·0123 grm. of strontium sulphate. One part of strontium sulphate dissolves consequently in 6895 parts of water.

b. In Water of 100°.

84·82 grm. of a solution prepared by boiling recently precipitated strontium sulphate several hours with water, left 0·0088 grm. Consequently 1 part of strontium sulphate dissolves in 9638 parts of boiling water.

22. Solubility of Strontium Sulphate in Water containing Hydrochloric Acid and Sulphuric Acid (to § 72, *a*).

a. 84·82 grm. of a solution prepared by 3 days' digestion, left 0·0077 grm. $SrSO_4$.

b. 42·41 grm. of a solution prepared by 4 days' digestion, left 0·0036 grm.

c. Pure strontium carbonate was dissolved in an excess of hydrochloric acid, and the solution precipitated with an excess of sulphuric acid and then allowed to stand in the cold for a fortnight. 84·82 grm. of the filtrate left 0·0066 grm.

In *a*. 1 part of $SrSO_4$ required			11016 parts.
b. 1	"	"	11780 "
c. 1	"	"	12791 "
	Mean.............		11862 parts.

23. Solubility of Strontium Sulphate in dilute Nitric Acid, Hydrochloric Acid, and Acetic Acid (to § 72, *a*).

a. Recently precipitated pure strontium sulphate was digested for 2 days in the cold with nitric acid of 4·8 per cent. 150 grm. of the filtrate left 0·3451 grm. One part of the salt required accordingly 435 parts of the dilute acid for its solution; in another experiment 1 part of the salt was found to require 429 parts of the dilute acid. Mean, 432 parts.

b. The same salt was digested for 2 days in the cold with hydrochloric acid of 8·5 per cent. 100 grm. left 0·2115, and in another experiment, 0·2104 grm. One part of the salt requires, accordingly, in the mean, 474 parts of hydrochloric acid of 8·5 per cent. for its solution.

c. The same salt was digested for 2 days in the cold with acetic acid of 15·6 per cent. $C_2H_4O_2$. 100 grm. left 0·0126, and in another experiment, 0·0129 grm. One part of the salt requires, accordingly, in the mean, 7843 parts of acetic acid of 15·6 per cent.

24. Solubility of Strontium Carbonate in Cold Water (to § 72, *b*).

Recently precipitated, thoroughly washed strontium carbonate was digested several days with cold distilled water, with frequent shaking. 84·82 grm. of the filtrate left, upon evaporation, a residue weighing, after ignition, 0·0047 grm. One part of strontium carbonate requires therefore 18045 parts of water for its solution.

25. Solubility of Strontium Carbonate in Water containing Ammonia and Ammonium Carbonate (to § 72, *b*).

Recently precipitated, thoroughly washed strontium carbonate was digested for 4 weeks with cold water containing ammonia and ammonium carbonate, with frequent shaking. 84·82 grm. of the filtrate left 0·0015 grm. $SrCO_3$. Consequently, 1 part of the salt requires 56545 parts of this fluid for its solution.

If solution of strontium chloride is precipitated with ammonium carbonate and ammonia as directed § 102, 2, *a*, sulphuric acid produces no turbidity in the filtrate, after addition of alcohol.

26. Solubility of Calcium Carbonate in Cold and in Boiling Water (to § 73, *b*).

a. A solution prepared by boiling as in 26, *b*, was digested in the cold for 4 weeks, with frequent agitation, with the undissolved precipitate. 84·82 grm. left 0·0080 $CaCO_3$. One part therefore required 10601 parts.

b. Recently precipitated calcium carbonate was boiled for some time with distilled water. 42·41 grm. of the filtrate left, upon evaporation and gentle ignition of the residue, 0·0048 $CaCO_3$. One part requires consequently 8834 parts of boiling water.

27. SOLUBILITY OF $CaCO_3$ IN WATER CONTAINING AMMONIA AND AMMONIUM CARBONATE (to § 73, *b*).

Pure dilute solution of calcium chloride was precipitated with ammonium carbonate and ammonia, allowed to stand 24 hours, and then filtered. 84·82 grm. left 0·0013 grm. Ca CO_3. One part requires consequently 65246 parts.

28. DEPORTMENT OF CALCIUM CARBONATE UPON IGNITION IN A PLATINUM CRUCIBLE (to § 73, *b*).

0·7955 grm. of perfectly dry calcium carbonate was exposed, in a small and thin platinum crucible, to the gradually increased and finally most intense heat of a good BERZELIUS' lamp. The crucible was open and placed obliquely. After the first 15 minutes the mass weighed 0·6482—after half an hour 0·6256—after one hour 0·5927, which latter weight remained unaltered after 15 minutes' additional heating. This corresponds to 74·5 per cent., whilst the proportion of CaO in the carbonate is calculated at 56 per cent.; there remained therefore evidently still a considerable amount of the carbonic acid.

29. COMPOSITION OF CALCIUM OXALATE DRIED AT 100° (to § 73, *c*).

0·8510 grm. of thoroughly dry pure calcium carbonate was dissolved in hydrochloric acid; the solution was precipitated with ammonium oxalate and ammonia, and the precipitate collected upon a weighed filter and dried at 100°, until the weight remained constant. The calcium oxalate so produced weighed 1·2461 grm. Calculating this as $CaC_2O_4 + H_2O$, the amount found contained 0·4772 CaO, which corresponds to 56·07 per cent. in the calcium carbonate; the calculated proportion of CaO in the latter is 56 per cent.

30. DEPORTMENT OF MAGNESIUM SULPHATE IN THE AIR AND UPON IGNITION (to § 74, *a*).

0·8135 grm. of perfectly pure anhydrous $MgSO_4$ in a covered platinum crucible acquired, on a fine and warm day in June, in half an hour, an increase of weight of 0·004 grm., and in the course of 12 hours, of 0·067 grm. The salt could not be accurately weighed in the open crucible, owing to continual increase of weight.

0·8135 grm., exposed for some time to a very moderate red heat, suffered no diminution of weight; but after 5 minutes' exposure to an intense red heat, the substance was found to have lost 0·0075 grm., and the residue gave no longer a clear solution with water. About 0·2 grm. of pure magnesium sulphate exposed in a small platinum crucible, for 15 to 20 minutes, to the heat of a powerful blast gas lamp, gave, with dilute hydrochloric acid, a solution in which barium chloride failed to produce the least turbidity.

31. SOLUBILITY OF AMMONIUM MAGNESIUM PHOSPHATE IN PURE WATER (to § 74, *b*).

a. Recently precipitated ammonium magnesium phosphate was thoroughly washed with water, then digested for 24 hours with water of about 15°, with frequent shaking.

84·42 grm. of the filtrate left.................................. 0·0047 grm. of magnesium pyrophosphate.

b. The same precipitate was digested in the same manner for 72 hours.

84·42 grm. of the filtrate left.................................. 0·0043 "

Mean.... .... 0·0045 "

which corresponds to 0·00552 grm. of the anhydrous double salt. One part of that salt dissolves therefore in 15293 parts of pure water.

The cold saturated solution gave, with ammonia, after the lapse of a short time, a distinctly perceptible crystalline precipitate;—on the addition of sodium phosphate, it remained perfectly clear, and even after the lapse of 2 days no precipitate had formed;—ammonium sodium phosphate produced a precipitate as large as that by ammonia.

32. Solubility of Ammonium Magnesium Phosphate in Water containing Ammonia (to § 74, *b*).

a. Pure ammonium magnesium phosphate was dissolved in the least possible amount of nitric acid; a large quantity of water was added to the solution, then ammonia in excess. The mixture was allowed to stand at rest for 24 hours, then filtered; its temperature was 14°. 84·42 grm. left 0·0015 magnesium pyrophosphate, which corresponds to 0·00184 of the anhydrous double salt. Consequently 1 part of the latter requires 45880 parts of ammoniated water for its solution.

b. Pure ammonium magnesium phosphate was digested for 4 weeks with ammoniated water, with frequent shaking; the fluid (temperature 14°) was then filtered off; 126·63 grm. left 0·0024 magnesium pyrophosphate, which corresponds to 0·00296 of the double salt. One part of it therefore dissolves in 42780 parts of ammoniated water. Taking the mean of *a* and *b*, 1 part of the double salt requires 44330 parts of ammoniated water for its solution.

33. Another Experiment on the same Subject (to § 74, *b*).

Recently precipitated ammonium magnesium phosphate, most carefully washed with water containing ammonia, was dissolved in water acidified with hydrochloric acid, ammonia added in excess, and allowed to stand in the cold for 24 hours. 169·64 grm. of the filtrate left 0·0031 magnesium pyrophosphate, corresponding to 0·0038 of anhydrous ammonium magnesium phosphate. One part of the double salt required therefore 44600 parts of the fluid.

34. Solubility of Ammonium Magnesium Phosphate in Water containing Ammonium Chloride (to § 74, *b*).

Recently precipitated, thoroughly washed ammonium magnesium phosphate was digested in the cold with a solution of 1 part of ammonium chloride in 5 parts of water. 18·4945 grm. of the filtrate left 0·002 magnesium pyrophosphate, which corresponds to 0·00245 of the double salt. One part of the salt dissolves therefore in 7548 parts of the fluid.

35. Solubility of Ammonium Magnesium Phosphate in Water containing Ammonia and Ammonium Chloride (to § 74, *b*).

Recently precipitated, thoroughly washed ammonium magnesium phosphate was digested in the cold with a solution of 1 part of ammonium chloride in 7 parts of ammoniated water. 23·1283 grm. of the filtrate left 0·0012 magnesium pyrophosphate, which corresponds to 0·00148 of the double salt. One part of the double salt requires consequently 15627 parts of the fluid for its solution.

36. Deportment of Acid Solutions of Magnesium Pyrophosphate with Ammonia (to § 74, *c*).

0·3985 grm. magnesium pyrophosphate was treated for several hours, at a high temperature, with concentrated sulphuric acid. This exercised no perceptible action. It was only after the addition of some water that the salt dissolved. The fluid, heated for some time, gave, upon addition of ammonia in excess, a crystalline precipitate, which was filtered off after 18 hours; the quantity of magnesium pyrophosphate obtained was 0·3805 grm., that is, 95·48 per cent. Sodium phosphate produced in the filtrate a trifling precipitate, which gave 0·0150 grm. of magnesium pyrophosphate, that is, 3·76 per cent.

0·3565 grm. magnesium pyrophosphate was dissolved in 3 grm. nitric acid, of 1·2 sp. gr.; the solution was heated, diluted, and precipitated with ammonia: the quantity of magnesium pyrophosphate obtained amounted to 0·3485 grm., that is, 98·42 per cent.; 0·4975 grm. was treated in the same manner with 7·6 grm. of the same nitric acid: the quantity re-obtained was 0·4935 grm., that is, 99·19 per cent.

0·786 grm. treated in the same manner with 16·2 grm. of nitric acid, gave 0·7765 grm., that is, 98·79 per cent.

The result of these experiments may be tabulated thus:

Proportion of Mg_2P_2O to nitric acid.	Re-obtained.	Loss.
1 : 9	98·42 per cent.	1·58
1 : 15	99·19 "	0·81
1 : 20	98·79 "	1·21

37. Solubility of pure Magnesia in Water (to § 74, *d*).

a. In Cold Water.

Perfectly pure well-crystallized magnesium sulphate was dissolved in water, and the solution precipitated with ammonium carbonate and caustic ammonia; the precipitate was thoroughly washed—in spite of which it still retained a perceptible trace of sulphuric acid—then dissolved in pure nitric acid, an excess of acid being carefully avoided. The solution was then re-precipitated with ammonium carbonate and caustic ammonia, and the precipitate thoroughly washed. The so-prepared perfectly pure magnesium carbonate was ignited in a platinum crucible until the weight remained constant. The residuary pure magnesia was then digested in the cold for 24 hours with distilled water, with frequent shaking. The distilled water used was perfectly free from chlorine, and left no fixed residue upon evaporation.

α. 84·82 grm. of the filtrate, cautiously evaporated in a platinum dish, left a residue weighing, after ignition, 0·0015 grm. One part of the pure magnesia dissolved therefore in.. 56546
parts of cold water.

The digestion was continued for 48 hours longer, when

β. 84·82 grm. left 0·0016 grm. One part required therefore....... 53012
γ. 84·82 grm. left 0·0015 grm. One part required................ 56546

Average.. 55368

The solution of magnesia prepared in the cold way has a feeble yet distinct alkaline reaction, which is most easily perceived upon the addition of very faintly reddened tincture of litmus; the alkaline reaction of the solution is perfectly manifest also with slightly reddened litmus paper, or with turmeric or dahlia paper, if these test-papers are left for some time in contact with the solution.

Alkali carbonates fail to render the solution turbid, even upon boiling.

Sodium phosphate also fails to impair the clearness of the solution, but if the fluid is mixed with a little ammonia and shaken, it speedily becomes turbid, and deposits after some time a perceptible precipitate of ammonium magnesium phosphate.

b. In Hot Water.

Upon boiling pure magnesia with water, a solution is obtained which comports itself in every respect like the cold-prepared solution of magnesia. A hot-prepared solution of magnesia does not become turbid upon cooling, nor does a cold-prepared solution upon boiling. 84·82 grm. of hot-prepared solution of magnesia left 0·0016 grm. MgO.

38. SOLUBILITY OF PURE MAGNESIA IN SOLUTIONS OF POTASSIUM CHLORIDE AND SODIUM CHLORIDE (to § 74, *d*).

3 flasks of equal size were charged as follows:—

1. With 1 grm. pure potassium chloride, 200 c.c. water and some perfectly pure magnesia.
2. With 1 grm. pure sodium chloride, 200 c.c. water and some pure magnesia.
3. With 200 c.c. water and some pure magnesia.

The contents of the 3 flasks were kept boiling for 40 minutes, then filtered, and the clear filtrates mixed with equal quantities of a mixture of sodium phosphate, ammonium chloride and ammonia. After 12 hours a very slight precipitation was visible in 3, and a considerably larger precipitation had taken place in 1 and 2.

39. PRECIPITATION OF ALUMINIUM BY AMMONIA, ETC. (to § 75, *a*).

a. Ammonia produces in neutral solutions of aluminium salts or of alum, as is well known, a gelatinous precipitate of aluminium hydroxide. Upon further addition of ammonia in considerable excess, the precipitate redissolves gradually, but not completely.

b. If a drop of a dilute solution of alum is added to a copious amount of ammonia, and the mixture shaken, the solution appears almost perfectly clear; however, after standing at rest for some time, slight flakes separate.

c. If a solution of aluminium, mixed with a large amount of ammonia, is filtered, and

α. The filtrate boiled for a considerable time, flakes of aluminium hydroxide separate gradually in proportion as the excess of ammonia escapes.

β. The filtrate mixed with solution of ammonium chloride, a very perceptible flocculent precipitate of aluminium hydroxide separates immediately; the whole of the aluminium present in the solution will thus separate if the ammonium chloride be added in sufficient quantity.

γ. The filtrate mixed with ammonium sesquicarbonate, the same reaction takes place as in *β*.

δ. The filtrate mixed with solution of sodium chloride or of potassium chloride, no precipitate separates, but, after several days' standing, slight flakes of aluminium hydroxide subside, owing to the loss of ammonia by evaporation.

d. If a neutral solution of aluminium is precipitated with ammonium carbonate, or if a solution strongly acidified with hydrochloric or nitric acid is precipitated with pure ammonia, or if to a neutral solution a sufficient amount of ammonium chloride is added besides the ammonia; even a considerable excess of the precipitants will fail to redissolve the precipitated aluminium hydroxide, as appears from the continued perfect clearness of the filtrates upon protracted boiling and evaporation.

40. PRECIPITATION OF ALUMINIUM BY AMMONIUM SULPHIDE (to § 75, *a*).

(*Experiments made by Mr.* J. FUCHS, *formerly Assistant in my Laboratory.*)

a. 50 c.c. of a solution of pure ammonium-alum, which contained 0·3939 Al_2O_3, were mixed with 50 c.c. water and 10 c.c. solution of ammonium sulphide, and filtered after ten minutes. The ignited precipitate weighed 0·3825 grm.

b. The same experiment was repeated with 100 c.c. water; the precipitate weighed 0·3759 grm.

c. The same experimeut was repeated with 200 c.c. water; the precipitate weighed 0·3642 grm.

41. PRECIPITATION OF CHROMIUM BY AMMONIA (to § 76, *a*).

Solutions of chromic chloride and of chrome-alum (concentrated and dilute, neutral and acidified with hydrochloric acid) were mixed with ammonia in excess. All the filtrates drawn off immediately after precipitation appeared red, but when filtered after ebullition, they all appeared colorless, if the ebullition had been sufficiently protracted.

42. SOLUBILITY OF THE BASIC ZINC CARBONATE IN WATER (to § 77, *a*).

Perfectly pure, recently (hot) precipitated basic zinc carbonate was gently heated with distilled water, and subsequently digested cold for many weeks, with frequent shaking. The clear solution gave no precipitate with ammonium sulphide, not even after long standing.

84·82 grm. left 0·0014 grm. zinc oxide, which corresponds to 0·0019 basic zinc carbonate (74 per cent. of ZnO being assumed in this salt). One part of the basic carbonate requires therefore 44642 parts of water for solution.

IN EACH OF THE THREE FOLLOWING NUMBERS THE SULPHIDE WAS PRECIPItated from the solution of the normal salt with addition of ammonium chloride by yellow ammonium sulphide, and allowed to stand in a closed vessel. After 24 hours the clear fluid was poured on to 6 filters of equal size, and the precipitate was then equally distributed among them. The washing was at once commenced and continued, without interruption, the following fluids being used:—

I. Pure water.
II. Water containing hydrogen sulphide.
III. Water containing ammonium sulphide.
IV. Water containing ammonium chloride, afterwards pure water.

V. Water containing hydrogen sulphide and ammonium chloride, afterwards water containing hydrogen sulphide.

VI. Water containing ammonium sulphide and ammonium chloride, afterwards water containing ammonium sulphide.

43. DEPORTMENT OF ZINC SULPHIDE ON WASHING (to § 77, *c*).

The filtrates were at first colorless and clear. On washing, the first three filtrates ran through turbid, the turbidity was strongest in II. and weakest in III.; the last three remained quite clear. On adding ammonium sulphide no change took place; the turbidity of the first three was not increased, the clearness of the last three was not impaired. Ammonium chloride therefore decidedly exercises a favorable action, and the water containing it may be displaced by water containing ammonium sulphide.

44. DEPORTMENT OF MANGANESE SULPHIDE ON WASHING (to § 78, *e*).

The filtrates were at first clear and colorless. But after the washing had been continued some time, I. appeared colorless, slightly opalescent; II. whitish and turbid; III. yellowish and turbid; IV. colorless, slightly turbid; V. slightly yellowish, nearly clear; VI. clear, yellowish. To obtain a filtrate that remains clear, therefore, the wash-water must at first contain ammonium chloride. Addition of ammonium sulphide also cannot be dispensed with, as all the filtrates obtained without this addition gave distinct precipitates of manganese sulphide when the reagent was subsequently added to them.

45. DEPORTMENT OF NICKEL SULPHIDE (ALSO OF COBALT SULPHIDE AND FERROUS SULPHIDE) ON WASHING (to § 79, *e*).

In the experiments with nickel sulphide the clear filtrates were put aside, and then the washing was proceeded with. The washings of the first 3 ran through turbid, of the last three clear. When the washing was finished, I. was colorless and clear; II. blackish and clear; III. dirty yellow and clear; IV. colorless and clear; V. slightly opalescent; VI. slightly brownish and opalescent. On addition of ammonium sulphide, I. became brown; II. remained unaltered; III. remained unaltered; IV. became black and opaque; V. became brown and clear; VI. became pure yellow and clear.

Cobalt sulphide and ferrous sulphide behaved in an exactly similar manner. It is plain that these sulphides oxidize more rapidly when the wash-water contains ammonium chloride, unless ammonium sulphide is also present. Hence it is necessary to wash with a fluid containing ammonium sulphide; and the addition of ammonium chloride at first is much to be recommended, as this diminishes the likelihood of our obtaining a muddy filtrate.

46. DEPORTMENT OF COBALTOUS HYDROXIDE PRECIPITATED BY ALKALIES (to § 80, *a*).

A solution of cobaltous chloride was precipitated boiling with solution of soda, and the precipitate washed with boiling water until the filtrate gave no longer the least indication of presence of chlorine. The dried and ignited residue, heated with water, manifested no alkaline reaction. It was reduced by ignition in hydrogen gas, and the metallic cobalt digested hot with water. The decanted water manifested no alkaline reaction, even after considerable concentration; but the metallic cobalt, brought into contact, moist, with turmeric paper, imparted to the latter a strong brown color.

47. Solubility of Lead Carbonate (to § 83, *a*).

a. In pure Water.

Recently precipitated and thoroughly washed pure lead carbonate was digested for 8 days with water at the common temperature, with frequent shaking. 84·42 grm. of the filtrate were evaporated with addition of some pure sulphuric acid; the residuary lead sulphate weighed 0·0019 grm., which corresponds to 0·00167 lead carbonate. One part of the latter salt dissolves therefore in 50551 parts of water. The solution, mixed with hydrogen sulphide water, remained perfectly colorless, not the least tint being detected in it, even upon looking through it from the top of the test-cylinder.

b. In Water containing a little Ammonium Acetate and also Ammonium Carbonate and Ammonia.

A highly dilute solution of pure lead acetate was mixed with ammonium carbonate and ammonia in excess, and the mixture gently heated and then allowed to stand at rest for several days. 84·42 grm. of the filtrate left, upon evaporation with a little sulphuric acid, 0·0041 grm. lead sulphate, which corresponds to 0·0036 of the carbonate. One part of the latter salt requires accordingly 23450 parts of the above fluid for solution. The solution was mixed with hydrogen sulphide water; when looking through the fluid from the top of the test-cylinder, a distinct coloration was visible; but when looking through the cylinder laterally, this coloration was hardly perceptible. Traces of lead sulphide separated after the lapse of some time.

c. In Water containing a large proportion of Ammonium Nitrate, together with Ammonium Carbonate and Caustic Ammonia.

A highly dilute solution of lead acetate was mixed with nitric acid, then with ammonium carbonate and ammonia in excess; the mixture was gently heated, and allowed to stand at rest for 8 days. The filtrate, mixed with hydrogen sulphide, exhibited a very distinct brownish color upon looking through it from the top of the cylinder; but this color appeared very slight only when looking through the cylinder laterally. The amount of lead dissolved was unquestionably more considerable than in *b*.

48. Solubility of Lead Oxalate (to § 83, *b*).

A dilute solution of lead acetate was precipitated with ammonium oxalate and ammonia, the mixture allowed to stand at rest for some time, and then filtered. The filtrate, mixed with hydrogen sulphide, comported itself exactly like the filtrate of No. 47, *b*. The same deportment was observed in another similar experiment, in which ammonium nitrate had been added to the solution.

49. Solubility of Lead Sulphate in Pure Water (to § 83, *d*).

Thoroughly washed and still moist lead sulphate was digested for 5 days with water, at 10—15°, with frequent shaking. 84·42 grm. of the filtrate (filtered off at 11°) left 0·0037 grm. lead sulphate. Consequently 1 part of this salt requires 22816 parts of pure water of 11° for solution.

The solution, mixed with hydrogen sulphide, exhibited a distinct brown color when viewed from the top of the cylinder, but this color appeared very slight upon looking through the cylinder laterally.

50. Solubility of Lead Sulphate in Water containing Sulphuric Acid (to § 83, *d*).

A highly dilute solution of lead acetate was mixed with an excess of dilute pure sulphuric acid; the mixture was very gently heated, and the precipitate allowed several days to subside. 80·31 grm. of the filtrate left 0·0022 grm. lead sulphate. One part of this salt dissolves therefore in 36504 parts of water containing sulphuric acid. The solution, mixed with hydrogen sulphide, appeared colorless to the eye looking through the cylinder laterally, and very little darker when viewed from the top of the cylinder.

51. Solubility of Lead Sulphate in Water containing Ammonium Salts and free Sulphuric Acid (to § 83, *d*).

A highly dilute solution of lead acetate was mixed with a tolerably large amount of ammonium nitrate, and sulphuric acid in excess added. After several days' standing, the mixture was filtered. The filtrate was nearly indifferent to hydrogen sulphide water; viewed from the top of the cylinder, it looked hardly perceptibly darker than pure water.

52. Deportment of Lead Sulphate upon Ignition (to § 83, *d*).

Speaking of the determination of the atomic weight of sulphur, Erdmann and Marchand* state that lead sulphate loses some sulphuric acid upon ignition. In order to inform myself of the extent of this loss, and to ascertain how far it might impair the accuracy of the method of determining lead as a sulphate, I heated 2·2151 grm. of absolutely pure $PbSo_4$ to the most intense redness, over a spirit-lamp with double draught. I could not perceive the slightest decrease of weight; at all events, the loss did not amount to 0·0001 grm.

53. Deportment of Lead Sulphide on Drying at 100° (to § 83, *f*).

Lead sulphide was precipitated from a solution of pure lead acetate with hydrogen sulphide, and when dry, kept for a considerable time at 100° and weighed occasionally. The following numbers represent the results of the several weighings:—

I. 0·8154. II. 0·8164. III. 0·8313. IV. 0·8460. V. 0·864

54. Deportment of Metallic Mercury at the Common Temperature and upon Ebullition with Water (to § 84, *a*).

To ascertain in what manner loss of metallic mercury occurs upon drying, and likewise upon boiling with water, and to determine which is the best method of drying, I made the following experiments:—

I treated 6·4418 grm. of perfectly pure mercury in a watch-glass, with distilled water, removed the water again as far as practicable (by decantation and finally by means of blotting-paper), and weighed. I now had 6·4412 grm. After several hours' exposure to the air, the mercury was reduced to 6·4411. I placed these 6·4411 grm. under a bell-jar over sulphuric acid, the temperature being about 17°. After the lapse of 24 hours the weight had not altered in the least. I introduced the 6·4411 grm. mercury into a flask, treated it with a copious quantity of distilled water, and boiled for 15 minutes violently. I then placed the mercury again upon the watch-glass, dried it most carefully with blotting-

* Journ. für Prakt. Chem. 31, 385.

paper, and weighed. The weight was now 6·4402 grm. Finding that a trace of mercury had adhered to the paper, I repeated the same experiment with the 6·4402 grm. After 15 minutes' boiling with water, the mercury had again lost 0·0004 grm. The remaining 6·4398 grm. were exposed to the air for 6 days (in summer, during very hot weather), after which they were found to have lost only 0·0005 grm.

55. DEPORTMENT OF MERCURIC SULPHIDE WITH SOLUTION OF POTASSA, AMMONIUM SULPHIDE, ETC. (to § 84, *c*).

a. If recently precipitated pure mercuric sulphide is boiled with pure solution of potassa, not a trace of it dissolves in that fluid; hydrochloric acid produces no precipitate, nor even the least coloration, in the filtrate.

b. If mercuric sulphide is boiled with solution of potassa, with addition of some hydrogen sulphide water, ammonium sulphide, or sulphur, complete solution is effected.

c. If freshly precipitated mercuric sulphide is digested in the cold with yellowish or very yellow ammonium sulphide, slight but distinctly perceptible traces are dissolved, while in the case of hot digestion scarcely any traces of mercury can be detected in the solution.*

d. Thoroughly washed mercuric sulphide, moistened with water, suffers no alteration upon exposure to the air; at least, the fluid which I obtained by washing mercuric sulphide which had been thus exposed for 24 hours, did not manifest acid reaction, nor did it contain mercury or sulphuric acid.

56. DEPORTMENT OF CUPRIC OXIDE UPON IGNITION (to § 85, *b*).

Pure cupric oxide (prepared from cupric nitrate) was ignited in a platinum crucible, then cooled under a bell-jar over sulphuric acid, and finally weighed. The weight was 3·542 grm. The oxide was then most intensely ignited for 5 minutes, over a BERZELIUS' lamp, and weighed as before, when the weight was found unaltered; the oxide was then once more ignited for 5 minutes, but with the same result.

57. DEPORTMENT OF CUPRIC OXIDE IN THE AIR (to § 85, *b*).

A platinum crucible containing 4·3921 grm. of gently ignited cupric oxide (prepared from the nitrate) stood for 10 minutes, covered with the lid, in a warm room (in winter); the weight of the oxide was found to have increased to 4·3939 grm.

The oxide was then intensely ignited over a spirit-lamp; after 10 minutes' standing in the covered crucible, the weight had not perceptibly increased; after 24 hours' it had increased by 0·0036 grm.

58. DEPORTMENT OF BISMUTH SULPHIDE UPON DRYING AT 100° (to § 86, *g*).

0 4558 grm. of bismuth sulphide prepared in the wet way were placed in the desiccator on a watch-glass, and allowed to stand at the common temperature. After 3 hours the weight was 0·4270, after 6 hours 0·4258, after 2 days the same.

0·3602 grm. of the bismuth sulphide so dried was put into a water-bath, in 15 minutes it weighed 0·3596, half an hour afterwards 0·3599, in half an hour

* Comp. my experiments in the Zeitschrift f. Anal. Chem. 3, 140.

more 0·3603, in two hours 0·3626. In a second experiment the drying was kept up for 4 days, and a continual increase of weight was observed.

0·5081 grm. of bismuth sulphide dried in the desiccator was heated in a boat in a stream of carbonic acid. After gentle ignition the weight was 0·5002, after repeated heating 0·4992. The bismuth sulphide was visibly volatilized on ignition in the current of carbonic acid.

59. DEPORTMENT OF CADMIUM SULPHIDE WITH AMMONIA, ETC. (to § 87, *c*).

Recently precipitated pure cadmium sulphide was diffused through water, and the following experiments were made with the mixture:

a. A portion was digested cold with ammonia in excess, and filtered. The filtrate remained perfectly clear upon addition of hydrochloric acid.

b. Another portion was digested hot with excess of ammonia, and filtered. This filtrate likewise remained perfectly clear upon addition of hydrochloric acid.

c. Another portion was digested for some time with solution of potassium cyanide, and filtered. This filtrate also remained perfectly clear upon addition of hydrochloric acid.

d. Another portion was digested with ammonium hydrosulphide, and filtered. The turbidity which hydrochloric acid imparted to this filtrate was pure white.

(A remark made by WACKENRODER, in BUCHNER'S Repertor. d. Pharm., xlvi. 226, induced me to make these experiments.)

60. DEPORTMENT OF PRECIPITATED ANTIMONIOUS SULPHIDE ON DRYING (to § 90, *a*).

0·2899 grm. of pure precipitated antimonious sulphide dried in the desiccator lost, when dried at 100°, 0·0007.

0·4457 grm. of the substance dried at 100° lost, when heated to blackening in a stream of carbonic acid, 0·0011 water.

0·1932 grm. of the substance dried at 100° gave up 0·0012, when heated to blackening in a stream of carbonic acid, and after stronger heating, during which fumes of antimony sulphide began to escape, the total loss amounted to 0·0022 grm.

0·1670 grm. of the substance dried at 100° lost 0·0005 grm. on being heated to blackening in a stream of carbonic acid.

61. AMOUNT OF WATER IN HYDRATED SILICA (to § 93, 9).

(*Experiments made by my assistant, Mr.* LIPPERT.)

A dilute solution of soluble glass was slowly dropped into hydrochloric acid, as long as the precipitate continued to dissolve rapidly, then the clear fluid was heated in the water-bath, till it set to a transparent jelly. This jelly was dried as far as possible with blotting paper, diffused in water, and washed by decantation till the fluid altogether ceased to give the chlorine reaction. It was then transferred to a filter, and the latter spread on blotting paper and exposed till a crumbly mass was left from the spontaneous evaporation of water. One half (I.) was dried for 8 weeks in the desiccator over sulphuric acid, with occasional trituration, the other half (II.) was dried under similar circumstances, but in a vacuum. Both were transferred to closed tubes and these were kept in the desiccator.

The weighing of the substance dried at 100° was effected between watch glasses. For the purpose of igniting the residue, it was allowed to satiate itself with aqueous vapor by exposure to the air, otherwise a considerable quantity of the substance would have been lost, then water was dropped upon it in the watch-glass, then it was rinsed into a platinum crucible, dried in a water-bath, and ignited, at first cautiously, towards the end intensely.

The substance I. contained	Expt. 1.	Expt. 2.
Water, escaping at or below 100°	4·19	9·28
" " above 100°	4·76	
Silica	91·05	90·72
	100·00	100·00

Consequently the hydrate dried at 100° consists of 4·97 water and 95·03 silica. In the substance dried in the desiccator the oxygen of the total water: the oxygen of the silica (SiO_2), according to the first experiment : : 1 : 6·1, according to the second experiment : : 1 : 5·86. And in the substance dried at 100° the oxygen of the water : the oxygen of the silica : : 1 : 11·5.

The substance II. contained	Expt. 1.	Expt. 2.	Expt. 3.
Water, escaping at or below 100°	4·75	4·71	9·95
" " above 100°	5·26	5·21	
Silica	89·99	90·08	90·05
	100·00	100·00	100·00

Consequently the hydrate dried at 100° consists on the average of 5·49 water and 94·51 silica. In the substance dried in a vacuum over sulphuric acid the oxygen of the total water : the oxygen of the silica—on an average : : 1 : 5·41. And in the substance dried at 100° the oxygen of the water : the oxygen of the silica : : 1 : 10·43.

62. DETERMINATION OF BARIUM BY PRECIPITATION WITH AMMONIUM CARBONATE (to § 101, 2, *a*).

0·7553 grm. pure ignited barium chloride precipitated after § 101, 2, *a*, gave 0·7142 $BaCO_3$, which corresponds to 0·554719 BaO = 73·44 per cent. (100 parts of $BaCl_2$ ought to have given 73·59 parts). The result accordingly was 99·79 instead of 100.

63. DETERMINATION OF BARIUM IN ORGANIC SALTS (to § 101, 2, *b*).

0·686 grm. barium racemate, treated according to § 101, 2, *b*, gave 0·408 barium carbonate = 0·3169 BaO = 46·20 per cent. (calculated 46·38 per cent.); *i.e.*, 99·61 instead of 100.

64. DETERMINATION OF STRONTIUM AS STRONTIUM SULPHATE (to § 102, 1, *a*).

a. An aqueous solution of 1·2398 grm. $SrCl_2$ was precipitated with sulphuric acid in excess, and the precipitated strontium sulphate washed with water. It weighed 1·4113, which corresponds to 0·795408 SrO = 64·15 per cent. (calculated 65·38 per cent.); *i.e.*, 98·12 instead of 100.

b. 1·1510 grm. $SrCo_3$ was dissolved in excess of hydrochloric acid, the solution diluted, and then precipitated with sulphuric acid; the precipitated $SrSO_4$ was washed with water; it weighed 1·4024 = 0·79039 SrO = 68·68 per cent. (calculated 70·07 per cent.); *i.e.*, 98.02 instead of 100.

65. DETERMINATION OF STRONTIUM AS SULPHATE, WITH CORRECTION (to § 102, 1, *a*).

The *filtrate* obtained in No. 64, *b*, weighed 190·84 grm. According to experiment No. 22, 11862 parts of water containing sulphuric acid dissolve 1 part of strontium sulphate; therefore, 190·84 grm. dissolve 0·0161. The *washings* weighed 63·61 grm. According to experiment No. 21, 6895 parts of water dissolve 1 part of $SrSO_4$; therefore, 63·61 grm. dissolve 0·0092 grm.

Adding 0·0161 and 0·0092 to the 1·4024 actually obtained, we find the total amount = 1·4277 grm., which corresponds to 0·80465 SrO = 69·91 per cent. in $SrCO_3$ (calculated 70·07 per cent.); *i.e.*, 99·77 instead of 100.

66. DETERMINATION OF STRONTIUM AS STRONTIUM CARBONATE (to § 102, 2).

1·3104 grm. strontium chloride, precipitated according to § 102, 2, gave 1·2204 $SrCO_3$, containing 0·8551831 SrO = 65·26 per cent. (calculated 65·38); *i.e.*, 99·82 instead of 100.

IN THE FOUR FOLLOWING EXPERIMENTS, AND ALSO IN No. 72, PURE AIR-dried calcium carbonate was used, in a portion of which the amount of anhydrous carbonate had been determined by very cautious heating. 0·7647 grm. left 0·7581 grm., which weight remained unaltered upon further (extremely gentle) ignition; *the air-dried carbonate contained accordingly* 55·516 *per cent. of lime.*

67. DETERMINATION OF CALCIUM AS CALCIUM SULPHATE BY PRECIPITATION (to § 103, 1, *a*).

1·186 grm. of "the air-dried calcium carbonate" was dissolved in hydrochloric acid, and the solution precipitated with sulphuric acid and alcohol, after § 103, 1, *a*. Obtained 1·5949 grm. $CaSO_4$, containing 0·65598 CaO, *i.e.*, 55·31 per cent. (calculated 55·51), which gives 99·64 instead of 100.

68. DETERMINATION OF CALCIUM AS $CaCO_3$, BY PRECIPITATION WITH AMMONIUM CARBONATE AND WASHING WITH PURE WATER (to § 103, 2, *a*).

A hydrochloric acid solution of 1·1437 grm. of "the air-dried calcium carbonate" gave upon precipitation as directed, 1·1243 grm. anhydrous calcium carbonate, corresponding to 0·629608 CaO = 55·05 per cent. (calculated 55·51 per cent.), which gives 99·17 instead of 100.

69. DETERMINATION OF CALCIUM AS $CaCO_3$, BY PRECIPITATION WITH AMMONIUM OXALATE FROM ALKALINE SOLUTION (to § 103, 2 *b*, *α*).

1·1734 grm. of "the air-dried calcium carbonate" dissolved in hydrochloric acid, and treated as directed § 103, 2, *b*, *a*, gave 1·1632 grm. $CaCO_3$ (reaction not alkaline), containing 0·651392 of CaO = 55·513 per cent. (calculated 55·516 per cent.), which gives 99·99 instead of 100.

70. DETERMINATION OF CALCIUM AS OXALATE (to § 103, 2, *b*, *α*).

0·857 grm. of "the air-dried calcium carbonate" were dissolved in hydrochloric acid; the solution was precipitated with ammonium oxalate and ammonia, the precipitate washed, and then dried at 100°, until the weight remained constant. The precipitate ($CaC_2O_4 + H_2O$) weighed 1·2461 grm. containing 0·477879 CaO = 55·76 per cent. (calculated 55·516 per cent.), which gives 100·45 instead of 100.

71. Volumetric Determination of Calcium Precipitated as Oxalate (to § 103, 2, *b*, *α*).

Six portions, of 10 c.c. each, were taken of a solution of pure calcium chloride; in 2 portions the calcium was determined in the gravimetric way (by precipitation with ammonium oxalate, and weighing $CaCO_3$); in two by the alkalimetric method (p. 236), and in two by precipitation with ammonium oxalate, and estimation of the oxalic acid in the precipitate by solution of potassium permanganate. The following were the results obtained:

a. In the gravimetric way.	*b* By the alkalimetric method.	*c.* By solution of potassium permanganate.
0·5617 $CaCO_3$	0·5614	0·5613
0·5620 "	0·5620	0·5620

72. Determination of Calcium as $CaCO_3$ by Precipitation as Calcium Oxalate from Acid Solution (to § 103, 2, *b*, *β*).

0·857 grm. of "the air-dried calcium carbonate" dissolved in hydrochloric acid and precipitated from this solution according to the directions of § 103, 2, *b*, *β*, gave 0·8476 calcium carbonate (which did not manifest alkaline reaction, and the weight of which did not vary in the least upon evaporation with ammonium carbonate), containing 0·474656 CaO = 55·39 per cent. (calculated 55·51), which gives 99·78 instead of 100.

73. Determination of Magnesium as $Mg_2P_2O_7$ (to § 104, 2).

a. A solution of 1·0587 grm. pure anhydrous magnesium sulphate in water, precipitated according to § 104, 2, gave 0·9834 magnesium pyrophosphate, containing 0·35438 MgO = 33·476 per cent. (calculated 33·33 per cent.), which gives 100·43 instead of 100.

b. 0·9672 $MgSO_4$ gave 0·8974 $Mg_2P_2O_7$ = 33·43 per cent. of MgO (calculated 33·33, which gives 100·30 instead of 100.

74. Precipitation of Zinc Acetate by Hydrogen Sulphide (to § 108, *b*).

a. A soluble of pure zinc acetate was treated with the gas in excess, allowed to stand at rest for some time, and then filtered. The filtrate was mixed with ammonia. It remained perfectly clear at first, and even after long standing a few hardly visible flakes only had separated.

b. A solution of zinc acetate to which a tolerably large amount of acetic acid had been added previously to the precipitation with hydrogen sulphide, showed exactly the same deportment.

75. Determination of Iron as Sulphide (to § 113, 2).

10 c.c. of a pure solution of ferric chloride was precipitated with ammonia; obtained 0·1453 Fe_2O_3 = 0·10171 Fe.

10 c.c. was precipitated with ammonia and ammonium sulphide, and treated after § 113, 2, obtained 0·1596 FeS = 0·10157 Fe.

10 c.c. again yielded 0.1605 FeS = 0·1021 Fe.

76. DETERMINATION OF LEAD AS CHROMATE (to § 116, 4).

1·0083 grm. pure lead nitrate were treated according to § 116, 4. The precipitate was collected on a weighed filter, and dried at 100°, obtained 0·9871 grm. = 0·67833 PbO. This gives 67·3 per cent. Calculation 67·4.

0·9814 lead nitrate again yielded 0·9625 chromate = 67.4 per cent.

77. DETERMINATION OF MERCURY IN THE METALLIC STATE, IN THE WET WAY, BY MEANS OF STANNOUS CHLORIDE (to § 118, 1, *b*).

2·01 grm. mercuric chloride gave 1·465 grm. metallic mercury = 72·88 per cent. (calculated 73·83 per cent.), which gives 98·71 instead of 100 (SCHAFFNER). The loss is not inherent in the method, *i.e.*, it does not arise from mercury evaporating during the ebullition and desiccation (Expt. No. 54); but its origin lies in the fact that one usually does not allow sufficient time for the mercury to settle quite completely, and in general is not careful enough in decanting, and drying with paper, &c.

78. DETERMINATION OF COPPER BY PRECIPITATION WITH ZINC IN A PLATINUM DISH (to § 119, 2).

30·882 grm. pure cupric sulphate were dissolved in water to 250 c.c.; 10 c.c. of the solution contained accordingly 0·31387 grm. metallic copper.

a. 10 c.c. precipitated with zinc in a platinum dish, gave 0·3140 = 100·06 per cent.

b. In a second experiment 10 c.c. gave 0·3138 = 100 per cent.

80. DETERMINATION OF COPPER AS CUPROUS SULPHOCYANATE (to § 119, 3, *b*).

0·5965 grm. of pure cupric sulphate was dissolved in a little water, and, after addition of an excess of sulphurous acid, precipitated with potassium sulphocyanate. The well-washed precipitate, dried at 100°, weighed 0·2893, corresponding to 0·1892 CuO = 31·72 per cent. As cupric sulphate contains 31·83 per cent., this gives 99·66 instead of 100.

81. DETERMINATION OF COPPER BY DE HAEN'S METHOD (to § 119, 4, *a*).

Four 10 c.c.'s of a solution of cupric sulphate, each 10 c.c., containing 0·0254 grm. Cu, were severally mixed with potassium iodide, then with 50 c.c. of a solution of sulphurous acid (50 c.c. corresponding to 12·94 c.c. iodine solution). After addition of starch paste, iodine solution was added until the fluid appeared blue.

This required,—

In *a*, 4·09
b, 3·95
c, 4·06
d, 3·95

As 100 c.c. of iodine solution contained 0·58043 grm. iodine, this gives—

For *a*, 0·0256 Cu instead of 0·0254
" *b*, 0·0260 " "
" *c*, 0·0257 " "
" *d*, 0·0260 " "

Another experiment, made with 100 c.c. of the same solution of cupric sul phate, gave 0·2606 instead of 0·254 of copper. Ammonium nitrate having been added to 10 c.c. of the solution of cupric sulphate, then some dilute hydrochloric acid, 3·4 and 3·5 c.c. of iodine solution were required instead of 4 c.c.—a proof that considerably more iodine had separated than corresponded to the copper.

83. Precipitation of Bismuth Nitrate by Ammonium Carbonate (to § 120, 1, *a*).

If a solution of bismuth nitrate, no matter whether containing much or little free nitric acid, is mixed with water, precipitated with ammonium carbonate and ammonia, and filtered without applying heat, the filtrate acquires, upon addition of hydrogen sulphide water, a blackish-brown color. But if the mixture before filtering is heated for a short time nearly to boiling, hydrogen sulphide fails to impart this color to the filtrate, or, at all events, the change of color is hardly visible to the eye looking through the full test-tube from the top.

84. Determination of Antimony as Sulphide (to § 125, 1).

0·559 grm. of pure air-dried tartar emetic, treated according to § 125, 1, gave 0·2902 grm. antimonious sulphide dried at 100° = ·2492 grm. or 44·58 per cent. of antimonious oxide. Heated to blackening in a current of carbonic acid, the precipitate lost 0·0079 grm. (reckoned from a part to the whole), leaving accordingly 0·2823 grm. of anhydrous antimonious sulphide, which corresponds to 0·24245 grm., or 43·37 per cent. of antimonious oxide. As the tartar emetic contains 43·70 per cent. of antimonious oxide, the process gives, if the precipitate is dried at 100°, 102·01; if heated to blackening, 99·22 instead of 100.

89. Determination of Phosphoric Acid as Magnesium Pyrophosphate (to § 134, *b*, *α*).

1·9159 and 2.0860 grm. pure crystallized sodium hydrogen phosphate, treated as directed § 134, *b*, *α*, gave 0·5941 and 0·6494 grm. of magnesium pyrophosphate respectively. These give 19·83 and 19·91 per cent. of P_2O_5 in sodium hydrogen phosphate, instead of 19·83 per cent.

90. Determination of Phosphoric Acid as Uranyl Pyrophosphate (to § 134, *c*).

30 c.c. of a solution of pure sodium hydrogen phosphate, treated with magnesium sulphate, ammonium chloride, and ammonia, as directed § 134, *b*, *α*, gave 0·3269 grm. of magnesium pyrophosphate. 10 c.c. contained accordingly 0·06982 grm. of phosphoric anhydride.

10 c.c. of the same solution were then precipitated with uranyl acetate as directed § 134, *c*. The ignited precipitate was treated with a little nitric acid, then again ignited; after cooling, it weighed 0·3478 grm. corresponding to 0·06954 grm. of phosphoric anhydride.

91. Determination of Free Hydrogen Sulphide by Means of Solution of Iodine (to § 148, I., *a*).

The experiments were made to settle the following questions:—

a. Does the quantity of iodine required remain the same for solutions of hydrogen sulphide of different degrees of dilution?

b. Does the equation $H_2S + I_2 = 2HS + S$ really represent the decomposition which takes place?

The hydrogen sulphide water was contained in a flask closed by a doubly-perforated cork; into one aperture a siphon with pinchcock was fitted, to draw off the fluid; into the other aperture a short open tube, which did not dip into the fluid.

Question *a*.

α. About 30 c.c. of iodine solution were introduced into a flask, which was then tared; hydrogen sulphide water was added until the yellow color had just disappeared. The flask was then closed, weighed, starch paste added, and then solution of iodine until the fluid appeared blue.

70·2 grm. H_2S water required 23·4 c.c. iodine solution, 100 accordingly 33·33 c.c.

68·4 grm. required 22·7 c.c. iodine solution, 100 accordingly 33·20 c.c.

β. Same process; but the fluid was diluted with water free from air.

61·5 grm. H_2S water + 200 grm. water required 20·7 c.c. iodine solution, 100 accordingly 33·65 c.c.

52·4 grm. + 400 grm. water required 17.7 c.c. iodine solution, 100 accordingly 33·77.

The iodine solution contained 0·00498 iodine in 1 c.c. Considering that addition of a larger volume of water necessarily involves a slight increase in the quantity of iodine solution, these results may be considered sufficiently corresponding.

Question *b*.

According to *a*, 100 grm. of the H_2S water contained 0·02215 grm. H_2S, assuming the proportion to be 100 : 33·2.

173·6 grm. of the same water were, immediately after the experiments in *a*, drawn off into a hydrochloric acid solution of arsenious acid; after 24 hours, the arsenious sulphide was filtered off, dried at 100°, and weighed. 0·0920 grm. were obtained, which corresponds to 0·03814 H_2S, or a percentage of 0·02197.

The second question also is therefore answered in the affirmative.

92. Solution of Magnesium Chloride dissolves Calcium Oxalate (to § 154, 6).

If some calcium chloride is added to a solution of magnesium chloride, then a little ammonium oxalate, no precipitate is formed at first; but upon slightly increasing the quantity of ammonium oxalate, a trifling precipitate gradually separates after some time.

If an excess of ammonium oxalate is added, the whole of the calcium is thrown down, but the precipitate contains also magnesium oxalate. This shows that to effect the separation of the two bases by ammonium oxalate, the reagent must be added in excess; whilst, on the other hand, in the presence of much magnesium, the operator must expect to precipitate some of the magnesium, as the following experiments (No. 93) clearly show.

93. Separation of Calcium from Magnesium (to § 154, 6).

The fluids employed in the following experiments were, a solution of calcium chloride, 10 c.c. of which corresponded to 0·5618 $CaCO_3$; a solution of magnesium chloride, containing 0·250 MgO in 10 c.c.; a solution of ammonium chloride (1 : 8); solution of ammonia, containing 10 per cent. NH_3; solution of ammonium oxalate (1 : 24); acetic acid, containing 30 per cent. $C_2H_4O_2$.

The precipitation was effected at the common temperature; the precipitate of calcium oxalate was filtered off after 20 hours.

a. Influence of the degree of dilution.

α. 10 c.c. $MgCl_2$, 10 c. c. $CaCl_2$, 10 c. c. NH_4Cl, 4 drops NH_4OH, 50 c.c. water, 20 c. c. $(NH_4)_2C_2O_4$. Result, 0·5705 $CaCO_3$.

β. Same as α, with 150 c.c. water instead of 50 c.c. Result, 0·5670 $CaCO_3$.

b. Influence of excess of ammonia.

Same as *a*, β + 10 c.c. NH_4OH. Result, 0·5614 grm. $CaCO_3$.

c. Influence of excess of ammonium chloride.

Same as *a*, β + 40 c.c. NH_4Cl. Result, 0·5652 grm.

d. Influence of excess of ammonia and ammonium chloride.

Same as *a*, β + 30 c.c. NH_4Cl + 10 c.c. NH_4OH. Result, 0·5613 grm.

e. Influence of free acetic acid.

Same as *a*, β, only with 6 drops $C_2H_4O_2$, instead of the 4 drops NH_4OH. Result, 0·5594 grm.

f. Influence of excess of ammonium oxalate in feebly alkaline solution.

Same as *a*, β + 20 c.c. $(NH_4)_2C_2O_4$. Result, 0·5644 grm. $CaCO_3$.

g. Influence of excess of ammonium oxalate in strongly alkaline solution.

Same as *a*, β, + 10 c.c. NH_4OH + 20 c.c. $(NH_4)_2C_2O_4$. Result, 0·5644.

h. Influence of excess of ammonium oxalate in presence of much NH_4Cl and NH_4OH.

Same as *a*, β, + 10 NH_4OH + 30 NH_4Cl + 20 $(NH_4)_2C_2O_4$. Result, 0·5709 grm.

i. Influence of excess of ammonium oxalate in solution slightly acidified with $C_2H_4O_2$.

Same as *a*, β, − 4 drops NH_4OH + 6 drops $C_2H_4O_2$ + 20 c.c. $(NH_4)_2C_2O_4$. Result, 0·5661 grm.

Consequently, when a notable amount of magnesium is present there is always a chance of magnesium oxalate, or ammonium magnesium oxalate precipitating along with the calcium oxalate.

Another series of experiments in which a solution of magnesium oxalate in hydrochloric acid was mixed with ammonia under varying circumstances, proved also that, in presence of a notable quantity of magnesium, magnesium oxalate, or magnesium ammonium oxalate, will always separate after standing for some time, no matter whether in a cold or a warm place.

In a third series of experiments, the separation was effected by double precipitation, in accordance with § 154, 28. The same solutions were employed as in the first series, with the exception of the magnesium chloride, for which a solution was substituted containing 0·2182 grm. MgO, in 10 c.c.

10 c.c. $CaCl_2$ + 30 c.c. $MgCl_2$ + 20 c.c. NH_4Cl, + 300 c.c. water, + 6 drops ammonia, + a sufficient excess of ammonium oxalate. Results, in two experiments, 0·5621 and 0·5652, mean 0·5636, instead of 0·5618 $CaCO_3$; also 0·6660 and 0·6489 MgO, mean 0·6574, instead of 0·6546.

94. Separation of Iodine from Chlorine by Pisani's Method.

0·2338 grm. potassium iodide, dissolved in water, + ½ c.c. of solution of iodide of starch, required 14 c.c. of decinormal silver solution = 0·2322 grm. potassium iodide.

0·3025 grm. potassium iodide, mixed with about double the quantity of sodium chloride, required 18·2 c.c. silver solution = 0·3021 KI.

0·2266 grm. potassium iodide, mixed with about 100 times as much sodium chloride, required 13·7 c.c. silver solution = 0·2272 KI.

95. Separation of Iodine from Bromine, by Pisani's Method.

0·3198 grm. potassium iodide, mixed with double the quantity of potassium bromide, required 19·2 c.c. of decinormal silver solution = 0·3187 KI.

99. Chlorimetrical Experiments (to § 199).

10 grm. of chloride of lime were rubbed up with water to one litre, with which the following experiments were made:

a. By Penot's method (§ 200); obtained 23·5 and 23·5 per cent.

b. By means of iron (§ 201, modification); obtained 23·6 per cent.

c. By Bunsen's method (§ 201); results, 23·6—23·6 per cent.

100. Drying of Manganese (to § 202, I.).

Four small pans, containing each 8 grm. of manganese of 53 per cent., were first heated in the water-bath. After 3 hours, I. had lost 0·145; after 6 hours, II. 0·15; after 9 hours, III. 0·15; after 12 hours, IV. 0·15 grm. I. and II. having been left standing, loosely covered, in the room for 12 hours, II. was found to weigh exactly as much as at first; I. wanted only 0·01 grm. of the original weight.

The four pans were now heated for 2 hours to 120°. After cooling, they were found to have lost each 0·180 of the original weight. I. and II. having been left standing, loosely covered, in the room for 60 hours, were found to have again acquired their original weight by attracting moisture. III. and IV. were heated for 2 hours to 150°. The loss of weight in both cases was 0·215 grm. Having been left standing, loosely covered, in the room for 72 hours, both were found to weigh 0·05 less than at first. Assuming the hygroscopic moisture expelled to be reabsorbed by standing in the air, this shows that at 150° a little chemically combined water escapes along with the moisture, and accordingly that the temperature must not exceed 120°.

My experiments will be found described in detail in Dingler's polyt. Journ., 135, 277 et seq.

CALCULATION OF ANALYSES.

THE calculation of the results obtained by an analysis presupposes, as an indispensable preliminary, a knowledge of the general laws of the combining proportions of bodies, on the one hand, and of the more simple rules of arithmetic on the other. It is a great error to suppose that the ability to make chemical calculations involves an extensive acquaintance with mathematics, a knowledge of decimal fractions and simple equations being for the most part sufficient. These remarks are not intended to dissuade students of chemistry from pursuing the highly important study of mathematics; but merely to encourage those who have had no opportunity of entering more deeply into this science, and who, as experience has shown me, are often afraid to venture upon chemical calculations. For this reason, I have made the whole of the calculations given in the following paragraphs, in the most intelligible manner possible, and without logarithms.

I. *Calculation of the Constituents sought from the Compound obtained in the Analytical Process, and exhibition of the Result in Per-cents.*

The bodies the weight of which it is intended to determine, are separated, as we have seen in Division I., treating of the "Execution of Analysis," either in the free state, or—and this most frequently—in combinations of known composition. The results are usually calculated upon 100 parts of the examined substance, since this gives a clearer and more intelligible view of the composition. In cases where the several constituents have been separated in the free state, the calculation may be made at once; but if the constituents have been separated in combination with other substances, they must first be calculated from the compounds obtained.

1. *Calculation of the Results into Per-cents by Weight, in Cases where the Substance sought has been separated in the Free State.*

a. Solid Bodies, Liquids, and Gases, which have been determined by Weight.

The calculation here is exceedingly simple.

Suppose you have analyzed mercurous chloride, and separated the mercury in the metallic state. 2·945 grm. mercurous chloride have given say 2·499 grm. metallic mercury.

$$2{\cdot}945 : 2{\cdot}499 :: 100 : x$$
$$x = 84{\cdot}85,$$

which means that your analysis shows 100 parts of mercurous chloride to contain 84·85 of mercury, and consequently 15·15 of chlorine.

Now as mercurous chloride is known to consist of 2 at. mercury and 2 at. chlorine, and as the atomic weights of both these elements are also known, the true percentage composition of the body may be readily calculated from these data. When analyzing substances of known composition for practice, the results theoretically calculated and those obtained by the analysis are usually placed in juxtaposition, as this enables the student at once to perceive the degree of accuracy with which the analysis has been performed.

Thus for instance—

	Found.	Calculated (compare § 84, b).
Mercury..........	84·85	84·94
Chlorine..........	15·15	15·06
	100·00	100·00

b. Gases which have been determined by Measure.

If a gas has been determined by measure, it is, of course, necessary first to ascertain the weight corresponding to the volume found, before the percentage by weight can be calculated.

But as the exact weights of a definite volume of the various gases have been severally determined by accurate experiments, this calculation also is a simple rule-of-three question, if the gas may be measured under the same circumstances to which the known relation of weight to volume refers. The circumstances to be taken into consideration here, are:

Temperature and Atmospheric Pressure.

Besides these, the

Tension of the Aqueous Vapor

may also claim consideration in cases where water is used as the confining fluid, or generally where the gas has been measured in the moist state,

The respective weights assigned in Table V.* to 1 litre of the gases there enumerated, refer to a temperature of 0°, and an atmospheric pressure of 0·76 metre of mercury. We have, therefore, in the first place, to consider the manner in which volumes of gas measured at another temperature and another height of the barometer, are to be reduced to 0° and 0·76 of the barometer.

α. Reduction of a Volume of Gas of any given Temperature to 0°, or any other Temperature between 0° and 100°.

The following propositions regarding the expansion of gases were formerly universally adopted:

1. All gases expand alike for an equal increase of temperature.

The expansion of one and the same gas for each degree of the thermometer is independent of its original density.

Although the correctness of these propositions has not been fully confirmed by the minute investigations of MAGNUS and REGNAULT, yet they may be safely followed in reductions of the temperature of those gases which are most frequently measured in the course of analytical processes, as the coefficients of expansion of these gases scarcely differ from each other, and as there is never any very considerable difference in the atmospheric pressure under which the gases are severally measured.

The investigations just alluded to have given

$$0{\cdot}3665$$

as the coefficient of the expansion of gases which comes nearest to the truth; in other words, as the extent to which gases expand when heated from the freezing to the boiling point of water. They expand, therefore, for every degree of the centigrade thermometer,

$$\frac{0{\cdot}3665}{100} = 0{\cdot}003665.$$

* See Tables at the end of the volume.

If we wish to ascertain how much space 1 c. c. of gas at 0° will occupy at 10°, we find

$$1 \times [1 + (10 \times 0{\cdot}003665)] = 1{\cdot}03665.$$

If we wish to ascertain how much space 100 c. c. at 0° will occupy at 10°, we find

$$100 \times [1 + (10 \times 0{\cdot}003665)]$$
$$= 100 \times 1{\cdot}03665 - 103{\cdot}665.$$

If we wish to know how much space 1 c. c. at 10° will occupy at 0°, we find

$$\frac{1}{1 + (10 \times 0{\cdot}003665)} = 0{\cdot}965.$$

How much space do 103·665 c. c. at 10° occupy at 0°?

$$\frac{103{\cdot}665}{1 + (10 \times 0{\cdot}003665)} = 100.$$

The general rule of these calculations may be expressed as follows:

To calculate the volume of a gas from a lower to a higher temperature, we have in the first place to find the expansion for the volume unit, which is done by adding to 1 the product of the multiplication of the thermometrical difference by 0·003665; and then to multiply this by the number of volume units found in the analytical process. On the other hand, to reduce the volume of a gas from a higher to a lower temperature, we have to divide the number of volume units found in the analytical process, by 1 + the product of the multiplication of the thermometrical difference by 0·003665.

β. Reduction of the Volume of a Gas of a certain given Density to ·76 Metre Barometric Pressure, or any other given Pressure.

According to the law of MARIOTTE, the volume of a gas is inversely as the pressure to which it is exposed; in accordance with this, a gas occupies the greater space the less the pressure upon it, and the less space the greater the pressure upon it.

Thus, supposing a gas to occupy a space of 10 c. c. at a pressure of 1 atmosphere, it will occupy 1 c. c. at a pressure of 10 atmospheres, and 100 c. c. at a pressure of $\frac{1}{10}$ atmosphere.

Nothing, therefore, can be more easy than the reduction of a gas of a certain given tension to 760 mm. bar. pressure, or any other given pressure, *e.g.*, 1000 mm., which is frequently used in the analysis of gases.

Supposing a gas to occupy 100 c. c. at 780 mm. bar., how much space will it occupy at 760 mm.?

$$760 : 780 :: 100 : x$$
$$x = 102{\cdot}63.$$

How much space will 100 c. c. at 750 mm. bar. occupy at 760 mm.?

$$760 : 750 :: 100 : x$$
$$x = 98{\cdot}68.$$

How much space will 150 c. c. at 760 mm. bar. occupy at 1000 mm.?

$$1000 : 760 :: 150 : x$$
$$x = 114.$$

γ. Reduction of the Volume of a Gas saturated with Aqueous Vapor, to its actual Volume in the Dry State.

It is a well-known fact that water has a tendency, at all temperatures, to assume the gaseous state. The degree of this tendency (the tension of the aqueous vapor)—which is dependent solely and exclusively upon the temperature, and not upon the circumstance of the water being *in vacuo* or in any gaseous atmosphere—is usually expressed by the height of a column of mercury counterbalancing it. The following table indicates the amount of tension for the various temperatures at which analyses are likely to be made.*

TABLE.

Temperature (in degrees C.)	Tension of the aqueous vapor expressed in millimetres.	Temperature (in degrees C.)	Tension of the aqueous vapor expressed in millimetres.
0	4·525	21	18·505
1	4·867	22	19·675
2	5·231	23	20·909
3	5·619	24	22·211
4	6·032	25	23·582
5	6·471	26	25·026
6	6·939	27	26·547
7	7·436	28	28·148
8	7·964	29	29·832
9	8·525	30	31·602
10	9·126	31	33·464
11	9·751	32	35·419
12	10·421	33	37·473
13	11·130	34	39·630
14	11·882	35	41·893
15	12·677	36	44·268
16	13·519	37	46·758
17	14·409	38	49·368
18	15·351	39	52·103
19	16·345	40	54·969
20	17·396		

Therefore, if a gas is confined over water, its volume is, *cæteris paribus*, always greater than if it were confined over mercury; since a quantity of aqueous vapor, proportional to the temperature of the water, mixes with the gas, and the tension of this partly counterbalances the column of air that presses upon the gas, and to that extent neutralizes the pressure. To ascertain the actual pressure upon the gas, we must therefore subtract from the apparent pressure so much as is neutralized by the tension of the aqueous vapor.

Suppose we had found a gas to measure 100 c.c. at 759 mm. bar., the temperature of the confining water being 15°; how much space would this volume of gas occupy in the dry state and at 760 mm. of the barometer?

Our table gives the tension of aqueous vapor at 15° = 12·677; the gas is consequently not under the apparent pressure of 759 mm., but under the actual pressure of 759 − 12·677 = 746·323 mm.

* Compare Magnus, Pogg. Annal. 61, 247.

The calculation is now very simple; it proceeds in the manner shown in β; we say,

$$760 : 746{\cdot}323 :: 100 : x$$
$$x = 98{\cdot}20.$$

When the volume of a gas has thus been adjusted by the calculations in α and β, or γ, to the thermometrical and barometrical conditions to which the data of Table V. refer, the percentage by weight may now be readily calculated by substituting the weight for the volume, and proceeding by simple rule of three.

What is the percentage by weight of nitrogen in an analyzed substance, of which 0·5 grm. have yielded 30 c.c. of dry nitrogen gas at 0°, and 760 mm. bar.?

In Table V. we find that 1 litre (1000 c.c.) of nitrogen gas at 0°, and 760 mm. bar., weighs 1·25456 grm.

We say accordingly:

$$1000 : 1{\cdot}25456 :: 30 : x$$
$$x = 0{\cdot}0376.$$

And then:

$$0{\cdot}5 : 0{\cdot}0376 :: 100 : x$$
$$x = 7{\cdot}52.$$

The analyzed substance contains consequently 7·52 per cent. by weight of nitrogen.

DR. GIBBS' *method of finding at once the total correction for temperature, pressure, and moisture in absolute determinations of nitrogen, or other gases:* *

"I take a graduated tube, which I fill with mercury, then displace about two-thirds of the mercury with air, and invert the tube into a cistern of mercury. Then I make four or five determinations of the volume of the included (moist) air in the usual manner, and find the volume of the air at 0° and 760 mm. as a mean of all the determinations. This tube I call the companion tube, and it always hangs in the little room I use for gas analyses. Suppose the volume of (dry) air at 0° and 760 mm. is 132·35 c. c.

"Now, in making an absolute nitrogen determination I collect the nitrogen moist over mercury in a graduated tube, and then suspend the measuring tube by the side of the companion tube. I then by a cord and pulley bring the level of the mercury in the two tubes to correspond exactly, and then read off the volume of air in the companion tube and the volume of nitrogen in the measuring tube. I ought to have stated that the two tubes hang in the same cistern of mercury. Suppose the volume of air in the companion tube to be 143 c. c.; then the total correction for temperature, pressure, and moisture will be 143 − 132·35 = 10·65 c.c. The correction for the nitrogen will then be found by rule of three. As the observed volume of air in the companion tube is to the observed volume of nitrogen, so is (in this case) 10·65 to the required correction. In this way, when the volume of air in the companion tube is once found, *no further observations of temperature, pressure, or height of mercury above the mercury in the cistern* are necessary. The companion tube lasts for an indefinite time. I have even used it filled with water, without any appreciable change in some weeks, but I prefer mercury. As the two tubes hang side by side, there is never an

* Private communication.

appreciable difference of temperature. My results are most satisfactory. Williamson & Russell have, as you know, used a companion tube for *equating pressures*, but not for finding the total value of the temperature and pressure correction at once; and I believe that my process is wholly new. Certainly it is wonderfully convenient, and saves all tables and labor of computation."

2. *Calculation of the Results into Per-cents by Weight, in Cases where the Body sought has been separated in Combination, or where a Compound has to be determined from one of its Constituents.*

If the body to be determined has not been weighed or measured in its own form, but in some other form, *e.g.*, carbonic acid as calcium carbonate, sulphur as barium sulphate, ammonia as nitrogen, chlorine by a standard solution of iodine, &c., its quantity must first be reckoned from that of the compound found before the calculation described in 1 can be made.

This may be accomplished either by rule of three or by some abridged method.

Suppose we have weighed hydrogen in the form of water, and have found 1 grm. of water; how much hydrogen does this contain?

A molecule of water consists of:

Hydrogen..........................	2 at. =	2 pts.
Oxygen..........................	1 at. =	16 "
		18 "

We say accordingly:

$$18 : 2 :: 1 : x$$
$$x = 0{\cdot}11111.$$

Or, expressed in general terms:

$$\textit{Water} \times 0{\cdot}11111 = \textit{Hydrogen}.$$

EXAMPLE.—

517 of water; how much hydrogen?
$517 \times 0{\cdot}11111 = 57{\cdot}444.$

The following equation results also from the above proportion:

$$\frac{18}{2} = \frac{1}{x}$$
$$18 = \frac{2}{x}$$
$$\therefore\ x = \frac{1}{9}$$

Or, expressed in general terms,

$$\textit{Water divided by } 9 = \textit{Hydrogen}.$$

EXAMPLE.—

517 of water, how much hydrogen?

$$\frac{517}{9} = 57{\cdot}444.$$

In this manner we may find for every compound constant numbers by which to multiply or divide the weight of the compound, in order to find the weight of the constituent sought (comp. Table III.*).

---

* See Tables at the end of the volume.

Thus, for instance, the nitrogen contained in ammonium platinic chloride may be obtained by multiplying the weight of the latter by 0·06296; thus the carbon may be calculated from carbonic acid by multiplying the weight of the latter by 0·2727, or dividing it by 3·666.

These numbers are by no means so simple, convenient, and easy to remember as in the case of hydrogen. It is therefore advisable, in the case of carbonic acid, for instance, to fix upon another general expression, viz.,

$$\frac{\textit{Carbonic acid} \times 3}{11} = \textit{Carbon};$$

12 parts in 44 ($= \frac{3}{11}$) in carbonic acid being carbon, as may be seen from the composition:

$$\begin{array}{ll} C & 12 \\ O_2 & 32 \\ \hline & 44 \end{array}$$

The object in view may also be attained in a very simple manner, by reference to Table IV.,* which gives the amount of the constituent sought for every number of the compound found, from 1 to 9; the operator need, therefore, simply add the several values together.

As regards hydrogen, for instance, we find:

TABLE.

Found.	Sought.	1	2	3	4	5	6	7	8	9
water	hydrogen	0·11111	0·22222	0·33333	0·44444	0·55555	0·66667	0·77778	0·88889	1·00000

From this table it is seen that 1 part of water contains 0·11111 of hydrogen, that 5 parts of water contain 0·55555 of hydrogen; 9 parts, 1·00000, &c.

Now if we wish to know, for instance, how much hydrogen is contained in 5·17 parts of water, we find this by adding the values for 5 parts, for $\frac{1}{10}$ part, and for $\frac{7}{100}$ parts, thus:

$$\begin{array}{l} 0{\cdot}55555 \\ 0{\cdot}011111 \\ 0{\cdot}0077778 \\ \hline 0{\cdot}5744388 \end{array}$$

Why the numbers are to be placed in this manner, and not as follows:

$$\begin{array}{l} 0{\cdot}55555 \\ 0{\cdot}11111 \\ 0{\cdot}77778 \\ \hline 1{\cdot}44444 \end{array}$$

is self-evident, since arranging them in the latter way would be adding the value for 5, for 1, and for 7 ($5 + 1 + 7 = 13$), and not for 5·17. This reflection shows

* See Tables at the end of the volume.

also that, to find the amount of hydrogen contained in 517 parts of water, the points must be transposed as follows:

$$\begin{array}{r} 55{\cdot}555 \\ 1{\cdot}1111 \\ 0{\cdot}77778 \\ \hline 57{\cdot}44388 \end{array}$$

3. *Calculation of the Results of Indirect Analyses into Per-cents by Weight.*

The import of the term "*indirect analysis*," as defined in § 151, p. 478 shows sufficiently that no universally applicable rules can be laid down for the calculations which have to be made in indirect analyses. The selection of the right way must be left in every special case to the intelligence of the analyst. I will here give the mode of calculating the results in the more important indirect separations described in Section V. They may serve as examples for other similar calculations.

a. Indirect Determination of Sodium and Potassium.

This is effected by determining the sum total of the chlorides, and the chlorine contained in them.

The calculation may be made as follows:

Suppose we have found 3 grm. of sodium and potassium chlorides, and in these 3 grm. 1·6877 of chlorine.

At. Chlorine.		Mol. KCl.		Chlorine found.
35·46	:	74·59	::	1·6877 : x
		x	=	3·55007.

If all the chlorine present were combined with potassium, the weight of the chloride would amount to 3·55007. As the chloride weighs less, sodium chloride is present, and this in a quantity proportional to the difference (*i.e.*, 3·55007 − 3 = ·55007), which is calculated as follows:

The difference between the mol. weight of KCl and that of NaCl (16·09) is to the mol. weight of NaCl (58·50), as the difference found is to the sodium chloride present:

$$16{\cdot}09 : 58{\cdot}50 :: {\cdot}55007 : x$$
$$x = 2 \text{ NaCl}$$
$$\text{and } 3 - 2 = 1 \text{ KCl}.$$

From this the following short rule is derived:

Multiply the quantity of chlorine in the mixture by 2·1035, deduct from the product the sum of the chlorides, and multiply the remainder by 3·6358; the product expresses the quantity of sodium chloride contained in the mixed chloride.

b. Indirect Determination of Strontium and Calcium.

This may be effected by determining the sum total of the carbonates, and the carbonic acid contained in them (§ 154, 31). Suppose we have found 2 grm. of mixed carbonate, and in these 2 grm. 0·7383 of carbonic acid.

Mol. CO_2	Mol. $SrCO_3$		CO_2 found.
44	147·50	::	0·7383 : x
	x	=	2·47498.

If, therefore, the whole of the carbonic acid were combined with strontia, the weight of the carbonate would amount to 2·47498 grm. The deficiency, = 0·47498, is proportional to the calcium carbonate present, which is calculated as follows:

The difference between the molecule of $SrCO_3$ and the molecule of $CaCO_3$ (47·50) is to the molecule of $CaCO_3$ (100) as the difference found is to the calcium carbonate contained in the mixed salt:

$$\therefore \quad 47{\cdot}5 : 100 :: 0{\cdot}47498 : x$$
$$x = 1.$$

The mixture, therefore, consists of 1 grm. calcium carbonate and 1 grm. strontium carbonate.

From this the following short rule is derived:

Multiply the carbonic acid found by 3·3523, deduct from the product the sum of the carbonates, and multiply the difference by 2·10526; the product expresses the quantity of the calcium carbonate.

c. Indirect Determination of Chlorine and Bromine (§ 169, 1).

Let us suppose the mixture of silver chloride and bromide to have weighed 2 grm., and the diminution of weight consequent upon the transmission of chlorine to have amounted to 0·1 grm. How much chlorine is there in the mixed salt, and how much bromine?

The decrease of weight here is simply the difference between the weight of the silver bromide originally present, and that of the silver chloride which has replaced it; if this is borne in mind, it is easy to understand the calculation which follows:

The difference between the molecules of silver bromide and silver chloride is to the molecule of silver bromide as the ascertained decrease of weight is to x, *i.e.*, to the silver bromide originally present in the mixture:

$$\therefore \quad 44{\cdot}49 : 187{\cdot}88 :: 0{\cdot}1 : x$$
$$x = 0{\cdot}422297.$$

The 2 grm. of the mixture therefore contained 0·422297 grm. silver bromide, and consequently 2 − 0·422297 = 1·577703 grm. silver chloride.

It results from the above, that we need simply multiply the ascertained decrease of weight by

$$\frac{187{\cdot}88}{44{\cdot}49} \text{ i.e., by } 4{\cdot}22297$$

to find the amount of silver bromide originally present in the analyzed mixture. And if we know this, we also know of course the amount of the silver chloride; and from these data we next calculate the quantities of chlorine and bromine in the ordinary way.

Supplement to I.

REMARKS ON LOSS AND EXCESS IN ANALYSES, AND ON TAKING THE AVERAGE.

If, in the analysis of a substance, one of the constituents is estimated from the loss, or, in other words, by subtracting from the original weight of the analyzed substance the ascertained united weight of the other constituents, it is evident that in the subsequent percentage calculation the sum total must invariably be 100. Every loss suffered or excess obtained in the determination of the several constituents will, of course, fall exclusively upon the one constituent which is estimated from the loss. Hence estimations of this kind cannot be considered accurate, unless the other constituents have been determined by good methods, and with the greatest care. The accuracy of the results will, of course, be the greater, the less the number of constituents determined in the direct way.

If, on the other hand, every constituent of the analyzed compound has been determined separately, it is obvious that, were the results absolutely accurate, the united weight of the several constituents must be exactly equal to the original weight of the analyzed substance. Since, however, as we have seen in § 96, certain inaccuracies attach to every analysis, without exception, the sum total of the results in the percentage calculation will sometimes exceed, and sometimes fall short of, 100.

In all cases of this description, the only proper way is to give the results as actually found.

Thus, for instance, PELOUZE found, in his analysis of chromate of potassium chloride,

Potassium	21·88
Chlorine	19·41
Chromic acid	58·21
	99·50

BERZELIUS, in his analysis of potassium uranate,

Potassa	12·8
Uranic oxide	86·8
	99·6

PLATTNER, in his analysis of pyrrhotite,

	Of Fahlun.	Of Brasil.
Iron	59·72	59·64
Sulphur	40·22	40·43
	99·94	100·07

It is altogether inadmissible to distribute any chance deficiency or excess proportionately among the several constituents of the analyzed compound, as such deficiency or excess of course never arises from the several estimations in the same measure; moreover, such "doctoring" of the analysis deprives other chemists of the power of judging of its accuracy. No one need be ashamed to confess having obtained somewhat too little or somewhat too much in an analysis, provided, of course, the deficiency or excess be confined within certain limits, which differ in different analyses, and which the experienced chemist always knows how to fix properly.

In cases where an analysis has been made twice, or several times, it is usual to take the mean as the most correct result. It is obvious that an average of the kind deserves the greater confidence the less the results of the several analyses differ. The results of the several analyses must, however, also be given, or, at all events, the maximum and minimum.

Since the accuracy of an analysis is not dependent upon the quantity of substance employed (provided always this quantity be not altogether too small), the average of the results of several analyses is to be taken quite independently of the quantities used; in other words, you must not add together the quantities used, on the one hand, and the weights obtained in the several analyses on the other, and deduce from these data the percentage amount; but you must calculate the latter from the results of each analysis separately, and then take the mean of the numbers so obtained.

Suppose a substance, which we will call AB, contains fifty per cent. of A; and suppose two analyses of this substance have given the following results:

(1) 2 grm. AB gave 0·99 grm. of A.
(2) 50 " " 24·00 "

From 1, it results that AB contains 49·50 per cent. of **A.**

" 2, " "	48·00 "
Total	97·50
Mean	48·75

It would be quite erroneous to say

$$2 + 50 = 52 \text{ of AB gave } 0{\cdot}99 + 24{\cdot}00 = 24{\cdot}99 \text{ of A,}$$

$$\text{therefore } 100 \text{ of AB contain } 48{\cdot}06 \text{ of A;}$$

for it will be readily seen that this way of calculating destroys nearly altogether the influence of the more accurate analysis (1) upon the average, on account of the proportionally small amount of substance used.

II.—DEDUCTION OF FORMULÆ.

1. *From the percentages of single elements in compounds.*

The process of deducing an empirical formula from the expression of the composition of a compound in parts per hundred of its constituents (*i.e.*, its percentage composition) will be readily understood by considering first the somewhat simpler reverse process of calculating percentage compositions from formulæ.

Applying this latter process to the formula, for instance, of mannite, $C_6H_{14}O_6$, we first compute from the relative number of atoms of the elements shown by the formula the relative quantities by weight of each, by means of their known atomic weights.

Carbon.....	6 at. × 12 =	72 pts. by weight.	
Hydrogen..	14 " × 1 =	14 " "	
Oxygen....	6 " × 16 =	96 " "	
		182 " "	of mannite.

Since 182 pts. of the compound contain 72 pts. of carbon, the number of pts. of carbon which 100 contain may be found by the rule of three:

$$182 : 100 :: 72 : x$$

$$\frac{100}{182} \times 72 = 39{\cdot}56 \text{ carbon.}$$

In like manner

$$\frac{100}{182} \times 14 = 7{\cdot}69 \text{ hydrogen.}$$

$$\frac{100}{182} \times 96 = 52{\cdot}75 \text{ oxygen.}$$

$$100{\cdot}00$$

Returning now to the first expression of the relative quantities, which was obtained by multiplying the relative number of atoms of carbon, oxygen, and hydrogen by their atomic weights, it is evident by dividing the relative quantities by the atomic weights, the relative number of atoms will again be obtained:

Parts of carbon........	$72 \div 12 =$	6 carbon atoms.
" " hydrogen......	$14 \div 1 =$	14 hydrogen "
" " oxygen........	$96 \div 16 =$	6 oxygen "

It is moreover evident that if numbers obtained by increasing or diminishing 72, 14, and 96 proportionally, be divided by 12, 1, and 16 respectively, the resulting quotients will express the atomic ratio also:

$$\begin{array}{llll}
\text{Carbon......} & 72 \times \frac{100}{182} = & 39{\cdot}56 \div 12 = & 3{\cdot}296 \text{ carbon atoms.} \\
\text{Hydrogen....} & 14 \times \frac{100}{182} = & 7{\cdot}69 \div 1 = & 7{\cdot}690 \text{ hydrogen "} \\
\text{Oxygen......} & 96 \times \frac{100}{182} = & 52{\cdot}75 \div 16 = & 3{\cdot}296 \text{ oxygen "} \\
 & \overline{182} & \overline{100{\cdot}00} &
\end{array}$$

The atomic ratio is found therefore by *dividing the percentages of the elements by their atomic weights.* In the present case, the formula $C_{3 \cdot 296}H_{7 \cdot 690}O_{3 \cdot 296}$ expresses the relative number of atoms.

It now remains to find the smallest whole numbers that express exactly, or approximately, the same atomic ratio as those directly obtained by such calculation. This is usually best done by dividing each number by the smallest, and multiplying, if necessary, the resulting quotients by some number that will wholly or nearly eliminate their fractional parts:

$$\begin{array}{l}
3{\cdot}296 \div 3{\cdot}296 = 1 \quad \times 3 = 3 \\
7{\cdot}690 \div 3{\cdot}296 = 2{\cdot}333 \times 3 = 6{\cdot}999 \\
3{\cdot}296 \div 3{\cdot}296 = 1 \quad \times 3 = 3.
\end{array}$$

It can now be seen that 3, 7, 3 are the smallest whole numbers which can express the relative number of atoms of carbon, hydrogen, and oxygen respectively, *i.e.*, that $C_3H_7O_3$ is the empirical formula.

When, as in the present example, the percentage composition is calculated from a formula, the empirical formula deduced from it will, of course, exhibit the same relative number of atoms as the original formula, except the slight variation arising from neglecting fractions in divisions. But when the empirical formula is deduced from a percentage composition found by analysis, it cannot be expected that the calculated atomic ratio can be expressed *exactly* by small whole numbers.

OPPERMANN found by actual analysis of mannite:

C..	39·31
H..	7·71
O..	52·98

which, calculated as above, gives $C_{3 \cdot 276}H_{7 \cdot 710}O_{3 \cdot 311}$ as the first formula. Dividing each number by the least, this becomes $C_1H_{2 \cdot 353}O_{1 \cdot 010}$, which multiplied by

3 gives $C_3H_{7.069}O_{3.030}$. These last numbers show that the carbon, hydrogen, and oxygen atoms found by analysis are so nearly in the proportion 3, 7, 3, that it is reasonable to believe that $C_3H_7O_3$ is a correct empirical formula, and that the slight differences from these numbers exhibited by the numbers actually obtained, are due to defects inherent in the method of analysis used. We can judge better whether such differences are greater than may be due to error in analysis by calculating from the deduced formula the percentage composition which it requires and comparing it with that found. The composition found may also be compared with that required by any other assumed formula which it indicates to be possible.

	Found.	Calculated for $C_3H_7O_3$.	For $C_4H_9O_4$.
Carbon........	39·31	39·56	39·67
Hydrogen.....	7·71	7·69	7·44
Oxygen........	52·98	52·75	52·89
	100·00	100·00	100·00

2. *From the percentages of groups of elements in compounds.*

a. When isomorphous constituents are not present.

In the analysis of oxygen salts, although data are obtained from which the percentage of each element or each radical present might be computed, it is far more convenient, and in fact customary, to calculate the percentage of oxides and water equivalent in quantity to the elements (see § 67, p. 131).

For example, the results of the analysis of sodium ammonium phosphate were presented in this form:

Na_2O	17·93
$(NH_4)_2O$	15·23
SO_3	46·00
H_2O	20·84
	100·00

From this statement the percentage of each element might first be calculated, and next the empirical formula by the method already described. The same end may be attained by the following shorter course.

Applying the term "molecule" to each group of elements here presented, it is evident that the relative number of molecules of sodium oxide, ammonium oxide, sulphuric anhydride and water can be found by dividing the quantity of each by its molecular weight—a process the same in principle as that employed for calculating atomic ratio (p. 848).

Relative quantities.			Molecular weights.		Relative number of molecules.				
Na_2O	17·93	÷	62·08	=	·2888	÷	·2888	=	1
$(NH_4)_2O$	15·23	÷	52·08	=	·2928	÷	·2888	=	1·01
SO_3	46·00	÷	80·	=	·5750	÷	·2888	=	1·99
H_2O	20·84	÷	18·	=	1·1577	÷	·2888	=	4·00

The numbers 1, 1·01, 1·99, 4, are so nearly in the same proportion as 1, 1, 2, 4, that there can be no doubt that $(Na_2O)_1(NH_4)_2O)_1(SO_3)_2(H_2O)_4$ is a correct formula. This formula shows necessarily the same grouping of elements that was used in stating the percentage composition. Rearranging the order in which the symbols of the elements stand, $(Na_2O)_1(NH_4)_2O)_1(SO_3)_2(H_2O)_4$ =

$Na_2N_2S_2H_{16}O_{12}$, and dividing by 2, we obtain the strictly empirical formula $NaNSH_8O_6$.

Rational formulæ.—Having obtained the empirical formula of a compound, any theoretical conclusion regarding its molecular weights may be expressed by increasing (if necessary) each atom an equal number of times; and any supposition, suggestion, or conclusion regarding its chemical constitution may be expressed by a conformable arrangement of the atoms. From the empirical formulæ of most oxygen salts rational formulæ may be readily deduced. In the above case, for instance (sodium ammonium sulphate), $NaNSH_8O_6 = NaNH_4SO_2H_4O_4$; a rational formula implying that the nitrogen exists in the form of ammonium and the sulphur in the form of the acid radical SO_2 (sulphuryl).

By inspection of the component parts of this formula it is seen that the sum of the quantivalence of the two basic radicals Na′ and $(NH_4)'$ equals that of the acid radical $(SO_2)''$. The salt must therefore be a normal salt, and none of the hydrogen can be in combination with either basic or acid radical. Two atoms of oxygen are required to unite the radicals, leaving $(H_2O)_2$. This leads to the conclusion that the 4 atoms of hydrogen exist in the form of water, which is in a state of combination called water of crystallization—

$$NaNH_4SO_2H_4O_4 = \genfrac{}{}{0pt}{}{Na\ -O}{NH_4-O}> SO_2 + 2H_2O.$$

b. When isomorphous constituents are present.

In deducing formulæ, it must be borne in mind that closely related elements or radicals, more especially the basic metals, may replace each other in all proportions. Elements of like quantivalence are oftenest found replacing each other, but in some cases equivalent amounts of elements having different valence appear to replace each other. The following example will illustrate the kind of formula and method of deducing it commonly used in such cases.

S. L. PENFIELD found by analysis of triphylite the following composition:

			Molecular weights.		Mol. ratio.				At. ratio.		
P_2O_5	44·76	÷	142	=	·315	× 2	=	P	·630		
FeO	26·40	÷	72	=	·366			Fe	·366	}	= R″ ·633
MnO	17·84	÷	71	=	·251			Mn	·251		
CaO	·24	÷	56	=	·004			Ca	·004		
MgO	·47	÷	40	=	·012			Mg	·012		
Li_2O	9·36	÷	30	=	·312	× 2	=	Li	·624	}	= R′ ·634
Na_2O	·35	÷	78·08	=	·005	× 2	=	Na	·010		
H_2O	·42							O	2·525		
	99·84										

Disregarding the small amount of water, the relative numbers of molecules of the oxides (mol. ratio) are first found by dividing quantities by molecular weights, as in the preceding example. Next the atoms contained by the molecules are written in another column (at. ratio). *This column, with the adjoined symbols, is the empirical formula.* It is apparent, or can be proved by trial, that the numbers of different atoms are not in any simple ratio. Such an atomic relation is to be expected when isomorphous constituents are present. It remains now to unite the atoms of such elements as are supposed to be capable of mutually replacing each other, and ascertain whether the numbers thus obtained are in any simple proportion. For this purpose let R″ represent one atom of any dyad basic metal and R′ one atom of any monad basic metal present. The sum of the dyad atoms is ·633; that of the monad atoms, ·634, as above shown. The atomic ratio thus obtained is expressed by the formula $R''_{633}R'_{634}P_{630}O_{2525}$; or simpler, dividing by 630, almost exactly by $R''R'PO_4$ which is equal to

$$(PO)''' \begin{cases} \begin{matrix} O \\ O \end{matrix} > R'' \\ O-R', \end{cases}$$

anhydrous normal lithium ferrous phosphate in which iron is partially replaced by manganese, magnesium, and calcium; and lithium to a slight extent by sodium.

It may be here observed that in presenting atomic ratios in connection with analyses of natural oxygen salts (minerals), computation and statement of oxygen atoms is often omitted, since they may be deduced from a formula showing the other constituents. Omitting oxygen in the above example we have $R''R'P$. By referring to the percentage composition it is seen that for two P five O must be present,—for two R′ one O,—for one R″ one O. Doubling $R''R'P$ and appending to each constituent the required oxygen atoms, we have: $R''_2O_2R'_2OP_2O_5 = R''_2R'_2P_2O_8$, and dividing by 2, $R''R'PO_4$, as before.

TABLES FOR THE CALCULATION OF ANALYSIS.

TABLE I.

ATOMIC WEIGHTS OF THE ELEMENTS CONSIDERED IN THE PRESENT WORK.*

Element	Symbol	Weight	Element	Symbol	Weight
Aluminium	Al†	27·50	Magnesium	Mg	24·00
Antimony	Sb†	122·00	Manganese	Mn	55·00
Arsenic	As	75·00	Mercury	Hg	200·00
Barium	Ba	137·00	Molybdenum	Mo	92·00
Bismuth	Bi	208·00	Nickel	Ni	59·00
Boron	B	11·00	Nitrogen	N	14·04
Bromine	Br	79·95	Oxygen	O	16·00
Cadmium	Cd	112·00	Palladium	Pd	106·58
Cæsium	Cs	133·00	Phosphorus	P	31·00
Calcium	Ca	40·00	Platinum	Pt	197·18
Carbon	C	12·00	Potassium	K	39·13
Chlorine	Cl	35·46	Rubidium	Rb	85·40
Chromium	Cr	52·48	Selenium	Se	79·00
Cobalt	Co	59·00	Silicon	Si	28·00
Copper	Cu	63·40	Silver	Ag	107·93
Fluorine	Fl	19·00	Sodium	Na	23·04
Gold	Au	196·71	Strontium	Sr	87 50
Hydrogen	H	1·00	Sulphur	S	32·00
Iodine	I	126·85	Tin	Sn	118·00
Iron	Fe	56·00	Titanium	Ti	50·00
Lead	Pb	207·00	Uranium	Ur	237·60
Lithium	Li	7·00	Zinc	Zn	65·06

TABLE II.

COMPOSITION OF THE BASIC AND ACID OXIDES.

GROUP I. *a.* BASIC OXIDES.

Cæsia	Cs_2	266·00	94·33
	O	16·00	5·67
	Cs_2O	282·00	100·00
Rubidia	Rb_2	170·80	91·43
	O	16·00	8·57
	Rb_2O	186.80	100.00

* [The numbers here given are based on the atomic weights used in the sixth German edition, the atomic weights of the "old system" being doubled when necessary.]

† Recent critical investigations—by J. P. COOKE, on the atomic weight of antimony—by J. W. MALLET, on that of aluminium, have conclusively shown that 120 and 27·02 respectively should be taken as the atomic weights of these elements. See Transactions of Am. Acad. Sci., 13, 15. Atomic Weight of Antimony, and Philosophical Transactions of the Royal Society (London), Revision of the Atomic Weight of Aluminium.

Oxide	Constituent	Weight	Per cent
Potassa	K_2	78·26	83·03
	O	16·00	16·97
	K_2O	94·26	100·00
Soda	Na_2	46·08	74·23
	O	16·00	25·77
	Na_2O	62·08	100·00
Lithia	Li_2	14·00	46·67
	O	16·00	53·33
	Li_2O	30·00	100·00
Ammonium oxide	$(NH_4)_2$	36·16	69·28
	O	16·00	30·72
	$(NH_4)_2O$	52·16	100·00
GROUP II.			
Baryta	Ba	137·00	89·54
	O	16·00	10·46
	BaO	153·00	100·00
Strontia	Sr	87·50	84·54
	O	16·00	15·46
	SrO	103·50	100·00
Lime	Ca	40·00	71·43
	O	16·00	28·57
	CaO	56·00	100·00
Magnesia	Mg	24·00	60·03
	O	16·00	39·97
	MgO	40·00	100·00
GROUP III.			
Alumina	Al_2	55·00	53·40
	O_3	48·00	46·60
	Al_2O_3	103·00	100·00
Chromic oxide	Cr_2	104·96	68·62
	O_3	48·00	31·38
	Cr_2O_3	152·96	100·00
GROUP IV.			
Zinc oxide	Zn	65·06	80·26
	O	16·00	19·74
	ZnO	81·06	100·00

Compound	Constituent	Parts	Per cent
Manganous oxide	Mn	55·00	77·46
	O	16·00	22·54
	MnO	71·00	100·00
Manganic oxide	Mn_2	110·00	69·62
	O_3	48·00	30·38
	Mn_2O_3	158·00	100·00
Nickelous oxide	Ni	59·00	78·67
	O	16·00	21·33
	NiO	75·00	100·00
Cobaltous oxide	Co	59·00	78·67
	O	16·00	21·33
	CoO	75·00	100·00
Cobaltic oxide	Co_2	118·00	71·08
	O_3	48·00	28·92
	Co_2O_3	166·00	100·00
Ferrous oxide	Fe	56·00	77·78
	O	16·00	22·22
	FeO	72·00	100·00
Ferric oxide	Fe_2	112·00	70·00
	O_3	48·00	30·00
	Fe_2O_3	160·00	100·00
GROUP V.			
Silver oxide	Ag_2	215·86	93·10
	O	16·00	6·90
	Ag_2O	231·86	100·00
Lead oxide	Pb	207·00	92·83
	O	16·00	7·17
	PbO	223·00	100·00
Mercurous oxide	Hg_2	400·00	96·15
	O	16·00	3·85
	Hg_2O	416·00	100·00
Mercuric oxide	Hg	200·00	92·59
	O	16·00	7·41
	HgO	216·00	100·00

Cuprous oxide	Cu_2	126·80	88·80
	O	16·00	11·20
	Cu_2O	142·80	100·00
Cupric oxide	Cu	63·40	79·85
	O	16·00	20·15
	CuO	79·40	100·00
Bismuth trioxide	Bi_2	416·00	89·66
	O_3	48·00	10·34
	Bi_2O_3	464·00	100·00
Cadmium oxide	Cd	112·00	87·50
	O	16·00	12·50
	CdO	128·00	100·00

GROUP VI.

Auric oxide	Au_2	392·00	89·09
	O_3	48·00	10·91
	Au_2O_3	440·00	100·00
Platinic oxide	Pt	197·18	86·04
	O_2	32·00	13·96
	PtO_2	229·18	100·00
Antimonious oxide	Sb_2	244·00	83·56
	O_3	48·00	16·44
	Sb_2O_3	292·00	100·00
Stannous oxide	Sn	118·00	88·06
	O	16·00	11·94
	SnO	134·00	100·00
Stannic oxide	Sn	118·00	78·67
	O_2	32·00	21·33
	SnO_2	150·00	100·00
Arsenious oxide	As_2	150·00	75·76
	O_3	48·00	24·24
	As_2O_3	198·00	100·00
Arsenic oxide	As_2	150·00	65·22
	O_5	80·00	34·78
	As_2O_5	230·00	100·00

b. ACID OXIDES (ANHYDRIDES).

Chromic anhydride	Cr	52·48	52·23
	O_3	48·00	47·77
	CrO_3	100·48	100·00
Sulphuric anhydride	S	32·00	40·00
	O_3	48·00	60·00
	SO_3	80·00	100·00
Phosphoric anhydride	P_2	62·00	43·66
	O_5	80·00	56·34
	P_2O_5	142·00	100·00
Boracic anhydride	B_2	22·00	31·43
	O_3	48·00	68·57
	B_2O_3	70·00	100·00
Oxalic anhydride	C_2	24·00	33·33
	O_3	48·00	66·67
	C_2O_3	72·00	100·00
Carbonic anhydride	C	12·00	27·27
	O_2	32·00	72·73
	CO_2	44·00	100·00
Silicic anhydride	Si	28·00	46·67
	O_2	32·00	53·33
	SiO_2	60·00	100·00
Nitric anhydride	N_2	28·08	25·98
	O_5	80·00	74·02
	N_2O_5	108·08	100·00
Chloric anhydride	Cl_2	70·92	46·99
	O_5	80·00	53·01
	Cl_2O_5	150·92	100·00

TABLE III.

REDUCTION OF COMPOUNDS FOUND TO CONSTITUENTS SOUGHT BY SIMPLE MULTIPLICATION OR DIVISION.

This Table contains only some of the more frequently occurring compounds the formulæ preceded by ! give absolutely accurate results.

For Inorganic Analysis.

CARBON DIOXIDE.

! Calcium Carbonate × 0.44 = Carbon dioxide.

CHLORINE.

Silver chloride × 0·2473 = Chlorine.

COPPER.

Cupric oxide × 0·79849 = Copper.

IRON.

! Ferric oxide × 0·7 = Iron.
! Ferric oxide × 0·9 = Ferrous oxide.

LEAD.

Lead oxide × 0·9283 = Lead.

MAGNESIA.

Magnesium pyrophosphate × 0·36036 = Magnesia.

MANGANESE.

Protosesquioxide of manganese × 0·72052 = Manganese.
Protosesquioxide of manganese × 0·93013 = Manganous oxide.

PHOSPHORIC ANHYDRIDE (P_2O_5).

Magnesium pyrophosphate × 0·6396 = Phosphoric acid.
Uranyl pyrophosphate ($(UO_2)_2P_2O_7$) × 0·1991 = P_2O_5.

POTASSIUM.

Potassium chloride × 0·5246 = Potassium.
Potassium sulphate × 0·54092 = Potassa.

Potassium platinic chloride × 0·30557, or $\frac{\text{Potassium platinic chloride}}{3·2725}$ = Potassium chloride.

Potassium platinic chloride × 0·19308, or $\frac{\text{Potassium platinic chloride}}{5·179}$ = Potassa.

SODA.

Sodium chloride × 0·5306 = Soda.
Sodium sulphate × 0·43694 = Soda.

SULPHUR.

Barium sulphate × 0·13734 = Sulphur.

SULPHURIC ACID.

Barium sulphate × 0·34335 = Sulphuric anhydride (SO_3).

For Organic Analysis.

CARBON.

$$\left.\begin{array}{c}\text{Carbon dioxide} \times 0{\cdot}2727 \\ \text{or} \\ \dfrac{\text{Carbon dioxide}}{3{\cdot}666} \\ \text{or} \\ \dfrac{\text{Carbon dioxide} \times 3}{11}\end{array}\right\} = \text{Carbon.}$$

HYDROGEN.

$$\left.\begin{array}{c}\text{Water} \times 0{\cdot}11111 \\ \text{or} \\ \dfrac{\text{Water}}{9}\end{array}\right\} = \text{Hydrogen.}$$

NITROGEN.

Ammonium platinic chloride × 0·06296 = Nitrogen.
Platinum × 0·1424 = Nitrogen.

TABLE

Showing the Amount of the

Number of the

Elements.	Found.	Sought.	1
Aluminium..	Alumina Al_2O_3	Aluminium Al	0.53398
(Ammonium).	Ammonium chloride NH_4Cl	Ammonia NH_3	0.31850
	Ammonium platinic chloride $(NH_4Cl)_2 \cdot PtCl_4$	Ammonium oxide $(NH_4)_2O$	0.11677
	Ammonium platinic chloride $(NH_4Cl)_2 \cdot PtCl_4$	Ammonia NH_3	0.07641
Antimony....	Antimonious oxide Sb_2O_3	Antimony Sb	0.83562
	Antimonious sulphide Sb_2S_3	Antimony Sb	0.71765
	Antimonious sulphide Sb_2S_3	Antimonious oxide Sb_2O_3	0.85882
	Antimony tetroxide Sb_2O_4	Antimonious oxide Sb_2O_3	0.94805
Arsenic......	Arsenious oxide As_2O_3	Arsenic As	0.75758
	Arsenic oxide As_2O_5	Arsenic As	0.65217
	Arsenic oxide As_2O_5	Arsenious oxide As_2O_3	0.86087
	Arsenious sulphide As_2S_3	Arsenious oxide As_2O_3	0.80488
	Arsenious sulphide As_2S_3	Arsenic oxide As_2O_5	0.93496
Barium......	Baryta BaO	Barium Ba	0.89542
	Barium sulphate $BaSO_4$	Baryta BaO	0.65665

IV.

Constituent sought for every

Compound found.

2	3	4	5	6	7	8	9
1.06796	1.60194	2.13592	2.66990	3.20389	3.73787	4.27185	4.80583
0.63701	0.95551	1.27402	1.59252	1.91103	2.22953	2.54804	2.86654
0.23353	0.35030	0.46706	0.58383	0.70060	0.81736	0.93413	1.05089
0.15282	0.22923	0.30564	0.38205	0.45846	0.53487	0.61128	0.68769
1.67123	2.50685	3.34247	4.17808	5.01370	5.84932	6.68194	7.52055
1.43529	2.15294	2.87059	3.58834	4.30588	5.02353	5.74118	6.45882
1.71765	2.57647	3.43530	4.29412	5.15294	6.01177	6.87059	7.72942
1.89610	2.84416	3.79221	4.74026	5.68831	6.63636	7.58442	8.53247
1.51516	2.27274	3.03032	3.78790	4.54548	5.30306	6.06064	6.81822
1.30435	1.95652	2.60870	3.26087	3.91304	4.56522	5.21739	5.86957
1.72174	2.58261	3.44348	4.30435	5.16521	6.02608	6.88695	7.74782
1.60975	2.41463	3.21951	4.02439	4.82927	5.63415	6.43902	7.24390
1.86992	2.80488	3.73984	4.67480	5.60975	6.54471	7.47967	8.41463
1.79085	2.68627	3.58170	4.47712	5.37255	6.26797	7.16340	8.05882
1.31330	1.96996	2.62661	3.28326	3.93991	4.59656	5.25322	5.90987

TABLE IV.

Elements.	Found.	Sought.	1
Barium......	Barium carbonate $BaCO_3$	Baryta BaO	0.77665
	Barium silico-fluoride $BaFl_2 \cdot SiFl_4$	Baryta BaO	0.54839
Bismuth.....	Bismuth trioxide Bi_2O_3	Bismuth Bi	0.89655
Boron.......	Boracic anhydride B_2O_3	Boron B	0.31429
Bromine.....	Silver bromide AgBr	Bromine Br	0.42554
Cadmium.....	Cadmium oxide CdO	Cadmium Cd	0.87500
Calcium......	Lime CaO	Calcium Ca	0.71429
	Calcium sulphate $CaSO_4$	Lime CaO	0.41176
	Calcium carbonate $CaCO_3$	Lime CaO	0.56000
Carbon......	Carbonic acid CO_2	Carbon C	0.27273
	Calcium carbonate $CaCO_3$	Carbonic acid CO_2	0.44000
Chlorine.....	Silver chloride AgCl	Chlorine Cl	0.24730
	Silver chloride AgCl	Hydrochloric acid HCl	0.25427
Chromium...	Chromic oxide Cr_2O_3	Chromium Cr	0.68619
	Chromic oxide Cr_2O_3	Chromic anhydride CrO_3	1.31381
	Lead chromate $PbCrO_4$	Chromic anhydride CrO_3	0.31062
Cobalt.......	Cobalt Co	Cobaltous oxide CoO	1.27119
	Cobaltous sulphate $CoSO_4$	Cobaltous oxide CoO	0.48387

(Continued).

2	3	4	5	6	7	8	9
1.55330	2.32995	3.10660	3.88325	4.65990	5.43655	6.21320	6.98985
1.09677	1.64516	2.19355	2.74194	3.29032	3.83871	4.38710	4.93548
1.79310	2.68965	3.58620	4.48275	5.37930	6.27586	7.17240	8.06895
0.62857	0.94286	1.25714	1.57143	1.88572	2.20000	2.51429	2.82857
0.85107	1.27661	1.70215	2.12768	2.55322	2.97876	3.40430	3.82983
1.75000	2.62500	3.50000	4.37500	5.25000	6.12500	7.00000	7.87500
1.42857	2.14286	2.85714	3.57143	4.28571	5.00000	5.71429	6.42857
0.82353	1.23529	1.64706	2.05882	2.47059	2.88235	3.29412	3.70588
1.12000	1.68000	2.24000	2.80000	3.36000	3.92000	4.48000	5.04000
0.54546	0.81818	1.09091	1.36364	1.63636	1.90909	2.18181	2.45455
0.88000	1.32000	1.76000	2.20000	2.64000	3.08000	3.52000	3.96000
0.49460	0.74188	0.98919	1.23649	1.48378	1.73108	1.97838	2.22568
0.50854	0.76281	1.01708	1.27135	1.52563	1.77990	2.03417	2.28844
1.37238	2.05858	2.74477	3.43096	4.11715	4.80334	5.48954	6.17573
2.62762	3.94142	5.25523	6.56904	7.88285	9.19666	10.51046	11.82427
0.62124	0.93187	1.24249	1.55311	1.86373	2.17435	2.48498	2.79560
2.54237	3.81356	5.08474	6.35593	7.62712	8.89830	10.16949	11.44067
0.96774	1.45161	1.93548	2.41935	2.90323	3.38710	3.87097	4.35484

TABLE IV.

Elements.	Found.	Sought.	1
Cobalt.......	Cobaltous sulphate + potassium sulphate $2(CoSO_4) + 3(K_2SO_4)$	Cobaltous oxide CoO	0.18012
	Cobaltous sulphate + potassium sulphate $2(CoSO_4) + 3(K_2SO_4)$	Cobalt Co	0.14170
Copper......	Cupric oxide CuO	Copper Cu	0.79849
	Cuprous sulphide Cu_2S.	Copper Cu	0.79849
Fluorine.....	Calcium fluoride $CaFl_2$	Fluorine Fl	0.48718
	Silicon fluoride $SiFl_4$	Fluorine Fl	0.73077
Hydrogen....	Water H_2O	Hydrogen H	0.11111
Iodine.......	Silver iodide AgI	Iodine I	0.54029
	Palladious iodide PdI_2	Iodine I	0.70417
Iron.........	Ferric oxide Fe_2O_3	Iron Fe	0.70000
	Ferric oxide Fe_2O_3	Ferrous oxide FeO	0.90000
	Ferrous sulphide FeS	Iron Fe	0.63636
Lead........	Lead oxide PbO	Lead Pb	0.92825
	Lead sulphate $PbSO_4$	Lead oxide PbO	0.73597
	Lead sulphate $PbSO_4$	Lead Pb	0.68317
	Lead sulphide PbS	Lead oxide PbO	0.93305
Lithium......	Lithium carbonate Li_2Co_3	Lithia Li_2O	0.40541

(*Continued*).

2	3	4	5	6	7	8	9
0.36024	0.54036	0.72048	0.90060	1.08072	1.26084	1.44096	1.62108
0.28339	0.42508	0.56676	0.70847	0.85016	0.99186	1.13355	1.27525
1.59698	2.39547	3.19396	3.99244	4.79093	5.58942	6.38791	7.18640
1.59698	2.39547	3.19396	3,99244	4.79093	5.58942	6.38791	7.18640
0.97436	1.46154	1.94872	2.43590	2.92307	3.41027	3.89743	4.38461
1.46154	2.19231	2.92308	3.65385	4.38461	5.11538	5.84615	6.57692
0.22222	0.33333	0.44444	0.55555	0.66667	0.77778	0.88889	1.00000
1.08059	1.62088	2.16118	2.70147	3.24176	3.78206	4.32235	4.86264
1.40835	2.11252	2.81670	3.52087	4.22505	4.92922	5.63340	6.33757
1.40000	2.10000	2.80000	3.50000	4.20000	4.90000	5.60000	6.30000
1.80000	2.70000	3.60000	4.50000	5.40000	6.30000	7.20000	8.10000
1.27273	1.90909	2.54546	3.18182	3.81818	4.45455	5.09091	5.72728
1.85650	2.78475	3.71300	4.64126	5.56951	6.49776	7.42601	8.35426
1.47195	2.20792	2.94390	3.67987	4.41584	5.15182	5.88779	6.62377
1.36634	2.04950	2.73267	3.41584	4.09901	4.78218	5.46534	6.14851
1.86611	2.79916	3.73222	4.66527	5.59832	6.53138	7.46443	8.39749
0.81081	1.21622	1.62162	2.02703	2.43243	2.83784	3.24324	3.64865

TABLE IV.

Elements.	Found.	Sought.	1
Lithium.....	Lithium sulphate Li_2SO_4	Lithia Li_2O	0.27273
	Lithium phosphate Li_3PO_4	Lithia Li_2O	0.38793
Magnesium...	Magnesia MgO	Magnesium Mg	0.60030
	Magnesium sulphate $MgSO_4$	Magnesia MgO	0.33350
	Magnesium pyrophosphate $Mg_2P_2O_7$	Magnesia MgO	0.36036
Manganese...	Manganous oxide MnO	Manganese Mn	0.77465
	Protosesquioxide of manganese $MnO + Mn_2O_3$	Manganese Mn	0.72052
	Manganic oxide Mn_2O_3	Manganese Mn	0.69620
	Manganous sulphate $MnSO_4$	Manganous oxide MnO	0.47020
	Manganous sulphide MnS	Manganous oxide MnO	0.81609
	Manganous sulphide MnS	Manganese Mn	0.63218
Mercury.....	Mercury Hg	Mercurous oxide Hg_2O	1.04000
	Mercury Hg	Mercuric oxide HgO	1.08000
	Mercurous chloride Hg_2Cl_2	Mercury Hg	0.84940
	Mercuric sulphide HgS	Mercury Hg	0.86207
Nickel.......	Nickelous oxide NiO	Nickel Ni	0.78667
Nitrogen.....	Ammonium platinic chloride $(NH_4Cl)_2, PtCl_4$	Nitrogen N	0.06296
	Platinum Pt	Nitrogen N	0.14241

(Continued).

2	3	4	5	6	7	8	9
0.54545	0.81818	1.09091	1.36364	1.63636	1.90909	2.18182	2.45454
0.77586	1.16379	1.55172	1.93966	2.32759	2.71552	3.10345	3.49138
1.20061	1.80091	2.40121	3.00151	3.60182	4.20212	4.80242	5.40273
0.66700	1.00051	1.33401	1.66751	2.00101	2.33451	2.66802	3.00152
0.72072	1.08108	1.44144	1.80180	2.16216	2.52252	2.88288	3.24324
1.54930	2.32394	3.09859	3.87324	4.64789	5.42254	6.19718	6.97183
1.44105	2.16157	2.88210	3.60262	4.32314	5.04367	5,76419	6.48472
1.39241	2.08861	2.78481	3.48102	4.17722	4.87342	5.56962	6.26583
0.94040	1.41060	1.88080	2.35099	2.82119	3.29139	3.76159	4.23179
1.63218	2.44828	3.26437	4.08046	4.89655	5.71264	6.52874	7.34483
1.26437	1.89655	2.52874	3.16092	3.79310	4.42529	5.05747	5.68966
2.08000	3.12000	4.16000	5.20000	6.24000	7.28000	8.32000	9.36000
2.16000	3.24000	4.32000	5.40000	6.48000	7.56000	8.64000	9.72000
1.69880	2.54820	3.39760	4.24701	5.09641	5.94581	6.79521	7.64461
1.72414	2.58621	3.44828	4.31034	5.17241	6.03448	6.89655	7.75862
1.57333	2.36000	3.14667	3.93333	4.72000	5.50667	6.29334	7.08000
0.12591	0.18887	0.25182	0.31478	0.37774	0.44069	0.50365	0.56660
0.28482	0.42722	0.56963	0.71204	0.85445	0.99686	1.13926	1.28167

TABLE IV.

Elements.	Found.	Sought.	1
Nitrogen.....	Silver cyanide $AgCN$	Cyanogen CN	0.19437
	Silver cyanide $AgCN$	Hydrocyanic acid HCN	0.20184
Oxygen......	Alumina Al_2O_3	Oxygen O	0.46602
	Antimonious oxide Sb_2O_3	Oxygen O	0.16438
	Arsenious oxide As_2O_3	Oxygen O	0.24242
	Arsenic oxide As_2O_5	Oxygen O	0.34783
	Baryta BaO	Oxygen O	0.10458
	Bismuth trioxide Bi_2O_3	Oxygen O	0.10345
	Cadmium oxide CdO	Oxygen O	0.12500
	Chromic oxide Cr_2O_3	Ogygen O	0.31381
	Cobaltous oxide CoO	Oxygen O	0.21333
	Cupric oxide CuO	Oxygen O	0.20151
	Ferrous oxide FeO	Oxygen O	0.22222
	Ferric oxide Fe_2O_3	Oxygen O	0.30000
	Lead oxide PbO	Oxygen O	0.07175
	Lime CaO	Oxygen O	0.28571
	Magnesia MgO	Oxygen O	0.39970
	Manganous oxide MnO	Oxygen O	0.22535

(Continued).

2	3	4	5	6	7	8	9
0.38874	0.58312	0.77749	0.97186	1.16623	1.36060	1.55498	1.74935
0.40367	0.60551	0.80734	1.00918	1.21102	1.41285	1.61469	1.81652
0.93204	1.39806	1.86408	2.33010	2.79611	3.26213	3.72815	4.19417
0.32877	0.49315	0.65754	0.82192	0.98630	1.15069	1.31507	1.47946
0.48484	0.72726	0.96968	1.21210	1.45452	1.69694	1.93936	2.18178
0.69565	1.04348	1.39130	1.73913	2.08696	2.43478	2.78261	3.13043
0.20915	0.31373	0.41830	0.52288	0.62745	0.73203	0.83660	0.94118
0.20690	0.31035	0.41380	0.51725	0.62070	0.72415	0.82760	0.93105
0.25000	0.37500	0.50000	0.62500	0.75000	0.87500	1.00000	1.12500
0.62762	0.94143	1.25524	1.56905	1.88286	2.19667	2.51048	2.82429
0.42667	0.64000	0.85333	1.06667	1.28000	1.49333	1.70666	1.92000
0.40302	0.60453	0.80604	1.00756	1.20907	1.41058	1.61209	1.81360
0.44444	0.66667	0.88889	1.11111	1.33333	1.55555	1.77778	2.00000
0.60000	0.90000	1.20000	1.50000	1.80000	2.10000	2.40000	2.70000
0.14350	0.21525	0.28700	0.35874	0.43049	0.50224	0.57399	0.64574
0.57143	0.85714	1.14286	1.42857	1.71429	2.00000	2.28571	2.57143
0.79939	1.19909	1.59879	1.99849	2.39818	2.79788	3.19758	3.59727
0.45070	0.67606	0.90141	1.12676	1.35211	1.57746	1.80282	2.02817

TABLE IV.

Elements.	Found.	Sought.	1
Oxygen......	Protosesquioxide of Manganese $MnO + Mn_2O_3$	Oxygen O	0.27947
	Manganic oxide Mn_2O_3	Oxygen O	0.30380
	Mercurous oxide Hg_2O	Oxygen O	0.03846
	Mercuric oxide HgO	Oxygen O	0.07407
	Nickelous Oxide NiO	Oxygen O	0.21333
	Potassa K_2O	Oxygen O	0.16974
	Silicic anhydride SiO_2	Oxygen O	0.53333
	Silver oxide Ag_2O	Oxygen O	0.06901
	Soda Na_2O	Oxygen O	0.25773
	Strontia SrO	Oxygen O	0.15459
	Stannic oxide SnO_2	Oxygen O	0.21333
	Water H_2O	Oxygen O	0.88889
	Zinc oxide ZnO	Oxygen O	0.19740
Phosphorus...	Phosphoric anhydride P_2O_5	Phosphorus P	0.43662
	Magnesium pyrophosphate $Mg_2P_2O_7$	Phosphoric anhydride P_2O_5	0.63964
	Ferric phosphate $FePO_4$	Phosphoric anhydride P_2O_5	0.47020
	Silver phosphate Ag_3PO_4	Phosphoric anhydride P_2O_5	0.16953
	Uranyl pyrophosphate $(UO_2)_2P_2O_7$	Phosphoric anhydride P_2O_5	0.19910

(Continued).

2	3	4	5	6	7	8	9
0.55895	0.83843	1.11790	1.39738	1.67686	1.95633	2.23581	2.51528
0.60759	0.91139	1.21519	1.51899	1.82278	2.12658	2.43038	2.73417
0.07692	0.11539	0.15385	0.19231	0.23077	0.26923	0.30770	0.34616
0.14815	0.22222	0.29630	0.37037	0.44444	0.51852	0.59259	0.66667
0.42667	0.64000	0.85333	1.06667	1.28000	1.49333	1.70667	1.92000
0.33949	0.50923	0.67897	0.84871	1.01846	1.18820	1.35794	1.52768
1.06667	1.60000	2.13333	2.66667	3.20000	3.73333	4.26667	4.80000
0.13801	0.20702	0.27603	0.34503	0.41404	0.48305	0.55206	0.62106
0.51546	0.77320	1.03093	1.28866	1.54639	1.80412	2.06186	2.31959
0.30918	0.46377	0.61836	0.77295	0.92753	1.08212	1.23671	1.39130
0.42667	0.64000	0.85333	1.06667	1.28000	1.49333	1.70667	1.92000
1.77778	2.66667	3.55556	4.44445	5.33333	6.22222	7.11111	8.00000
0.39480	0.59220	0.78960	0.98700	1.18440	1.38180	1.57920	1.77660
0.87324	1.30986	1.74648	2.18309	2.61971	3.05633	3.49295	3.92957
1.27928	1.91892	2.55856	3.19820	3.83784	4.47748	5.11712	5.75676
0.94040	1.41060	1.88080	2.35099	2.82119	3.29139	3.76159	4.23179
0.33907	0.50860	0.67814	0.84767	1.01721	1.18674	1.35628	1.52581
0.39821	0.59731	0.79641	0.99551	1.19462	1.39372	1.59282	1.79192

TABLE IV.

Elements.	Found.	Sought.	1
Potassium...	Potassa K_2O	Potassium K	0.83026
	Potassium sulphate K_2SO_4	Potassa K_2O	0.54091
	Potassium chloride KCl	Potassium K	0.52460
	Potassium chloride KCl	Potassa K_2O	0.63185
	Potassium platinic chloride $(KCl)_2PtCl_4$	Potassa K_2O	0.19308
	Potassium platinic chloride $(KCl)_2PtCl_4$	Potassium chloride KCl	0.30557
Silicon.......	Silicic anhydride SiO_2	Silicon Si	0.46667
Silver........	Silver chloride AgCl	Silver Ag	0.75270
	Silver chloride AgCl	Silver oxide Ag_2O	0.80849
Sodium......	Soda Na_2O	Sodium Na	0.74227
	Sodium sulphate Na_2SO_4	Soda Na_2O	0.43694
	Sodium chloride NaCl	Soda Na_2O	0.53060
	Sodium chloride NaCl	Sodium Na	0.39384
	Sodium carbonate Na_2CO_3	Soda Na_2O	0.58522
Strontium....	Strontia SrO	Strontium Sr	0.84541
	Strontium sulphate $SrSO_4$	Strontia SrO	0.56403
	Strontium carbonate $SrCO_3$	Strontia SrO	0.70169
Sulphur	Barium sulphate $BaSO_4$	Sulphur S	0.13734

(Continued).

2	3	4	5	6	7	8	9
1.66051	2.49077	3.32103	4.15128	4.98154	5.81180	6.64206	7.47231
1.08183	1.62274	2.16366	2.70457	3.24549	3.78640	4.32732	4.86823
1.04920	1.57380	2.09840	2.62300	3.14761	3.67221	4.19681	4.72141
1.26371	1.89556	2.52742	3.15927	3.79112	4.42298	5.05483	5.68669
0.38615	0.57923	0.77230	0.96538	1.15846	1.35153	1.54461	1.73768
0.61114	0.91671	1.22228	1.52785	1.83343	2.13900	2.44457	2.75014
0.93333	1.40001	1.86667	2.33333	2.80000	3.26667	3.73333	4.20000
1.50540	2.25811	3.01081	3.76351	4.51621	5.26891	6.02162	6 77432
1.61700	2.42548	3.23398	4.04247	4.85096	5.65946	6.46795	7.27645
1.48454	2.22680	2.96907	3.71134	4.45361	5.19588	5.93814	6.68041
0.87387	1.31081	1.74775	2.18468	2.62162	3.05856	3.49550	3.93243
1.06120	1.59179	2.12239	2.65299	3.18359	3.71419	4.24478	4.77538
0.78769	1.18154	1.57538	1.96923	2.36308	2.75692	3.15077	3.54461
1.17044	1.75566	2.34088	2.92610	3.51132	4.09654	4.68176	5.26698
1.69082	2.53623	3.38164	4.22705	5.07247	5.91788	6.76329	7.60870
1.12807	1.69210	2.25613	2.82017	3.38420	3.94823	4.51226	5.07630
1.40339	2.10508	2.80678	3.50848	4.21017	4.91186	5.61356	6.31526
0.27468	0.41202	0.54936	0.68670	0.82403	0.96137	1.09871	1.23605

TABLE IV.

Elements.	Found.	Sought.	1
Sulphur......	Arsenious sulphide As_2S_3	Sulphur S	0.39024
	Barium sulphate $BaSO_4$	Sulphuric anhydride SO_3	0.34335
Tin..........	Stannic oxide SnO_2	Tin Sn	0.78667
	Stannic oxide SnO_2	Stannous oxide SnO	0.89333
Zinc.........	Zinc oxide ZnO	Zinc Zn	0.80260
	Zinc sulphide ZnS	Zinc oxide ZnO	0.83515
	Zinc sulphide ZnS	Zinc Zn	0.67031

(*Continued*).

2	3	4	5	6	7	8	9
0.78049	1.17073	1.56097	1.95122	2.34146	2.73170	3.12194	3.51219
0.68670	1.03004	1.37339	1.71674	2.06009	2.40344	2.74678	3.09013
1.57333	2.36000	3.14667	3.93333	4.72000	5.50667	6.29334	7.08000
1.78667	2.68000	3.57333	4.46667	5.36000	6.25333	7.14666	8.04000
1.60520	2.40780	3.21040	4.01300	4.81560	5.61820	6.42080	7.22340
1.67031	2.50546	3.34062	4.17577	5.01092	5.84608	6.68123	7.51639
1.34061	2.01092	2.68123	3.35154	4.02184	4.69215	5.36246	6.03276

TABLE V.

SPECIFIC GRAVITY AND ABSOLUTE WEIGHT OF SEVERAL GASES.

	Specific gravity, atmospheric air = 1·0000.	1 litre (1000 cubic centimetres) of gas at 0° and 0·76 metre bar. pressure weighs grammes.
Atmospheric air	1·0000	1·29366
Oxygen	1·10832	1·43379
Hydrogen	0·06927	0·08961
Water, vapor of	0·62343	0·80651
Carbon, vapor of	0·83124	1·07534
Carbon dioxide	1·52394	1·97146
Carbon monoxide	0·96978	1·25456
Marsh gas	0·55416	0·71689
Elayl gas	0·96978	1·25456
Phosphorus, vapor of	4·29474	5·55593
Sulphur, vapor of	6·64992	8·60273
Hydrosulphuric acid	1·17759	1·52340
Iodine, vapor of	8·78898	11·36995
Bromine, vapor of	5·53952	7·16625
Chlorine	2·45631	3·17763
Nitrogen	0·96978	1·25456
Ammonia	0·58879	0·76169
Cyanogen	1·80102	2·32991

TABLE VI.

COMPARISON OF THE DEGREES OF THE MERCURIAL THERMOMETER WITH THOSE OF THE AIR THERMOMETER.

According to Magnus.

Degrees of the mercurial thermometer.	Degrees of the air thermometer.
100	100·00
150	148·74
200	197·49
250	245·39
300	294·51
330	320·92

ALPHABETICAL INDEX.

Lightning Source UK Ltd.
Milton Keynes UK
UKHW010305190119
335762UK00007B/653/P